Monographs in Mathematics
Vol. 102

Managing Editors:
H. Amann
Universität Zürich, Switzerland
J.-P. Bourguignon
IHES, Bures-sur-Yvette, France
K. Grove
University of Maryland, College Park, USA
P.-L. Lions
Université de Paris-Dauphine, France

Associate Editors:
H. Araki, Kyoto University
F. Brezzi, Università di Pavia
K.C. Chang, Peking University
N. Hitchin, University of Warwick
H. Hofer, Courant Institute, New York
H. Knörrer, ETH Zürich
K. Masuda, University of Tokyo
D. Zagier, Max-Planck-Institut Bonn

Alexander Brudnyi • Yuri Brudnyi

Methods of Geometric Analysis in Extension and Trace Problems

Volume 1

Alexander Brudnyi
Department of Mathematics & Statistics
University of Calgary
2500 University Dr. NW
Calgary, Alberta, Canada, T2N 1N4
abrudnyi@ucalgary.ca

Yuri Brudnyi
Mathematics Department
Technion - Israel Institute of Technology
Haifa 32000
Israel
ybrudnyi@math.technion.ac.il

2010 Mathematics Subject Classification: 26A16, 26B35, 46B85, 46B70, 51H25, 52A07, 53C23, 54E35, 54E40

ISBN 978-3-0348-0208-6   e-ISBN 978-3-0348-0209-3
DOI 10.1007/978-3-0348-0209-3

Library of Congress Control Number: 2011939996

© Springer Basel AG 2012
This work is subject to copyright. All rights are reserved, whether the whole or part of the material is concerned, specifically the rights of translation, reprinting, re-use of illustrations, recitation, broadcasting, reproduction on microfilms or in other ways, and storage in data banks. For any kind of use permission of the copyright owner must be obtained.

Printed on acid-free paper

Springer Basel AG is part of Springer Science+Business Media

www.birkhauser-science.com

Contents

Preface — xi

Basic Terms and Notation — xvii

I Classical Extension-Trace Theorems and Related Results — 1

1 Continuous and Lipschitz Functions — 5
    Continuous Functions — 5
    1.1 Notation and definitions — 6
    1.2 Extension and trace problems: formulations and examples — 7
        1.2.1 Example: Continuous functions — 9
        1.2.2 Example: Uniformly continuous functions — 9
        1.2.3 Example: Continuously differentiable functions on $\mathbb{R}^n$ — 10
        1.2.4 Example: BMO and Sobolev spaces — 11
    1.3 Continuous selections — 12
    1.4 Simultaneous continuous extensions — 14
    1.5 Extensions of continuous maps acting between metric spaces — 16
    1.6 Absolute metric retracts — 17
    Lipschitz Functions — 20
    1.7 Notation and definitions — 20
    1.8 Trace and extension problems for Lipschitz functions — 22
    1.9 Lipschitz selection problem — 23
        1.9.1 Counterexample — 23
        1.9.2 Combinatorial–geometric selection results — 25
    1.10 Extensions preserving Lipschitz constants — 27
        1.10.1 Banach-valued Lipschitz functions — 27
        1.10.2 Extension and the intersection property of balls — 30
        1.10.3 Proof of Theorem 1.26 — 33
        1.10.4 Lipschitz maps acting in spaces of constant curvature — 33
    1.11 Lipschitz extensions — 42
    1.12 Simultaneous Lipschitz extensions — 53

1.13		Simultaneous Lipschitz selection problem	55
Comments			56
Appendices			61
A.		Topological dimension and continuous extensions of maps into $\mathbb{S}^n$	61
B.		Helly's topological theorem	64
	B.1.	The Classical Helly theorem and related results	64
	B.2.	Cohomology theory – a computational aspect	67
	B.3.	Helly's topological theorem	71
C.		Sperner's lemma and its consequences	73
D.		Contractions of n-spheres	78

2 Smooth Functions on Subsets of $\mathbb{R}^n$ 83

2.1		Classical function spaces: notation and definitions	84
	2.1.1	Differentiable functions	84
	2.1.2	k-jets	86
	2.1.3	Lipschitz functions of higher order	87
	2.1.4	Extension and trace problems for classical function spaces	92
2.2		Whitney's extension theorem	93
2.3		Divided differences, local approximation and differentiability	101
2.4		Trace and extension problems for univariate C^k functions	120
	2.4.1	Whitney's theorem	120
	2.4.2	Reformulation of Whitney's theorem	129
	2.4.3	Finiteness and linearity	130
	2.4.4	Basic conjectures	133
2.5		Restricted main problem for some classes of domains in $\mathbb{R}^n$	134
	2.5.1	Quasiconvex domains	134
	2.5.2	Lipschitz domains	149
2.6		Sobolev spaces: selected trace and extension results	155
	2.6.1	P. Jones' theorem and related results	155
	2.6.2	Peetre's nonexistence theorem	160
Comments			165
Appendices			170
E.		Difference identities	170
	E.1.	Kemperman's identity	170
	E.2.	Marchaud's identity	176
F.		Local polynomial approximation and moduli of continuity	178
	F.1.	Degree of local polynomial approximation	178
	F.2.	Whitney's constants	186
	F.3.	Conjectures	188
G.		Local inequalities for polynomials	188

II Topics in Geometry of and Analysis on Metric Spaces — 197

3 Topics in Metric Space Theory — 201
3.1 Principal concepts and related facts — 201
- 3.1.1 Pseudometrics, metrics and quasimetrics — 201
- 3.1.2 Metric and quasimetric spaces — 203
- 3.1.3 Paracompactness and continuous partitions of unity — 212
- 3.1.4 Compact and precompact metric spaces — 215
- 3.1.5 Proper metric spaces — 219
- 3.1.6 Doubling metric spaces — 222
- 3.1.7 Metric length structure — 226
- 3.1.8 Basic metric constructions — 237

3.2 Measures on metric spaces — 252
- 3.2.1 Measure theory — 252
- 3.2.2 Integration — 254
- 3.2.3 Measurable selections — 255
- 3.2.4 Hausdorff measures — 256
- 3.2.5 Doubling measures — 263
- 3.2.6 Families of pointwise doubling measures — 266

3.3 Basic classes of metric spaces — 274
- 3.3.1 Ultrametric spaces — 274
- 3.3.2 Spaces of bounded geometry — 278
- 3.3.3 Riemannian manifolds as metric spaces — 280
- 3.3.4 Gromov hyperbolic spaces — 286
- 3.3.5 Sub-Riemannian manifolds — 291
- 3.3.6 Metric graphs — 292
- 3.3.7 Metric groups — 301

Comments — 312

4 Selected Topics in Analysis on Metric Spaces — 317
4.1 Dvoretsky type theorem for finite metric spaces — 318
4.2 Covering metric invariants — 326
- 4.2.1 Metric dimension — 326
- 4.2.2 Hausdorff dimension — 329
- 4.2.3 Hausdorff dimension of doubling metric spaces — 334
- 4.2.4 Nagata dimension — 341

4.3 Existence of doubling measures — 357
- 4.3.1 Finite metric spaces — 358
- 4.3.2 Compact metric spaces — 367
- 4.3.3 Complete metric spaces — 369
- 4.3.4 Dyn'kin conjecture — 369
- 4.3.5 Concluding remarks — 371

4.4 Space of balls — 372
- 4.4.1 $\mathcal{B}(\mathcal{M})$ as a length space — 373

	4.4.2	$\mathcal{B}(\mathbb{R}^n)$ as a space of pointwise homogeneous type	379
	4.4.3	Generalized hyperbolic spaces $\mathbb{H}_\omega^{n+1}$	384
4.5	Differentiability of Lipschitz functions		386
	4.5.1	Lipschitz functions on $\mathbb{R}^n$	386
	4.5.2	Lipschitz functions on metric spaces	389
4.6	Lipschitz spaces		394
	4.6.1	Modulus of continuity	394
	4.6.2	Real interpolation of Lipschitz spaces	397
	4.6.3	Duality theorem	405
Comments			413

5 Lipschitz Embedding and Selections   417

5.1	Embedding of metric spaces into the space forms of nonpositive curvature		418
	5.1.1	Finite metric spaces	418
	5.1.2	Infinite metric trees	423
	5.1.3	Doubling metric spaces	438
	5.1.4	Gromov hyperbolic spaces	447
5.2	Roughly similar embeddings of Gromov hyperbolic spaces		454
	5.2.1	Coarse Geometry, a survey	454
	5.2.2	Coarse geometry of $\mathbb{H}^n$	458
	5.2.3	The Bonk–Schramm theorem	461
5.3	Lipschitz selections		470
	5.3.1	Barycenter and Steiner selectors	470
	5.3.2	Helly type result: a conjecture	476
	5.3.3	A Sylvester type selection result	483
5.4	Simultaneous Lipschitz selections		493
	5.4.1	The problem	493
	5.4.2	Formulation of the main theorem	494
	5.4.3	Auxiliary results	495
	5.4.4	Proof of Theorem 5.66	497
	5.4.5	Proof of Proposition 5.68	512
	5.4.6	Proof of Proposition 5.69	519
Comments			523

Bibliography   **527**

Index   **557**

Contents of Volume II

Preface

Basic Terms and Notation

III Lipschitz Extensions from Subsets of Metric Spaces

6 Extensions of Lipschitz Maps
- 6.1 Lipschitz n-connectedness
- 6.2 Whitney covers
- 6.3 Main extension theorem
- 6.4 Corollaries of the main extension theorem
- 6.5 Nonlinear Lipschitz extension constants
- Comments

7 Simultaneous Lipschitz Extensions
- 7.1 Characterization of simultaneous Lipschitz extension spaces
- 7.2 Main extension result
- 7.3 Locally doubling metric spaces with uniform lattices
- 7.4 Spaces with the universal linear Lipschitz extension property
- Comments

8 Linearity and Nonlinearity
- 8.1 Snowflake stability of Lipschitz extension properties
- 8.2 Relation between linear and nonlinear extension constants
- 8.3 Metric spaces without simultaneous Lipschitz extension property
- Comments

IV Smooth Extension and Trace Problems for Functions on Subsets of $\mathbb{R}^n$

9 Traces to Closed Subsets: Criteria, Applications

9.1 Traces to closed subsets: criteria . . . .
9.2 Traces to Markov sets . . . .
9.3 Simultaneous extensions from uniform domains . . . .
Comments . . . .

10 Whitney Problems
10.1 Formulation of the problems . . . .
10.2 Trace and extension problems for Markov sets . . . .
10.3 $C^{k,\omega}(\mathbb{R}^n)$ spaces: finiteness and linearity . . . .
10.4 Fefferman's solution to the classical Whitney problems . . . . . .
10.5 Jet space $J^\ell \Lambda^{2,\omega}(\mathbb{R}^n)$: finiteness and linearity . . . .
Comments . . . .

Bibliography

Index

Preface

The title refers to extension and trace problems for Lipschitz functions on metric spaces and smooth functions on subsets of $\mathbb{R}^n$, linked by a unified geometric analysis approach. While methods of Geometric Analysis are clearly relevant to the study of functions on metric spaces, the smooth extension and trace problems are of pure analytic origin and seem to require other tools for their study. An approach allowing us to involve Geometric Analysis in this study will be briefly considered below; to begin we discuss some features of the problems in question.

Seemingly very specific, mostly intended for applications, this topic has been, from its very beginning, a powerful source of ideas, concepts and methods that essentially influenced and sometimes even unified considerable parts of Analysis. These include Complex and Differential Analysis, PDE and Image Processing (to name but a few) but the most spectacular example is Algebraic Topology. Here the problem of existence (or nonexistence) of continuous extensions that arose about one hundred years ago (Lebesgue, Brower) has eventually turned into a main object of study (see, in particular, the Princeton Colloquium Lectures by N. Steenrod).

This amazing research power of the topic may apparently be explained by the nature of human knowledge that, citing philosopher W. V. Quine "... is a man-made fabric which impinges on experience only along the edges".

In the case of metric spaces (whose study forms a considerable part of the book), continuous maps are naturally replaced by their metric equivalents, Lipschitz and rough Lipschitz maps and the like. Moreover, together with the analysis of continuity, compactness and so forth, one should use concepts and tools of (interpreted broadly) Geometric Analysis such as geometric measures and probabilities, metric graphs, metric analogs of dimension and related methods and results.

In accordance with a diversity of geometric and analytic objects joined by the concept of metric space, the fundamental extension problem of topology is now divided into several components. The most natural analog asks about the existence of a Lipschitz extension for a map from a subset of one metric space to another. For real-valued functions a positive answer is given by a remarkably simple nonlinear operator (McShane, 1934) while sufficient conditions for the general problem were found only a few years ago (U. Lang-Schlichenmaier, 2005). The answer includes several concepts (Nagata dimension, Lipschitz n-connectivity etc.) belonging to a new developing area that may be naturally called Lipschitz Topology.

If the image of a Lipschitz map is a normed linear space, one can study a linear (*simultaneous*) Lipschitz extension of the map and the norm evaluation of the corresponding linear extension operator using geometric characteristics of the domain. Unlike the continuous case where a positive solution is due to Borsuk (1933) for separable metric spaces and Dugundji (1951) for the general case, the answer is now known to be negative even for real-valued Lipschitz functions (Pelczynski, 1960).

A wide class of metric spaces admitting simultaneous Lipschitz extensions was discovered simultaneously by three different approaches that exploit, respectively, tools of Lipschitz Topology (Lang-Schlichenmaier), Geometric Measure Theory (the authors) and Probabilistic Combinatorics (Lee-Naor). The first approach estimates the corresponding norms by unspecified constants while the two others give effective upper bounds.

A class of metric spaces with the required extension property may be essentially enlarged by using bi-Lipschitz embeddings into universal "nice" spaces such as the classical metric forms $\mathbb{R}^n$, $\mathbb{S}^n$ or $\mathbb{H}^n$. To successfully apply this approach one should really know much more about universal spaces for given classes of metric spaces, quasi-isometric invariants and the like. As model cases one may point to Urysohn's universal space (1928) containing isometric copies of all separable metric spaces, the Bonk-Schramm result (2000) on the universal property of $\mathbb{H}^n$ with respect to the class of Gromov hyperbolic spaces of bounded geometry and results of Lipschitz Topology (metric, Hausdorff and Nagata dimensions and so forth). Nevertheless, even the results proved so far have been applied to obtain extension theorems of considerable value.

The final part of the book is devoted to the extension and trace problems for spaces of multivariate differentiable and smooth functions[1] and related jet spaces. Given such a *smoothness space* X on $\mathbb{R}^n$ and a function f on a subset $S \subset \mathbb{R}^n$ the following problems will be studied:

Qualitative trace problem. Does there exist a function $F \in X$ whose trace $F|_S$ agrees with f?

Quantitative trace problem. Find an effective two-sided estimate for the trace norm of f, i.e., for $\inf\{\|F\|_X \,;\, f = F|_S\}$.

Simultaneous extension problem. Does there exist a linear bounded extension operator from the trace space $X|_S$ into X?

These problems go back to the two classic papers that Whitney published in 1934 where the first solves all these problems for the jet space generated by functions from the space $C_b^k(\mathbb{R}^n)$ of all bounded k-times continuously differentiable functions with bounded higher derivatives. The second, less well known paper,

---

[1] i.e., with controlled moduli of continuity of given order of their higher derivatives

solves all of them for the space $C_b^k(\mathbb{R})$ of univariate functions. The indication of number I in the title of the second paper may be seen as Whitney's intention (or appeal?) to proceed with the multivariate case. However, considerable progress in this direction did not appear for more than fifty years.

To illustrate the seemingly unsurmountable complexity of the multivariate problem one can compare the two simplest cases, namely, the spaces $C_b^1(\mathbb{R}^n)$ for the known case $n=1$ and the recently studied case $n>1$.

For the quantitative trace problem, the Whitney theorem characterizes the extendability of a univariate function $f: S \to \mathbb{R}$ to a function of $C_b^1(\mathbb{R})$ by behaviour of f on 3-point subsets of S and the number 3 of points is sharp (in the seemingly analogous case of Lipschitz functions, the number of required points is 2 for every n). However, the corresponding number of points for the multivariate case is of exponential growth in the dimension!

For the second trace problem Whitney's theorem gives an elegant explicit formula evaluating the trace norm via three point divided differences. In contrast, a result of this kind is unknown for the multivariate case.

Finally, a simultaneous extension for $n=1$ is given by Whitney's (linear) extension construction developed in his first paper. Since this construction is, in a sense, universal, it also solves positively the simultaneous extension problem for the space $C_u^1(\mathbb{R})$ consisting of functions with uniformly continuous derivatives. However, this latter result does not hold for $n>1$.

It is worth noting that Whitney's problem, as special as it may seen, is, in fact, one of the most challenging topics of a vast, intensively developing area, that studies problems with incomplete data. This, in particular, includes Differential Analysis on sets without differential structure (large finite subsets of $\mathbb{R}^n$, metric spaces etc.), inverse and incorrect problems of Mathematical Physics and such fields of Applied Mathematics as image restoration, mathematical tomography and computer graphics. The concepts and methods that have been and will be developed for the study of Whitney's problem will doubtless play a considerable role in the development of this area.

The road to applying geometric methods to the pure analytic issue in question is opened by Local Approximation Theory developed by the second named author in the 1960s. For a wide class of smoothness spaces including Sobolev and Besov spaces over $L_\infty(\mathbb{R}^n)$ and the associated jet spaces, the theory gives a complete description of their trace spaces to an arbitrary subset of $\mathbb{R}^n$. The sufficiency part of these criteria is equivalently reformulated as existence of Lipschitz selections for set-valued functions on specially constructed metric spaces or more involved spaces with values in the set $\mathcal{C}(\mathbb{R}^n)$ of nonempty convex subsets of $\mathbb{R}^n$.

A very attractive example of this kind is the next conjecture proved in several special cases.

Let $\varphi: \mathcal{M} \to \mathcal{C}(\mathbb{R}^n)$ be a set-valued map on a metric space $\mathcal{M}$. Assume that its trace to every 2^n point subset admits a 1-Lipschitz selection. Then φ itself admits a selection with Lipschitz constant depending only on n.

The first breakthrough in the realization of this program was due to Shvartsman who in his 1984 PhD thesis proved the aforementioned conjecture for the set of affine subspaces of $\mathbb{R}^n$ and derived from here a solution to the qualitative trace problem for Zygmund spaces[2]. He also constructed a very ingenious example showing that the cardinality of subsets involved in the solution is sharp (hence, the number 2^n in the Lipschitz selection conjecture is also sharp).

The next important result is a solution to the linear (simultaneous) extension problem for Zygmund spaces (Yu. Brudnyi and Shvartsman, 1985). The involved set-valued map now depends *linearly* on a functional parameter and the required Lipschitz selection should preserve this dependence.

Subsequent work in this direction was started only at the end of the previous century. New results obtained concern, in particular, spaces $C^{1,\omega}(\mathbb{R}^n)$ (the second of the only known sharp results) and the generalization of the Whitney-Glaeser theorem to the spaces of jets whose higher components satisfy the Zygmund condition. The research potential of the discussed method is by no means exhausted by these briefly discussed results. A challenging task justifying this claim is the proof of the trace-extension problem for the space of multivariate C^k functions whose higher derivatives satisfy the Zygmund condition.[3]

A new powerful method for the study of the classical Whitney problem (i.e., for spaces $C_b^k(\mathbb{R}^n)$ and $C^{k,\omega}(\mathbb{R}^n)$) was invented by Ch. Fefferman. In a series of papers (2003–2009) he solved for this case the qualitative trace and simultaneous extension problems and made important advancements (in partial collaboration with Klartag) in the solution of the quantitative trace problem. This breakthrough work is only surveyed in this book, since even a superficial explanation of its proofs requires over a hundred pages of an extremely complicated text. However, a comprehensive understanding of Fefferman's proofs and ideas beyond them is one of the concrete aims of the theory that we can believe will lead to new discoveries in the area.

Now, we briefly discuss the contents of the book. More information can be found in the introduction to each chapter.

The book is divided into two volumes each consisting of two parts. Volume II is devoted to the study of the main themes, extension and trace problems for Lipschitz and smooth functions, respectively; see Preface to this volume for more information. Part 1 of Volume I (Chapters 1, 2) is introductory and gives background material and the important initial results that motivated and shaped the area. These include classical results of Lebesgue, Brouwer, Whitney, Bernstein and Valentine that have never appeared in book form.

Appendices to each chapter intend to familiarize the beginners with some facts and concepts used in the subsequent text. Those to Chapter 1 contain, in

---

[2] defined by the norm $\sup\limits_{\mathbb{R}^n}|f| + \sup\limits_{x\neq y}\dfrac{|f(x) - 2f(\frac{x+y}{2}) + f(y)|}{\omega(\|x-y\|)}$

[3] Note that Taylor polynomials of order $k+1$ may nowhere exist for functions of this space. This explains the inapplicability to this case of the Fefferman approach that we discuss below.

particular, the basic material on covering dimension and its relation to continuous extensions and Helly's topological generalization of his classical convex body result. This includes some topics in Algebraic Topology used in the proofs there and in the subsequent parts of the book.

Appendices to Chapter 2 contain, in particular, two fundamental facts of Local Polynomial Approximation Theory: the multivariate Whitney and Remez type inequalities.

Finally, Part 2 of Volume I (Chapters 3–5) contains concepts and results of Metric Space Theory used in the subsequent parts of the book. To avoid fragmenting the text into a discontinuous string of theorems, we add some background and several classical results that turn this part into a specifically oriented course on Metric Space Theory. Most of the basic results there were proved not long ago and have never appeared in book form. They, along with several classical results, are accompanied by detailed proofs.

Comments to each chapter discuss generalizations and related results. The latter are presented partly in a historical context; we believe that the reader may learn something important from such a presentation.

Reader's Guide

The reader whose main interest is Lipschitz Analysis may begin with Chapter 1 (Volume I) and its appendices presenting topological background and motivations for Lipschitz extension results studied in Chapters 6–8 (Volume II). Geometric Analysis results used for this study are contained in Sections 3.1, 3.2, 4.2, 4.3, 4.6, 5.1 and 5.2 of Volume I.

The reader interested in Whitney's problems may begin with Chapter 2 of Volume I and its appendices presenting background material on Multivariate Differential Analysis, the aforementioned classical Whitney theorems and related conjectures and several basic facts of Local Approximation Theory. This material forms an introduction to the intensive study of Whitney's extension and trace problems presented in Chapters 9, 10 of Volume II. The Geometric Analysis results used in this study are contained in Sections 3.2, 4.4, 5.3 and 5.4 of Volume I.

The prerequisites for reading this book include the material covered by first year graduate study (in particular, Linear Algebra, Real Analysis, Functional Analysis and some General Topology). More specialized topics are carefully presented in survey or more extended form. We either give a complete proof or else a detailed outline of the proof for very recent results. Many of these proofs are based on rather complicated geometric constructions; their study may be essentially facilitated by using appropriate geometric presentations. We give some of them in the book and strongly recommend the readers to draw their own geometric sketches in every such proof.

The vast majority of the main results presented in the book appear for the first time in book form. This holds not only for recent results but also for some of the deep classical results mentioned above.

Acknowledgments. It is a pleasure to express our thanks to several colleagues who helped us in preparation of this book.

Len Bos' suggestions and remarks allowed us to essentially improve the exposition of Chapters 1–3 and 9 while Peter Zwengrowski helped us to improve the exposition of Chapter 5. Charles Fefferman prepared for us a survey of his extension results and provided us with preprints of his papers. Our special thanks go to Pavel Shvartsman whose PhD Thesis and collaboration with us are reflected in many deep results of this book.

We thank also Ms. Galya Khanin whose marvelous work transformed our rough handwritten text into an excellent manuscript.

Basic Terms and Notation

Set-theoretic operations

$\in$ membership

$\cup$ union

$\cap$ intersection

$\setminus$ set theoretic difference

$\subset$ embedding (not necessarily proper)

$\oplus$ direct sum (also known as direct or cartesian product)

Sets and subsets

Let $T \subset S$ be sets.

- $T := \{x \in S \,;\, \mathcal{P}\}$: all elements of the subset T have the property $\mathcal{P}$

  Note the *figured brackets* and *semicolon* designed for the notation of sets. Hereafter the symbol $:=$ means that the left-hand side is defined or denoted by the right-hand side.

- $T^c := \{x \in S \,;\, x \notin T\}$: complement of T in S; it is usually clear from the context with respect to which larger set S the complement is taken

- 2^S: collection of subsets in S

- Let $\mathcal{F} := \{S_j\}_{j \in J} \subset 2^S$ be a *family* (indexed set). Then

$$\cup \mathcal{F} := \bigcup_{j \in J} S_j, \qquad \cap \mathcal{F} := \bigcap_{j \in J} S_j,$$

  $\sqcup \mathcal{F}$: disjoint union (($S_j \cap S_{j'} = \emptyset$ if $j \neq j'$ in this case)

- $\mathcal{F}$ is a *cover* of S if $\cup \mathcal{F} = S$

- a cover $\mathcal{F}'$ is a *refinement* of $\mathcal{F}$ if every $S' \in \mathcal{F}'$ is a subset of some $S \in \mathcal{F}$

Functions

Let S, S' be sets and $T \subset S$, $T' \subset S'$.

- $f : S \to S'$: function (map, transform) acting from S into S'
- $x \mapsto f(x)$: the alternative notation of f whenever S, S' are clear from the context
- $\operatorname{Im} f := \{f(x) \in S'\,;\, x \in S\}$: image (range) of f
- $f(S)$: the alternative notation of $\operatorname{Im} f$
- $f^{-1}(T') := \{x \in S\,;\, f(x) \in T'\}$: coimage of $T' \subset S'$
- $f^{-1} : \operatorname{Im} f \to 2^S$: inverse to f given by $x' \mapsto f^{-1}(\{x'\})$
- $f|_T : T \to S'$: trace (restriction) of f to $T \subset S$

Let $F : S \to 2^{S'} \setminus \{\emptyset\}$ be a set-valued (multivalued) map.

- $f : S \to S'$: *selection* of F if $f(x) \in F(x)$ for all x
- $\operatorname{card} : 2^S \to [0, +\infty]$: cardinality (number of points)
- $\operatorname{ord} \mathcal{F}$: *order (multiplicity)* of $\mathcal{F}$, i.e., $\operatorname{ord} \mathcal{F} := \sup_{x \in S}\left(\operatorname{card}\{j \in J\}\,;\, S_j \ni x\}\right)$
- $\mathbf{1}_T : S \to \{0, 1\}$: indicator (characteristic function) of $T \subset S$

Numbers and related vector spaces

- $\mathbb{N}, \mathbb{Z}, \mathbb{Q}, \mathbb{R}$: natural, integer, rational and real numbers
- $\mathbb{Z}_+, \mathbb{R}_+$: nonnegative integers and real numbers
- $(a, b), [a, b]$: open and closed intervals with endpoints $a, b \in \mathbb{R} \cup \{-\infty, \infty\}$
- $\{e_j\}_{1 \leq j \leq n}$: the standard basis of $\mathbb{R}^n$
- $\langle x, y \rangle := \sum_{i=1}^{n} x_i y_i$: the standard scalar product (sometimes also denoted by $x \cdot y$)
- $x \mapsto \|x\|_p := \left\{\sum_{i=1}^{n} |x_i|^p\right\}^{1/p}$: ℓ_p-norm (quasinorm if $0 < p < 1$)
- $\ell_p^n := (\mathbb{R}^n, \|\cdot\|_p)$

Subsets of $\mathbb{R}^n$

- $\mathbb{Z}^n := \{x \in \mathbb{R}^n\,;\, x_i \in \mathbb{Z}\text{ for all }i\}$
- $\mathbb{Z}_+^n := \{x \in \mathbb{Z}^n\,;\, x_i \geq 0 \text{ for all }i\}$

Basic Terms and Notation

- α, β, γ: elements of $\mathbb{Z}_+^n$
- $\mathbb{S}^{n-1} := \{x \in \mathbb{R}^n \,;\, \|x\|_2 = 1\}$: the unit sphere
- $(x, y), [x, y]$: open and closed intervals with endpoints $x, y \in \mathbb{R}^n$
- $Lin(\mathbb{R}^n)$, $\mathrm{Aff}(\mathbb{R}^n)$: the sets of linear and affine subspaces in $\mathbb{R}^n$
- $\mathcal{C}(\mathbb{R}^n)$: the set of nonempty bounded convex sets in $\mathbb{R}^n$
- hull: *linear hull* (span, envelope)
- aff: *affine hull*
- conv: *convex hull*
- $Q_r(x)$ (briefly Q, Q' etc.): closed cube (ℓ_∞^n ball) in $\mathbb{R}^n$ of center x and radius $r > 0$
- c_Q, r_Q: the center and radius of a cube Q
- $\mathcal{K}(\mathbb{R}^n)$: the set of closed cubes in $\mathbb{R}^n$
- $\mathcal{K}_S := \{Q_r(x) \in \mathcal{K}(\mathbb{R}^n) \,;\, x \in S \text{ and } 0 < r \le 2\,\mathrm{diam}\,S\}$
- $\mathcal{W}_S$: Whitney cover of S^c for a closed subset $S \subset \mathbb{R}^n$

Polynomials, derivatives, differences

Let $\alpha, \beta \in \mathbb{Z}_+^n$.

- $x \mapsto x^\alpha := \prod_{i=1}^n x_i^{\alpha_i}$, $x \in \mathbb{R}^n$: α-monomial (stipulation: $0^0 := 1$)
- $|\alpha| := \sum_{i=1}^n \alpha_i$, $\quad \alpha! := \prod_{i=1}^n \alpha_i!$, $\quad \binom{\alpha}{\beta} := \dfrac{\alpha!}{(\alpha - \beta)!\,\beta!}$
- $\mathcal{P}_{k,n}$: the space of polynomials in $x \in \mathbb{R}^n$ of degree k, the linear hull of α-monomials with $|\alpha| \le k$
- $D_i := \dfrac{\partial}{\partial x_i}$, $1 \le i \le n$: the i-th partial derivative
- $D_x := \sum_{i=1}^n x_i D_i$: derivative in direction $x \in \mathbb{S}^{n-1}$
- $\nabla := (D_1, \ldots, D_n)$: gradient
- $D^\alpha := \prod_{i=1}^n D_i^{\alpha_i}$: mixed α-derivative

Let f be k-times differentiable at $x \in \mathbb{R}^n$.

- $T_x^k f := \sum_{|\alpha| \leq k} \frac{(\cdot - x)^\alpha}{\alpha!} D^\alpha f(x)$: Taylor's polynomial at x of degree k

Let $f : \mathbb{R}^n \to \mathbb{R}$ and $h \in \mathbb{R}^n$.

- $\tau_h f := f(\cdot + h)$: h-shift

- $\Delta_h^k := (\tau_h - 1)^k = \sum_{j=0}^{k} (-1)^{k-j} \binom{k}{j} \tau_{jh}$: k-difference of step h

- $\Delta_h^\alpha := \prod_{i=1}^{n} \Delta_{h_i e_i}^{\alpha_i}$: (mixed) α-difference

Topological and metric spaces

Let S be a subset of a Hausdorff topological space.

- $\bar{S}$: closure, S^o: interior, $\partial S := \bar{S} \cap \bar{S^c}$: boundary

- $(\mathcal{M}, d)$: metric space with underlying set $\mathcal{M}$ and metric d (briefly, $\mathcal{M}$ whenever d is clear from the context)

  Throughout the book $\mathcal{M}$ is assumed to be nontrivial, i.e., card $\mathcal{M} > 1$.

- m, m' etc.: points of $\mathcal{M}$

- $S \subset (\mathcal{M}, d)$ (briefly, $S \subset \mathcal{M}$): a metric subspace of $(\mathcal{M}, d)$

- $(\mathcal{M}, m_0, d)$: punctured metric space ($m_0 \in \mathcal{M}$)

- $S \subset (\mathcal{M}, m_0, d)$: a metric subspace of the punctured metric space (i.e., $m_0 \in S$)

- $B_r(m_0)$, $\bar{B}_r(m_0)$: open and closed balls in $\mathcal{M}$ of *center* x_0 and radius $r > 0$;

  $\bar{B}_r(m_0) := \{m \in \mathcal{M} \,;\, d(m, m_0) \leq r\}$ does not, in general, coincide with the closure $\overline{B_r(m_0)}$ of $B_r(m_0) := \{m \in \mathcal{M} \,;\, d(m, m_0) < r\}$

Let S, S' be subsets of $(\mathcal{M}, d)$.

- $S_\varepsilon := \cup\{B_\varepsilon(m) \subset \mathcal{M} \,;\, m \in S\}$: ε-neghborhood of S

- $d(m, S) := \inf\{d(m, m') \,;\, m' \in S\}$: distance from m to S

- $Pr_S : \mathcal{M} \to 2^S$: metric projection onto S, i.e.,

$$Pr_S(m) := \{m' \in S \,;\, d(m, m') = d(m, S)\}$$

- $d(S, S') := \inf\{d(m, m') \,;\, (m, m') \in S \oplus S'\}$: distance between S and S'

Basic Terms and Notation   xxi

- $d_\mathcal{H}(S, S') := \inf\{\varepsilon > 0\,;\, S \subset S'_\varepsilon,\, S' \subset S_\varepsilon\}$: Hausdorff distance between S and S'

Lipschitz functions

Let $f : (\mathcal{M}, d) \to (\mathcal{M}', d')$.

- $L(f; \mathcal{M}, \mathcal{M}') := \sup\limits_{m_1 \neq m_2} \dfrac{d'(f(m_1), f)m_2))}{d(m_1, m_2)}$: Lipschitz constant (briefly, $L(f)$)
- $|f|_{Lip(\mathcal{M}, \mathcal{M}')}$: alternative notation for $L(f)$
- f is C-*Lipschitz*, if $L(f) \leq C$ and f is *Lipschitz* if $L(f)$ is finite
- f is C-*bi-Lipschitz embedding*, if f^{-1} exists and its *distortion* $D(f)$ satisfies $D(f) := \max\{L(f), L(f^{-1})\} \leq C$
- f is a C-*isometry* (isometry for $C = 1$) if f is a bijection with $D(f) \leq C$
- f is a *bi-Lipschitz homeomorphism* if f is a C-isometry for some C

Continuous and Lipschitz spaces

Let $\mathcal{M}, \mathcal{M}'$ be metric spaces.

- $C(\mathcal{M})$: the space of real continuous functions
- $C_u(\mathcal{M})$: the space of real uniformly continuous functions
- $C_b(\mathcal{M})$: the space of real bounded continuous functions equipped with the uniform norm
- $Lip(\mathcal{M}, \mathcal{M}')$: the space of Lipschitz maps from $\mathcal{M}$ into $\mathcal{M}'$ equipped with the seminorm $f \mapsto L(f)$
- $Lip(\mathcal{M}) := Lip(\mathcal{M}, \mathbb{R})$
- $Lip(\mathcal{M}, m_0, \mathbb{R}^n) := \{f \in Lip(\mathcal{M}, \mathbb{R}^n)\,;\, f(m_0) = 0\}$ (briefly, $Lip_0(\mathcal{M}, \mathbb{R}^n)$ if the choice of m_0 is clear)

Let $G \subset \mathbb{R}^n$ be a *domain* (open connected set) and ω belongs to the class of k-*majorant* Ω_k:

- $t \mapsto \omega_k(t; f)_G$, $t > 0$: k-modulus of continuity of $f : G \to \mathbb{R}$ (the subindex G is omitted for $G = \mathbb{R}^n$)
- $\dot{\Lambda}^{k,\omega}(G)$: the "homogeneous" space of k-Lipschitz functions on G equipped with the seminorm $f \mapsto |f|_{\Lambda^{k,\omega}(G)} := \sup\limits_{t>0} \dfrac{\omega_k(t; f)_G}{\omega(t)}$
- $\Lambda^{k,\omega}(G) \subset \dot{\Lambda}^{k,\omega}(G)$: the "nonhomogeneous" space of k-Lipschitz functions on G equipped with the norm $f \mapsto \|f\|_{\Lambda^{k,\omega}(G)} := \sup\limits_{G} |f| + |f|_{\Lambda^{k,\omega}(G)}$

- $S \mapsto E_k(S; f)$, $S \subset \mathbb{R}^n$: (local) best approximation of f by polynomials of degree $k-1$
- $\dot{\mathcal{E}}^{k,\omega}(S)$: the space of real functions on $S \subset \mathbb{R}^n$ equipped with the seminorm
$$f \mapsto \sup \left\{ \frac{E_k(S \cap Q; f)}{\omega(r_Q)} ; Q \in \mathcal{K}_S \right\}$$

Spaces of differentiable and smooth functions

- $C^\ell(G)$: the space of ℓ-times continuously differentiable real functions on a domain G
- $\dot{C}_b^\ell(G)$: the subspace of $C^\ell(G)$ defined by finiteness of the seminorm
$$f \mapsto |f|_{\dot{C}_b^\ell(G)} := \max_{|\alpha|=\ell} \sup_G |D^\alpha f|$$
- $\dot{C}_u^\ell(G)$: the subspace of $\dot{C}_b^\ell(G)$ consisting of functions with uniformly continuous higher derivatives
- $C_b^\ell(G)$: the subspace of $\dot{C}_b^\ell(G)$ defined by finiteness of the norm
$$f \mapsto \|f\|_{C_b^\ell(G)} := \sup_G |f| + |f|_{\dot{C}_b^\ell(G)}$$
- $C^\ell \dot{\Lambda}^{k,\omega}(G)$: the subspace of $C^\ell(G)$ consisting of functions whose higher derivatives belong to $\dot{\Lambda}^{k,\omega}(G)$ equipped by the seminorm
$$f \mapsto |f|_{C^\ell \dot{\Lambda}^{k,\omega}(G)} := \max_{|\alpha|=\ell} |D^\alpha f|_{\Lambda^{k,\omega}(G)}$$
- $C^\ell \Lambda^{k,\omega}(G)$: the subspace of $C^k \dot{\Lambda}^{k,\omega}(G)$ defined by the finiteness of the norm
$$f \mapsto \|f\|_{C^\ell \Lambda^{k,\omega}(G)} := \sup_G |f| + |f|_{C^\ell \Lambda^{k,\omega}(G)}$$
- $J^\ell \dot{\Lambda}^{k,\omega}(G)$ the space of ℓ-jets $\vec{f} := \{f_\alpha\}_{|\alpha| \leq \ell}$ on G defined by finiteness of the seminorm
$$\vec{f} \mapsto |\vec{f}|_{J^\ell \dot{\Lambda}^{k,\omega}(G)} := \max_{|\alpha|=\ell} |f_\alpha|_{\Lambda^{k,\omega}(G)}$$
- $J^\ell \Lambda^{k,\omega}(G)$: the subspace of $J^\ell \dot{\Lambda}^{k,\omega}(G)$ defined by finiteness of the norm
$$\vec{f} \mapsto \|\vec{f}\|_{J^\ell \lambda^{k,\omega}(G)} := \max_{|\alpha| \leq \ell} |f_\alpha| + |\vec{f}|_{J^\ell \Lambda^{k,\omega}(G)}$$

Let X be one of the above introduced functions spaces on $\mathbb{R}^n$; let $S \subset \mathbb{R}^n$ be closed.

- $X|_S := \{f : S \to \mathbb{R} \,;\, f = g|_S \text{ for some } g \in X\}$: the trace of X to S equipped with the *trace seminorm*

$$f \mapsto |f|_{X|_S} := \inf\{|g|_X \,;\, g|_S = f\}$$

if X is seminormed and the analogous trace norm if X is normed

Extension constants

Let $\mathcal{M}, \mathcal{M}'$ be metric spaces and S be a metric subspace of $\mathcal{M}$.

- $L_{ext}(f)$: Lipschitz extension constant for $f : S \to \mathcal{M}'$ (the trace norm of f in $Lip(\mathcal{M}, \mathcal{M}')|_S$)
- $\Lambda(S, \mathcal{M}; \mathcal{M}') := \sup\left\{\dfrac{L_{ext}(f)}{L(f)} \,;\, f \in Lip(S, \mathcal{M}')\right\}$: (local) Lipschitz extension constant
- $\Lambda(\mathcal{M}, \mathcal{M}') := \sup\{\Lambda(S, \mathcal{M}, \mathcal{M}') \,;\, S \subset \mathcal{M}\}$: (global) Lipschitz extension constant

Let X be a Banach space.

- $Ext(S, X)$: the space of all bounded linear extension operators (*simultaneous extensions*) from $Lip(S, X)$ into $Lip(\mathcal{M}, X)$
- $\lambda(S, \mathcal{M}; X) := \inf\{\|E\| \,;\, E \in Ext(S, X)\}$: (local) linear Lipschitz extension constant
- $\lambda(\mathcal{M}, X) := \sup\{\lambda(S, \mathcal{M}; X) \,;\, S \subset \mathcal{M}\}$: (global) linear Lipschitz extension constant
- $\lambda(\mathcal{M}) := \lambda(\mathcal{M}, \mathbb{R})$

Let X be one of the above introduced spaces of differentiable or smooth functions on $\mathbb{R}^n$ (*smoothness spaces*) and Σ be a class of subsets in $\mathbb{R}^n$.

- $\delta_N(f; S; X) := \sup\{|f|_{X|_{S'}} \,;\, S' \subset S \text{ and card } S' \leq N\}$: (the seminorm here and below is replaced by the norm if X is normed)
- $\mathcal{F}_\Sigma(X)$: the *finiteness constant* of X with respect to Σ (the minimal N for which $X|_S$ coincides with linear space $\{g \in C(S) \,;\, \delta_N(g; S; X) < \infty\}$)
- $\mathcal{FP}(\Sigma)$: the class of all smoothness spaces with $\mathcal{F}_\Sigma(X) < \infty$
- $\gamma_\Sigma(X)$: the extension constant for X with respect to Σ equals

$$\sup\{|g|_{X|_S} \,;\, \delta_N(g; S; X) \leq 1,\ S \in \Sigma\}$$

- $\mathcal{FP}_u(\Sigma)$: the subclass of $\mathcal{FP}(\Sigma)$ with $\gamma_\Sigma(X) < \infty$

For Σ being the class of all nonempty closed subsets the symbol Σ in all these notations is omitted.

Part I

Classical Extension-Trace Theorems and Related Results

> ... when one has lost himself in the flower gardens of abstract algebra or topology, as so many of us do nowadays, one becomes aware here once more, perhaps with some surprise, of how mighty and fruitbearing an orchard is classical analysis.
>
> Hermann Weyl

Chapter 1

Continuous and Lipschitz Functions

Chapter 1 is intended to give the reader a fairly clear introduction to Part 3 of the book (Chapters 5–8) devoted to extensions of Lipschitz maps between metric spaces. This will be done on the basis of methods and results regarded as "classical" in this rather young field of modern analysis. The matter on continuous functions presented is used to motivate the problems and methods that will be developed in Part 3. For instance, topological theory relating (covering) dimension and continuous extensions of maps into spheres (Theorem 1.9 and Appendix A) has a metric counterpart (Chapter 6) with Nagata dimension substituted for the covering one.

In the process, certain preliminary concepts and terminology will be introduced. The reader is referred to the list of notations for the standard symbols used in this and consequent parts.

Although we have in mind an audience with prior exposure in modern analysis, a rather good knowledge of point-set topology, linear algebra and calculus should suffice. The more advanced results (when needed) will be carefully stated within the text itself and discussed and referred to in the comments and appendices that will appear at the end of the chapter. So the reader should be able to read it (as well as the entire book) with minimal prerequisites.

Continuous Functions

This half of the chapter is devoted to extension theorems for continuous functions defined on closed subsets of metric spaces. The first result of this type was the classical Lebesgue–Brouwer theorem asserting that every bounded continuous function on a closed subset of $\mathbb{R}^n$ can be extended to a continuous function on $\mathbb{R}^n$ without increasing its uniform norm. This statement was generalized by

Tietze to the case of metric spaces; then Urysohn proved the analogous theorem for normal topological spaces[1]. The latter result cannot be sharpened, since every Hausdorff topological space possessing this extension property is normal. The Urysohn method is based on his celebrated lemma and subsequent elegant iteration procedure completing the extension construction. The basic elements of this construction can be seen in the proofs of the continuous selection theorems by Bartle–Graves and by E. Michael presented below. To formulate these and subsequent results we have to introduce several concepts and notations related to metric spaces. Most of this information is well known to the readers and does not require complete and thorough presentation.

1.1 Notation and definitions

Throughout the book the symbol $(\mathcal{M}, d)$ will designate a nontrivial *metric space* with the *underlying set* $\mathcal{M}$ of card $\mathcal{M} > 1$ and *metric* $d : \mathcal{M} \times \mathcal{M} \to \mathbb{R}_+$. Thus, d is a nonnegative and symmetric function satisfying the *triangle inequality* and vanishing exactly at points of the set $\Delta(\mathcal{M}) := \{(m, m)\,;\, m \in \mathcal{M}\}$. Hereafter we will write $\mathcal{M}$ instead of $(\mathcal{M}, d)$, if the metric can be recovered from the context.

Given a set $S \subset \mathcal{M}$, we will use the standard notation $\overline{S}, S^\circ, \partial S$ and S^c for *the closure, interior, boundary* and *complement* of S in $\mathcal{M}$. In particular,

$$\partial S = \overline{S} \cap \overline{S^c} \quad \text{and} \quad S^\circ = \overline{S} \setminus \partial S.$$

Further, an *open ball* $B_r(m_0)$ of radius $r > 0$ and center $m_0 \in \mathcal{M}$ is defined by

$$B_r(m_0) := \{m \in \mathcal{M}\,;\, d(m, m_0) < r\},$$

and the *closed ball* of the same center and radius is given by

$$\overline{B}_r(m_0) := \{m \in \mathcal{M}\,;\, d(m, m_0) \leq r\}.$$

Note that this set is not, generally speaking, the closure of $B_r(m_0)$ as well as the *sphere*

$$S_r(m_0) := \{m \in \mathcal{M}\,;\, d(m, m_0) = r\}$$

is not necessarily the boundary of $\overline{B}_r(m_0)$. Finally, the distance $d(m_0, S)$ from a point m_0 to a subset S in $\mathcal{M}$ is defined by

$$d(m_0, S) := \inf\{d(m_0, m)\,;\, m \in S\}.$$

Now let $(\mathcal{M}', d')$ be another metric space, and $F := F(\mathcal{M}, \mathcal{M}')$ be a set of maps from $\mathcal{M}$ to $\mathcal{M}'$. Given a subset $S \subset \mathcal{M}$, we define the *trace* $F|_S$ of F to S by

$$F|_S := \{f|_S\,;\, f \in F\}. \tag{1.1}$$

---

[1] A topological space is *normal* if all its single point subsets are closed and every pair of disjoint closed sets can be separated by disjoint open neighborhoods.

Hereafter $f|_S$ is the *restriction* of f to S, i.e., a map from S to $\mathcal{M}'$ given by

$$(f|_S)(m) := f(m), \quad m \in S. \tag{1.2}$$

If F is a normed linear space, we equip $F|_S$ with the *trace norm*

$$\|f\|_{F|_S} := \inf\{\|g\|_F \; ; \; g|_S = f\}. \tag{1.3}$$

In this case $F|_S$ is called a *trace space*.

The function spaces F of this section are introduced by

Definition 1.1. (a) Let X be a Banach space; then $C_b(\mathcal{M}, X)$ is the Banach space of bounded continuous maps $f : \mathcal{M} \to X$ equipped with the norm

$$\|f\| := \sup\{\|f(m)\|_X \; ; \; m \in \mathcal{M}\}. \tag{1.4}$$

(b) $C_u(\mathcal{M}, X)$ is the closed subspace of $C_b(\mathcal{M}, X)$ consisting of bounded uniformly continuous maps $f : \mathcal{M} \to X$.

If X is the real line endowed with the standard metric, we use the notation $C_b(\mathcal{M})$ and $C_u(\mathcal{M})$.

Remark 1.2. (a) By the Heine–Cantor theorem,

$$C_u(\mathcal{M}, X) = C_b(\mathcal{M}, X)$$

provided that $\mathcal{M}$ is compact.

(b) It is well known and readily seen that the trace spaces introduced above are Banach spaces.

1.2 Extension and trace problems: formulations and examples

The purpose of this section is to outline the main problems studied in this book. We use general formulations covering the variety of these problems; then accompanying instructive examples are introduced and briefly discussed. Every example is a special case of the respective basic result presented here and in one of the subsequent parts of the book; at this stage the reader may concentrate on these examples.

Let $F := F(\mathcal{M}, \mathcal{M}')$ be a class of maps between metric spaces $\mathcal{M}$ and $\mathcal{M}'$, and S be a subset of $\mathcal{M}$.

Main Problem. *Find a complete description of the trace of F to S.*

In other words, we are looking for necessary and sufficient conditions on a map $f : S \to \mathcal{M}'$ to belong to $F|_S$.

This problem can be divided into two subproblems of independent interest.

Trace Problem. *Find a characterization of the elements of $F|_S$.*

That is to say, we now use characteristics inherited from F in order to give as complete a description of $F|_S$ as possible. This, in particular, requires more detailed information about F; the results obtained in this direction are sometimes even more important than the final solution. At this stage we obtain only necessary conditions for a map to belong to $F|_S$. Let us denote the set of maps satisfying these necessary conditions by $F(S)$; then we have

$$F|_S \subset F(S). \tag{1.5}$$

This might solve positively the Main Problem, if we verified that the embedding (1.5) is in fact an equality. The latter can be established by solving the next problem. For its formulation we use some classes of maps, say $G(S)$ and G, acting from S to $\mathcal{M}'$ and from $\mathcal{M}$ to $\mathcal{M}'$, respectively. These classes may be different from $F(S)$ and F.

Extension Problem. *Prove that every $f \in G(S)$ admits an extension to the whole of $\mathcal{M}$ such that the extended function belongs to G.*

To solve this problem we must either construct an extension algorithm or apply a suitable existence theorem, e.g., Brouwer's fix point theorem or Helly's theorem. The first approach is usually more difficult but yields much more information and has an essentially wider range of applications. If, in particular, the classes introduced coincide, respectively, with $F(S)$ and F, then the solution to the Extension Problem leads to equality in (1.5) and therefore solves the Main Problem.

In many cases, however, such extensions exist only for maps from a subclass smaller than $F(S)$. Preserving for this subclass the same notation $G(S)$ we then have

$$G(S) \subset F|_S \subset F(S). \tag{1.6}$$

These embeddings could be seen as an approximate solution to the Main Problem. However, the methods developed for their derivation and results obtained in the process are frequently of great value for rather distant areas of analysis.

One version of the Extension Problem concerns the case of $F = F(\mathcal{M}, X)$ where X is a Banach space. Now $F|_S$ is also Banach and the Bartle–Graves theorem (see Theorem 1.6 below) states that for a closed S there exists a *nonlinear* continuous extension operator $E : F|_S \to F$. In this setting, it is natural to ask the following question:

Simultaneous Extension Problem. *Does there exist a linear bounded operator $E : F|_S \to F$ such that $(Ef)|_S = f$?*

We now illustrate all these problems by examples.

1.2.1 Example: Continuous functions

Suppose that the class F is the space $C_b(\mathcal{M})$. Then for any $S \subset \mathcal{M}$,

$$C_b(\mathcal{M})\big|_S \subset C_b(S)$$

but these spaces are in general distinct. If S is closed, the Tietze–Urysohn extension theorem yields the inverse embedding. Hence, for closed subsets the solution to the Main Problem is as follows.

A function $f : S \to \mathbb{R}$ belongs to the trace space $C_b(\mathcal{M})\big|_S$ if and only if f is continuous.

However, for unclosed subsets the solution requires more involved characteristics. Invoking again the Tietze–Urysohn theorem we easily derive the required description in the following form.

A function $f : S \to \mathbb{R}$ belongs to $C_b(\mathcal{M})\big|_S$ if and only if it satisfies the conditions:

(i) *At every point m of the closure $\overline{S}$ there exists the limit*

$$\bar{f}(m) := \lim_{m' \to m} f(m').$$

(ii) *The function $\bar{f} : \overline{S} \to \mathbb{R}$ is continuous.*

Note that Urysohn's extension operator is nonlinear. But the required extension can be achieved by linear bounded operators; this clearly solves the Simultaneous Extension Problem for this case (see Theorem 1.8 by Borsuk–Dugundji).

Since the Urysohn operator preserves the supremum norm, the space $C_b(\mathcal{M})$ in the above result can be replaced by an arbitrary closed ball of $C_b(\mathcal{M})$.

For sets of continuous maps the situation is more complicated. We confine ourselves to the case of continuous maps from a separable metric space $\mathcal{M}$ to the unit sphere $\mathbb{S}^n$. The classical Hurewicz theorem states that, for a closed subset $T \subset \mathcal{M}$,

$$C(\mathcal{M}, \mathbb{S}^n)\big|_T = C(T, \mathbb{S}^n),$$

if and only if T is of dimension at most n, see Section 1.5.

1.2.2 Example: Uniformly continuous functions

Suppose F is now the space $C_u(\mathcal{M})$ of uniformly continuous bounded functions. It is evident that

$$C_u(\mathcal{M})\big|_S \subset C_u(S).$$

In fact, these spaces coincide and $C_u(S)$ is the required description of the trace space for S. We explain this fact later, see Corollary 1.29.

It is worth noting that $C_u(\mathcal{M})\big|_S$ is a Banach algebra. The reader versed in Banach algebras could invoke this theory to describe its space of maximal ideals and so on.

1.2.3 Example: Continuously differentiable functions on $\mathbb{R}^n$

Let us consider first the space $C_b^1(\mathbb{R})$ of continuously differentiable functions $f : \mathbb{R} \to \mathbb{R}$ such that

$$|f|_{C_b^1(\mathbb{R})} := \sup\{|f'(x)|\ ;\ x \in \mathbb{R}\}$$

is finite. In this case it suffices to consider only closed $S \subset \mathbb{R}$; then a function $f : S \to \mathbb{R}$ from the trace space $C_b^1(\mathbb{R})\big|_S$ inherits the following property.

For every limit point of S the limit

$$f^{[1]}(x) := \lim_{x' \to x} \frac{f(x') - f(x)}{x' - x}$$

exists and $f^{[1]}$ is bounded on the set of limit points.

This property determines the space of functions on S, which we denote temporarily by $C_b^{[1]}(S)$. Hence,

$$C_b^1(\mathbb{R})\big|_S \subset C_b^{[1]}(S)$$

but these spaces are in general distinct. The Whitney extension theorem for $C^k(\mathbb{R})$ (see Theorem 2.47 below) implies the desired description of the trace space in the following way. Let $f[x_1, x_2, x_3]$ be the second divided difference of f over the subset $\{x_1, x_2, x_3\} \subset \mathbb{R}$. By the Mean Value Theorem,

$$\lim_{x_2, x_3 \to x_1} (x_3 - x_2) f[x_1, x_2, x_3] = 0 \qquad (1.7)$$

for every $f \in C_b^1(\mathbb{R})$. Denoting by $C_b^1(S)$ the space of functions $f : S \to \mathbb{R}$ satisfying condition (1.7) with the points x_i in S, $i = 1, 2, 3$, we therefore have

$$C_b^1(\mathbb{R})\big|_S \subset C_b^1(S).$$

In turn, Whitney's construction gives a *linear* extension operator from $C_b^1(S)$ to $C_b^1(\mathbb{R})$. This implies the equality

$$C_b^1(\mathbb{R})\big|_S = C_b^1(S)$$

and solves the Simultaneous Extension Problem for this case.

Let us now turn to functions of n variables (the reader is referred to the last chapter of Volume II for the proofs of the results outlined below). Since the definition of the second divided difference uses three valuations, i.e., functionals of the form $\delta_x(f) := f(x)$, it is natural to find out how many valuations may be involved in the description of $C_b^1(\mathbb{R}^n)\big|_S$. The striking answer is $3 \cdot 2^{n-1}$. Therefore it seems to be unrealistic to find a simple constructive characterization of $C_b^1(\mathbb{R}^n)\big|_S$ with n sufficiently large.

1.2. Extension and trace problems: formulations and examples

There is another striking difference between the one-dimensional and multidimensional cases. Actually, Whitney's extension operator $W : C_b^1(\mathbb{R})\big|_S \to C_b^1(\mathbb{R})$ is not only linear but also *universal* in a sense. In particular, it linearly extends functions f from $C_u^1(\mathbb{R})\big|_S$ as well; here $C_u^1(\mathbb{R})$ is the space of functions with uniformly continuous first derivatives. But for $n = 2$ there exists a closed subset $S \subset \mathbb{R}^2$ such that there are *no* linear extension operators from $C_u^1(\mathbb{R}^2)\big|_S$ to $C_u^1(\mathbb{R}^2)$. On the other hand, the Glaeser theorem states that a simultaneous extension operator from $C_b^1(\mathbb{R}^n)\big|_S$ to $C_b^1(\mathbb{R}^n)$ does exist!

1.2.4 Example: BMO and Sobolev spaces

Suppose F is the space of functions of *bounded mean oscillation* (BMO). Recall that this space consists of locally integrable functions $f : \mathbb{R}^n \to \mathbb{R}$ such that

$$|f|_* := \sup_Q \frac{1}{|Q|} \int_Q |f - f_Q| dx \tag{1.8}$$

is finite. Here Q denotes a cube of $\mathbb{R}^n$ and $f_Q := \frac{1}{|Q|} \int_Q f dx$, where $|Q|$ stands for the volume of Q.

The trace space $\mathrm{BMO}\,|_S$ is nontrivial only for S of positive Lebesgue measure. For an arbitrary open subset S of $\mathbb{R}^n$ we can define the space $\mathrm{BMO}(S)$ using (1.8) with cubes Q contained in S. It is obvious that $\mathrm{BMO}(\mathbb{R}^n)\big|_S \subset \mathrm{BMO}(S)$ for such S. The first natural problem is as follows.

Characterize open subsets S such that

$$\mathrm{BMO}(\mathbb{R}^n)\big|_S = \mathrm{BMO}(S). \tag{1.9}$$

The answer is given by the P. Jones theorem [Jon-1980]. We consider here only a special case concerning simply connected open subsets $S \subset \mathbb{R}^2$. For such S, equality (1.9) holds if and only if ∂S is a *quasicircle* (the image of the unit circle under a quasiconformal map). A quasicircle may be of a very complicated structure, e.g., a set of Hausdorff dimension more than $\dim \partial S\, (= 1)$.

We obtain the very same answer for the Sobolev space $W_2^1(\mathbb{R}^2)$ consisting of square integrable functions on the plane whose distributional first derivatives are also square integrable (the Goldstein–Vodop'yanov theorem, see subsection 2.6.1 below). Since, by definition, Sobolev spaces are quite different from BMO, this similarity seems to be very unusual. In fact, there exists a definition of all these spaces revealing their inner resemblance.

Finally, let us mention Th. Wolff's deep theorem [Wo-1980] giving, in particular, a complete characterization of the trace space $\mathrm{BMO}(\mathbb{R}^n)\big|_S$ for arbitrary S of positive measure. This result connects the Trace Problem for BMO with the theory of A_p-weights.

1.3 Continuous selections

There is a natural generalization of the continuous extension problem which can be motivated as follows.

Consider the trivial "extension" of a continuous function $f : S \to \mathbb{R}$ to a function $\bar{f}$ from the ambient metric space $\mathcal{M}$ to convex subsets of $\mathbb{R}$ given by

$$\bar{f}(m) = \begin{cases} \{f(m)\} & \text{if } m \in S, \\ \mathbb{R} & \text{if } m \in \mathcal{M} \setminus S. \end{cases}$$

If we could select from every set $\bar{f}(m)$ a number, say $\varphi(m)$, such that the map $m \mapsto \varphi(m)$ were continuous, this would yield the required continuous extension of f.

Now, further generalization is clear. In order to formulate the corresponding result we consider a map (called a *multivalued or set-valued* function) F from a metric space $\mathcal{M}$ to nonempty subsets of a Banach space Y. This map is said to be *lower semicontinuous* if for every open set $U \subset Y$ its *preimage*

$$F^{-1}(U) := \{m \in \mathcal{M} \; ; \; F(m) \cap U \neq \phi\}$$

is open.

Theorem 1.3 (E. Michael). *Assume that F is a lower semicontinuous map from $\mathcal{M}$ to nonempty closed convex subsets of Y. Then F admits a continuous selection, i.e., a continuous function $f : \mathcal{M} \to Y$ such that*

$$f(m) \in F(m) \quad \text{for all} \quad m \in \mathcal{M}.$$

Proof. Given $r > 0$, introduce the open cover $\{F^{-1}(B_r(y))\}_{y \in \mathcal{M}}$ of $\mathcal{M}$ by the preimages of open balls in Y of radius r. Let $\{p_\alpha\}_{\alpha \in A}$ be a *continuous partition* of unity subordinate to this cover, see subsection 3.1.3 below for its definition. Pick for each α a point $y(\alpha) \in Y$ such that the support of p_α, i.e., the closed set

$$\operatorname{supp} p_\alpha := \overline{\{m \in \mathcal{M} \; ; \; p_\alpha(m) \neq 0\}}, \tag{1.10}$$

is contained in $F^{-1}(B_r(y(\alpha)))$. Now let $F_r : \mathcal{M} \to Y$ be the continuous function given by

$$f_r = \sum_\alpha p_\alpha y(\alpha).$$

Since $p_\alpha \geq 0$ and $\sum_\alpha p_\alpha = 1$, and every $m \in \mathcal{M}$ is covered by only a finite number of the supports (1.10), $f_r(m)$ is a convex combination of finitely many points $y(\alpha)$. Moreover, all these points lie in the r-neighborhood of $F(m)$, i.e., in the convex set $\{y \in Y \; ; \; d(y, F(m)) < r\}$. Therefore $f_r(m)$ is also a point of this convex set. In other words, f_r is an *r-approximation* to the required continuous selection (briefly, *r-selection*).

1.3. Continuous selections

Given now this r-selection, we construct an $\frac{r}{2}$-selection that differs from f_r by at most $\frac{3}{2}r$. To this end define a new multivalued function $\widetilde{F}$ by

$$\widetilde{F}(m) := F(m) \cap \bar{B}_r(f_r(m)), \quad m \in \mathcal{M}. \tag{1.11}$$

Then $\widetilde{F}$ meets the conditions of the theorem. In fact, $\widetilde{F}(m)$ is nonempty by the definition of f_r, convex and closed. To show that $\widetilde{F}$ is lower semicontinuous, one should check that the preimage $V := \widetilde{F}^{-1}(U)$ is open for every open set $U \subset Y$. To accomplish this, note that $F(m) \cap B_r(f(m))$ is a convex nonempty set whose closure coincides with $\widetilde{F}(m)$, see (1.11). Therefore the point m belongs to V, if and only if

$$F(m) \cap B_r(f_r(m)) \cap U \neq \emptyset.$$

In other words, $\widetilde{F}^{-1}(U) = F^{-1}(U \cap B_r(f_r(m)))$. Since F is lower semicontinuous, this implies the openness of V.

Now let $f_{\frac{r}{2}}$ be an $\frac{r}{2}$-selection of $\widetilde{F}$. By the definition of $\widetilde{F}$, the point $f_{\frac{r}{2}}(m)$ lies in the ball of radius $\frac{3r}{2}$ centered at $f_r(m)$ and, in addition, $f_{\frac{r}{2}}$ is an $\frac{r}{2}$-selection of F, see (1.11).

Finally, let $\{f_r, f_{\frac{r}{2}}\}$ be the introduced pair of approximate selections. Taking successively $r = 1, 2^{-1}, 2^{-2}, \ldots$, one defines the Cauchy sequence $\{f_{2^{-i}}\}_{i \in \mathbb{Z}}$ of continuous functions from $\mathcal{M}$ to Y such that $f_{2^{-i}}(m)$ lies in the $3 \cdot 2^{-i}$-neighborhood of $F(m)$. The limit of this sequence is clearly the required continuous selection of F. $\square$

We now present two important consequences of Michael's result; the first of them is a generalization of the Tietze–Urysohn theorem for the case of vector-valued functions

Theorem 1.4. *Let S be a closed subspace of a metric space $\mathcal{M}$ and Y be a Banach space. Then for every $f \in C_b(S, Y)$ there is a continuous extension $\tilde{f} \in C_b(\mathcal{M}, Y)$ such that*

$$\tilde{f}(\mathcal{M}) \subset \overline{\operatorname{conv} f(S)}.$$

Proof. Set, for $m \in \mathcal{M}$,

$$F(m) := \begin{cases} \{f(m)\} & \text{if } m \in S, \\ \overline{\operatorname{conv} f(S)} & \text{otherwise.} \end{cases} \tag{1.12}$$

Then F maps points of $\mathcal{M}$ to nonempty convex closed subsets of Y. We now check that F is lower semicontinuous. For this, pick an open set $U \subset Y$ and consider its preimage $V := \{m \in \mathcal{M} \,;\, F(m) \cap U \neq \emptyset\}$. By (1.12) this set is empty whenever $U \cap \overline{\operatorname{conv} f(S)} = \emptyset$, and equals $f^{-1}(U) \cup (\mathcal{M} \setminus S)$, otherwise. Since f is continuous and S is closed, V is open in both cases. Hence F meets the conditions of Theorem 1.3 and therefore there exists a continuous selection $\tilde{f} : \mathcal{M} \to Y$ for F. By the definition of F, this $\tilde{f}$ is the desired extension of f. $\square$

The following version of the Tietze–Urysohn theorem is of use in many applications.

Corollary 1.5. *Let f be a continuous map from a closed subspace S of $\mathcal{M}$ into another metric space $\mathcal{M}'$. Assume that $\mathcal{M}'$ is homeomorphic[2] to a closed convex subset of a Banach space. Then f admits a continuous extension to $\mathcal{M}$.*

Proof. Let φ be a homeomorphism of $\mathcal{M}'$ onto a closed convex subset C of a Banach space Y. Then $\varphi \circ f \in C(S, Y)$ and, by Theorem 1.4, there exists $\tilde{f} \in C(\mathcal{M}, Y)$ that agrees with $\varphi \circ f$ on S and such that $\tilde{f}(\mathcal{M}) \subset C$. Then $\varphi^{-1} \circ \tilde{f} : \mathcal{M} \to \mathcal{M}'$ is clearly the required extension of f. □

The second consequence was published earlier than Theorem 1.3.

Theorem 1.6 (Bartle–Graves). *Suppose that $t : Y \to X$ is a continuous linear surjection of a Banach space Y onto a Banach space X. Then there exists a continuous map $f : X \to Y$ such that the composition $t \circ f$ is the identity map.*

Proof. Define the multivalued function F by

$$F(x) := t^{-1}(\{x\}), \quad x \in X.$$

Then $F(x)$ is a closed and convex (even affine) subset of Y. Moreover, for every open set $U \subset Y$ the preimage $F^{-1}(U)$ equals, by definition, $t(U)$, while the latter set is open by the Banach open map theorem, see [DS-1958, Sec. II.2]. Hence, F meets the assumptions of Theorem 1.3, and therefore admits a continuous selection $f : X \to Y$. By the definition of F, this selection satisfies $t \circ f = id_X$. □

Remark 1.7. The map F of this proof can be regarded as a surjection of X onto the quotient space $Y/t^{-1}(0)$. Therefore it is natural to ask whether the selection f could be chosen to be linear. The answer is negative as the next consideration shows.

Let Y be a closed uncomplemented subspace of a Banach space X. One can take, e.g., $X = \ell^\infty$, the space of all bounded sequences $\{x_n\}_{1 \leq n \leq \infty}$ with norm $\sup_n |x_n|$, and $Y = c_0$, the subspace of ℓ_∞ containing sequences converging to zero, see [Ph-1940]. Let $\pi : X \to X/Y$ be the canonical surjection on the factor space. If the linear version of Theorem 1.6 is true, there exists a linear continuous operator $\ell : X/Y \to X$ such that composition $\pi \circ \ell = id_{X/Y}$. Then the map $id_X - \ell \circ \pi$ is a continuous projection of X onto Y, a contradiction.

1.4 Simultaneous continuous extensions

Unlike the Bartle–Graves result, the Banach-valued version of the Tietze–Urysohn Theorem admits a linearization. Namely, the question raised in the Si-

---

[2] Metric spaces $\mathcal{M}_1$ and $\mathcal{M}_2$ are *homeomorphic* if there exists a bijection $\varphi : \mathcal{M}_1 \to \mathcal{M}_2$ such that φ and φ^{-1} are continuous.

1.4. Simultaneous continuous extensions

multaneous Extension Problem, see Section 1.2, can be answered affirmatively for this case. The corresponding result is presented as

Theorem 1.8 (Borsuk–Dugundji). *Let S be a closed subspace of a metric space $\mathcal{M}$ and Y be a Banach space. There exists a linear operator E from $C_b(S,Y)$ to $C_b(\mathcal{M},Y)$ of norm 1, so that for every $f \in C_b(S,Y)$ the map Ef is an extension of f to the whole of $\mathcal{M}$. Moreover,*

$$(Ef)(\mathcal{M}) \subset \operatorname{conv} f(S).$$

Proof. Cover $S^c := \mathcal{M}\setminus S$ by open balls B_m, $m \in \mathcal{M}\setminus S$, where

$$B_m := B_{r_m}(m) \quad \text{and} \quad r_m := \frac{1}{3} d(m, S). \tag{1.13}$$

Let $\{p_\alpha\}_{\alpha \in A}$ be a partition of unity subordinate to this cover, see subsection 3.1.3 for its definition. For each α pick points $m_i(\alpha)$, $i = 1, 2$, so that

$$m_1(\alpha) \in S \quad \text{and} \quad m_2(\alpha) \in \operatorname{supp} p_\alpha$$

and

$$d\bigl(m_1(\alpha), m_2(\alpha)\bigr) < 2d\bigl(m_2(\alpha), S\bigr). \tag{1.14}$$

Such points do exist, since $\operatorname{supp} p_\alpha$ is contained in some ball B_m contained in S^c, see (1.13).

Introduce now the required operator by setting

$$Ef := \begin{cases} f & \text{on } S, \\ \sum_{\alpha \in A} f(m_1(\alpha)) p_\alpha & \text{on } S^c. \end{cases}$$

The function Ef is clearly continuous on S and on S^c. We now show that it is also continuous at each point of the boundary ∂S. Let $m \in \partial S$ and U_m be a convex open neighborhood of $f(m)$ (in Y). Pick $\delta > 0$ so that

$$f\bigl(S \cap B_\delta(m)\bigr) \subset U_m. \tag{1.15}$$

We check that for $m' \in S^c$,

$$d(m, m') < \frac{\delta}{6} \Rightarrow (Ef)(m') \in U_m. \tag{1.16}$$

This clearly implies the continuity of Ef at points of ∂S and, hence, on all of $\mathcal{M}$.

To establish (1.16), we first note that $(Ef)(m')$ is a convex combination of the points $f(m_1(\alpha))$ with α running over a finite subset $A_0 \subset A$ given by

$$A_0 := \{\alpha \in A \ ; \ m' \in \operatorname{supp} p_\alpha\}.$$

Hence, it suffices to prove that under the assumption of (1.16)
$$f(m_1(\alpha)) \in U_m \text{ for } \alpha \in A_0.$$

To this end, given $\alpha \in A_0$, choose $m(\alpha) \in S^c$ so that $\operatorname{supp} p_\alpha \subset B_{m(\alpha)}$. Then $m' \in B_{m(\alpha)}$ and this inclusion, together with (1.13) and (1.16), imply that

$$d(m(\alpha),S) \leq d(m(\alpha),m) \leq d(m(\alpha),m') + d(m',m) \leq \frac{1}{3} d(m(\alpha),S) + \frac{\delta}{6}.$$

These inequalities, in turn, imply that $d(m(\alpha),S) < \frac{\delta}{4}$ and $d(m(\alpha),m) < \frac{\delta}{4}$. The latter inequality and (1.13) then yield

$$d(m_2(\alpha),m) \leq d(m_2(\alpha),m(\alpha)) + d(m(\alpha),m) < \frac{1}{3} d(m(\alpha),S) + \frac{\delta}{4} < \frac{\delta}{3}.$$

Finally, from here and (1.14) one obtains

$$d(m_1(\alpha),m) \leq d(m_1(\alpha),m_2(\alpha)) + d(m_2(\alpha),m)$$
$$< 2d(m_2(\alpha),S) + \frac{\delta}{3} < 2d(m_2(\alpha),m) + \frac{\delta}{3} < \delta.$$

Hence, the point $m_1(\alpha)$ belongs to $B_\delta(m) \cap S$, and therefore $f(m_1(\alpha))$ is in U_m, as it is required.

Consequently, E is a linear extension operator from $C_b(S,Y)$ to $C_b(\mathcal{M},Y)$ such that $(Ef)(\mathcal{M}) \subset \operatorname{conv}(f(S))$. Since $(Ef)(1) = 1$, its norm is clearly 1. □

1.5 Extensions of continuous maps acting between metric spaces

For continuous maps with topologically nontrivial target spaces, the extension problem is much more complicated (and interesting). The simplest target space of this kind is the unit n-sphere

$$\mathbb{S}^n := \left\{ x \in \mathbb{R}^{n+1} \; ; \; \sum_{i=1}^{n+1} x_i^2 = 1 \right\}.$$

In general, continuous maps to $\mathbb{S}^n$ are not extendable from a closed subset to the whole metric space. The classical example is the identity map from the boundary of the unit ball B^{n+1} of $\mathbb{R}^{n+1}$ to $\mathbb{S}^n$ (identifying with ∂B^{n+1}) which has no such an extension. This fact which is equivalent to the Brouwer fixed point theorem will be discussed in example (c) following Definition 1.10.

The next classical result shows that the only obstacle to desired extensions is the topological dimension of the set to which a map has to be extended.

1.6. Absolute metric retracts

Theorem 1.9 (Hurewicz). *Let S be a closed subspace of a separable metric space $\mathcal{M}$. Then the following is true.*

(i) *If $\dim(S^c) \leq n$ $(n \geq 0)$, then*

$$C_b(\mathcal{M}, \mathbb{S}^n)\big|_S = C_b(S, \mathbb{S}^n). \tag{1.17}$$

(ii) *Equality in (1.17) holds for every nonempty closed S if and only if $\dim \mathcal{M} \leq n$.*

For more information on topological dimension and its relation to continuous extensions of maps to $\mathbb{S}^n$ see Appendix A.

1.6 Absolute metric retracts

We now introduce two basic concepts of topology adapted to the metric space context and discuss their relations to the extension problems studied in this part of Chapter 1.

Definition 1.10 (Borsuk). A subspace $\mathcal{M}_0 \subset \mathcal{M}$ is said to be a *retract* of $\mathcal{M}$, if there is a continuous map $r : \mathcal{M} \to \mathcal{M}_0$ such that $r|_{\mathcal{M}_0}$ is the identity map. This map is called a *retraction* of $\mathcal{M}$ onto $\mathcal{M}_0$.

The next facts illustrate this concept:

(a) Let X be a complemented subspace of a Banach space Y, i.e., there is a continuous linear projection $p : Y \to X$. This subspace is clearly a retract of Y. Thus, a retract is a topological counterpart of the notion of a complemented subspace in Banach Space Theory.

(b) However, by the Borsuk–Dugundji theorem *every* closed subspace is a retract of its ambient Banach space. An interesting example is the space c_0 regarded as a closed subspace of ℓ_∞. By the above mentioned Phillips theorem, c_0 is uncomplemented in ℓ_∞ while it is a retract of ℓ_∞. The latter can be shown straightforwardly using a retraction map $r : \ell_\infty \to c_0$ given by

$$r(x)_n := \left(|x_n| - \overline{\lim_{n \to \infty}} |x_n|\right)_+ \operatorname{sgn} x_n,$$

where $x_+ := \max(x, 0)$ for $x \in \mathbb{R}$.

It is obvious that $r(x) = x$ for $x \in c_0$; moreover, this map is continuous (in fact, Lipschitz), since

$$\|r(x) - r(y)\|_{\ell_\infty} \leq 2 \sup_n |x_n - y_n| \leq 2\|x - y\|_{\ell_\infty}.$$

(c) The n-sphere $\mathbb{S}^n$ regarded as the boundary of the unit ball B^{n+1} of $\mathbb{R}^{n+1}$ is not a retract of this ball. We present the proof of this result based on several facts of Algebraic Topology; an elementary proof is presented in Appendix C.

Assume, on the contrary, that a retraction $r : B^{n+1} \to \mathbb{S}^{n+1}$ does exist. Denote by i the embedding map $\mathbb{S}^{n+1} \subset B^{n+1}$; then $r \circ i = id_{\mathbb{S}_n}$. Passing to the n-th homology groups of the spaces involved, one obtains the commutative diagram

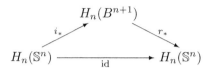

where i_* and r_* are homomorphisms induced by i and r. Hence, the identity homomorphism id of the group $H_n(\mathbb{S}^n) = \mathbb{Z}$ is factored through the group $H_n(B^{n+1}) = \{0\}$; this yields a contradiction.

In the same vein one can use cohomology groups H^n for this proof; their basic properties will be discussed in Appendix B.

Definition 1.11 (Borsuk). (a) A metric space $\mathcal{M}$ is an *absolute retract*, if it is a retract of every metric space containing $\mathcal{M}$ as a closed subspace.

(b) $\mathcal{M}$ is an *absolute neighborhood retract*, if for each of its ambient metric spaces $\mathcal{M}'$ there is a $\delta > 0$ such that $\mathcal{M}$ is a retract of the open δ-neighborhood of $\mathcal{M}$ in $\mathcal{M}'$.

The following examples and results explain the role of these concepts for continuous extension problems.

(a) Let $\ell_\infty(\Gamma)$ be the Banach space of bounded functions $x : \Gamma \to \mathbb{R}$ defined on a set Γ and equipped with the norm
$$\|x\|_\infty := \sup\{|x(\gamma)| \;;\; \gamma \in \Gamma\}.$$
Then its closed unit ball
$$\overline{B}_\infty(\Gamma) := \{x \in \ell_\infty(\Gamma) \;;\; \sup_\gamma |x(y)| \leq 1\}$$
is an absolute retract.

To show this, consider a metric space $\mathcal{M}$ containing this ball as a closed subspace. Let $c_\gamma : \overline{B}_\infty(\Gamma) \to \mathbb{R}$ be a function given by $c_\gamma(x) := x(\gamma)$, $\gamma \in \Gamma$. By the Tietze–Urysohn theorem, there is a continuous extension, say $\tilde{c}_\gamma$, of c_γ to all of $\mathcal{M}$ such that
$$\sup\{|\tilde{c}_\gamma(m)| \;;\; m \in \mathcal{M}\} = \sup\{c_\gamma(x) \;;\; x \in \overline{B}_\infty(\Gamma)\}.$$
Since the latter supremum is at most one, a map $r : \mathcal{M} \to \overline{B}_\infty(\Gamma)$ given by
$$r(m) := \bigl(\tilde{c}_\gamma(m)\bigr)_{\gamma \in \Gamma}$$
yields the required retraction of $\mathcal{M}$ onto $\overline{B}_\infty(\Gamma)$.

1.6. Absolute metric retracts

(b) $\mathbb{S}^n$ is not an absolute retract by Theorem 1.9. Nevertheless, it is an absolute neighborhood retract. To show this, one uses the retraction map $r : \mathcal{M} \to B_\infty(\Gamma)$ from (a) with $\Gamma := \{1, \ldots, n+1\}$ and $\mathcal{M}$ being an ambient metric space for $\mathbb{S}^n$. That is to say, $r := (\tilde{c}_i)_{i=1}^{n+1}$, where $\tilde{c}_i$ is a supremum norm preserving continuous extension of $c_i : \mathbb{S}^n \to \mathbb{R}$ to all of $\mathcal{M}$, and c_i is defined by $c_i(x) := x_i$ for $x \in \mathbb{R}^{n+1}$.

Let U be the open set of $\mathcal{M}$ given by

$$U := \left\{ m \in \mathcal{M} \; ; \; \sum_{i=1}^{n+1} \tilde{c}_i(m)^2 > 0 \right\}.$$

Since for $m \in \mathbb{S}^n$ the sum in the brackets equals 1, $\mathbb{S}^n$ is contained in U. Therefore a map $r : U \to \mathbb{S}^n$ given for $m \in U$ by

$$r(m)_i := \tilde{c}_i(m) \bigg/ \sum_{i=1}^{n+1} \tilde{c}_i(m)^2, \quad 1 \leq i \leq n+1,$$

yields the required retraction of U onto $\mathbb{S}^n$.

The next two facts relate the concept of absolute retract to continuous extensions problems.

Proposition 1.12. *$\mathcal{M}_0$ is an absolute retract if and only if it has the following property:*

For every metric space $\mathcal{M}$ and its closed subset S each continuous map from S to $\mathcal{M}_0$ can be continuously extended to all of $\mathcal{M}$.

Proof. Sufficiency trivially follows from the definition of absolute retract. To prove necessity, one needs the following

Lemma 1.13. *Every metric space $\mathcal{M}$ is isometric to a closed subset of some space $\ell_\infty(\Gamma)$.*

Recall the proof of this well-known result. Fix $m_0 \in \mathcal{M}$ and define Γ to be the set $\mathcal{M}$. Introduce the Kuratowsky map $I := \{I_\gamma\}_{\gamma \in \mathcal{M}}$ of $\mathcal{M}$ by

$$I_\gamma(m) := d(\gamma, m) - d(\gamma, m_0), \quad \gamma, \, m \in \mathcal{M}.$$

Then one has, by the triangle inequality,

$$\|I(m) - I(m')\|_{\ell_\infty(\Gamma)} := \sup_\gamma |d(\gamma, m) - d(\gamma, m')| = d(m, m').$$

Hence I is the required isometry of $\mathcal{M}$ into $\ell_\infty(\Gamma)$. $\square$

Now let S be a closed subspace of a metric space $\mathcal{M}$, and $f : S \to \mathcal{M}_0$ be a continuous map. By the lemma, $\mathcal{M}_0$ can be identified with a closed subset of some

$\ell_\infty(\Gamma)$. Since $\mathcal{M}_0$ is an absolute retract, there is a retraction map $r : \ell_\infty(\Gamma) \to \mathcal{M}_0$. Moreover, the continuous map $f : S \to \mathcal{M}_0$ can be seen as a continuous map from S to the Banach space $\ell_\infty(\Gamma)$. By Theorem 1.4, there is a continuous extension $\tilde{f} : \mathcal{M} \to \ell_\infty(\Gamma)$ of f to all of $\mathcal{M}$. Then a continuous extension of f to the whole $\mathcal{M}$ is given by the composition $r \circ \tilde{f}$. □

Remark 1.14. In view of the fact established in Proposition 1.12, one can restate Theorem 1.4 as follows.

Every Banach space is an absolute retract.

The remaining important result tells us that the existence of continuous extensions for a map, whose target space is an absolute neighborhood retract, depends only on the *homotopy class* of this map. To formulate this let us recall that the maps f_0, f_1 from S to $\mathcal{M}$ are *homotopic* if there exists a *continuous* map (*homotopy*) $F : X \times [0,1] \to \mathcal{M}$ so that

$$F(\cdot, 0) = f_0 \quad \text{and} \quad F(\cdot, 1) = f_1.$$

Theorem 1.15 (Borsuk). *If f_0, f_1 are homotopic maps from $S \subset \mathcal{M}$ to an absolute neighborhood retract $\mathcal{M}_0$ and f_0 admits a continuous extension to all of $\mathcal{M}$, then f_1 also does.*

For the proof see, e.g., Section VI.3 of the book [HW-1941].

Lipschitz Functions

Now we turn to a new area of research which is rather distant from pure topology. It is devoted to extension problems for Lipschitz maps of metric spaces and is a considerable part of Geometric Analysis.

We follow the line of the previous part presenting either straightforward "Lipschitz" analogs of continuous extension results or counterexamples to such claims. The first results in this area were obtained by McShane for scalar Lipschitz functions and by Kirszbraun for Lipschitz maps from subsets of a Euclidean space into itself. We begin with these results and then proceed with Valentine's beautiful extension of the latter result to the remaining classical space forms, hyperbolic spaces and Euclidean spheres. Discussion of counterexamples is scattered within this part of the chapter. It is specifically aimed at motivating the problems studied in Part 3 and illuminating the results presented there.

1.7 Notation and definitions

We first recall several basic concepts of Geometric Analysis; the notation introduced below is consistent with that of Section 1.1.

1.7. Notation and definitions

Suppose f is a map from a metric space $(\mathcal{M}, d)$ into $(\widetilde{\mathcal{M}}, \tilde{d})$. The *Lipschitz constant* of f is defined (for a nontrivial $\mathcal{M}$) by

$$L(f) := \sup_{m \neq m'} \left\{ \frac{\tilde{d}(f(m), f(m'))}{d(m, m')} \; ; \; m, m' \in \mathcal{M} \right\}. \tag{1.18}$$

We also write $L(f; \mathcal{M}, \widetilde{\mathcal{M}})$, if this additional information is needed.

The map f is said to be *C-Lipschitz*, if $L(f) \leq C$, and *Lipschitz*, if its constant is finite.

For the set of such maps we use the notation

$$\operatorname{Lip}(\mathcal{M}, \widetilde{\mathcal{M}}) := \{ f : \mathcal{M} \to \widetilde{\mathcal{M}} \; ; \; L(f) < \infty \} \tag{1.19}$$

and sometimes write $|f|_{\operatorname{Lip}(\mathcal{M}, \widetilde{\mathcal{M}})}$ instead of $L(f)$. This notation is simplified to $\operatorname{Lip}(\mathcal{M})$ for $\widetilde{\mathcal{M}}$ being the real line $\mathbb{R}$ (with the standard metric). The space of scalar Lipschitz functions $\operatorname{Lip}(\mathcal{M})$ becomes a Banach space after factorization by the subspace consisting of constant functions. The same is true for the factor space

$$\operatorname{Lip}(\mathcal{M}, X)/X \tag{1.20}$$

with the Banach target space X. In both cases $L(f)$ becomes a Banach norm.

There is another way of turning $\operatorname{Lip}(\mathcal{M}, X)$ into a Banach space. Fix a point $m^* \in \mathcal{M}$ and set

$$\operatorname{Lip}(\mathcal{M}, m^*, X) := \{ f \in \operatorname{Lip}(\mathcal{M}, X) \; ; \; f(m^*) = 0 \}. \tag{1.21}$$

Then $L(f)$ is a Banach norm of this space. In the sequel, for the space (1.21), we use the notation $\operatorname{Lip}_0(\mathcal{M}, X)$ or $\operatorname{Lip}_0(\mathcal{M})$, if m^* can be restored from the context.

Finally, we recall the following basic notion.

Definition 1.16. A map $f : \mathcal{M} \to \widetilde{\mathcal{M}}$ is called a *C-bi-Lipschitz embedding*, if f and its inverse f^{-1} (defined of the subspace $f(\mathcal{M})$ of $\widetilde{\mathcal{M}}$) are C-Lipschitz maps.

If, in particular, this map is a bijection onto $\widetilde{\mathcal{M}}$, f is called a *C-isometry* (the name *C-bi-Lipschitz homeomorphism* is also in use). In this case we also define the *distortion* of f by

$$D(f) := L(f) L(f^{-1}). \tag{1.22}$$

In what follows we simply call the map of the above definition a *bi-Lipschitz embedding* whenever the constant C is unspecified. Unfortunately, this simplification cannot be used for C-isometries, so a *bi-Lipschitz homeomorphism* is the unique name for this case.

Finally, like in Definitions 1.10 and 1.11 for the continuous case, one can introduce the concepts of a *Lipschitz retract* and an *absolute Lipschitz retract*, see Section 1.11 for these definitions.

1.8 Trace and extension problems for Lipschitz functions

In accordance with the notions and notation introduced in Section 1.2, we define the *trace* of $\mathrm{Lip}(\mathcal{M}, \widetilde{\mathcal{M}})$ to a subset $S \subset \mathcal{M}$ by

$$\mathrm{Lip}(\mathcal{M}, \widetilde{\mathcal{M}})|_S := \{f|_S \, ; \, f \in \mathrm{Lip}(\mathcal{M}, \widetilde{\mathcal{M}})\}; \tag{1.23}$$

then we define the *Lipschitz trace constant* for $f|_S$ by

$$L_S(f) := \inf\{L(g) \, ; \, g \in \mathrm{Lip}(\mathcal{M}, \widetilde{\mathcal{M}}) \text{ and } g|_S = f|_S\}. \tag{1.24}$$

If the target space $\widetilde{\mathcal{M}}$ is a Banach space, say X, then (1.24) defines a Banach norm of the factor space

$$\mathrm{Lip}(\mathcal{M}, X) \big/ \{f \in \mathrm{Lip}(\mathcal{M}, X) \, ; \, f|_S = 0\}.$$

Definition 1.17. A map $g \colon \mathcal{M} \to \widetilde{\mathcal{M}}$ is called a *Lipschitz extension* of $f \colon S \to \widetilde{\mathcal{M}}$, if $g|_S = f$ and g is Lipschitz.

The *extension (Lipschitz) constant* of f is then introduced by

$$L_{\mathrm{ext}}(f) := \inf\{L(g) \, ; \, g|_S = f\}. \tag{1.25}$$

This constant is conventionally equal to infinity, if there is no Lipschitz extension of f. On the other hand,

$$L_S(f) = L_{\mathrm{ext}}(f) \tag{1.26}$$

provided that $f \in \mathrm{Lip}(\mathcal{M}, \widetilde{\mathcal{M}})|_S$.

In view of these definitions the formulation of the *Main Problem for Lipschitz maps* is now clear. Just as for that of Section 1.2, this problem is divided into two subproblems. We present the second of them in the following form.

Extension Problem for Lipschitz Maps. *Under what conditions on $f \colon S \to \mathcal{M}'$ does this map admit an extension to a Lipschitz map between $\mathcal{M}$ and $\mathcal{M}'$?*

A quantitative version of this problem includes suitable estimates of the *Lipschitz extension constants* introduced by

$$\Lambda(S, \mathcal{M}; \widetilde{\mathcal{M}}) := \sup\left\{\frac{L_{\mathrm{ext}}(f)}{L(f)} \, ; \, f \in \mathrm{Lip}(S, \widetilde{\mathcal{M}})\right\} \tag{1.27}$$

and, accordingly, by

$$\Lambda(\mathcal{M}; \widetilde{\mathcal{M}}) := \sup\{\Lambda(S, \mathcal{M}; \widetilde{\mathcal{M}}) \, ; \, S \subset \mathcal{M}\}. \tag{1.28}$$

We say that an extension method is *universal*, if it can be applied to an arbitrary S and the corresponding extension constants are controlled by (1.28).

Trace Problem. *Characterize maps from* $\mathrm{Lip}(\mathcal{M}, \widetilde{\mathcal{M}})\big|_S$.

Characteristics obtained in the study of this problem determine a class that contains $\mathrm{Lip}(\mathcal{M}, \widetilde{\mathcal{M}})\big|_S$. The simplest result of this kind, sufficient for achieving our goal, is given by the embedding

$$\mathrm{Lip}(\mathcal{M}, \widetilde{\mathcal{M}})\big|_S \subset \mathrm{Lip}(S, \widetilde{\mathcal{M}}),$$

which is clearly true. The finiteness of the Lipschitz extension constant (1.27) immediately implies the following solution to the Main Problem.

It is true that

$$\mathrm{Lip}(\mathcal{M}, \widetilde{\mathcal{M}})\big|_S = \mathrm{Lip}(S, \widetilde{\mathcal{M}}). \tag{1.29}$$

Moreover, for f in the right-side space it is true that

$$L(f) \leq L_S(f) \leq CL(f) \tag{1.30}$$

with $C := \Lambda(S, \mathcal{M}; \widetilde{\mathcal{M}})$.

Let us note that for the continuous extension problem the result analogous to (1.29) is true only for closed subsets S.

1.9 Lipschitz selection problem

The goal of this section is twofold. First, we show that the Lipschitz analog of Michael's selection theorem fails to be true for infinite-dimensional target spaces. Secondly, we reformulate the classical Helly and Sylvester combinatorial–geometric theorems as selection results for multivariate (set-valued) maps into convex subsets of $\mathbb{R}^n$ and prove a similar result for multivalued maps into closed intervals of $\mathbb{R}^n$. These facts will motivate a general Lipschitz selection conjecture and related results presented in subsections 5.3.2 and 5.3.3, which, in turn, will play a decisive role in the solution of the Trace Problem for Lipschitz functions of higher order and continuously differentiable functions on $\mathbb{R}^n$.

1.9.1 Counterexample

Let us recall the definition of *Hausdorff distance*. For subsets of a metric space $(\mathcal{M}, d)$ it is defined by

$$d_{\mathcal{H}}(S_0, S_1) := \max_{i=0,1} \sup \{d(m_i, S_{1-i}) \, ; \, m_i \in S_i\}. \tag{1.31}$$

The restriction of $d_{\mathcal{H}}$ to the set of all closed nonempty subsets in $\mathcal{M}$ is well known to be a metric, see, e.g., [Had-1957].

Now let $\mathcal{M}$ be a Banach space (denoted by X), and

$$\mathcal{C}(X) := \{S \subset X \, ; \, S \text{ closed convex bounded nonempty}\}. \tag{1.32}$$

We equip this subspace with the Hausdorff metric and denote it by $\mathcal{C}_{\mathcal{H}}(X)$.

Definition 1.18. A map $f : \mathcal{M} \to X$ is called a *Lipschitz selection* for the multi-valued function $F : \mathcal{M} \to \mathcal{C}(X)$, if f is Lipschitz and

$$f(m) \in F(m) \text{ for every } m \in \mathcal{M}.$$

The following claim is the natural analog of Michael's selection theorem for Lipschitz functions.

Conjecture 1.19. *Every F from $\mathrm{Lip}(\mathcal{M}, \mathcal{C}_\mathcal{H}(X))$ admits a Lipschitz selection.*

It was proved independently by Przesławski and D. Yost [PY-1989] and by Aubin and Frankowska [AF-1990] that this is true for finite-dimensional X.

For $\dim X = \infty$ the previous authors also established the following counterexample to Conjecture 1.19.

Theorem 1.20. *For every infinite-dimensional Banach space X there exist a metric space and a map from this space into $\mathcal{C}_\mathcal{H}(X)$ that does not admit a Lipschitz selection.*

Proof. Let us choose the required metric space to be $\mathcal{C}_\mathcal{H}(X)$ itself. The identity map $I : \mathcal{C}_\mathcal{H}(X) \to \mathcal{C}_\mathcal{H}(X)$ is clearly Lipschitz. We show that there is no Lipschitz selection of I regarded as a multivalued function defined on the metric space $\mathcal{C}_\mathcal{H}(X)$ whose values $I(x)$, $x \in \mathcal{C}_\mathcal{H}(X)$, are considered as convex subsets of X. Assume the converse and suppose that $f : \mathcal{C}_\mathcal{H}(X) \to X$ is a Lipschitz selection of I. Identify X with a metric subspace of $\mathcal{C}_\mathcal{H}(X)$ in the following way. Let $x \mapsto \{x\}$ be the (trivial) map from X into $\mathcal{C}_\mathcal{H}(X)$. Since $d_\mathcal{H}(\{x\}, \{y\}) = \|x - y\|_X$, this map isometrically embeds X into $\mathcal{C}_\mathcal{H}(X)$; then we identify X with its image.

After this identification f becomes a Lipschitz retraction of $\mathcal{C}_\mathcal{H}(X)$ onto X. In fact, $f|_X$ is the identity map of X and $L(f) < \infty$.

Now let Y be a (closed) subspace of X. Then $f_Y := f|_{\mathcal{C}_\mathcal{H}(Y)}$ is clearly a retraction of $\mathcal{C}_\mathcal{H}(Y)$ onto Y and

$$L(f_Y) \leq L(f) < \infty. \tag{1.33}$$

To obtain a contradiction we then use a version of the following classical result.

Theorem 1.21 (Dvoretsky). *If $\dim X = \infty$, then for every n greater than some n_0 there exists a subspace $Y_n \subset X$ that is linearly isomorphic and 2-isometric to the n-dimensional Euclidean space E_n.*

Applying (1.33) to the subspace Y_n and then replacing it by E_n we obtain a retraction $r_n : \mathcal{C}_\mathcal{H}(E_n) \to E_n$ such that

$$L(r_n) \leq 4L(f_{Y_n}) \leq 4L(f) < \infty.$$

However, it is established by Posicelski [Pos-1971] that $L(r_n) \to \infty$ as $n \to \infty$.

This contradiction completes the proof. □

1.9.2 Combinatorial–geometric selection results

We begin with a simply formulated (but deep!) case of constant selections. As before, we deal with a multivalued function from a metric space $\mathcal{M}$ into the class $\mathcal{C}(X)$ of nonempty bounded convex subsets of a Banach space X. In accordance with the previous counterexample we assume that

$$n := \dim X < \infty.$$

We formulate now as a selection result the classical Helly theorem in a form adjusted to our setting.

Let C be a collection of convex subsets in X satisfying one of the following conditions:

(a) C is finite;

(b) there exist n subsets of C whose intersection is compact.

Theorem 1.22 (Helly). *A map $F : \mathcal{M} \to C$ admits a constant selection whenever its restriction to every $(n+1)$-point subset of $\mathcal{M}$ does.*

In particular, the result holds for set-valued maps into the set $\mathcal{C}(X)$.

To explain the equivalence of this result to the classical one, let us restate our assumption as follows:

Every subfamily of the family $\{F(m)\}_{m \in \mathcal{M}}$ consisting of $n+1$ subsets has a nonempty intersection.

Then the Helly theorem asserts that all $F(m)$ have a common point, say x_0. Putting $f(m) := x_0$ for all $m \in \mathcal{M}$ we get the required selection. (See Appendix C for a detailed discussion of this topic.)

In the sequel we will need a generalization of Helly's theorem providing existence of *nonconstant* Lipschitz selections. Of course, the cardinality of the subfamilies involved may be more than $n+1$. Another generalization concerns multivalued functions assigning values being noncompact convex subset. We present below two simple results revealing the basic characteristics of such generalizations.

To formulate the first of them, the so-called Sylvester's theorem, we let $\mathrm{Aff}_n(X)$ denote the class of affine sets of dimension n (hyperplanes) in a Banach space X of dimension $n+1$.

Theorem 1.23 (Sylvester–Gallai). *A map $F : \mathcal{M} \to \mathrm{Aff}_n(X)$ admits a constant selection whenever its restriction to every $(n+1)$-point subset of $\mathcal{M}$ does.*

We again use here the reformulation which is relevant to our setting. The classical version deals with the existence of a common point for hyperplanes $F(m)$, $m \in \mathcal{M}$.

Our next result may be considered as a model case for the required n-dimensional Lipschitz selection theorem.

Proposition 1.24. *A map $F : \mathcal{M} \to \mathcal{C}(\mathbb{R})$ into compact intervals of $\mathbb{R}$ admits a C-Lipschitz selection, if its restriction to every 2-point subset of $\mathcal{M}$ does.*

Proof. It suffices to prove the result for $C = 1$. Set $F(m) := [a(m), b(m)]$ and introduce the required selection of F by the formula

$$f(m) := \sup\{a(m') - d(m, m') \; ; \; m' \in \mathcal{M}\}, \quad m \in \mathcal{M}. \qquad (1.34)$$

To check that $f(m) \in F(m)$ we first note that, by the definition,

$$f(m) \geq a(m) - d(m, m) = a(m).$$

To prove the remaining inequality $f(m) \leq b(m)$, we denote by $f_{mm'}$ a 1-Lipschitz selection of the restriction of F to $\{m, m'\}$. Since $a(m') \leq f_{mm'}(m')$ and $f_{mm'}(m) \leq b(m)$, we have

$$f(m) := \sup\{a(m') - d(m, m'); m' \in \mathcal{M}\}$$
$$\leq \sup\{a(m') + f_{mm'}(m) - f_{mm'}(m') ; m' \in \mathcal{M}\} \leq f_{mm'}(m) \leq b(m).$$

This implies that f is a selection of F.

Let us show that f is 1-Lipschitz. Given $m \in \mathcal{M}$ and $\varepsilon > 0$, there exists a point $m_\varepsilon \in \mathcal{M}$ such that

$$f(m) \leq a(m_\varepsilon) - d(m, m_\varepsilon) + \varepsilon, \qquad (1.35)$$

see definition (1.34). By the same definition,

$$f(m') \geq a(m_\varepsilon) - d(m', m_\varepsilon).$$

Extracting this from (1.35) we get

$$f(m) - f(m') \leq d(m', m_\varepsilon) - d(m, m_\varepsilon) + \varepsilon \leq d(m, m') + \varepsilon.$$

Hence, f is 1-Lipschitz.

The proof is complete. $\square$

It is easily seen that the result holds also for pseudometrics, i.e., for $d(m, m')$ being zero for some $m \neq m'$. In applications we need a more general version of the proposition given by

Corollary 1.25. *A map F from a pseudometric space $(\mathcal{M}, d)$, with d which may assign the value $+\infty$ into the set of closed balls of $\ell_\infty(\Gamma)$, admits a C-Lipschitz selection if the restriction of F to every 2-point subset does.*

Proof. It suffices to prove the result for Γ consisting of a single point, i.e., for $\ell_\infty(\Gamma) = \mathbb{R}$. In fact, every bounded closed ball of $\ell_\infty(\Gamma)$ is the direct product of compact intervals, and every unbounded ball is the product of compact intervals and the real line. The coordinatewise application of the selection result for $\mathbb{R}$ gives then the required assertion.

To prove the result for $F: \mathcal{M} \to \mathcal{C}(\mathbb{R})$, we introduce an equivalence relation on $\mathcal{M}$ writing $m \sim m'$ if $d(m, m') < \infty$. If we prove existence of a C-Lipschitz selection for the restriction of F to every equivalence class, then the union of these selections gives the required selection for F. Hence, without loss of generality, we assume that d is finite on $\mathcal{M}$.

Further, let
$$\mathcal{M}_0 := \{m \in \mathcal{M} \,;\, F(m) \text{ is compact}\}.$$

Applying Proposition 1.24 we find a C-Lipschitz selection f of the trace $F|_{\mathcal{M}_0}$. Then we extend f, using the McShane Theorem 1.27 presented below, to a C-Lipschitz function on $\mathcal{M}$. Since $F(m) = \mathbb{R}$ for $m \in \mathcal{M} \setminus \mathcal{M}_0$, this extension is the required C-Lipschitz selection of F. $\square$

1.10 Extensions preserving Lipschitz constants

We present here several classical results including those of Kirszbraun and Valentine. They deal with such basic objects as Euclidean spaces, and spherical and hyperbolic spaces with the geodesic metrics. To formulate the corresponding theorems, the following notion is of use.

We say that the pair of metric spaces $(\mathcal{M}, \mathcal{M}')$ has the *strong extension property*, if, for every $S \subset \mathcal{M}$, any map $f \in \text{Lip}(S, \mathcal{M}')$ with $L(f) \leq 1$ admits an extension $\tilde{f}$ to all of $\mathcal{M}$ so that $L(\tilde{f}) \leq 1$.

In this case we write $(\mathcal{M}, \mathcal{M}') \in \mathcal{SE}$.

We begin the account with a relatively simple case of Banach-valued functions. Unlike the situation for Lipschitz selections, see Theorem 1.20, there does exist a large class of infinite-dimensional Banach spaces X, for which the pair $(\mathcal{M}, X)$ belongs to $\mathcal{SE}$ for every $\mathcal{M}$. Its description is presented in the next subsection. Then we give two different proofs of Kirszbraun's theorem and use the introduced methods to obtain several other extension results.

1.10.1 Banach-valued Lipschitz functions

We are looking for characteristics of a Banach space X such that pairs $(\mathcal{M}, X)$ belong to $\mathcal{SE}$ for every $\mathcal{M}$. Two different descriptions for such X, of functional analytic and geometric nature, respectively, were discovered by Nachbin, Goodner, Kelly and Hasumi. To remain in the realm of Geometric Analysis, we present only the second one.

Theorem 1.26. *The following conditions on a Banach space X are equivalent:*

(i) *For every metric space $\mathcal{M}$ the pair $(\mathcal{M}, X) \in \mathcal{SE}$.*

(ii) *Every collection of mutually intersecting closed balls of X has a common point.*

We will prove this result in due course, including its proof in a wider geometric context. For now we consider only several consequences of Theorem 1.26.

Let $\{[a_i, b_i]\}_{i \in I}$ be a family of mutually intersecting compact intervals of the real line. Then $\sup a_i \leq \inf b_i$, and every x lying in between is a common point of all $[a_i, b_i]$. Hence, by Theorem 1.26, $(\mathcal{M}, \mathbb{R}) \in \mathcal{SE}$ for every $\mathcal{M}$. In this setting, however, there exists a simple extension operator; it is introduced by the following classical result.

Theorem 1.27 (McShane). *Let $f : S \to \mathbb{R}$ be a C-Lipschitz function on a subset S of a pseudometric space $\mathcal{M}$. Then the function $f_- : \mathcal{M} \to \mathbb{R}$ defined for $m \in \mathcal{M}$ by*

$$f_-(m) := \sup\{f(m') - Cd(m, m') \; ; \; m' \in S\} \tag{1.36}$$

is a C-Lipschitz extension of f.

Proof. Without loss of generality, set $C = 1$. Then $f(m') \leq f(m) + d(m, m')$ for $m, m' \in S$ and therefore $f_-(m) = f(m)$, if $m \in S$. We now check that the extension f_- is 1-Lipschitz on $\mathcal{M}$. Since $|\sup F - \sup G| \leq \sup |F - G|$, we have for $m_1, m_2 \in \mathcal{M}$,

$$|f_-(m_1) - f_-(m_2)| \leq \sup\{|d(m_1, m') - d(m_2, m')| \; ; \; m' \in S\} \leq d(m_1, m_2). \quad \square$$

Remark 1.28. (a) Let ω be a *subadditive function* on $(0, +\infty)$, i.e., ω satisfies the condition

$$\omega(t_1 + t_2) \leq \omega(t_1) + \omega(t_2) \quad \text{for} \quad t_1, t_2 > 0. \tag{1.37}$$

Assume also that $\omega(0+) := \lim_{t \to 0} \omega(t)$ is zero. Then ω is nonnegative and continuous and admits a continuous extension on $\mathbb{R}_+$ by setting $\omega(0) := 0$. In this case the function

$$d_\omega := \omega \circ d \tag{1.38}$$

is a metric on $\mathcal{M}$ (if d is).

The class of Lipschitz functions on $\mathcal{M}$ with the metric d_ω is denoted by $\text{Lip}^\omega(\mathcal{M})$. The original version of the McShane theorem concerns functions of the class $\text{Lip}^\omega(\mathcal{M})$ and states that the operator (1.36) with Cd replaced with d_ω is an extension of $f \in \text{Lip}^\omega(S)$ to all of $\mathcal{M}$ preserving the seminorm

$$|f|_{\text{Lip}^\omega(S)} := \sup\left\{\frac{|f(m) - f(m')|}{\omega(d(m, m'))} \; ; \; m, m' \in S\right\}. \tag{1.39}$$

(b) Another extension operator of this type is given by the formula

$$f_+(m) := \inf\{f(m') + Cd(m, m') \; ; \; m' \in S\}.$$

1.10. Extensions preserving Lipschitz constants

It is easy to check that every C-Lipschitz extension $\tilde{f}$ of f satisfies the inequality
$$f_- \leq \tilde{f} \leq f_+.$$

As a consequence of this theorem we present an extension result for uniformly continuous functions mentioned in Example 1.2.2. Recall that $C_u(\mathcal{M})$ is the Banach space of bounded uniformly continuous real-valued functions on the metric space $(\mathcal{M}, d)$, see Definition 1.1.

Corollary 1.29. $C_u(\mathcal{M})\big|_S = C_u(S)$.

Proof. We must only prove that every function $f \in C_u(S)$ admits an extension to all of $\mathcal{M}$ as a bounded uniformly continuous function. To establish this, we use the *modulus of continuity* for $f : S \to \mathbb{R}$, i.e., a function on $(0, +\infty)$, given for $t > 0$ by
$$\omega(t;f) := \sup\{f(m) - f(m') \,;\, d(m,m') \leq t \text{ and } m, m' \in S\}. \tag{1.40}$$

Now we find a subadditive function ω such that
$$\omega(t;f) \leq \omega(t), \quad 0 \leq t \leq 1, \text{ and } \omega(0+) = 0. \tag{1.41}$$

To this end, consider the function $\varphi(t) := t + \omega(t;f)$, $t > 0$. This function is strictly increasing and therefore the inverse function φ^{-1} is well defined and increasing. Hence the integral $\phi(t) := \int_0^t \varphi^{-1}(s)ds$, $t \geq 0$, is a convex strictly increasing function that equals zero at zero. Moreover, $\varphi^{-1}(t) \leq \phi(t)$ for $0 \leq t \leq 1$ and therefore
$$t + \omega(t;f) \leq \phi^{-1}(t) =: \omega(t), \quad 0 \leq t \leq 1.$$

Since ϕ is convex, ω is concave and equals zero at zero. Hence, it remains to show that ω is subadditive. But the slope $\frac{\omega(t)}{t}$ of the chord connecting the points $(0,0)$ and $(t, \omega(t))$ of the graph of ω clearly increases as $t \to 0$. Therefore
$$\frac{\omega(t_1 + t_2)}{t_1 + t_2} \leq \frac{\omega(t_i)}{t_i}, \quad i = 1, 2,$$
and subadditivity follows.

Now replace the metric d with a new metric $\tilde{d}$ given by
$$\tilde{d} := \min\{\omega \circ d, 2\|f\|_{C_b(S)}\}.$$

By (1.41) and the trivial estimate $\omega(t;f) \leq 2\|f\|_{C_b(S)}$, we get
$$|f(m) - f(m')| \leq \tilde{d}(m,m') \quad \text{for} \quad m, m' \in S.$$

Applying McShane's extension operator (1.36) with d replaced with $\tilde{d}$, we obtain the extension $f_- : \mathcal{M} \to \mathbb{R}$ of f satisfying, for all $m, m' \in \mathcal{M}$, the inequality
$$|f_-(m) - f_-(m')| \leq \tilde{d}(m,m').$$

Since $\tilde{d}(m,m') \to 0$ as $d(m,m') \to 0$, the function f_- is uniformly continuous on $\mathcal{M}$. Moreover, $|f_-|$ is clearly bounded by

$$2\|f\|_{C_b(S)} + \sup\{|f_-(m')| \,;\, m' \in S\} = 3\|f\|_{C_b(S)}.$$

Hence, $f_- \in C_u(\mathcal{M})$, as required. $\square$

The space $\ell_\infty(\Gamma)$ is another example of a Banach space possessing the *binary intersection property* of Theorem 1.26. This fact follows from the similar property of $\mathbb{R}$, since a closed ball of $\ell_\infty(\Gamma)$ is the direct product of compact intervals.

Remark 1.30. The aforementioned functional analytic description of the Banach spaces X under consideration is as follows.

For every $\mathcal{M}$ the pair $(\mathcal{M}, X) \in \mathcal{SE}$ if and only if X is isometrically isomorphic to the Banach space $C(K)$ of continuous functions on a totally disconnected[3] compact Hausdorff space K.

This result has been obtained by the efforts of several mathematicians including Nachbin, Goodner and Klee, see, e.g., [Da-1955] and references therein.

In particular, $\ell_\infty(\Gamma)$ is isometrically isomorphic to the space $C(\hat{\Gamma})$ of this kind where $\hat{\Gamma}$ is the Stone-Čech compactification of Γ equipped with the discrete topology, see, e.g., [DS-1958, Thm. IV.6.22]. Another example of the same type is the space $L_\infty(0,1)$.

1.10.2 Extension and the intersection property of balls

The proofs of Theorem 1.26 and the consequent results of this section are based on a geometric criterion, which is now introduced. Its formulation requires the following notion.

Suppose $\mathcal{B} := \{B_\alpha\}_{\alpha \in A}$ and $\mathcal{B}' := \{B'_\alpha\}_{\alpha \in A}$ are families of closed balls from metric spaces $(\mathcal{M}, d)$ and $(\mathcal{M}', d')$, respectively, indexed by the same set A. Let r_α and c_α denote the radius and center of B_α; similar notation with primes is used for B'_α.

We will say that $\mathcal{B}$ *dominates* $\mathcal{B}'$ and write $\mathcal{B} \succ \mathcal{B}'$, if for all $\alpha, \beta \in A$,

$$r_\alpha = r'_\alpha \quad \text{and} \quad d(c_\alpha, c_\beta) \geq d'(c'_\alpha, c'_\beta). \tag{1.42}$$

In what follows $\cap \mathcal{B}$ and $\cup \mathcal{B}$ stand for the intersection and union of the balls of $\mathcal{B}$.

Proposition 1.31. *The following conditions are equivalent:*

(i) *The pair $(\mathcal{M}, \mathcal{M}')$ belongs to $\mathcal{SE}$.*

---

[3] i.e., the closure of every open set of K is open.

1.10. Extensions preserving Lipschitz constants

(ii) *For every family $\mathcal{B}, \mathcal{B}'$ of closed balls from $\mathcal{M}$ and $\mathcal{M}'$, respectively, the conditions*

$$\mathcal{B} \succ \mathcal{B}' \quad \text{and} \quad \cap \mathcal{B} \neq \emptyset \tag{1.43}$$

imply that

$$\cap \mathcal{B}' \neq \emptyset. \tag{1.44}$$

Proof. (The implication (ii)$\Rightarrow$(i)). According to the definition of $\mathcal{SE}$ the extending map $f : S \to \mathcal{M}'$ is a *contraction*, i.e., satisfies the condition

$$L(f) \leq 1. \tag{1.45}$$

A standard application of the Zorn lemma allows us to consider only an extension of f to one more point outside of S, say m_0. To find such an extension we introduce families of closed balls $\mathcal{B} := \{B_m\}_{m \in \mathcal{M}}$ and $\mathcal{B}' := \{B'_m\}_{m \in \mathcal{M}}$ from $\mathcal{M}$ and $\mathcal{M}'$, respectively, in the following fashion. The radii r_m and r'_m of B_m and B'_m are determined by

$$r_m = r'_m := d(m, m_0)$$

and their centers c_m and c'_m are given by

$$c_m := m \quad \text{and} \quad c'_m := f(m).$$

Because of (1.45),

$$d'(c'_m, c'_n) \leq d(c_m, c_n) \quad \text{for all} \quad n, m \in \mathcal{M}.$$

Hence, $\mathcal{B} \succ \mathcal{B}'$; moreover, m_0 belongs to every B_m by the definition of r_m, i.e., $\cap \mathcal{B} \neq \emptyset$. Thus condition (ii) holds for $\mathcal{B}$ and $\mathcal{B}'$; therefore the balls of $\mathcal{B}'$ have a common point, say m'_0. Extend f to the set $S \cup \{m_0\}$ by letting $\tilde{f}(m_0) := m'_0$. By the choice of m'_0 we have

$$d'(\tilde{f}(m_0), \tilde{f}(m)) := d'(m'_0, f(m)) := d(m'_0, c'_m) \leq r'_m.$$

Since the last quantity equals $d(m, m_0)$, this extension is a contraction, and this part of the proof is complete.

We now prove the implication (i)$\Rightarrow$(ii). Let $\mathcal{B}$ and $\mathcal{B}'$ be families of balls subject to conditions (1.42). Assume that $\cap \mathcal{B} \neq \emptyset$ and derive from here and assumption (i) that $\cap \mathcal{B}' \neq \emptyset$. To this end we use the sets $S := \{c_\alpha\}_{\alpha \in A}$ and $S' := \{c'_\alpha\}_{\alpha \in A}$ of the centers of these balls to introduce a map $f : S \to \mathcal{M}'$ by

$$f(c_\alpha) := c'_\alpha, \quad \alpha \in A.$$

Since $\mathcal{B} \succ \mathcal{B}'$, this map satisfies (1.45) and, due to (i), admits an extension which is a contraction $\tilde{f} : \mathcal{M} \to \mathcal{M}'$. As $\tilde{f}$ sends the center of B_α to the center of B'_α and the radii of these balls are equal, we get

$$\tilde{f}(B_\alpha) \subset B'_\alpha \quad \text{for all} \quad \alpha.$$

Consequently, we obtain
$$\emptyset \neq \tilde{f}(\cap \mathcal{B}) \subset \cap \mathcal{B}',$$
and therefore the latter intersection is not empty.

The proof of Proposition 1.31 is complete. □

Applications of the criterion just proved can be made substantially easier by the following characteristic of the target metric space $\mathcal{M}'$.

Definition 1.32. A metric space has the *ball intersection property*, if, for some integer $n > 1$, the following is true:

Every family of closed balls of this space has a common point whenever each of its subfamilies of cardinality n has.

The minimal n of this definition is called the *Helly index* (written $i_H(\mathcal{M})$). We illustrate this concept by examples. As has been mentioned before,

$$i_H(\mathbb{R}) = i_H(\ell_\infty(\Gamma)) = 2.$$

The Helly theorem, in turn, implies that

$$i_H(\mathbb{R}^n) = n + 1. \tag{1.46}$$

We will explain in Appendix C that the same equality holds for an open hemisphere of the n-sphere $\mathbb{S}^n$ and for the hyperbolic space $\mathbb{H}^n$.

Based on this notion we now introduce another version of Proposition 1.31 which is more suitable for applications.

Corollary 1.33. *Assume that $i_H(\mathcal{M}') = n$. Then the following conditions are equivalent:*

(i) *The pair $(\mathcal{M}, \mathcal{M}')$ belongs to $\mathcal{SE}$.*

(ii) *The implication*

$$\cap \mathcal{B} \neq \emptyset \Rightarrow \cap \mathcal{B}' \neq \emptyset \tag{1.47}$$

is true for all families $\mathcal{B} \succ \mathcal{B}'$ containing at most n balls.

Proof. We must only check that (ii) implies the same implication for all families $\mathcal{B} \succ \mathcal{B}'$ consisting of more than n balls. But in this case every subfamily of $\mathcal{B}$ of cardinality n has a nonempty intersection; due to (1.47) the corresponding subfamily of $\mathcal{B}'$ also has. Since $i_H(\mathcal{M}') = n$, this immediately implies that $\cap \mathcal{B}' \neq \emptyset$. □

1.10.3 Proof of Theorem 1.26

Let the Banach space X satisfy the binary intersection property of this theorem. In our terms this means that $i_H(X) = 2$. But for $\mathcal{B} \succ \mathcal{B}'$, each consisting of two balls, the condition (1.47) is trivially true. Then, due to Corollary 1.33, every contraction (a 1-Lipschitz map) from $S \subset \mathcal{M}$ to X admits an extension to a contraction determined on all of $\mathcal{M}$. Notice that the metric of X is positive homogeneous, and therefore the same is true for the corresponding Lipschitz constant $L(f)$.

In the opposite direction we must show that a family of mutually intersecting closed balls $\{\mathcal{B}_\alpha\}$ in X has nonempty intersection assuming that $(\mathcal{M}, X) \in \mathcal{SE}$ for every $\mathcal{M}$. To this end, we identify X with a metric subspace, say $\widehat{X}$, of $\ell_\infty(\Gamma)$ (see Lemma 1.13) and regard the B_α as subsets of $\ell_\infty(\Gamma)$. Let $\widehat{B}_\alpha$ be a closed ball of $\ell_\infty(\Gamma)$ whose center and radius are the same as those of B_α. Then $B_\alpha \subset \widehat{B}_\alpha$ and the balls $\widehat{B}_\alpha$ are mutually intersecting as well. Since $i_H(\ell_\infty(\Gamma)) = 2$, there exists a point, say $\hat{x}$, common to all the $\widehat{B}_\alpha$. To derive from here that $\cap B_\alpha \neq \emptyset$, consider the identity map $id_X : X \to X$. It gives rise to an isometry of $\widehat{X} \subset \ell_\infty(\Gamma)$ into X. By condition (i) of Theorem 1.26 this isometry admits an extension to a 1-Lipschitz map $\varphi : \ell_\infty(\Gamma) \to X$. By the definition of the balls $\widehat{B}_\alpha$ and B_α, we then have $\varphi(\widehat{B}_\alpha) \subset B_\alpha$. Therefore $\varphi(\hat{x})$ is the required common point of the balls B_α.

This completes the proof. □

1.10.4 Lipschitz maps acting in spaces of constant curvature

We now apply the criterion introduced in Corollary 1.33 to the proof of the classical Kirszbraun and Valentine extension theorems. They concern Lipschitz maps acting in Euclidean, spherical and hyperbolic spaces. We begin with the Euclidean case; in view of further applications two different proofs of this result are presented below. In its formulation, E_n denotes an n-dimensional Euclidean space whose norm and scalar product are denoted by $|x|$ and $x \cdot y$.

Theorem 1.34 (Kirszbraun). *The pair (E_n, E_n) belongs to $\mathcal{SE}$.*

First proof (Valentine). Since $i_H(E_n) = n + 1$, we can apply Corollary 1.33. Hence, the following must be established.

Let $\mathcal{B} := \{B_i\}_{i=1}^{k+1}$ and $\mathcal{B}' := \{B_i\}_{i=1}^{k+1}$ be families of closed balls in E_n such that

$$\mathcal{B} \succ \mathcal{B}', \ \cap \mathcal{B} \neq \emptyset \ \text{and} \ 1 \leq k \leq n. \tag{1.48}$$

Then it is true that

$$\cap \mathcal{B}' \neq \emptyset. \tag{1.49}$$

This is derived from a result whose proof (presented in Appendix D) is based on the celebrated Sperner lemma.

Lemma 1.35. *Let $\sigma := \mathrm{conv}\{v_i\}_{i=1}^{k+1}$ be a k-simplex in $\mathbb{R}^n$ (degenerate or nondegenerate). Assume that $\{F_i\}_{i=1}^{k+1}$ is a family of closed subsets of $\mathbb{R}^n$ such that every proper face $\sigma_I := \mathrm{conv}\{v_i\}_{i \in I}$ of σ is covered by the subfamily $\{F_i\}_{i \in I}$. Then*

$$\cap F_i \neq \emptyset.$$

We apply this lemma to the case of a simplex $\Delta' := \mathrm{conv}\{c_i'\}_{i=1}^{k+1}$ formed by the centers of the closed balls $B_i' \in \mathcal{B}'$ and to the family $\mathcal{B}'$. We show that in this setting the assumptions of Lemma 1.35 are fulfilled; then the claim (1.49) would be true and imply the statement of Theorem 1.34.

Assume, on the contrary, that a face $\Delta_I' := \mathrm{conv}\{c_i'\}_{i \in I}$ of Δ' is not covered by the family $\mathcal{B}_I' := \{B_i'\}_{i \in I}$, where I is a proper subset of $\{1, \ldots, k+1\}$. Then there exists a point x' such that

$$x' \in \Delta_I' \setminus \cup \mathcal{B}_I'. \tag{1.50}$$

On the other hand, by (1.48), there is a point x such that

$$x \in \cap \mathcal{B}_I,$$

where $\mathcal{B}_I := \{B_i\}_{i \in I}$. Together with (1.50) this will lead to a contradiction. In view of the subsequent proofs for the hyperbolic and spherical cases, we prefer, however, a more complicated derivation using the inclusion

$$x \in \cap \mathcal{B}_I \cap \Delta_I; \tag{1.51}$$

here $\Delta_I := \mathrm{conv}\{c_i\}_{i \in I}$ is a face of the simplex $\Delta := \mathrm{conv}\{c_i\}_{i=1}^{k+1}$ formed by the centers of $B_i \in \mathcal{B}$.

The latter implication will be proved below, see Lemma 1.36, while for the time being we derive from it and (1.50) the desired contradiction.

Since conditions (1.48) do not change after translations of $\mathcal{B}$ and $\mathcal{B}'$ by different vectors, we assume without loss of generality that

$$x = x' = 0; \tag{1.52}$$

from here and (1.50) and (1.51), it follows that

$$|c_i| \leq r_i = r_i' < |c_i'|, \quad 1 \leq i \leq k+1.$$

This means that

$$c_i' \cdot c_i' > c_i \cdot c_i \quad \text{for} \quad 1 \leq i \leq k+1. \tag{1.53}$$

In turn, the second condition in (1.42) gives

$$c_i \cdot c_i + c_j \cdot c_j - 2c_i \cdot c_j \geq c_i' \cdot c_i' + c_j' \cdot c_j' - 2c_i' \cdot c_j';$$

1.10. Extensions preserving Lipschitz constants

along with (1.53) this implies

$$c'_i \cdot c'_j > c_i \cdot c_j \quad \text{for} \quad 1 \le i, j \le k+1. \tag{1.54}$$

Finally, by (1.52) and the inclusions $x \in \Delta_I$ and $x' \in \Delta'_I$, there exist real numbers α_i and α'_i satisfying

$$\alpha_i \ge 0, \quad \alpha'_i \ge 0, \quad \sum_{i \in I} \alpha_i = 1, \quad \sum_{i \in I} \alpha'_i = 1,$$

and such that

$$\sum_{i \in I} \alpha_i c_i = 0, \quad \sum_{i \in I} \alpha'_i c'_i = 0. \tag{1.55}$$

Multiplying (1.54) by $\alpha_i \cdot \alpha'_j$ and summing over $i, j \in I$, we obtain

$$\left(\sum \alpha_i c'_i\right) \cdot \left(\sum \alpha'_j c'_j\right) > \left(\sum \alpha_i c_i\right) \cdot \left(\sum \alpha'_j c_i\right).$$

In view of (1.55) this inequality is a contradiction.

It remains to prove (1.51), which is a consequence of the following fact.

Lemma 1.36. *Under condition (1.48) and with the previous notation, we have*

$$\cap \mathcal{B}_I \cap \Delta_I \neq \emptyset. \tag{1.56}$$

Proof. Without loss of generality we set $I := \{1, \ldots, k+1\}$ so that $\Delta_I = \Delta$ and $\mathcal{B}_I = \mathcal{B}$. Note that for $k = 1$ the result is trivial, since in this case $\Delta = [c_1, c_2]$. We now assume that the result is true for a family of k balls satisfying (1.48) and prove it for a family of $k+1$ balls. We also assume that $\Delta := \text{conv}\{c_i\}_{i=1}^{k+1}$ is of dimension k; then the case of degenerate Δ is derived from this by taking small perturbations.

Let A be the affine hull of Δ. Consider a collection of $k+2$ convex sets in A comprising $\Delta \cap A$ and k-dimensional balls $B_i \cap A$, $1 \le i \le k+1$. Then every $(k+1)$-element subcollection has a common point. Indeed, this is true for $\{B_i \cap A\}_{i=1}^{k+1}$ by (1.48), and follows for other such subcollections by the induction hypothesis. Hence, the collection $\{\Delta \cap A, B_i \cap A; 1 \le i \le k+1\}$ satisfies the assumption of the Helly theorem and therefore

$$\Delta \cap (\cap \mathcal{B}) \cap A \neq \emptyset.$$

This completes the induction. □

Thus, the Kirszbraun theorem is proved by the Valentine geometric method. We now present an analytic proof of this fact.

Second proof (Mickle). As in the previous proof we must show that the balls B'_i, $1 \le i \le k+1$, have a common point, say x', assuming that the conditions of (1.48) are true.

Let, as before, x be a point of $\bigcap_{i=1}^{k+1} B_i$. The dominance condition makes the following claim plausible:

there exists a point x' so that

$$|x' - c'_i| \leq |x - c_i|, \quad 1 \leq i \leq k+1. \tag{1.57}$$

Since $x \in \cap \mathcal{B}$, the right-hand side is at most r_i. By (1.42), we have $r_i = r'_i$ and therefore (1.57) implies that x' is a common point of the balls B'_i. Hence, it suffices to find a point x' satisfying (1.57).

We find the desired point as a solution to the following extremal problem. Define a function $\varphi : E_n \to \mathbb{R}$ by

$$\varphi(y) := \max_{1 \leq i \leq n} \frac{|y - c'_i|}{|x - c_i|}, \quad y \in \mathbb{R}^n;$$

here we assume, without loss of generality, that $x \neq c_i$ for all i. As φ becomes infinity at infinity, there exists y_0 so that

$$\varphi(y_0) = \min \varphi.$$

If we show that $\varphi(y_0) \leq 1$, then the inequalities in (1.57) will be satisfied for $x' = y_0$. To evaluate this minimum, denote by I the set of $1 \leq i \leq k+1$ for which

$$|y_0 - c'_i| = \varphi(y_0)|x - c_i|; \tag{1.58}$$

then, for $i \notin I$, we get

$$|y_0 - c'_i| < \varphi(y_0)|x - c_i|. \tag{1.59}$$

We show that y_0 belongs to $\mathrm{conv}\{c'_i\}_{i \in I}$. If this is not true, y_0 is separated from the convex hull by a hyperplane, say H. Let ℓ be the straight line orthogonal to H and passing through y_0. Suppose p_i is the orthogonal projection of c'_i onto ℓ, and $p_{\min}$ is the closest to the point y_0 among the points p_i, $i \in I$. Then for all $0 \leq t < 1$ and every $i \in I$ we have

$$\left|(1-t)p_i + ty_0 - c'_i\right| \leq \left|(1-t)p_{\min} + ty_0 - c'_i\right| < \left|y_0 - c'_i\right|.$$

Setting now $y_t := (1-t)p_{\min} + ty_0$ and choosing t sufficiently close to 1, we obtain from here, (1.58) and (1.59), the inequalities

$$|y_t - c'_i| < \varphi(y_0)|x - c_i|, \quad 1 \leq i \leq k+1.$$

Since they contradict the minimality of $\varphi(y_0)$, the point y_0 belongs to $\mathrm{conv}\{c'_i\}_{i \in I}$. This implies

$$\sum_{i \in I} \alpha_i (y_0 - c'_i) = 0 \tag{1.60}$$

for some scalars $\alpha_i \geq 0$ satisfying

$$\sum_{i \in I} \alpha_i = 1.$$

From here we derive the desired inequality

$$\mu := \varphi(y_0) \leq 1. \tag{1.61}$$

To this end, set

$$y_i := x - c_i \quad \text{and} \quad y'_i := y_0 - c'_i.$$

Then $|y'_i| = \mu |y_i|$ for $i \in I$, and therefore (1.42) implies for $i, j \in I$,

$$|y_i - y_j|^2 \geq |y'_i - y'_j|^2 = \mu^2 \big(|y_i|^2 + |y_j|\big)^2 - 2 y'_i y'_j.$$

Rewriting this in the form

$$2 y'_i \cdot y'_j \geq (\mu^2 - 1)\big(|y_i|^2 + |y_j|^2\big) + 2 y_i \cdot y_j,$$

multiplying by the nonnegative products $\alpha_i \alpha_j$ and summing over $i, j \in I$ we then get

$$2 \bigg|\sum_{i \in I} \alpha_i y'_i\bigg|^2 \geq (\mu^2 - 1) \sum_{i,j \in I} \alpha_i \alpha_j \big(|y_i|^2 + |y_j|^2\big) + 2 \bigg|\sum_{i \in I} \alpha_i y_i\bigg|^2.$$

If the α_i are the nonnegative numbers from (1.60), then the left-hand side is zero. The inequality obtained in this way holds only for $\mu^2 \leq 1$.

This proves (1.61) and the theorem. $\square$

Corollary 1.37. *Every pair of Hilbert spaces belongs to* $\mathcal{SE}$.

Proof. Let $\mathcal{B} := \{B_\alpha\}_{\alpha \in A}$ and $\mathcal{B}' := \{B'_\alpha\}_{\alpha \in A}$ be families of closed balls from, respectively, Hilbert spaces H and H'. We must prove the implication

$$\mathcal{B} \succ \mathcal{B}' \quad \text{and} \quad \cap \mathcal{B} \neq \emptyset \Rightarrow \cap \mathcal{B}' \neq \emptyset. \tag{1.62}$$

If $\dim H' < \infty$, this clearly follows from Theorem 1.34. For the case of $\dim H' = \infty$, we use the *finite intersection property* of H', meaning that a family of closed balls has a nonempty intersection, if every finite subfamily has. The result follows from the similar property for a family of compact subsets in a topological space and the weak compactness of closed balls of a Hilbert space, see [DS-1958, Sec. V.4]. So it suffices to prove (1.62) for finite families of balls. Intersecting the balls of such a family by finite-dimensional subspaces of H and H', respectively, passing through the centers of these balls, we reduce the problem to the case of finite-dimensional Hilbert spaces and then use once again Theorem 1.34. $\square$

We use now the method of the first proof for Theorem 1.34 to establish a similar result for hyperbolic and spherical spaces. We begin with the n-dimensional hyperbolic space $\mathbb{H}^n$; the *hyperbolic model* of this space is the most suitable for the consequent proof (see, e.g., [Rat-1994, Ch. 3] for the geometric fact used below). Namely, $\mathbb{H}^n$ is identified with the upper sheet of a hyperboloid in $\mathbb{R}^{n+1}$ endowed with the geodesic metric. In more detail,

$$\mathbb{H}^n := \{x \in \mathbb{R}^{n+1} \ ; \ b(x,x) = 1, \ x_{n+1} > 0\},$$

where the bilinear form b is given by

$$b(x,y) := \sum_{i=1}^{n} x_i y_i - x_{n+1} y_{n+1}, \quad x, y \in \mathbb{R}^{n+1},$$

and the geodesic distance $d(x,y)$ between $x, y \in \mathbb{H}^n$ is defined to be the length of the *geodesic segment* $[x,y]$ (the image of a curve of the shortest length joining x and y). This geodesic segment is unique and lies in the intersection of $\mathbb{H}^n$ with the 2-dimensional subspace of $\mathbb{R}^{n+1}$ determined by its endpoints. Finally, the distance d is calculated by the formula

$$\cosh d(x,y) = b(x,y), \quad x, y \in \mathbb{H}^n.$$

Since the hyperbolic cosine is strictly increasing on the positive half-line $\mathbb{R}_+$, we get the implication

$$d(x,y) > d(x',y') \iff b(x,y) > b(x',y'). \tag{1.63}$$

Based on this information we now prove the following

Theorem 1.38 (Valentine). $(\mathbb{H}^n, \mathbb{H}^n) \in \mathcal{SE}$.

Proof. As before, we should prove the implication (1.62) but now for families of closed balls in $\mathbb{H}^n$. Since $i_H(\mathbb{H}^n) = n+1$ (see Appendix B), each of these families contains at most $k+1$ balls with $1 \le k \le n$. To establish (1.62) under these conditions, we follow the line of the first proof of Theorem 1.34 indicating simple changes which are required in our setting.

We first introduce an analog of Lemma 1.35. Keeping the notation of this lemma, consider a simplex $\sigma \subset \mathbb{H}^n$ with vertices v_i, $1 \le i \le k+1$, and a collection of closed subsets $\{F_i\}_{i=1}^{k+1}$ from $\mathbb{H}^n$. Here the following definitions are used.

A set $C \subset \mathbb{H}^n$ is said to be *convex*, if the geodesic segment joining every pair of points of this set lies in C. In turn, a simplex of vertices v_i, $1 \le i \le k+1$, is defined as the intersection of all convex sets containing $\{v_i\}_{i=1}^{k+1}$.

It is known and easily verified that a simplex is contained in the smallest convex cone with apex at 0 containing all its vertices. To check that the analog of Lemma 1.35 is true for $\mathbb{H}^n$, we orthogonally project $\mathbb{H}^n$ onto the hyperplane $\sum_{i=1}^{n+1} x_i = 0$. Since this projection is a homeomorphism, and the image of σ is a

1.10. Extensions preserving Lipschitz constants

Euclidean k-simplex, the result for the hyperbolic case immediately follows from the Euclidean one.

Second, Lemma 1.36 also holds for the hyperbolic case. In fact, its proof uses only the Helly theorem. Therefore, it suffices to note that the balls and the simplices of $\mathbb{H}^n$ are convex in the sense of the above introduced notion of convexity, and that the Helly theorem is true for families of such convex sets (see Appendix B).

Finally, the second condition of dominance in (1.42) and assertion (1.63) imply that

$$b(c_i, c_j) \geq b(c'_i, c'_j), \quad 1 \leq i, j \leq k+1, \tag{1.64}$$

for the families of the centers of $\mathcal{B}$ and $\mathcal{B}'$.

Using now the notation of the proof of Theorem 1.34 we find, using (1.62) and the analog of Lemma 1.36, points $x, x' \in \mathbb{H}^n$ such that

$$x \in \cap \mathcal{B}_I \cap \Delta_I \quad \text{and} \quad x' \in \Delta'_I \setminus \cup \mathcal{B}'_I, \tag{1.65}$$

where Δ and Δ' are the simplices of $\mathbb{H}^n$ with vertices c_i and c'_i, $1 \leq i \leq k+1$, respectively.

Now (1.63) and the first conditions in (1.65) and (1.62) imply

$$b(x', c'_i) > b(x, c_i), \quad i \in I. \tag{1.66}$$

Since $x \in \Delta_I$, this point lies in the smallest convex cone with apex at 0 containing the points c_i, $i \in I$. The corresponding assertion with primes holds for x'. Hence there exist real numbers α_i and α'_i satisfying

$$\alpha_i \geq 0, \quad \alpha'_i \geq 0, \quad \sum_{i \in I} \alpha_i \neq 0, \quad \sum_{i \in I} \alpha'_i \neq 0$$

and such that

$$x = \sum_{i \in I} \alpha_i c_i, \quad x' = \sum_{i \in I} \alpha'_i c'_i. \tag{1.67}$$

Multiplying (1.64) by $\alpha_i \alpha'_j$ and summing over $i, j \in I$, we obtain

$$b\left(\sum \alpha_i c_i, \sum \alpha'_j c_j\right) \geq b\left(\sum \alpha_i c'_i, \sum \alpha'_j c'_j\right),$$

whence by (1.67),

$$b\left(x, \sum \alpha'_j c_j\right) \geq b\left(\sum \alpha_i c'_i, x'\right).$$

Similarly, multiplying (1.66) by α_i and summing over $i \in I$, we get

$$b\left(x', \sum a_i c'_i\right) > b(x, x).$$

The last two inequalities imply that

$$b\left(x, \sum \alpha'_j c_j\right) > b(x, x).$$

However, multiplying (1.66) by α'_i and summing over $i \in I$, we get

$$b(x', x') > b\left(x, \sum \alpha'_j c_j\right).$$

Since x, x' are points of $\mathbb{H}^n$, we have $b(x, x) = b(x', x') = 1$ and the above inequality is a contradiction.

Thus (1.65) is not true and therefore Δ_I is covered by the family $\mathcal{B}'_I$. To complete the proof, it remains to apply the analog of Lemma 1.35. $\square$

Remark 1.39. The assertion of Theorem 1.38 is also true for the case of the pair $(\mathbb{S}^n, \mathbb{S}^n_+)$, where

$$\mathbb{S}^n_+ := \{x \in \mathbb{R}^{n+1} \; ; \; x \cdot x = 1 \text{ and } x_{n+1} > 0\}.$$

Here $\mathbb{S}^n$ is equipped with the geodesic metric defined by geodesic segments. In this case a geodesic segment lies in a great circle of $\mathbb{S}^n$, given by its intersection with a 2-dimensional subspace of $\mathbb{R}^{n+1}$. The geodesic distance d is calculated by the formula

$$\cos(d(x,y)) = x \cdot y, \quad 0 \leq d(x,y) \leq \pi, \quad x, y \in \mathbb{S}^n.$$

Since cosine is strictly decreasing on $[0, \pi]$, this implies the implication

$$d(x,y) > d(x', y') \iff x \cdot y < x' \cdot y'. \tag{1.68}$$

Now convex sets and simplices are introduced for $\mathbb{S}^n$ similarly to those in $\mathbb{H}^n$. Unfortunately, the Helly index for $\mathbb{S}^n$ is greater than $n+1$ (consider three mutually intersecting arcs covering the circle). But it is true that

$$i_H(\mathbb{S}^n_+) = n + 1$$

and, moreover, the Helly theorem holds for convex subsets of the open hemisphere (see Appendix C). Therefore we can proceed as in the proof of Theorem 1.38. The only difference is the signs in inequalities (1.63) and (1.68). This leads to the uniform change in the direction of the inequality signs, but does not change the final conclusion. Hence the following is true:

$$(\mathbb{S}^n, \mathbb{S}^n_+) \in \mathcal{SE}. \tag{1.69}$$

We use (1.69) to prove the last result of this subsection concerning spherical spaces. In this case, the method of the proof for hyperbolic spaces breaks down as $i_H(\mathbb{S}^n) > n+1$. This obstacle will be overcome by using a property of contractions of the n-sphere, see Proposition 1.41 below.

1.10. Extensions preserving Lipschitz constants

Theorem 1.40 (Valentine). $(\mathbb{S}^n, \mathbb{S}^n) \in \mathcal{SE}$.

Proof. The main step of the proof is the following striking fact whose derivation is presented in Appendix D.

Proposition 1.41. *Let f be a 1-Lipschitz map from a subset of $\mathbb{S}^n$ into $\mathbb{S}^n$. If the image of f does not lie in any closed hemisphere, then f is an isometry.*

Now, the proof of the theorem falls naturally into two parts.

Case I. Suppose that $f : \Sigma \to \mathbb{S}^n$ is a 1-Lipschitz map of the subset $\Sigma \subset \mathbb{S}^n$, and $f(\Sigma)$ lies in a closed hemisphere H_n. Let $H_n^\varepsilon \subset H_n$ be the open ball of $\mathbb{S}^n$ having the same center as H_n and radius $\frac{\pi}{2} - \varepsilon$. Set $\Sigma_\varepsilon := f^{-1}(H_n^\varepsilon) \subset \Sigma$ and consider the 1-Lipschitz map $f_\varepsilon := f|_{\Sigma_\varepsilon}$ whose image is contained in H_n^ε. According to (1.69) f_ε admits a 1-Lipschitz extension to a map $\tilde{f}_\varepsilon : \mathbb{S}^n \to \mathbb{S}^n$. By the Arcela–Ascoli theorem, there exists a limit point, say $\tilde{f}$, of the set $\{\tilde{f}_\varepsilon\}_{\varepsilon>0}$. This $\tilde{f}$ is clearly 1-Lipschitz and agrees with f on Σ. Hence $\tilde{f}$ is the required extension.

Case II. Suppose now that $f(\Sigma)$ does not lie in any closed hemisphere. Then by Proposition 1.41, f is an isometry of Σ onto $f(\Sigma)$. We show that f admits an extension to an isometry (orthogonal transform of $\mathbb{R}^{n+1}$) of $\mathbb{S}^n$.

Let $\{x'_1, \ldots, x'_m\}$, $1 \le m \le n+1$, be a *maximal* linearly independent set in $f(\Sigma)$. We claim that $m = n+1$; in fact, $f(\Sigma)$ lies in the intersection of $\mathbb{S}^n$ and the affine hull of the set $\{x'_i\}$, that is, in a sphere of dimension $m-1$. Since $f(\Sigma)$ does not lie in any closed hemisphere, m must be equal to n.

Consider now the set of preimages $x_i := f^{-1}(x'_i)$, $1 \le i \le n+1$. Since f is an isometry, $d(x_i, x_j) = d(x'_i, x'_j)$ for all i, j. By the definition of d, see Remark 1.39,

$$x_i \cdot x_j = x'_i \cdot x'_j, \quad 1 \le i, j \le n+1. \tag{1.70}$$

This implies the existence of an isometry ρ of $\mathbb{S}^n$ such that

$$\rho(x_i) = x'_i \ (:= f(x_i)), \quad 1 \le i \le n+1.$$

We claim that, for every $x \in \Sigma \setminus \{x_i\}$,

$$f(x) = \rho(x).$$

In fact, the isometries f and ρ agree on the set $\{x_i\}_{i=1}^{n+1}$. Therefore

$$d(\rho(x), x'_i) = d(x, x_i) = d(f(x), x'_i), \quad 1 \le i \le n+1.$$

As in the derivation of (1.70), this implies that

$$\rho(x) \cdot x'_i = f(x) \cdot x'_i, \quad 1 \le i \le n+1. \tag{1.71}$$

Now let $\rho(x) = \sum \alpha_i x'_i$ and $f(x) = \sum \beta_i x'_i$ for some scalars α_i, β_i. From here and (1.71) we obtain the homogeneous system of linear equations

$$\sum_{i=1}^{n+1} (\alpha_i - \beta_i) x'_i \cdot x'_j = 0, \quad 1 \le j \le n+1.$$

Since the determinant (Grammian) of this system is not zero, $\alpha_i = \beta_i$, and hence $f(x) = \rho(x)$. This shows that $f|_\Sigma = \rho$, that is, ρ is the required extension of f to all of $\mathbb{S}^n$. $\square$

Corollary 1.42. (i) *For all $m, n \geq 1$,*
$$(\mathbb{H}^m, \mathbb{H}^n) \in \mathcal{SE}.$$

(ii) *For all $1 \leq m \leq n$,*
$$(\mathbb{S}^m, \mathbb{S}^n) \in \mathcal{SE}.$$

Proof. (i) If $m \leq n$, we consider $\mathbb{H}^m$ as a metric subspace of $\mathbb{H}^n$ given by
$$\mathbb{H}^m := \{x \in \mathbb{H}^n \,;\, x_1 = x_2 = \cdots = x_{n-m} = 0\}.$$

Then $f : \mathcal{S} \to \mathbb{H}^n$ can be regarded as a map from a subset of $\mathbb{H}^n$. Its extension to all of $\mathbb{H}^n$ and then the restriction of this extension to $\mathbb{H}^m$ yields the desired result.

Now let $m > n$, and f map a subset $\mathcal{S}$ of $\mathbb{H}^m$ into $\mathbb{H}^n$. Using the canonical embedding $\mathbb{H}^n \subset \mathbb{H}^m$, we first extend f to a 1-Lipschitz map from $\mathbb{H}^m$ to $\mathbb{H}^m$, say, $\tilde{f}$. Note that the canonical projection $p : \mathbb{H}^m \to \mathbb{H}^n$ is 1-Lipschitz, and therefore $p \circ \tilde{f}$ is the required extension.

(ii) In this case $m \leq n$ and the canonical embedding $\mathbb{S}^m \subset \mathbb{S}^n$ yields the required result. $\square$

Remark 1.43. The second assertion of this corollary is not true for $m > n$, by the topological argument presented in Section 1.6, see example (c) after Definition 1.10.

1.11 Lipschitz extensions

Definition 1.44. A pair of metric spaces $(\mathcal{M}, \mathcal{M}')$ has the Lipschitz extension property (briefly, belongs to $\mathcal{LE}$), if its Lipschitz constant $\Lambda(\mathcal{M}, \mathcal{M}')$ is finite.

In other words, for some constant $C = C(\mathcal{M}, \mathcal{M}') \geq 1$ every Lipschitz map f from a subset $\mathcal{S} \subset \mathcal{M}$ into $\mathcal{M}'$ admits an extension $\tilde{f} : \mathcal{M} \to \mathcal{M}'$ such that
$$L(\tilde{f}) \leq C L(f). \tag{1.72}$$

Let us recall, see (1.28), that
$$\Lambda(\mathcal{M}, \mathcal{M}') := \inf C.$$

The simplest example of an $\mathcal{LE}$-pair is $(\mathcal{M}, X)$, where X is a finite-dimensional Banach space. In fact, coordinatewise application of McShane's extension Theorem 1.27 gives
$$\Lambda(\mathcal{M}, X) \leq \dim X.$$

1.11. Lipschitz extensions

It follows from Theorem 1.49 below and Kadets-Snobar result [KS-1971] that the right-hand side can be replaced by $\sqrt{\dim X}$ while the results of Section 1.10 show that this Lipschitz constant may attain the minimal value 1. In particular,

$$\Lambda(\ell_p^n, \ell_q^n) = 1 \tag{1.73}$$

whenever $q = \infty$ or $(p, q) = (2, 2)$; Theorem 1.46 below implies that it is also true for $0 < p \leq \frac{1}{2}$ and $q = 2$.

Let us recall that the metric of ℓ_p^n is given at $x, y \in \mathbb{R}^n$ for $1 \leq p \leq \infty$ by

$$\|x - y\|_p := \left(\sum_{i=1}^n |x_i - y_i|^p \right)^{1/p}$$

and for $0 < p < 1$ by

$$\|x - y\|_p := \sum_{i=1}^n |x_i - y_i|^p.$$

Remark 1.45. Except these cases the Lipschitz constant in (1.73) is apparently greater than 1. To confirm this claim one considers the case of $p = \infty$ and $q = 2$ (the reader may check that the same derivation works for $1 \leq q < \infty$ and then modify it for $1 < p < 2$).

Consider a family $\mathcal{B}$ of $n+1$ closed balls (cubes) in ℓ_∞^n of radius 1 centered at the points $c_i := -e_i + \sum_{j \neq i} e_j$, $1 \leq i \leq n$, and $c_{n+1} = \sum_{j=1}^n e_j$; here $\{e_i\}_{i=1}^n$ is the standard orthogonal basis of $\mathbb{R}^n$. Note that $\|c_i - c_j\|_\infty = 2$ for all $i \neq j$.

Now let $\{c_i'\}_{i=1}^{n+1}$ be the vertices of a regular n-simplex in ℓ_2^n whose edges are of length 2. Consider the family $\mathcal{B}'$ of $n+1$ closed balls in ℓ_2^n of radius 1 centered at c_i', $1 \leq i \leq n+1$. By our construction

$$\mathcal{B} \succ \mathcal{B}' \quad \text{and} \quad \cap \mathcal{B} = \{0\},$$

but $\cap \mathcal{B}' = \emptyset$. Indeed, every two balls in $\mathcal{B}'$ are separated by the hyperplane tangent to each of them at a point of the boundary of the simplex. Due to Proposition 1.31 a 1-Lipschitz map sending c_i into c_i' cannot be extended to the point $0 \in \ell_\infty^n$ preserving the same Lipschitz constant. Calculating the radius of the circumscribed Euclidean ball of the regular simplex of ℓ_2^n, we then get

$$\Lambda(\ell_\infty^n, \ell_2^n) \geq \prod_{j=2}^n \left(1 - \frac{1}{j^2} \right)^{-\frac{1}{2}} > 1 \quad \text{(and tends to } \sqrt{2} \text{ as } n \to \infty\text{)}.$$

There are also infinite-dimensional pairs of Banach spaces belonging to $\mathcal{LE}$. The simplest example is (L_p, L_∞), $0 < p \leq \infty$ (in this case the Lipschitz constant is 1, see Remark 1.30). Here $L_p = L_p[0, 1]$ for $1 \leq p \leq \infty$ and

$$\|f - g\|_p := \int_0^1 |f - g|^p \, dx \quad \text{for} \quad 0 < p < 1.$$

Essentially more interesting (but very difficult) results concern pairs (L_p, L_q) with $q < \infty$. The results presented in subsection 6.5.2 of Volume II, in particular, imply that $(L_p, L_q) \in \mathcal{LE}$ for $2 \leq p < \infty$ and $1 < q \leq 2$, and $(L_p, L_q) \notin \mathcal{LE}$ for $1 \leq p < 2$. The long-standing conjecture of Johnson and Lindenstrauss claims that $(L_p, L_2) \in \mathcal{LE}$ for $2 \leq p < \infty$. For the time being, even a fairly plausible conjecture has not formulated for other pairs $p, q \in [1, \infty)$.

The essential part of the book, see, in particular, Chapter 6 (Volume II) and related material in Chapters 4 and 5, is devoted to the study of two problems concerning the class $\mathcal{LE}$. The first one is

Dominance Problem. *Given a metric space $\mathcal{M}$ find all or a sufficiently large class of metric spaces $\mathcal{M}'$ such that*

$$\Lambda(\mathcal{M}', \mathcal{M}) < \infty.$$

The solutions to the problem presented in Chapter 6 include also direct constructions of extension operators and estimates of the corresponding Lipschitz constants in terms of geometric and analytic characteristics of the involved metric spaces.

At this stage we illustrate solutions of such a kind choosing a Hilbert space H as a target space.

Theorem 1.46 (Minti [Min-1970]). *Assume that a metric space $(\mathcal{M}, d)$ is such that d^p is, for some $2 \leq p < \infty$, also a metric on $\mathcal{M}$. Then we have*

$$\Lambda(\mathcal{M}, H) = 1.$$

Proof. Because of the inequality $(a+b)^{2/p} \leq a^{2/p} + b^{2/p}$, $a, b > 0$, $p \geq 2$, the assumption of the theorem holds for $p = 2$ if it does for $p > 2$. Hence we may assume that d^2 is a metric, i.e., for $m', m', m'' \in \mathcal{M}$,

$$d^2(m, m') \leq d^2(m, m'') + d^2(m', m''). \tag{1.74}$$

Now let f be a 1-Lipschitz map from a subset $S \subset \mathcal{M}$ into H (the case of C-Lipschitz maps clearly reduces to this one). As in the proof of Theorem 1.34, the required assertion is derived from the following geometric fact.

Let $\mathcal{B} := \{B_i\}_{i=1}^n$ and $\mathcal{B}' := \{B_i'\}_{i=1}^n$ be arbitrary finite subsets of closed balls in $\mathcal{M}$ and H, respectively. Assume that (see (1.43))

$$\mathcal{B} \succ \mathcal{B}' \quad \text{and} \quad \cap \mathcal{B} \neq \emptyset.$$

Then $\cap \mathcal{B}' \neq \emptyset$.

The claim holds if some center of balls B_i, say c_{i_0}, belongs to $\cap \mathcal{B}$, since due to (1.42) $c_{i_0}' \in \cap \mathcal{B}'$.

Now let $x \in \cap \mathcal{B}$ and differ from all c_i. Following the argument of the second proof of Theorem 1.34 we define a function $\varphi : H \to \mathbb{R}$ given by

$$\varphi(y) := \max_{0 \leq i \leq n} \frac{\|y - c_i'\|_H}{d(x, c_i)}.$$

1.11. Lipschitz extensions

Since φ is infinity at infinity and the closed sphere of H is weakly compact, there is a minimum point y_0 for φ.

Let I be the set of indices satisfying

$$\|y_0 - c_i'\|_H = \varphi(y_0) d(x, c_i). \tag{1.75}$$

Setting $y_i' := y_0 - c_i'$, $i \in I$, and arguing as in the second proof of Theorem 1.34, one finds nonnegative numbers α_i such that

$$\sum_{i \in I} \alpha_i y_i' = 0, \quad \sum_{i \in I} \alpha_i = 1. \tag{1.76}$$

Moreover, the condition $\mathcal{B} \succ \mathcal{B}'$ implies that

$$\|y_i' - y_j'\|_H \leq d(c_i, c_j).$$

Squaring this inequality, multiplying by $\alpha_i \alpha_j$ and then summing over $i, j \in I$ and applying (1.75) and (1.76), we obtain for $\mu := \varphi(y_0)$,

$$0 = \sum_{i,j \in I} \alpha_i \alpha_j \left(\|y_i'\|_H^2 + \|y_j'\|_H^2 - \|y_i' - y_j'\|_H^2 \right)$$

$$\geq \mu^2 \sum_{i,j \in I} \alpha_i \alpha_j \left(d(x, c_i)^2 + d(x, c_j)^2 - d(c_i, c_j)^2 \right)$$

$$+ (\mu^2 - 1) \sum_{i,j \in I} \alpha_i \alpha_j d(c_i, c_j)^2.$$

Both sums on the right-hand side are nonnegative, since $\alpha_i \geq 0$ and d^2 satisfies (1.74). Hence, the inequality can be true only if $\mu := \varphi(y_0) \leq 1$. This implies the existence of a common point for the family $\{B_i'\}$. □

If a target space $\mathcal{M}$ is not homotopically trivial, e.g., $\mathcal{M}$ is the Euclidean n-sphere, the situation is much more complicated. In fact, in this case there exist obstructions even to existence of continuous extensions (for $\mathbb{S}^n$ such an obstruction is its dimension, see the proof of Brower's fixed point theorem in Section 1.6). In Chapter 6, we present an extension theorem which implies a Lipschitz analog of Hurewicz's Theorem 1.9 with topological dimension replaced by one of its metric substitutes (Nagata dimension).

However, a Lipschitz analog of topological obstruction theory has not been developed for the time being. The solution to the following problem might be considered as a starting point for developing such a theory.

Problem. *For which pairs $\mathcal{M}', \mathcal{M}$ the following principle is valid.*

Assume that for every closed subset $S \subset \mathcal{M}'$ and Lipschitz $f : S \to \mathcal{M}$ there exists a continuous extension of f to $\mathcal{M}'$. Then there exists a Lipschitz extension of f to $\mathcal{M}'$.

Some concepts and results in this direction will be discussed in Chapter 6 of Volume II, see, in particular, Theorem 6.12.

The second main problem is formulated as follows.

Universal Target Space Problem. *Find metric spaces $\mathcal{M}$ for which*

$$\Lambda(\mathcal{M}', \mathcal{M}) < \infty$$

for every *metric space $\mathcal{M}'$.*

We call such $\mathcal{M}$ a *universal Lipschitz target space* (briefly, *ULT*-space).

The subsequent result presents the basic properties of *ULT*-spaces. The second of them allows us to reformulate the problem in a quantitative form.

Proposition 1.47. *Let $(\mathcal{M}, d)$ be a ULT-space. Then the following holds:*

(a) *$\mathcal{M}$ is complete;*

(b) *The universal Lipschitz constant of $\mathcal{M}$ given by*

$$\Lambda(\mathcal{M}) := \sup_{\mathcal{M}'} \Lambda(\mathcal{M}', \mathcal{M}) \tag{1.77}$$

is finite.

Proof. (a) Let $\{m_j\}$ be a Cauchy sequence in $\mathcal{M}$ and $\bar{m}$ be its limit in the completion $\overline{\mathcal{M}}$ of $\mathcal{M}$. The identity map $m_j \mapsto m_j$ regarded as a map from a subset of $\overline{\mathcal{M}}$ into $\mathcal{M}$ admits a Lipschitz extension, say f, on the whole of $\overline{\mathcal{M}}$. Clearly $f(\bar{m})$ is the limit of the sequence $f(\{m_j\}) := \{m_j\}$ in $\mathcal{M}$.

(b) If, on the contrary, $\Lambda(\mathcal{M}) = \infty$, then for every integer $n \geq 1$ there exists a metric space $(\mathcal{M}_n, d_n)$ such that

$$\Lambda(\mathcal{M}_n, \mathcal{M}) \geq n. \tag{1.78}$$

Let us define a metric space $(\widetilde{\mathcal{M}}, \widetilde{d})$ with the underlying set consisting of those points $(m_n)_{n \in \mathbb{N}}$ of the direct product $\prod_{n \in \mathbb{N}} \mathcal{M}_n$ that satisfy

$$\sup_n d_n(m_n, m_n^o) < \infty;$$

here $(m_n^o)_{n \in \mathbb{N}}$ is a fixed point of the direct product. The metric $\widetilde{d}$ is given at $m := (m_n)_{n \in \mathbb{N}}$, $m' := (m_n')_{n \in \mathbb{N}}$ by the formula

$$\widetilde{d}(m, m') := \sup_n d_n(m_n, m_n').$$

By the assumption there exists a constant $C \geq 1$ such that

$$\Lambda(\widetilde{\mathcal{M}}, \mathcal{M}) \leq C. \tag{1.79}$$

1.11. Lipschitz extensions

Choosing $N > C$ we will derive from (1.79) that

$$\Lambda(\mathcal{M}_N, \mathcal{M}) \leq C \qquad (1.80)$$

in contradiction with (1.78).

To this end given a set $S \subset \mathcal{M}_N$ and a Lipschitz map $f \in \mathrm{Lip}(S, \mathcal{M})$ we find an extension $\hat{f} : \mathcal{M}_N \to \mathcal{M}$ such that $L(\hat{f}) \leq CL(f)$. This will clearly mean that (1.80) holds.

Let $\widetilde{S} := \{(m_n)_{n \in \mathbb{N}} \, ; \, m_n = m_n^o \text{ for } n \neq N \text{ and } m_N \in S\}$; clearly, $\widetilde{S} \subset \widetilde{\mathcal{M}}$. Further, a map $\tilde{f} : \widetilde{S} \to \mathcal{M}$ is given by

$$\tilde{f}(m) := f(m_N), \quad m \in \widetilde{S}.$$

Due to (1.79) there exists an extension $F : \widetilde{\mathcal{M}} \to \mathcal{M}$ of $\tilde{f}$ such that

$$L(F) \leq CL(\tilde{f}),$$

that is, for every pair $m, m' \in \widetilde{\mathcal{M}}$,

$$d(F(m), F(m')) \leq CL(\tilde{f}) \sup_n d_n(m_n, m_n').$$

Taking here m, m' with $m_n = m_n' = m_n^o$ for $n \neq N$ and denoting these points by $\hat{m}, \hat{m}'$ we get

$$d(F(\hat{m}), F(\hat{m}')) \leq CL(\tilde{f}) d_N(m_N, m_N'). \qquad (1.81)$$

Moreover, since $\tilde{d}(m, m') = d_N(m_N, m_N')$ and $\tilde{f}(m) = f(m_N)$, $\tilde{f}(m') = f(m_N')$ for $m, m' \in \widetilde{S}$, we get

$$L(\tilde{f}) = \sup_{m_N \neq m_N'} \frac{d(f(m_N), f(m_N'))}{d_N(m_N, m_N')} = L(f).$$

Finally, the function $\hat{f} : m_N \mapsto F(\hat{m})$ is clearly an extension of f to $\mathcal{M}_N$ which due to (1.81) satisfies

$$L(\hat{f}) \leq CL(f).$$

This proves (1.80) and the proposition. □

One of the first examples of an ULT-space is given in Section 1.10 where it was proved that $\Lambda(\mathcal{M}, \ell_\infty(\Gamma)) = 1$ for every metric space $\mathcal{M}$. In other words,

$$\Lambda(\ell_\infty(\Gamma)) = 1. \qquad (1.82)$$

We essentially enlarge the number of such examples using the relation of the before introduced concept with Lipschitz and Banach space analogs of absolute metric retract, see Definitions 1.10 and 1.11. To define the first we simply replace in those definitions continuous maps by Lipschitz ones. In particular, a *Lipschitz retraction* of a metric space $\mathcal{M}$ onto its subspace S is a Lipschitz map $\rho : \mathcal{M} \to S$ such that $\rho|_S$ is the identity map of S. The subspace S is called a *Lipschitz retract* of $\mathcal{M}$.

In turn, $\mathcal{M}$ is called an *absolute Lipschitz retract* (briefly, *ALR*) if $\mathcal{M}$ is a Lipschitz retract of every metric space containing $\mathcal{M}$ isometrically.

As in the case of $\Lambda(\mathcal{M})$ the last notion has an equivalent quantitative definition. It exploits an *absolute Lipschitz retraction constant* denoted by $R(\mathcal{M})$ and given by

$$R(\mathcal{M}) := \sup_{\mathcal{M}' \supset \mathcal{M}} \left\{ \inf_{\rho : \mathcal{M}' \to \mathcal{M}} L(\rho) \right\},$$

where ρ runs over all Lipschitz retractions onto $\mathcal{M}$ of a metric space $\mathcal{M}'$ containing $\mathcal{M}$ isometrically.

Then, $\Lambda(\mathcal{M})$ is an absolute Lipschitz retract if and only if $R(\mathcal{M}) < \infty$.

Since every metric space is isometrically embedded into some $\ell_\infty(\Gamma)$, see Lemma 1.13, the argument of Proposition 1.47 can be easily adapted to prove equivalence of these two definitions of *ALR*.

Now the following result gives the aforementioned relation between *ULT*- and *ALR*-spaces.

Proposition 1.48. $\Lambda(\mathcal{M}) = R(\mathcal{M})$. *In particular, the classes of ULT- and ALR-spaces coincide.*

Proof. Let $\Lambda(\mathcal{M}) < \infty$ and $\mathcal{M}'$ contains $\mathcal{M}$ isometrically. Then the identity map of $\mathcal{M}$ regarded as a subspace of $\mathcal{M}'$ can be for every $\varepsilon > 0$ extended to $\mathcal{M}'$ with Lipschitz constant at most $(1+\varepsilon)\Lambda(\mathcal{M})$. This implies the inequality $R(\mathcal{M}) \leq \Lambda(\mathcal{M})$ that trivially holds also for $\Lambda(\mathcal{M}) = \infty$.

Conversely, let $R(\mathcal{M}) < \infty$ and f be a Lipschitz map from a subset of a metric space $\mathcal{M}'$ into $\mathcal{M}$. We should show that for every $\varepsilon > 0$ the f admits an extension $\widetilde{f} : \mathcal{M}' \to \mathcal{M}$ with $L(\widetilde{f}) \leq (1+\varepsilon)R(\mathcal{M})$. To this end we regard $\mathcal{M}$ as a metric subspace of some $\ell_\infty(\Gamma)$, see Lemma 1.13. Then there exists a Lipschitz retraction $\rho : \ell_\infty(\Gamma) \to \mathcal{M}$ with $L(\rho) \leq (1+\varepsilon)R(\mathcal{M})$. Further, considering f as a function from S into $\ell_\infty(\Gamma)$ we extend it to a function $\widehat{f} : \mathcal{M}' \to \ell_\infty(\Gamma)$ with $L(\widehat{f}) = L(f)$, see the text above Remark 1.30. Then composition $\rho \circ \widehat{f} : \mathcal{M}' \to \mathcal{M}$ is clearly an extension of f with Lipschitz constant at most $(1+\varepsilon)R(\mathcal{M})$.

This proves the converse inequality $\Lambda(\mathcal{M}) \leq R(\mathcal{M})$ that trivially holds also for $R(\mathcal{M}) = \infty$. □

As a typical application of this result we indicate the following:

$$\Lambda(c_0) \leq 2. \tag{1.83}$$

In fact, the map $r : \ell_\infty \to c_0$ used in Section 1.6 (see example (b) there) to prove that c_0 is a metric retract of ℓ_∞ is evidently 2-Lipschitz. This implies that $R(c_0) \leq 2$ and the result follows.

1.11. Lipschitz extensions

The reader may find several interesting examples of *ALR*- (hence, *ULT*-) spaces in the book [BL-2000, Ch.2] by Benyamini and Lindenstrauss. We present here only one concerning the metric space $\mathcal{C}(X)$ of convex bounded nonempty closed subsets of a Banach space X equipped with the Hausdorff metric, see subsection 1.9.1 for details. Then Theorem 2.7 of the cited book implies that

$$\Lambda(\mathcal{C}(X)) \leq 8.$$

Now we prove a Banach analog of Proposition 1.48 where the role of the *ALR* constant plays its substitute, the *projective constant* of a Banach space X denoted by $\Pi(X)$. It is given by

$$\Pi(X) := \sup_Y \inf_P \|P\|,$$

where P runs over all linear projections of a Banach space Y onto its subspace linearly isometric to X and Y runs over all Banach spaces containing such subspaces.

To formulate the result we also need the following notion.

A Banach space X is said to be *constrained in its bidual* if X is the range of a norm one linear projection when canonically embedded in its bidual X^{**}.

It is well known (see Dixmier [Di-1948]) that an L_1-space or a dual Banach space X (i.e., $X = Y^*$ for a Banach space Y) meets this condition. Also, it is easy to see that a Banach space X is constrained in its bidual iff X is isometric to the range of a norm one linear projection in some dual space.

Theorem 1.49. $\Lambda(X) \leq \Pi(X)$ *with equality holding for X being constrained in its bidual.*

Proof. To prove the inequality it suffices to check that for every $\varepsilon > 0$, $S \subset \mathcal{M}$ and $f : S \to X$ with $L(f) \leq 1$ there exists an extension $\widehat{f} : \mathcal{M} \to X$ of f such that

$$L(\widehat{f}) \leq (1+\varepsilon)\Pi(X).$$

To do this we first note that since every Banach space admits a linear isometric embedding into some $\ell_\infty(\Gamma)$, it suffices in the definition of $\Pi(X)$ to take supremum only over all Banach spaces $\ell_\infty(\Gamma)$ containing X isometrically as a subspace, see, e.g., Day [Da-1955]. We may therefore pick a space $\ell_\infty(\Gamma)$ containing X and a linear projection $P : \ell_\infty(\Gamma) \to X$ so that

$$\|P\| \leq (1+\varepsilon)\Pi(X).$$

Since $X \subset \ell_\infty(\Gamma)$, the f may be regarded as a map from S into $\ell_\infty(\Gamma)$. But every such f admits an extension $\widetilde{f} : \mathcal{M} \to \ell_\infty(\Gamma)$ such that $L(\widetilde{f}) = L(f) \leq 1$. Therefore composition $\widehat{f} := P \circ \widetilde{f} : \mathcal{M} \to X$ is an extension of f to $\mathcal{M}$ with

$$L(\widehat{f}) \leq L(\widetilde{f})\|P\| \leq (1+\varepsilon)\Pi(X).$$

Hence, $\hat{f}$ is the required Lipschitz extension of f and the inequality $\Lambda(X) \leq \Pi(X)$ is proved.

Now, let X be constrained in its bidual. We should prove that

$$\Pi(X) \leq \Lambda(X) \, (= R(X)).$$

Due to the definition of $R(X)$

$$R(X) \geq \widehat{R}(X) := \sup_{\rho} L(\rho),$$

where now the supremum is taken over all Lipschitz retracts onto X of *Banach spaces* Y containing X linearly and isometrically.

Now we show that every retract ρ of a Banach space Y onto its subspace X gives rise to a linear projection P_ρ of Y onto X such that $L(\rho) \geq \|P_\rho\|$. This will clearly imply that

$$(R(X) \geq) \, \widehat{R}(X) \geq \sup_{\rho} \|P_\rho\|$$

and the latter is by definition more than or equal to $\Pi(X)$ as required.

To construct the correspondence $\rho \mapsto P_\rho$ we use the *invariant mean* $\mathfrak{M} = \int_X \ldots dx$ on the space $\ell_\infty(X)$. Recall that the invariant mean is a linear functional on $\ell_\infty(X)$ satisfying

$$\left| \int_X f(x) dx \right| \leq \sup_X |f|; \tag{1.84}$$

$$\int_X dx = 1; \tag{1.85}$$

and

$$\int_X f(x + x') dx = \int_X f(x) dx \tag{1.86}$$

for all $x' \in X$.

Define now an operator $S : \mathrm{Lip}(X) \to \mathrm{Lip}(Y)$ by

$$(Sf)(z) := \int_Y \left\{ \int_X [(f \circ \rho)(x + y + z) - (f \circ \rho)(x + y)] dx \right\} dy,$$

where $f \in \mathrm{Lip}(X)$ and $z \in Y$.

Since the function within [ ] is bounded for every fixed z (recall that $f \circ \rho \in \mathrm{Lip}(Y)$), the definition is consistent. Moreover, from (1.84) and (1.85) it follows that

$$|Sf|_{\mathrm{Lip}(Y)} \leq \|L(\rho)\| \, |f|_{\mathrm{Lip}(X)}.$$

In particular, Sf is continuous.

1.11. Lipschitz extensions

Now we check that Sf is an element of the dual space Y^*. Clearly, it suffices to establish additivity of Sf. Applying the translation invariance of $\int_X \ldots dx$ and $\int_Y \ldots dy$ we have

$$(Sf)(z_1+z_2) = \int_Y \left\{ \int_X [(f\circ\rho)(x+y+z_1+z_2) - (f\circ\rho)(x+y+z_2)]\,dx \right\} dy$$
$$+ \int_Y \left\{ \int_X [(f\circ\rho)(x+y+z_2) - (f\circ\rho)(x+y)]\,dx \right\} dy =: (Sf)(z_1) + (Sf)(z_2).$$

Thus, S maps $\mathrm{Lip}(X)(\supset X^*)$ into Y^*.

Set now $T := S|_{X^*}$. Then T is a linear bounded operator from X^* into Y^*. Let us establish the extension property of T, i.e., prove that $(Tf)(z) = f(z)$ for $f \in X^*$, $z \in X$. To this end we rewrite $(Tf)(z)$, $z \in X$, as follows:

$$(Tf)(z) = \int_Y \left\{ \int_X [(f\circ\rho)(x+y+z) - (f\circ\rho)(y+z)]\,dx \right\} dy$$
$$+ \int_Y \left\{ \int_X [(f\circ\rho)(y+z) - (f\circ\rho)(x+y)]\,dx \right\} dy.$$

Since $\int_X \ldots dx$ is shift invariant and z belongs to X, the element z can be omitted in the first term on the right. Moreover, $(f\circ\rho)(x) = f(x)$ for $f \in X^* \subset \mathrm{Lip}(X)$. Thus, the right-hand side is equal to

$$\int_Y \left\{ \int_X [(f\circ\rho)(x+y) - f(y) + (f\circ\rho)(y+z) - (f\circ\rho)(x+y)]\,dx \right\} dy.$$

Since $(f\circ\rho)(y+z) = f(y+z) = f(y) + f(z)$, this integral equals

$$f(z) \int_Y dy \int_X dx = f(z),$$

as required.

Thus, T is a linear operator from X^* into Y^* with norm

$$\|T\| \leq L(\rho)$$

such that for every linear functional $x^* \in X^*$ its image Tx^* is a linear extension of x^* from X to Y.

Now we are in a position to define the desired projection P_ρ of Y onto X. To this end we exploit the conjugate operator $T^* : Y^{**} \to X^{**}$ whose norm $\|T^*\| = \|T\|$. Because of the extension property of T the trace $T^*|_Y$ is a projection of Y into X^{**}. Here we use the canonical linear isometric embedding of a Banach space into its second dual regarding X, Y as linear closed subspaces of X^{**} and Y^{**}, respectively. To check that $T^*|_Y$ is a projection we should show that $T^*x = x$ for every $x \in X \subset X^{**}$. But we get for $x^* \in X^*$,

$$\langle T^*x, x^* \rangle = \langle x, Tx^* \rangle = \langle x, x^* \rangle,$$

since Tx^* is an extension of x^* from X to Y. (Here the angle brackets denote the canonical bilinear form on a Banach space and its dual.) Hence, $T^*x = x$ for every $x \in X$ as required.

Finally, by the assumption on X there exists a linear projection $P : X^{**} \to X$ of norm 1. Setting
$$P_\rho := P(T^*|_X)$$
we obtain the desired projection from Y onto X of norm bounded by $\|T\| \leq L(\rho)$. The result has been proved. $\square$

We use Theorem 1.49 to evaluate Lipschitz constants for several classes of Banach spaces constrained in their biduals. For every such space X we have
$$\Lambda(X) = \Pi(X).$$
Thus using well-known sharp or asymptotically sharp results for projection constants we obtain the corresponding results for universal Lipschitz constants.

We begin with the family $\{\ell_p^n\}_{1 \leq p < \infty}$ where the results for projection constants are due to Grünnbaum [G-1960] ($p = 1$ and the sharp upper bound for $p = 2$), Rutovitz [Rut-1965] (the sharp lower bound for $p = 2$ and two-sided estimates for $2 < p < \infty$) and König, Schütt and Tomczak-Jaegermann [KST-1999] (sharp asymptotics for $1 < p < 2$). In the next formulations the symbol $a_n \sim b_n$, $n \to \infty$, means that $\lim_{n \to \infty} \frac{a_n}{b_n} = 1$.

Corollary 1.50. *The following relations are true:*

(a)
$$\Lambda(\ell_1^n) = \frac{(2n-1)\Gamma(n-\tfrac{1}{2})}{\sqrt{\pi}\,\Gamma(n)} \sim \sqrt{\frac{2n}{\pi}}, \quad n \to \infty;$$

(b)
$$\Lambda(\ell_2^n) = \frac{n\Gamma(\tfrac{1}{2}n)}{\sqrt{\pi}\,\Gamma(\tfrac{n+1}{2})} \sim \sqrt{\frac{2n}{\pi}}, \quad n \to \infty;$$

(c) *for $1 < p < 2$,*
$$\Lambda(\ell_p^n) \sim \sqrt{\frac{2n}{\pi}}, \quad n \to \infty;$$

(d) *for $2 < p < \infty$,*
$$\sqrt{\frac{2}{\pi}}\, n^{1/p} < \Lambda(\ell_p^n) \leq n^{1/p}.$$

Comparison of the last result with the sharp ones for $p = 2$ and $p = \infty$ (i.e., $\Lambda(\ell_\infty^n) = 1$) shows that these inequalities are far from being asymptotically sharp. Therefore the following question may be of interest.

Problem. *Find the sharp asymptotic of $\Lambda(\ell_p^n)$, $2 < p < \infty$, as $n \to \infty$.*

Finally, we present several results concerning infinite-dimensional Banach spaces exploiting the classical results on projection constants due to Grothendieck [Gro-1953] (separable and reflexive Banach spaces) and Pełczyński [Pe-1960] (spaces not containing isomorphic copies of c_0). In all these cases, the corresponding projection constants equal infinity. Hence, we get

Corollary 1.51. *Let X be an infinite-dimensional Banach space. Then the following is true.*

(a) *If X is reflexive (hence, dual), then $\Lambda(X) = \infty$.*
   In particular, this holds for infinite-dimensional L_p-spaces, $1 < p < \infty$.

(b) *The same holds, if X is constrained in its bidual and either separable or not containing an isomorphic copy of c_0.*
   In particular this is valid for infinite-dimensional L_1-spaces.

Remark 1.52. (a) For nondual separable Banach spaces the result may be wrong. As a simple example one singles out the inequality

$$\Lambda(c_0) \leq 2 < \Pi(c_0) = \infty \tag{1.87}$$

following from (1.83) and separability of c_0 (this, by the way, shows that the inequality $\Lambda(X) \leq \Pi(X)$ of Theorem 1.49 may be strict).

(b) There is a generalization of inequality (1.83) due to Kalton [Kal-2007]. It states that

$$\Lambda(C(\mathcal{M})) \leq 2$$

for every metric compact $\mathcal{M}$ and, moreover, for some such $\mathcal{M}$ the left-hand side is strictly greater than 1.
Since $C(\mathcal{M})$ is, in this setting, separable, $\Pi(C(\mathcal{M})) = \infty$.

(c) The only known examples of infinite-dimensional Banach spaces with finite Lipschitz constants are $C(S)$ with S being either a compact metric space or totally disconnected and compact topological space (the latter follows from the equality $\Pi(C(S)) = 1$, see, e.g., Day [Da-1955], which clearly implies that $\Lambda(C(S)) = 1$). It might be conjectured that the aforementioned Banach spaces are the only ones (up to Banach isomorphisms) having finite Lipschitz constants.

1.12 Simultaneous Lipschitz extensions

In this section we discuss Lipschitz analogs of the Borsuk–Dugundji theorem on simultaneous extension, see Section 1.4. We will show that such an analog does not hold even for scalar functions. The problem now is formulated as follows:

Given a subset S of a metric space $\mathcal{M}$ and a Banach space X, find a linear bounded operator E acting from $\mathrm{Lip}(S,X)$ into $\mathrm{Lip}(\mathcal{M},X)$ such that
$$Ef|_S = f.$$

We denote the set of these operators by $\mathrm{Ext}(S,X)$ and set

$$\lambda(S,\mathcal{M};X) := \inf\{\|E\|\ ;\ E \in \mathrm{Ext}(S,X)\}. \qquad (1.88)$$

We also set

$$\lambda(\mathcal{M},X) := \sup\{\lambda(S,\mathcal{M};X)\ ;\ S \subset \mathcal{M}\} \qquad (1.89)$$

shortening this notation to $\lambda(\mathcal{M})$ for scalar functions; so

$$\lambda(\mathcal{M}) := \lambda(\mathcal{M},\mathbb{R}). \qquad (1.90)$$

We will see in Chapter 8 of Volume II that $\lambda(\mathcal{M})$ may be infinity for rather simple metric spaces, e.g., two-dimensional Riemannian manifolds or Hilbert spaces. On the other hand, we will describe in Chapter 7 of that volume a lot of interesting geometric objects with finite $\lambda(\mathcal{M})$ and $\lambda(\mathcal{M},X)$ for an arbitrary X. These include Gromov hyperbolic spaces of bounded geometry, the direct product of metric trees, Carnot–Carathéodory spaces and many others. Unlike Lipschitz extensions results, simultaneous extensions were barely studied in the 20th century. We represent here two known results, the first of which being a consequence of the classical Whitney extension theorem described in Chapter 2.

Theorem 1.53. *For an arbitrary n-dimensional Banach space X,*
$$\lambda(X) \leq c^n,$$
where c is a numerical constant (≤ 5).

Later we will present results of several authors showing that $\lambda(X)$ is of linear growth in n. Note that the sharp order of growth is unknown even for $X := \ell_p^n$.

The next result is a counterexample to the Lipschitz version of the Borsuk–Dugundji theorem. For its presentation we need the following deep fact.

Theorem 1.54 (Lindenstrauss and Tzafriri [LT-1971]). *Let X be a Banach space linearly nonisomorphic to a Hilbert space. Then X contains a noncomplemented subspace.*

Suppose now that X is a *reflexive* Banach space and Y is its *uncomplemented* subspace. In this setting the following result holds.

Theorem 1.55. *There is no linear bounded extension operator from the trace space $\mathrm{Lip}(X)\big|_Y$ into $\mathrm{Lip}(X)$.*

Proof. By Theorem 1.27, $\mathrm{Lip}(X)\big|_Y$ is isometric to $\mathrm{Lip}(Y)$. Thus we should prove that there is no linear bounded extension operator from $\mathrm{Lip}(Y)$ into $\mathrm{Lip}(X)$. Suppose, conversely, that there does exist such an operator $L : \mathrm{Lip}(Y) \to \mathrm{Lip}(X)$. From this fact we derive the existence of a linear bounded extension operator $T : Y^* \to X^*$, i.e., T extends linearly and continuously every linear and continuous functional on Y to one on X. Assuming that the latter is true, the proof is completed as follows.

The conjugate operator T^* acts continuously from $X^{**} = X$ into $Y^{**} = Y$. We prove that T^* is actually a projection onto Y, from which we get the desired contradiction. To this end, take an arbitrary y^* from Y^* and y from Y. Identifying y with an element of X^{**} we obtain

$$\langle T^* y, y^* \rangle = \langle y, T y^* \rangle = (T y^*)(y) = \langle y, y^* \rangle.$$

The last equality is true, since T is an extension operator. As y^* is arbitrary, $T^* y = y$ for every $y \in Y$.

Thus it remains to construct the extension operator T using the given extension operator L. To this end we use the linearization procedure from the proof of Theorem 1.49 and define an operator $S : \mathrm{Lip}(Y) \to \mathrm{Lip}(X)$ by

$$(Sf)(z) := \int_X \left\{ \int_Y \big[(Lf)(x+y+z) - (Lf)(x+y) \big] dy \right\} dx,$$

where $f \in \mathrm{Lip}(Y)$ and $z \in X$.

Repeating line by line the argument of this proof we conclude that $T := S|_{Y^*}$ is a linear operator from Y^* into X^* with $\|T\| \leq \|L\|$ and, moreover, $(Tf)(z) = f(z)$ for every $f \in Y^*$, $z \in Y$.

Hence, T is the desired extension operator. $\square$

1.13 Simultaneous Lipschitz selection problem

The problem is a linearized version of that discussed in Section 1.3. We are now given a fixed map F from a metric space $(\mathcal{M}, d)$ into the set of closed convex centrally symmetric subsets of $\mathbb{R}^n$ and an associate linear space $\Sigma_F(\mathcal{M}; \mathbb{R}^n)$ of all maps $f : \mathcal{M} \to \mathbb{R}^n$ such that $f + F$ has a Lipschitz selection. Setting

$$|f|_{\Sigma_F(\mathcal{M};\mathbb{R}^n)} := \inf\{|s|_{\mathrm{Lip}(\mathcal{M},\mathbb{R}^n)}\,;\, s(m) \in f(m) + F(m),\, m \in \mathcal{M}\}$$

we are asking for conditions on the metric space $\mathcal{M}$ that provide existence of a linear operator $T_F : \Sigma_F(\mathcal{M}; \mathbb{R}^n) \to \mathrm{Lip}(\mathcal{M}, \mathbb{R}^n)$ such that

(i) $T_F f$ is a Lipschitz selection of $f + F$ for every f;

(ii) $\|T_F\|$ is bounded by a constant depending only on $\mathcal{M}$ and n.

In other words, T_F is a linearization of a nonlinear operator assigning an (almost) optimal Lipschitz selection s of $f + F$ to every $f \in \Sigma_F(\mathcal{M}; \mathbb{R}^n)$.

The solution to this problem was due to Yu. Brudnyi and Shvartsman [BSh-1999, Thm. 4.15]. It turns out that necessary and sufficient condition for $\mathcal{M}$ is finiteness of its linear Lipschitz extension constant $\lambda(\mathcal{M})$ introduced in the previous section, see (1.90). In particular, the required linear selection operator may not exist in the case of Theorem 1.55.

In Section 5.4, we will present the solution to the problem for the special case of multivalued maps into the set of linear subspaces of $\mathbb{R}^n$; a highly nontirivial passage from here to the general result is presented in the aforementioned paper [BSh-1999].

In fact, the result discussed plays a decisive role in solutions to the linear extension problems for a large class of smoothness spaces. This will be studied in Chapter 10 (Volume II) while now we only single out a relation to the problem of the previous section.

Let $\mathcal{M}'$ be a metric subspace of $\mathcal{M}$ and F be a multivalued map from $\mathcal{M}$ into the set $\{\{0\}, \mathbb{R}\}$ of all linear subspaces of $\mathbb{R}$ given by

$$F(m) = \begin{cases} \{0\} & \text{if} \quad m \in \mathcal{M}', \\ \mathbb{R} & \text{if} \quad m \in \mathcal{M} \setminus \mathcal{M}'. \end{cases}$$

Then any Lipschitz selection of $f + F$ coincides with f on $\mathcal{M}'$ and extends $f|_{\mathcal{M}'}$ as a Lipschitz function to $\mathcal{M}$. Hence, the required linear operator T_F is in this case a linear extension operator from $\mathrm{Lip}(\mathcal{M}')$ into $\mathrm{Lip}(\mathcal{M})$. Since $\|T_F\|$ is finite, we obtain for the linear extension constant (1.90) the inequality

$$\lambda(\mathcal{M}) \leq \|T_F\| < \infty.$$

The converse statement is also true, that is, if $\lambda(\mathcal{M}) < \infty$, then the required operator T_F exists. The general result of this type forms the main content of the proof presented in Section 5.4.

The intimate relation between the class of metric spaces with $\lambda(\mathcal{M}) < \infty$ and several important extension problems of Differential Analysis partially motivates detailed study of this class in Chapters 6 and 7 (Volume II).

Comments

The first continuous extension theorem, for functions of two variables, was proved by Lebesgue in his 1908 paper [Le-1907, pp. 99–100]. His account is too sketchy; moreover, as most prominent French mathematicians of that time, he liked words more than symbols. Apparently, that is why the Lebesgue result fell into oblivion soon after its publication. However, the Lebesgue extension method is of value and deserves to be discussed. The following account of Lebesgue's text may be regarded as a fairly precise description of the method.

Let $f \in C_b(E)$, where E is a closed subset of the plane[4]. We will construct a function $F \in C_b(\mathbb{R}^2)$ that agrees with f on E.

To this end, we subdivide the plane into squares of side length 1 by lines which are parallel to the axes. This division is denoted by D_1. The subdivision D_2 is obtained by dividing each square of D_1 into four congruent squares and so forth. However, we do not divide those squares of a subdivision which do not contain in the interior and/or boundary some points of E. In this way we obtain a decomposition of $E^c := \mathbb{R}^2 \setminus E$ into squares:

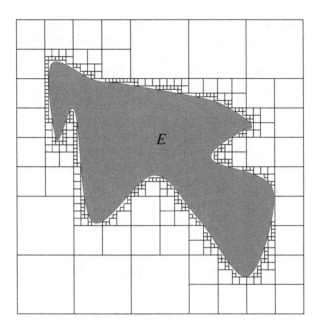

Figure 1.1: Decomposition of E^c.

At a vertex V of one of these squares, we take the function F to be equal to the infumum of values assigned by f at the points of E which are closest to V. Now on every edge of the squares, F is defined on a finite set of points. On these edges we extend F linearly between the points where its values are known. Hence, F varies linearly on the perimeter, except at certain points. Through these points we draw lines parallel to the axes which divide the square into subrectangles. On the lines which are parallel to the x-axis and not on the perimeter we take F to vary linearly in x (using the values at the endpoints). Finally, in every subrectangle, we define F to be a polynomial of a form $a_0 + a_1 x + a_2 y + a_3 xy$ that interpolates F at the vertices of the subrectangle. Since the restriction of the polynomial to

---

[4] Lebesgue imposed some additional restrictions on E, e.g., simple connectedness, which he did not use in the proof.

the boundary of the rectangle varies linearly on every edge, the polynomial agrees with F on the lines which are parallel to the x-axis.

F is now defined on E^c and gives the required bounded continuous extension of f from E to the whole $\mathbb{R}^2$.

It is clear from the construction but was not stated explicitly in the Lebesgue paper that $\sup |F| = \sup |f|$. It is also clear that the Lebesgue method can be easily applied to functions of more than two variables. But the corresponding result appeared only about ten years later without an explicit connection to Lebesgue's work. It was done independently and in different ways by Brouwer [Bro-1918] and H. Bohr, see pages 541–543 of the book [Ca-1918] by Carathéodory. Brouwer referred to his earlier paper [Bro-1912] where the result was contained but not formulated explicitly.

The Tietze generalization of the Lebesgue–Brouwer theorem to metric spaces is presented in his paper [Ti-1915]. It clearly contains as a simple consequence the aforementioned result for $\mathbb{R}^n$ which was published later. Apparently, the novelty of the metric space concept at that time and the war conditions obscured this fact. The Tietze method of continuous extension is as follows.

Let S be a closed subset of a metric space $\mathcal{M}$, and $f \in C_b(S)$. Without loss of generality we assume that $f \geq \lambda > 0$ for some number λ. Then the required continuous extension F is given at a point $m \in S^c$ by

$$F(m) := \sup_{m' \in S} \frac{f(m')}{(1 + d(m, m'))^\sigma},$$

where $\sigma := 1/d(m, S)$.

This bizarre formula was essentially simplified by Hausdorff [Hau-1919] who proposed the function

$$\widetilde{F}(m) := \inf_{m' \in \mathcal{S}} \left(f(m') + \frac{d(m, m')}{d(m, \mathcal{S})} - 1 \right) \quad \text{for} \quad m \in \mathcal{S}^c$$

as the required extension.

The Urysohn characterization of normal topological spaces through the continuous extendability property is presented in Section 28 of his fundamental paper [Ur-1925]. The Urysohn extension method was then developed in the work that followed in this field. Some ingredients of the method are, in particular, used in the proof of the first continuous selection theorem due to Bartle and Graves [BG-1952], see Theorem 1.4, and in the proof of Theorem 1.3 by E. Michael [Mi-1956]. In fact, Michael's theorem concerns a more general result dealing with paracompact spaces. Moreover, the continuous selection property of this theorem completely characterizes the class of paracompact topological spaces (see subsection 3.1.3 below for the definition of paracompactness).

The facts of Functional Analysis, namely, the Banach open map theorem and existence of uncomplemented subspaces used in Section 1.3, can be found, for example, in the treatise [DS-1958] by Dunford and Schwartz.

Comments

The simultaneous extension theorem of Section 1.4 was proved by Borsuk [Bor-1933a] for scalar functions under the additional assumption of the separability of the metric space $\mathcal{M}$. He used for the extension the integral mean of the distance function with respect to a certain positive Borel measure on $\mathcal{M}$. Then Kakutani [Ka-1940] proved this result in another way assuming only that $\mathcal{M}$ is locally separable. Finally, Dugundji [Du-1951] removed the separability assumption and proved the theorem for Banach-valued functions; his proof is reproduced in Section 1.4.

The Hurewicz extension theorem from Section 1.5 was proved in [Hu-1935]. The definition and certain properties of topological dimension and its relation to the continuous extension property will be discussed in Appendix A.

The basic notions of absolute and neighborhood metric retracts presented in Section 1.6 were introduced and studied for the general case of topological spaces by Borsuk [Bor-1931]. The role of these notions in Algebraic Topology and their relation to continuous extensions is discussed, e.g., in the Steenrod colloquium lectures [St-1957]; there the reader can find several remarkable results similar to that in example (c) following Definition 1.10.

A lot of interesting examples of Lipschitz retracts for subsets of Banach spaces similar to those in Section 1.6 can be found in Chapters 2 and 3 of the book [BL-2000] by Y. Benyamini and J. Lindenstrauss.

The classical Dvoretzky theorem used in the proof of Proposition 1.20 appears in [Dv-1961].

The celebrated Helly's Theorem 1.22 was discovered in 1913 but published only ten years later in [He-1923] because of World War I. This theorem and its topological version will be discussed in Appendix B.

The so-called Sylvester–Gallai theorem was posed by Sylvester in 1893 as a student problem. It concerns a finite set of lines on the plane, and the solution was given by Gallai [Grünwald] some 40 years later. The result was the starting point of many investigations in the field devoted to the study of the structure (and cardinality) of incidences between a set of points and a set of algebraic submanifolds with fixed degrees of freedom in $\mathbb{R}^n$ (planes, spheres, etc.), see, e.g., Chapter 13 of the book [GR-2000].

The last result of subsection 1.9.2, Proposition 1.24, appeared in a weaker form and with another proof in the paper [Li-1964] by Lindenstrauss.

The first strong Lipschitz extension result, Theorem 1.27, was due to McShane [McSh-1934] but the nonlinear extension operator (1.36) had already appeared in the paper [Hau-1919] by Hausdorff; he attributed this formula to M. Pach. The criterion of strong Lipschitz extendability of Proposition 1.31 was due to Kirszbraun [Ki-1934] for the Euclidean case and then was formulated in the general form by Valentine [Va-1945]. Both authors combine the criterion with Helly's theorem to obtain the desired extensions.

The intersection ball property and the related Theorem 1.26 appeared in the paper [AP-1956] by Aronszajn and Panitchpakdi.

The classical Kirszbraun Theorem 1.34 was proved by a rather intricate method in his paper [Ki-1934]. The proofs of the theorem presented in subsection 1.10.4 were given by Valentine [Va-1945] and by Mickle [Mic-1949], respectively. Mickle's ingenious idea has been used in many subsequent works in this field whereas the Valentine method has been applied only to prove the extension results for Euclidean, spherical and hyperbolic spaces, see [Va-1945] and [Va-1944], where Theorems 1.34, 1.38 and 1.40 were proved. However, we present here the Valentine method, only in part due to its beauty. We believe that the geometric approach used will find a new domain of applications (see Gromov's claim below which apparently confirms this belief).

A new proof of the Kirszbraun theorem appeared in Gromov's paper [Gr-1987]. The proof is based on the following *volume intersection property*.

Let $\mathcal{B}$ and $\mathcal{B}'$ be families of closed Euclidean balls in $\mathbb{R}^n$ of cardinality $n+2$. Then $\mathcal{B} \succ \mathcal{B}'$ implies

$$\mathrm{vol}_n(\cap\mathcal{B}) \leq \mathrm{vol}_n(\cap\mathcal{B}'). \tag{1.91}$$

This clearly leads to $\cap\mathcal{B}'$ being nonempty provided that $\cap\mathcal{B} \neq \emptyset$, and therefore proves Theorem 1.40. The Kirszbraun theorem can be also easily derived from this property. Unfortunately, the proof of (1.91) presented in [Gr-1987] does not work for the hyperbolic case.

A far reaching generalization of the Kirszbraun and Valentine theorems was obtained by Lang and Schroeder [LSch-1997]. Their nice result will be discussed in subsection 6.5.3 of Volume II and now we only formulate a special case.

Let $\mathcal{M}_\pm$ be two complete Riemannian manifolds regarded as metric spaces with respect to their geodesic metrics. Assume that $\mathcal{M}_-$ is simply connected and sectional curvatures $\mathcal{K}_\pm$ of $\mathcal{M}_\pm$ satisfy $\mathcal{K}_+ \geq 0 \geq \mathcal{K}_-$. Then it is true that

$$(\mathcal{M}_+, \mathcal{M}_-) \in \mathcal{SE}.$$

It is worth noting that the Valentine sphere theorem does not follow even from the general Lang–Schroeder result. Moreover, Gromov claims in [Gr-2000, p. 21–22] that the latter fact may be deduced from the Valentine sphere contraction result given in Proposition 1.41.

Theorem 1.46 was proved in the paper [Min-1970] in another way. Other extension results mentioned in Section 1.11 will be discussed in Chapter 6 of Volume II.

The linearization procedure used in the proofs of Theorems 1.49 and 1.55 was invented by Lindenstrauss [Li-1964]. Its application to extension problems for uniformly continuous functions is outlined in the Remarks to §2 of the book [Pe-1968] by Pełczyński. Existence of the invariant mean used in these proofs follows from amenability of abelian groups, see, e.g., [HR-1963].

Theorem 1.49 for finite-dimensional Banach spaces was due to Rieffel [Rie-2006]. He used this result to prove the following elegant equality.

Let $M_n(\mathbb{C})^{sa}$ be the set of $n \times n$ complex self-adjoint matrices equipped with the metric

$$d(A, B) := \|A - B\|;$$

here $\|A\|$ is the operator norm of an $n \times n$ complex matrix A regarded as a linear operator acting in the Hermitian space $\mathbb{C}^n$.

Then the following is true:

$$\Lambda(M_n(\mathbb{C})^{sa}) = 2n\left(\frac{n}{n+1}\right)^{n-1} - 1 \sim \frac{2n}{e}, \quad n \to \infty.$$

Appendices

A. Topological dimension and continuous extensions of maps into $\mathbb{S}^n$

Since ancient times dimension has been considered as one of the most intuitively clear concepts of geometry. But the discoveries of such objects as Peano curves filling out the unit square and fractal sets (the middle third Cantor set, von Koch's curve, Sierpinski's gasket, etc.) convinced mathematicians of the necessity of a precise definition.

The dimension of topological spaces was introduced and comprehensively studied by Urysohn, Menger and their followers; prior ideas and results were due to Lebesgue, Poincaré and Brouwer. The part of the theory concerning separable metric spaces was essentially completed in the period from 1920 till 1940; the reader can find an excellent account in the classical book [HW-1941] by Hurewicz and Wallmen.

Here we present only those facts of the theory which are required for the continuous extension theorem from Section 1.5. We begin with the definition of *covering (Čech–Lebesgue) dimension*. Let us note that in the realm of separable metric spaces this notion coincides with that of *inductive (Menger–Urysohn) dimension*, see [Hu-1927b].

Definition A.1. The covering dimension of a Hausdorff topological space S is said to be at most n, if for every open covering of S there exists an open refinement of order $n+1$ (i.e., no more than $n+1$ subsets of the refinement has nonempty intersection).

The covering dimension denoted by $\dim S$ is then the smallest integer n such that S has covering dimension $\leq n$.

The first fundamental fact of the theory is

Theorem A.2 (Lebesgue–Brouwer).

$$\dim \mathbb{R}^n = n.$$

Since dim is a topological invariant [5], n-simplices, n-cubes or n-balls of $\mathbb{R}^n$ have dimension n as well. The reader can use the Heine–Borel theorem to show that

---
[5] i.e., homeomorphic spaces have the same covering dimension.

$\dim[0,1] = 1$ and then try to prove the result for $n = 2$ (the proof of Theorem A.2 will be presented in Appendix C). Another exercise for better understanding the definition is to prove that $\dim \mathbb{Q}^n = 0$; here the set $\mathbb{Q}^n \subset \mathbb{R}^n$ of rational points in $\mathbb{R}^n$ is endowed by the induced topology. In contrast, the set $\mathbb{Q}^\infty$ of rational points in the Hilbert space ℓ_2 has dimension 1 (!), see [Er-1940].

Now we introduce two basic results of the theory, the so-called *separation theorems*, in a form suitable to our purpose. The first of these results was proved by Menger [Men-1926] for the compact case and then extended by Hurewicz [Hu-1927b] to the general situation. In what follows $\mathcal{M}$ is assumed to be a *separable* metric space.

Theorem A.3. *Let A_i be closed subsets of $\mathcal{M}$, $i = 1, 2$, such that*

$$\dim(\mathcal{M} \setminus \cup A_i) \leq n.$$

Then there exist closed subsets V_i of $\mathcal{M}$ satisfying the conditions

$$\mathcal{M} = \cup V_i, \quad A_i \subset V_i \quad \text{and} \quad A_i \cap V_j = A_1 \cap A_2 \tag{A.1}$$

for $i \neq j$ and $i, j \in \{1, 2\}$. Moreover,

$$\dim(\cap V_i \setminus \cap A_i) \leq n - 1. \tag{A.2}$$

In fact, the Menger–Hurewicz theorem has a special (but equivalent) case with *disjoint* closed A_i. This special case admits a more elegant formulation based on the notion of a *partition* which is introduced as follows.

A subset $P \subset \mathcal{M}$ is a partition between subsets A_1, A_2 of $\mathcal{M}$, if there are disjoint open sets W_i such that

$$A_i \subset W_i \quad \text{and} \quad \mathcal{M} \setminus P = \cup W_i.$$

The original version of the discussed result simply asserts that for the case $A_1 \cap A_2 = \emptyset$ there exists a partition P of these subsets such that the dimension of $(A_1 \cup A_2) \cap P$ is at most $n - 1$.

The next separation theorem characterizes the dimension of $\mathcal{M}$ in terms of partitions.

Theorem A.4. *$\mathcal{M}$ satisfies the condition*

$$\dim \mathcal{M} \leq n$$

for some $n \geq 0$ if and only if for every collection $\{(A_i, B_i) \ ; \ i = 1, \ldots, n+1\}$ of pairs of disjoint and closed subsets there exist $n+1$ closed subsets P_i such that P_i is a partition between A_i and B_i, $1 \leq i \leq n+1$, and $\cap P_i = \emptyset$.

Appendix A. Topological dimension and continuous extension of maps 63

For $n = 0$ the theorem simply asserts that every pair of disjoint closed subsets is partitioned by $P = \emptyset$. To include this special case in the general context we will use here and below the convention

$$\dim \mathcal{M} = -1 \iff \mathcal{M} = \emptyset.$$

Theorem A.4 was proved by Eilenberg and Otto [EO-1938].

Using these results we now outline the proof of Hurewicz's continuous extension theorem (see Theorem 1.9 of Section 1.5) following the derivation presented in Section 1.9 of the book [En-1978] by Engelking. Hurewicz's original proof is given in section VI.2 of the above cited classical book [HW-1941].

The first part of Theorem 1.9 asserts that a map $f \in C(\mathcal{M}_0, \mathbb{S}^n)$ admits a continuous extension to all of $\mathcal{M}$ provided that

$$\dim(\mathcal{M} \backslash \mathcal{M}_0) \leq n.$$

Let us recall that $\mathcal{M}_0$ is assumed to be closed.

The proof begins with $n = 0$ and then proceeds by induction on n. So at the first step f is a continuous map from $\mathcal{M}_0$ into $\mathbb{S}^0 = \{-1, 1\}$ and

$$\dim(\mathcal{M} \backslash \mathcal{M}_0) = 0.$$

Set $A_i := f^{-1}(i)$, $i = \pm 1$; then the A_i are closed and, moreover,

$$\mathcal{M}_0 = \cup A_i \quad \text{and} \quad \cap A_i = \emptyset.$$

By Theorem A.3, there exist closed subsets V_i of $\mathcal{M}$ so that $A_i \subset V_i$, $\mathcal{M} = \cup V_i$ and $\dim(\cap V_i) = -1$. The last condition means that the V_i are disjoint and closed. So, letting

$$\hat{f} := i \quad \text{on} \quad V_i \quad \text{for} \quad i = \pm 1,$$

we define a continuous map from $\mathcal{M}$ to $\mathbb{S}^0$ that agrees with f on $\mathcal{M}_0$.

Assume now that the result is true for dimension $n - 1$ with $n \geq 1$. To prove it for $\dim(\mathcal{M} \backslash \mathcal{M}_0) = n$, consider a map $f \in C(\mathcal{M}_0, \mathbb{S}^n)$ and define sets A_i by

$$A_1 := f^{-1}(\mathbb{S}^n_+) \quad \text{and} \quad A_2 := f^{-1}(\mathbb{S}^n_-).$$

Then the A_i are closed and

$$\mathcal{M}_0 = \cup A_i \quad \text{and} \quad f(\cap A_i) \subset \mathbb{S}^{n-1}.$$

By Theorem A.3, there exist closed subsets V_i of $\mathcal{M}$, $i = 1, 2$, satisfying conditions (A.1) and (A.2). The latter condition allows us to apply the induction hypothesis to the restriction $f|_{\cap A_i}$ mapping $\cap A_i$ into $\mathbb{S}^{n-1}$. Then one gets a continuous extension $\hat{f} : \cap V_i \to \mathbb{S}^{n-1}$ of this map. Now conditions (A.1) imply that the formula

$$f_i := \begin{cases} f & \text{on } A_i, \\ \hat{f} & \text{on } \cap V_i, \end{cases}$$

defines a continuous map on $A_i \cup (\cap V_i)$ with values in $\mathbb{S}^n_+$ for $i=1$ and in $\mathbb{S}^n_-$ for $i=2$.

Since hemispheres are homeomorphic to the unit cube of $\mathbb{R}^n$, the Tietze–Urysohn theorem (see Corollary 1.5) implies that f_i admits a continuous extension g_i of f_i to the subspace V_i, $i=1,2$. Finally, letting

$$\tilde{f} := g_i \quad \text{on} \quad V_i, \quad i=1,2,$$

we get a continuous map from $\cup V_i = \mathcal{M}$ into $\mathbb{S}^n$ that agrees with f on $\mathcal{M}_0$.

The second part of Theorem 1.9 asserts that if for every closed $\mathcal{M}_0 \subset \mathcal{M}$ and every $f \in C(\mathcal{M}_0, \mathbb{S}^n)$ there exists a continuous extension to all of $\mathcal{M}$, then

$$\dim \mathcal{M} \leq n. \tag{A.3}$$

To prove this inequality we will apply Theorem A.4 dealing with a partition of $n+1$ pairs of disjoint closed subsets. Let $\{(A_i, B_i) \, ; \, i=1,\ldots,n+1\}$ be a family of pairs of disjoint closed subsets in $\mathcal{M}$. Consider continuous functions $f_i : A_i \cup B_i \to [0,1]$ such that $f_i(A_i) = 0$ and $f_i(B_i) = 1$, $i=1,\ldots,n+1$. Letting $f := (f_i)_{i=1}^{n+1}$ we define a continuous map from $\bigcup (A_i \cup B_i)$ to the boundary ∂Q^{n+1} of the unit cube $Q^{n+1} = [0,1]^{n+1} \subset \mathbb{R}^{n+1}$. Since this boundary is homeomorphic to $\mathbb{S}^n$, there is an extension $\hat{f}$ of f to all of $\mathcal{M}$. Let $\hat{f} = (\hat{f}_i)_{i=1}^{n+1}$; then $\hat{f}_i$ is a continuous extension of the $f_i : \cup (A_i \cup B_i) \to [0,1]$. Hence the closed set $P_i := \hat{f}_i^{-1}(\frac{1}{2})$ is a partition between A_i and B_i, $1 \leq i \leq n+1$, and $\cap P_i = \emptyset$. Now we apply Theorem A.4 to get (A.3).

B. Helly's topological theorem

B.1. The Classical Helly theorem and related results

In 1913 Helly proved the following celebrated result.

Theorem B.1. *Let $\mathcal{F}$ be a finite family of convex sets of $\mathbb{R}^n$. If every $n+1$ members of $\mathcal{F}$ have a point in common, then $\cap \mathcal{F} \neq \emptyset$.*

The theorem easily implies the similar assertion for infinite families of *compact* convex subsets. This generalization is formulated (in an equivalent form) in Section 1.9, see Theorem 1.22.

Helly's theorem has profoundly influenced the development of the field which is now known as Combinatorial Geometry. Numerous generalizations and variants of Theorem B.1 can be found in the surveys [DGK-1963] and [Eck-1993]. We present below only one of them (Theorem B.2) revealing an unexpected connection between two seemingly distinct classical theorems of Combinatorial Geometry. This deep and elegant theorem was discovered by Dol'nikov in 1981, then announced in 1987 and published in 1993, see [Dol-1993] and references therein.

Let us recall that an m-dimensional affine manifold (m-plane) L of $\mathbb{R}^n$ is called an *m-transversal for a family of subsets in $\mathbb{R}^n$*, if L intersects all members

Appendix B. Helly's topological theorem 65

of the family. Here $0 \leq m \leq n$; in particular, a 0-transversal is a point in common for all members of the family.

Now let $\mathcal{F}_i$, $1 \leq i \leq m$, be families of convex sets of $\mathbb{R}^n$ and $m \leq n$. Assume that each family $\mathcal{F}_i$ is either finite or contains a compact subset.

Theorem B.2. *If every $n - m + 2$ members of each $\mathcal{F}_i$ has a point in common, then the family $\mathcal{F} := \bigcup_{i=1}^{m} \mathcal{F}_i$ has an $(m-1)$-transversal.*

So, for $m = 1$, i.e., for one family $\mathcal{F}$, this coincides with the Helly theorem. Another extreme case, $m = n$, implies the following classical result due to Borsuk [Bor-1933b].

Theorem B.3. *If $\{F_i\}_{i=1}^{n}$ is a cover of $\mathbb{S}^{n-1}$ by n closed sets, then at least one of them contains two antipodal points[6].*

Proof. If, on the contrary, every F_i does not contain antipodal points, then its diameter (in $\mathbb{R}^n$) is strictly less than $\operatorname{diam} \mathbb{S}^{n-1} = 2$. In other words, for some $r \in (0, 1)$,

$$\operatorname{diam} F_i \leq 2r \ (< 2), \quad i = 1, \ldots, n.$$

Hence, every two closed balls of radius r and centers in F_i have a point in common. In other words, the families

$$\mathcal{F}_i := \{\overline{B}_r(m) \ ; \ m \in F_i\}, \quad i = 1, \ldots, n,$$

satisfy the assumptions of Theorem B.2 with $m = n$. Therefore there exists an $(n-1)$-transversal L that intersects every ball of family $\mathcal{F} := \bigcup_{i=1}^{n} \mathcal{F}_i$. This, in particular, implies that the distance from the center of $B_r(m) \in \mathcal{F}$ to L is at most $r < 1$. But $\{F_i\}$ is a cover of $\mathbb{S}^{n-1}$ and therefore the same is true for *arbitrary* $m \in \mathbb{S}^n$. This means that $\operatorname{diam} \mathbb{S}^{n-1}$ is at most $2r < 2$, a contradiction. □

Helly's theorem has also important applications to optimization problems in analysis. Its use allows one to reduce an infinite-dimensional problem to a finite-dimensional problem of the same type. We have encountered this kind of reduction in Section 1.10. Now we present an application of another type dealing with an optimization problem originating from Approximation Theory.

Let S be a compact topological space, and L_n be an n-dimensional linear space. Assume that a function $f : S \times L_n \to \mathbb{R}$ is upper semicontinuous in $s \in S$ and convex in $\ell \in L_n$. Set, for a subset $\widehat{S} \subset S$,

$$\mu(\widehat{S}) := \inf_{\ell \in L_n} \sup_{s \in \widehat{S}} f(s, \ell).$$

---
[6] i.e., points x, y of $\mathbb{S}^{n-1}$ such that $x = -y$.

Theorem B.4. *There exists a point $\hat{\ell} \in L_n$ and a subset $\widehat{S} \subset S$ of at most $n+1$ points such that*
$$\mu(S) = \mu(\widehat{S}) = \sup_{s \in \widehat{S}} f(s, \hat{\ell}).$$

Hence, the minimax problem with infinite number of parameters is reduced to that with at most $2n + 1$ parameters.

We outline the remarkable proof of this result based on Helly's theorem. It was due to L. Schnirelman [Shn-1938] and later rediscovered by Rademacher and Schoenberg [RSch-1950]. These authors considered a special (but basic) case of a function $(s, \ell) \mapsto |f(s) - \ell(s)|$, $(s, \ell) \in S \times L_n$, where f is a continuous scalar function and $L_n \subset C(S)$. In this case, which is important for Approximation Theory, Theorem B.4 was first proved (in another way) in Remez's 1935 book (in Ukrainian), see [Rem-1957] and references therein.

To explain Schnirelman's proof one sets
$$\mu_n := \sup\{\mu(\widehat{S}) \,;\, \text{card } \widehat{S} \le n+1\}.$$

It is clear that $\mu_n \le \mu(S)$. We show that, in fact,
$$\mu_n = \mu(S). \tag{B.1}$$

To this end one introduces a family $\{V(s)\}_{s \in S}$ of convex subsets in L_n determined by
$$V(s) := \{\ell \in L_n \,;\, f(s, \ell) \le \mu_n\}.$$

This set is convex because of the convexity of the function $\ell \mapsto f(s, \ell)$. Moreover, the Helly intersection property holds for this family, that is, for an arbitrary set $\{s_0, \ldots, s_n\}$ of distinct points of S,
$$\bigcap_{i=0}^{n} V(s_i) \ne \emptyset. \tag{B.2}$$

In fact, the function $\ell \mapsto \max_{0 \le i \le n} f(s_i, \ell)$ is continuous, being the maximum of a finite set of convex (and, hence, continuous) functions. By virtue of the Weierstrass theorem we therefore can find $\ell^* \in L_n$ such that
$$\max_{0 \le i \le n} f(s_i, \ell^*) = \inf\left\{\max_{0 \le i \le n} f(s_i, \ell) \,;\, \ell \in L_n\right\}.$$

For this ℓ^* we have
$$f(s_i, \ell^*) \le \mu_n, \quad i = 0, 1, \ldots, n,$$
that is to say, $\ell^* \in \bigcap_{i=0}^{n} V(s_i)$. So (B.2) is true; however, the Helly theorem cannot be applied to $\{V(s)\}$ as its subsets are noncompact. Fortunately, there is a variant of Helly's theorem designed for this situation. It suffices only to prove that a

Appendix B. Helly's topological theorem 67

finite subfamily of $\{V(s)\}$ has compact intersection. This can be derived from the compactness of S and the assumptions on f. Applying then this version of Helly's theorem, we find a point $\hat{\ell} \in L_n$ belonging to every $V(s)$ with $s \in S$. This simply means that
$$f(s, \hat{\ell}) \leq \mu_n \quad \text{for all} \quad s \in S,$$
whence $\mu(S) \leq \mu_n$.

So, (B.1) is true. Now note that the function $(s_0, \ldots, s_n) \mapsto \mu(\{s_0, \ldots, s_n\})$ is upper semicontinuous being the infimum of upper semicontinuous functions $\varphi_\ell(s_0, \ldots, s_n) := \max_{0 \leq i \leq n} f(s_i, \ell)$, $\ell \in L_n$. Since S is compact, there exists a set $\widehat{S}$ of at most $n+1$ points such that $\mu(\widehat{S}) = \mu_n$. Together with (B.1) this proves the first assertion of Theorem B.4; taking the point $\hat{\ell}$ that has already appeared in the proof, we easily establish the second assertion as well.

B.2. Cohomology theory – a computational aspect

The basic point of the proof for the topological Helly's theorem is to distinguish two metric spaces of a rather complicated geometric nature. We will resolve the problem using for their discrimination certain topological invariants. Unfortunately, commonly known numerical invariants, such as the dimension or the number of connected components, do not work in this case. Therefore we have to apply more involved tools, namely, (Čech) cohomology groups [7] whose definition and properties can be found in any book on Algebraic Topology, see, e.g., [Mun-1984]. Here we devote a few lines to the general properties of these objects with special emphasis on computational aspects of the theory.

In the fancy language of Homological Algebra, the n-th cohomological functor H^n ($n \geq 0$) associates with every paracompact topological space S an abelian group $H^n(S)$ and with every continuous map $f : S_1 \to S_2$ an (induced) homomorphism $H^n(f) : H^n(S_2) \to H^n(S_1)$ such that H^n reverses composition and preserves the identity:
$$H^n(f \circ g) = H^n(g) \circ H^n(f), \quad H^n(id_S) = id_{H^n(S)}.$$

This, in particular, implies that homeomorphic spaces have the same cohomologies. In fact, the following stronger assertion is true.

(Homotopy Invariance Principle) Homotopic maps induce the same map in cohomology.

Recall that a *homotopy* between continuous maps $f, g : S_1 \to S_2$ is a continuous map $F : S_1 \times [0, 1]$ such that for $s \in S_1$
$$F(s, t) = \begin{cases} f(s), & \text{if } t = 0, \\ g(s), & \text{if } t = 1. \end{cases}$$

---

[7] We restrict our consideration to the cohomology groups with the coefficients in the group $\mathbb{Z}$ of integer numbers.

In turn, the spaces [8] S_1 and S_2 are homotopic, if there is a *homotopic right invertible map* from S_1 onto S_2. In other words, there are surjections $f : S_1 \to S_2$ and $g : S_2 \to S_1$ such that $g \circ f$ is homotopic to the identity id_{S_1}.

Example B.5. A space S is said to be *contractible*, if it is homotopic to some point $s_0 \in S$. By virtue of the Homotopy Invariance Principle, we get in this case, for all $n \geq 0$ and some $s_0 \in S$,

$$H^n(S) = H^n(\{s_0\}).$$

The latter is known to be equal to $\{0\}$ for $n \geq 1$ and $\mathbb{Z}$ for $n = 0$.

In particular, $\mathbb{R}^m$ or its open or closed balls, cubes and simplices are clearly contractible. Hence, all of them are *cohomologically trivial*, i.e., their cohomologies of order $n \geq 1$ are trivial.

There exists a relatively simple algorithm of finding $H^n(S)$ for S having *simplicial structure*. Bearing in mind the consequent applications we include this notion in a more general context related to some basic concepts of Convex Geometry, see, e.g., [KK-1994] for a detailed discussion.

Recall that a *polyhedral cell* C is the convex hull conv P of a *finite* set of points P in a real linear space L. The *dimension* of C is the dimension of the smallest affine subspace containing C. The *interior* of C (written C^0) is the interior of C as a subset of this affine subspace.

A hyperplane [9] $H \subset L$ determines the closed half-spaces H_+ and H_- such that $H_+ \cap H_- = H$. If the cell C lies in one of these half-spaces and $F := C \cap H \neq \phi$, then F is called a *face* of C.

The *dimension* of F is the dimension of the smallest affine subspace containing F, and F° is the interior of F as a subset of this subspace.

A face of dimension 0 is called a *vertex* of C. The set of vertices V is a subset of the generating set P and $C = \text{conv}\, V$. If the number of vertices for C equals $m := 1 + \dim C$, then C is called a *simplex* (or m-simplex).

It can be easily established that every face of C is a polyhedral cell, and the intersection of two faces is a face or an empty set.

Definition B.6. A polyhedral complex K is a finite set of cells C in a real linear space L such that

(i) if $C \in K$, then all its faces are also in K;

(ii) for every pair $C_1, C_2 \in K$ its intersection is either empty or a face both of C_1 and of C_2.

The *dimension* of K is given by

$$\dim K := \sup\{\dim C\,;\, C \in K\},$$

---

[8] In what follows we write 'space' instead of 'paracompact topological space'.
[9] i.e., a proper maximal subspace of L.

Appendix B. Helly's topological theorem

and the *underlying* set of K is the set
$$[K] := \bigcup_{C \in K} C.$$

If all cells of K are simplices, K is called a *simplicial complex*. There exists an important operation that turns a polyhedral complex into a simplicial one. For its definition we recall that the *barycenter* b_C of a polyhedral cell C is a point of L given by
$$b_C := \frac{1}{\operatorname{card} V} \sum_{v \in V} v,$$
where V is the set of vertices for C. Now the (first) *barycentric subdivision* of C is a simplicial complex (denoted by $\widehat{C}$) defined as follows.

Let $F_0 \subset F_1 \subset \ldots \subset F_k$ be a strictly ascending sequence of faces of C; the convex hull of the barycenters of these faces determines a (unique) simplex. The nonempty intersection of two such simplices is again such a simplex. So, the collection of all these simplices forms the simplicial complex $\widehat{C}$.

If now K is a polyhedral complex, then the family $\{\widehat{C}\,;\,C \in K\}$ is a simplicial complex called the barycentric subdivision of K and is denoted by $\widehat{K}$.

Finally, we recall the notion of the *star* for a vertex v from a polyhedral complex K. This is given by
$$\operatorname{St}(v) := \bigcup_{v \in C \in K} C^\circ.$$

The underlying set $[K] \subset L$ of a polyhedral complex K can be equipped with a topology named the *weak topology*, in the following way. Given a Hausdorff topological space, we consider the family of functions $f : [K] \to X$ such that $f|_C$ is continuous for every cell $C \in K$. Then the weak topology is the weakest topology on $[K]$ such that for all X, every function of this family is continuous, see, e.g., [Mun-1984].

We can also identify $[K]$ with a subset of the Hilbert space $\ell^2(\mathcal{V})$, where $\mathcal{V}$ is the set of vertices of K. Namely, let S be a (unique) simplex of minimal dimension containing a point $p \in [K]$. Then this point can be uniquely presented as
$$p = \sum_{v \in \mathcal{V}} \beta_v(p) v,$$
where $v \mapsto \beta_v(p)$ is a nonnegative function supported by the set of vertices for S and such that $\sum_{v \in \mathcal{V}} \beta_v(p) = 1$. Hence, this function belongs to $\ell_2(v)$, and we can define the distance between points p_1, p_2 of $[K]$ by
$$d(p_1, p_2) := \left\{ \sum_{v \in \mathcal{V}} (\beta_v(p_1) - \beta_v(p_2))^2 \right\}^{\frac{1}{2}}.$$

The metric topology defined by this distance coincides with the weak topology if and only if K is finite.

In the remaining part of this section we will consider only simplicial complexes K.

We will say that a space S has a *simplicial structure*, if it is homeomorphic to some $[K]$. In this case $H^n(S)$ is clearly equal to $H^n([K])$; the latter is much easier to evaluate.

A simple way to relate a simplicial structure to a space goes through the important notion of a *nerve* introduced by P. Aleksandroff [Al-1928]. For its definition one considers a locally finite open cover $\mathcal{U} := \{U_\alpha\}_{\alpha \in A}$ of a space S. A nerve $N(\mathcal{U})$ associated with the cover is an (abstract) simplicial complex whose vertices are $\{U_\alpha\}$ and where $[U_{\alpha_0}, \ldots, U_{\alpha_n}]$ is an n-simplex, if $\bigcap_{i=0}^{n} U_{\alpha_i} \neq \emptyset$. If this cover is finite (the only case we will use below), its nerve $N(\mathcal{U})$ may and will be understood as a finite simplicial complex in the space $\mathbb{R}^n$ with $n := \operatorname{card} A$ spanned by linearly independent vectors in a fixed one-to-one correspondence with $U_\alpha \in \mathcal{U}$. In this case the underlying set $[N(\mathcal{U})]$ is a subset of $\mathbb{R}^n$. In general, $[N(\mathcal{U})]$ can be also realized as a subset in $\ell_2(A)$ using the inductive limit of the ordered set of finite simplicial complexes associated with finite subcovers of $\mathcal{U}$, see, e.g., [Mun-1984]. Finally, one defines a *canonical $\mathcal{U}$-map* φ from the space S to $[N(\mathcal{U})]$ as follows.

Let $\{\varphi_\alpha\}_{\alpha \in A}$ be a continuous partition of unity subordinate to $\mathcal{U}$, see Proposition 3.17 below. Since $\varphi_\alpha \geq 0$ and $\Sigma \varphi_\alpha = 1$, the finite set of nonzero numbers $\varphi_\alpha(s)$ can be regarded as the barycentric coordinates of a point in a Euclidean simplex determined by the set $\{\alpha \in A \,;\, \varphi_\alpha(s) > 0\}$. Therefore the correspondence $s \mapsto \{\varphi_\alpha(s)\}$ is a continuous map from S to $[N(\mathcal{U})]$, the desired *canonical $\mathcal{U}$-map*.

The next result connecting all the concepts introduced above is the classical Leray–Weil theorem, see [We-1952]. Our formulation is restricted to a special case required for the proof of the topological Helly's theorem.

Theorem B.7. *Let $\mathcal{U}$ be a locally finite open cover of a space S. Assume that all finite intersections of sets from $\mathcal{U}$ are cohomologically trivial. Then the canonical $\mathcal{U}$-map $\varphi : S \to [N(\mathcal{U})]$ is homotopic right invertible. In particular, S and $[N(\mathcal{U})]$ are homotopic.*

Remark B.8. The result remains true for locally finite *closed* covers. In this case the canonical $\mathcal{U}$-map can be defined in a similar fashion.

Theorem B.7 and the Homotopy Invariance Principle allow us to reduce the evaluation of cohomologies for S with a "good" cover $\mathcal{U}$ to the corresponding problem for the simplicial complex $N(\mathcal{U})$. The latter evaluation can be accomplished in many cases. We use, in particular, the following two results.

Let Δ^{n+1} be an $(n+1)$-dimensional simplex in $\mathbb{R}^{n+1}$. Then for its boundary $\partial \Delta^{n+1}$ we get

$$H^n(\partial \Delta^{n+1}) = \mathbb{Z}. \tag{B.3}$$

Note that this boundary is homeomorphic to $\mathbb{S}^n$.

Now let K be a finite simplicial complex in $\mathbb{R}^n$. Then
$$H^n([K]) = \{0\}.$$
This is also true for a more general case. Namely, let S be a closed subset of $\mathbb{R}^n$. Then
$$H^n(S) = \{0\}. \tag{B.4}$$
This can be derived from the previous formula by a limit procedure named simplicial approximation, see [Mun-1984] for details.

B.3. Helly's topological theorem

In 1930, Helly [He-1930] proved a topological version of his famous intersection theorem. We present below a variant of his result adapted to the needs of the Valentine extension theorems of Section 1.10.

Let $\mathcal{F}$ be a cover of a metric space $\mathcal{M}$ by compact sets. Assume that $\mathcal{M}$ is homeomorphic to $\mathbb{R}^n$, and the sets of the cover and their finite nonempty intersections are cohomologically trivial. Under these conditions the following is true.

Theorem B.9. *If the intersection of $n+1$ members of $\mathcal{F}$ is nonempty, then $\cap \mathcal{F} \neq \emptyset$.*

Proof. We begin with the case
$$\operatorname{card} \mathcal{F} = n + 2.$$
Assume, on the contrary, that
$$\cap \mathcal{F} = \emptyset. \tag{B.5}$$

To come to a contradiction, let us regard $\mathcal{F}$ as a cover of the compact set $\cup \mathcal{F}$. Then its nerve $N(\mathcal{F})$ consists of all open faces of an $(n+1)$-simplex $\Delta^{n+1} \subset \mathbb{R}^{n+2}$ with the vertices identified with F_i, $1 \leq i \leq n+2$. Indeed, every proper subfamily of $\mathcal{F}$ has nonempty intersection by assumption, and these intersections determine the corresponding faces. However, the interior of Δ^{n+1} does not belong to the nerve because of (B.5). Hence, $[N(\mathcal{F})]$ coincides with the boundary $\partial \Delta^{n+1}$, and by Theorem B.7 and (B.3) we have
$$H^n(\cup \mathcal{F}) = H^n(\partial \Delta^{n+1}) = \mathbb{Z}.$$

On the other hand, $\cup \mathcal{F}$ is a compact set in the space $\mathcal{M}$, homeomorphic to $\mathbb{R}^n$. Therefore, the Homotopy Invariance Principle and (B.4) give
$$H^n(\cup \mathcal{F}) = \{0\}$$
in contradiction with the previous equality.

We proceed with the proof by induction on $m := \operatorname{card} \mathcal{F}$. Assume that the result is true for $m \geq n+2$ and prove it for $m+1$. Let $\mathcal{F} = \{F_i\}_{1 \leq i \leq m+1}$ and $\widehat{\mathcal{F}} := \{F_i \cap F_{m+1}\}_{1 \leq i \leq m}$. By the assumption of the theorem, all nonempty intersections of members from $\widehat{\mathcal{F}}$ are cohomologically trivial. Moreover, every $n+1$ members have nonempty intersections. In fact, consider a subfamily $\widetilde{\mathcal{F}}$ of $n+1$ members $\{F_i \cap F_{m+1}\}_I$ with some $I \subset \{1, \ldots, m\}$. It was proved that the subfamily of $n+2$ members consisting of F_{n+1} and of these F_i has nonempty intersection. Hence, $\cap \widetilde{\mathcal{F}} = \cap \{F_i \cap F_{n+1}\} \neq \emptyset$. Now the induction hypothesis applied to the family $\widehat{\mathcal{F}} := \{F_i \cap F_{n+1}\}_{1 \leq i \leq m}$ implies that $\cap \widehat{\mathcal{F}} = \cap \mathcal{F} \neq \emptyset$.

Thus, the result is true for all finite $\mathcal{F}$. Now let $\mathcal{F}$ be an infinite family satisfying the conditions of the theorem. We show that $\cap \mathcal{F} \neq \emptyset$. If, on the contrary, the intersection is empty, then, by compactness of members of $\mathcal{F}$, there exists a finite subfamily $\mathcal{F}' \subset \mathcal{F}$ such that $\cap \mathcal{F}' = \emptyset$. But $\mathcal{F}'$ clearly satisfies the conditions of the theorem and is finite. For such $\mathcal{F}'$, it has been proved that $\cap \mathcal{F}' \neq \emptyset$.

This contradiction completes the proof of the general case. □

Remark B.10. The compactness of all the F_α is used only for the case of infinite families $\mathcal{F}$. This assumption can be weakened as follows.

Let all the F_α in Theorem B.9 be closed and some finite subfamily $\widehat{\mathcal{F}} \subset \mathcal{F}$ have nonempty compact intersection. Keeping the assumption of cohomological triviality we then obtain $\cap \mathcal{F} \neq \emptyset$. To prove this, it suffices to consider a new family consisting of the sets $(\cap \widehat{\mathcal{F}}) \cap F_\alpha$ with $F_\alpha \notin \widehat{\mathcal{F}}$ and apply the argument of the second part of the proof.

Note that closed balls of E^n, $\mathbb{H}^n$ and $\mathbb{S}^n_+$, the n-dimensional Euclidean and hyperbolic spaces and open n-hemispheres, are contractible and therefore cohomologically trivial; the same is true for their finite intersections. Moreover, E^n, $\mathbb{H}^n$ and $\mathbb{S}^n_+$ are homeomorphic to $\mathbb{R}^n$. Hence, Theorem B.7 can be applied to families of closed balls in each of these spaces. According to Definition 1.32 of Helly's index, this implies that

$$i_H(\mathbb{R}^n) = i_H(\mathbb{H}^n) = i_H(\mathbb{S}^n_+) = n+1,$$

as was stated in Section 1.10.

Finally, we formulate a Helly type theorem for the n-sphere proved by Robinson [Rob-1942]. In this case the intersection of two closed balls of $\mathbb{S}^n$ may be noncontractible as the example of two closed hemispheres with common equator shows. Hence, Theorem B.7 cannot be applied to families of closed balls. Moreover, the Helly index of $\mathbb{S}^n$ cannot be less than $2n+2$ as the following example shows. Consider a family of $2n+2$ closed hemispheres of $\mathbb{S}^n$ with centers at the points $\pm e_i$, $1 \leq i \leq n+1$, (recall that $\{e_i\}$ is the standard basis of $\mathbb{R}^{n+1}$). Then every $2n+1$ members of this family clearly have a point in common, but the intersection of all $2n+2$ of them is empty.

To formulate a general Helly type theorem for this case we need the following notion.

A (closed) subset of $\mathbb{S}^n$ is *spherically convex*, if it is the intersection of a family of closed hemispheres.

It is clear that the intersection of a family of spherically convex sets also has this property. Moreover, a spherically convex set, other than an m-subsphere

Appendix C. Sperner's lemma and its consequences

$(0 \leq m \leq n)$, is contractible. The exceptional role of m-spheres affects the form of Robinson's theorem presented now.

Theorem B.11. *A family of more than $2n+2$ spherically convex subsets of $\mathbb{S}^n$ has a point in common provided that every $2n+1$ of its members has nonempty intersection.*

The example presented above shows that the theorem is not true for families of $2n+2$ subsets. This example and the above theorem show that the Helly index of $\mathbb{S}^n$ equals $2n+2$.

C. Sperner's lemma and its consequences

A single combinatorial result discovered by Sperner [Spe-1928] not only leads to surprisingly simple proofs of profound topological theorems, but is in itself a remarkable fact of Combinatorial Geometry. We present here the proofs of this result and several of its corollaries (including Lemma 1.35 from Section 1.10) as well as the Lebesgue–Brouwer theorem mentioned in Appendix A.

In order to introduce these results one recalls that a *triangulation* of a nondegenerate n-simplex $\Delta_n \subset \mathbb{R}^n$ is a simplicial complex K such that the union $[K]$ of its simplices coincides with Δ_n, see Definition B.6 of a simplicial complex. We assume that K contains also "small" n-simplices (different from Δ_n). Then K gives rise to a decomposition of Δ_n into a finite number of n-simplices such that two intersecting simplices touch each other at a unique $(n-1)$-simplex. We will denote this decomposition by the same letter K while K^0 will stand for the set of vertices (0-faces) of K.

A *coloring* of K^0 with *colors* $0, 1, \ldots, n$ is a function γ from K^0 to $\{0, 1, \ldots, n\}$.

Theorem C.1. (Sperner's Lemma). *Let a coloring γ of K^0 meet the following conditions:*

(i) *if $v_0, v_1, \ldots, v_n \in K^0$ are the vertices of Δ_n, then*

$$\gamma(v_j) = j, \quad 0 \leq j \leq n;$$

(ii) *if $v \in K^0$ and belongs to a $(n-1)$-face $\Delta_I := \operatorname{conv}\{v_i\}_{i \in I}$ of Δ_n, then $\gamma(v) \in I$;*

(iii) *interior vertices of K^0 are colored arbitrarily with $0, 1, \ldots, n$.*

Then there exists a small n-simplex of K for which all $n+1$ vertices have different colors.

Proof. A stronger fact will be established by induction on n:

The number of such colored small n-simplices is *odd*.

For $n=1$, K^0 is a finite set of points in $\Delta_1 = [v_0, v_1]$, say, $v_0 < v^1 < \cdots < v^N < v_1$. In this case $\gamma(v_i) = i$, $i = 0, 1$, and $\gamma(v^j) \in \{0,1\}$. Note that the number

$\varepsilon_j := \gamma(v^j) - j(v^{j-1})$ is odd (equals ± 1) if and only if the 1-simplex $[v^{j-1}, v^j]$ is properly colored and is zero otherwise. Setting $v^0 := v_0$ and $v^{N+1} := v_1$ we then get $\pm 1 = \gamma(v_1) - \gamma(v_0) = \sum_{i-1}^{N+1} \varepsilon_i$. Hence the number of odd ε_i is odd.

Assume now that the result has been proved for all nondegenerate $(n-1)$-simplices, and prove it for $\Delta_n = \mathrm{conv}\{v_1, \ldots, v_n\}$ with triangulation K and coloring γ, subject to the conditions of the theorem. For this goal, introduce a graph $\Gamma = \Gamma(K, \gamma)$ whose vertices are small n-simplices of K and one more vertex, $\Delta_n^c := \mathbb{R}^n \setminus \Delta_n$. Now introduce the set of edges for Γ as follows. Two distinct vertices (small n-simplices) of K are joined by an edge, if they have a (unique) common $(n-1)$-face colored with all the colors $1, 2, \ldots, n$. In turn, Δ_n^c is connected by edges with those small n-simplices which have $(n-1)$-faces contained in $\Delta_{n-1} := \mathrm{conv}\{v_1, \ldots, v_n\}$ and are properly colored with respect to Δ_{n-1}. According to conditions (i) and (ii) this means that these $(n-1)$-faces are colored by n different colors $1, 2, \ldots, n$. By the induction hypothesis, the number of these faces is odd. Hence the *degree* $\deg \Delta_n^c$ of vertex Δ_n^c, i.e., the number of edges emanating from the vertex, is odd, say,

$$\deg \Delta_n^c = 2d + 1 \quad \text{for some} \quad c \in \mathbb{Z}_+.$$

On the other hand, $\deg \Delta$ for $\Delta \in K$ is 1, if Δ is properly colored and is 2, otherwise. In fact, in the latter case only two vertices of Δ have the same color. Hence, the sum of all degrees

$$\deg \Delta_n^c + \sum_{\Delta \in K} \deg \Delta = a + 2b + (2d+1),$$

where a is the number of properly colored small n-simplices and $b \geq 0$ is some integer.

Count then this sum in another way noting that every edge of Γ has two endpoints (incident vertices). Hence, the sum of the degrees of all vertices equals the doubled number of edges between them, say $2e$, for some $d \in \mathbb{Z}_+$. Hence, $2e = a + 2b + (2d+1)$ and therefore a, the number of properly colored small simplices, must be odd. □

In order to derive from here the aforementioned fundamental topological theorems we need two auxiliary results due to Knaster, Kuratowski and Mazurkiewicz [KKM-1929]. The first of them was formulated in Section 1.10 as Lemma 1.35.

In what follows $\Delta = \mathrm{conv}\{v_i\}_{0 \leq i \leq n}$ is a nondegenerate n-simplex in $\mathbb{R}^n$. If I is a proper subset of $\{0, 1, \ldots, n\}$, then Δ_I is the face of Δ with vertices v_i, $i \in I$. This Δ_I is *opposite to a vertex* v_i, if $I = \{0, 1, \ldots, n\} \setminus \{i\}$; in this case we use the notation Δ^i.

Theorem C.2. *Let $\{F_i\}_{0 \leq i \leq n}$ be a family of closed subsets in $\mathbb{R}^n$ such that for every proper subset $I \subset \{0, 1, \ldots, n\}$*

$$\Delta_I \subset \bigcup_{i \in I} F_i. \tag{C.1}$$

Then the intersection of all the F_i is nonempty.

Proof. Let K_ε be a triangulation of Δ with subsimplices of diameter at most ε. Color vertices $v \in K_\varepsilon^0$ with colors $0, 1, \ldots, n$, giving v a color i, if v belongs to F_i. (We choose one of these i's arbitrarily, if v belongs to several F_i.) Because of (C.1) the coloring satisfies conditions (i)–(iii) of Theorem C.1. Hence, there exists a properly colored n-simplex of K_ε, say, Δ_ε. In other words, Δ_ε intersects all of F_i. Now let $m_0 \in \Delta$ be a limit point of the family of points $m_\varepsilon \in \Delta_\varepsilon$ as $\varepsilon \to 0$. Then every open neighborhood of m_0 contains some Δ_ε and therefore contains points of every F_i. Because F_i is closed this point belongs to every F_i, i.e., $m_0 \in \bigcap_{i=0}^{n} F_i$. □

Corollary C.3. *Let $\{F_i\}_{i=0}^n$ be a cover of Δ by closed subsets such that for every vertex v_i of Δ,*

$$v_i \in F_i, \tag{C.2}$$

and, moreover, for every $(n-1)$-face Δ^i opposite to v_i,

$$F_i \cap \Delta^i = \emptyset. \tag{C.3}$$

Then $\bigcap_{i=0}^{n} F_i \neq \emptyset$.

Proof. It suffices to show that conditions (C.2) and (C.3) imply (C.1). Let Δ_I be a proper face of Δ and $i \notin I$. Then Δ_I is clearly a subset of Δ^i. Since it is true for every $i \notin I$ and $\{F_i\}$ covers Δ, Δ_I must be covered by the F_i with $i \in I$, by (C.3). □

Remark C.4. Both of the above results are true also for a degenerate n-simplex Δ. In fact, choose an ε-neighborhood F_i^ε of every F_i with ε sufficiently small and use a small perturbation to convert Δ into a nondegenerate n-simplex Δ_ε preserving either condition (C.1) or conditions (C.2) and (C.3). Then there is a point m_ε in Δ_ε belonging to $\bigcap_{i=0}^{n} F_i^\varepsilon$. Using a sequence of such Δ_ε converging to Δ, take a limit point m_0 of sequence $\{m_\varepsilon\}$. Clearly, m_0 belongs to $\bigcap_{i=0}^{n} F_i$.

Note that Lemma 1.35 is formulated for a Δ which may be degenerate.

Now we prove the famous Lebesgue–Brouwer theorem on the dimension of $\mathbb{R}^n$, see [Le-1911] and [Bro-1912].

Theorem C.5. $\dim \mathbb{R}^n = n$.

Proof. By Definition A.1 of the dimension, $\dim A \leq \dim B$ provided that $A \subset B \subset \mathbb{R}^n$. Since, in addition, the interior Δ^0 of a closed nondegenerate n-simplex $\Delta \subset \mathbb{R}^n$ is homeomorphic to $\mathbb{R}^n$, we get

$$\dim \mathbb{R}^n = \dim \Delta^0 \leq \dim \Delta \leq \dim \mathbb{R}^n.$$

Hence, it suffices to prove that

$$\dim \Delta = n. \tag{C.4}$$

Invoking again Definition A.1, one divides (C.4) into the next two statements.

Statement I. Every locally finite open cover $\mathcal{U}$ of Δ admits an open refinement of order $\leq n+1$.

Statement II. Every locally finite open cover $\mathcal{U}$ of Δ whose subsets are sufficiently small has order $\geq n+1$.

Proof of I.. Because of the compactness of Δ, the cover $\mathcal{U}$ is assumed to be finite. Therefore the required refinement of $\mathcal{U}$ can be formed by closed subsets of Δ. We construct this refinement as follows.

Introduce, first, a cover $\mathcal{V}_\varepsilon$ of $\mathbb{R}^n$ by closed subsets of diameter $< \varepsilon$. If its order is at most $n+1$, then $\mathcal{V}_\varepsilon$ gives rise to the closed cover of Δ, say $\hat{\mathcal{V}}_\varepsilon$ of order $\leq n+1$. If ε is sufficiently small, then $\hat{\mathcal{V}}_\varepsilon$ is a refinement of $\mathcal{U}$. This will be deduced from

Lemma C.6 (Lebesgue). *Let $\mathcal{U} = \{\mathcal{U}_\alpha\}$ be a locally finite open cover of a compact metric space $\mathcal{M}$. Then there exists a number $\lambda = \lambda(U) > 0$ such that every subset of diameter $< \lambda$ is contained in one of the $\mathcal{U}_\alpha$.*

Proof. Introduce a function $\varphi : \mathcal{M} \to \mathbb{R}_+$ given at $m \in \mathcal{M}$ by

$$\varphi(m) := \sup\{r \; ; \; B_r(m) \subset \mathcal{U}_\alpha \text{ for some } \alpha\}.$$

Since $\mathcal{U}$ is locally finite and open, this function is continuous (even 1-Lipschitz) and strictly positive. Then φ attains its minimum at some $m_0 \in \mathcal{M}$ and one can set $\lambda := \varphi(m_0) > 0$. $\square$

Hence, $\hat{\mathcal{V}}_\varepsilon$ is a refinement of $\mathcal{U}$, if $\varepsilon < \lambda$. It remains to construct the closed cover $\mathcal{V}_\varepsilon$ of $\mathbb{R}^n$ (of order $\leq n+1$). For this goal, decompose $\mathbb{R}^n$ into congruent dyadic n-cubes of diameter $< \varepsilon$. Then triangulate each cube into n-simplices having a common vertex at the center of the cube and the other n vertices at the centers of the k-faces of the cube with $k = 0, 1, \ldots, n-1$ (one face in each dimension). In this way the cube is triangulated into $2^n n!$ congruent n-simplices. Denote the triangulation formed by all these simplices by K_ε.

Now let $\widetilde{\Delta} := \mathrm{conv}\{v_0, \ldots, v_n\} \in K_\varepsilon$. Define for each i a hyperplane L_i passing through the barycenter $\frac{1}{n+1}\sum_{i=0}^{n} v_i$ of $\widetilde{\Delta}$ and parallel to the $(n-1)$-face opposite to v_i of $\widetilde{\Delta}$. Then L_i cuts off from $\widetilde{\Delta}$ a simplex containing v_i; denote it by $F_{\widetilde{\Delta}}^i$, $0 \leq i \leq n$.

For a vertex $v \in K_\varepsilon^0$ of the constructed triangulation, consider the set

$$F_v := \bigcup_{\widetilde{\Delta} \ni v} F_{\widetilde{\Delta}}^i.$$

Appendix C. Sperner's lemma and its consequences

Since v belongs to at most $2^n n!$ of $\widetilde{\Delta}$, this set is closed. The family $\{F_v\}_{v \in K_\varepsilon^0}$, clearly, covers $\mathbb{R}^n$, and it remains to estimate its order.

Pick a point $m \in \mathbb{R}^n$. Then $m \in \text{conv}\{v_0, \ldots, v_r\}$ for some $v_i \in K_\varepsilon^0$ and $r \leq n$. Then m can be contained only in the sets F_{v_i}, $0 \leq i \leq r$. Hence, the order of $\{F_v\}$ is $n+1$, and Statement I is proved. □

Now let us prove Statement II. Let $\mathcal{U} = \{U_\alpha\}$ be a cover of a given nondegenerate n-simplex $\Delta = \text{conv}\{v_0, \ldots, v_n\}$ by open subsets of diameter $< \varepsilon$. Choose $\varepsilon > 0$ so small that the closure $\overline{U}_\alpha$ of U_α containing a vertex v_i does not intersect the $(n-1)$-face Δ^i opposite to v_i. We show that the order of $\mathcal{U}$ is $\geq n+1$. Clearly, it suffices to prove this for closed cover $\overline{\mathcal{U}} := \{\overline{U}_\alpha\}$. Moreover, because of the compactness of Δ, we can deal with a finite family $\overline{\mathcal{U}} = \{\overline{U}_0, \ldots, \overline{U}_m\}$. Because of the choice of $\varepsilon > 0$, this cover must contain at least $n+1$ elements, so $m \geq n$. We assume that $\overline{\mathcal{U}}$ is ordered so that $v_i \in \overline{U}_i$, $0 \leq i \leq n$.

Assume that $m > n$ and $n < j \leq m$. Then $\overline{U}_j$ contains some v_i and does not intersect the $(n-1)$-face opposite to v_i. Then we replace $\overline{U}_i$ by a new closed set $\overline{U}_i^{(1)} := \overline{U}_i \cup \overline{U}_j$ and in this way obtain a new cover $\mathcal{U}^{(1)}$ which consists of $m-1$ subsets.

Iterating these procedure we finally obtain a cover $\mathcal{U}^{(p)} = \{\overline{U}_0^{(p)}, \ldots, \overline{U}_n^{(p)}\}$ of Δ consisting of $n+1$ closed subsets, such that every $\overline{U}_i^{(p)}$ contains v_i and does not intersect Δ^i. Then the family $\overline{\mathcal{U}}^{(p)}$ satisfies the conditions of Corollary C.3 and therefore $\cap \overline{\mathcal{U}}^p \neq \emptyset$. Hence, the order of $\overline{\mathcal{U}}^{(p)}$ is $n+1$. Since the above replacement procedure does not increase the order of the cover, the initial cover $\mathcal{U}$ has order at least $n+1$.

This proves Statement II and the theorem. □

Finally, we present a remarkable simple proof of the most famous and applicable theorem of Topology, Brouwer's fixed point theorem. In its formulation a *closed n-cell* means a topological space homeomorphic to the closed unit ball of $\mathbb{R}^n$.

Theorem C.7. *Every continuous map of a closed n-cell into itself has a fixed point.*

Proof. ([KKM-1929]) Since the unit n-ball is homeomorphic to a nondegenerate n-simplex $\Delta := \text{conv}\{v_0, \ldots, v_n\} \subset \mathbb{R}^n$ and the property of having a fixed point is preserved under a homeomorphism, it suffices to prove the result for a continuous map $f : \Delta \to \Delta$. Let $\beta_i(m)$, $i = 0, 1, \ldots, n$, be the barycentric coordinates of a point $m \in \Delta$. Hence, they satisfy

$$\sum_{i=0}^{m} \beta_i(m) = 1 \quad \text{and} \quad \beta_i(m) \geq 0,$$

and are such that $m = \sum_{i=0}^{n} \beta_i(m) v_i$. Introduce sets $F_i \subset \Delta$ by letting

$$F_i := \{m \in \Delta \; ; \; \beta_i(f(m)) \leq \beta_i(m)\}, \quad 0 \leq i \leq n.$$

Since the coordinate functions are continuous, these sets are closed. Finally, for m belonging to a proper face $\Delta_I = \operatorname{conv}\{v_i\}_{i \in I}$ of Δ, we get

$$\sum_{i \in I} \beta_i(m) = 1 \geq \sum_{i \in I} \beta_i(f(m)).$$

Therefore, for at least one $i \in I$, we get $\beta_i(m) \geq \beta_i(f(m))$, i.e., $m \in F_i \subset \bigcup_{i \in I} F_i$. Hence, Δ_I is covered by $\bigcup_{i \in I} F_i$, i.e., the conditions of Theorem C.2 are valid for the F_i's. By virtue of this theorem we get a point $m_0 \in \bigcap_{i=0}^{n} F_i$; in other words, this point satisfies the inequalities

$$\beta_i(m_0) \geq \beta_i(f(m_0)) \quad \text{for} \quad 0 \leq i \leq n.$$

But $\sum \beta_i(m_0) = \sum \beta_i(f(m_0)) = 1$, and these inequalities can be true only if $\beta_i(m_0) = \beta_i(f(m_0))$, $0 \leq i \leq n$. Hence $f(m_0) = m_0$. □

D. Contractions of n-spheres

We present here a proof of Proposition 1.41 from Section 1.10. The result appeared in the paper of Valentine [Va-1945], see Lemmas 2 and 3 there. Its role for Lipschitz extensions of maps between Riemannian manifolds was recognized by Gromov, see [Gr-1987, pp. 21-22]

Let us recall the formulation of the Valentine theorem:

Assume that $f : \Sigma \to \mathbb{S}^n$ is a contraction (1-Lipschitz map) of a compact set $\Sigma \subset \mathbb{S}^n$ satisfying the following condition:

$f(\Sigma)$ is not contained in any closed hemisphere of $\mathbb{S}^n$.

Then f is an isometry.

Proof. The proof breaks up into the following alternative cases:

I. The set of all antipodal points of $f(\Sigma)$ is contained in the closure of $f(\Sigma)$.

Since the center 0 of $\mathbb{S}^n$ is the origin of $\mathbb{R}^{n+1}$, the antipodal point of $s \in \mathbb{S}^n$ is $-s$. Hence, in this case

$$-f(\Sigma) \subset \overline{f(\Sigma)}. \tag{D.1}$$

We derive from (D.1) that $\Sigma = -\Sigma$ and

$$f(-s) = -f(s) \quad \text{for all} \quad s \in \Sigma. \tag{D.2}$$

The latter suffices to show for points $-f(s)$ from $f(\Sigma)$. Then the case $-f(s) \in \overline{f(\Sigma)} \setminus f(\Sigma)$ follows from the previous one by a compactness argument.

Appendix D. Contractions of n-spheres

Thus, let $-f(s) = f(s')$ for some $s' \in \Sigma$. Since f is a contraction and the spherical distance d of $\mathbb{S}^n$ is bounded by π, we have

$$\pi = d\big(f(s), -f(s)\big) = d\big(f(s), f(s')\big) \leq d(s, s') \leq \pi.$$

Hence, $d(s, s') = \pi$ and $s' = -s \in \Sigma$.

Now let $s_1, s_2 \in \Sigma$. By (D.2) and the definition of d, we conclude that

$$d(s_1, s_2) + d(s_1, -s_2) = \pi, \quad d\big(f(s_1), f(s_2)\big) + d(f(s_1), f(-s_2)) = \pi.$$

Moreover, $d(f(s_1), f(\pm s_2)) \leq d(s_1, \pm s_2)$, and these inequalities together with the previous equalities lead to the required relation $d(f(s_1), f(s_2)) = d(s_1, s_2)$.

This proves the theorem for this case.

II. There exists $s_0 \in \Sigma$ such that

$$-f(s_0) \notin \overline{f(\Sigma)}. \tag{D.3}$$

We derive from (D.3) the following

Lemma D.1. *There exist points $s_1, \ldots, s_{j+1}$ of $\mathbb{S}^n$ with $1 \leq j \leq n$ such that the simplex*

$$\Delta := \mathrm{conv}\{f(s_0), f(s_1), \ldots, f(s_{j+1})\}$$

contains the center 0 of $\mathbb{S}^n$ in its interior.

Proof. Consider the convex hull $\mathrm{conv}\, f(\Sigma)$ of $f(\Sigma)$. Since $f(\Sigma)$ is not contained in any closed hemisphere, $\mathrm{conv}\, f(\Sigma)$ contains the center 0 of $\mathbb{S}^n$ in its *interior*.

Now let $x_0 \in \mathbb{R}^{n+1}$ be the (unique) point of the ray $\{-\lambda f(s_0)\,;\, \lambda \geq 0\}$ lying in the boundary of $\mathrm{conv}\, f(\Sigma)$. By the definition of $\mathrm{conv}\, f(\Sigma)$, each of its boundary points is a convex combination of at most $n+1$ points of $f(\Sigma)$; hence,

$$x_0 = \sum_{i=1}^{n+1} \alpha_i f(s_i)$$

for some $s_i \in \Sigma$ and $\alpha_i \geq 0$ such that $\sum \alpha_i = 1$.

Note that the number of α_i that > 0 is at least 2; otherwise, $x_0 (= -\lambda_0 f(s_0)$ for some $\lambda_0 > 0$) would be in $f(\Sigma)$, in contradiction with (D.3). Hence, x_0 is an interior point of the simplex $\mathrm{conv}\{f(s_1), \ldots, f(s_{j+1})\}$ with some $s_i \in \Sigma$ and $1 \leq j \leq n$ (renumerating, if necessary, the vertices). But the center 0 belongs to the open interval $(f(s_0), x_0)$ and therefore it is contained in the interior of the simplex $\Delta := \mathrm{conv}\{f(s_0), f(s_1), \ldots, f(s_{j+1})\}$. $\square$

For the simplex $\Delta = \mathrm{conv}\{f(s_0), \ldots, f(s_{j+1})\}$ of the above lemma, the following is true.

Lemma D.2. *For every pair $0 \leq i \neq k \leq j+1$, we have*

$$d(f(s_i), f(s_k)) = d(s_i, s_k). \tag{D.4}$$

Proof. Assume, on the contrary, that (D.4) is not true for, say, $i = 0$ and $k = 1$:
$$d(f(s_0), f(s_1)) < d(s_0, s_1). \tag{D.5}$$

As before, we get for antipodal pairs
$$d(s_0, s_1) + d(s_i, -s_0) = \pi, \quad d(f(s_0), f(s_i)) + d(f(s_i), -f(s_0)) = \pi,$$

where $i = 1, \ldots, j+1$. Moreover, $d(f(s_0), f(s_i)) \leq d(s_0, s_i)$ and therefore these equalities and (D.5) imply that
$$d(f(s_i), -f(s_0)) \geq d(s_i, -s_0), \quad i = 2, \ldots, j+1,$$
$$d(f(s_1), -f(s_0)) > d(s_1, -s_0).$$

Using the property of d presented in (1.68) of Section 1.10, we equivalently rewrite these inequalities:
$$\begin{aligned} -s_0 \cdot s_i &\geq -f(s_0) \cdot f(s_i), \quad i = 2, \ldots, j+1, \\ -s_0 \cdot s_1 &> -f(s_0) \cdot f(s_1). \end{aligned} \tag{D.6}$$

Similarly, the inequalities $d(f(s_i), f(s_k)) \leq d(s_i, s_k)$ imply that
$$f(s_i) \cdot f(s_k) \geq s_i \cdot s_k, \quad 1 \leq i, k \leq j+1. \tag{D.7}$$

In turn, 0 is in the interior of the simplex $\Delta = \mathrm{conv}\{f(s_0), \ldots, f(s_{j+1})\}$, and therefore $-f(s_0)$ is an interior point of the minimal cone containing all $f(s_i)$ with $i \neq 0$. Hence, there exist constants $\lambda_i > 0$ such that
$$-f(s_0) = \sum_{i=1}^{j+1} \lambda_i f(s_i). \tag{D.8}$$

Now, multiplying inequalities (D.6) by λ_i and summing on $i \geq 1$, then multiplying inequalities (D.7) by $\lambda_i \lambda_k$ and summing on $i, k \geq 1$, and, finally, applying (D.8), we derive the contradictory inequalities
$$-s_0 \cdot \sum_{i \geq 1} \lambda_i s_i > (-f(s_0)) \cdot (-f(s_0)) = 1,$$
$$\left| \sum_{i \geq 1} \lambda_i s_i \right|^2 \leq (-f(s_0)) \cdot (-f(s_0)) = 1.$$

Actually, by the Schwartz inequality, we get from here that
$$1 < |-s_0| \cdot \left| \sum_{i \geq 1} \lambda_i s_i \right| \leq 1.$$

The result is verified. □

Appendix D. Contractions of n-spheres

To complete the derivation we must prove, for arbitrary pair $s', s'' \in \Sigma$, the equality
$$d(f(s'), f(s'')) = d(s', s''). \tag{D.9}$$
Using an appropriate rotation of $\mathbb{S}^n$ we may and will assume that s' coincides with the point s_0 in (D.3). Further, using an appropriate isometry ρ of $\mathbb{S}^n$ (orthogonal transformation of $\mathbb{R}^{n+1}$) we map the simplex $\Delta = \text{conv}\{f(s_0), \ldots, f(s_{j+1})\}$ onto $\text{conv}\{s_0, \ldots, s_{j+1}\}$. This is possible, since the corresponding edges of these simplices are equal by Lemma D.2. If we prove that $\tilde{f} := \rho \circ f$ is an isometry, the same holds for f. To simplify notation, we will now write f instead of $\tilde{f}$. Hence, the following now holds:
$$s_i = f(s_i), \quad 0 \le i \le j+1.$$
We must prove that
$$d(f(s_0), f(s)) = d(s_0, s)$$
for arbitrary $s \in \Sigma$ distinct from points s_i.

If this is not true, that is, $d(f(s_0), f(s)) < d(s_0, s)$, then, for the antipodal points, we get, as before, the opposite inequality $d(f(s), -f(s_0)) > d(s, -s_0)$. This, in turn, gives
$$-s_0 \cdot s > -f(s_0) \cdot f(s) = -s \cdot f(s). \tag{D.10}$$
On the other hand, inequalities $d(f(s_i), f(s)) \le d(s_i, s)$ imply that
$$f(s_i) \cdot f(s) \ge s_i \cdot s, \quad i = 1, \ldots, j+1,$$
whence
$$\sum_{i \ge 1} \lambda_i f(s_i) \cdot f(s) \ge \sum_{i \ge 1} \lambda_i s_i \cdot s.$$
But $f(s_i) = s_i$ and $\sum_{i \ge 1} \lambda_i f(s_i) = -f(s_0)(= -s_0)$, see (D.8). Hence, we can rewrite the above inequality as
$$-s_0 \cdot f(s) \ge -s_0 \cdot s,$$
and this contradicts inequality (D.10). Hence, (D.9) is true and f is an isometry.
$\square$

Chapter 2

Smooth Functions on Subsets of $\mathbb{R}^n$

This chapter is an introduction to the final part of the book (Chapters 9–11) that studies the problems of a field which may be regarded as a part of Global Differential Analysis on subsets of $\mathbb{R}^n$. Such subsets may be of a very complicated geometrical nature (i.e., fractals) and, in general, have no naturally associated differential structure. The presence of such different structures – rigid in the sense of Differential Calculus and possibly chaotic in terms of the underlying set – makes the extension–trace problems for smooth functions one of the most difficult problems of Analysis. Therefore the Whitney solution to the problem for univariate C^k functions appears as a miracle. The title of Whitney's paper [Wh-1934b] includes number I which, probably, indicates his intention to proceed further with this study. However, significant multivariate results appeared only about fifty years after Whitney's seminal papers [Wh-1934a] and [Wh-1934b]. The extension theorem from the first paper is one of the basic results of Global Differential Analysis; its generalizations and ramifications have been presented in many papers and books, see, e.g., the books [Hor-1983], [Mal-1966] and [Ste-1970]. Therefore it would be natural to present this material in as brief a manner as possible. However, Whitney's extension method plays an important role in almost all approaches to the problems under consideration. In Section 2.2 we will give it a rather detailed exposition. The reader is referred to the aforementioned books for further details.

On the other hand, Whitney's beautiful result [Wh-1934b] has never appeared in book form. We present (in Section 2.4) its complete proof but not only for this reason. In fact, an appropriate reformulation of this result leads to several general conjectures for multivariate functions whose study appears to be one of the main trends in the field. We discuss the results obtained in this direction within the last two decades in the final part of the book.

Difference characteristics of C^k functions in the univariate case and local approximation by Taylor polynomials in the multivariate case play an essential role in Whitney's proofs. The results of this kind which are not commonly known can be found in Section 2.3.

Turning to the extension and trace problems for Lipschitz functions of higher order (see Section 2.1 for the corresponding definitions) one immediately notices that Taylor approximation is inefficient in this setting. For instance, the Weierstrass nowhere differentiable function is a Lipschitz function of second order. A new tool, *local polynomial best approximation*, which will be used for this case, will first appear in Section 2.3 in connection with S. Bernstein's classical theorem [Ber-1940]. This result gives a complete description of the space $C^k(\mathbb{R})$ via local best approximation.

In Chapters 9 and 10 (Volume II), we will show that local polynomial approximation theory gives powerful tools for the study of extension–trace problems for Lipschitz functions of higher order. In fact, the range of its applications is much broader and includes the study of basic properties of the classical spaces of smooth functions.

The results obtained are then applied in Section 2.5 to solve the restricted Main Problem for several classes of domains in $\mathbb{R}^n$ (quasiconvex and Lipschitz).

The final section presents selected trace and extension problems for weakly differentiable functions, in particular, the P. Jones extension theorem [Jon-1981] and the Peetre theorem [Peet-1979] on the nonexistence of a simultaneous extension for the Sobolev space $W_1^1(\mathbb{R}^n_+)$.

2.1 Classical function spaces: notation and definitions

2.1.1 Differentiable functions

Throughout the book $C^k(\mathbb{R}^n)$ denotes the space of k-times continuously differentiable functions on $\mathbb{R}^n$ equipped with the topology of uniform convergence of the functions and their derivatives on compact subsets of $\mathbb{R}^n$; here k may be infinity. Hence, $C^k(\mathbb{R}^n)$ is a Fréchet space with topology determined by the collection of seminorms

$$|f|_m^K := \sum_{|\alpha|\leq m} \sup_K |D^\alpha f|, \qquad (2.1)$$

where $K \subset \mathbb{R}^n$ is compact and $0 \leq m \leq k$.

Hereafter we will use the standard notation of Differential Analysis. In particular, $\alpha = (\alpha_1, \ldots, \alpha_n), \beta, \ldots$ are multi-indices, i.e., vectors from $\mathbb{Z}^n_+$, and $|\alpha| := \sum_{i=1}^n \alpha_i$. Moreover, for $x \in \mathbb{R}^n$,

$$x^\alpha := \prod_{i=1}^n x_i^{\alpha_i} \quad \text{and} \quad D^\alpha := \prod_{i=1}^n D_i^{\alpha_i}, \qquad (2.2)$$

2.1. Classical function spaces: notation and definitions

where $D_i := \frac{\partial}{\partial x_i}$.

For example, the Taylor polynomial of degree k at a point $y \in \mathbb{R}^n$ may be written as

$$T_y^k F(x) := \sum_{|\alpha| \leq k} D^\alpha F(y) \frac{(x-y)^\alpha}{\alpha!}, \qquad (2.3)$$

where $\alpha! := \prod_{i=1}^{n} (\alpha_i!)$.

If now G is a *domain* (connected open subset) of $\mathbb{R}^n$, then the Fréchet space $C^k(G)$ is defined similarly to the case $G = \mathbb{R}^n$ by the collection of seminorms (2.1), but with the subsets K being compactly embedded in G. Such an embedding is denoted by $K \Subset G$.

Definition 2.1. $C_b^k(G)$ is a Banach subspace of $C^k(G)$ defined by the norm

$$\|f\|_{C_b^k(G)} := \sum_{|\alpha| \leq k} \sup_G |D^\alpha f|. \qquad (2.4)$$

A "homogeneous" C^k space is defined by the seminorm

$$|f|_{C_b^k(G)} := \sum_{|\alpha| = k} \sup_G |D^\alpha f| \qquad (2.5)$$

and is denoted by $\dot{C}_b^k(G)$ (with a dot!).

Another subspace of $C^k(G)$ consists of functions with prescribed behavior of the moduli of continuity for their higher derivatives; in its definition presented now k is finite and ω denotes a nonnegative function on $\mathbb{R}_+ := [0, +\infty)$ satisfying the following conditions:

(i) $\omega(t)$ and $\frac{t}{\omega(t)}$ are nondecreasing as t increases to $+\infty$;

(ii) $\omega(0+) = 0$.

Definition 2.2. $C_b^{k,\omega}(G)$ is a Banach subspace of $C^k(G)$ defined by the norm

$$\|f\|_{C_b^{k,\omega}(G)} := \|f\|_{C_b^k(G)} + |f|_{C_b^{k,\omega}(G)}, \qquad (2.6)$$

where the seminorm on the right-hand side is given by

$$|f|_{C_b^{k,\omega}(G)} := \sum_{|\alpha|=k} \sup_{[x,y] \subset G} \frac{|D^\alpha f(x) - D^\alpha f(y)|}{\omega(\|x-y\|)}. \qquad (2.7)$$

Here $[x, y]$ is the closed segment with the endpoints x and y, and $\|\cdot\|$ is the Euclidean norm of $\mathbb{R}^n$.

As before, $\dot{C}_b^{k,\omega}(G)$ will denote the homogeneous $C_b^{k,\omega}$-space defined by seminorm (2.7).

For $G = \mathbb{R}^n$, in particular, we have the spaces $C_b^k(\mathbb{R}^n)$, $\dot{C}_b^k(\mathbb{R}^n)$ and so on.

Remark 2.3. If the function $\omega : \mathbb{R}_+ \to \mathbb{R}_+$ satisfies the above condition (i), then ω is *subadditive*, i.e., for $0 < t_1, t_2 < \infty$,

$$\omega(t_1 + t_2) \leq \omega(t_1) + \omega(t_2). \tag{2.8}$$

This and (ii) imply that $d_\omega(x, y) := \omega(\|x - y\|)$, $x, y \in \mathbb{R}^n$, is a metric on $\mathbb{R}^n$ and $\dot{C}^{0,\omega}(\mathbb{R}^n) = \mathrm{Lip}(\mathbb{R}^n, d_\omega)$.

Finally, $\dot{C}^k_u(G)$ is the closed linear subspace of $\dot{C}^k_b(G)$ consisting of functions with *uniformly continuous* higher derivatives:

$$\dot{C}^k_u(G) := \{f \in \dot{C}^k_b(G); \ D^\alpha f, \ |\alpha| = k, \ \text{are uniformly continuous on } G\}. \tag{2.9}$$

2.1.2 k-jets

Let F be a linear space of k-times differentiable functions on a *domain* $G \subset \mathbb{R}^n$. One defines a space of *k-jets* $J^k F$ as the space of all vector functions (k-jets) $\vec{f} := \{f_\alpha\}_{|\alpha| \leq k}$ on G with values in $\mathbb{R}^N$ such that for some $f \in F$ and all $|\alpha| \leq k$,

$$D^\alpha f = f_\alpha.$$

Here $N = N(k, n) := \mathrm{card}\{\alpha \in \mathbb{Z}^n_+ ; \ |\alpha| \leq k\}$.

If $\mathcal{F}$ is a (semi-) normed space, $J^k F$ is equipped with the (semi-) norm

$$\|\vec{f}\|_{J^k F} := \inf\{\|f\|_F ; \ D^\alpha f = f_\alpha, \ |\alpha| \leq k\}.$$

Specifically, choosing F being equal to $C^k(G), C^k_b(G)$, etc., we will simply write $J^k(G), J^k_b(G), \ldots$ instead of $J^k F$. Since G is open, all f_α with $\alpha \neq 0$ are uniquely defined by f_0, e.g., the norm of $\vec{f} \in J^k_b(G)$ satisfies

$$\|\vec{f}\|_{J^k_b(G)} = \|f_0\|_{C^k_b(G)} := \sup_{|\alpha| \leq k} \|f_\alpha\|_{C_b(G)}. \tag{2.10}$$

In particular, the linear projection $\vec{f} \mapsto f_0$ maps isometrically $J^k_b(G)$ onto $C^k_b(G)$; sometimes these spaces will be identified in the sequel.

In turn, the Fréchet topology on $J^k(G)$ (k may be infinity) is defined by the uniform convergence on compact subsets of G, i.e., this topology is determined by the family of seminorms

$$|f|_{m,C} := \sup_{|\alpha| \leq m} \sup_{x \in C} |f_\alpha(x)|,$$

where $0 \leq m \leq k$ and $C \subset G$ runs over all compact subsets of G.

2.1.3 Lipschitz functions of higher order

The definition of this family is based on the notion of *k-modulus of continuity* characterizing smoothness of functions in more detail than do derivatives and Taylor polynomials.

We first recall the definition of the *k-th difference*. Given $h \in \mathbb{R}^n$ and $k \in \mathbb{N}$, the k-th difference is a linear operator on functions on $\mathbb{R}^n$ given by

$$\Delta_h^k := (\tau_h - 1)^k,$$

where $\tau_h : f \mapsto f(\cdot + h)$ is the shift operator.

Hence, for $f : \mathbb{R}^n \to \mathbb{R}$,

$$\Delta_h^k f(x) := \sum_{j=0}^{k} (-1)^{k-j} \binom{k}{j} f(x + jh), \quad x \in \mathbb{R}^n. \tag{2.11}$$

Definition 2.4. *k-modulus of continuity* is the function on $\ell_\infty^{\text{loc}}(\mathbb{R}^n) \times (0, +\infty)$ with range in $\mathbb{R}_+ \cup \{+\infty\}$ given by

$$\omega_k(t\,;f) := \sup_{\|h\| \leq t} \left\| \Delta_h^k f \right\|_{\ell_\infty(\mathbb{R}^n)}. \tag{2.12}$$

Here $\ell_\infty^{\text{loc}}(\mathbb{R}^n)$ is the (Fréchet) space of locally bounded functions f on $\mathbb{R}^n$ equipped with the collection of seminorms $\left\{ \sup_C |f| \right\}$ where C runs over the family of compact subsets of $\mathbb{R}^n$.

The following fact whose proof may be found in Appendix F to this chapter explains the necessity of local boundedness in this definition.

Theorem 2.5. (a) *Let f be bounded on a compact subset of $\mathbb{R}^n$. Then*

$$\omega_k(\cdot\,;f) = 0$$

if and only if f is a polynomial of degree $k - 1$.

(b) *Let $f \in \ell_\infty^{\text{loc}}(\mathbb{R}^n)$ and $\omega_k(t\,;f) < \infty$ for some (and therefore for all) $t > 0$. Then $f = f_0 + p$ where f_0 is bounded on $\mathbb{R}^n$ and p is a polynomial of degree $k - 1$.*

Remark 2.6. In assertion (a), boundedness may be replaced by measurability on $\mathbb{R}^n$. The classical Hamel example of a function H which is nonmeasurable and unbounded on every open subset and satisfies

$$\Delta_h^k H = 0 \quad \text{for all} \quad h \in \mathbb{R}^n \quad \text{and} \quad k \geq 2$$

explains the necessity of having at most one of these assumptions.

In the next theorem we formulate the basic properties of k-modulus of continuity and outline proofs for some of them; see any book on Approximation Theory, e.g., [DeVL-1993] or [Tim-1963] for the proofs of the remaining facts.

Theorem 2.7. *Let $\omega_k(t_0\,;f) < \infty$ for some $t_0 > 0$. Then the following is true:*

(a) *ω_k is nondecreasing and continuous in $t \in (0, +\infty)$ and, moreover,*
$$\omega_k(0+\,;f) = 0$$
if and only if f is uniformly continuous on $\mathbb{R}^n$.

(b) *For every integer $N \geq 1$ and $t > 0$,*
$$\omega_k(Nt\,;f) \leq N^k \omega_k(t\,;f).$$

(c) *There is a function $\widehat{\omega} : (0, +\infty) \to \mathbb{R}_+$ satisfying the conditions*
  (i) $2^{-k}\widehat{\omega} \leq \omega_k(\cdot\,;f) \leq \widehat{\omega}$;
  (ii) $\widehat{\omega}(t)$ and $t^k/\widehat{\omega}(t)$ are nondecreasing as t increases to $+\infty$;

(d) *For $0 \leq \ell < k$,*
$$\omega_k(\cdot\,;f) \leq 2^{k-\ell} \omega_\ell(\cdot\,;f);$$
here $\omega_0(\cdot\,;f) := \|f\|_{\ell_\infty(\mathbb{R}^n)}$.

Conversely, there is a polynomial p of degree $k-1$ such that for some $c > 0$ depending only on k and all $t > 0$,
$$\omega_\ell(t\,;f-p) \leq ct^\ell \int_t^\infty \frac{\omega_k(s\,;f)}{s^{\ell+1}}\,ds. \tag{2.13}$$

(e) *If f has locally bounded derivatives of order ℓ on $\mathbb{R}^n$ where $\ell < k$, then for some constant $c > 0$ and all $t > 0$,*
$$\omega_\ell(t\,;f) \leq ct^\ell \max_{|\alpha|=\ell} \omega_{k-\ell}(t\,;D^\alpha f);$$
here c depends only on k and n.

Conversely, for some $c = c(k,n) > 0$ and all $t > 0$,
$$\max_{|\alpha|=\ell} \omega_{k-\ell}(t\,;D^\alpha f) \leq c \int_0^t \frac{\omega_k(s\,;f)}{s^{\ell+1}}\,ds. \tag{2.14}$$

Proof (outline). (b) follows from the identity
$$\Delta_{Nh}^k = (\tau_h^N - 1)^k = \left(\sum_{j=0}^{N-1} \tau_{jh}\right)^k \Delta_h^k$$

2.1. Classical function spaces: notation and definitions

and the equality $\|T_h\| = 1$.

Assertions (d) and (e) are usually called Marchaud type inequalities after Marchaud [Mar-1927] who proved these results for continuous univariate functions (in a less precise form for (d)). The multivariate case was proved in the paper [BSha-1973] by Yu. Brudnyi and V. Shalashov. For the reader's convenience we will present the proof of (d) in Appendix F.

Finally, the regularization $\widehat{\omega}$ in (c) may be defined by

$$\widehat{\omega}(t) := t^k \sup_{s \geq t} \frac{\omega_k(s;f)}{s^k}.$$

Then (ii) is clear while (i) follows from (b) which implies that

$$\frac{\omega_k(s;f)}{s^k} \leq 2^k \frac{\omega_k(t;f)}{t^k} \quad \text{for} \quad s > t. \qquad \square$$

Property (c) of this theorem motivates

Definition 2.8. A function $\omega : (0, +\infty) \to \mathbb{R}_+$ belongs to the class Ω_k if it satisfies the conditions

(a) ω is nondecreasing, continuous and

$$\omega(0+) = 0;$$

(b) for all $0 < t \leq s$,

$$\frac{\omega(s)}{s^k} \leq \frac{\omega(t)}{t^k}.$$

In the sequel the functions of Ω_k will be called *k-majorants*.

Using the notions introduced we now define the desired family of function spaces.

Definition 2.9. Let $\omega \in \Omega_k$. The homogeneous Lipschitz space $\dot{\Lambda}^{k,\omega}(\mathbb{R}^n)$ consists of locally bounded on $\mathbb{R}^n$ functions f satisfying

$$|f|_{\dot{\Lambda}^{k,\omega}(\mathbb{R}^n)} := \sup_{t>0} \frac{\omega_k(t;f)}{\omega(t)} < \infty. \qquad (2.15)$$

We also define the Banach space $\Lambda^{k,\omega}(\mathbb{R}^n)$ of Lipschitz functions of order k by

$$|f|_{\Lambda^{k,\omega}(\mathbb{R}^n)} := \sup_{\mathbb{R}^n} |f| + |f|_{\dot{\Lambda}^{k,\omega}(\mathbb{R}^n)}. \qquad (2.16)$$

Finally, we introduce the spaces of *smooth functions* combining definitions from both of the subsections. For instance, the space $C_b^k \Lambda^{s,\omega}(\mathbb{R}^n)$, where $\omega \in \Omega_s$, is the linear subspace of $C_b^k(\mathbb{R}^n)$ given by the norm

$$\|f\|_{C_b^k \Lambda^{s,\omega}(\mathbb{R}^n)} := \|f\|_{C_b^k(\mathbb{R}^n)} + \sup_{|\alpha|=k} |D^\alpha f|_{\Lambda^{s,\omega}(\mathbb{R}^n)} \qquad (2.17)$$

while the corresponding homogeneous space $C^k \dot{\Lambda}^{s,\omega}(\mathbb{R}^n)$ is defined by the seminorm

$$|f|_{C^k \Lambda^{s,\omega}(\mathbb{R}^n)} := \max_{|\alpha|=k} |D^\alpha f|_{\Lambda^{s,\omega}(\mathbb{R}^n)}. \tag{2.18}$$

Similarly we define other combinations; e.g., $J^k \Lambda^{s,\omega}(\mathbb{R}^n)$ is the space of k-jets $\vec{f} = \{f_\alpha\}_{|\alpha|\leq k}$ from $J_b^k(\mathbb{R}^n)$, equipped with the norm

$$\|\vec{f}\|_{J^k \Lambda^{s,\omega}(\mathbb{R}^n)} := \max_{|\alpha|\leq k} \sup_{\mathbb{R}^n} |f_\alpha| + \sup_{|\alpha|=k} |f_\alpha|_{\Lambda^{s,\omega}(\mathbb{R}^n)}. \tag{2.19}$$

The properties of the k-modulus of continuity presented in Theorem 2.7 immediately imply the following relations between some of these spaces.

Theorem 2.10. *Let $\omega \in \Omega_k$ and $0 < s < k$ be an integer. Then the following is true:*

(a) *If the function $\widetilde{\omega} : (0, +\infty) \to \mathbb{R}_+$ given by*

$$\widetilde{\omega}(t) := \int_0^t \frac{\omega(u)}{u^{s+1}} \, du$$

is finite, then

$$\Lambda^{k,\omega}(\mathbb{R}^n) \subset C^s \Lambda^{k-s,\widetilde{\omega}}(\mathbb{R}^n) \tag{2.20}$$

and the embedding constant[1] is bounded by some $C = C(k, n)$.[2]

(b) *If the function $\widehat{\omega} : (0, +\infty) \to \mathbb{R}_+$ given by*

$$\widehat{\omega}(t) := t^s \int_t^\infty \frac{\omega(u)}{u^{s+1}} \, du$$

is finite, then

$$\Lambda^{k,\omega}(\mathbb{R}^n) \subset \Lambda^{s,\widehat{\omega}}(\mathbb{R}^n) \tag{2.21}$$

and the embedding constant is bounded by $C = C(k, n)$.

In particular, let $\omega(t) := t^\sigma$, $0 < \sigma \leq k$, and s be the largest integer less than σ. Then the previous result yields the following equality:

$$\Lambda^{k,\omega}(\mathbb{R}^n) = C^s \Lambda^{2,\overline{\omega}}(\mathbb{R}^n), \tag{2.22}$$

where $\overline{\omega}(t) := t^{\sigma-s}$ and the corresponding norms are equivalent.

---

[1] that is, the norm of the linear embedding operator.
[2] We write $C = C(k, \ell, \dots)$ etc. to indicate dependence *only* on the arguments in the brackets.

2.1. Classical function spaces: notation and definitions

Note that if σ is not integer or $\sigma = k$, one can replace the right-hand side by $C^s \Lambda^{1,\overline{\omega}}(\mathbb{R}^n)$. Otherwise $s = \sigma - 1$ and the space in the right-hand side is defined by the norm

$$\|f\|_{B^\sigma_\infty(\mathbb{R}^n)} := \|f\|_{\ell_\infty(\mathbb{R}^n)} + \max_{|\alpha|=\sigma-1} \sup_{t>0} \frac{\omega_2(t; D^\alpha f)}{t}. \qquad (2.23)$$

For $s = 0$ (i.e., $\sigma = 1$) this space is sometimes called the Zygmund space after A. Zygmund who was the first to discover its role in Approximation Theory and Harmonic Analysis [Z-1945]. He also coined the term "smooth function".

In the modern literature the space $\Lambda^{k,\omega}(\mathbb{R}^n)$ with $\omega(t) := t^\sigma$, $0 < \sigma \leq k$, is denoted by $B^\sigma_\infty(\mathbb{R}^n)$, since it is a member of the important family of the so-called Besov spaces. This explains the notation in (2.23); in the sequel we adopt this term and notation.

Remark 2.11. The embeddings of Theorem 2.10 also hold for the corresponding homogeneous spaces. E.g., using Theorem 2.7 (d) we have

$$\dot{\Lambda}^{k,\omega}(\mathbb{R}^n) \subset \dot{\Lambda}^{s,\widehat{\omega}}(\mathbb{R}^n)/\mathcal{P}_{k-1,n},$$

where $\mathcal{P}_{k,n} \subset \mathbb{R}[x_1, \ldots, x_n]$ hereafter stands for the space of polynomials of degree k on $\mathbb{R}^n$.

Now we define similar families of spaces over a domain $G \subsetneq \mathbb{R}^n$. In this case, $f \in \ell^{\text{loc}}_\infty(G)$ and the function $x \mapsto \Delta^k_h f(x)$ is defined on the set

$$G_{k,h} := \{x \,;\, x + jh \in G, \quad j = 0, 1, \ldots, k\}.$$

We remark that the definition commonly used in the literature utilizes the set

$$G_{kh} := \{x \,;\, [x, x + kh] \subset G\}$$

which is smaller than the previous if G is not convex.

Note the distinction between these two sets. The former may be disconnected with infinitely many connected components (as, e.g., for a domain bounded by two spirals in the plane twisted infinitely many times around the origin) while the latter is connected.

Now the definition of k-modulus of continuity for a function $f \in \ell^{\text{loc}}_\infty(G)$ is given by

$$\omega_k(t; f)_G := \sup_{\|h\| \leq t} \|\Delta^k_h f\|_{\ell_\infty(G_{kh})}, \quad t > 0. \qquad (2.24)$$

All of the properties of k-modulus of continuity formulated above for $\mathbb{R}^n$ remain to be true for *convex* domains G but most of them do not hold for general domains.

We leave to the reader to formulate the corresponding results and to define the spaces of Lipschitz functions $\dot{\Lambda}^{k,\omega}(G)$ and of smooth functions $C^k \dot{\Lambda}^{s,\omega}(G)$ and $\dot{B}^\lambda_\infty(G)$ and their nonhomogeneous counterparts.

Remark 2.12. (a) In accordance with the above introduced notation the spaces $\dot{C}^{k,\omega}(G), \dot{J}^{k,\omega}(G)$ will be denoted by $C^k \dot{\Lambda}^{1,\omega}(G), J^k \dot{\Lambda}^{1,\omega}(G)$.

(b) We emphasize the distinction between two Lipschitz spaces over an open set G, the space $\mathrm{Lip}^\omega(G)$ defined by the seminorm

$$|f|_{\mathrm{Lip}^\omega(G)} := \sup\left\{\frac{|f(x) - f(y)|}{\omega(\|x - y\|)}\, ;\, x, y \in G\right\}$$

and the space $\dot{\Lambda}^{1,\omega}(G)$ where in both cases ω is a 1-majorant. Clearly,

$$\dot{\Lambda}^{1,\omega}(G) \supset \mathrm{Lip}^\omega(G),$$

but the second space may be essentially smaller for nonconvex G.

Let, e.g., G be the union of open balls (disks) $B_i := B_{r_i}(i, 0) \subset \mathbb{R}^2$ where $r_i := \frac{1}{2} - \frac{1}{4i}$, $i \in \mathbb{N}$.

Define a function $f : G \to \mathbb{R}$ by letting f on B_i to be equal $(-1)^i$, $i \in \mathbb{N}$. Then, by (2.24),

$$\omega_1(t; f)_G = \sup_{\|h\| \le t} \|\Delta_h^1 f\|_{\ell^\infty(G_h)} = 0,$$

since every component of G_h is contained in some B_i.

On the other hand,

$$|f|_{\mathrm{Lip}^\omega(G)} \ge 2 \sup_i \frac{1}{\omega((4i)^{-2})} = \infty.$$

2.1.4 Extension and trace problems for classical function spaces

Now let S be an arbitrary subset of $\mathbb{R}^n$, and F be one of the normed, seminormed or Fréchet spaces introduced above. In accordance with the definitions of Section 1.1, see (1.1)–(1.3) there, we define the trace space $F|_S$ and the corresponding trace norm or seminorm, etc. For example, for $F := \dot{C}_b^k(\mathbb{R}^n)$ the trace seminorm is given by

$$|f| := \inf\left\{\sum_{|\alpha|=k} \sup_{\mathbb{R}^n} |D^\alpha g|\, ;\, g|_S = f\right\}.$$

For these spaces we then pose the problems formulated in Section 1.2. In particular, the *Main Problem* for $\dot{C}_b^k(\mathbb{R}^n)$ asks for a complete characterization of the trace space $\dot{C}_b^k(\mathbb{R}^n)\big|_S$. This is naturally divided into two subproblems, the *Trace Problem* and the *Extension Problem*. Since all the spaces are linear, we may also ask about the existence of a linear extension operator (the *Simultaneous Extension Problem*).

Another possibility is to consider these problems only for special classes of subsets instead of the class of all closed ones. We will consider only one such case related to bounded domains in $\mathbb{R}^n$. For the space $C_b^k(\mathbb{R}^n)$ this, for instance, leads to the following modification of the Main Problem:

Restricted Main Problem. *Characterize bounded domains $G \subset \mathbb{R}^n$ such that*

$$C_b^k(\mathbb{R}^n)\big|_G = C_u^k(G). \tag{2.25}$$

Note that the restrictions of the higher derivatives of a function $f \in C_b^k(\mathbb{R}^n)$ to a bounded domain $G \subset \mathbb{R}^n$ are uniformly continuous on the closure $\overline{G}$. This explains the appearance of $C_u^k(G)$ in (2.25). In the same fashion, the Restricted Trace and Extension and Simultaneous Extension Problems are formulated for $C_b^k(\mathbb{R}^n)$ and bounded domains.

2.2 Whitney's extension theorem

We present a brief exposition of this classical result with emphasis on the Whitney extension method. The reader could restore the omitted details of the proof by consulting one of the books mentioned at the beginning of the chapter.

Our goal is to characterize the trace space $J^k(\mathbb{R}^n)\big|_S$ for an arbitrary subset $S \subset \mathbb{R}^n$ which in this settings may be assumed to be *closed*. The following remark motivates the appearance of *Taylor chains* in the formulation of Whitney's result.

Suppose that f is a function in $C^k(\mathbb{R}^n)$ and $T_y^k F$ is its Taylor polynomial, see (2.3). Then

$$f = T_y^k f + R_k$$

where R_k, the *remainder*, is a k-times continuously differentiable function in $x \in \mathbb{R}^n$ and continuous in y. Its mixed α-derivative in x with $|\alpha| \leq k$,

$$D_x^\alpha R_k(x, y) = D^\alpha f(x) - \sum_{|\beta| \leq k - |\alpha|} D^{\alpha+\beta} f(y) \cdot \frac{(x-y)^\beta}{\beta!}, \tag{2.26}$$

is clearly the remainder of order $k - |\alpha|$ for $D^\alpha F$ and by the Taylor–Peano theorem,

$$\big|D_x^\alpha R_k(x, y)\big| = o\big(\|x - y\|^{k-|\alpha|}\big) \quad \text{as} \quad y \to x.$$

The *reduced remainder*

$$r_\alpha(x, y) := D_x^\alpha R_k(x, y) / \|x - y\|^{k-|\alpha|}$$

is a continuous function on $(\mathbb{R}^n \times \mathbb{R}^n) \setminus \Delta$, where $\Delta := \{(x, y)\,;\, x = y\}$, and can be continuously extended by zero to $\mathbb{R}^n \times \mathbb{R}^n$.

Thus, we have the following

Taylor chain condition. *Assume that $\vec{f} = \{f_\alpha\}_{|\alpha| \le k}$ is a k-jet on the closed set S generated by a function $f \in C^k(\mathbb{R}^n)$[3]. Then all its reduced remainders*

$$r_\alpha(\vec{f}; x, y) := \left| f_\alpha(x) - \sum_{|\beta| \le k - |\alpha|} f_{\alpha+\beta}(y) \frac{(x-y)^\beta}{\beta!} \right| \Big/ \|x - y\|^{k-|\alpha|} \quad (2.27)$$

are continuous on $(S \times S) \setminus \Delta$ and can be continuously extended by zero to all of $S \times S$.

In other words, if $\vec{f}$ belongs to the trace space $J^k(\mathbb{R}^n)\big|_S$, then $\vec{f}$ satisfies the Taylor chain condition. A theorem of Whitney states that this condition is also sufficient for $\vec{f}$ to be in $J^k(\mathbb{R}^n)\big|_S$.

Theorem 2.13 (Whitney). *A k-jet $\vec{f} = \{f_\alpha\}_{|\alpha| \le k}$ defined on a closed set $S \subset \mathbb{R}^n$ belongs to $J^k(\mathbb{R}^n)\big|_S$ if and only if $\vec{f}$ satisfies the Taylor chain condition.*

We single out the basic ingredients of Whitney's extension method and briefly explain how they work in the proof.

Whitney's covering lemma

Let $\mathcal{K} := \mathcal{K}(\mathbb{R}^n)$ stand for the collection of n-cubes in $\mathbb{R}^n$ homothetic to the cube

$$Q_0 := [-1, 1]^n. \quad (2.28)$$

We regard such a cube as the closed ball of ℓ_∞^n; so

$$Q_r(x) := \{ y \in \mathbb{R}^n \; ; \; \|x - y\|_\infty := \max_{1 \le i \le n} |y_i - x_i| \le r \}$$

is the ball of center x and radius r.

In the sequel, r_Q and c_Q stand for radius and center of $Q \in \mathcal{K}$. Further, by λQ with $\lambda > 0$ we denote the cube of center c_Q and radius λr_Q. Finally, we set

$$\mathcal{K}(S) := \{ Q \in \mathcal{K} \; ; \; Q \subset S \};$$

this collection may be empty.

Lemma 2.14. *There is a cover of the open set $S^c := \mathbb{R}^n \setminus S$ by cubes of $\mathcal{K}(S^c)$ denoted hereafter by $\mathcal{W}_S$ such that*

(a) *interiors of distinct cubes do not intersect,*

(b) *for every $Q \in \mathcal{W}_S$,*

$$\frac{1}{5} r_Q \le \operatorname{dist}(Q, S) \le 5 r_Q,$$

where the distance is measured in the ℓ_∞-norm, i.e.,

$$d(Q, S) := \inf \{ \max_{1 \le i \le n} |x_i - y_i| \; ; \; x \in Q, \; y \in S \}.$$

---

[3]i.e., $f_\alpha = D^\alpha F|_S$ for all $|\alpha| \le k$.

2.2. Whitney's extension theorem

Let us explain how to find such a cover (cf. Lebesgue's decomposition in Figure 1.1 in Comments to Chapter 1). A cube $Q \in \mathcal{K}$ is said to be *dyadic*, if it has a form
$$Q = 2^{-j}(Q_0 + k)$$
for some $j \in \mathbb{Z}$ and $k \in \mathbb{Z}^n$.

So, the dyadic cube has center $2^{-j}k$ and radius 2^{-j}.

Now we define $\mathcal{W}_S$ as follows. Cover S^c by dyadic cubes Q satisfying the condition
$$2Q \in \mathcal{K}(S^c) \quad \text{but} \quad 5Q \notin \mathcal{K}(S^c). \tag{2.29}$$

For every $x \in S^c$ take among these cubes the biggest containing x in its interior, say Q^x, and discard all smaller dyadic cubes contained in Q^x. Then take a point y from the open set $S^c \setminus Q^x$ and find for it the cube Q^y and so on. Since the interiors of two dyadic cubes either lie one in another or are mutually disjoint, this procedure yields a cover $\mathcal{W}_S$ of S^c satisfying condition (a) of the lemma. The second condition easily follows from (2.29).

The cover $\mathcal{W}_S$ has several additional properties which will be used in the sequel. We present them in the next statement where we use the notation
$$Q^* := \lambda Q \quad \text{with} \quad \lambda := \frac{9}{8}. \tag{2.30}$$

Because of the choice of λ and the estimate $\mathrm{dist}(Q, S) > \frac{1}{5} r_Q$, see Lemma 2.14, every Q^* is contained in the open set S^c, i.e.,
$$S^c = \cup\{Q^* \,;\, Q \in \mathcal{W}_S\}.$$

Corollary 2.15. *Let Q, K be cubes in $\mathcal{W}_S$. Then the following is true:*

(a) $Q \cap K \neq \emptyset$ *if and only if* $Q^* \cap K^* \neq \emptyset$. *Moreover, for some* $c = c(n) > 0$ *and every such pair* Q^*, K^*, *it is true that*
$$|Q^* \cap K^*| \geq c \min\{|Q^*|, |K^*|\}.$$

Here $|\cdot|$ stands for the Lebesgue measure in $\mathbb{R}^n$.

(b) *If* $Q \cap K \neq \emptyset$, *then*
$$\frac{1}{4} r_Q \leq r_K \leq 4 r_Q.$$

(c) *The order (multiplicity) of cover* $\mathcal{W}_S^* := \{Q^* \,;\, Q \in \mathcal{W}_S\}$, *i.e., the quantity*
$$\mathrm{ord}\,\mathcal{W}_S^* := \sup_{x \in S^c} \mathrm{card}\{Q^* \in \mathcal{W}_S^* \,;\, A^* \ni x\}$$

is bounded by a constant depending only on n.

Smooth partition of unity

Let φ be a C^∞ function supported by the cube $Q_0^*\,(=\frac{9}{8}[-1,1]^n)$ which equals 1 on Q_0. By scaling we define the function

$$\tilde{\varphi}_Q(x) := \varphi\left(\frac{x-c_Q}{r_Q}\right), \quad x \in \mathbb{R}^n, \quad Q \in \mathcal{K}.$$

Note that
$$\tilde{\varphi}_Q = 1 \text{ on } Q \text{ and } \operatorname{supp}\varphi_Q \subset Q^*.$$

Further, we define the C^∞ function φ_Q on S^c by

$$\varphi_Q := \tilde{\varphi}_Q \Big/ \sum_{Q \in \mathcal{W}_S} \tilde{\varphi}_Q$$

and extend it to $\mathbb{R}^n$ by $\varphi_Q := 0$ on S.

Since $\mathcal{W}_S$ is a locally finite cover of S^c, φ_Q is well defined. The collection $\{\varphi_Q\,;\,Q \in \mathcal{W}_S\}$ is the required *smooth partition of unity subordinate to the cover* $\mathcal{W}_S^*$.

From its definition and Corollary 2.15 we get

Lemma 2.16. (i) *For every* $x \in S^c$,

$$\sum_{Q \in \mathcal{W}_S} \varphi_Q(x) = 1;$$

(ii) *for each* $\alpha \in \mathbb{Z}_+^n$,

$$\sup_{\mathbb{R}^n} |D^\alpha \varphi_Q| \leq C(\alpha, n) r_Q^{-|\alpha|}.$$

Linear extension operator

Let $\vec{f} = \{f_\alpha\}_{|\alpha| \leq k}$ be a continuous vector function on a closed set $S \subset \mathbb{R}^n$. We introduce an analog of Taylor's polynomial for $\vec{f}$ by

$$T_x^k \vec{f}(y) := \sum_{|\alpha| \leq k} f_\alpha(x) \frac{(y-x)^\alpha}{\alpha!}.$$

To define the required extension operator, we pick a point s_Q of S such that

$$d(s_Q, Q) = d(S, Q).$$

Now we define an extension operator E_k^S given for $\vec{f}$ by the formula

$$E_k^S \vec{f} := \begin{cases} \displaystyle\sum_{Q \in \mathcal{W}_S} \varphi_Q T_{s_Q}^k \vec{f} & \text{on } S^c, \\ f_0 & \text{on } S. \end{cases}$$

2.2. Whitney's extension theorem

This is clearly a linear operator; moreover, $E_k^S \vec{f}$ is a C^∞ function on S^c. The basic two facts related to the extension operator are as follows.

Lemma 2.17. *There is a constant $\lambda = \lambda(n) \in (0,1]$ such that for every cube Q with center at S,*

$$E_k^S \vec{f} = E_k^{S \cap Q} \vec{f} \quad \text{on} \quad \lambda Q. \tag{2.31}$$

Further, let the reduced remainders r_α, $|\alpha| \le k$, see (2.27), satisfy the inequality

$$r_\alpha(\vec{f}; x, y) \le \omega(\|x - y\|), \quad x, y \in S, \tag{2.32}$$

for some continuous concave function $\omega : \mathbb{R}_+ \to \mathbb{R}_+$ which equals zero at zero (i.e., for $\omega \in \Omega_1$).

Lemma 2.18. *Under condition (2.32) the following is true:*

(a) *For every α with $|\alpha| \le k + 1$ and all points $x \in S$ and $y \in S^c$,*

$$\left| D_y^\alpha (E_k^S - T_x^k) \vec{f}(y) \right| \le C \omega(\|x - y\|) \tag{2.33}$$

for some $C = C(k, n)$;

(b) *for every cube Q with center at S there is a cube $\widetilde{Q} = \lambda Q$ with $0 < \lambda < 1$ such that*

$$\sum_{|\alpha| \le k} \sup_{Q \setminus S} \left| D^\alpha E_k^S \vec{f} \right| \le C \sum_{|\alpha| \le k} \sup_{\widetilde{Q} \cap S} |f_\alpha|; \tag{2.34}$$

here $\lambda = \lambda(n)$ and $C = C(S, k, n)$.

Outline of the proof for Whitney's Theorem

Let $\vec{f} = \{f_\alpha\}_{|\alpha| \le k}$ be a vector function on S satisfying the Taylor chain condition of Theorem 2.13. First, we must prove the existence and continuity of the derivatives for $E_k^S \vec{f}$ of order at most k on the whole of $\mathbb{R}^n$. It suffices to do this only for an open neighborhood of S. In turn, Lemma 2.17 allows us to work with $Q \cap S$ instead of S. Hence, in this part of the proof we can and will assume that S is compact.

Since every r_α is continuous on the compact set $S \times S$, $r_\alpha(x, y) \to 0$ as $\|x - y\| \to 0$, and this convergence is *uniform*. Therefore every r_α satisfies condition (2.32) with some function ω depending, clearly, on S and $\vec{f}$. Accordingly, inequality (2.33) holds for this ω. For $\alpha = 0$ this yields the continuity of $E_k^S \vec{f}$ on $\mathbb{R}^n$. Proceeding by induction on $|\alpha|$ we establish, with the help of (2.33), the existence and continuity of $D^\alpha E_k^S \vec{f}$ for all α with $|\alpha| \le k$. Hence, E_k^S is a linear extension operator acting from $J^k(\mathbb{R}^n)\big|_S$ to $C^k(\mathbb{R}^n)$. Continuity of the operator in the Fréchet topologies of these spaces follows from (2.34).

Whitney's theorem for other jet spaces

We present here variants of Theorem 2.13 for the spaces $\dot{J}_b^{k,\omega}(\mathbb{R}^n)$ and $J_b^{k,\omega}(\mathbb{R}^n)$. According to our notation, the former (with a dot!) consists of all vector functions $\vec{f} = \{f_\alpha\}_{|\alpha| \le k}$ that are continuous on $\mathbb{R}^n$ and such that the seminorm

$$|\vec{f}|_{k,\omega} := \sum_{|\alpha|=k} \sup_{x \ne y} \frac{|f_\alpha(x) - f_\alpha(y)|}{\omega(\|x-y\|)} < \infty.$$

In turn, $J_b^{k,\omega}(\mathbb{R}^n)$ is defined by the norm

$$\|\vec{f}\|_{k,\omega} := |\vec{f}|_{k,\omega} + \sum_{|\alpha| \le k} \sup_{\mathbb{R}^n} |f_\alpha|.$$

To justify the corresponding results, it is useful to notice that the Taylor chain condition for $f \in C_b^{k,\omega}(\mathbb{R}^n)$ is bounded as follows:

$$\left| D^\alpha (f - T_y^k f)(x) \right| \le C \|x - y\|^{k-|\alpha|} \omega(\|x-y\|)$$

where $C = C(k, n)$.

Hence the necessity condition for the reduced remainders $r_\alpha(\vec{f})$, see (2.27), is

$$r_\alpha(\vec{f}\,; x, y) \le C\omega(\|x-y\|), \quad x, y \in S, \tag{2.35}$$

and this inequality coincides with (2.32). Therefore part (a) of Lemma 2.18 may be applied and this leads to the corresponding result for the homogeneous space $\dot{J}_b^{k,\omega}(\mathbb{R}^n)$. But we cannot estimate the uniform norms of the derivatives $D^\alpha E_k^S \vec{f}$ on $\mathbb{R}^n$ using part (b) of Lemma 2.18 (in fact, they may be unbounded). Therefore, we should modify the definition of the extension operator; namely, we set

$$\widehat{E}_k^S \vec{f} := \sum_{Q \in \widehat{\mathcal{W}}_S} \varphi_Q T_{s_Q}^k \vec{f} \quad \text{on} \quad S^c$$

and define $\widehat{E}_k^S \vec{f}$ to agree with f_0 on S. Here $\widehat{\mathcal{W}}_S$ is a part of the cover $\mathcal{W}_S$ containing only the cubes of radius at most 1. In other words, in the definition of $E_k^S \vec{f}$ we substitute $T_{s_Q}^k \vec{f}$ for zero, if $r_Q > 1$.

Estimate (2.33) remains to be true for this modification, but inequality (2.34) is replaced by the stronger inequality

$$\sum_{|\alpha| \le k} \sup_{S^c} \left| D^\alpha \widehat{E}_k^S \vec{f} \right| \le C(n, k) \sum_{|\alpha| \le k} \sup_S |f_\alpha|.$$

Using these facts one can derive the following result due to G. Glaser [Gl-1958].

2.2. Whitney's extension theorem

Theorem 2.19. (a) *Let $\vec{f} = \{f_\alpha\}_{|\alpha| \le k}$ be a vector function defined on a closed set $S \subset \mathbb{R}^n$. Then $\vec{f}$ belongs to the trace space $\dot{J}^{k,\omega}(\mathbb{R}^n)\big|_S$ if and only if $\vec{f}$ satisfies (2.32).*

(b) *$\vec{f}$ belongs to the trace space $J_b^{k,\omega}(\mathbb{R}^n)\big|_S$ if and only if $\vec{f}$ is bounded on S and satisfies (2.32).*

In both cases, there exists a linear continuous extension operator.

Remark 2.20. For $k = 0$ and $\omega(t) := t$, $t > 0$, Theorem 2.19 yields the following:

For every closed $S \subset \mathbb{R}^n$, there is a linear extension operator $E : \text{Lip}(\mathbb{R}^n)\big|_S \to \text{Lip}(\mathbb{R}^n)$ whose norm is bounded by a constant depending only on n.

Since all Banach norms of $\mathbb{R}^n$ are equivalent, the same is true for $\text{Lip}(X)$, where X is an n-dimensional Banach space.

It is worth noting that the extension operator $\widehat{E}_k^S$ is *universal* in the sense that it does not depend on the majorant ω. On the other hand, we have

$$\dot{J}_u^k(\mathbb{R}^n) = \bigcup_\omega \dot{J}_b^{k,\omega}(\mathbb{R}^n),$$

where ω runs over the set of 1-majorants. Let us recall that the space on the left-hand side consists of all vector functions $\vec{f} = \{f_\alpha\}_{|\alpha| \le k}$ generated by C^k functions whose higher derivatives are bounded and uniformly continuous on $\mathbb{R}^n$.

These facts lead to

Theorem 2.21. *A vector function $\vec{f} = \{f_\alpha\}_{|\alpha| \le k}$ defined and bounded on a closed set $S \subset \mathbb{R}^n$ belongs to the trace space $\dot{J}_u^k(\mathbb{R}^n)\big|_S$ if and only if the reduced remainders $r_\alpha(\vec{f})$ with $|\alpha| = k$ are uniformly continuous on $S \times S$.*

Moreover, $\widehat{E}_k^S$ is a linear bounded extension operator from the trace space into $\dot{C}_u^k(\mathbb{R}^n)$.

Whitney's theorem for C^∞ functions

Hestens [Hes-1941] modified the Whitney extension method to adapt it to C^∞ functions. In this case we deal with the space $J^\infty(\mathbb{R}^n)$ of vector functions $\vec{f} = \{f_\alpha\}_{|\alpha| < \infty}$ on $\mathbb{R}^n$ generated by C^∞ functions, i.e., for every such $\vec{f}$ there is a function $f \in C^\infty(\mathbb{R}^n)$ such that $f_\alpha = D^\alpha f$ for all α.

Given now $\vec{f} = \{f_\alpha\}_{|\alpha| < \infty}$ defined on a closed subset $S \subset \mathbb{R}^n$, we introduce as before the family $\{r_\alpha(\vec{f})\}_{|\alpha|,\infty}$ of reduced remainders. For this case the Taylor chain condition is:

For all $\alpha \in \mathbb{Z}_n^+$,

$$r_\alpha(\vec{f}\,; x, y) \to 0 \quad \text{as} \quad y \to x, \; y, x \in S. \tag{2.36}$$

Then the following is true.

Theorem 2.22. *A vector function $\vec{f} = \{f_\alpha\}_{|\alpha|<\infty}$ defined on a closed set $S \subset \mathbb{R}^n$ belongs to the trace space $J^\infty(\mathbb{R}^n)|_S$ if an only if $\vec{f}$ satisfies the Taylor chain condition* (2.36).

To prove sufficiency of this condition, Hestens used the following *nonlinear* extension operator.

Given $\vec{f} = \{f_\alpha\}_{|\alpha|<\infty}$ defined on S and a cube Q from Whitney's cover $\mathcal{W}_S$, a nondecreasing sequence of numbers $\{N_j\}_{j=1}^\infty$ is chosen so that

$$\left| \sum_{|\alpha|=j} \frac{f_\alpha(x)}{\alpha!} y^\alpha \right| < N_j$$

for every y with $\|y\|_\infty = 1$ and every x in $Q_j(s_Q) \cap S$. Recall that s_Q is a point of S nearest to Q.

Selecting constants $\delta_j > 0$ such that $\delta_j < \frac{1}{N_j}$ and $\delta_{j+1} < \delta_j$, $j = 1, 2, \ldots$, we set

$$k_Q := \begin{cases} j, & \text{if } \delta_{j+1} \leq d(Q, S) < \delta_j, \\ 0, & \text{if } d(Q, S) \geq \delta_j. \end{cases}$$

Finally, we define an extension operator E_∞^S on $\vec{f} = \{f_\alpha\}_{|\alpha|<\infty}$ by

$$E_\infty^S \vec{f} := \begin{cases} \sum_{Q \in \mathcal{W}_S} \varphi_Q T_{s_Q}^{k_Q} \vec{f} & \text{on } S^c, \\ f_0 & \text{on } S. \end{cases}$$

Using the estimates of Lemma 2.18 one then proves that $E_\infty^S \vec{f}$ belongs to $C^\infty(\mathbb{R}^n)$.

The Whitney–Hestens Theorem 2.22 is a far reaching generalization of the classical E. Borel Theorem [EBo-1895] concerning univariate functions and one-point sets $S = \{x_0\} \subset \mathbb{R}$. In this case, $J^\infty(\mathbb{R})|_S$ is the linear space of all sequences $\{c_i\}_{i \in \mathbb{Z}_+} \subset \mathbb{R}$, and the Taylor chain condition clearly holds. Therefore Theorem 2.22 states that, for every sequence $\{c_i\}_{i \in \mathbb{Z}_+} \subset \mathbb{R}$, there is a C^∞ function f on $\mathbb{R}$ such that

$$f^{(i)}(x_0) = c_i \quad \text{for every} \quad i \in \mathbb{Z}_+,$$

(Borel's theorem).

Remark 2.23. In fact, a Borel type result is true for every *countable* subset $S \subset \mathbb{R}^n$. Specifically, let $\{s_\alpha\,; \alpha \in \mathbb{Z}_+^n\}$ and $S := \{x_\alpha\,; \alpha \in \mathbb{Z}_+^N\}$ be arbitrary families of real numbers and points of $\mathbb{R}^n$, respectively. Then there is a function $f \in C^\infty(\mathbb{R}^n)$ such that for every $\alpha \in \mathbb{Z}_+^n$,

$$D^\alpha f(x_\alpha) = s_\alpha.$$

The result is a direct consequence of the classical Eidelheit theorem [Ei-1936] on the solvability of infinite linear systems on Fréchet spaces.

The Borel theorem gives a remarkably simple example of a closed subset $S \subset \mathbb{R}$ (a singleton $\{x_0\}$) for which there is no *linear* continuous extension operator from $J^\infty(\mathbb{R}^n)\big|_S$ to $J^\infty(\mathbb{R}^n)$. The result was due to Mitiagin [Mit-1961]; its proof presented now follows the argument from the paper [LyT-1969] by Yu. Lyubich and Tkachenko.

Theorem 2.24. *There is no linear continuous extension operator from $J^\infty(\mathbb{R}^n)\big|_{\{x_0\}}$ to $J^\infty(\mathbb{R}^n)$.*

Proof. Assume, on the contrary, that such an operator, say E, exists. Let $\{e_\alpha\}_{\alpha \in \mathbb{Z}_+^n}$ be the standard basis of the Fréchet space of all families $\{s_\alpha\}_{\alpha \in \mathbb{Z}_+^n}$, i.e., the coordinate $(e_\alpha)_\beta = 1$ if $\beta = \alpha$ and $(e_\alpha)_\beta = 0$ if $\alpha \neq \beta$. By the continuity of E,

$$E(\{s_\alpha\}) = \sum_\alpha s_\alpha E e_\alpha,$$

where the series converges in $C^\infty(\mathbb{R}^n)$ (identified with $J^\infty(\mathbb{R}^n)$) and therefore uniformly converges on $Q_0 := [-1, 1]^n$. This implies that

$$\sup_\alpha |s_\alpha| \, \|Ee_\alpha\|_{C(Q_0)} < \infty.$$

Since this is true for an arbitrary family $\{s_\alpha\}$, there is only a *finite* number of α for which $Ee_\alpha \neq 0$. Hence, the image of E is a finite-dimensional subspace of $J^\infty(\mathbb{R}^n)$. On the other hand, the composition $R \circ E$, where $Rf := f\big|_{\{x_0\}}$ is the identity map, a contradiction. □

2.3 Divided differences, local approximation and differentiability

Turning to the trace and extension problems for C^k functions, we immediately notice their essential difference from those for k-jets. In the latter case, the following information is available:

(i) the traces to S of all the derivatives of the function up to order k;

(ii) a collection of inequalities relating the values of these traces for every *two-point* subset of S.

However, in the case of C^k functions, information is much more restricted: we know only the trace of a function to S. The deficiency in data should therefore be completed by a more expanded set of relations between the values of the trace. In Section 2.4, we prove the aforementioned Whitney theorem describing the complete set of these relations for univariate C^k functions. Every such relation includes at most $k + 2$ distinct values and this number cannot be diminished. In Chapter 10, we present several results of this kind concerning different smoothness spaces in n variables; the number of values in all these cases is bounded by a constant depending only on smoothness order and n. Apparently, there exists a deep reason for this phenomenon called below the *Finiteness Property*, but for the time being we have only some heuristic argument for its explanation.

It is natural to begin the study with the simplest case of C^k functions defined on the whole of $\mathbb{R}^n$. The results presented below will provide several characteristics of C^k functions based on their behavior on subsets of cardinality $k+1$ and on that of their *local polynomial approximation*. The first group of results describes relations between k-differentiability of multivariate functions and behavior of its difference characteristics. For their formulations we need a few notions and facts concerning such characteristics.

Divided and mixed differences

Let $f : \mathbb{R} \to \mathbb{R}$ and let S be a finite subset of $\mathbb{R}$. The *divided difference of f on S* (denoted by $f[S]$) is described by induction on card S starting from $S := \{x_0\}$ as follows. Set
$$f[\{x_0\}] := f(x_0);$$
if $f[S]$ is now defined on all k-point subsets, then, for $S := \{x_0, \ldots, x_k\}$, set
$$f[S] := \frac{f[\{x_0, \ldots, x_{k-1}\}] - f[\{x_1, \ldots, x_k\}]}{x_0 - x_k}.$$

Proposition 2.25. *Let $S \subset \mathbb{R}$ and card $S = k+1$. Set $\omega_S(t) := \prod_{s \in S}(t-s)$. Then the following is true:*

(a)
$$f[S] = \sum_{s \in S} \frac{f(s)}{\omega'_S(s)}.$$

In particular, $f[S]$ is a symmetric function in the arguments $s \in S$.

(b) *Let $L_S(f)$ be the Lagrange interpolation polynomial for f with the nodes $s \in S$. Then*
$$L_S^{(k)}(f) = f[S].$$

(c) *If f belongs to C^k in some open interval I containing S, then*
$$f[S] = f^{(k)}(c)$$

for a point c from $(\operatorname{conv} S)^0$.

These classical results are well known, see, e.g., the book [deB-2001] by de Boor.

2.3. Divided differences, local approximation and differentiability

Now let the points of $S := \{x_0, \ldots, x_k\}$ be equally spaced, i.e., $x_{j+1} - x_j = h$ for all $0 \leq j < k$ and some $h \neq 0$. Then it is easy to see that

$$f[S] = \frac{k!}{h^k} \sum_{j=0}^{k} (-1)^{k-j} \binom{k}{j} f(x_0 + jh). \tag{2.37}$$

Let us recall that for a multivariate function $f : \mathbb{R}^n \to \mathbb{R}$ and a vector $h \in \mathbb{R}^n$ the k-th-difference of f is given by the similar formula:

$$\Delta_h^k f(x) := \sum_{j=0}^{k} (-1)^{k-j} \binom{k}{j} f(x + jh). \tag{2.38}$$

A variant of this notion, the k-th *difference of f in a direction* $e \in \mathbb{S}^{n-1}$, is given by

$$\Delta^k(te)f := \Delta_{te}^k f, \quad t \in \mathbb{R}.$$

Finally, we introduce *(mixed) α-difference* Δ_h^α by

$$\Delta_h^\alpha := \prod_{i=1}^{n} \Delta^{\alpha_i}(h_i e_i); \tag{2.39}$$

here $h = (h_1, \ldots, h_n) \in \mathbb{R}^n$ and $\{e_i\}_{i=1}^n$ is the standard basis of $\mathbb{R}^n$.

Our first result goes back to Brouwer [Bro-1908] who proved it for $n = 1$.

Theorem 2.26. *A continuous function $f : \mathbb{R}^n \to \mathbb{R}$ belongs to $C^k(\mathbb{R}^n)$ if and only if for every $e \in \mathbb{S}^{n-1}$ and every cube Q the function $t^{-k}\Delta^k(te)f$ converges uniformly on Q as t tends to 0 through an arbitrary sequence $\{t_j\}$.*

Proof. (Necessity) If $f \in C^k(\mathbb{R}^n)$, then by Proposition 2.25 (c) and (2.37) we have

$$t^{-k} \Delta^k(te) f(x) = D_e^k f(c), \tag{2.40}$$

here c is a point in the interval $(x, x + kte)$ and the directional derivative D_e^k is defined by the iteration of D_e where

$$D_e f(x) := \lim_{t \to 0} t^{-1} \big(f(x + te) - f(x) \big).$$

Therefore $D_e = \sum_{i=1}^n e_i D_i$ and

$$t^{-k} \Delta^k(te) f(x) = \sum_{|\alpha|=k} e^\alpha D^\alpha f(c).$$

Since $f \in C^k(\mathbb{R}^n)$, the right-hand side is continuous in c (and therefore uniformly continuous) on every closed cube of $\mathbb{R}^n$. Hence, the left-hand side of (2.40) tends to $D_e^k f(x)$ uniformly in x as $t \to 0$.

(Sufficiency) We begin with

Proposition 2.27. *The assertion of Theorem 2.26 is true for $n = 1$.*

Proof. By the assumption of Theorem 2.26 for every closed interval $I := [a, b] \subset \mathbb{R}$ there is a function $\varphi_I \in C(I)$ such that

$$\max_I \left| t_j^{-k} \Delta_{t_j}^k f - \varphi_I \right| \to 0 \quad \text{as} \quad j \to \infty.$$

Let ψ be a C^∞ function compactly supported in $I^0 = (a, b)$. Then for a sufficiently small t,

$$\int_I (\Delta_t^k f) \psi \, dx = \int_I f(\Delta_{-t}^k \psi) \, dx.$$

Dividing by t^{-k} and passing to the limit as $t = t_j \to 0$, one then obtains

$$\int_I \varphi_I \cdot \psi \, dx = (-1)^k \int_I f \psi^{(k)} \, dx.$$

Finally, integrating by parts k-times we get

$$\int_I \left(f - \int_a^x \frac{(x-t)^{k-1}}{(k-1)!} \varphi_I(t) \, dt \right) \psi^{(k)} \, dx = 0.$$

Since this is true for an arbitrary ψ, the classical Dubois–Reymond lemma implies that the term in brackets is a polynomial of degree $k - 1$. Hence,

$$f(x) = \sum_{j=0}^{k-1} a_j x^j + \int_a^x \frac{(x-t)^{k-1}}{(k-1)!} \varphi_I(t) \, dt \in C^k(\mathbb{R}). \tag{2.41}$$

□

Applying Proposition 2.27 to the restriction of $f \in C(\mathbb{R}^n)$ to a straight line in a direction $e \in \mathbb{S}^n$, we immediately get

Corollary 2.28. *Under the condition of Theorem 2.26 f has k-th continuous directional derivatives for any $e \in \mathbb{S}^n$.*

We derive from here the existence and continuity of all mixed derivatives for f up to order k. For this aim we need an algebraic identity concerning the difference operators, see Appendix E to this chapter for the proof.

In this identity, τ_v, $v \in \mathbb{R}^n$, denotes the shift operator given by $\tau_v f := f(\cdot + v)$, where f is a function on $\mathbb{R}^n$. We also set $\Delta(v) := \tau_v - 1$ for the difference in the direction v.

Let V be a collection of N vectors in $\mathbb{R}^n$ and $\varphi : V \to \{1, \frac{1}{2}, \ldots, \frac{1}{N}\}$ be a bijection. Then

$$\prod_{v \in V} \Delta(v) = \sum_{\omega \subset V} (-1)^{\operatorname{card} \omega} \tau_{w_\omega} \Delta^N(v_\omega), \tag{2.42}$$

2.3. Divided differences, local approximation and differentiability

where the sum is taken over all nonempty subsets of V and

$$v_\omega := \sum_{v \in \omega} \varphi(v) v, \quad w_\omega := \sum_{v \notin \omega} v.$$

By standard convention, $w_V = 0$.

Let V_α, $|\alpha| = k$, be determined by

$$\prod_{v \in V_\alpha} \Delta(v) = \Delta_h^\alpha \left(:= \prod_{j=1}^n \Delta^{\alpha_j}(h_j e_j) \right).$$

Then (2.42) yields

$$\Delta_h^\alpha f(x) = \sum_{\omega \subset V_\alpha} (-1)^{\text{card } \omega} \Delta_{v_\omega}^k f(x + w_\omega). \tag{2.43}$$

If here $h_j = t$ for $1 \leq j \leq n$ and some $t > 0$, then

$$v_\omega = t\hat{v}_\omega \quad \text{and} \quad w_\omega = t\hat{w}_\omega,$$

where the vectors with the hat are independent of t. Therefore the condition of Theorem 2.26 and (2.43) imply that $|h|^{-k} \Delta_h^\alpha f$ converges uniformly on compact subsets to a continuous on $\mathbb{R}^n$ function as the vector $h := t \sum_{j=1}^n e_j$ tends to zero.

Hence we establish

Corollary 2.29. *Under the condition of Theorem 2.26, $\|h\|^{-k} \Delta_h^\alpha f$ tends uniformly on every cube Q to a continuous function φ_Q as h tends to zero along the ray $\left\{ t \sum_{i=1}^n e_i ; t > 0 \right\}$.*

Now we apply the argument of Proposition 2.27 to each variable x_i to get the following

Proposition 2.30. *If the assertion of Corollary 2.29 holds for a continuous function f and a cube $Q := \prod_{j=1}^n [a_j, b_j]$, then for every $x \in Q$,*

$$f(x) = p_Q^\alpha(x) + \left(\prod_{j=1}^n I_j^{\alpha_j} \right)(\varphi_Q \, ; \, x),$$

where the operator I_j denotes integration with respect to the j-th variable from a_j to x_j and p_Q^α is a continuous function on Q given by

$$p_Q^\alpha(x) = \sum_{j=1}^n \sum_{s=0}^{\alpha_j - 1} \varphi_{sj}(x) x_j^s \tag{2.44}$$

with φ_{sj} being a continuous function independent of x_j.

Now we can complete the proof of Theorem 2.26. By Corollary 2.28 f has first continuous derivatives. Proceeding by induction, assume that f has continuous mixed derivatives of order s with $1 \leq s < k$ and prove the same for $s+1$. Let $|\alpha| = s+1$; then

$$\alpha = \beta + e_i \quad \text{for some} \quad |\beta| = s \text{ and } 1 \leq i \leq n. \tag{2.45}$$

Applying Proposition 2.30 for this α to have, for a cube Q,

$$f = p_Q^\alpha + \hat{f}_Q,$$

where p_Q^α is given by (2.44). Hence, $D^\gamma \hat{f}_Q$ exists and is continuous on Q for every multi-index γ with $\gamma_i \leq \alpha_i$, $1 \leq i \leq n$. Moreover, by the induction hypothesis, $D^\beta f$ exists and is continuous; hence, the same is true for $D^\beta p_Q^\alpha$. Now write

$$\Delta(te_i) D^\beta f = \Delta(te_i) D^\beta p_Q^\alpha + \Delta(te_i) D^\beta \hat{f}_Q.$$

Since $D^\beta p_Q^\alpha = s! \varphi_{si}$, i.e., is independent of x_i, the first term in the right-hand side is zero. Dividing then by $t \to 0$, we conclude that $\lim_{t \to 0} t^{-1} \Delta(te_i) D^\beta f(x)$ exists and is equal on Q to the continuous function $D^\alpha \hat{f}_Q$, see (2.45) and (2.44).

Hence, $D^\alpha f$ exists and is continuous for every α with $|\alpha| = s+1$. □

The theorem proved leads to a similar difference characteristic of the space $\dot{C}^{k,\omega}(\mathbb{R}^n)$; see (2.7) for its definition.

Theorem 2.31. *A continuous function $f : \mathbb{R}^n \to \mathbb{R}$ belongs to $\dot{C}^{k,\omega}(\mathbb{R}^n)$ if and only if for some constant $C > 0$ and all sufficiently small $h, g \in \mathbb{R}^N$, the inequality*

$$\|h\|^{-k} |\Delta_g \Delta_h^k f(x)| \leq C \omega(\|g\|) \tag{2.46}$$

holds for all $x \in \mathbb{R}^n$. Moreover, the equivalence[4]

$$\inf C \approx |f|_{\dot{C}^{k,\omega}(\mathbb{R}^n)} \tag{2.47}$$

holds (with constants independent of f).

Proof. By (2.46), the ratio $|h|^{-k} \Delta_h^k f$ converges uniformly as $h \to 0$ (recall that $\omega(0+) = 0$). The application of Theorem 2.26 then implies that $f \in C^k(\mathbb{R}^n)$. Passing to the limit in (2.46) we therefore obtain

$$|\Delta_g D_e^k f(x)| \leq C \omega(\|g\|),$$

where $e \in \mathbb{S}^{n-1}$ and $x \in \mathbb{R}^n$. This and identity (2.43) yield

$$|\Delta_g D^\alpha f(x)| \leq a C \omega(\|g\|), \quad |\alpha| = k,$$

---

[4] Hereafter we write $F \approx G$, if these functions satisfy a two-sided inequality $C_1 F \leq G \leq C_2 F$ with constants $C_1, C_2 > 0$ independent of the arguments of F, G.

2.3. Divided differences, local approximation and differentiability

for some constant $a = a(n,k)$, whence $|f|_{C^{k,\omega}(\mathbb{R}^n)} \leq bC$ with $b = b(k,n)$.

In the opposite direction, it follows from Proposition 2.25 (c) and (2.38) that

$$\left|\Delta_g \Delta_h^k f(x)\right| \leq \|h\|^k \left|\Delta_g D_e^k f(y)\right|,$$

provided that $f \in \dot{C}^{k,\omega}(\mathbb{R}^n)$; here $y \in (x, x + kh)$ and $e := \frac{h}{\|h\|}$.

Estimating the right-hand side as in Theorem 2.26 we then have

$$\|h\|^k \sum_{|\alpha|=k} \left|e^\alpha \Delta_g D^\alpha f(g)\right| \leq C(n) |f|_{C^{k,\omega}(\mathbb{R}^n)} \|h\|^k \omega(\|g\|).$$

This completes the proof. $\qquad\square$

For the case of $\omega(t) = t^\lambda$, $t \in \mathbb{R}_+$, with $0 < \lambda \leq 1$ the criterion of Theorem 2.31 can be further simplified. In the consequent formulation, we use the notation

$$\dot{C}^{k,\omega} := \dot{C}^{k,\lambda} \quad \text{for} \quad \omega(t) := t^\lambda,\; t \in \mathbb{R}_+. \tag{2.48}$$

Theorem 2.32. *A continuous function $f : \mathbb{R}^n \to \mathbb{R}$ belongs to the space $\dot{C}^{k,\lambda}(\mathbb{R}^n)$ if and only if*

$$\mathcal{M}(f) := \sup_h \|h\|^{-k-\lambda} \left\|\Delta_h^{k+1} f\right\|_{C(\mathbb{R}^n)} < \infty.$$

Moreover, $\mathcal{M}(f) \approx |f|_{C^{k,\lambda}(\mathbb{R}^n)}$.

The result is a consequence of Theorem 2.10 (a). The case of $\lambda = 1$ is true under the weaker assumption:

$$m(f) := \lim_{h \to 0} \|h\|^{-k-1} \left\|\Delta_h^{k+1} f\right\|_{C(\mathbb{R}^n)} < \infty. \tag{2.49}$$

We outline the proof of this result. Necessity may be proved in the same fashion as in Theorem 2.26 and leads to the inequality

$$\|h\|^{-k-1} \left\|\Delta_h^{k+1} f\right\|_{C(\mathbb{R}^n)} \leq a(n,k) \|h\|^{-1} \sum_{|\alpha|=k} \left\|\Delta_h D^\alpha f\right\|_{C(\mathbb{R}^n)}.$$

This implies that

$$m(f) \leq a(n,k) |f|_{C^{k,1}(\mathbb{R}^n)}.$$

The converse may be derived from the corresponding univariate case using the argument of Theorem 2.26 based on identity (2.42). Therefore it remains only to prove the following one-dimensional result:

If $f \in C(\mathbb{R})$ and $m(f) < \infty$, then $f \in C^{k,1}(\mathbb{R})$ and

$$|f|_{C^{k,1}(\mathbb{R})} \leq a m(f)$$

with some constant a independent of f. As in the proof of Proposition 2.27, we pick a C^∞ function ψ compactly supported on $I = (a, b) \subset \mathbb{R}$. Then for a sufficiently small $t > 0$, we have

$$\int_I \left(t^{-k-1}\Delta_t^{k+1}f\right)\psi\, dx = \int_I f\left(t^{-k-1}\Delta_{-t}^{k+1}\psi\right) dx. \tag{2.50}$$

If $t \to 0$, the right-hand side becomes $(-1)^{k+1}\int_I f\psi^{(k+1)}dx$. On the other hand, for an appropriate sequence $t_j \to 0$, the sequence of functions $f_j := t_j^{-k-1}\Delta_{t_j}^{k+1}f$, $j = 1, 2, \ldots$, is bounded in the L_∞-norm by (2.49). Since $L_\infty(I)$ is conjugate to $L_1(I)$, there exists a subsequence of $\{f_j\}$ converging in the weak* topology of $L_\infty(I)$ to some function φ_I, see, e.g., Dunford and Schwartz [DS-1958]. Passing to the limit in (2.50) as $t_j \to \infty$, we get

$$\int_I \varphi_I \psi\, dx = (-1)^{k+1}\int_I f\psi^{(k+1)}\, dx. \tag{2.51}$$

Moreover, by semicontinuity of the norm with respect to the weak* convergence,

$$\|\varphi_I\|_{L_\infty(I)} \leq \varliminf_{j\to\infty} t_j^{-k-1}\|\Delta_t^{k+1}\|_{C(\mathbb{R})} =: m(f).$$

Integrating (2.51) by parts and using the Dubois–Reymond lemma we then have

$$f(x) = \sum_{j=0}^k a_j x^j + \int_a^x \frac{(x-t)^k}{k!}\varphi_I(t)dt.$$

This immediately implies that

$$\left|f^{(k)}(x+h) - f^{(k)}(x)\right| \leq \int_x^{x+h}|\varphi_I(t)|\, dt \leq |h|m(f).$$

Hence, $f \in \dot{C}^{k,1}(\mathbb{R})$ and its seminorm in this space is bounded by $m(f)$.

Remark 2.33. (a) This proof also yields the following useful fact:

$$\dot{C}^{k,1}(\mathbb{R}^n) = \dot{W}_\infty^{k+1}(\mathbb{R}^n) \tag{2.52}$$

with equivalence of the seminorms.

(b) The analog of Theorem 2.32 with a similar proof based on duality, holds for the homogeneous Sobolev space $\dot{W}_p^{k+1}(\mathbb{R}^n)$ with $1 < p < \infty$.

Let us recall that the (homogeneous) Sobolev space $\dot{W}_p^{k+1}(\mathbb{R}^n), 1 \leq p \leq \infty$, is defined by the seminorm

$$|f|_{W_p^{k+1}(\mathbb{R}^n)} := \sum_{|\alpha|=k+1}\|D^\alpha f\|_{L_p(\mathbb{R}^n)}.$$

Here $D^\alpha f$ are the distributional derivatives of $f \in L_p^{\mathrm{loc}}(\mathbb{R}^n)$ which are assumed to belong to $L_p(\mathbb{R}^n)$.

Local polynomial approximation and derivatives

The Whitney extension theorem relates the differentiable characteristics of a function to its local approximation by Taylor polynomials. Now we present other results of this kind which concern local Taylor approximation and also best approximation. The approximation of the latter type plays an essential role in the proofs of the extension theorems for Lipschitz functions of higher order, see Section 2.5 and Chapters 9 and 10. These proofs are fairly complicated and therefore it would be natural to consider this approach in the relatively simple situation of Theorem 2.38 below.

We begin with approximation by Taylor polynomials.

Theorem 2.34. *Let $\{T_y\}_{y \in \mathbb{R}^n}$ be a family of polynomials in $x \in \mathbb{R}^n$ of degree k, and f be a continuous function on $\mathbb{R}^n$. Assume that*

$$|f(x) - T_y(x)|/\|x - y\|^k \to 0 \quad as \quad y \to x \tag{2.53}$$

uniformly on every closed cube. Then f belongs to $C^k(\mathbb{R}^n)$.

Proof. Write

$$T_y(x) = \sum_{|\alpha| \le k} f_\alpha(y) \frac{(x-y)^\alpha}{\alpha!}$$

and consider the k-jet

$$\vec{f} := \{f_\alpha\}_{|\alpha| \le k}.$$

By (2.53), $f_0 = f$ and we must show that $f_0 \in C^k(\mathbb{R}^n)$. We derive this from Whitney's Theorem 2.13 by checking that $\vec{f}$ satisfies the Taylor chain condition on every closed cube Q, see (2.27). Then Whitney's extension theorem implies that $f_0|_Q$ is the trace of a $C^k(\mathbb{R}^n)$ function and the result will follow.

To establish the required property of $\vec{f}$ we need

Lemma 2.35 (Markov's type inequality). *Let P be a polynomial on $\mathbb{R}^n$ of degree k. Then for every closed cube C of side length r and $|\alpha| \le k$, the inequality*

$$\max_C |D^\alpha P| \le \mu r^{-|\alpha|} \max_C |P| \tag{2.54}$$

holds with a constant μ depending only on k and n.

Proof. Since the space $\mathcal{P}_{k,n}$ of polynomials on $\mathbb{R}^n$ of degree k is affine-invariant, one can use scaling to replace C in (2.54) by the unit cube $Q_0 := [0,1]^n$. Then we should prove that

$$\max_{Q_0} |D^\alpha P| \le \mu(k,n) \max_{Q_0} |P|. \tag{2.55}$$

But every linear operator acting in a finite-dimensional Banach space is bounded. Applying this to the space $\mathcal{P}_{k,n}\big|_{Q_0}$ equipped with the uniform norm and to the operator D^α, we immediately get (2.55). □

We now introduce the polynomial in z,
$$P(z) := T_x(z) - T_y(z) \quad \text{with} \quad x, y \in Q$$
and apply Markov's inequality to P and the cube $C := \overline{Q}_{|x-y|}(x)$. Then we get
$$|D_z^\alpha P|\big|_{z=x} \leq \mu(k,n)\|x-y\|^{-|\alpha|} \max_C |P|.$$

The left-hand side equals
$$R_\alpha(\vec{f}\,;\,x,y) := f_\alpha(x) - \sum_{|\beta|\leq k-|\alpha|} f_{\alpha+\beta}(y) \frac{(x-y)^\beta}{\beta!}.$$

On the other hand, by the definition of the cube C and (2.53),
$$\max_C |P| \leq \max_C \left|f(z) - T_x(z)\right| + \max_C \left|f(z) - T_y(z)\right| \leq \varepsilon_Q(|x-y|)|x-y|^k,$$
where $\varepsilon_Q(t) \to 0$ as $t \to 0^+$. Combining this with the previous two relations we obtain
$$\left| f_\alpha(x) - \sum_{|\beta|\leq k-|\alpha|} f_{\alpha+\beta}(y) \frac{(x-y)^\beta}{\beta!} \right| \leq \mu(k,n)\varepsilon_Q(\|x-y\|)\|x-y\|^{k-|\alpha|}$$
for all $x, y \in Q$ and $|\alpha| \leq k$.

Hence, $\vec{f}$ satisfies the Taylor chain condition on Q. $\square$

Another characterization of C^k functions is based on the notion of *local polynomial (best) approximation*. In its definition, $\mathcal{B}(\mathbb{R}^n)$ stands for the family of *bounded* subsets in $\mathbb{R}^n$.

Definition 2.36. The local polynomial (best) approximation of order k is a function $E_k : \ell_\infty^{\mathrm{loc}}(\mathbb{R}^n) \times \mathcal{B}(\mathbb{R}^n) \to \mathbb{R}_+$ given by
$$E_k(S\,;f) := \inf\{\|f-P\|_{\ell_\infty(S)}\,;\, P \in \mathcal{P}_{k-1,n}\}. \tag{2.56}$$

Notice that the order k differs by 1 from the corresponding degree of the approximating polynomials.

We will discuss the basic properties of this set-function later; for now we only need the next profound fact whose univariate case was due to Whitney [Wh-1957] and the multivariate one was proved by Yu. Brudnyi [Br-1970a]. For the formulation of the latter we define the k-*oscillation* of $f : \mathbb{R}^n \to \mathbb{R}$ on a set $S \subset \mathbb{R}^n$ by
$$\omega_k(S\,;f) := \sup\{|\Delta_h^k f(x)|\,;\, x+jh \in S,\ j = 0, 1, \ldots, k\}. \tag{2.57}$$

2.3. Divided differences, local approximation and differentiability

The set $\bigcap_{j=0}^{k}(S+jh)$ may be empty for every $h \neq 0$; in this case the k-oscillation is assumed to be infinity. However, for S of positive Lebesgue measure this intersection is nonempty for a set of h of positive measure, a consequence of the general Ruziewicz theorem [Ruz-1925].

Theorem 2.37. *Let f be a locally bounded function on a convex subset C of $\mathbb{R}^n$ (which may be unbounded). Then there is a constant $w_k(C)$ such that*

$$E_k(C;f) \leq w_k(C)\omega_k(C;f). \tag{2.58}$$

Moreover, the number

$$w_{k,n} := \sup_C w_k(C)$$

is finite.

We discuss this result in Appendix F to this chapter, while for now we apply it to the following characterization of the space $\dot{C}^{k,\lambda}(\mathbb{R}^n)$, $0 < \lambda \leq 1$.

Theorem 2.38. *A locally bounded function $f : \mathbb{R}^n \to \mathbb{R}$ belongs to $\dot{C}^{k,\lambda}(\mathbb{R}^n)$ if and only if*

$$M_{k,\lambda}(f) = \sup_Q \frac{E_{k+1}(Q;f)}{|Q|^{\frac{k+\lambda}{n}}}$$

is finite; here the supremum is taken over all closed cubes of $\mathbb{R}^n$. Moreover, the equivalence

$$M_{k,\lambda}(f) \approx |f|_{C^{k,\lambda}(\mathbb{R}^n)}$$

holds with constants depending only on k and n.

Proof. (Necessity) Let $f \in \dot{C}^{k,\lambda}(\mathbb{R}^n)$. By virtue of (2.38) and Proposition 2.25 (c),

$$\omega_{k+1}(Q;f) \leq \sup\{|\Delta_h D_e^k f(x)| \cdot \|h\|^k \; ; \; e \in \mathbb{S}^n, \; x, x+h \in Q\}.$$

As in the proof of necessity in Theorem 2.26, we then bound this upper estimate by

$$\sum_{|\alpha|=k} \sup\{|\Delta_h D^\alpha f(x)| \cdot \|h\|^k \; ; \; x, x+h \in Q\}$$
$$\leq |f|_{C^{k,\lambda}(\mathbb{R}^n)} \sup\{\|h\|^{k+\lambda} \; ; \; x, x+h \in Q\}$$
$$\leq \left(\sqrt{n}|Q|^{\frac{1}{n}}\right)^{k+\lambda} |f|_{C^{k,\lambda}(\mathbb{R}^n)}.$$

Applying now (2.58), we get the required inequality

$$E_{k+1}(Q;f) \leq c(k,n)|Q|^{\frac{k+\lambda}{n}} |f|_{C^{k,\lambda}(\mathbb{R}^n)}.$$

(Sufficiency) Suppose that for every cube Q,

$$E_{k+1}(Q;f) \leq M|Q|^{\frac{k+\lambda}{n}}. \tag{2.59}$$

Since $\Delta_h^{k+1} P = 0$ for $P \in \mathcal{P}_{k,n}$ and, moreover,

$$|\Delta_h^{k+1} f(x)| \leq 2^{k+1} \max_{0 \leq j \leq k} |f(x+jh)|,$$

(see (2.38)), we therefore have

$$|\Delta_h^{k+1} f(x)| = |\Delta_h^{k+1}(f-P)(x)| \leq 2^{k+1} \|f-P\|_{C(Q)},$$

where P is a polynomial from $\mathcal{P}_{k,n}$ and Q denotes the cube of minimal volume containing x and $x + kh$. Taking here the infimum over all polynomials P and using (2.59) we then obtain

$$|\Delta_h^{k+1} f(x)| \leq cM \|h\|^{k+\lambda}$$

for all $x, h \in \mathbb{R}^n$ and some constant c depending only on k and n.

Hence, the assumption of Theorem 2.32 holds for f and therefore $f \in \dot{C}^{k,\lambda}(\mathbb{R}^n)$ and $|f|_{C^{k,\lambda}(\mathbb{R}^n)} \leq cM$ with some constant $c = c(k,n)$. This and the first part of the proof also give the required equivalence

$$M_{k,\lambda}(f) \approx |f|_{C^{k,\lambda}(\mathbb{R}^n)}.$$

The proof is complete. □

Remark 2.39. Using Theorem 2.37 and the inequality

$$2^{-k} \sup\{|\Delta_h^k f(x)| \; ; \; x, x+kh \in Q\} \leq E_k(f \; ; \; Q),$$

(see the argument of the proof of (2.59)), we obtain in the very same way the equivalence

$$|f|_{\Lambda^{k,\omega}(\mathbb{R}^n)} \approx \sup_Q \frac{E_k(Q \; ; f)}{\omega(|Q|^{\frac{1}{n}})}, \tag{2.60}$$

where the constants of equivalence are independent of f.

A characterization of C^k functions via local polynomial approximation has been proved only for the univariate case (S. Bernstein [Ber-1940]). The result is hardly known and its proof is only outlined in Bernstein's note. For this reason we present a complete proof of this remarkable theorem. We refer the reader to the books [Tim-1963] by Timan and [DeVL-1993] by DeVore and Lorentz for the results of classical Approximation Theory which will be used (but not proved) in the proof.

Theorem 2.40. *A continuous function $f : [-1,1] \to \mathbb{R}$ is k-times continuously differentiable in $[-1,1]$ if and only if the limit*

$$\lambda(f \; ; x) := \lim_{Q \to x} \frac{E_k(Q \; ; f)}{|Q|^k} \tag{2.61}$$

2.3. Divided differences, local approximation and differentiability

exists and converges uniformly in $[-1,1]$ [5]. Moreover, in this case, for every x

$$\lambda(f;x) = \frac{|f^{(k)}(x)|}{2^{2k-1} \cdot k!}. \tag{2.62}$$

Proof. (Necessity) Given $f \in C^k[-1,1]$ and $\{x\}$, $Q \subset [-1,1]$, we write for $y \in Q$,

$$f(y) = T_x^{k-1}(y) + \frac{f^{(k)}(x)}{k!}(y-x)^k + R_k,$$

where the remainder R_k of the Taylor formula (in Cauchy's form) is bounded by $\frac{|y-x|^k}{k!}\omega_1(|y-x|;f^{(k)})$. Since $f^{(k)}$ is uniformly continuous in $[-1,1]$,

$$\lim_{Q \to x} |R_k| \cdot |x-y|^{-k} = 0$$

uniformly for $y \in Q$.

Hence, for given $\varepsilon > 0$ and all $|x| \leq 1$, there is $\delta > 0$ such that

$$\left| E_k(Q;f) - E_k\left(Q; T_x^{k-1} + \frac{f^{(k)}(x)}{k!}(\cdot - x)^k\right) \right| < \varepsilon |Q|^k,$$

provided that $|Q| + d(x,Q) < \delta$.

By the definition of E_k, the second term equals

$$\frac{|f^{(k)}(x)|}{k!} \min_{\{a_i\}} \max_{y \in Q} \left| x^k - \sum_{i=0}^{k-1} a_i y^i \right| = \frac{|f^{(k)}(x)|}{k!} \cdot 2\left(\frac{|Q|}{4}\right)^k,$$

where we use here the classical Chebyshev theorem on the polynomial of least deviation. Combining all these facts we get

$$\left| |Q|^{-k} E_k(Q;f) - \frac{|f^{(k)}(x)|}{2^{2k-1} \cdot k!} \right| < \varepsilon$$

for all x satisfying $|Q| + d(x,Q) < \delta$.

This part of the result is proved.

(Sufficiency) We need some properties of best polynomial approximation which are collected in

Proposition 2.41. *Let $f \in C(Q)$ and differ from a polynomial of degree $k-1$, $k \geq 1$. Then the following facts are true.*

(a) *There exists a unique polynomial (of best approximation) of degree $k-1$ denoted by $P_Q(f)$ such that*

$$E_k(Q;f) = \|f - P_Q(f)\|_{C(Q)}.$$

---

[5] $Q \to x$ means that the endpoints of the interval Q tend to x and (2.61) converges uniformly in Q.

(b) *There is a collection of (equioscillation) points* $x_0 < x_1 < \cdots < x_k$ *in* Q *such that*

$$|f(x_i) - P_Q(f)(x_i)| = \max_Q |f - P_Q(f)| \left(= E_k(Q;f)\right), \quad i = 0, \ldots, k,$$

and, moreover, for $\varepsilon_i := \operatorname{sgn}(f - P_Q(f)(x_i))$,

$$\varepsilon_i \varepsilon_{i+1} < 0, \quad i = 0, 1, \ldots, k-1.$$

Conversely, if for a polynomial P of degree $k-1$ the difference $f - P$ has $k+1$ points of equioscillation in Q, then $P = P_Q(f)$.

(c) *If $f_\kappa \in C(Q)$ continuously depends on a parameter κ ranging in a Hausdorff topological space, then $E_k(Q; f_\kappa)$ and $P_Q(f_\kappa)$ are continuous in κ.*

Assertion (c) and (2.61) imply that

Lemma 2.42. *The function λ is continuous.*

Proof. Let $T_Q : \mathbb{R} \to \mathbb{R}$ be an affine transform mapping $[-1, 1]$ onto Q. Then for $f_Q := f \circ T_Q$ we have, by the definition of E_k,

$$E_k(Q; f) = E_k([-1, 1]; f_Q).$$

Hence, the function $Q \mapsto |Q|^{-k} E_k(Q; f)$ continuously depends on Q and therefore the uniform limit in (2.61) is continuous. □

The function λ is therefore bounded in $[-1, 1]$ and the condition of Theorem 2.38 holds for $n = 1$ and the exponent $(k-1) + 1$. Hence, f *belongs to the space* $C^{k-1,1}[-1, 1]$.

Since the Lipschitz function $f^{(k-1)}$ is absolutely continuous, $f^{(k)}$ exists almost everywhere, and for all $x \in Q$,

$$f(x) = T_a(x) + \frac{1}{(k-1)!} \int_a^x (x-t)^{k-1} f^{(k)}(t) dt, \qquad (2.63)$$

where T_a is the (Taylor) polynomial of degree $k-1$ at a fixed point a in Q.

Set $S_f := \{x \in Q ; f^{(k)}(x) \text{ exists }\}$. By the Taylor formula with the remainder in Peano form we have, for $x \in S_f$ and $y \in Q$,

$$f(y) = T_x^{k-1}(f; y) + \frac{f^{(k)}(x)}{k!}(y-x)^k + \varepsilon_k \cdot (y-x)^k$$

where $\varepsilon_k \to 0$ as $y \to x$.

The argument used for the proof of necessity then implies that

$$\lambda(x) := \lim_{Q \to x} \frac{E_k(Q; f)}{|Q|^k} = \frac{|f^{(k)}(x)|}{k! 2^{2k-1}}.$$

2.3. Divided differences, local approximation and differentiability

Hence, (2.63) may be written as

$$f(x) = T_a(x) + k \cdot 2^{2k-1} \int_a^x (x-t)^{k-1} \lambda(t) \sigma(t) dt, \qquad (2.64)$$

where we set

$$\sigma := \operatorname{sgn} f^{(k)}.$$

Now let

$$G := \{x \in Q\,;\, \lambda(x) > 0\}.$$

This set is open, since λ is continuous. We will show below that σ admits a continuous extension from $G \cap S_f$ to G. In other words, we will prove that $f^{(k)}$ preserves its sign in any connected component (interval) of the open set G.

Let us assume for the time being that $\bar{\sigma}$ be this extension. Replacing σ by $\bar{\sigma}$ in (2.64) and differentiating k times the Riemann integral so obtained, we conclude that $f^{(k)}$ exists and is continuous in $[-1, 1]$, as required.

In turn, we derive the above stated extension result from the following

Claim. Given $\varepsilon > 0$ and a closed interval I from the open set

$$G_\varepsilon := \{x \in Q\,;\, \lambda(x) > \varepsilon\},$$

the function σ preserves its sign in $I \cap S(f)$.

We prove this by contradiction. Assume, on the contrary, that there are points z_1, z_2 in $I \cap S(f)$ such that

$$\sigma(z_1)\sigma(z_2) := \operatorname{sgn} f^{(k)}(z_1) \cdot \operatorname{sgn} f^{(k)}(z_2) < 0. \qquad (2.65)$$

To proceed, we reformulate this condition using equioscillation points, see Proposition 2.41 (b). To this end, we fix a closed interval $J \subset Q$, denote by $x_0 < x_1 < \cdots < x_k$ the set of equioscillation points for $f - P_J(f)$ in J and set

$$\varepsilon_J(f) := \operatorname{sgn}(f(x_k) - P_J(f)(x_k)).$$

Lemma 2.43. *Let $f \in C^{k-1,1}(J)$ and $f^{(k)}(t_0) > 0$ for some $t_0 \in J \cap S_f$. Then there is $\delta > 0$ such that for every interval $\Delta \subset J$ of length $\leq \delta$ which contains t_0, we have*

$$\varepsilon_\Delta(f) > 0.$$

Proof. Assume, on the contrary, that there is a sequence of intervals $\{\Delta_n\}_{n \geq 1}$ such that

$$t_0 \in \Delta_n \subset J,\ \varepsilon_{\Delta_n}(f) < 0\ \text{for all}\ n\ \text{and}\ |\Delta_n| \to 0\ \text{as}\ n \to \infty. \qquad (2.66)$$

Let Δ be one of these intervals and $x_0 < x_1 < \cdots < x_k$ be the equioscillation points in Δ for $f - P_\Delta(f)$. Due to (2.66) and Proposition 2.41, there is a point $y_{k-1} \in (x_{k-1}, x_k)$ such that

$$(f - P_\Delta(f))'(y_{k-1}) < 0.$$

Similarly, between x_{k-2} and x_{k-1} there is a point y_{k-2} such that
$$(f - P_\Delta(f))'(y_{k-2}) > 0,$$
and so forth.

Applying this argument to $(f - P_\Delta)'$, then to $(f - P_\Delta)''$ and so on, we finally find two points $w_0 < w_1$ in Δ such that
$$\operatorname{sgn}(f - P_\Delta)^{(k-1)}(w_i) = (-1)^i, \quad i = 0, 1.$$

Since $P_\Delta^{(k-1)}$ is a constant, this implies that
$$\frac{f^{(k-1)}(w_1) - f^{(k-1)}(w_0)}{w_1 - w_0} < 0.$$

Sending $|\Delta|$ to zero through the sequence $\{\Delta_n\}$ we obtain the inequality $f^{(k)}(t_0) \leq 0$ which contradicts the assumption of the lemma. $\square$

From the lemma, we now immediately obtain

Corollary 2.44. *Assume that condition (2.65) holds for $z_1, z_2 \in I \cap S(f)$. Then there is $\delta > 0$ such that for arbitrary subintervals $\Delta_i \ni z_i$, $i = 1, 2$, in I of equal length less than δ we have*
$$\varepsilon_{\Delta_1}(f) \varepsilon_{\Delta_2}(f) < 0. \tag{2.67}$$

Now we proceed as follows. Let $q_\Delta(f) := \gamma_\Delta x^k + \cdots$ be a polynomial of degree k closest to $f|_\Delta$ in $C(\Delta)$. It will be shown that $\varepsilon_\Delta(f) = \operatorname{sgn} \gamma_\Delta$. Therefore, shifting Δ_1 toward Δ_2 and using (2.67) and the continuous dependence of γ_Δ on Δ, see Lemma 2.42, we find an intermediate segment Δ such that $\gamma_\Delta = 0$. But q_Δ has $k + 2$ equioscillation points in Δ, say, $x_0 < x_1 < \cdots < x_{k+1}$, and the segment $\Delta' := [x_0, x_k] \subset \Delta$ contains $k + 1$ of them. By Proposition 2.41 this implies
$$E_k(\Delta; f) = E_k(\Delta'; f). \tag{2.68}$$

Further, $\Delta \subset G_\varepsilon$, i.e., $\lambda(x) > \varepsilon$ on Δ, and $f^{(k-1)}$ satisfies the Lipschitz condition with the constant $\gamma(k) \max_Q \lambda$. This allows us to estimate the distance between adjacent equioscillation points and, hence, the size of Δ' by the inequality
$$\frac{|\Delta'|}{|\Delta|} \leq q(\varepsilon, k, f) < 1. \tag{2.69}$$

Assuming this to be true for the moment we may derive from here the desired contradiction as follows. Let $\{\Delta_i^n\}_{n \in \mathbb{N}}$ be sequences of intervals in I such that $|\Delta_i^n| \to 0$ as $n \to \infty$ and $z_i \in \Delta_i^n$, $i = 1, 2$. Find for sufficiently large n, the aforementioned subintervals $\Delta_n' \subset \Delta_n$ related to the pairs Δ_1^n, Δ_2^n. Then we get for all n greater than some n_0,

2.3. Divided differences, local approximation and differentiability

$$E_k(\Delta'_n; f) = E_k(\Delta_n; f) \quad \text{and} \quad \frac{|\Delta'_n|}{|\Delta_n|} \leq q < 1.$$

By compactness we may assume that Δ_n tends to some point x as $n \to \infty$. For this x we then have

$$\lambda(x) = \lim_{\Delta_n \to x} \frac{E_k(f; \Delta_n)}{|\Delta_n|^k} \leq q^k \lim_{\Delta'_n \to x} \frac{E_k(f; \Delta'_n)}{|\Delta'_n|^k} < \lambda(x),$$

a contradiction.

Thus, it remains to prove (2.68) and (2.69). For this purpose we need the next

Lemma 2.45. *Suppose that the function f_λ is continuous on a closed interval I and depends continuously on $\lambda \in [\lambda_0, \lambda_1]$. Assume that the sign $\varepsilon_I(f_\lambda)$ satisfies*

$$\varepsilon_I(f_{\lambda_0}) \varepsilon_I(f_{\lambda_1}) < 0. \tag{2.70}$$

Then there is a $\lambda \in (\lambda_0, \lambda_1)$ such that $f_\lambda - P_I(f_\lambda)$ has one more equioscillation point[6].

Proof. We denote the polynomial of degree k closest to f_λ in $C(I)$ by $q_I(f_\lambda)$. We will show that for some $\lambda \in (\lambda_0, \lambda_1)$ this equals $P_I(f_\lambda)$. If this will be proved, then $f - P_I(f_\lambda)$ has the same number of equioscillation points as $f - q_I(f_\lambda)$, i.e., $k+2$.

Assume, on the contrary, that for all $\lambda \in (\lambda_0, \lambda_1)$,

$$P_I(f_\lambda) \neq q_I(f_\lambda).$$

Then the difference

$$q_I(f_\lambda) - P_I(f_\lambda) = \gamma(\lambda) x^k + \cdots \tag{2.71}$$

has at every equioscillation point of $f_\lambda - P_I(f_\lambda)$ a sign that equals that of $f_\lambda - P_I(f_\lambda)$. Actually, at such a point x, we get

$$|f_\lambda(x) - q_I(f_\lambda)(x)| \leq E_{k+1}(J; f_\lambda) < E_k(J; f_\lambda) = |f_\lambda(x) - P_I(f_\lambda)(x)|,$$

and the result follows from the equality

$$q_I(f_\lambda) - P_I(f_\lambda) = (f_\lambda - P_I(f_\lambda)) - (f_\lambda - q_I(f_\lambda)).$$

Thus, the polynomial in (2.71) changes sign at the $(k+1)$ equioscillation points. Let $x_I(f_\lambda)$ denote the largest of these points. Then the polynomial in (2.71) has k zeros less than $x_I(f_\lambda)$ and therefore

$$\text{sgn}(q_I(f_\lambda) - P_I(f_\lambda))(x_I(f_\lambda)) = \text{sgn}\, \gamma(\lambda).$$

---
[6] i.e., $k+2$ points, since $\deg P_I = k - 1$, see Proposition 2.41.

Since, as we have proved,
$$\varepsilon_I(f_\lambda) := \mathrm{sgn}(f_\lambda - P_I(f_\lambda))(x_I(f_\lambda)) = \mathrm{sgn}(q_I(f_\lambda) - P_I(f_\lambda))(x_I(f_\lambda)),$$
we get
$$\varepsilon_I(f_\lambda) = \mathrm{sgn}\,\gamma(\lambda).$$
By Proposition 2.41 (c), the function γ is continuous on $[\lambda_0, \lambda_1]$ while (2.70) shows that γ changes its sign at the endpoints. Hence there is $\lambda \in (\lambda_0, \lambda_1)$ such that $\gamma(\lambda) = 0$ and $P_I(f_\lambda) = q_I(f_\lambda)$ for this λ. $\square$

Now we show how to find the intervals $\Delta' \subset \Delta \subset \mathrm{conv}(\Delta_1, \Delta_2)$ in (2.68). Set $\Delta_i := [a_i, b_i]$, $i = 1, 2$, and $\ell := b_1 - a_1 \,(= b_2 - a_2)$. Assuming that $a_1 < a_2$, we define the function $f_\lambda \in C[0, \ell]$, where $\lambda \in [a_1, a_2]$, by
$$f_\lambda(x) := f(x + \lambda), \quad x \in [0, \ell].$$
Applying the previous lemma to the function f_λ and (2.70) we find $\lambda \in [a_1, a_2]$ such that $f_\lambda - P_{[0,\ell]}(f_\lambda)$ has $k+2$ equioscillation points in $[0, \ell]$. Translating by ℓ we then obtain the same for $f - P_\Delta(f)$ where $\Delta := [\lambda, \ell + \lambda] \subset \mathrm{conv}(\Delta_1 \cup \Delta_2)$. Let $x_0 < x_1 < \cdots < x_{k+1}$ be the equioscillation points of this difference in Δ. According to Proposition 2.41 (b), the polynomial P_Δ is of best approximation to f in $C[x_0, x_k]$. Setting
$$\Delta' := [x_0, x_k] \subset \Delta,$$
we then get
$$E_k(\Delta'; f) = E_k(\Delta; f),$$
and it remains to show that for this Δ',
$$\frac{|\Delta'|}{|\Delta|} \le q = q(k, \varepsilon, f) < 1; \tag{2.72}$$
this will prove (2.70) and the theorem.

In turn, (2.72) is a consequence of

Lemma 2.46. *Let the function $f \in C^{k-1,1}(I)$, where I is a closed interval in $\mathbb{R}$, satisfy the conditions*
$$\left|f^{(k-1)}(x) - f^{(k-1)}(y)\right| \le M_0 |x - y|, \quad x, y \in I, \tag{2.73}$$
$$E_k(I; f) \ge M_1 |I|^k. \tag{2.74}$$
Then for every pair of adjacent equioscillation points x_i, x_{i+1} for $f - P_I(f)$, it is true that
$$|x_{i+1} - x_i| \ge c \frac{M_1}{M_0} |I|, \tag{2.75}$$
where $c > 0$ depends only on k and f.

2.3. Divided differences, local approximation and differentiability

We first derive from here (2.72). Let $\Delta_1, \Delta_2 \subset I$ be as in Corollary 2.44, where the closed interval $I \subset G_\varepsilon := \{x \in Q \,;\, \lambda(x) > \varepsilon\}$. By the uniform convergence in (2.61), we get for some $\delta_1 > 0$ and every subinterval $\widetilde{\Delta} \subset I$ of length $< \delta_1$,

$$\frac{E_k(\widetilde{\Delta}\,;f)}{|\widetilde{\Delta}|^k} \geq \frac{1}{2} \inf_{x \in \widetilde{\Delta}} \lambda(x) > \frac{1}{2}\varepsilon.$$

Choosing an interval $\Delta \subset \mathrm{conv}(\Delta_1 \cup \Delta_2)$ such that $|\Delta| < \delta_1$, we then get

$$E_k(\Delta\,;f) \geq \frac{1}{2}\varepsilon|\Delta|^k.$$

Applying now Lemma 2.46 with $M_0 := |f|_{C^{k-1,1}(\Delta)}$, $M_1 := \frac{1}{2}\varepsilon$, and $I := \Delta$ we obtain

$$|\Delta'| := [x_0, x_k] \leq |\Delta| - |x_{k+1} - x_k| \leq |\Delta| - \frac{\varepsilon c(k)}{2|f|_{C^{k-1,1}(\Delta)}}|\Delta|.$$

This gives the required inequality

$$\frac{|\Delta'|}{|\Delta|} \leq q = q(k,\varepsilon,f) < 1.$$

So, it remains to establish Lemma 2.46.

Proof. Let first $k = 1$. Then $P_I(f)$ is a constant and there are two adjacent equioscillation points x_0, x_1 in I. By the definition of these points, (2.74) and (2.73) we have

$$2M_1|I| \leq 2E_1(I\,;f) = \left|(f - P_I(f))(x_0) - (f - P_I(f))(x_1)\right| \leq M_0|f|_{C^{0,1}(I)}|x_0 - x_1|$$

and (2.75) for $k = 1$ is established with $c := \frac{2}{|f|_{C^{0,1}(I)}}$.

Now let $k \geq 2$ and $x_0 < x_1 < \cdots < x_k$ be the equioscillation points of $f - P_I$ in I. As before,

$$\begin{aligned} 2M_1|I|^k \leq 2E_k(I\,;f) &= \left|(f - P_I(f))(x_i) - (f - P_I(f))(x_{i+1})\right| \\ &\leq |x_i - x_{i+1}| \max_I |(f - P_I(f))'|. \end{aligned} \tag{2.76}$$

To estimate this maximum we use the Taylor formula and (2.73) to write

$$\max_I |f - T(f)| \leq \frac{M_0}{(k-1)!}|I|^k,$$

where $T(f)$ is the Taylor polynomial for f of degree $k - 1$ at the left endpoint of I.

The same argument also gets

$$\max_I |(f - T(f))'| \leq \frac{M_0}{(k-2)!} |I|^{k-1}. \tag{2.77}$$

Using the inequality

$$\max_I |(P_I(f) - T(f))| \leq E_k(I;f) + \max_I |f - T(f)| \leq 2 \max_I |f - T(f)|$$

we derive from here that

$$\max_I |P_I(f) - T(f)| \leq \frac{2M_0}{(k-1)!} |I|^k.$$

This and the Markov inequality, see Lemma 2.35 or Appendix G, give

$$\max_I |(P_I(f) - T(f))'| \leq \frac{4M_0(k-1)^2}{(k-1)!} |I|^{k-1}.$$

Together with (2.77) this leads to the estimate

$$\max_I |(f - P_I)'| \leq c(k) M_0 |I|^{k-1}.$$

Combining this last inequality with (2.76), we prove the lemma. □

As it was explained above this gives the proof of Theorem 2.40. □

2.4 Trace and extension problems for univariate C^k functions

2.4.1 Whitney's theorem

The Whitney solution to the problem is based on a detailed study of the combinatorial properties of closed subsets of the real line and a subsequent use of such analytic tools as Lagrange interpolation and divided differences. The Lagrange interpolation polynomial which coincides with f on a $(k+1)$-point set X is denoted by $L(f;X)$; its degree equals k and the k-th derivative satisfies

$$L^{(k)}(f;X) = k! f[X]. \tag{2.78}$$

Now let S be a closed subset of $\mathbb{R}$, and let $S^{\langle k \rangle}$ denote the class of all k-point subsets of S. The main role in the solution will play a class of functions on S satisfying the condition

$$\lim_{X \to x} \{|f[X]| \cdot \operatorname{diam} X \,;\, X \in S^{\langle k \rangle}\} = 0 \tag{2.79}$$

at every point $x \in S$; here $X \to x$ means that $d(\{x\}, X) \to 0$.

The linear space of continuous functions $f : S \to \mathbb{R}$ satisfying (2.79) is denoted by $D^k(S)$.

2.4. Trace and extension problems for univariate C^k functions

Theorem 2.47 (Whitney [Wh-1934b]). (a) *It is true that*

$$C^k(\mathbb{R})\big|_S = D^{k+2}(S). \tag{2.80}$$

(b) *There exists a linear extension operator from the trace space into $C^k(\mathbb{R})$.*

Proof. Assume that $f = F|_S$ where $F \in C^k(\mathbb{R})$. If $X = \{x_0, \ldots, x_{k+1}\} \subset S$ and $x_i < x_{i+1}$, then for some c_1, c_2 from the interval $[x_0, x_{k+1}]$,

$$f[X] = F[X] = k! \, \frac{F^{(k)}(c_1) - F^{(k)}(c_2)}{x_0 - x_{k+1}}.$$

Since $F^{(k)}$ is continuous, the condition

$$\lim_{X \to x} \left\{ |f[X]| \cdot \operatorname{diam} X \; ; \; X \in S^{\langle k+2 \rangle} \right\} = 0$$

holds at every $x \in X$. Hence,

$$D^{k+2}(S) \supset C^k(\mathbb{R})\big|_S. \qquad \square$$

The proof of the converse embedding is essentially more complicated. We begin with a result whose proof will be postponed until the end of the section.

Lemma 2.48 (Combinatorial). *There exists a map $\phi : S \to S^{\langle 1 \rangle} \cup S^{\langle k+1 \rangle}$ such that*

(i) $x \in \phi(x)$;

(ii) *if x is a limit point, then*

$$\phi(x) = \{x\};$$

(iii) *if x is an isolated point, then*

$$\operatorname{card} \phi(x) = k + 1;$$

(iv) *if either of x_1, x_2 is a limit point, then*

$$\operatorname{diam} \phi(x_1) + \operatorname{diam} \phi(x_2) \leq C|x_1 - x_2|$$

for some constant $C = C(k)$.

This is also true for isolated points if $\phi(x_1) \neq \phi(x_2)$.

Now let P be a univariate polynomial of degree k with k mutually distinct real roots situated in an interval I of length ℓ.

Lemma 2.49. *For every $j \leq k$,*

$$\max_I |P^{(j)}| \leq \frac{|P^{(k)}|}{j!(k-j)!} \ell^{k-j}.$$

Proof. Let $Z = \{z_1, \ldots, z_k\} \subset I$ be the set of zeros. Then
$$P(x) = \frac{a}{k!} \prod_{z \in Z} (x - z),$$
where the constant $a := P^{(k)}$. Therefore,
$$P^{(j)}(x) = \frac{a}{k!} \sum_{Y \in Z^{\langle k-j \rangle}} \prod_{y \in Y} (x - y),$$
whence
$$\max_I |P^{(j)}| \leq \frac{|P^{(k)}|}{k!} \binom{k}{k-j} \ell^{k-j}. \qquad \square$$

Now let $X_1, X_2 \in S^{\langle k+1 \rangle}$ and $X_1 \neq X_2$. Suppose that I is an interval of length ℓ in $\mathbb{R}$ such that
$$X := X_1 \cup X_2 \subset I.$$
Then for the Lagrange polynomials $L(f; X_i)$ the following holds.

Lemma 2.50. *For every $j \leq k$ and $x \in I$,*
$$\big|(L(X_1, f) - L(X_2, f))^{(j)}(x)\big|$$
$$\leq c(k) \ell^{k-j} \cdot \max\{|f[Y]| \cdot \operatorname{diam} Y \,;\, Y \in X^{\langle k+2 \rangle}\}. \tag{2.81}$$

Proof. Connect X_1 with X_2 by a chain of $(k+1)$-point subsets $Y_i \in X^{\langle k+1 \rangle}$, $i \leq m := k+2$, such that $Y_1 = X_1$, $Y_m = X_2$ and
$$\operatorname{card}(Y_i \cap Y_{i+1}) = k \quad \text{for every} \quad i \leq m - 1. \tag{2.82}$$
Setting $P_i := L(Y_i, f)$ we may bound the left-hand side of (2.81) by the sum $\sum_{i < m} \max_I |(P_i - P_{i+1})^{(j)}|$ and then bound the i-th term by Lemma 2.49 with $P := P_i - P_{i+1}$, a polynomial of degree k whose set of roots $Y_i \cap Y_{i+1}$ contains k points of the interval I.

Hence, the sum is bounded by
$$\frac{\ell^{k-j}}{j!(k-j)!} \sum_{i=1}^{m} |(P_i - P_{i+1})^{(k)}| = \binom{k}{j} \ell^{k-j} |f[Y_i] - f[Y_{i+1}]|,$$
see (2.78). Finally, by the definition of the divided difference and (2.82), we get
$$|f[Y_i] - f[Y_{i+1}]| \leq |f[Y_i \cup Y_{i+1}]| \operatorname{diam}(Y_1 \cup Y_2)$$
$$\leq \max\{|f[Y]| \operatorname{diam} Y \,;\, Y \in X^{\langle k+2 \rangle}\}.$$
Putting all these estimates together we establish (2.81). $\qquad \square$

Now we relate the space $D^{k+2}(S)$ to the trace space $J^k(\mathbb{R})\big|_S$. In this setting, $J^k(\mathbb{R})$ consists of vector functions $\{f_j\}_{j=0}^k$ generated by functions $f \in C^k(\mathbb{R})$ such that $f_j = f^{(j)}$.

2.4. Trace and extension problems for univariate C^k functions

Lemma 2.51. *There exists a linear operator* $W : D^{k+2}(S) \to J^k(\mathbb{R})|_S$ *such that*

$$(Wf)_0 = f.$$

Proof. By Theorem 2.13 the space $J^k(\mathbb{R})|_S$ consists of vector functions $\vec{f} = \{f_j\}_{j=0}^k$ continuous on S and satisfying the following condition:
For every $j \leq k$ and $x, y \in S$,

$$r_j(\vec{f}; x, y) \to 0 \quad \text{as} \quad y \to x. \tag{2.83}$$

Let us recall that (in this case) the reduced remainders are defined by

$$r_j(\vec{f}; x, y) := \left| f_j(x) - \sum_{i=0}^{k-j} f_{i+j}(y) \frac{(x-y)^i}{i!} \right| \bigg/ |x-y|^{k-j}.$$

Hence, given $f \in D^{k+2}(S)$, we must find a vector function $\vec{f} = \{f_j\}_{j=0}^k$ satisfying (2.83) that depends linearly on f. We define the function f_j as the limit of the j-th derivatives for the corresponding interpolation polynomials.

We begin with the value of f_j at a limit point $x \in S$. In this case, the set

$$T_\varepsilon(x) := \{X \in S^{\langle k+2 \rangle} \; ; \; \text{diam}(\{x\} \cup X) < \varepsilon\}$$

contains for sufficiently small ε, say, $\varepsilon < \varepsilon_0$, infinitely many points. This allows us to define f_j by

$$f_j(x) := \lim_{Y \to x} \{L^{(j)}(Y, f)(x) \; ; \; Y \in S^{\langle k+1 \rangle}\}; \tag{2.84}$$

the limit has the same meaning as that in (2.79).

To prove that the limit exists, we apply Lemma 2.50 to subsets $X_1, X_2 \in T_\varepsilon(x) \cap S^{\langle k+1 \rangle}$ where $\varepsilon < \frac{\varepsilon_0}{2}$. This yields

$$\left|((L(X_1, f) - L(X_2, f))^{(j)}(x)\right| \leq c(k) \max\{|f[Y]| \, \text{diam}\, Y \; ; \; Y \in T_{2\varepsilon}(x) \cap S^{\langle k+2 \rangle}\}.$$

Since $f \in D^{k+2}(S)$, the right-hand side tends to zero as $Y \to x$, see (2.79). Hence, the limit in (2.84) exists for all $j \leq k$. Moreover, for such x,

$$f_0(x) = f(x),$$

since $L(X, f)(x) = f(x)$ as $x \in X$.

Further, we define $\vec{f}(x)$ at an *isolated* point $x \in X$ using the set $\phi(x)$ of Combinatorial Lemma 2.48. In this case, $\phi(x)$ consists of precisely $k+1$ points and $L(\phi(x), f)$ is a polynomial of degree k. We introduce now $f_j(x)$ by

$$f_j(x) := \frac{d^j}{dt^j} L(\phi(x), f)(t) \bigg|_{t=x} := L^{(j)}(\phi(x), f)(x) \tag{2.85}$$

noting that $x \in \phi(x)$ and therefore $f_0(x) = f(x)$ at isolated points as well.

Thus, $\vec{f}$ is now determined at all points of S. The operator $W : f \mapsto \vec{f}$ is clearly linear and satisfies

$$(Wf)_0 = f.$$

Let us show that $\vec{f} \in J^k(\mathbb{R})\big|_S$, that is, the reduced remainders r_j satisfy condition (2.83). To this end we choose, for a given $x \in S$, a sequence of sets $\{X_i(x)\}$ from $S^{\langle k+1 \rangle}$ such that

$$\operatorname{diam} X_i(x) \to \operatorname{diam} \phi(x) \quad \text{as} \quad i \to \infty, \tag{2.86}$$

and, moreover, for $0 \leq j \leq k$,

$$f_j(x) = \lim_{i \to \infty} L^{(j)}(X_i(x), f)(x). \tag{2.87}$$

The existence of such a sequence is obvious for a limit point x; for an isolated x we simply set

$$X_i(x) := \phi(x), \quad i = 1, 2, \ldots. \tag{2.88}$$

Since $L(\phi(x), f)$ is a polynomial of degree k, the remainder $r_j(\vec{f}; x, y)$ may be written in the form

$$r_j(\vec{f}; x, y) = \lim_{i \to \infty} \left| (P_{i,x} - P_{i,y})^{(j)}(x) \right| \big/ |x - y|^{k-j}, \tag{2.89}$$

where we set for $x \in S$ and $z \in \mathbb{R}$,

$$P_{i,x}^{(j)}(z) := \frac{d^j}{dz^j} L\big(X_i(x), f\big)(z),$$

see (2.85) and (2.87).

It remains to estimate the j-th derivative in (2.89).

First, let x be a limit point. By Lemma 2.48,

$$\operatorname{diam} \phi(x) + \operatorname{diam} \phi(y) \leq c(k)|x - y|.$$

This and (2.86) yield the estimate for the upper limit of lengths ℓ_i of intervals $J_i(x, y) := \operatorname{conv}\big(X_i(x) \cup X_i(y)\big)$:

$$\varlimsup_{i \to \infty} \ell_i \leq \lim_{i \to \infty} \big(\operatorname{diam} X_i(x) + \operatorname{diam} X_i(y) + |x - y|\big) \leq c(k)|x - y|.$$

Using this and Lemma 2.50 we bound the limit in (2.89) by

$$c(k)|x-y|^{k+j} \varlimsup_{i \to \infty} \left[\ell_i^{k-j} \max\{|f[X]| \operatorname{diam} X \,;\, X \in \big(X_i(x) \cup X_i(y)\big)^{\langle k+2 \rangle} \} \right]$$

$$\leq c(k) \varlimsup_{i \to \infty} \max\{|f[X]| \operatorname{diam} X \,;\, X \in \big(X_i(x) \cup X_i(y)\big)^{\langle k+2 \rangle} \}.$$

2.4. Trace and extension problems for univariate C^k functions

Since $f \in D^{k+2}(S)$, see (2.79), and $\ell_i = |J_i(x,y)| \to 0$ uniformly in i as $y \to x$, the right-hand side tends to zero as $y \to x$ and therefore

$$r_j(f\,;x,y) \to 0 \quad \text{as} \quad y \to x.$$

Now, let $x \in S$ be an isolated point. Then $y = x$ if $y \in S$ is sufficiently close to x and therefore, for such y,

$$r_j(f\,;x,y) = r_j(f\,;x,x) = 0. \qquad \square$$

Lemma 2.51 implies the required inverse embedding

$$D^{k+2}(S) \subset C^k(\mathbb{R})\big|_S,$$

since by Theorem 2.13, $(Wf)_0 = F\big|_S$ for some $F \in C^k(\mathbb{R})$ while $(Wf)_0 = f$.

Thus, part (a) of Theorem 2.47 is proved.

Finally, the required linear extension operator $\mathcal{E}_k^X$ from $D^{k+2}(S)$ to $C^k(\mathbb{R})$ is defined by composing Whitney's extension operator E_k^S acting from $J^k(\mathbb{R})\big|_S$ to $J^k(\mathbb{R})$ with W and the projection $P\vec{f} := f_0$, $\vec{f} \in J^k(\mathbb{R})$. In other words,

$$\mathcal{E}_k^S := PE_k^S W. \qquad (2.90)$$

This completes the proof of Theorem 2.47. $\qquad \square$

We apply now the Whitney extension method to obtain a version of Theorem 2.47 concerning the space $\dot{C}^{k,\omega}(\mathbb{R})$ defined by the seminorm (2.7); recall that the continuous function $\omega : \mathbb{R}_+ \to \mathbb{R}_+$ satisfies the conditions

$$\omega(t) \text{ and } \omega^*(t) := \frac{t}{\omega(t)} \text{ increase as } t \to +\infty \text{ and } \omega(0+) = 0. \qquad (2.91)$$

To formulate the result due to Merrien [Mer-1966] we introduce a space of functions $f : S \to \mathbb{R}$ by the condition

$$m_{k,\omega}(f) := \sup\{|f[X]|\omega^*(\operatorname{diam} X)\,;\, X \in S^{\langle k+2 \rangle}\} < \infty. \qquad (2.92)$$

We denote this space by $\dot{D}_\omega^{k+2}(S)$.

Theorem 2.52. (a) $\dot{D}_\omega^{k+2}(S) = \dot{C}^{k,\omega}(\mathbb{R})\big|_S$ and $m_{k,\omega}(f)$ is equivalent to the trace norm of f with constants depending only on k.

(b) The restriction of the extension operator $\mathcal{E}_k^S$ to $\dot{C}^{k,\omega}(\mathbb{R})\big|_S$ is a linear continuous extension operator from this space into $\dot{C}^{k,\omega}(\mathbb{R})$.

Proof. Let $f = F|_S$ for some $F \in \dot{C}^{k,\omega}(\mathbb{R})$. Then we have, for $X \in S^{\langle k+2 \rangle}$,

$$|f[X]| = k!\,\frac{|F^{(k)}(c_1) - F^{(k)}(c_2)|}{\operatorname{diam} X},$$

where $c_1, c_2 \in \operatorname{conv} X$.

This immediately yields

$$\bigl|f[X]\bigr|\omega^*(\operatorname{diam} X) \leq k!|F|_{C^{k,\omega}(\mathbb{R})},$$

whence

$$m_{k,\omega}(f) \leq k!|f|_{C^{k,\omega}(\mathbb{R})|_S} \tag{2.93}$$

by the definition of the seminorm involved.

To prove the converse result, we first state that

$$\bigl|f[X]\bigr|\operatorname{diam} X \leq m_{k,\omega}(f)\omega(\operatorname{diam} X), \tag{2.94}$$

see (2.92). But $\omega(0+) = 0$ and therefore finiteness of $m_{k,\omega}(f)$ implies that f belongs to $D^{k+2}(S)$. Hence, the operator $\mathcal{E}_k^S$ is defined on the space $\dot{D}_\omega^{k+2}(S)$ and it remains to estimate the $C^{k,\omega}$-seminorm of $\mathcal{E}_k^S f$ for $f \in \dot{D}_\omega^{k+2}(S)$. To this end we first apply the inequality established for the remainder $r_j(\vec{f}; x, y)$ with $x, y \in S$, see (2.89), and then use (2.94) to bound the right-hand side of this inequality by

$$C(k)m_{k,\omega}(f) \varlimsup_{i \to \infty} \bigl\{\omega(\operatorname{diam} X) \; ; \; X \in J_i(x,y)^{\langle k+2 \rangle}\bigr\}$$
$$\leq C(k)m_{k,\omega}(f)\omega\bigl(C_1(k)|x-y|\bigr);$$

here $\vec{f}$ is the k-jet generated by f, i.e., $\vec{f} = Wf$, see Lemma 2.51.

Since ω^* is nondecreasing, $\omega(ct) \leq \max(1,c)\omega(t)$ and therefore,

$$r_j(\vec{f}; x, y) \leq C(k)m_{k,\omega}(f)\omega(|x-y|)$$

provided that $x, y \in S$ and $0 \leq j \leq k$. This means that $\vec{f}$ satisfies the Taylor condition of Theorem 2.22, see (2.36), which then implies that the Whitney operator E_k^S extends $\vec{f}$ to the k-jet $E_k^S \vec{f}$ from $\dot{J}^{k,\omega}(\mathbb{R})$. Since, in addition, this operator is linear and continuous, the operator $\mathcal{E}_k^S := PE_k^S W$ (see (2.90)) extends functions from $\dot{D}_\omega^{k+2}(S)$ to $\dot{C}^{k,\omega}(\mathbb{R})$ and

$$|\mathcal{E}_k^S f|_{C^{k,\omega}(\mathbb{R})} \leq C(k)m_{k,\omega}(f).$$

Together with (2.93) this completes the proof. $\square$

Remark 2.53. The same proof can be applied to the case of the Banach space $C_b^{k,\omega}(\mathbb{R})$ with norm

$$\|f\|_{C_b^{k,\omega}(\mathbb{R})} := \sup_\mathbb{R} |f| + |f|_{C^{k,\omega}(\mathbb{R})}$$

We leave to the reader to verify that up to the equivalence of the norms

$$C_b^{k,\omega}(\mathbb{R})\bigr|_S = D_\omega^{k+2}(S),$$

2.4. Trace and extension problems for univariate C^k functions

where the space of the right-hand side is defined by the norm $\sup_S |f| + m_{k,\omega}(f)$.

We now use *universality* of the extension operator $\mathcal{E}_k^S$ (meaning independence of ω) to obtain an extension result for the subspace $\dot{C}_u^k(\mathbb{R})$ of $C^k(\mathbb{R})$ consisting of functions whose higher derivatives are uniformly continuous on the real line.

Corollary 2.54. *The restriction of $\mathcal{E}_k^S$ to the space $\dot{C}_u^k(\mathbb{R})\big|_S$ is a linear continuous extension operator from this space into $\dot{C}_u^k(\mathbb{R})$.*

Proof. If $f \in \dot{C}_u^k(\mathbb{R})$, then the *modulus of continuity* of f defined for $f > 0$ by

$$\omega(t; f^{(k)}) := \sup_{|x-y|\leq t} \left| f^{(k)}(x) - f^{(k)}(y) \right|$$

tends to zero as $t \to 0$.

Since modulus of continuity is a subadditive function, there exists a function $\widetilde{\omega} : \mathbb{R}_+ \to \mathbb{R}_+$ satisfying conditions (2.91) and such that

$$\omega \leq \widetilde{\omega} \leq 2\omega;$$

for instance, one can choose $\widetilde{\omega}(t) := t \sup_{s \geq t} \frac{\omega(s)}{s}$. This implies the equality

$$\dot{C}_u^k(\mathbb{R}) = \bigcup_\omega \dot{C}^{k,\omega}(\mathbb{R}),$$

where the union is taken over all ω satisfying (2.91).

Then assertion (b) of Theorem 2.52 gives the required result. $\square$

Proof of Combinatorial Lemma 2.48. We begin the definition of the required set function $\phi : S \to S^{\langle 1 \rangle} \cup S^{\langle k+1 \rangle}$ by setting

$$\phi(x) := \{x\}$$

for $x \in S$ being a limit point of S.

Now let x be an isolated point of S. Then we set

$$\phi(x) := \{x_0, x_1, \ldots, x_k\},$$

where $x_0 := x$ and the remaining points are defined as follows.

Define x_1 by the condition

$$|x_0 - x_1| = d(\{x_0\}, S\setminus\{x_0\}), \quad x_1 \in S\setminus\{x_0\}$$

(recall that $d(S_1, S_2) := \inf\{|s_1 - s_2| \; ; \; s_i \in S_i, \; i = 1, 2\}$). If there are two such points, choose x_1 to be the one to the right.

If now x_1 is a limit point of the closed set $S\setminus\{x_0\}$, we choose distinct points x_j from $S\setminus\{x_0, x_1\}$, $2 \leq j \leq k$, such that

(R) $\quad |x_j - x_1| = d(x_j, \{x_0, x_1, \ldots, x_{j-1}\}) \leq d(x_{j-1}, \{x_0, x_1, \ldots, x_{j-2}\}).$

Since x_1 is a limit point of $S\setminus\{x_0\}$, such a choice is possible. Otherwise $S\setminus\{x_0,x_1\}$ is closed and we define x_2 by the condition

$$d(x_2,\{x_0,x_1\}) = d(\{x_0,x_1\},S\setminus\{x_0,x_1\}), \quad x_2 \in S\setminus\{x_0,x_1\};$$

as above, the point x_2 is to be the one to the right if the choice is not unique. Again, if x_2 is a limit point of $S\setminus\{x_0,x_1\}$, then we choose points $x_j \in S\setminus\{x_0,x_1,x_2\}$ by the above formulated rule (R) with x_2 replacing x_1.

Proceeding in this way, we define $\phi(x) := \{x_0,\ldots,x_k\}$.

Claim. If $\phi(x) \neq \phi(y)$, then

$$|x_i - x_j| \leq \max\{i,j\}|x-y|, \quad 0 \leq i,j \leq k, \tag{2.95}$$

and the same holds for $|y_i - y_j|$.

To establish this, we first show that for $0 \leq i \leq k-1$,

$$\begin{aligned} d(x_{i+1},\{x_0,\ldots,x_i\}) &\leq |x_0 - y_0|, \\ d(y_{i+1},\{y_0,\ldots,y_i\}) &\leq |x_0 - y_0|. \end{aligned} \tag{2.96}$$

These inequalities will be proved by induction on i. The result is trivial for $i=0$. Now let (2.96) hold for $i-1 < k-1$; we will prove it for i. By symmetry, it suffices to establish the first of these inequalities. Without loss of generality we may assume that the points $x_0,\ldots,x_i$ are isolated; for otherwise the required inequality follows from the rule (R) and the induction hypothesis. It follows from the condition $\phi(x) \neq \phi(y)$ that there are two possibilities:

A. $\{y_0,\ldots,y_i\}$ is a *proper* subset of $\{x_0,\ldots,x_i\}$.
B. There is $j \leq i$ such that

$$y_j \notin \{x_0,\ldots,x_i\}. \tag{2.97}$$

Case A is impossible, since all points $y_0, y_1, \ldots, y_i$ are pairwise distinct.

In case B, let j be the minimal index satisfying (2.97). By the extremal property of x_{i+1}, we then have

$$d(x_{i+1},\{x_0,\ldots,x_i\}) \leq d(y_j,\{x_0,\ldots,x_i\}).$$

If now $j=0$, then the right-hand side of the inequality is clearly bounded by $|x_0 - y_0|$, as required. Otherwise, $\{y_0,\ldots,y_{j-1}\}$ is a nonempty subset of $\{x_0,\ldots,x_i\}$. Hence, in this case,

$$d(y_j,\{x_0,\ldots,x_i\}) \leq d(y_j,\{y_0,\ldots,y_{j-1}\}),$$

and it follows from the induction hypothesis that this is bounded by $|x_0 - y_0|$.

Using now inequalities (2.96) we prove (2.95) by induction on $\ell := \max\{i,j\}$.

2.4. Trace and extension problems for univariate C^k functions

For $\ell = 0$ this is trivial. For $\ell > 0$, we may assume that $i > j$. Let $i' < i$ be such that $|x_i - x_{i'}| = d(x_i, \{x_0, \ldots, x_{i-1}\}) \leq |x_0 - y_0|$. By the induction hypothesis, $|x_{i'} - x_j| \leq \max\{i', j\}|x_0 - y_0| \leq (\ell - 1)|x_0 - y_0|$. Hence,

$$|x_i - x_j| \leq |x_i - x_{i'}| + |x_{i'} - x_j| \leq \ell|x_0 - y_0|.$$

The proof of (2.95) is complete.

Now we are ready to prove the Combinatorial Lemma. By the definition of ϕ, the point x belongs to $\phi(x)$. Moreover, $\phi(x) = \{x\}$ at a limit point and card $\phi(x) = k + 1$ at an isolated point.

We claim that if $x \neq y$ and one of these points is a limit point, then

$$\operatorname{diam} \phi(x) + \operatorname{diam} \phi(y) \leq 2k|x - y|.$$

In fact, in this case $\phi(x) \neq \phi(y)$ and, by (2.95), the diameters of $\phi(x)$ and $\phi(y)$ are at most $k|x - y|$. The same argument can be applied to the case of isolated points x, y such that $\phi(x) \neq \phi(y)$ and so the lemma has been proved. □

2.4.2 Reformulation of Whitney's theorem

In an attempt to generalize the Whitney theorem to $C^k(\mathbb{R}^n)$, we immediately enter the following difficulty. Unlike the univariate case we cannot use for the definition of divided differences the linear order of the real line or the connection with Lagrange interpolation as in Proposition 2.25. In the multivariate case, it would be natural to define the divided difference $f[\Sigma]$ for a finite set $\Sigma \subset \mathbb{R}^n$ as a collection of derivatives $\{D^\alpha L_\Sigma(f); |\alpha| = k\}$, where $L_\Sigma(f)$ is a polynomial of *minimal* degree interpolating the function f at points of Σ. Such a polynomial is not, in general, unique and one must select one of them. Unfortunately, such a choice cannot be done arbitrarily, since, given f being the trace of a C^k function to S, the chosen family of interpolation polynomials for f must satisfy quite strong restrictions. This leads to a very difficult *selection problem* whose geometric counterpart will be discussed in Section 5.3.

Therefore it seems to be reasonable to find an equivalent formulation of Whitney's theorem which eliminates divided differences (and related Lagrange interpolation) and then to extract from there the basic features of the multivariate extension problems. To achieve this purpose we need a new concept, the *Finiteness Property*, already mentioned in Section 2.3. For its motivation we first note that due to Theorem 2.52 for a $(k + 2)$-point set $\Sigma \subset \mathbb{R}$ and a function $g : \Sigma \to \mathbb{R}$,

$$|g|_{C^{k,\omega}|_\Sigma} \approx |g[\Sigma]\omega^*(\operatorname{diam} \Sigma)|,$$

where $\omega^*(t) := t/\omega(t)$, $t > 0$, and the constants of equivalence are independent of g and Σ.

Therefore Theorem 2.52 may be restated as follows.

Theorem 2.55. *Let S be a closed subset of $\mathbb{R}$ containing at least $k + 2$ points.*

(a) A function $f : S \to \mathbb{R}$ belongs to the trace space $\dot{C}^{k,\omega}\big|_S$ if and only if its restrictions $f|_\Sigma$ to subsets Σ from $S^{\langle k+2 \rangle}$ satisfy

$$\sup\left\{|f|_\Sigma|_{C^{k,\omega}|_\Sigma} \, ; \, \Sigma \in S^{\langle k+2 \rangle}\right\} < \infty.$$

Moreover, the supremum is equivalent to $|f|_{C^{k,\omega}|_S}$.

(b) There is a linear continuous extension operator from $\dot{C}^{k,\omega}\big|_S$ to $\dot{C}^{k,\omega}(\mathbb{R})$ independent of ω.

In order to reformulate Whitney's Theorem 2.47, we note that the entity $|f[\Sigma]\operatorname{diam}\Sigma|$, where $\Sigma := \{x_0, \ldots, x_{k+1}\} \subset \mathbb{R}$ may be rewritten up to the multiplier $k!$ as $\big|L_{\Sigma_0}(f) - L_{\Sigma_1}(f)\big|_{C^k(\mathbb{R})}$ for $\Sigma_i := \{x_i, \ldots, x_{k+i}\}$, $i = 0, 1$, see Proposition 2.25 (b). Moreover, the Lagrange polynomial (of degree k) $L_{\Sigma_i}(f)$ is an extension of the trace $f|_{\Sigma_i}$ to a function of $\dot{C}^k(\mathbb{R})$.

These facts lead to the following form of Whitney's Theorem 2.47.

Theorem 2.56. *Let S be a closed subset of $\mathbb{R}$ containing at least $k+2$ points.*

(a) *A function $f : S \to \mathbb{R}$ belongs to the trace space $\dot{C}^k\big|_S$ if and only if every trace $f|_\Sigma$ to a $(k+1)$-point subset $\Sigma \subset S$ admits an extension to a function F_Σ of $\dot{C}^k(\mathbb{R})$ such that for every $x \in S$,*

$$|F_\Sigma - F_{\Sigma'}|_{C^k(\mathbb{R})} \to 0 \quad \text{as} \quad \Sigma, \Sigma' \to \{x\}.$$

(b) *There exists a linear continuous extension operator $\mathcal{E}_k^S : \dot{C}^k\big|_S \to \dot{C}^k(\mathbb{R})$ such that for every 1-majorant ω,*

$$\mathcal{E}_k^S\left(\dot{C}^{k,\omega}\big|_S\right) \subset \dot{C}^{k,\omega}(\mathbb{R}).$$

In the case of the space $C_b^k(\mathbb{R})$ the family $\{F_\Sigma\}_{\Sigma \in S^{\langle k+2 \rangle}}$ may be taken to be uniformly bounded on $\mathbb{R}$. This fact and condition (a) allow us to apply the Arcelá-Ascoli theorem which leads to the following version of Whitney's result.

Theorem 2.57. *There exists a constant $c(k) > 0$ such that a function $f : S \to \mathbb{R}$ belongs to $C_b^k(\mathbb{R})|_S$ if and only if some family of $C_b^k(\mathbb{R})$ extensions $\{F_\Sigma\}_{\Sigma \in S^{\langle k+2 \rangle}}$ is contained in a compact subset of a ball in $C_b^k(\mathbb{R})$ centered at 0 and of radius at most $c(k) \sup_S |f|$.*

2.4.3 Finiteness and linearity

Now we are ready to introduce these general concepts and formulate the basic conjectures. An extensive study of this area will be presented in Chapter 10 (Volume II).

Let X be a space of smooth functions on $\mathbb{R}^n$ equipped with a seminorm $|\cdot|$, e.g., $\dot{C}^{k,\omega}(\mathbb{R}^n)$ or $\dot{\Lambda}^{k,\omega}(\mathbb{R}^n)$, or norm $f \mapsto \sup_{\mathbb{R}^n} |f| + |f|_X$. For a closed subset S

2.4. Trace and extension problems for univariate C^k functions

of $\mathbb{R}^n$ we, for brevity, denote by $X(S)$ the trace space $X|_S$, and by $|\cdot|_{X(S)}$ the corresponding trace seminorm, i.e.,

$$|f|_{X(S)} := \inf\{|g|_X \,;\, g|_S = f\}.$$

Given an integer $N > 1$ and a closed subset $S \subset \mathbb{R}^n$ containing at most N points, we now introduce a functional $\delta_N(\cdot\,;\, S\,;\, X)$ defined on functions $f : S \to \mathbb{R}$ by

$$\delta_N(f\,;\, S\,;\, X) := \sup\{|f|_{X(\Sigma)} \,;\, \Sigma \in S^{\langle N \rangle}\}. \tag{2.98}$$

Using this we introduce the basic

Definition 2.58. (a) The space X possesses the *finiteness property* on a closed subset $S \subset \mathbb{R}^n$ if for some integer N the following is true.

A function $f : S \to \mathbb{R}$ belongs to the trace space $X(S)$ if

$$\delta_N(f\,;\, S\,;\, X) < \infty.$$

Moreover, the equivalence

$$\delta_N(f\,;\, S\,;\, X) \approx |f|_{X(S)}, \tag{2.99}$$

holds with constants independent of f.

(b) X has the *(uniform) finiteness property* if (a) holds for every closed S with the constants in (2.99) independent of f and S.

The minimal N here is denoted by $\mathcal{F}_S(X)$ and is called the *local finiteness constant* of X.

In turn, the *(uniform) finiteness constant* of X is given by

$$\mathcal{F}(X) := \sup\{\mathcal{F}_S(X)\,;\, S \text{ is closed}\}. \tag{2.100}$$

In these terms, Theorem 2.55 (a) simply asserts that

$$\mathcal{F}(\dot{C}^{k,\omega}(\mathbb{R})) = k + 2. \tag{2.101}$$

To formulate in the same manner Theorems 2.56 and 2.57 we need a new concept.

Definition 2.59. The space X possesses the *strong finiteness property* if the following is true:

A function $f : S \to \mathbb{R}$ belongs to the trace space $X(S)$ if

(a) for the seminormed space X, some family $\{F_\Sigma\}_{\Sigma \in S^{\langle N \rangle}}$ satisfies

$$|F_\Sigma - F_{\Sigma'}|_X \to 0 \quad \text{if} \quad \Sigma, \Sigma' \to \{x\}$$

for every $x \in S$;

(b) for the normed space X, some family $\{F_\Sigma\}_{\Sigma \in S^{\langle N \rangle}}$ is contained in a compact subset of a ball in X centered at 0 of radius at most $c(n,k)\sup_S |f|$.

The minimal N of this definition is said to be the *strong finiteness constant* of X denoted by $\mathcal{F}_{\mathrm{str}}(X)$.

In these terms, Theorem 2.56 (a) simply states that

$$\mathcal{F}_{\mathrm{str}}\bigl(\dot{C}^k(\mathbb{R})\bigr) = k+1. \qquad (2.102)$$

One more illustration of these properties yields the Shevchuk theorem [She-1992, Thm.11.1] generalizing Theorem 2.55 to the space $C^k \dot\Lambda^{s,\omega}(\mathbb{R})$.

Let us recall that this space (over $\mathbb{R}^n$) is defined by the seminorm

$$|f|_{C^k \dot\Lambda^{s,\omega}(\mathbb{R}^n)} := \sup_{|\alpha|=k}\sup_{t>0}\frac{\omega_s(t\,;D^\alpha f)}{\omega(t)},$$

where ω is an s-majorant, see Section 2.1 for details.

The aforementioned theorem asserts that $f : X \to \mathbb{R}$ belongs to the trace space $C^k \dot\Lambda^{s,\omega}(\mathbb{R})\big|_S$ if

$$\sup\bigl\{|f[\Sigma]| \cdot \Psi_{k,s}(\omega,\Sigma)\,;\, \Sigma \in S^{\langle k+s+1\rangle}\bigr\} \qquad (2.103)$$

is finite. Here $\Psi_{k,s}(\omega,\Sigma)$ is a rather complicated function of ω and Σ for which an analytic expression can be given only in few cases; for instance,

$$\Psi_{k,1}(\omega,\Sigma) \approx \omega^*(\mathrm{diam}\,\Sigma) \qquad (2.104)$$

so that (2.103) implies Theorem 2.55 (a).

Another interesting case concerns the Besov space $\dot{B}^\sigma_\infty(\mathbb{R})$, see (2.23). For noninteger σ we have $\dot{B}^\sigma_\infty(\mathbb{R}) = \dot{C}^{k,\omega}(\mathbb{R})$ with $\omega(t):=t^{\sigma-k}$, $0 < \sigma - k < 1$; hence the corresponding result is given by (2.104) with $\omega(t):=t^\sigma$, $k := \lfloor\sigma\rfloor$ and $s=1$. However, for integer σ we obtain a new result; in this case, $k = \sigma - 1$, $s = 2$ and $\omega(t) := t$, and for this k, s, ω,

$$\Psi_{k,s}(\omega,\Sigma) \approx \frac{\mathrm{diam}\,\Sigma}{\bigl|\log\bigl(\frac{\mathrm{diam}\,\Sigma_1}{\mathrm{diam}\,\Sigma_0}\bigr)\bigr| + 1},$$

where $\Sigma := \{x_0 < x_1 < \cdots < x_{k+s}\}$ and $\Sigma_i := \{x_i, \ldots, x_{k+i}\}$, $i = 0, 1$.

Hence, $f : S \to \mathbb{R}$ belongs to the trace space $\dot{B}^\sigma_\infty(\mathbb{R})\big|_S$, where σ is an integer, if (and only if)

$$\sup\biggl\{|f[\Sigma]|\frac{\mathrm{diam}\,\Sigma}{\bigl|\log\bigl(\frac{\mathrm{diam}\,\Sigma_1}{\mathrm{diam}\,\Sigma_0}\bigr)\bigr|+1}\,;\,\Sigma \in S^{\langle \sigma\rangle}\biggr\}$$

is finite.

For $\sigma = 1$ this result was firstly proved by A. Jonsson [Jon-1980].

2.4. Trace and extension problems for univariate C^k functions

The final case concerns the space $\dot{\Lambda}^{2,\omega}(\mathbb{R})$ with arbitrary 2-majorant ω. In this setting,

$$\Psi_{0,2}(\omega, \Sigma) \approx (\operatorname{diam} \Sigma) \left(\int_{\delta(\Sigma)}^{\operatorname{diam} \Sigma} \frac{\omega(t)}{t^2} \, dt \right)^{-1},$$

where $\delta(\Sigma) := \min\{x_{i+1} - x_i \, ; \, i = 0, 1\}$ provided that $\Sigma := \{x_0 < x_1 < x_2\}$.

Then (2.103) yields the trace description result established independently by Shvartsman [Shv-1982] and Dziadik and Shevchuk [DShe-1983].

We now clarify the Shevchuk theorem using the finiteness constants. The result is naturally divided into two parts the first of which simply asserts that

$$\mathcal{F}\left(C^k \dot{\Lambda}^{s,\omega}(\mathbb{R})\right) = k + s + 1. \tag{2.105}$$

The second, the computational part of the theorem, evaluates the trace norm of $f : \Sigma \to \mathbb{R}$ for subsets Σ of cardinality $k + s + 1$. The Shevchuk result yields for such subsets

$$|f|_{C^k \dot{\Lambda}^{s,\omega}(\mathbb{R})|_\Sigma} \approx \Psi_{k,s}(\omega, \Sigma).$$

2.4.4 Basic conjectures

The previous discussion forms a basis for several conjectures for multivariate functions.

We begin with

Finiteness Conjecture 2.60. (a) *The space $C^k \dot{\Lambda}^{s,\omega}(\mathbb{R}^n)$ has the finiteness property.*

(b) *The space $\dot{C}^k(\mathbb{R}^n)$ has the strong finiteness property.*

Linearity Conjecture 2.61. *Let $X(\mathbb{R}^n)$ be one of the spaces of the previous conjecture. Then for every closed subset $S \subset \mathbb{R}^n$ there is a linear bounded extension operator from the trace space $X|_S$ into $X(\mathbb{R}^n)$. In other words, the trace space admits a simultaneous extension.*

Until the early 1980s the only result was Whitney's Theorem 2.56, confirming the first conjecture for $n = 1$. Since then the situation has essentially improved due to the works of Yu. Brudnyi, Shvartsman, Ch. Fefferman, Bierstone and P. Milman. We discuss their results in the final part of the book; for now we only single out the next striking fact established by Shvartsman [Shv-1987]:

$$\mathcal{F}\left(\dot{\Lambda}^{2,\omega}(\mathbb{R}^n)\right) = 3 \cdot 2^{n-1}. \tag{2.106}$$

Remark 2.62. (a) Unlike the one-dimensional case the extension operator for $n > 1$ depends on the majorant ω. This follows from the following fact:

There is a subset $S \subset \mathbb{R}^2$ such that $\dot{C}_u^1(\mathbb{R}^2)\big|_S$ does not admit a simultaneous extension to $\dot{C}_u^1(\mathbb{R}^2)$.

This result due to Yu. Brudnyi and P. Shvartsman [BSh-1999] will be also discussed in the final part of the book.

(b) "Interpolating" the Shevchuk and Shvartsman results (2.105) and (2.106) one may guess that
$$\mathcal{F}(C^k\dot{\Lambda}^{s,\omega}(\mathbb{R}^n)) = (k+s+1)2^{\gamma(k,n)}$$
where $\gamma(k,n) := -1 + \binom{n+s-2}{s-1}$. Note that the binomial coefficient is the dimension of the space of $(s-1)$-homogeneous polynomial in $x_1, \ldots, x_n$.

This quantitative version of Conjecture 2.60 (a) seems to be extremely difficult to prove (if true).

(c) The local extension constant $\mathcal{F}_S(X)$ for "massive" subsets $S \subset \mathbb{R}^n$ may have a strictly lesser rate of growth in n. For example, let S be n-regular, i.e., for some $c_0 > 0$ and $r_0 > 0$ and all cubes $Q_r(x)$ with $r \le r_0$ and $x \in S$,
$$|S \cap Q_r(x)| \ge c_0 r^n.$$
Then the extension theorem from the paper by Yu. Brudnyi [Br-1970b] and the Remez–Shnirelman finiteness principle, see Appendix B of Chapter 1, yield for these S and power k-majorants ω the inequality
$$\mathcal{F}_S(\dot{C}^{k,\omega}(\mathbb{R}^n)) \le 1 + \dim \mathcal{P}_{k,n} = 1 + \binom{n+k}{k}.$$
A general result of this kind concerning the space $\dot{\Lambda}^{k,\omega}(\mathbb{R}^n)$ and the so-called Markov subsets S will be presented in Section 9.2 (Volume II).

(d) In all these formulations we may replace seminorms of X by its normed counterpart defined by the norm $\|f\|_{C_b(\mathbb{R}^n)} + |f|_X$. The known results cited above remain to be true for the normed spaces as well.

2.5 Restricted main problem for some classes of domains in $\mathbb{R}^n$

2.5.1 Quasiconvex domains

We derive from Theorems 2.13 and 2.19 trace and extension results for C^k functions on open sets. We restrict our consideration to the space $\dot{C}^{k,\omega}(G)$, where hereafter $G \subset \mathbb{R}^n$ stands for a domain. Let us recall that this space is defined by the seminorm
$$|f|_{C^{k,\omega}(G)} := \sum_{|\alpha|=k} \sup_{[x,y] \subset G} \frac{|D^\alpha f(x) - D^\alpha f(y)|}{\omega(\|x-y\|)}, \qquad (2.107)$$

2.5. Restricted main problem for some classes of domains in $\mathbb{R}^n$

where ω is a nondecreasing nonnegative function on $\mathbb{R}_+$ such that $\omega(t)/t$ is nonincreasing and $\omega(0+) = 0$.

In particular, ω is *subadditive*, i.e., for all $t_1, t_2 > 0$,

$$\omega(t_1 + t_2) \leq \omega(t_1) + \omega(t_2); \tag{2.108}$$

indeed, $\frac{\omega(t_1+t_2)}{t_1+t_2} \leq \frac{\omega(t_i)}{t_i}$, and multiplying this by t_i and summing on i, one gets (2.108).

In the case under consideration, the Restricted Main Problem is formulated as follows:

Problem. (a) *Under what geometric characteristics of a domain G is it true that*

$$\dot{C}^{k,\omega}(G) = \dot{C}^{k,\omega}(\mathbb{R}^n)\Big|_G ? \tag{2.109}$$

(b) *Under what conditions on G is the equality (2.109) a linear isomorphism, i.e., there exists a linear bounded extension operator from the trace space into $\dot{C}^{k,\omega}(\mathbb{R}^n)$?*

We present here only a partial solution to the problem, namely, we describe a subclass of domains which satisfies (2.109). Since the proof of the result will be based on Whitney's extension theorems, the answer to question (b) is automatically affirmative.

To introduce the subclass in question, we need several geometric notions.

Suppose that $\gamma : [0,1] \to \mathbb{R}^n$ is a *curve* (continuous map). Recall that its *length* is defined by

$$\ell(\gamma) := \sup \sum_{i=0}^{n} \|\gamma(t_{i+1}) - \gamma(t_i)\|, \tag{2.110}$$

where the supremum is taken over all *polygonal lines* with associated segments $[\gamma(t_i), \gamma(t_{i+1})]$, $0 \leq t_1 < \cdots < t_n \leq 1$, and all n.

A curve γ is *rectifiable* if $\ell(\gamma) < \infty$.

Further, fixing ω one defines the ω-*length* of a *polygonal line* $\gamma : [0,1] \to \mathbb{R}^n$ with the segments $[\gamma(t_i), \gamma(t_{i+1})]$, where $t_0 = 0 < t_1 < \cdots < t_n = 1$, by

$$\ell_\omega(\gamma) := \sum_{i=0}^{n-1} \omega(\|\gamma(t_{i+1}) - \gamma(t_i)\|). \tag{2.111}$$

Now let G be a domain in $\mathbb{R}^n$. One defines the *(geodesic) ω-distance* in G setting for $x, y \in G$,

$$d_\omega(x, y) := \inf \ell_\omega(\gamma), \tag{2.112}$$

where γ runs over all polygonal lines γ joining x, y in G (i.e., $\gamma : [0,1] \to G$ and $\gamma(0) = x$, $\gamma(1) = y$).

Due to (2.108), $d_\omega(x,y) = \omega(\|x-y\|)$ if the segment $[x,y]$ lies in G and therefore d_ω is a metric on $\mathbb{R}^n$. It will be shown below, see the proof of Proposition 2.68, that d_ω is a metric in G as well.

Definition 2.63. A domain $G \subset \mathbb{R}^n$ is said to be (C,ω)-convex, if for every two points x,y in G there is a polygonal line γ joining them in G and such that

$$\ell_\omega(\gamma) \leq C\omega(\|x-y\|).$$

The optimal C will be denoted by $C_\omega(G)$.

If $\omega(t) = t$, $t \in \mathbb{R}_+$, then such a domain is said to be C-*quasiconvex* (*quasiconvex* if the constant does not matter).

A 1-quasiconvex domain is clearly convex. Moreover, for C close to 1, C-quasiconvex domains inherit some basic features of convex ones. In particular, such a domain is contractible if $C < \frac{\pi}{2}$, and *simply connected*[7], if $C < \frac{\pi\sqrt{2}}{2}$, see Gromov [Gr-2000, pp. 11–12]. On the other hand, the geometry of (C,ω)-convex domains may be rather complicated even for C close to 1. As an elementary example we consider a planar domain G_λ, $\lambda \geq 1$, in the open unit disk $\mathbb{D}$ centered at $(0,0)$ given by $G_\lambda := \{(x,y) \in \mathbb{D}\,;\, x \leq 0 \text{ or } |y| > x^\lambda\}$:

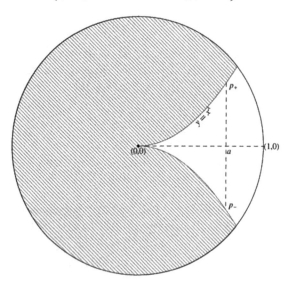

Figure 2.1: An example of a (C,ω)-convex domain.

This domain is (C,ω)-convex for $\omega(t) := t^{\frac{1}{\lambda}}$. In fact, it suffices to check the condition of Definition 2.63 only near the vertex $(0,0)$ of the cusp of G_λ.

---
[7] i.e., every closed curve into the domain is homotopic to a constant map.

2.5. Restricted main problem for some classes of domains in $\mathbb{R}^n$

In this case, fix a sufficiently small $a > 0$ and let γ_a be a polygonal line with the endpoints $p_\pm := (a, \pm a^\lambda)$ and with very small segments inscribed in the arc $|y| = x^\lambda$, $0 \leq x \leq a$. Then

$$\ell_\omega(\gamma_a) \approx 2 \int_0^a \left(1 + \left(\frac{d}{dt}\omega(t^\lambda)\right)^2\right)^{\frac{1}{2}} dt \approx a,$$

while

$$\omega(\|p_+ - p_-\|) = (2a^\lambda)^{\frac{1}{\lambda}} \approx a$$

and the required condition follows.

G_λ has a single inner cusp at 0; it is easy to construct a (C, ω)-domain with infinitely many inner and outer cusps.

Now we formulate and prove the basic result of this section following, in essence, Whitney's paper [Wh-1934d], where a simultaneous extension for $\dot{C}_u^k(G)$ in the case of a quasiconvex bounded domain is studied.

Theorem 2.64. *Assume that $G \subset \mathbb{R}^n$ is a (C, ω)-convex domain. Then for every k equality (2.109) holds up to equivalence of seminorms. Moreover, there exists a linear bounded extension operator from $\dot{C}^{k,\omega}(G)$ into $\dot{C}^{k,\omega}(\mathbb{R}^n)$.*

Proof. Since the restriction of $f \in \dot{C}^{k,\omega}(\mathbb{R}^n)$ to G obviously belongs to $\dot{C}^{k,\omega}(G)$, the embedding

$$\dot{C}^{k,\omega}(\mathbb{R}^n)\big|_G \subset \dot{C}^{k,\omega}(G) \tag{2.113}$$

holds with the embedding constant 1.

We prove the converse embedding using the Whitney extension operator E_k^G, see Section 2.2. To this end, given a function $f \in \dot{C}^{k,\omega}(G)$, we introduce a k-jet $\vec{f} := \{f_\alpha\}_{|\alpha| \leq k}$ defined on G by

$$f_\alpha := D^\alpha f.$$

We claim that under the assumption on G, the continuous extension of k-jet to $\overline{G}$ satisfies the Whitney–Glaeser Theorem 2.19. This can be done if we prove for the reduced remainders

$$r_\alpha(\vec{f}\,;\,x,y) := \left| D^\alpha f(x) - \sum_{|\beta| \leq k-|\alpha|} \frac{D^{\alpha+\beta} f(y)}{\beta!}(x-y)^\beta \right| \bigg/ \|x-y\|^{k-|\alpha|}$$

the inequality

$$r_\alpha(\vec{f}\,;\,x,y) \leq C|f|_{\dot{C}^{k,\omega}(G)}\,\omega(\|x-y\|) \tag{2.114}$$

with C depending only on k, n and the constant $C_\omega(G)$.

To establish (2.114), set

$$R_\alpha(f\,;\,x,y) := D^\alpha(f - T_y^k)(x),$$

where $T_y^k f$ is the k-th Taylor polynomial of f at point y.

Lemma 2.65. Let $f \in \dot{C}^{k,\omega}(G)$ and $[x, y] \subset G$. Then for every α and $z \in \mathbb{R}^n$,

$$R_\alpha(f\,;x,y) \le c|f|_{C^{k,\omega}(G)}\big(\|z-x\|^{k-|\alpha|} + \|z-y\|^{k-|\alpha|}\big) \cdot \omega(\|x-y\|), \quad (2.115)$$

where $c = c(k, n)$.

Proof. Applying the Taylor formula with the remainder in the integral form we represent $R_\alpha(f\,;x,y)$ with $|\alpha| < k$ in the form

$$R_\alpha(f\,;x,y) = (k - |\alpha|) \sum_{|\beta|=k-|\alpha|} \frac{(x-y)^\beta}{\beta!} \int_0^1 (1-t)^{k-|\alpha|-1}$$
$$\times \big[D^{\alpha+\beta}f(y + t(x-y)) - D^{\alpha+\beta}f(y)\big]\,dt.$$

Since $y + t(x-y) \in [x,y] \subset G$, the expression in the square brackets does not exceed in absolute value $|f|_{C^{k,\omega}(G)} \cdot \omega(\|x-y\|)$. Hence, we have

$$\big|R_\alpha(f\,;x,y)\big| \le c(k,n)|f|_{C^{k,\omega}(G)}\,\|x-y\|^{k-|\alpha|}\,\omega(\|x-y\|).$$

This estimate is true also for $|\alpha| = k$, because in this case

$$\big|R_\alpha(f\,;x,y)\big| = \big|D^\alpha f(x) - D^\alpha f(y)\big|.$$

Apply now these estimates to the identity

$$D^\alpha\big[T_x^k f - T_y^k f\big](z) = \sum_{|\beta|\le k-|\alpha|} \frac{1}{\beta!} R_{\alpha+\beta}(f\,;x,y)(z-x)^\beta.$$

This gives for the absolute value of the right-hand side the upper bound

$$c(k,n)|f|_{C^{k,\omega}(G)}\omega(\|x-y\|) \sum_{|\beta|\le k-|\alpha|} \frac{1}{\beta!} \|x-y\|^{k-|\alpha+\beta|}\|z-x\|^{|\beta|}.$$

The sum is clearly bounded by $c(k)\big(\|x-z\|^{k-|\alpha|} + \|y-z\|^{k-|\alpha|}\big)$, and the result follows. □

Going back to the proof of (2.114), choose a polygonal curve $\gamma : [0,1] \to G$ connecting points $x, y \in G$ and such that

$$\ell_\omega(\gamma) \le 2C_\omega(G)\omega(\|x-y\|).$$

Let $\{y_i\}_{0 \le i \le m} \subset \gamma([0,1])$ be the vertices of γ, so that $y_0 = x$, $y_m = y$, $[y_i, y_{i+1}] \subset G$; then

$$\ell_\omega(\gamma) \ge \sum_{i=0}^{m-1} \omega(\|y_{i+1} - y_i\|). \quad (2.116)$$

2.5. Restricted main problem for some classes of domains in $\mathbb{R}^n$

Applying Lemma 2.65 we get

$$|R_\alpha(f;x,y)| = |D^\alpha(T_x^k f - T_y^k f)(x)| \le \sum_{i=0}^{m-1} |D^\alpha(T_{y_{i+1}}^k f - T_{y_i}^k f)(x)|$$

$$\le C(k,n)|f|_{C^{k,\omega}(G)} \sum_{i=0}^{m-1} \omega(\|y_{i+1} - y_i\|)(\|x - y_i\|^{k-|\alpha|} + \|x - y_{i+1}\|^{k-|\alpha|}).$$

Setting $\mu := \sup_i (\|x - y_i\|^{k-|\alpha|} + \|x - y_{i+1}\|^{k-|\alpha|})$ we therefore have

$$|R_\alpha(f;x,y)| \le c(k,n)\mu |f|_{C^{k,\omega}(G)} \ell_\omega(\gamma). \tag{2.117}$$

We now show that

$$\mu \le c(k)[C_\omega(G)\|x-y\|]^{k-|\alpha|}. \tag{2.118}$$

Together with (2.117) and (2.112) this would yield the desired inequality (2.114) with

$$\widetilde{c} = c(k,n)c(k)C_\omega(G)^k. \tag{2.119}$$

To establish (2.118) we use

Lemma 2.66. *Let $\gamma : [0,1] \to \mathbb{R}^n$ be a curve with endpoints $\gamma(0) := x$ and $\gamma(1) := y$. Then*

$$\ell(\gamma) \le \frac{\|x-y\|}{\omega(\|x-y\|)} \ell_\omega(\gamma).$$

Proof. Since the map $t \mapsto \ell(\gamma|_{[0,t]})$ is uniformly continuous on $[0,1]$, one can choose $\{y_i\} \in \gamma([0,1])$ such that $\|y_{i+1} - y_i\| := \|\gamma(t_{i+1}) - \gamma(t_i)\| \le \|x-y\|$ for all i. Since the function $t \mapsto \frac{t}{\omega(t)}$ is nondecreasing, we then have

$$\|\gamma(t_{i+1}) - \gamma(t_i)\| \le \frac{\|x-y\|}{\omega(\|x-y\|)} \omega(\|\gamma(t_{i+1}) - \gamma(t_i)\|).$$

Summing these inequalities over i and taking the corresponding supremum we then obtain

$$\ell(\gamma) \le \frac{\|x-y\|}{\omega(\|x-y\|)} \sum \omega(\|\gamma(t_{i+1}) - \gamma(t_i)\|) \le \frac{\|x-y\|}{\omega(\|x-y\|)} \ell_\omega(\gamma).$$

The result is thus established. $\square$

To apply the result just proved to estimate μ, we write

$$\|x - y_i\|^{k-|\alpha|} + \|x - y_{i+1}\|^{k-|\alpha|} \le c(k)(\|x - y_i\| + \|x - y_{i+1}\|)^{k-|\alpha|}$$
$$\le 2^k c(k) \ell(\gamma)^{k-|\alpha|}.$$

Estimating here $\ell(\gamma)$ by Lemma 2.66 and recalling the definition of μ we obtain the inequality

$$\mu \leq 2^k c(k) \left(\frac{\|x-y\|}{\omega(\|x-y\|)} \ell_\omega(\gamma) \right)^{k-|\alpha|}$$

$$\leq 2^k c(k) \left(\frac{\|x-y\|}{\omega(\|x-y\|)} \cdot C_\omega(G)\omega(\|x-y\|) \right)^{k-|\alpha|}.$$

This proves (2.118) and (2.114).

Now (2.114) with $|\alpha|=k$ implies that

$$|D^\alpha f(x) - D^\alpha f(y)| \leq O(1) |f|_{C^{k,\omega}(G)} \, \omega(\|x-y\|).$$

Hence all derivatives $D^\alpha f$ with $|\alpha|=k$ are uniformly continuous on G. This and (2.114) with $|\alpha|=k-1$ then imply the uniform continuity of all $D^\alpha f$ with $|\alpha|=k-1$. Proceeding this way we prove that $D^\alpha f$ for all $|\alpha| \leq k$ are uniformly continuous on G. Extending these derivatives by continuity to the closure $\overline{G}$ we obtain the k-jet satisfying condition (2.114) on the closed set.

It remains to apply the Whitney–Glaeser Theorem 2.19 to complete the proof. □

Remark 2.67. Estimates (2.114) and (2.119) yield the upper bound $\tilde{c}(k,n)C_\omega(G)^k$ for the extension operator of Theorem 2.64. It can be shown that the estimate is asymptotically sharp (up to a constant) as $C_\omega(G)$ tends to 1.

Now we return to the Restricted Main Problem, see (2.109). With this in mind, we reformulate Theorem 2.64 as follows.

Let $\mathcal{D}_{\text{ext}}(k,\omega)$ denote the class of domains in $\mathbb{R}^n$ satisfying (2.109). Then Theorem 2.64 asserts that ω-convexity of a domain is sufficient for being a member of $\mathcal{D}_{\text{ext}}(k,\omega)$.

This condition is, in general, not necessary. Nevertheless, this is the case for a few situations presented below.

Proposition 2.68. *A domain $G \subset \mathbb{R}^n$ belongs to the class $\mathcal{D}_{\text{ext}}(0,\omega)$ if and only if G is ω-convex.*

Proof. According to Theorem 2.64 only ω-convexity of G should be established. For this purpose we first show that the distance d_ω introduced by (2.112) is a metric in G. It is easy to verify the metric axioms for d_ω, except that of finiteness (i.e., $d_\omega < \infty$). To show the finiteness of d_ω, we fix $x \in G$ and denote by $G(x)$ the set of all $y \in G$ that can be connected with x by a polygonal curve $\gamma : [0,1] \to G$ with $\ell_\omega(\gamma) < \infty$. Finiteness of d_ω means that $G = G(x)$ for every $x \in G$. But $G(x)$ is clearly open and it can be easily checked that it is also closed in G. Since G is connected, the result follows.

Fix now a point $x \in G$ and define a function $f_0 : G \to \mathbb{R}$ by

$$f_0(y) := d_\omega(x,y), \quad y \in G. \tag{2.120}$$

2.5. Restricted main problem for some classes of domains in $\mathbb{R}^n$

Let $[y, y'] \subset G$. The triangle inequality and the definition of d_ω then yield

$$|f_0(y) - f_0(y')| \leq d_\omega(y, y') = \omega(\|y - y'\|).$$

Hence $f_0 \in \dot{C}^{0,\omega}(G)$ and $|f|_{C^{0,\omega}(G)} \leq 1$. Since (2.109) with $k = 0$ holds for the domain G, every $f \in \dot{C}^{0,\omega}(G)$ admits an extension $F \in \dot{C}^{0,\omega}(\mathbb{R}^n)$ such that

$$|F|_{C^{0,\omega}(\mathbb{R}^n)} \leq C|f|_{C^{0,\omega}(G)}$$

with a constant C independent of f. Applying this to the function f_0 and denoting its extension by F_0, we obtain, for $x, y \in G$, the inequality

$$d_\omega(x, y) = |f_0(x) - f_0(y)| = |F_0(x) - F_0(y)|$$
$$\leq |F_0|_{C^{0,\omega}(\mathbb{R}^n)} \omega(\|x - y\|) \leq C\omega(\|x - y\|).$$

That is to say, for every pair $x, y \in G$, there is a curve $\gamma : [0, 1] \to G$ connecting them and such that $\ell_\omega(\gamma) \leq C\omega(\|x - y\|)$.

Hence, G is (C, ω)-convex. □

The next result, proved by Zobin [Zo-1999], shows that ω-convexity is a necessary condition for the space $\dot{C}^{k,1}(G)$, i.e., $\dot{C}^{k,\omega}(G)$ with $\omega(t) = t$, $t > 0$, and some class of domains $G \subset \mathbb{R}^2$ to satisfy (2.109). Let us recall that a ω-convex domain with this ω is called quasiconvex.

Theorem 2.69. (a) *Assume that G is a simply connected planar domain belonging to $\mathcal{D}_{\text{ext}}(k, \omega)$ with $\omega(t) := t$, $t > 0$, i.e.,*

$$\dot{C}^{k,1}(G) = \dot{C}^{k,1}(\mathbb{R}^2)\big|_G. \tag{2.121}$$

Then G is quasiconvex.

(b) *The same is true for finitely connected bounded planar domains of the class $\mathcal{D}_{\text{ext}}(k, \omega)$ with $\omega(t) := t$, $t > 0$.*

(c) *There is a bounded planar domain of this class which is not quasiconvex.*

Proof. We present the proof of assertion (a) and outline the proof of assertion (b) referring to Zobin's paper [Zo-1998] for a counterexample proving assertion (c).

(a) By the assumption, for every $f \in \dot{C}^{k,1}(G)$ there exists an extension $F \in \dot{C}^{k,1}(\mathbb{R}^2)$ such that

$$|F|_{C^{k,1}(\mathbb{R}^2)} \leq A|f|_{C^{k,1}(G)} \tag{2.122}$$

with A independent of f. To derive from here the quasiconvexity of G, we construct, for every $x \in G$, a function $f_x \in \dot{C}^{k,1}(G)$ such that its seminorm satisfies

$$|f_x|_{C^{k,1}(G)} \leq \sqrt{2}, \tag{2.123}$$

and, moreover, for every $y \in G$, the inequality

$$d_\omega(x,y) \leq \sum_{|\alpha|=k} |D^\alpha f_x(x) - D^\alpha f_x(y)| \qquad (2.124)$$

is true with $\omega(t) := t$.

We first show that this function can be used to complete the proof. Let $F_x \in \dot{C}^{k,1}(\mathbb{R}^2)$ be an extension of f_x satisfying (2.122). Then (2.122) and (2.124) yield the inequality

$$d_\omega(x,y) \leq \sum_{|\alpha|=k} |D^\alpha F_x(x) - D^\alpha F_x(y)| \leq |F_x|_{C^{k,1}(\mathbb{R}^2)} \|x-y\|$$

$$\leq A|f_x|_{C^{k,1}(G)} \|x-y\| \leq \sqrt{2}\,A\|x-y\|.$$

By the definition of d_ω (with the linear ω) this estimate means that every pair $x, y \in G$ is connected within G by a curve of length at most $C(k)A\|x-y\|$, and the result follows.

It remains to find the function $f_x \in \dot{C}^{k,1}(G)$. For this purpose, we denote by $\mathcal{L}(x,y)$ the set of piecewise linear curves in G connecting x and y and having segments parallel either to the x_1-axis (x_1-segments) or to the x_2-axis (x_2-segments). For such a curve, say β, one sets

$$\ell_i(\beta) := \text{sum of lengths of } x_i\text{-segments of } \beta.$$

Now define a pseudometric $d_G^i : G \times G \to \mathbb{R}_+$ by setting

$$d_G^i(x,y) := \inf\{\ell_i(\beta)\,;\, \beta \in \mathcal{L}(x,y)\}.$$

For the segment $[x,y] \subset G$ the Pythagoras theorem gives

$$\|x-y\| = \sqrt{d_G^1(x,y)^2 + d_G^2(x,y)^2}.$$

This implies the inequality

$$d_G \geq \frac{1}{\sqrt{2}}(d_G^1 + d_G^2), \qquad (2.125)$$

where one sets $d_G := d_\omega$ for $\omega(t) := t$.

It is important that for simply connected planar domains inequality (2.125) can be conversed. This is based on the following geometric fact.

Lemma 2.70. *Let β_1, β_2 be piecewise linear curves from the class $\mathcal{L}(x,y)$. Then there exists a piecewise linear curve β from this class whose length satisfies the inequality*

$$\ell(\beta) \leq \ell_1(\beta_1) + \ell_2(\beta_2). \qquad (2.126)$$

2.5. Restricted main problem for some classes of domains in $\mathbb{R}^n$

We postpone the proof of the lemma to the final part of the derivation. For now, we only conclude from (2.126) that

$$\inf \ell(\beta) \leq d_G^1(x,y) + d_G^2(x,y),$$

where the infimum is taken over all $\beta \in \mathcal{L}(x,y)$. Evidently, this infimum equals $d_G(x,y)$, see (2.112) with $\omega(t) = t$. Hence, the desired converse inequality

$$d_G \leq d_G^1 + d_G^2 \qquad (2.127)$$

is true.

Define now the required function $f_x : G \to \mathbb{R}$, where $x \in G$, by

$$f_x(y) := \frac{1}{(k-1)!} \int_\beta \sum_{i=1}^2 d_G^i(z,x)(y_i - z_i)^{k-1}\, dz, \qquad (2.128)$$

where β is an arbitrary curve from $\mathcal{L}(x,y)$.

This f_x is well defined, i.e., the integral does not depend on the choice of β. By the Green formula for such Lipschitz differential 1-forms, see, e.g., Federer [Fe-1969], and simply connectedness of G, this claim will follow from the identity

$$\frac{\partial}{\partial z_1} P_2(z) = \frac{\partial}{\partial z_2} P_1(z), \quad z \in G, \qquad (2.129)$$

where one sets $P_i(z) := d_G^i(z,x)(y_i - z_i)^{k-1}$.

Further, by definition, the function $z \mapsto d_G^1(z,x)$ is constant on every interval parallel to the x_2-axis, and the similar assertion holds for d_G^2 with respect to x_1-axis. Hence, $\frac{\partial}{\partial z_1} P_2(x) = \frac{\partial}{\partial z_2} P_1(z) = 0$ and f_x is well defined.

The same argument works for the evaluation of mixed derivatives of f_x. To compute $D_i f_x \left(:= \frac{\partial}{\partial z_i} f_x \right)$, choose curves $\beta \in \mathcal{L}(x,y)$ and $\beta' \in \mathcal{L}(x, y+he_i)$, where $e_i = (\delta_j^i)_{j=1,2}$ and $h \neq 0$, and use them in (2.128) to obtain

$$\frac{1}{h}\left(f_x(y+he_i) - f_x(y)\right)$$
$$= \frac{1}{(k-1)!} \left\{ \int_\beta d_G^i(z,x) \cdot \frac{(y_i + h - z_i)^{k-1} - (y_i - z_i)^{k-1}}{h}\, dz_i \right.$$
$$\left. + \int_{[y, y+he_i]} \frac{d_G^i(z,x)(y_i - z_i)^{k-1}}{2}\, dz_i \right\}.$$

Then passing to the limit as $h \to 0$ we have

$$D_i f_x(y) = \frac{1}{(k-2)!} \int_\beta d_G^i(z,x)(y_i - z_i)^{k-2} dz_i \quad \text{for} \quad k \geq 2$$

and, moreover,

$$D_i f_x(y) = d_G^i(x,y) \quad \text{for} \quad k = 1.$$

Continuing this way we obtain for all $y \in G$ and $0 < m < k$,

$$D_i^m f_x(y) = \frac{1}{(k-1-m)!} \int_\beta d_G^i(z,x)(y_i - z_i)^{k-1-m} \, dy_i$$

and, for $m = k$,

$$D_i^k f_x(y) = d_G^i(x,y). \tag{2.130}$$

In addition to this, we also have

$$D^\alpha f_x = 0 \quad \text{for} \quad |\alpha| := \alpha_1 + \alpha_2 \leq k \quad \text{and} \quad \alpha_1, \alpha_2 > 0. \tag{2.131}$$

This follows from the fact that the derivatives in (2.129) are zero. These calculations and (2.125) lead to the following estimate:

$$|f_x|_{C^{k,1}(G)} := \sum_{i=1}^2 \sup_{[y,y'] \subset G} \frac{|D_i^k f_x(y) - D_i^k f_x(y')|}{\|y - y'\|}$$

$$\leq \sum_{i=1}^2 \sup_{[y,y'] \subset G} \frac{d_G^i(y,y')}{\|y - y'\|} \leq \sqrt{2}.$$

Hence, f_x satisfies inequality (2.123). Moreover, inequality (2.127) and equalities (2.130) and (2.131) yield

$$d_G(x,y) \leq d_G^1(x,y) + d_G^2(x,y) = \sum_{|\alpha|=k} |D^\alpha f_x(x) - D^\alpha f_x(y)|,$$

and inequality (2.124) follows as well.

It remains to prove Lemma 2.70. We derive it from an elementary geometric result whose proof is postponed to the end of this subsection. In its formulation we use the following notion. A non-self-intersecting polygonal curve in $\mathbb{R}^2$ with segments parallel to the coordinate axes is called a *bolt*. If β is a *closed bolt* (i.e., homeomorphic to a circle) then $\mathcal{U}(\beta)$ designates the open (simply connected) polygon with boundary β. Clearly, all angles of this polygon equal $\pm \frac{\pi}{2}$.

Geometric Lemma. *Let x, y be distinct points of a closed bolt β, and let β_1, β_2 be bolts connecting x and y and such that*

$$\beta = \beta_1 \cup \beta_2 \quad \text{and} \quad \beta_1 \cap \beta_2 = \{x, y\}.$$

Then there is a bolt β_3 with endpoints x, y such that $\beta_3 \subset \overline{\mathcal{U}(\beta)}$ and its length satisfies the inequality

$$\ell(\beta_3) \leq \ell_1(\beta_1) + \ell_2(\beta_2).$$

2.5. Restricted main problem for some classes of domains in $\mathbb{R}^n$

Let us recall that $\ell_i(\beta)$ is the sum of the x_i-segments lengths of β.

Now we apply the lemma to prove inequality (2.126). Let $\beta_1, \beta_2 \in \mathcal{L}(x,y)$; we should find $\beta_3 \in \mathcal{L}(x,y)$ whose length satisfies the inequality

$$\ell(\beta_3) \leq \ell_1(\beta_1) + \ell_2(\beta_2).$$

Obviously, the piecewise linear curves β_1 and β_2 intersect in a finite number of points, say, $x = x^0, x^1, \ldots, x^k = y$. Then the parts β_1^i, β_2^i of β_1, β_2 from x^i to x^{i+1} form a closed bolt which is denoted by β^i. By the Geometric Lemma, there is a bolt β_3^i in $\overline{\mathcal{U}(\beta^i)}$ such that β_3^i connects x^i and x^{i+1} and such that

$$\ell(\beta_3^i) \leq \ell_1(\beta_1^i) + \ell_2(\beta_2^i).$$

Then $\beta_3 := \bigcup_{i=0}^{k-1} \beta_3^i$ belongs to $\mathcal{L}(x,y)$; summing the above inequalities over i we establish inequality (2.126) for β_3.

Part (a) of Theorem 2.69 has been proved.

(b) Now let G be a bounded finitely connected domain in $\mathbb{R}^2$ such that

$$\dot{C}^{k,1}(G) = \dot{C}^{k,1}(\mathbb{R}^2)\big|_G, \quad k \geq 1.$$

We must prove that G is quasiconvex. To this end, we first note that for *bounded* domains quasiconvexity is equivalent to *local quasiconvexity*. The latter means that there is a constant $\delta > 0$ such that

$$\sup\left\{\frac{d_G(x,y)}{\|x-y\|}\,;\, x,y \in G,\, \|x-y\| < \delta\right\} < \infty.$$

This fact may be established by a standard compactness argument.

Secondly, we may choose a sufficiently small $\varepsilon > 0$ such that every open disk $B_\varepsilon(x)$ in $\mathbb{R}^2$ of radius ε intersects at most one connected component of the complement $G^c := \mathbb{R}\setminus G$ and such that the intersection of this disk with G is simply connected or empty. This can be done, since G^c contains a finite number of connected components.

Finally, for a nonempty simply connected set $G \cap B_\varepsilon(x)$ where $x \in G$, we may, as in part (a), find a function $F \in \dot{C}^{k,1}(G)$ such that for every $y \in B_\varepsilon(x)$,

$$\min\left\{d_G(x,y), \frac{\varepsilon}{2}\right\} \leq 2 \sum_{|\alpha|=k} |D^\alpha F(x) - D^\alpha F(y)|$$

and, moreover, for some constant $C = C(k)$,

$$|F|_{\dot{C}^{k,1}(G)} \leq C(k).$$

As in the proof of part (a) we conclude from those inequalities that for every $y \in B_\varepsilon(x)$,

$$d_G(x,y) \leq C(\varepsilon, k)\|x-y\|.$$

Notice that if $B_\varepsilon(x)$ does not intersect G^c, then $d_G(x,y)$ simply equals $\|x-y\|$ and the above inequality trivially holds as well.

By definition, this means that G is locally (and therefore globally) quasiconvex, as required.

Proof of Geometric Lemma. (Igonin and Yanishevski [IYa-1998]) Let β be a closed bolt with fixed points $x, y \in \beta$. The parts of β joining x and y are denoted by β_1 and β_2. We must find a bolt β_3 joining x and y inside the closure of the domain $U(\beta)$ bounded by β, and such that its length satisfies

$$\ell(\beta) \leq \ell_1(\beta_1) + \ell_2(\beta_2).$$

Recall that $\ell_i(\beta)$ is the sum of lengths of x_i-segments [8] in β.

This is trivial if the closed polygon $\overline{U(\beta)}$ is a rectangle (put, e.g., $\beta_3 := \beta_1$). In the sequel we assume that β has at least six segments. In particular, β has at least one *outer angle* vertex meaning that the angle of $\overline{U(\beta)}$ associated to this vertex equals $-\frac{\pi}{2}$ and we call a vertex of $\overline{U(\beta)}$ an *inner angle* one if the associated angle equals $\frac{\pi}{2}$.

Now let $\lambda_i = [v_i, v_{i+1}]$, $i=1,2,3$, be the sequential segments of β, i.e., vertex v_{i+1} is common for λ_i and λ_{i+1}, $i=1,2$. We say that $(\lambda_1, \lambda_2, \lambda_3)$ forms a *marked triple* if the following holds:

(a) v_2 and v_3 are inner angle vertices of β.

(b) The orthogonal projection of v_1 onto the straight line containing λ_3 lies in λ_3.

The simplest example of this object is any triple of sides of a rectangle.

Claim. For every segment λ of β there exists a marked triple $(\lambda_1, \lambda_2, \lambda_3) \subset \beta$ such that $\lambda \neq \lambda_i$, $i=1,2,3$.

We use induction on the number of segments. If $U(\beta)$ is a rectangle, the result is evident. Otherwise, this number is at least six and therefore β has an outer vertex. Let v be such a vertex with the maximal vertical coordinate. We extend the horizontal segment ending at v inside $\overline{U(\beta)}$ and denote by w the closest to v vertex of the intersection of this extension and β. The intersection denoted by $[w, w']$ is a horizontal segment or a point, where $w' = w$ in the latter case. Then the points v, w divide β into two bolts denoted by β_1 and β_2. Since v has the maximal x_2-coordinate, one of the polygons $P_i := \overline{U}(\beta_i \cup [v,w])$, $i=1,2$, say, P_1, is a rectangle, see Figure 2.2.

First, let the chosen segment λ belong to β_2, and λ_i be a side of P_1 that does not contain $[v,w]$, $i \in \{1,2,3\}$. If $w' \neq w$, then $(\lambda_1, \lambda_2, \lambda_3)$ is the required marked triple. Otherwise, we extend the side ending at w (say, λ_3) down to the closest

---

[8] In the sequel, an x_i-segment is said to be *horizontal* if $i=1$ and *vertical* if $i=2$.

2.5. Restricted main problem for some classes of domains in $\mathbb{R}^n$

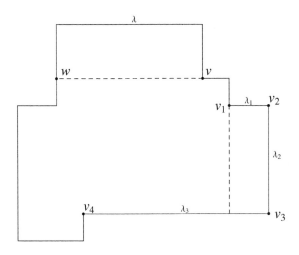

Figure 2.2: Marked triple $(\lambda_1, \lambda_2, \lambda_3)$.

to w vertex w' of β. The triple $(\lambda_1, \lambda_2, \lambda_3 \cup [w, w'])$ is then the required marked triple.

Now, let λ belong to β_1. The closed bolt $\widetilde{\beta}_2 := \beta_2 \cup [v, w]$ has clearly less segments than β. By the induction hypothesis there exists a marked triple $(\lambda_1, \lambda_2, \lambda_3)$ of the bolt $\widetilde{\beta}_2$ which does not contain its segment determined by $[v, w]$. If $w' \neq w$, then $(\lambda_1, \lambda_2, \lambda_3)$ is a marked triple of β which does not contain λ. Otherwise, we extend the segment containing w, say λ_3, up to the closest to w vertex w'' of β. Then the triple $(\lambda_1, \lambda_2, \lambda_3 \cup [w, w''])$ is required and the result follows.

Now we complete the proof of the lemma using again induction on the number of segments. Let β have at least six segments and $\lambda_i = [v_i, v_{i+1}]$, $i = 1, 2, 3$, be a marked triple of β which does not contain the given point x. We assume, for definiteness, that $\ell(\lambda_1) \leq \ell(\lambda_3)$ and denote by v_1' the projection of v_1 onto λ_3. Then we replace $\bigcup_{i=1}^{3} \lambda_i$ by $[v_1, v_1'] \cup [v_1', v_4]$ and denote by β' the closed bolt obtained in this way. Further, we consider two cases, see Figure 2.3 below.

Assume first that the point $y \in \beta'$, and denote by β_1' and β_2' the parts of β' determined by x and y. Clearly, β' has less segments than β. Therefore, by the induction hypothesis, there exists a bolt $\beta_3 \subset \overline{U(\beta')}$ joining x and y and such that

$$\ell(\beta_3) \leq \ell_1(\beta_1') + \ell_2(\beta_2').$$

Since $\ell_i(\beta_i') \leq \ell_i(\beta_i)$ and $\overline{U(\beta')} \subset \overline{U(\beta)}$, β_3 is the required curve.

Secondly, let $y \notin \beta'$ and therefore belongs to $\lambda_1 \cup \lambda_2 \cup [v_3, v_1']$. Without loss of generality we may assume that $v_1 \in \beta_1$, $v_1' \in \beta_2$ and λ_1 is horizontal. The points x and v_1' divide the bolt β' into parts denoted by β_1' and β_2'. We choose this notation so that in a neighborhood of x the polygonal curve β_i' coincides with β_i, $i=1,2$. It is easily seen that $\ell_i(\beta_i') \leq \ell_i(\beta_i)$, $i = 1, 2$.

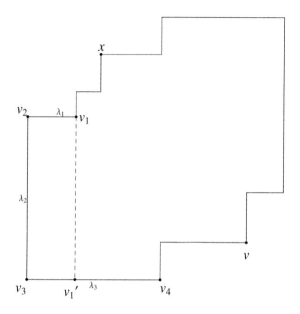

Figure 2.3: First case: $y = v$. Second case: $y = v_2$ or v_3.

We construct the required bolt β_3 by the juxtaposition of two bolts β_3' and β_3'' defined as follows. The former joins v_1 and x, contained in $\overline{U(\beta')}$ and satisfies the inequality

$$\ell(\beta_3') \leq \ell_1(\beta_1') + \ell_2(\beta_2').$$

This exists by the induction hypothesis.

Further, β_3'' is the smallest bolt joining v_1 and y inside the closed rectangle with vertices v_1, v_2, v_3, v_1'. In accordance with this definition

$$\ell(\beta_3'') \leq \ell_1(\beta_1) - \ell_1(\beta_1') + \ell_2(\beta_2) - \ell_2(\beta_2').$$

Then $\beta_2 := \beta_3' \cup \beta_3''$ satisfies the required inequality

$$\ell(\beta_3) \leq \ell_1(\beta_1) + \ell_2(\beta_2),$$

joins x and y and is contained in the polygon $\overline{U(\beta)}$, as required. □

The proof of Theorem 2.69 is complete. □

2.5.2 Lipschitz domains

For this subclass of the class of quasiconvex domains Theorem 2.64 may be strengthened in two respects. Firstly, the extension theorem can be proved for all spaces $\dot{\Lambda}^{k,\omega}(G)$ (without any restriction on ω) and the same may be done for $C^k \dot{\Lambda}^{s,\omega}(G)$. Secondly, the corresponding extension operator is "universal" in the sense that it extends simultaneously the family of all Lipschitz spaces $\dot{\Lambda}^{k,\omega}(G)$ with $1 \leq k \leq N$, where N is arbitrary (but the seminorm for $k = n$ tends to infinity as $N \to \infty$).

We discuss an approach leading to these results and formulate the main facts obtained. The corresponding proofs will only be outlined.

We introduce the required class of domains using the following more general concept, see, e.g., Stein [Ste-1970, Sec. VI.3.3].

Definition 2.71. *Let $G \subset \mathbb{R}^n$ be an open set. We say that its boundary ∂G is minimally smooth if there exist numbers ε, $L > 0$, integers $N, M \geq 1$ and a sequence of open sets G_j, $j \in \mathbb{N}$, such that*

(a) *the ε-neighborhoods $(G_j)_\varepsilon$, $j \in \mathbb{N}$, cover the boundary ∂G;*

(b) *the order of the cover $\{G_j\}_{j \in \mathbb{N}}$ is at most M;*

(c) *for every j, there is a special Lipschitz domain D_j with the Lipschitz constant at most L such that*
$$G_j \cap G = G_j \cap D_j.$$

A domain $D \subset \mathbb{R}^n$ is *special Lipschitz with the Lipschitz constant $L(D)$* if D is a subgraph of a Lipschitz function f with $L(f) \leq L(D)$ defined on a hypersubspace of $\mathbb{R}^n$. In other words, in a suitable coordinate system, f is defined on $\mathbb{R}^{n-1}$ and $D := \{(x, x_n) \in \mathbb{R}^{n-1} \times \mathbb{R}; \, x_n < f(x)\}$.

Definition 2.72. *A domain $G \subset \mathbb{R}^n$ is said to be Lipschitz if G is bounded and has a minimally smooth boundary as in Definition 2.71.*

This class of Lipschitz domains is denoted by $\mathcal{L}ip$.

The boundary of such a domain is *locally Lipschitz*, i.e., can be locally represented as the graph of a Lipschitz function defined on some open ball of $\mathbb{R}^{n-1}$. This follows from a simple fact noticed by Gagliardo [Ga-1958]. For its formulation we denote by $K := K(e, r, \varphi)$ the cone with axis in the direction $e \in \mathbb{S}^{n-1}$, apex at 0, height $r > 0$ and angle $0 < \varphi < \frac{\pi}{2}$; that is,
$$K := \{x \in \mathbb{R}^n; \, \|x\| \cos \varphi < x \cdot e < r\}.$$

Recall that $x \cdot y$ and $\|x\|$ stand for the standard scalar product and norm in $\mathbb{R}^n$, respectively.

Proposition 2.73. *A domain $G \subset \mathbb{R}^n$ belongs to the class $\mathcal{L}ip$ if and only if, for some integer N, there are open balls B_i centered at ∂G and cones K_i, $1 \leq i \leq N$, such that*

(a) $\{B_i\}$ covers ∂G;

(b) *for every* $1 \leq i \leq N$,
$$(B_i \cap \partial G) + K_i \subset G.$$

Proof. Because of boundedness of G we may take only a finite number of sets G_i and D_i in Definition 2.71 and then note that for every special domain D_i there is a cone K_i satisfying $D_i + K \subset D_i$. □

The locally Lipschitz structure of ∂G may be derived from here as follows. The surface of the shifted cone $x + K_i$ is clearly the graph of a Lipschitz function f_x with Lipschitz constant $c = c(K_i) > 0$ which is defined on a ball in $\mathbb{R}^{n+1}$. Extending f_x to $\mathbb{R}^{n-1}$ with the same Lipschitz constant and preserving the notation f_x for the extended function, we represent the part $B_i \cap \partial G$ of the boundary as the graph of the function $f := \sup\{f_x\,;\, x \in B_i \cap \partial G\}$. But f is Lipschitz as the supremum of Lipschitz functions with uniformly bounded Lipschitz constants.

From Proposition 2.73 we also easily derive that
$$\mathcal{L}ip \subset \mathcal{C} \tag{2.132}$$

where $\mathcal{C}$ stands for the class of quasiconcave domains (see Definition 2.63). However, there are bounded quasiconcave domains which are not Lipschitz. As an example we point out the domain G_λ following Definition 2.63.

Now we discuss the aforementioned extension results for $\dot{\Lambda}^{k,\omega}(G)$ with $G \in \mathcal{L}ip$ and an arbitrary k-majorant ω. To describe the main features of the extension method we begin with the case of a special Lipschitz domain $G \subset \mathbb{R}^n$. So let
$$G := \{(x, x_n) \in \mathbb{R}^{n-1} \times \mathbb{R}\,;\, x_n < \varphi(x)\}, \tag{2.133}$$

where $\varphi \in \mathrm{Lip}(\mathbb{R}^{n-1})$.

We use for such G the so-called "mirror reflection" method originated by L. Lichtenstein [Lich-1929] and developed to full extent by Hestens [Hes-1941]. This gives the following result (a special case of the Calderón–Stein theorem, see [Ste-1970, Sec. VI.3.2.1]).

Theorem 2.74. *Let G be given by* (2.133) *and $N \geq 1$ be a fixed integer. There is a linear extension operator $T_N : C_b(G) \to C_b(\mathbb{R}^n)$ such that for every $k \in \{0, 1, \ldots, N\}$,*
$$T_N(C_b^k(G)) \subset C_b^k(\mathbb{R}^n)$$
and, moreover, for some $c = c(N, G) > 0$,
$$\max_{|\alpha|=k} \sup_{\mathbb{R}^n} |D^\alpha T_N f| \leq c \max_{|\alpha|=k} \sup_G |D^\alpha f|. \tag{2.134}$$

For G an open half-space given by (2.133) with $\varphi = 0$, the Lichtenstein–Hestens operator T_N for $(x, x_n) \in \mathbb{R}^{n-1} \times (0, \infty)$ is given by
$$T_N f(x, x_n) := \int_0^\infty f(x - \lambda x_n)\psi(\lambda)d\lambda, \tag{2.135}$$

where $\psi : (0, \infty) \to \mathbb{R}$ is a compactly supported function with moments

$$\int_0^\infty \lambda^k \psi(\lambda) d\lambda = \begin{cases} 1 & \text{if } k = 0, \\ 0 & \text{if } 1 \leq k \leq N. \end{cases}$$

For a detailed proof for special Lipschitz domains see the above cited section of Stein's book.

Remark 2.75. In his proof, Stein exploits a function ψ decreasing at infinity faster than any power λ^{-N}, $N \geq 1$, and having zero moments for all $k = 1, 2, \ldots$. Using such a ψ he obtains a linear extension operator T_∞ such that

$$T_\infty\big(C_b^\infty(G)\big) \subset T_\infty\big(C_b^\infty(\mathbb{R}^n)\big).$$

Here $C_b^\infty(G) := \bigcap_{k \geq 0} C_b^k(G)$.

The second basic ingredient of the extension method is the equivalence of the so-called K-functional of the pair $C_b(G), \dot{C}_b^s(G)$ and the s-modulus of continuity $\omega_s(\cdot\,;f)_G$. Let us recall the definition of the former concept introduced by J. Peetre [Peet-1963].

Let X_0, X_1 be a *Banach couple*, i.e., a pair of Banach spaces linearly and continuously embedded into a topological vector space. This embedding allows us to define the sums $x_0 + x_1$ with $x_i \in X_i$.

Definition 2.76. The K-functional of a Banach couple X_0, X_1 is a function on $(X_0 + X_1) \times (0, +\infty)$ given by

$$K(t\,;x\,;X_0,X_1) := \inf_{x=x_0+x_1} \big\{\|x_0\|_{X_0} + t\|x_1\|_{X_1}\big\}. \qquad (2.136)$$

We will also use a modification of this definition when one or both spaces X_i are only seminormed (but complete).

The function K is a Banach norm on $X_0 + X_1$ for every fixed $t > 0$, and is a nondecreasing concave function in t, see, e.g., Bergh and Löfström [BLo-1976, Sec. 3.3.1].

Now let $X_0 := C_b(G)$ and $X_1 := \dot{C}_b^s(G)$; in this case, we use the notation $K_s(t\,;f\,;G)$, i.e.,

$$K_s(t\,;f\,;G) := \inf_{f=f_0+f_1} \Big\{\sup_G |f| + t \max_{|\alpha|=s} \sup_G |D^\alpha f|\Big\}.$$

The following result was proved in an equivalent form by J. L. Lions and Peetre [LP-1964] for $G := \mathbb{R}^n$. The proof presented below is based on the method of the paper [Br-1964] by Yu. Brudnyi.

Theorem 2.77. *Let $G \subset \mathbb{R}^n$ be a special Lipschitz domain and $f \in C_b(G)$. Then, for every $s \geq 1$,*

$$K_s(t^s\,;f\,;G) \approx \omega_s(t\,;f)_G, \quad t > 0, \qquad (2.137)$$

with constants of equivalence independent of f and t.

Proof. If G is given by (2.133), then there exists an *infinite* cone $K_\infty := K(e_0, \infty, \psi)$ such that

$$G + K_\infty \subset G. \tag{2.138}$$

Now let e be a unit vector in K_∞ and $t > 0$. We define an operator $S_t(e)$ on functions $f \in C_b(G)$ by

$$S_t(e)f := \frac{1}{t^s} \int_0^t \cdots \int_0^t \sum_{j=1}^s c(j) f(\cdot + j\tau e) d\tau_1 \ldots d\tau_s, \tag{2.139}$$

where $c(j) := (-1)^{s-j} \binom{s}{j}$ and $\tau := \tau_1 + \cdots + \tau_s$. Due to (2.138) and the choice of e, this operator acts continuously from $C_b(G)$ into $C_b(G)$.

Transforming the multiple integral, we rewrite (2.139) as

$$S_t(e)f(x) = \int_0^1 \left(f(x) - \Delta_{t\tau e}^s f(x)\right) \vartheta(\tau) d\tau, \tag{2.140}$$

where ϑ is a certain bounded function.

Moreover, $S_t(e)f$ is clearly s-times continuously differentiable on G and its derivative D_e^s in direction e equals

$$D_e^s S_t(e)f = \left(\sum_{j=1}^s d(j) \Delta_{js^{-1}te}^s f\right) \Big/ t^s, \tag{2.141}$$

where $d(j) := (sj)^{-1} c(j)$.

Choosing now $N := \binom{h+s+1}{s}$ vectors e_j in the cone K_∞ we set

$$S_t := \prod_{1 \leq j \leq N} S_t(e_j).$$

Then $S_t : C_b(G) \to C_b(G)$ and

$$f - S_t f = \sum_{j=1}^N \left(\prod_{i \geq j+1} S_t(e_i)\right) \int_0^1 (\Delta_{t\tau e_j}^s f) \vartheta(\tau) d\tau,$$

where the product is assumed to be the identity operator if $j + 1 > N$.

This immediately leads to the inequality

$$\sup_G |f - S_t f| \leq O(1) \omega_s(t; f)_G. \tag{2.142}$$

Now we use identity (2.43) for mixed derivatives D^α with $|\alpha| = s$ which gives

$$D^\alpha = \sum_{j=1}^N a_j D_{e_j}^s,$$

2.5. Restricted main problem for some classes of domains in $\mathbb{R}^n$

where a_j are some constants. Using this and (2.141) we have

$$|D^\alpha(S_t f)(x)| \leq \sum_{j=1}^{N} |a_j| \left| \left(\prod_{i \neq j} S_t(e_i)\right) D_{e_j}^s (S_t(e_j)f)(x) \right|$$

$$\leq O(1) t^{-s} \sup_{1 \leq i \leq s} \left| \Delta_{is^{-1}te}^s f(x) \right| \leq O(1) t^{-s} \omega_s(t; f)_G.$$

Combining this with (2.142) we obtain

$$K_s(t^s; f; G) \leq \sup_G |f - S_t f| + t^s \max_{|\alpha|=s} \sup_G |D^\alpha S_t f| \leq O(1) \omega_s(t; f)_G.$$

The converse inequality is easy. In fact, if $f = f_0 + f_1$ and $f_1 \in \dot{C}_b^s(G)$, then the known properties of ω_s imply

$$\omega_s(t; f)_G \leq \omega_s(t; f_0)_G + \omega_s(t; f_1)_G \leq 2^s \sup_G |f_0| + t^s \max_{|\alpha|=s} \sup_G |D^\alpha f_1|.$$

Taking here the infimum over all decompositions $f = f_0 + f_1$ we get the desired inequality

$$\omega_s(t; f)_G \leq O(1) K_s(t^s; f; G). \tag{2.143}$$

$\square$

Remark 2.78. (a) In fact, we have shown that the couple $(C_b(G), \dot{C}_b^s(G))$ is *K-linearized* meaning that there is the family $\{S_t\}_{t>0}$ of *linear* operators realizing nearly optimal decompositions for $K_s(t^s; f; G)$:

$$f = (1 - S_t)f + S_t f.$$

In general, a Banach couple (X_0, X_1) is said to be *K-linearized* if there exist families of linear operators $\{S_t^i\}_{t>0}$ acting from $X_0 + X_1$ into X_i, $i = 0, 1$, such that $S_t^0 + S_t^1 = Id_{X_0+X_1}$ for each $t > 0$, and for every $x \in X_0 + X_1$ and $t > 0$,

$$K(x; t; X_0, X_1) \approx \|S_t^0 x\|_{X_0} + t \|S_t^1 x\|_{X_1}$$

with constants independent of x and t.

(b) Inequality (2.143) is true for *any* open set G. The converse inequality does not hold even for domains with good extension properties, e.g., for bounded uniform domains, see Definition 2.81 below.

As a consequence of Theorems 2.74 and 2.77 we now obtain the desired extension result.

Theorem 2.79. Let $G \subset \mathbb{R}^n$ be a special Lipschitz domain and $N \geq 1$ be an integer. Then the linear extension operator T_N of Theorem 2.74 maps every homogeneous Lipschitz space $\dot{\Lambda}^{s,\omega}(G)$ into $\dot{\Lambda}^{s,\omega}(\mathbb{R}^n)$, $1 \leq s \leq N$. Moreover, for some $c = c(N, G) > 0$ and every $f \in \dot{\Lambda}^{s,\omega}(G)$,

$$|T_N f|_{\Lambda^{s,\omega}(\mathbb{R}^n)} \leq c|f|_{\Lambda^{s,\omega}(G)}.$$

Proof. Let $f \in C_b(G)$ and $f = f_0 + f_1$ be such a decomposition that

$$\sup_G |f_0| + t^s \max_{|\alpha|=s} \sup_G |D^\alpha f_1| \leq O(1)\omega_s(t; f)_G, \qquad (2.144)$$

see Theorem 2.77. By the definition of ω_s, we have for the function $T_N f = T_N f_0 + T_N f_1$,

$$\omega_s(t; T_N f) \leq O(1)\left\{\sup_{\mathbb{R}^n} |T_N f_0| + t^s \max_{|\alpha|=s} \sup_{\mathbb{R}^n} |D^\alpha T_N f_1|\right\}.$$

Estimating the right-hand side first by (2.134) and then by (2.144) we get

$$\omega_s(t; T_N f) \leq O(1)\omega_s(t; f)_G.$$

Dividing this by an s-majorant ω and taking the supremum over $t > 0$ we obtain the required inequality

$$|T_N f|_{\Lambda^{s,\omega}(\mathbb{R}^n)} \leq O(1)|f|_{\Lambda^{s,\omega}(G)}. \qquad \square$$

Remark 2.80. In other words, this theorem asserts that for some $c = c(N, G)$ and every $f \in C_b(G)$, $t > 0$ and $1 \leq s \leq N$,

$$\omega_s(t; T_N f) \leq c\omega(t; f)_G.$$

It is highly probable that this result is true for $N = \infty$.

The situation with Lipschitz domains is more complicated. According to the Calderón–Stein theorem, the extension operator analogous to that of Theorem 2.74 exists in this case even for open sets with minimally smooth boundaries, see [Ste-1970, Sec. VI.3.1]. However, inequality (2.134) is replaced by

$$\max_{|\alpha|=s} \sup_{\mathbb{R}^n} |D^\alpha T_N f| \leq O(1)\left\{\max_{|\alpha|=s} \sup_G |D^\alpha f| + \sup_G |f|\right\}, \qquad (2.145)$$

where $0 \leq s \leq N$. We might eliminate the second term here by assuming that T_N preserves polynomials of degree $N - 1$. Applying, in this case, (2.145) to the function $f - p$, where p is a polynomial of degree $s - 1$, and taking the infimum over all such p, we replace the second summand in (2.145) by

$$E_s(G; f) := \inf\left\{\sup_G |f - p| \,;\, p \in \mathcal{P}_{s-1,n}\right\}.$$

This, in turn, is bounded above, for a bounded Lipschitz domain G, by $O(1)(\operatorname{diam} G)^s \max_{|\alpha|=s} \sup_G |D^\alpha f|$, see, e.g., the paper [BrH-1970] by Bramble and Hilbert. Therefore, (2.145) can be rewritten as

$$\max_{|\alpha|=s} \sup_{\mathbb{R}^n} |D^\alpha T_N f| \leq O(1) \max_{|\alpha|=s} \sup_G |D^\alpha f|.$$

Further, Theorem 2.77 is also true for Lipschitz domains, see the papers [Br-1976] by Yu. Brudnyi and [JSch-1977] by H. Johnen and K. Scherer.

Thus, the version of Theorem 2.79 for bounded Lipschitz domains would be true, if the extension operator T_N preserved polynomials of degree $N-1$. Unfortunately, the Calderón–Stein operator does not possess this property. Their construction is modified as required in the paper [Br-1980] by Yu. Brudnyi where Theorem 2.79 is proved also for bounded Lipschitz domains.

2.6 Sobolev spaces: selected trace and extension results

2.6.1 P. Jones' theorem and related results

The local polynomial approximation methods partially discussed in Section 2.3 have a much wider range of applications. In particular, they may be applied to the study of trace and extension problems for spaces of weakly differentiable functions such as Sobolev or Besov spaces. Unfortunately, the corresponding results remained in manuscript form or were published in almost inaccessible journals. In Chapter 9, we present several applications of this approach to the extension and trace problems for Lipschitz functions of higher order, while in the Comments to that chapter we formulate several results contained in the aforementioned sources. In particular, we prove the extension theorem for the space $\dot{\Lambda}^{k,\omega}(G)$, where ω is a quasipower majorant and $G \subset \mathbb{R}^n$ is a so-called *uniform domain*. The class $\mathcal{U}$ of these domains plays a considerable role in Analysis; for now we only formulate the corresponding definition leaving a detailed discussion to Chapter 9.

Let $\gamma : [0,1] \to \mathbb{R}^n$ be a rectifiable curve with the endpoints x, y. A *(curvilinear) cone with axis γ and parameter $c > 0$* is the open set

$$Kn(\gamma; c) := \bigcup \{ B_{c\rho(z)}(z)\, ; \, z \in \gamma([0,1]) \}, \tag{2.146}$$

where

$$\rho(z) := \min\{\|x-z\|, \|y-z\|\}. \tag{2.147}$$

Such a cone is also defined for the case of $y = \infty$ (i.e., for γ being unbounded and locally rectifiable).

Definition 2.81. A domain $G \subset \mathbb{R}^n$ is said to be (c_0, c_1)-*uniform* if for every pair $x, y \in G$ satisfying

$$\|x - y\| \leq c_0$$

there is a curve γ joining x and y in G such that the length of γ satisfies

$$\ell(\gamma) \leq c_1 \|x - y\| \tag{2.148}$$

and for the associated cone it is true that

$$Kn(\gamma; c_1) \subset G. \tag{2.149}$$

Due to (2.148) G is locally quasiconvex, but Example 2.79 (c) below shows that the class $\mathcal{U}$ is a proper subclass of the class of locally quasiconvex domains.

The aforementioned extension result for $\dot{\Lambda}^{k,\omega}(G)$, see Theorem 9.51 of Volume II, and the argument of Theorem 2.32, see also Remark 2.33, lead for $p = \infty$ to the isomorphism

$$W_p^k(G) = W_p^k(\mathbb{R}^n)\big|_G, \tag{2.150}$$

provided that G is a uniform domain.

Let us recall that the Sobolev space $W_p^k(G)$ is defined by the norm

$$\|f\|_{W_p^k(G)} := \max_{|\alpha| \leq k} \|D^\alpha f\|_{L_p(G)}. \tag{2.151}$$

Here the mixed weak derivative $D^\alpha f$ is an L_p-function such that for every C^∞ function φ supported on a compact subset of G,

$$\int_G f \cdot D^\alpha \varphi \, dx = (-1)^{|\alpha|} \int_G D^\alpha f \cdot \varphi \, dx.$$

The following fundamental fact was due to P. Jones [Jon-1981].

Theorem 2.82. *Let $G \subset \mathbb{R}^n$ be a (c_0, c_1)-uniform domain. Then there is a linear continuous extension operator from $W_p^k(G)$ to $W_p^k(\mathbb{R}^n)$.*

The P. Jones theorem leads to the isomorphism (2.150) for all $1 \leq p \leq \infty$.

There are examples due to Maz'ya [Maz-1981] showing that the uniformity of G is not necessary for validity of this result. However, for planar simply connected domains and $kp = 2$ the uniformity of G is also necessary. This result was due to V. Gol'dshtein and Vodop'yanov for $k = 1$ and $p = 2$, see [GV-1980] and [GV-1981]; a generalization including the cases $k = 2$ and $p = 1$ was then proved by Christ [Chr-1984].

Theorem 2.83. *Let $G \subset \mathbb{R}^2$ be a simply connected domain. Then for $kp = 2$ equality (2.150) holds up to the equivalence of the seminorms if and only if G is a uniform domain.*

Proof. In view of Theorem 2.82, it suffices to prove that if (2.150) holds for $n = 2$ and $kp = 2$, then G is uniform. To establish this we need the following characteristics of uniform domains, see Gehring [Ge-1982, Thm. 3.6].

2.6. Sobolev spaces: selected trace and extension results

Proposition 2.84. *Assume that $G \subset \mathbb{R}^2$ is uniform. Then there is a constant $\lambda \geq 1$ such that for every $x \in \mathbb{R}^2$ and $r \in (0, \infty)$ the following holds:*

(a) *every pair of points of $G \cap B_r(x)$ can be joined in $G \cap \overline{B}_{\lambda r}(x)$;*

(b) *every pair of points of $G \setminus B_r(x)$ can be joined in $G \setminus \overline{B}_{\frac{r}{\lambda}}(x)$.*

Here "joined" means joined by a rectifiable curve lying within the specified set.

Assume now that (2.151) holds with the specified k and p but G is not uniform. Then according to Proposition 2.84 there are closed disks denoted by $D_r(x^0)$ and $D_R(x^0)$ with an arbitrary large ratio $\frac{R}{r}$ of their radii satisfying one of the following conditions:

(i) $G \cap D_R(x^0)$ contains two different connected components G_0, G_1 for which
$$D_r(x^0) \cap G_j \neq \emptyset, \quad j = 0, 1.$$

(ii) $G \setminus D_r(x^0)$ contains two different connected components G_0, G_1 for which
$$(\mathbb{R}^2 \setminus D_R(x^0)) \cap G_j \neq \emptyset, \quad j = 0, 1.$$

To derive a contradiction we need

Proposition 2.85. *Suppose that a domain G satisfies either (i) or (ii) with a fixed $r < \frac{R}{4}$. Suppose also that for every $f \in \dot{W}_p^k(G)$, where $kp = 2$, there exists an extension $F \in \dot{W}_p^k(\mathbb{R}^2)$ of f satisfying*

$$|F|_{W_p^k(\mathbb{R}^2)} \leq A|f|_{W_p^k(G)}. \tag{2.152}$$

Then the inequality

$$C\left(\log \frac{R}{r}\right)^{\frac{1}{p}} \leq A \tag{2.153}$$

holds with a numerical constant $C > 0$.

Before proving this, we derive the theorem from (2.153). If (2.150) holds for the domain G with $kp = 2$, then (2.152) does as well. But since G is non-uniform, one of the two conditions (i), (ii) holds for an arbitrary large $\frac{R}{r}$. This contradicts (2.153).

It remains to prove Proposition 2.85. We begin with the following auxiliary result.

Lemma 2.86. *Let $f \in C^\infty(G)$ and let γ be a rectifiable curve of length ℓ joining fixed points x and y in the domain G. Then for the Taylor polynomial $T_y^{k-1} f$ it is true that*

$$|f(x) - T_y^{k-1} f(x)| \leq C(k,n) \ell^{k - \frac{1}{p}} \left(\sum_{|\alpha|=k} \int_\gamma |D^\alpha f|^p ds \right)^{\frac{1}{p}}; \tag{2.154}$$

here $1 \leq p \leq \infty$.

Proof. It suffices to prove the inequality for a polygonal curve $\gamma \in G$. We apply to each segment $[a_i, a_{i+1}]$ of γ the inequality

$$\left|(T_u^{k-1}f - T_v^{k-1}f)(z)\right| \leq C(k,n)\bigl(\|u-z\| + \|z-v\|\bigr)^{k-1} \sum_{|\alpha|=k} \int_{[u,v]} |D^\alpha f|\,ds;$$

here $[u,v] \subset G$ and $z \in \mathbb{R}^n$. Its proof is a simple modification of the proof of Lemma 2.65 and can be left to the reader. Taking $u = a_i$, $v = a_{i+1}$ and summing the inequalities obtained, we get

$$\left|f(x) - T_y^{k-1}f(x)\right| \leq C(k,n) \sum_i \bigl(\|x-a_i\| + \|x-a_{i+1}\|\bigr)^{k-1} \sum_{|\alpha|=k} \int_{[a_i,a_{i+1}]} |D^\alpha f|\,ds.$$

Since $\|x - a_i\| + \|x - a_{i+1}\| \leq 2\ell$, the right-hand side does not exceed

$$2^{k-1} \cdot C(k,n) \ell^{k-1} \sum_{|\alpha|=k} \int_\gamma |D^\alpha f|\,ds.$$

Estimating the integral in the right-hand side by the Hölder inequality we obtain the desired result. □

Now let G_0, G_1 be domains in $\mathbb{R}^2$ satisfying the following condition: For every $t \in (r, R)$ the circle $C_t := \partial D_t(0)$ satisfies

$$G_j \cap C_t \neq \emptyset, \quad j = 0, 1. \tag{2.155}$$

Denote by $A_{r,R}$ the circular annulus $\{x \in \mathbb{R}^2\,;\, r < \|x\| < R\}$. In this setting, the following holds.

Lemma 2.87. *Let $f \in \dot{W}_p^k(\mathbb{R}^2)$ and $kp = 2$. Assume that*

$$f\big|_{G_j \cap A_{r,R}} = j, \quad j = 0, 1. \tag{2.156}$$

Then there is a constant $C > 0$ such that

$$|f|_{W_p^k(\mathbb{R}^2)} \geq C\left(\log \frac{R}{r}\right)^{\frac{1}{p}}.$$

Proof. Since $p < \infty$, the set $C_0^\infty(\mathbb{R}^2)$ is dense in $\dot{W}_p^k(\mathbb{R}^2)$, see, e.g., Maz'ya [Maz-1985], we can assume that $f \in C_0^\infty(\mathbb{R}^2)$. By the Fubini theorem,

$$\int_{A_{r,R}} \left(\sum_{|\alpha|=k} |D^\alpha f|^p\right) dx = \int_r^R \left(\sum_{|\alpha|=k} \int_{C_t} |D^\alpha f|^p\,ds\right) dt.$$

2.6. Sobolev spaces: selected trace and extension results

According to (2.155), there exist points $x_j(t) \in G_j \cap C_t$, $j = 0, 1$. Since $f = 0$ in the neighborhood of $x_0(t)$, the Taylor polynomial $T_{x_0(t)}^{k-1} f$ equals 0. Moreover, $f(x_1(t)) = 1$, see (2.156). Hence, by (2.154),

$$1 = \left| f(x_1(t)) - T_{x_0(t)}^{k-1} f(x_1(t)) \right|$$

$$\leq C_1 \ell(C_t)^{k-\frac{1}{p}} \left(\sum_{|\alpha|=k} \int_{C_t} |D^\alpha f|^p \, ds \right)^{\frac{1}{p}} = C_1 (2\pi t)^{\frac{1}{p}} \left(\sum_{|\alpha|=k} \int_{C_t} |D^\alpha f|^p ds \right)^{\frac{1}{p}}.$$

Combining this inequality with the previous identity we get

$$|f|_{W_p^k(\mathbb{R}^2)} \geq \left(\int_r^R \frac{1}{C_1^p 2\pi t} \, dt \right)^{\frac{1}{p}} = C_2 \left(\log \frac{R}{r} \right)^{\frac{1}{p}},$$

and the result follows. $\square$

Now we are ready to prove Proposition 2.85. Let $\varphi \in C_0^\infty(\mathbb{R}^2)$ be a test-function satisfying

$$\varphi|_{D_{\frac{1}{2}}(0)} = 1, \quad \operatorname{supp}(\varphi) \subset D_{\frac{3}{4}}(0).$$

Consider first the case of the connected component of $G \cap D_R(x^0)$ for which the corresponding G_0, G_1 satisfy condition (i) with $r < \frac{R}{4}$. Let us define a function $f : G \to \mathbb{R}$ by

$$f(x) := \begin{cases} \varphi\left(\dfrac{x - x^0}{R}\right) & \text{for } x \in G_1, \\ 0 & \text{for } x \in G \setminus G_1. \end{cases}$$

Since $\operatorname{supp} \varphi\left(\dfrac{\cdot - x^0}{R}\right) \subset D_{\frac{3}{4}R}(x^0)$, the function f is well defined and belongs to $C_0^\infty(G)$. By its definition and the equality $kp = 2$,

$$|f|_{W_p^k(G)} \leq \left| \varphi\left(\frac{\cdot - x^0}{R} \right) \right|_{W_p^k(\mathbb{R}^2)} = R^{\frac{2}{p}-k} |\varphi|_{W_p^k(\mathbb{R}^2)} = |\varphi|_{W_p^k(\mathbb{R}^2)}.$$

In the case of condition (ii), we define f as above but replace 0 by 1. Since $\varphi\left(\dfrac{x - x^0}{2r}\right) = 1$ for $x \in D_r(x^0)$, the function f is well defined and belongs to $C_0^\infty(G)$. In this case, we have a similar estimate:

$$|f|_{W_p^k(G)} = |f|_{W_p^k(G \setminus G_1)} = (2r)^{\frac{2}{p}-k} |\varphi|_{W_p^k(\mathbb{R}^2)} = |\varphi|_{W_p^k(\mathbb{R}^2)}.$$

Since $f \in \dot{W}_p^k(G)$, for its extension $F \in \dot{W}_p^k(\mathbb{R}^2)$ we get by (2.152)

$$|F|_{W_p^k(\mathbb{R}^2)} \leq A |\varphi|_{W_p^k(\mathbb{R}^2)}.$$

Applying now Lemma 2.87 to the function F and the annulus $x_0 + A_{r,\frac{R}{2}}$ in case (i), and the annulus $x_0 + A_{2r,R}$ in case (ii), we get

$$A|\varphi|_{W_p^k(\mathbb{R}^2)} \geq |F|_{W_p^k(\mathbb{R}^2)} \geq C\left(\log \frac{R}{2r}\right)^{\frac{1}{p}}.$$

This is clearly equivalent to the required inequality (2.153).

The proof of Proposition 2.85 is complete. □

2.6.2 Peetre's nonexistence theorem

We first formulate a special case of the trace theorems for Sobolev spaces. Let $x = (x', x_n) \subset \mathbb{R}_+^n$, where $x' \in \mathbb{R}^{n-1}$ and $x_n \geq 0$. Since C^∞ functions are dense in $W_p^k(\mathbb{R}_+^n)$ with $1 \leq p < \infty$, the trace operator $\mathrm{Tr} : f(x) \to f(x', 0)$ is defined on a dense subset of $W_p^k(\mathbb{R}_+^n)$ and can be continuously extended to the whole of this space.

The following classical result was due to Gagliardo [Ga-1957].

Theorem 2.88. *Tr is a linear continuous operator mapping $W_1^1(\mathbb{R}_+^n)$ onto $L_1(\mathbb{R}^{n-1})$. In other words,*

$$W_1^1(\mathbb{R}_+^n)\big|_{\mathbb{R}^{n-1}} = L_1(\mathbb{R}^{n-1}),$$

where the trace space is defined as the image of the trace operator.

Remark 2.89. Gagliardo's proof also gives the following result which will be used below.

If $f \in W_1^1(\mathbb{R}_+^n)$, then the function $x' \mapsto f(x', x_n)$ belongs to $L_1(\mathbb{R}^{n-1})$ for almost all x_n and the limit

$$f(x', 0) := \lim_{x_n \to 0} f(x', x_n) \quad (\text{in } L_1(\mathbb{R}^{n-1})) \tag{2.157}$$

exists and coincides with $\mathrm{Tr}\, f(x')$ almost everywhere on $\mathbb{R}^{n-1}$.

The extension operator used in the Gagliardo proof is nonlinear. The following result obtained by Peetre [Peet-1979] shows that the simultaneous extension problem is unsolvable in this case.

Theorem 2.90. *There is no linear continuous extension operator from $L_1(\mathbb{R}^{n-1})$ to $W_1^1(\mathbb{R}_+^n)$ for $n \geq 2$.*

Proof. Assume, on the contrary, that there is a linear continuous extension operator $E : L_1(\mathbb{R}^{n-1}) \to W_1^1(\mathbb{R}_+^n)$. As an element of $W_1^1(\mathbb{R}_+^n)$ the function Ef satisfies

$$\lim_{x_n \to \infty} Ef(\cdot, x_n) = 0 \quad (\text{convergence in } L_1(\mathbb{R}^{n-1})) \tag{2.158}$$

2.6. Sobolev spaces: selected trace and extension results

for every $f \in L_1(\mathbb{R}^{n-1})$, see, e.g., Maz'ya [Maz-1985]. This and (2.157) imply by integration by parts that

$$f = -\int_0^\infty (D_n Ef)(\cdot, x_n) dx_n, \tag{2.159}$$

where the integral is defined as the limit of $\int_\varepsilon^N \ldots dx_n$ as $\varepsilon \to 0$ and $N \to \infty$ (convergence in $L_1(\mathbb{R}^{n-1})$).

We use this representation to find a sequence of "smoothing" operators $\{E_j\}_{j\in\mathbb{Z}}$ acting from $L_1(\mathbb{R}^{n-1})$ to $W_1^1(\mathbb{R}^{n-1})$ whose properties are given by

Lemma 2.91. (a) *For every $f \in L_1(\mathbb{R}^{n-1})$*

$$f = \sum_j E_j f \quad (convergence \ in \ L_1(\mathbb{R}^{n-1})). \tag{2.160}$$

(b) *For some constant $C > 0$ and every f,*

$$\sum_j \|E_j f\|_{L_1(\mathbb{R}^{n-1})} \leq C\|f\|_{L_1(\mathbb{R}^{n-1})}; \tag{2.161}$$

$$\sum_j 2^j |E_j f|_{W_1^1(\mathbb{R}^{n-1})} \leq C\|f\|_{L_1(\mathbb{R}^{n-1})}. \tag{2.162}$$

Proof. We define the required sequence using test-functions $\chi_j : \mathbb{R}_+ \to \mathbb{R}$, $j \in \mathbb{Z}$, satisfying the conditions:

$$\operatorname{supp} \chi_j \subset S_j := [2^{j-2}, 2^{j+1}] \quad \text{and} \quad \sum_j \chi_j = 1; \tag{2.163}$$

$$\sup |\chi_j| \leq C \quad \text{and} \quad \sup |\chi_j'| \leq 2^{-j} C, \quad j \in \mathbb{Z}, \tag{2.164}$$

where $C > 0$ is a constant independent of j.

Then the required operator E_j is given for $f \in L_1(\mathbb{R}^{n-1})$ by

$$E_j f := -\int_{\mathbb{R}_+} \chi_j(x_n) D_n Ef(\cdot, x_n) dx_n. \tag{2.165}$$

The identity (2.160) follows from (2.159) and (2.163). The assertions in (b) are proved by the same argument; we derive only (2.162). Integrating by parts in (2.165) and applying (2.163) and (2.164), we obtain

$$|E_j f|_{W_1^1(\mathbb{R}^{n-1})} \leq C \cdot 2^{-j} \int_{S_j} \sum_{i=1}^{n-1} \|D_i Ef(\cdot, x_n)\|_{L_1(\mathbb{R}^{n-1})} \leq C \cdot 2^{-j} |Ef|_{W_1^1(\mathbb{R}^{n-1} \times S_j)}.$$

Since the family $\{\mathbb{R}^{n-1} \times S_j\}_{j\in\mathbb{Z}}$ covers $\mathbb{R}_+^n$ with order (multiplicity) ≤ 4, we obtain (2.162) by summing over j. $\square$

We further transform the sequence $\{E_j\}_{j\in\mathbb{Z}}$ into a sequence $\{\widehat{E}_j\}_{j\in\mathbb{Z}}$ of *translation invariant* linear operators on $L_1(\mathbb{R}^{n-1})$ satisfying (2.160)–(2.162). To accomplish this, we use the *invariant mean* $\mathcal{M}$ on the space $\mathcal{B}(\mathbb{R}^{n-1})$ of all functions bounded on $\mathbb{R}^{n-1}$ equipped with the uniform norm. Let us recall, see, e.g., [HR-1963], that $\mathcal{M}$ is a linear functional on $\mathcal{B}(\mathbb{R}^{n-1})$ satisfying

$$|\mathcal{M}[f]| \leq \sup_{\mathbb{R}^{n-1}} |f|; \tag{2.166}$$

$$\mathcal{M}[1] = 1; \tag{2.167}$$

$$\mathcal{M}[f(\cdot + y)] = \mathcal{M}[f] \tag{2.168}$$

for all $y \in \mathbb{R}^{n-1}$.

Now let $C_0(\mathbb{R}^{n-1})$ be the space of continuous functions vanishing at infinity, i.e., for every $\varepsilon > 0$ and a function f from this space, there is a compact set $S_\varepsilon \subset \mathbb{R}^{n-1}$ such that $|f(x)| < \varepsilon$ for $x \notin S_\varepsilon$.

Given functions $\varphi \in C_0(\mathbb{R}^{n-1})$ and $f \in L_1(\mathbb{R}^{n-1})$ we define a function $\psi_j = \psi_j(f, \varphi) : \mathbb{R}^{n-1} \to \mathbb{R}$ by

$$\psi_j : h \longmapsto \int_{\mathbb{R}^{n-1}} \tau_{-h}(E_j \tau_h f) \varphi \, dx, \tag{2.169}$$

where $\tau_h g := g(\cdot + h)$, $h \in \mathbb{R}^{n-1}$.

Since by (2.161)

$$\sup_h |\psi_j(h)| \leq C \|f\|_{L_1(\mathbb{R}^{n-1})} \|\varphi\|_{L_\infty(\mathbb{R}^{n-1})}, \tag{2.170}$$

the function ψ_j belongs to $\mathcal{B}(\mathbb{R}^{n-1})$.

Applying the invariant mean $\mathcal{M}$ to ψ_j, we define the bilinear functional

$$(f, \varphi) \to \mathcal{M}[\psi_j(f, \varphi)].$$

Due to (2.166) and (2.170), this functional is continuous in φ for each $f \in L_1(\mathbb{R}^{n-1})$, and its norm is bounded by $C\|f\|_{L_1(\mathbb{R}^{n-1})}$.

By the Riesz representation theorem there is a bounded Borel measure $\mu := \mu_{f,j}$ on $\mathbb{R}^{n-1}$ such that

$$\mathcal{M}[\psi_j(f,\varphi)] = \int_{\mathbb{R}^{n-1}} \varphi \, d\mu.$$

By inequality (2.170),

$$\|\mu\| \leq C\|f\|_{L_1(\mathbb{R}^{n-1})},$$

where $\|\mu\|$ stands for the variation of μ over $\mathbb{R}^{n-1}$, i.e., $\|\mu\| := \sup_{S \subset \mathbb{R}^{n-1}} |\mu(S)|$.

Thus, the correspondence $f \to \mu_{f,j}$ defines a linear bounded operator from $L_1(\mathbb{R}^{n-1})$ into the Banach space $M(\mathbb{R}^{n-1})$ of bounded Borel measures on $\mathbb{R}^{n-1}$. We denote this operator by $\widehat{E}_j$ and show that it satisfies conditions similar to those in Lemma 2.91.

2.6. Sobolev spaces: selected trace and extension results

Lemma 2.92. (a) *For every* $f \in L_1(\mathbb{R}^{n-1})$ *and the measure* μ_f *given by* $d\mu_f := f dx$, *it is true that*

$$\mu_f = \sum_j \widehat{E}_j f \quad (\text{convergence in } M(\mathbb{R}^{n-1})). \tag{2.171}$$

(b) *For some constant* $C > 0$ *and every* $f \in L_1(\mathbb{R}^{n-1})$,

$$\sum_j \|\widehat{E}_j f\|_{M(\mathbb{R}^{n-1})} \leq C \|f\|_{L_1(\mathbb{R}^{n-1})}. \tag{2.172}$$

(c) *For every* $h \in \mathbb{R}^{n-1}$,

$$\tau_h \widehat{E}_j = \widehat{E}_j \tau_h,$$

i.e., $\widehat{E}_j$ *is translation invariant.*[9]

Proof. Assertions (a) and (b) immediately follow from the corresponding ones of Lemma 2.91 and the definition of $\widehat{E}_j$.

To prove (c), we use (2.169) to conclude that for $\varphi \in C_0(\mathbb{R}^{n-1})$ and $g, h \in \mathbb{R}^{n-1}$,

$$\langle \widehat{E}_j \tau_g f, \varphi \rangle = \mathcal{M}[\langle \tau_{-h} E_j \tau_h \tau_g f, \varphi \rangle]$$
$$= \mathcal{M}[\langle \tau_{-h-g} E_j \tau_{h+g} f, \tau_{-g}\varphi \rangle] = \mathcal{M}[\tau_g(\psi_j(f, \tau_{-g}\varphi))].$$

Due to (2.168) and the translation invariance of $\mathcal{M}$, the last term of the above identity equals

$$\mathcal{M}[\psi_j(f, \tau_{-g}\varphi)] = \langle \widehat{E}_j f, \tau_{-g}\varphi \rangle := \langle \tau_g \widehat{E}_j f, \varphi \rangle.$$

Combining these identities we get

$$\langle \widehat{E}_j \tau_g f, \varphi \rangle = \langle \tau_g \widehat{E}_j f, \varphi \rangle,$$

as required. □

By virtue of the equivariance of $\widehat{E}_j$ and the Riesz representation theorem, there is a measure $\mu_j \in M(\mathbb{R}^{n-1})$ such that for all $f \in L_1(\mathbb{R}^{n-1})$,

$$\widehat{E}_j f = \mu_j * f := \int_{\mathbb{R}^{n-1}} f(\cdot + y) d\mu_j(y) \quad (\text{equality in } M(\mathbb{R}^{n-1})).$$

Inserting this in (2.172) we get

$$\sum_j \|\mu_j * f\|_{M(\mathbb{R}^{n-1})} \leq C \|f\|_{L_1(\mathbb{R}^{n-1})}.$$

---

[9] Recall that translation $\tau_h \mu$ of a measure μ is a linear functional defined by $\langle \tau_h \mu, \varphi \rangle := \int \tau_{-h} \varphi d\mu$, $\varphi \in C_0(\mathbb{R}^{n-1})$.

Now let $\{f_N\}_{N\in\mathbb{N}}$ be an approximate identity in $C_0(\mathbb{R}^{n-1})$, i.e., for every N and $\varepsilon > 0$,

$$\int_{\mathbb{R}^{n-1}} f_N\, dx = 1, \quad f_N \geq 0 \quad \text{and} \quad \max_{|x|>\varepsilon} f_N \to 0 \quad \text{as} \quad N \to \infty.$$

Then $\mu_j * f_N \to \mu_j$ in $M(\mathbb{R}^{n-1})$. This and the previous inequality imply that

$$\sum_j \|\mu_j\|_{M(\mathbb{R}^{n-1})} \leq C. \tag{2.173}$$

Lemma 2.93. *For each j the measure μ_j is absolutely continuous with respect to the Lebesgue $(n-1)$-measure. In other words, there is a function $g_j \in L_1(\mathbb{R}^{n-1})$ such that*

$$g_j * f = \mu_j * f \quad \text{and} \quad \|\mu_j\|_{M(\mathbb{R}^{n-1})} = \|g_j\|_{L_1(\mathbb{R}^{n-1})}. \tag{2.174}$$

Proof. First, we show that the first-order distributional derivatives $D_i \mu_j$, $1 \leq i \leq n-1$, (tempered distributions over the Schvartz space $\mathcal{S}(\mathbb{R}^{n-1})$) are bounded Borel measures over $\mathbb{R}^{n-1}$.

Using the definition of $\widehat{E}_j$ and (2.163), we obtain, for $\varphi \in \mathcal{S}(\mathbb{R}^{n-1})$,

$$\langle D_i \mu_j * f, \varphi \rangle := \langle D_i \widehat{E}_j f, \varphi \rangle := -\langle \widehat{E}_j f, D_i \varphi \rangle$$
$$:= \mathcal{M}\big[\langle \tau_{-h} E_j \tau_h f, D_i \varphi \rangle\big] = \mathcal{M}\big[\langle \tau_{-h} D_i(E_j \tau_h f), \varphi \rangle\big].$$

The last equality is proved by integrating by parts inside the square brackets using the fact that $E_j T_h f$ belongs to $W_1^1(\mathbb{R}^{n-1})$. Since $D_i E_j T_h f$ belongs to $L_1(\mathbb{R}^{n-1})$ and $S(\mathbb{R}^{n-1})$ is dense in $C_0(\mathbb{R}^{n-1})$, the last term of the above equality can be extended to a linear continuous functional on $C_0(\mathbb{R}^{n-1})$. This means, see, e.g., Stein [Ste-1970, Ch. 2], that $D_i \mu_j * f$ is a bounded measure on $\mathbb{R}^{n-1}$ for every $f \in L_1(\mathbb{R}^{n-1})$; therefore $D_i \mu_j \in M(\mathbb{R}^{n-1})$, as required.

We conclude from here that μ_j is absolutely continuous as follows. Using a C^∞-approximate identity $\{f_N\}_{n\in\mathbb{N}}$ we define the differentiable function $F_N := f_N * \mu_j$, $N \in \mathbb{N}$. By the Gagliardo embedding theorem [Ga-1957]

$$\|F_N\|_{L_p(\mathbb{R}^{n-1})} \leq C|F_N|_{W_1^1(\mathbb{R}^{n-1})}, \quad N \in \mathbb{N},$$

where $\frac{1}{p} := 1 - \frac{1}{n-1}$ and $C = C(n) > 0$.

The right-hand side is bounded by $C \sum_{i=1}^{n-1} \|D_i \mu_j\|_{M(\mathbb{R}^{n-1})}$. Since, by definition, $1 < p < \infty$ if $n > 2$ and $p = \infty$ if $n = 2$, the space $L_p(\mathbb{R}^{n-1})$ is reflexive for $n > 2$ and dual to $L_1(\mathbb{R}^{n-1})$ for $n = 2$. Hence in both cases there is a subsequence of $\{F_N\}$ (which, without loss of generality, may be identified with $\{F_N\}$) and a function, say g_j, such that for every $f \in L_q(\mathbb{R}^{n-1})$, where $\frac{1}{p} + \frac{1}{q} = 1$, we have

$$\int F_N \cdot f\, dx \to \int g_j \cdot f\, dx \quad \text{as} \quad N \to \infty.$$

But $F_N \to \mu_j$ in $M(\mathbb{R}^{n-1})$ as $N \to \infty$. Hence $\int_{\mathbb{R}^{n-1}} f d\mu_j = \int_{\mathbb{R}^{n-1}} f g_j \, dx$ for every $f \in C_0(\mathbb{R}^{n-1})$ and therefore $d\mu_j = g_j dx$.

This proves (2.174). □

Now we derive the desired contradiction to the existence of the linear extension operator $E : L_1(\mathbb{R}^{n-1}) \to W_1^1(\mathbb{R}_+^n)$.

Combining (2.174) and (2.173) we have

$$\sum_j \|g_j\|_{L_1(\mathbb{R}^{n-1})} \leq C.$$

Then the sum $g := \sum_j g_j$ is an element of $L_1(\mathbb{R}^{n-1})$. On the other hand, $f = g * f$ for every $f \in L_1(\mathbb{R}^{n-1})$ by (2.171). Hence g is the δ-measure which clearly does not belong to $L_1(\mathbb{R}^{n-1})$.

The proof is complete. □

Comments

Spaces of C^k functions whose higher derivatives satisfy the Hölder or Zygmund conditions and the associated spaces of k-jets are customary objects of modern analysis. The general class of Lipschitz spaces of higher order introduced in Section 2.1 was first introduced in Quantitative Approximation Theory originated by S. Bernstein, D. Jackson, de la Vallée–Poussin and A. Zygmund, see, e.g., comments to Chapter 4 of the book [TB-2004] by Trigub and Belinsky.

The notion of k-modulus of continuity was due to Lebesgue ($k = 1$) and S. Bernstein [Ber-1912] ($k > 1$). Its basic properties for univariate continuous functions were studied by Marchaud [Mar-1927].

Another approach to Lipschitz spaces of higher order and the associated spaces of smooth functions $C^k \Lambda^{s,\omega}$ is based on a generalized modulus of continuity; various definitions of this notion were given by A. Calderón [Cal-1964], Yu. Brudnyi [Br-1965b] and H. Shapiro, see his book [Sha-1971] and references therein. For our goal the following definition is the most suitable.

Let μ be a bounded Borel measure on $\mathbb{R}^n$ *orthogonal to the space of polynomials* $\mathcal{P}_{k-1,n}$. Then the μ-modulus of continuity for a function $f \in L_p^{\text{loc}}(\mathbb{R}^n)$ and $t > 0$ is given by

$$\omega_\mu(t; f; L_p) := \sup_{0 < s \leq t} \left\| \int_{\mathbb{R}^n} f(\cdot + sy) d\mu(y) \right\|_p.$$

For compactly supported μ this definition can be extended to functions $f \in L_p(G)$ where G is a domain in $\mathbb{R}^n$. In particular, choosing $\mu := \sum_{j=0}^k c_{kj} \delta_{kj}$ where $c_{kj} := (-1)^{k-j} \binom{k}{j}$, we obtain k-modulus of continuity $\omega_k(\cdot; f; L_p)_G$.

Theorem. (a) *For a convex domain $G \subset \mathbb{R}^n$ and $f \in L_p(G)$, $1 \leq p \leq \infty$,*

$$\omega_\mu(\,\cdot\,;\,f;\,L_p)_G \leq c(k,n)|\mu|\omega_k(\,\cdot\,;\,f;\,L_p)_G.$$

(b) *If the Fourier transform of μ does not vanish identically on any ray emanating from the origin, then for $t > 0$,*

$$\omega_k(t;\,f;\,L_p) \leq c(k,n,\mu) \int_{\mathbb{R}_+} \min\left[\left(\frac{t}{u}\right)^k, 1\right] \frac{\omega_\mu(u;\,f;\,L_p)}{u}\,du.$$

If μ is compactly supported, the inequality holds for $f \in L_p(G)$ and $0 < t \leq \operatorname{diam} G$.

Part (a) of the theorem was proved by Yu. Brudnyi [Br-1965a], see also [Br-1970a]. Part (b) for $d\mu = w dx$, where w is a spherically symmetric compactly supported C^∞ function, was, in essence, due to A. Calderón [Cal-1964]; in general, assertion (b) follows from the general Boman and Shapiro *comparison theorem* estimating ω_μ by an integral transform of another μ-modulus of continuity, see the book by H. Shapiro [Sha-1971] and the consequent Boman's paper [Bom-1977].

The Whitney–Hestens extension Theorem 2.22 for the jet space $J^\infty(\mathbb{R}^n)$ naturally poses the following problem.

We say that a closed subset $S \subset \mathbb{R}^n$ has the C^∞ *simultaneous extension property* if there is a linear continuous extension operator from the trace space $J^\infty(\mathbb{R}^n)\big|_S$ to $J^\infty(\mathbb{R}^n)$.

Let us recall that $J^\infty(\mathbb{R}^n)$ is a Frechet space consisting of ∞-jets associated to C^∞ functions; its topology is defined by the family of seminorms given on elements $\vec{f} := \{f_\alpha\}_{\alpha \in \mathbb{Z}_+^n}$ of $J^\infty(\mathbb{R}^n)$ by

$$\|\vec{f}\|_k := \max_{|\alpha| \leq k} \sup_{\mathbb{R}^n} |f_\alpha|, \quad k \in \mathbb{Z}_+.$$

In turn, the Frechet topology of the trace space $J^\infty(\mathbb{R}^n)\big|_S$ is defined by the family of seminorms

$$\|\vec{f}\|_k^S := \inf_F \big\|\{D^\alpha F\}_{\alpha \in \mathbb{Z}_+^n}\big\|_k, \quad k \in \mathbb{Z}_+,$$

where F runs over all C^∞ functions on $\mathbb{R}^n$ satisfying $D^\alpha F\big|_S = f_\alpha$, $\alpha \in \mathbb{Z}_+^n$.

Due to the Whitney–Hestens Theorem 2.19 the topology of the trace space may be equivalently defined by the reduced Taylor remainders $r_\alpha(\vec{f})$, $\alpha \in \mathbb{Z}_+^n$.

Problem. *Characterize closed subsets of $\mathbb{R}^n$ possessing the C^∞ simultaneous extension property.*

By virtue of Theorem 2.24 a set S from the class defined by this property (briefly, $S \in \operatorname{SEP}_\infty$) has no isolated points. The next criterion was due to Tidten [Tid-1979] and Vogt [Vogt-1983] who generalized the one-dimensional Mitiagin result [Mit-1961]. For its formulation let us denote by s the space of rapidly decreasing sequences $x := (x_j)_{j \in \mathbb{N}} \subset \mathbb{R}$ whose Frechet topology is defined by the sequence of seminorms $\|x\|_k$ given by

$$\|x\|_k := \sum_{j \in \mathbb{N}} |x_j| j^k, \quad k \in \mathbb{Z}_+.$$

Theorem. *Let S be a compact set in $\mathbb{R}^n$ such that $\overline{S^o} = S$. Then $S \in \mathrm{SEP}_\infty$ if and only if the trace space $J^\infty(\mathbb{R}^n)\big|_S$ is isomorphic to the space s.*

In general, the trace space is isomorphic to some factor space of the space s, see the papers [Tid-1979] by Tidten and [BS-1983] by Bierstone and G. Schwartz. The following example from the first paper demonstrates that $J^\infty(\mathbb{R}^n)\big|_S$ may be nonisomorphic to s.

The set $S := \{(x,y) \in \mathbb{R}^2 \,;\, x \geq 0,\, |y| \leq e^{-\frac{1}{x}}\}$ *does not belong to* SEP_∞ (but $\overline{\mathbb{R}^2 \setminus S}$ does!).

The condition for a compact subset $S \subset \mathbb{R}^n$ to be the closure of its interior is not necessary. For example, the classical Cantor set has empty interior but belongs to SEP_∞ (Tidten [Tid-1983]).

A considerable number of papers has been devoted to the study of other subclasses of the class SEP_∞: Nash subanalytic sets (Bierstone and P. Milman [BM-1991]), sets with real analytic boundaries having polynomial cusps (Pawlucki and Pleśniak [PP-1988]), sets for which a Markov type inequality is valid (Pleśniak [Pl-1990]).

There are some variants of the extension problem in question which consider spaces of subanalytic functions (Kurdyka and Pawlucki [KP-1997]) and spaces of ultradifferentiable functions $\mathcal{E}_\omega(\mathbb{R}^n)$ in the Beurling–Bjork sense, see, e.g., the paper by Meise and B. Taylor [MT-1989] and references therein. Note that $\mathcal{E}_\omega(\mathbb{R}^n) = C^\infty(\mathbb{R}^n)$ for $\omega(t) := \log(1+t)$, $t > 0$; other choices of the weight give the Denjoy–Carleman classes of quasianalytic functions and (nonquasianalytic) Gevrey classes.

Theorems 2.26 and 2.31 were proved by Brouwer for univariant continuous functions in his almost forgotten paper [Bro-1908]. The multivariate generalizations presented in Section 2.3 were proved in Yu. Brudnyi's papers [Br-1965b] and [Br-1967].

Identity (2.43) is due to Kemperman; the problem of finding such identities was formulated by A. Timan and solved in different ways by Yu. Brudnyi [Br-1965b] (see [Br-1970a] for the proof) and M. Timan [MTim-1969]. The discussion of these results is presented in Appendix E.

Theorem 2.32 for $C(\mathbb{R}_+)$-functions was due to Marchaud [Mar-1927]; the argument for the proof of the multivariate case is taken from [Br-1967].

Theorem 2.34 for univariate continuous functions was proved by Brouwer [Bro-1908] and rediscovered by Whitney [Wh-1934c].

Theorem 2.37 for bounded functions on intervals of the real line was due to Whitney [Wh-1957, Wh-1959]. For convex domains in $\mathbb{R}^n$ and integrable functions the result was proved by Yu. Brudnyi in [Br-1970a]; see Appendix F for the corresponding proofs and other references.

Raikov [Rai-1939] was the first to characterize differential properties of functions by behavior of their local best approximation. His paper contains Theorem 2.38 for univariate continuous functions; the result is an easy consequence of the aforementioned Marchaud theorem. A new approach based on Whitney's inequality from [Wh-1957] and on a local Markov inequality was proposed by Yu. Brudnyi and Gopengauz [BGo-1961], see also [BGo-1963] for the proofs. In particular, the (generalized) Marchaud theorem is a consequence of their results. These ideas were then developed in a series of papers by Yu. Brudnyi's and his students' and collaborators' papers beginning with [Br-1965a], see the bibliography to this book. In these papers, the basic properties of the classical spaces of smooth multivariate functions were studied using the local approximation methods. Some results of this kind will be discussed and proved in Chapter 9.

The n-dimensional generalization of the Raikov theorem, Theorem 2.38, was due to Yu. Brudnyi [Br-1965a] (see [Br-1971] for the proof). A generalization to a wide class of subsets in $\mathbb{R}^n$ including Lipschitz domains and fractals was due to A. and Yu. Brudnyis [BB-2007a]; the result will be presented in subsection 9.2.3 (Volume II).

Theorem 2.38 raises the following extremal

Problem. *Find the sharp constant $\gamma = \gamma(k, \lambda, n, p)$ for the inequality*

$$M_{k,\lambda}(f) \leq \gamma |f|_{C^{k,\lambda}(G)}.$$

The only known result concerns univariate functions and $\lambda = 1$; in this case $\gamma = \frac{1}{2^{2k+1}(k+1)!}$.

The proof of Bernstein's Theorem 2.40 was outlined in his note [Ber-1940]; the derivation in the book follows his basic ideas.

In the multivariate case, the following claim might be true.

Conjecture. *A function $f : G \to \mathbb{R}$, where G is a convex domain of $\mathbb{R}^n$, belongs to the space $C_u^k(G)$ if and only if the limit*

$$\lim_{Q \to x} \frac{E_k(Q \cap G; f)}{|Q|^{\frac{k}{n}}}$$

exists at every $x \in G$ and convergence is uniform.

Here $C_u^k(G)$ stands for the space of k-times continuously differentiable on G functions whose higher derivatives are uniformly continuous.

The variant of Whitney's extension Theorem 2.47 for the space $\dot{C}^{k,\omega}(\mathbb{R})$, i.e., Theorem 2.52, is due to Merrien [Mer-1966]. The derivations of both theorems follow the Whitney argument [Wh-1934b] but the proof of combinatorial Lemma 2.48 is taken from the book [KM-1997] by Kriegl and Michor.

Theorem 2.64 is motivated by the Whitney extension theorem [Wh-1934d] for the space of k-times continuously differentiable functions on $G \subset \mathbb{R}^n$ with uniformly continuous higher derivatives; in this case G is quasiconvex.

The proof of Zobin's Theorem 2.69 (a) is based on his argument but was essentially simplified by Shvartsman in [BSh-2001a]; he, in particular, formulated the Geometrical Lemma which then was proved by Igonin and Yanishevski.

All the results of subsection 2.5.2 are also true for p-integrable functions with $1 \leq p \leq \infty$; the proofs, up to trivial changes, are the same. In connection with the Yu. Brudnyi result [Br-1980] mentioned in Remark 2.80, the following conjecture seems to be valid.

Conjecture. *Let $f \in L_p(G)$, where $G \subset \mathbb{R}^n$ is a special Lipschitz domain. There exists a linear continuous operator $T : L_p(G) \to L_p(\mathbb{R}^n)$ such that, for all $k \geq 0$ and some $c(k,n) > 0$,*

$$\omega_k(\cdot\,;Tf)_{L_p(\mathbb{R}^n)} \leq c(k,n)\omega_k(\cdot\,;f)_{L_p(G)}.$$

The local approximation approach to Jones type extension theorems (see subsection 2.6.1) will be used in Chapter 9 for general spaces of smooth functions. The results of subsection 2.6.1 yield the following (very difficult)

Problem. *Find a geometric characteristic of domains $G \subset \mathbb{R}^n$ which admit a simultaneous extension from $W_p^k(G)$ into $W_p^k(\mathbb{R}^n)$.*

The proof of the nonexistence theorem of subsection 2.6.2 was outlined in Peetre's note [Peet-1979]; the proof presented here is due to Yu. Brudnyi in [BSh-2001a]. Another variant of this proof was then proposed by Pełczyński and Wojciechowski [PW-2002] who used this result to show that the Sobolev space $W_1^k(\mathbb{R}^n)$ for $n \geq 2$ is not isomorphic to any Banach lattice (unlike the case $n = 1$, see Borsuk [Bor-1933a]).

It is worth noting that a Peetre type nonexistence result holds only for the embedding in $L_p(\mathbb{R}^{n-1})$ with $p = 1$; a simultaneous extension does exist for $p > 1$. This, clearly, means that the set of linear continuous operators acting in L_1-spaces is very small. Actually, the L_1-space is a "border point" between L_p spaces with $p < 1$, where there is no nontrivial linear continuous operators, and L_p spaces with $p > 1$, where the set of these operators is as large as in L_2.

Another nonexistence phenomenon was discovered by Burenkov and Goldman [BuGo-1979]. In particular, they proved that for the embedding operator

$$B_p^{\sigma,1}(\mathbb{R}^n) \subset L_p(\mathbb{R}^m)$$

there is no inverse linear extension operator.

Here $1 \leq p < \infty$, $1 \leq m < n$, $\sigma = \frac{n-m}{p}$ and $\mathbb{R}^m$ is identified with an m-dimensional coordinate subspace of $\mathbb{R}^n$; for the definition of the Besov space standing in the left-hand side see, e.g., Triebel [Tri-1992].

Appendices

E. Difference identities

E.1. Kemperman's identity

We present the identity for mixed differences formulated in Section 2.3, see (2.43), and also a certain modification which will be more relevant for applications. The former was due to J. H. B. Kemperman and was first published in [JSch-1977]. Previous identities of this kind were found by Yu. Brudnyi [Br-1965b] and M. Timan [MTim-1969]. All of them may be written in a form which uses the following notation.

By $\tau(h)$, $h \in \mathbb{R}^n$, we denote the shift by h acting on functions $f : \mathbb{R}^n \to \mathbb{R}$ as $\tau(h)f := f(\cdot + h)$. Then the kth difference operator is written as

$$\Delta^k(h) := \Delta(h)^k = \sum_{j=0}^{k}(-1)^{k-j}\binom{k}{j}\tau(jh),$$

where $\Delta(h) := \tau(h) - \tau(0) \; (:= \tau(h) - 1)$, and the α-difference operator, $\alpha \in \mathbb{Z}_+^n$, as

$$\Delta^\alpha(h) := \prod_{i=1}^{n} \Delta^{\alpha_i}(h_i e_i);$$

here $\{e_i\}_{1 \leq i \leq n}$ is the standard basis of $\mathbb{R}^n$.

We deviate here from the notation of Section 2.3, where $\Delta^k(h)$ is denoted by Δ_h^k, etc.

Now the desired identity looks as follows:

$$\Delta^\alpha(h) = \int_{\mathbb{R} \times \mathrm{Mat}_n(\mathbb{R})} \Delta^k(th)\tau(Mh)\,d\lambda(t)\,d\mu(M), \tag{E.1}$$

where $|\alpha| = k$ and λ and μ are compactly supported Borel measures on $\mathbb{R}$ and on $\mathrm{Mat}_n(\mathbb{R})$, respectively; the latter stands for the space of real $n \times n$ matrices.

Since $\Delta^\alpha(h)$ and $\Delta^k(th)$ are finitely supported (as linear combinations of δ-measures), it will be natural to look for finitely supported measures λ and μ in this identity.

This is the case of the Kemperman and M. Timan identities; in the proof of Yu. Brudnyi, the measures are unspecified but the identity has some additional property which the previous ones do not possess. Specifically, in this identity, for every number $t \in \mathrm{supp}\,\lambda$ and every matrix $M \in \mathrm{supp}\,\mu$,

$$Mh + \mathrm{conv}(\mathrm{supp}\,\Delta^k(th)) \subset \mathrm{conv}(\mathrm{supp}\,\Delta_h^\alpha). \tag{E.2}$$

Appendix E. Difference identities 171

Let us note that $\operatorname{conv}(\operatorname{supp} \Delta^k(h))$ is the closed h-interval $[0, kh] := \operatorname{conv}\{0, kh\}$ while the right-hand side is the rectangular box $\prod_{i=1}^{h} [0, \alpha_i h_i]$; here we adopt the convention $[0, a] := [a, 0]$ if $a < 0$.

Now we formulate and prove a modification of Kemperman's identity which also satisfies (E.2).

Theorem E.1. *There are finitely supported measures λ and μ such that (E.1) and (E.2) hold.*

Proof. We begin with the proof of Kemperman's identity.

Proposition E.2. *Let V be a collection of k vectors in $\mathbb{R}^n$ and $\varphi : V \to \{1, \frac{1}{2}, \ldots, \frac{1}{k}\}$ be a bijection. Then*

$$\prod_{v \in V} \Delta(v) = \sum_{\omega \in V} (-1)^{\operatorname{card} \omega} \tau(w_\omega) \Delta^k(v_\omega), \qquad (E.3)$$

where the sum is taken over all nonempty subsets ω of V and

$$v_\omega := \sum_{v \in \omega} \varphi(v) v, \quad w_\omega := \sum_{v \notin \omega} v.$$

Note that $w_V = 0$, since $\{v \notin V\} = \emptyset$.

Proof. We enumerate the vectors of V such that $v = v_i$ if $\varphi(v) = \frac{1}{i}$. Then we identify V with the set $\{1, \ldots, k\}$ and regard the ω's in (E.3) as nonempty subsets of the latter set. For arbitrary $0 \le i \le k$ we have

$$\prod_{j=1}^{k} \Delta((j-i)v_j) = \prod_{j=1}^{k} [\tau((j-i)v_j) - 1]$$

$$= \sum_{\omega} (-1)^{k-|\omega|} \tau\left(\sum_{j \in \omega} j v_j\right) \tau\left(-i \sum_{j \in \omega} v_j\right).$$

Here ω runs over *all* subsets of $\{1, \ldots, k\}$; the term with $\omega = \emptyset$ equals $(-1)^k \tau(0)$.

The left-hand side is not zero only for $i = 0$. Therefore, multiplying both sides by $(-1)^{k-i} \binom{k}{i}$ and summing over i, we obtain

$$(-1)^k \prod_{j=1}^{k} \Delta(jv_j) = \sum_{\omega} (-1)^{k-|\omega|} \tau\left(\sum_{j \in \omega} j v_j\right) \Delta^k\left(-\sum_{j \in \omega} v_j\right),$$

where we may exclude $\omega = \emptyset$, since the corresponding term is zero.

Next we replace v_j by $-v_j$ and apply to the right-hand side the formula

$$\Delta(-jv_j) = -\tau(-jv_j)\Delta(jv_j). \qquad (E.4)$$

This leads to the equality

$$\tau\left(-\sum_{j=1}^{k} jv_j\right) \prod_{j=1}^{k} \Delta(jv_j) = \sum_{\omega}(-1)^{k-|\omega|}\tau\left(-\sum_{j\in\omega} jv_j\right)\Delta^k\left(\sum_{j\in\omega} v_j\right).$$

Multiplying both sides by $\tau\left(\sum_{j=1}^{k} jv_j\right)$ and then substituting v_j for jv_j one gets (E.3). □

Now we choose here the set V in such a way that $\prod_{v\in V}\Delta(v)$ becomes the mixed α-difference $\Delta^\alpha(h) := \prod_{i=1}^{n}\Delta^{\alpha_i}(h_i e_i)$. Hence, k equals $|\alpha|$ and $V = \{v_1, \ldots, v_k\}$, where $v_j := h_1 e_1$ if $1 \le j \le \alpha_1$, $v_j := h_2 e_2$ if $\alpha_1 < j \le \alpha_1 + \alpha_2$, etc. To write the corresponding vectors v_ω and w_ω in (E.3), we define a surjection $\psi : \{1, \ldots, k\} \to \{1, \ldots, n\}$ by

$$\psi(j) := i \quad \text{if} \quad v_j = h_i e_i.$$

Then, in accordance with (E.3), we have

$$\begin{aligned} v_\omega := v_\omega(h) &= \sum_{j\in\omega} j^{-1} h_{\psi(j)} e_{\psi(j)}, \\ w_\omega := w_\omega(h) &= \sum_{j\notin\omega} h_{\psi(j)} e_{\psi(j)}. \end{aligned} \quad (E.5)$$

Here the ω is regarded as a subset of $\{1, 2, \ldots, k\}$. In particular, for $\omega := \{1, \ldots, k\}$, $v_\omega(h) = \sum_{i=1}^{n} c_i \alpha_i h_i e_i$ and $w_\omega(h) = 0$, where $c_i := \sum_{\alpha_{i-1} < j \le \alpha_i} j^{-1}$.

Now condition (E.2) requires that for every nonempty ω the n-interval

$$I_\omega(h) := \operatorname{conv}\{w_\omega(h),\ w_\omega(h) + kv_\omega(h)\}$$

be contained in the rectangular box $\Pi_\alpha(h) := \prod_{i=1}^{n}[0, \alpha_i h_i]$. That is to say, for every nonempty ω we must have

$$w_\omega(h) \quad \text{and} \quad w_\omega(h) + kv_\omega(h) \in \Pi_\alpha(h). \quad (E.6)$$

It is readily seen that this is not yet the case for $|\alpha| = k > 1$.

We now modify the Kemperman identity to get rid of this shortage. To this end we fix an integer $N > 1$ and apply the Multinomial Theorem to the identity

$$\Delta^k(Nh) := (\tau_h^N - 1)^k = (\tau_h - 1)^k \left(\sum_{j=0}^{N-1} \tau_h^j\right)^k.$$

Appendix E. Difference identities

This yields
$$\Delta^k(Nh) = \sum_{\beta} \tau(|\beta|h)\Delta_h^k,$$

where the vectors $(\beta_0, \ldots, \beta_{N-1})$ run over the lattice
$$L(k, N) := \mathbb{Z}_+^N \cap [0, (k-1)N]^N.$$

Substituting k for α_i and h for $\frac{1}{N}h_i e_i$ and multiplying the identities so obtained over $1 \le i \le n$ we then get
$$\Delta^\alpha(h) = \sum_{\beta} \tau(N^{-1}u_\beta(h))\Delta^\alpha(N^{-1}h), \tag{E.7}$$

where $\beta := (\beta^1, \ldots, \beta^n)$, vector β^i runs over the lattice $L(\alpha_i, N)$ and
$$u_\beta(h) := \sum_{i=1}^n |\beta^i|h_i e_i.$$

Note that all points $N^{-1}u_\beta(h)$ satisfy
$$N^{-1}u_\beta(h) \in \Pi_\alpha(h). \tag{E.8}$$

Now we associate to each β a signature $\varepsilon_\beta \in \{-1, 1\}^n \in \mathbb{Z}^n$ as follows. We divide $\Pi_\alpha(h)$ into 2^n congruent subboxes and fix an arbitrary one, say Π. There exists a unique vertex of Π common to $\Pi_\alpha(h)$, and every vertex of $\Pi_\alpha(h)$ has the form $\frac{1}{2}\sum_{i=1}^n(1+\varepsilon_i)\alpha_i h_i e_i$, where $\varepsilon := (\varepsilon_1, \ldots, \varepsilon_n)$ is a signature. We denote the signature related to the common vertex by $\varepsilon(\Pi)$.

Because of (E.8) the point $N^{-1}u_\beta(h)$ belongs to one of such Π and we set
$$\varepsilon_\beta := -\varepsilon(\Pi). \tag{E.9}$$

If $N^{-1}u_\beta(h)$ lies on the boundary of Π and therefore belongs also to other subboxes of the subdivision, we choose in (E.9) one of them arbitrarily.

Now we modify each term in (E.7) in the following way. Fix a signature and set for $h \in \mathbb{R}^n$,
$$h_\varepsilon := \sum_{i=1}^n \varepsilon_i h_i e_i.$$

Raising identity (E.4) to the power α_i we get
$$\Delta^{\alpha_i}(h_i e_i) = (-1)^{\alpha_i}\tau(-\alpha_i h_i e_i)\Delta^\alpha(-h_i e_i).$$

Multiplying these equalities together for those i where $\varepsilon_i = -1$ we further obtain
$$\Delta^\alpha(h) = \pm\tau(z_\alpha(h_\varepsilon))\Delta^\alpha(h_\varepsilon), \tag{E.10}$$

where

$$z_\alpha(h_\varepsilon) := \sum_{i=1}^n \min(0, \varepsilon_i) \alpha_i h_i e_i.$$

Replacing finally h by $N^{-1}h$ and choosing $\varepsilon := \varepsilon_\beta$ we express the β-summand of (E.7) in the form

$$\tau\big(N^{-1}u_\beta(h) + N^{-1}z_\alpha(h_\varepsilon)\big)\Delta^\alpha(N^{-1}h_\varepsilon), \quad \text{where} \quad \varepsilon = \varepsilon_\beta.$$

Then we apply the Kemperman identity to $\Delta^\alpha(N^{-1}h_{\varepsilon_\beta})$ and insert the result into (E.7). This gives an identity of the type (E.1) and we should only explain why condition (E.2) holds for this case.

According to (E.5) and (E.6) this condition is equivalent to the belonging of the points $x_\omega := N^{-1}u_\beta(h) + N^{-1}\big(z_\alpha(h_{\varepsilon_\beta}) + w_\omega(h_{\varepsilon_\beta})\big)$ and $y_\omega := x_\omega + N^{-1}v_\omega(h_{\varepsilon_\beta})$ to the parallelotope $\Pi_\alpha(h)$. Setting $N := 2k$ we establish this directly computing the coordinates of x_ω and y_ω.

The result has been proved. $\square$

To formulate a consequence of identity (E.1) we introduce several notions. Let $x, y \in \mathbb{R}^n$. The *pointwise multiplication* of these vectors is given by

$$x \cdot y := \sum_{i=1}^n x_i y_i e_i. \tag{E.11}$$

Further, the *ordered interval* determined by x and y is the rectangular box defined by

$$\Pi[x,y] := \{z \in \mathbb{R}^n \,;\, z_i \in [x_i, y_i],\ i = 1,\ldots,n\}. \tag{E.12}$$

Now let G be a domain in $\mathbb{R}^n$. We say that G is *orderly convex* if G along with every two points x, y contains the ordered interval $\Pi[x,y]$.

Now we introduce an analog of the k-th modulus of continuity, see (2.24), based on mixed differences.

Definition E.3. Let G be a domain in $\mathbb{R}^n$. Then the α-modulus of continuity, where $\alpha \in \mathbb{Z}_+^n$, is a function $\omega_\alpha : C(G) \times (0, +\infty)^n \to \mathbb{R}_+$ given by

$$\omega_\alpha(h;f)_G := \sup\{\|\Delta_t^\alpha f\|_{C(G_{\alpha \cdot t})} \,;\, t \in \Pi[0,h]\},$$

where $G_y := \{x \in G \,;\, \Pi[x, x+y] \subset G\}$.

Using the notions now introduced we derive from identity (E.1)

Corollary E.4. *Let $G \subset \mathbb{R}^n$ be orderly convex. Then for every $f \in C(G)$ and $t > 0$,*

$$\omega_k(t;f)_G \approx \sup_{|\alpha|=k} \omega_\alpha(te;f)_G, \tag{E.13}$$

where the constants of equivalence depend only on n and k.

Here $e := \sum_{i=1}^n e_i = (1, \ldots, 1)$.

Appendix E. Difference identities

Proof. Identity (E.1) and condition (E.2) immediately imply, for $|\alpha| = k$, $h := te$ and the orderly convex domain G the inequality

$$\omega_\alpha(te; f)_G \leq c(k,n)\omega_k(t; f)_G.$$

The converse follows from an identity expressing Δ_h^k via the corresponding mixed differences. In fact, by raising the identity

$$\Delta(h) = \sum_{i=1}^{n} \tau\left(\sum_{j=1}^{i-1} h_j e_i\right) \Delta(h_i e_i)$$

to the k-th power we get

$$\Delta^k(h) = \sum_{|\alpha|=k} \frac{k!}{\alpha!} \tau(h_\alpha) \Delta_h^\alpha, \qquad (E.14)$$

where $h_\alpha := \sum_{i=1}^{n-1} \alpha_i \left(\sum_{j=1}^{i-1} h_j e_j\right)$. This and the orderly convexity of G immediately imply that

$$\omega_k(t; f)_G \leq n^k \sup_{|\alpha|=k} \omega_\alpha(te; f)_G.$$

□

This result can be easily extended to a wide class of (quasi-)norms and arbitrary bases in $\mathbb{R}^n$. We begin with

Definition E.5. Suppose that the linear space X of measurable (classes of) functions on $\mathbb{R}^n$ is equipped with the quasinorm [10] $\|\cdot\|_X$. This space is said to be a quasi-Banach translation-invariant lattice if the following conditions hold:

(a) (Translation-invariance) For every $h \in \mathbb{R}^n$,

$$\|\tau(h)f\|_X = \|f\|_X.$$

(b) (Monotonicity) If $|f| \leq |g|$ almost everywhere and $g \in X$, then $f \in X$ and

$$\|f\|_X \leq \|g\|_X.$$

(c) (Completeness) $(X, \|\cdot\|)$ is complete.

Property (b) allows us to define for every measurable subset $S \subset \mathbb{R}^n$ a quasinorm $\|\cdot\|_{X(S)}$ given for measurable $f: S \to \mathbb{R}$ by

$$\|f\|_{X(S)} := \|\bar{f}\|_X, \qquad (E.15)$$

---
[10] i.e., the triangle inequality holds in a weaker form: for a fixed $c > 1$ and all $f, g \in X$ $\|f + g\|_X \leq c\{\|f\|_X + \|g\|_X\}$.

where $\bar{f}$ is the extension of f to $\mathbb{R}^n$ by zero.

It can be easily checked that the linear space $X(S)$ defined by this quasinorm is complete and the quasinorm is monotone.

Now let G be a domain in $\mathbb{R}^n$ and $f \in X(G)$. Then similarly to the case of continuous functions we define the k-modulus of continuity for f by

$$\omega_k(t;f)_{X(G)} := \sup_{\|h\|\leq t} \|\Delta^k(h)f\|_{X(G_{kh})}; \qquad (\text{E.16})$$

recall that $G_y := \{x \in G\,;\, [x,y] \subset G\}$.

Next, let $\mathcal{B} := \{b_1,\ldots,b_n\}$ be a basis of $\mathbb{R}^n$. Replacing the standard orthonormal basis $\{e_1,\ldots,e_n\}$ by $\mathcal{B}$ in all the above definitions we introduce orderly convex domains and mixed moduli of continuity *with respect to the basis $\mathcal{B}$*. For instance, the latter notion is given for $f \in X(G)$ and $h \in \mathbb{R}^n$ by

$$\omega_\alpha^\mathcal{B}(h;f)_{X(G)} := \sup\{\|\Delta_t^\alpha f\|_{X(G_{\alpha\cdot t})}\,;\, t \in \Pi_\mathcal{B}[0,h]\}, \qquad (\text{E.17})$$

where $x \cdot y$ and $\Pi_\mathcal{B}[x,y]$ are now defined by (E.11) and (E.12) with $\mathcal{B}$ substituted for $\{e_1,\ldots,e_n\}$.

The applications of identities (E.1) and (E.14) immediately lead to

Corollary E.6. *Let $G \subset \mathbb{R}^n$ be an orderly convex domain with respect to a basis $\mathcal{B}$ of $\mathbb{R}^n$. Then for every $f \in X(G)$ and $h \in \mathbb{R}^n$,*

$$\omega_k(t;f)_{X(G)} \approx \sup_{|\alpha|=k} \omega_\alpha^\mathcal{B}(te;f)_{X(G)}. \qquad (\text{E.18})$$

Example E.7. Let $G \subset \mathbb{R}^n$ be a special Lipschitz domain, i.e., the subgraph of a Lipschitz function, say $\varphi : \mathbb{R}^{n-1} \to \mathbb{R}$,

$$G := \{(x, x_n) \in \mathbb{R}^{n-1} \times \mathbb{R}\,;\, x_n < \varphi(x)\}.$$

Then there is an infinite (circular) cone K such that $\partial G + K \subset G$. Choose a basis $\mathcal{B} := \{b_1,\ldots,b_n\}$ such that all the b_i are in K. It can be then easily checked that G is orderly convex with respect to this $\mathcal{B}$, and we may apply (E.18) to this setting.

E.2. Marchaud's identity

This identity expresses the k-th difference via differences of bigger orders in the following fashion.

Proposition E.8. *Let $k, s \geq 1$ be integers and $h \in \mathbb{R}^n$. Then the following holds:*

$$\Delta^k(h) = 2^{-sk}\Delta^k(2^s h) + \sum_{j=0}^{s-1} 2^{-jk} T_k(2^j h)\Delta^{k+1}(2^j h), \qquad (\text{E.19})$$

Appendix E. Difference identities

where $T_k(h) := \sum_{i=0}^{k-1} c_k(i)\tau(ih)$ and the numbers $c_k(i) \geq 0$ and satisfy

$$\sum_{i=1}^{k-1} c_k(i) = \frac{k}{2}.$$

Proof. Using the identity preceding (E.7) for $N = 2$ we get

$$\Delta^k(2h) - 2^k\Delta^k(h) = \sum_{j=0}^{k} \binom{k}{j}[\tau_{jh}\Delta_h^k - \Delta_h^k] = \sum_{j=1}^{k} \binom{k}{j}\sum_{i=0}^{j-1} \tau(ih)\Delta^{k+1}(h).$$

Further, we set $c_k(i) := 2^{-k}\sum_{j=i+1}^{k}\binom{k}{j}$, $0 \leq i \leq k-1$, and define $T_k(h)$ using these numbers. Then $\sum_i c_k(i) = \frac{k}{2}$ and the previous identity yields

$$2^{-k}\Delta^k(2h) - \Delta^k(h) = T_k(h)\Delta^{k+1}(h).$$

Substituting here h for $2^j h$, dividing by 2^{-jk} and summing over $j = 0, 1, \ldots, s-1$ we then get

$$2^{-sk}\Delta^k(2^s h) - \Delta^k(h) = \sum_{j=0}^{s-1} 2^{-jk} T_k(2^j h)\Delta^{k+1}(2^j h),$$

as required. $\square$

As a consequence we obtain the following Marchaud inequality, see [Mar-1927] for $n = 1$.

Theorem E.9. *Let X be a translation-invariant Banach lattice on $\mathbb{R}^n$, and $G \subset \mathbb{R}^n$ be a convex domain. Given integers $0 \leq k < \ell$, there is a constant $c(\ell) > 0$ such that for every $f \in X(G)$ and $0 < t \leq \frac{1}{\ell}\operatorname{diam} G$*

$$\omega_k(t; f)_{X(G)} \leq c(\ell) t^k \left\{ \int_t^d \frac{\omega_\ell(u; f)_{X(G)}}{u^{k+1}}\, du + \frac{\|f\|_{X(G)}}{(\operatorname{diam} G)^k} \right\}.$$

Here $d := \frac{1}{\ell}\operatorname{diam} G$ and the second term in the sum is zero if $\operatorname{diam} G = \infty$.

Proof. It suffices to consider the case of $\ell = k+1$ and then iterate the inequality obtained. Applying identity (E.19) with this ℓ we get

$$\|\Delta^k(h)f\|_{X(G_{kh})} \leq \frac{k}{2}\sum_{j=0}^{s-1} 2^{-jk}\omega_{k+1}(\|h\|; f)_{X(G)} + 2^{-sk} \cdot 2^{k+1}\|f\|_{X(G)}. \quad (E.20)$$

Now let $\|h\| \leq t$ and s satisfies the condition

$$\frac{\operatorname{diam} G}{2\ell} \leq 2^{st} < \frac{\operatorname{diam} G}{\ell};$$

in the case $\operatorname{diam} G = \infty$, we replace $\operatorname{diam} G$ by an integer N and let it tend to infinity in the final inequality. Applying inequality (E.20) and using the monotonicity of ω_{k+1} we obtain for this s,

$$\omega_k(t;f)_{X(G)} \leq c(k)\left\{\int_t^d \frac{\omega_{k+1}(u;f)_{X(G)}}{u^{k+1}}\,du + \frac{\|f\|_{X(G)}}{(\operatorname{diam} G)^k}\right\}.$$

The result has been proved. □

Remark E.10. Now let X be a translation-invariant quasi-Banach lattice on $\mathbb{R}^n$. In this setting, inequality (E.20) should be multiplied by c^s, where $c > 1$ is the constant in the triangle inequality for X. This clearly destroys the proof. To avoid this obstacle one uses the Aoki–Rolewica theorem, see, e.g., [BLo-1976, Lemma 3.10.2], which states that for some $\rho = \rho(X) \in (0,1)$,

$$\left\|\sum_{i=1}^\infty f_i\right\|_X \leq 2\left\{\sum_{i=1}^\infty \|f_i\|_X^\rho\right\}^{\frac{1}{\rho}}. \tag{E.21}$$

Using this replacement of the triangle inequality we derive the Marchaud inequality for a quasi-Banach X in the form

$$\omega_k(t;f)_{X(G)} \leq c(\ell,p)t^k\left\{\left(\int_t^d \left[\frac{\omega_\ell(u;f)_{X(G)}}{u^k}\right]^p \frac{du}{u}\right)^{\frac{1}{p}} + \frac{\|f\|_{X(G)}}{(\operatorname{diam} G)^k}\right\}.$$

Let, e.g., $X = L_p$ where $0 < p < 1$. Then $\rho(L_p) = p$ and (E.21) holds without the factor 2 because of concavity of the function $t \mapsto t^p$, $t > 0$.

F. Local polynomial approximation and moduli of continuity

We will prove Theorem 2.37 which was formulated and widely used in Section 2.3 and will be intensively exploited in Chapter 9, and then present several related results.

F.1. Degree of local polynomial approximation

We first prove Theorem 2.37 for continuous functions. Let us recall (in an equivalent form) its formulation.

Appendix F. Local polynomial approximation and moduli of continuity

Theorem F.1. *Let $V \subset \mathbb{R}^n$ be a closed convex body[11]. Let $f : V \to \mathbb{R}$ be continuous and satisfy the condition, for $x, x + kh \in V$,*

$$|\Delta_h^k f(x)| \leq 1. \tag{F.1}$$

Then there is a constant $w = w(k, n)$ and a polynomial p of degree $k - 1$ such that

$$\sup_V |f - p| \leq w(k, n). \tag{F.2}$$

Remark F.2. Local best approximation of order 1 may be evaluated by

$$E_1(f\,;\,C) = \frac{1}{2}\sup\{|f(x) - f(y)|\,;\,x, y \in C\}.$$

Hence, (F.2) holds with $w(1, n) = \frac{1}{2}$ and we may assume in the sequel that $k \geq 2$.

Proof. We begin with V being a cube. Using scaling we reduce the proof to the case

$$V = Q := [0, 1]^n.$$

Lemma F.3. *Let $Q_k := [0, 1 - \frac{1}{k}]^n$. For every $f \in C(Q)$ there exists a function $f_k \in C^{k-1,1}(Q_k)$ such that*

$$\begin{aligned}\|f - f_k\|_{C(Q_k)} &\leq c(k, n)\omega_k(Q\,;\,f),\\ \sup_{|\alpha|=k}\|D^\alpha f_k\|_{C(Q_k)} &\leq c(k, n)\omega_k(Q\,;\,f).\end{aligned} \tag{F.3}$$

Here $\omega_k(S\,;\,f)$ is the k-oscillation of f on $S \subset \mathbb{R}^n$, i.e., $\omega_k\left(\frac{\operatorname{diam} S}{k}\,;\,f\right)_S$, see (2.57).

Proof. We use the operator $S_t := \prod_{1 \leq j \leq N} S_t(e_j)$ introduced in the proof of Theorem 2.77 with all the unit vectors e_j contained in the cube Q. According to its definition, S_t is an integral operator whose kernel is supported on $\operatorname{conv}\{te_j\}_{1 \leq j \leq n}$. Choosing an appropriate $t = t(k, n)$ we obtain for every $x \in Q_k$,

$$x + \operatorname{conv}\{te_j\}_{1 \leq j \leq N} \subset Q.$$

Under this choice of t and e_j the function $f_k := S_t f$ is defined on a set containing Q_k. In turn, inequalities (F.3) follow from those proved in Theorem 2.77, see (2.142) and the subsequent inequality for $D^\alpha S_t f$. $\square$

Lemma F.4. *Theorem F.1 is true for $V = Q$.*

Proof. Define the desired polynomial $p \in \mathcal{P}_{k-1,n}$ as the Taylor polynomial of order $k - 1$ for f_k at 0, i.e., $p := T_0^{k-1} f_k$. Then for $x \in Q_k$,

$$|(f - p)(x)| = |(f_k - T_0^{k-1} f_k)(x)| \leq c(k, n)|f_k|_{C^k(Q_k)}$$

---
[11] i.e., the interior of V is nonempty.

by the Taylor formula.

Combining this with (F.3) we get

$$\|f - p\|_{C(Q_k)} \leq c(k,n)\omega_k(Q;f). \tag{F.4}$$

Now let $x \in Q \setminus Q_k$. Then for an appropriate h,

$$x - jh \in Q_k, \quad j = 1, \ldots, k.$$

Moreover, $\Delta_h^k p = 0$ and therefore

$$(f - p)(x) = \Delta_h^k f(x) - \sum_{j=1}^{k} (-1)^{k-j} \binom{k}{j} (f - p)(x - jh).$$

Applying (F.4) to each term of the sum we then have

$$|(f - p)(x)| \leq \omega_k(Q;f) + (2^k - 1)c(k,n)\omega_k(Q;f) \leq 2^k c(k,n).$$

Together with (F.4) this proves the result. □

Lemma F.5. *Theorem* F.1 *holds for Euclidean balls.*

Proof. Without loss of generality we may assume that f is continuous and satisfies (F.1) on the closed unit ball B centered at 0. Let Q be a cube of maximal volume contained in B. Then Q is of center 0 and side length $\frac{2}{\sqrt{n}}$. Restricting f to Q we use Lemma F.4 to find a polynomial $p \in \mathcal{P}_{k-1,n}$ with

$$|f(x) - p(x)| \leq c(k,n) \quad \text{for} \quad x \in Q.$$

Fix any $x \in B$. By the one-dimensional variant of Lemma F.4, there is a polynomial $\varphi \in \mathcal{P}_{k-1,1}$ such that

$$|f(tx) - \varphi(t)| \leq c(k,1)$$

for $|t| \leq 1$. Hence for $|t| \leq n^{-\frac{1}{2}}$ we have

$$|p(tx) - \varphi(t)| \leq c(k,1) + c(k,n).$$

Now we apply the Remez type inequality, see Corollary G.2 of Appendix G, to extend this to all of $t \in [-1,1]$. This gives for these t,

$$|p(tx) - \varphi(t)| \leq \left(\frac{4}{n^{-\frac{1}{2}}}\right)^{k-1} (c(k,1) + c(k,n)) =: \tilde{c}(k,n).$$

Combining this with the inequality for $f(tx)$ and choosing $t = 1$ we finally get

$$|f(x) - p(x)| \leq c(k,1) + \tilde{c}(k,n) \quad \text{for any} \quad x \in B. \quad \square$$

Appendix F. Local polynomial approximation and moduli of continuity

We now establish the result for *bounded* convex bodies. To this end, we need the classical F. John result [Jo-1948].

Lemma F.6. *Let V be a bounded closed convex body in $\mathbb{R}^n$. There is an ellipsoid E containing V and such that the homothety $\frac{1}{n} E$ of E with respect to its center is contained in V.*

Lemma F.7. *Theorem F.1 holds for bounded closed convex bodies.*

Proof. The assertion of Theorem F.1 may be reformulated as follows.
There is a constant $c > 0$ such that for any $f \in C(V)$,
$$E_k(f\,;V) \le c\omega_k(V\,;f).$$
Set $w(k,V) := \inf c$. It is easily seen that this constant is affine invariant. Using an appropriate affine transform we may then assume that the John ellipsoid for V is the closed unit ball B centered at 0. Then
$$\frac{1}{n} B \subset V \subset B.$$
Now let $f \in C(V)$. Restricting f to the ball $\frac{1}{n} B$ and applying Lemma F.5 we find a polynomial $p \in \mathcal{P}_{k-1,n}$ such that for $x \in \frac{1}{n} B$,
$$|f(x) - p(x)| \le w_k\left(\frac{1}{n} B\,;f\right) = w(B\,;f) =: \tilde{c}(k,n).$$
Repeating then the argument of the proof of Lemma F.5 we extend this inequality to all points of B as follows:
$$|f(x) - p(x)| \le c(k,1) + (4n)^k \big(c(k,1) + \tilde{c}(k,n)\big). \qquad \square$$

At the final stage V is *unbounded*. Let V_N denote the intersection of V with the closed ball of radius N centered at a fixed point of V. Suppose that $f \in C(V)$ and satisfies (F.1). Then there is a polynomial $p_N \in \mathcal{P}_{k-1,n}$ such that
$$|f(x) - p_N(x)| \le c(k,n) \quad \text{for} \quad x \in V_N. \tag{F.5}$$
The sequence $\{p_N\}_{N \ge 1}$ is then uniformly bounded on the compact set V_1; therefore there is a subsequence which converges uniformly on V_1 to some polynomial p of degree $k-1$. Since $\mathcal{P}_{k-1,n}$ is finite-dimensional, this subsequence converges to p on any compact subset of $\mathbb{R}^n$. Restricting (F.5) to the set V_{N_0} with fixed $N_0 \le N$ and passing to the limit as $N \to \infty$, we get
$$|f(x) - p(x)| \le c(k,n) \quad \text{for} \quad x \in V_{N_0}.$$
Since N_0 is arbitrary, this proves the result for unbounded bodies. $\qquad \square$

A generalization of Theorem F.1 is also true for bounded (maybe, non-measurable) functions and for measurable functions.

Let $B(V)$ be the Banach space of bounded on V functions equipped with the uniform norm. As in the proof of Theorem F.1, the main point is the derivation of the result for $Q := [0,1]^n$. According to Whitney's theorem [Wh-1959] there is a constant w_k such that for every $f \in B([0,1])$ the Lagrange polynomial interpolating f at k equally distributed points of $[0,1]$ satisfies

$$\|f - L(f)\|_B \leq w_k \sup\{|\Delta_h^k f(x)|\,;\, 0 \leq x \leq x + kh \leq 1\}. \qquad \text{(F.6)}$$

Let $f \in B(Q)$ and $L_i(f)$ be defined by applying the interpolation operator L to the function $x_i \mapsto f(x)$, $0 \leq x_i \leq 1$. Then $L_i(f)$ is a polynomial of degree $k-1$ in x_i with the coefficients being bounded functions on Q independent of x_i. By their definition, the operators L_i mutually commute. Therefore the function $\widehat{L}(f) := (L_1 \cdots L_n)(f)$ is a polynomial of degree $k-1$ in each variable. Moreover, (F.6) implies that

$$\|f - L_i(f)\|_{B(Q)} \leq w_k \sup\{|\Delta_{he_i}^k f(x)|\,;\, [x, x + khe_i] \subset Q\}.$$

In particular, the norm of L_i is bounded by $(2^k + 1)w_k$. This implies that for $f \in B(Q)$,

$$\|f - \widehat{L}(f)\|_{B(Q)} \leq \sum_{i=1}^n \left(\prod_{j \neq i} \|L_j\|\right) \|f - L_i f\|_{B(Q)} \leq c(k,n) \omega_k(Q\,;f). \qquad \text{(F.7)}$$

Now we derive from here the required inequality

$$E_k(Q\,;f) \leq c(k,n) \omega_k(Q\,;f).$$

The left-hand side is a Banach norm on the factor space $B(Q)/\mathcal{P}_{k-1,n}$. From inequality (F.7) we conclude that the functional $f \mapsto \omega_k(Q\,;f)$ is also a norm on this factor space. In fact, it suffices to check that $\omega_k(Q\,;f) = 0$ implies that $f \in \mathcal{P}_{k-1,n}$. But due to (F.7) f is a polynomial. Moreover, $\Delta_h^k f = 0$ for every h and therefore all its derivatives of order k equal zero. Hence, f is a polynomial of degree $k-1$.

Further, the norm $\omega_k(Q\,;\cdot)$ is Banach, see Lemma 6 of the paper [Br-1970a] by Yu. Brudnyi. Finally, for every $p \in \mathcal{P}_{k-1,n}$,

$$\omega_k(Q\,;f) = \omega_k(Q\,;f - p) \leq 2^k \|f - p\|_{B(Q)}$$

and therefore

$$\omega_k(Q\,;f) \leq 2^k E_k(Q\,;f).$$

By the Banach open mapping theorem, see, e.g., [DS-1958], this implies the inverse inequality

$$E_k(Q\,;f) \leq c\,\omega_k(Q\,;f)$$

with $c > 0$ independent of f. That is to say, the required result for cubes is true, and Theorem F.1 for bounded functions then follows.

Appendix F. Local polynomial approximation and moduli of continuity

The theorem for measurable functions may be reduced to the case of bounded functions. To this end it suffices to prove that a measurable function $f : [0,1] \to \mathbb{R}$ satisfying the inequality

$$|\Delta_h^k f(x)| \leq 1 \tag{F.8}$$

for $x, x + kh \in [0,1]$, is bounded on $[0,1]$. This allows us to use for f Whitney's theorem (F.6) and to complete the proof as above.

Let $f : [0,1] \to \mathbb{R}$ be measurable, and let (F.8) hold. By measurability, for any $\varepsilon > 0$ there is $N_\varepsilon > 0$ such that the measure of the set

$$S_\varepsilon := \{x \in [0,1]\,;\, |f(x)| > N_\varepsilon\}$$

is bounded by ε.

We then fix an $x \in \left[0, 1 - \frac{1}{k}\right]$ and consider the set $\{h > 0\,;\, x + ih \in S_\varepsilon\}$ for $1 \leq i \leq k$. The measure of this set is bounded by $\left|\frac{1}{i}(S_\varepsilon - x)\right| = \frac{1}{i}|S_\varepsilon| < \frac{\varepsilon}{i}$; here λS is the λ-homothety of $S \subset [0,1]$ with respect to zero. Then the measure of the set of all $h > 0$ satisfying $x + ih \in S_\varepsilon$ for some $1 \leq i \leq k$ is at most $\left(\sum_{i=1}^{k} \frac{1}{i}\right)\varepsilon$. Choose ε such that this number becomes less than $\frac{1}{k}$. Then there is $h \in \left(0, \frac{1}{k}\right)$ such that all points $x + ih$ belong to $[0,1]\setminus S_\varepsilon$.

Finally, we write for these x and h,

$$|f(x)| \leq |\Delta_h^k f(x)| + \sum_{i=1}^{k} \binom{k}{i} |f(k+ih)| \leq 1 + (2^k - 1) N_\varepsilon.$$

Hence, f is bounded on $\left[0, 1 - \frac{1}{k}\right]$.

Applying the same argument (with negative h) to $x \in \left[\frac{1}{k}, 1\right]$ we prove boundedness of f on $[0,1]$.

As a consequence, we now derive the precise form of Marchaud's inequality formulated in Theorem 2.7 (d).

Corollary F.8. *Let f be a locally bounded function on the closure of a convex domain $G \subset \mathbb{R}^n$. Assume that $\omega_k(f\,;G) < \infty$ for a fixed integer $k \geq 1$. Then there exists a polynomial p of degree $k-1$ and a constant $c = c(n,k) > 0$ such that for every integer $0 \leq \ell < k$ the inequality*

$$\omega_\ell(t\,;f-p)_G \leq c t^\ell \int_t^{2d} \frac{\omega_k(s\,;f)_G}{s^{\ell+1}}\, ds$$

holds for all $t \leq d$; here $d := \frac{1}{\ell}\operatorname{diam} G$.

Proof. First let G be bounded and $p \in \mathcal{P}_{k-1,n}$ be such that

$$\sup_G |f - p| \leq c\,\omega_k(G\,;f) \left(:= c\,\omega_k\!\left(\frac{\operatorname{diam} G}{k}\,;f\right)_G\right) \tag{F.9}$$

for some $c = c(k,n)$. Applying Theorem E.8 to $f - p$ we obtain, for $0 < t \leq d$,

$$\omega_\ell(t; f - p)_G \leq c_1(k,n) t^\ell \left(\int_t^d \frac{\omega_k(s; f)_G}{s^{\ell+1}} ds + \frac{\sup_G |f - p|}{(\operatorname{diam} G)^\ell} \right). \tag{F.10}$$

Since ω_k is equivalent to a k-majorant, see Theorem 2.7, we get for $t \leq s \leq 2d$,

$$\frac{\omega_k(s; f)_G}{s^{\ell+1}} \geq 2^{-k} \frac{\omega_k(G; f)}{(\operatorname{diam} G)^k} s^{k-\ell-1}.$$

Integrating this over $[t, 2d]$ and noting that for some $c(k) > 0$ and $t \leq d$,

$$\int_t^{2d} s^{k-\ell-1} ds \geq c(k) (\operatorname{diam} G)^{k-\ell},$$

we finally obtain for this t,

$$\int_t^{2d} \frac{\omega_k(s; f)_G}{s^{\ell+1}} ds \geq 2^{-k} c(k) \frac{\omega_k(G; f)}{(\operatorname{diam} G)^\ell},$$

i.e., the second term in the sum of (F.10) is absorbed by the first.

Now let G be unbounded and G_R denote the intersection of G with a Euclidean ball of radius R centered at a fixed point of G. Let $p \in \mathcal{P}_{k-1,n}$ be chosen as in (F.9). Then applying Marchaud's inequality to $f - p$ and G_R, we get

$$\omega_\ell(t; f - p)_{G_R} \leq c(k,n) t^\ell \left(\int_t^\infty \frac{\omega_k(s; f)_G}{s^{\ell+1}} ds + \frac{\omega_k(G; f)}{(\operatorname{diam} G_R)^\ell} \right).$$

Sending R to infinity we prove the corollary for this case. $\square$

Remark F.9.

(a) Let $S := \overline{G}$ be the closure of a bounded domain in $\mathbb{R}^n$. Assume that S is *starlike*, i.e., there is a point $x_0 \in S$ such that for every line L passing through x_0 the set $S \cap L$ is a nontrivial line segment and, moreover, G contains a ball B centered at x_0. Applying the argument of Lemma F.7 to the closed set $S := \overline{G}$ we get the inequality

$$E_k(S; f) \leq c \omega_k(S; f) \tag{F.11}$$

with the constant c depending on n, k and on the star-like coefficient

$$\sigma(S) := \sup \left\{ \frac{\operatorname{diam} S}{\operatorname{diam} B} \, ; \, B \subset S \right\},$$

where B runs over all balls in S centered at x_0.

Because of the last dependence, inequality (F.10) may be incorrect for unbounded star-like domains.

Appendix F. Local polynomial approximation and moduli of continuity

(b) Another extension of the class of bounded convex bodies can be obtained by applying the extension theorems of subsection 2.6.3. This gives inequality (F.8) for (bounded) Lipschitz domains and (unbounded) special Lipschitz domains, see Definitions 2.71 and 2.72.

Example F.10.

(a) Unlike Lipschitz domains the boundary of star-like domains may be far from any kind of regular. The classical von Koch snowflake, see, e.g., Falconer [Fal-1999, p. XV and Example 9.5], bounds a star-like domain whose boundary is a "thick" curve (its Hausdorff dimension[12] is strictly greater than 1). In particular, any arc of this curve is of infinite length.

(b) The following example mentioned in the paper [BI-1985] by Binev and Ivanov demonstrates the role of cusps for the validity of Theorem F.1.

The set $S := \left\{ x \in \mathbb{R}^2 \,;\, \frac{1}{2} x_1^2 \leq x_2 \leq x_1^2, \; 0 \leq x_1 \leq 1 \right\}$ has a cusp at $(0,0)$:

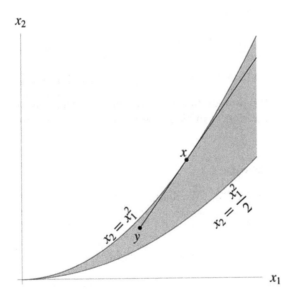

Figure 2.4: The role of cusps for the validity of Theorem F.1.

We define the function $f_\varepsilon : S \to \mathbb{R}$ by

$$f_\varepsilon(x) := \sin\left(\varepsilon \log \frac{1}{x_1}\right).$$

---

[12] see subsection 3.2.4 below for the definition of this concept.

Since f_ε assumes values ± 1 infinitely many times and $|f_\varepsilon| \leq 1$, we have

$$E_k(S\,;f_\varepsilon) = 1.$$

On the other hand,

$$\omega_k(S\,;f_\varepsilon) \leq 2^{k-1}\,\omega_1(S\,;f_\varepsilon) \to 0 \quad \text{as} \quad \varepsilon \to 0. \tag{F.12}$$

In fact, by definition,

$$\omega_1(S\,;f_\varepsilon) = \sup\bigl|f_\varepsilon(x_1) - f_\varepsilon(y_1)\bigr|,$$

where the supremum is taken over all x, y such that the segment $[x,y] \subset S$. Note that this interval with a fixed x is of maximal length if y is a point of the lower parabola $x_2 = \frac{1}{2} x_1^2$ and $[x,y]$ is tangent to the upper parabola $x_2 = x_1^2$. A direct computation shows that in this case $y_1 \leq c_0 x_1$ with some numerical constant $c_0 > 1$.

Finally, by the Mean Value Theorem,

$$\bigl|f_\varepsilon(x_1) - f_\varepsilon(y_1)\bigr| \leq \varepsilon \left|\log \frac{x_1}{y_1}\right| \leq \varepsilon \log c_0.$$

Hence, (F.12) holds and Theorem F.1 is not true for this setting.

However, the domain obtained by replacing the lower parabola by $x_2 = -\frac{1}{2} x_1^2$ is star-like. Hence, in this case the cusp does not prevent the validity of (F.8). □

F.2. Whitney constants

Let S be a closed subset of $\mathbb{R}^n$. We define the *Whitney constant* $w_k(S)$ by

$$w_k(S) := \sup\{E_k(S\,;f)\,;\, f \in C(S) \quad \text{and} \quad \omega_k(S\,;f) \leq 1\}. \tag{F.13}$$

We also define the *global Whitney constant* $w_k(n)$ by

$$w_k(n) := \sup\{w_k(S)\,;\, S \subset \mathbb{R}^n \quad \text{bounded and convex}\}. \tag{F.14}$$

In the spirit of Whitney's paper [Wh-1957] who considered the case of dimension one [13], let us consider also the constants $w_k^*(n)$ and $w_k^{**}(n)$ defined by (F.13) with $S := \mathbb{R}_+^n$ and $S := \mathbb{R}^n$. Using the argument of Beurling, see [Wh-1957], it is easy to prove the following estimates:

$$w_k^*(n) \leq 2, \quad w_k^{**}(n) \leq \min_{1 \leq j \leq n} 1 \Big/ \binom{n}{j}.$$

---

[13] In this case $w_k(1) = w_k([0,1])$.

Appendix F. Local polynomial approximation and moduli of continuity 187

In contrast, the sharp upper bound for $w_k(n)$ depends on the dimension, and in fact, $\lim_{n\to\infty} w_k(n) = \infty$ if $k \geq 2$. We discuss this situation below following the paper [BKa-2000] by Yu. Brudnyi and Kalton, but first consider the one-dimensional Whitney constant $\omega_k := \omega_k(1)$.

It is easy to show that $\omega_2 = \frac{1}{2}$ (H. Burkill [Bu-1952]) but the value of ω_3 is still unknown. In the above mentioned paper, Whitney proved that $\frac{8}{15} \leq \omega_3 \leq \frac{7}{10}$ and $\omega_k < \infty$ for all k. The latter result is essentially improved by Sendov [Sen-1987] whose method (after a slight modification) leads to the inequality $\omega_k \leq 3$. Further improvement was due to Giliwicz, Kryakin and Shevchuk [GKS-2002]; the result states that $\omega_k \leq 2 + e^{-2}$. It was conjectured by Sendov for all of k and proved for $k \leq 7$ by Kryakin and Zhelnov that $\omega_k \leq 1$, see the survey by Kryakin [Kry-2002] for more information.

Now we present several conjectures and results of the aforementioned paper by Yu. Brudnyi and Kalton.

There is a fairly precise estimate for $w_2(n)$, i.e.,

$$\frac{1}{2}\log_2\left(\left\lfloor\frac{n}{2}\right\rfloor + 1\right) \leq w_2(n) \leq \frac{1}{2}\lfloor\log_2 n\rfloor + \frac{5}{4}.$$

Curiously enough, $w_2(n)$ is almost attained not for the unit n-simplex S^n as it may be thought, but for $S^n \oplus S^n \subset \mathbb{R}^{2n}$. Meanwhile for S^n the precise asymptotic is given by

$$\lim_{n\to\infty} \frac{w_2(S^n)}{\log_2 n} = \frac{1}{4}.$$

In the sequel we will write $w_k(\ell_p^n)$ instead of $w_k(S)$ when S is the closed unit ball of ℓ_p^n. Then $w_2(\ell_1^n) \approx \log n$ while $w_2(\ell_p^n)$ with $1 < p \leq \infty$ is equivalent, up to a logarithmic factor, to $(p-1)^{-1}$ as $p \to 1$. This striking difference in asymptotic behavior is explained by Theorem 3.12 of [BKa-2000] which gives the upper bound of $w_2(B_X)$, where B_X is the unit ball of a finite-dimensional Banach space X, in terms of the p-type constant of X.

Now let $w_k^{(\text{sym})}(n)$ be defined as in (F.14) but for centrally symmetric convex bodies. Then for some numerical constants $c_1, c_2 > 0$,

$$c_1\sqrt{n} \leq w_3^{(\text{sym})}(n) \leq c_2\sqrt{n}\log(n+1).$$

As in the linear approximation case, this result can be improved for $w_3(\ell_p^n)$. For example, $w_3(\ell_2^n) \approx \log(n+1)$ and

$$c_1\log(n+1) \leq w_3(\ell_\infty^n) \leq c_2\bigl(\log(n+1)\bigr)^2.$$

There are also a few estimates for $k \geq 4$. In particular,

$$w_k^{(\text{sym})}(X) \leq cn^{\frac{k}{2}-1}\log(n+1),$$

while

$$w_k(\ell_p^n) \leq cn^{\frac{(k-3)}{2}}\log(n+1)$$

for $2 \leq p \leq \infty$ and $w_k(\ell_1^n) \approx \log(n+1)$.

F.3. Conjectures

(a) If $k \geq 2$, then
$$w_k(n) \approx w_k^{(\text{sym})}(n) \approx n^{\frac{k}{2}-1} \log(n+1)$$
as $n \to \infty$.

This is proved for $k = 2$ while the upper estimate for $w_k^{(\text{sym})}(n)$ is established for $k \geq 2$. As for the lower bound, we only have $w_k(n) \geq w_k^{(\text{sym})}(n) \geq c\sqrt{n}$ for $k \geq 3$.

(b) If $k \geq 3$ and $1 \leq p < \infty$, then
$$w_k(\ell_p^n) \approx \log(n+1)$$
as $n \to \infty$.

The result is established for $p = 1$ and all $k \geq 2$ and for $k = 3$ and $2 \leq p < \infty$, while the lower bound is established for all $k \geq 3$. It is quite possible, that it is way off the mark when $k \geq 4$.

(c) $w_2(\ell_\infty^n)$ is "small", say, $w_2(\ell_\infty^n) \leq 2$. The only known results are $w_2(\ell_\infty^1) = \frac{1}{2}$, and $w_2(\ell_\infty^2) = 1$, and $w_2(\ell_\infty^n) \leq 802$ for $n \geq 3$. If the conjecture held, then for every convex function f on an n-cube Q we would have the inequality
$$E_2(Q;f) \leq \omega_2(Q;f).$$

(d) If X is an infinite-dimensional Banach space, then $w_3(X) = \infty$.

G. Local inequalities for polynomials

We will prove several inequalities estimating the uniform or L_p-norm of a polynomial on a convex body via that on a subset of positive measure. The results have their origin in the classical Chebyshev inequality, see, e.g., [Tim-1963, pp. 67–68]. It asserts that a polynomial $p \in \mathcal{P}_{k,1}$ satisfying the inequality
$$\max_I |p| \leq M$$
on the interval $I := [a,b] \subset \mathbb{R}$ grows outside the interval as
$$|p(x)| \leq M t_k^I(x), \quad x \notin I.$$
Here t_k^I is the normalized k-th *Chebyshev polynomial*, i.e.,
$$t_k^I(x) := t_k\left(\frac{2x-a-b}{b-a}\right),$$
where $t_k(x) := \cos(k \arccos x)$.

Appendix G. Local inequalities for polynomials

If, in particular, $[a, b] \subset [0, 1]$ then this implies the inequality

$$\max_{[0,1]} |p| \leq t_k\left(\frac{2-\ell}{\ell}\right) \max_I |p|, \qquad (G.1)$$

where $\ell := b - a$.

In a little known and hardly available paper [Rem-1936], Remez generalizes (G.1) by replacing $[a, b]$ by an arbitrary measurable subset $S \subset \mathbb{R}$ of measure ℓ. This result has been rediscovered several times, see, e.g., Dudley and Randall [DR-1962] and Yu. Brudnyi and Ganzburg [BrG-1973, Lemma 2] but the Remez proof presented below remains the most simple and elegant.

A multidimensional result of this kind is due to Yu. Brudnyi and Ganzburg [BrG-1973] presented now.

Theorem G.1. *Let $V \subset \mathbb{R}^n$ be a compact convex body and $S \subset V$ be a subset of relative Lebesgue measure*

$$\ell := \frac{|S|}{|V|} > 0.$$

Then for every polynomial $p \in \mathcal{P}_{k,n}$ the sharp inequality

$$\max_V |p| \leq t_k\left(\frac{1 + \sqrt[n]{1-\ell}}{1 - \sqrt[n]{1-\ell}}\right) \sup_S |p| \qquad (G.2)$$

holds.

Proof. We begin with the aforementioned Remez result (and proof).

Claim I. (G.2) is true for $V := [a, b] \subset \mathbb{R}$ and a subset $S \subset V$ of relative Lebesgue measure $\ell > 0$.

We may assume that $V = [0, 1]$; then $S \subset [0, 1]$ is of measure ℓ. Without loss of generality, we may also assume that S is closed and has no isolated points. Moreover, for S being an interval, the result follows from (G.1). Otherwise, $[0, 1] \setminus S$ contains at least one open interval, say, (b, c), and therefore

$$S \subset [a, b] \cup [c, d],$$

where $a := \min S$ and $d := \max S$. Further, let $\xi \in [0, 1]$ be an extreme point for a polynomial $p \in \mathcal{P}_{k,1}$, i.e.,

$$|p(\xi)| = \max_{[0,1]} |p|.$$

As the case $\xi \in S$ is trivial, we should consider only three possibilities:

(a) $\xi \in [0, a]$ or $\xi \in [d, 1]$;

(b) $\xi \in (b, c)$.

First let $\xi \in [0, a]$ and t_k^I be the normalized Chebyshev polynomial for $I := [a, a + \ell]$, and let
$$a =: x_1 < x_2 < \cdots < x_{k+1} := a + \ell$$
be points where t_k^I assumes alternatively the values ± 1. Then we choose points $\hat{x}_1 < \hat{x}_2 < \cdots < \hat{x}_{k+1}$ in S as follows. Set $\hat{x}_1 := a$ and determine $\hat{x}_i$ for $2 \leq i \leq k+1$ by the condition
$$\left| S \cap [\hat{x}_1, \hat{x}_i] \right| = x_i - x_1.$$
These points are correctly defined, since $|S| = \ell = x_{k+1} - x_1$. Due to this choice,
$$a + \ell - x_j \geq |\xi - \hat{x}_j| \quad \text{and} \quad |\hat{x}_i - \hat{x}_j| \geq |x_i - x_j|.$$
Further, $\hat{x}_i \in S$ and therefore $|p(\hat{x}_i)| \leq \sup_S |p|$.

Using Lagrange interpolation we then have
$$|p(\xi)| \leq \sum_{i=1}^{k+1} \left(\prod_{i \neq j} \frac{|\xi - \hat{x}_j|}{|\hat{x}_i - \hat{x}_j|} \right) |p(\hat{x}_j)|$$
$$\leq \left(\sum_{i=1}^{k+1} \left(\prod_{j \neq i} \frac{a + \ell - x_j}{|x_i - x_j|} \right) |t_k^I(x_i)| \right) \sup_S |p|.$$

The sum in the right-hand side equals
$$\left| \sum_{i=1}^{k+1} \left(\prod_{j \neq i} \frac{a + \ell - x_j}{x_i - x_j} \right) t_k^I(x_i) \right| = |t_k^I(a + \ell)| = t_k\left(\frac{2 - \ell}{\ell} \right),$$
and the result follows.

The case of $\xi \in [d, 1]$ is considered similarly.

Finally, let $\xi \in (b, c)$. Then at least one of the fractions
$$\ell_1 := \frac{|[0, \xi] \cap S|}{\xi}, \quad \ell_2 := \frac{|[\xi, 1] \cap S|}{1 - \xi}$$
is greater than or equal to
$$\ell = \frac{|[0, \xi] \cap S| + |[\xi, 1] \cap S|}{\xi + (1 - \xi)}.$$

If, e.g., $\ell_1 \geq \ell$, then we apply the just proved result to the interval $I_1 := [0, \xi]$ and the subset $S_1 := S \cap [0, \xi]$ to have
$$\max_{[0,1]} |p| = |p(\xi)| \leq t_k\left(\frac{2 - \ell_1}{\ell_1} \right) \max_{S_1} |p| \leq t_k\left(\frac{2 - \ell}{\ell} \right) \max_S |p|.$$

Claim I is proved.

Appendix G. Local inequalities for polynomials

To proceed we need the following geometric fact.

Let x_0 be an interior point of the body V and $0 < \ell \leq 1$. Let R stand for a ray emanating from x_0. Consider the extreme problem

$$\gamma(\ell) := \sup_S \left(\underset{R}{\mathrm{ess\,inf}} \, \frac{|V \cap R|_1}{|S \cap R|_1} \right),$$

where S runs over all subsets of V of relative measure $\frac{|S|}{|V|} \geq \ell$. Here $|\cdot|_1$ is the Lebesgue 1-measure.

Claim II. The following is true:

$$\gamma(\ell) = \frac{1}{1 - \sqrt[n]{1 - \ell}}. \tag{G.3}$$

To prove this we use the spherical coordinate system $(r, \varphi) = (r, \varphi_1, \ldots, \varphi_{n-1})$ with the origin at x_0. Let

$$r = H(\varphi) = H(\varphi_1, \ldots, \varphi_{n-1})$$

be the equation of the surface ∂V.

Then we define the subset $\widetilde{S} \subset V$ by the inequalities

$$(\sqrt[n]{1 - \ell}) H(\varphi) \leq r \leq H(\varphi).$$

Comparing the n-volume of V and the Lebesgue n-measure of $\widetilde{S}$, we get

$$|\widetilde{S}| = \ell |V|.$$

Moreover, a similar computation gives

$$\frac{|V \cap R|_1}{|\widetilde{S} \cap R|_1} = \frac{1}{1 - \sqrt[n]{1 - \ell}} \tag{G.4}$$

for every ray R from x_0.

It remains to show that for a subset $S \subset V$ satisfying $\frac{|S|}{|V|} \geq \ell$ the following inequality holds:

$$\sup_S \left(\underset{R}{\mathrm{ess\,inf}} \, \frac{|V \cap R|_1}{|S \cap R|_1} \right) \leq \frac{1}{1 - \sqrt[n]{1 - \ell}}. \tag{G.5}$$

Suppose, on the contrary, that for some $S_0 \subset V$ with $\frac{|S_0|}{|V|} \geq \ell$, the converse inequality holds. This and (G.4) then imply that for almost all R,

$$|S_0 \cap R|_1 < |\widetilde{S} \cap R|_1. \tag{G.6}$$

Further, let $I_0 \subset V \cap R$ be the closed interval of length $|S_0 \cap R|_1$ with right endpoint $|V \cap R|_1$. By monotonicity of the power function,

$$\int_{S_0 \cap R} r^{n-1} dr \leq \int_{I_0} r^{n-1} dr. \tag{G.7}$$

Due to (G.6), I_0 is a proper subinterval of an interval $\widetilde{I}$ which has the same right endpoint $|V \cap R|_1$ and is of length $|\widetilde{S} \cap R|_1$. Substituting in (G.7) $\widetilde{I}$ for I_0, then multiplying by $\prod_{i=1}^{n-1} \cos^{i-1} \varphi_i$ and integrating over φ we have

$$|S_0| = \int_{S_0 \cap R} dv < \int_{\widetilde{S}} dv = |\widetilde{S}| = \ell|V|.$$

This contradicts the inequality $|S_0| \geq \ell|V|$ for the chosen subset S_0.

The result has been proved.

Now we complete the proof of (G.2). Let $S \subset V$ and $|S| \geq \ell|V|$ for some $0 < \ell \leq 1$. Let $p \in \mathcal{P}_{k,n}$ and $x_0 \in V$ be such that

$$\max_V |p| = |p(x_0)|.$$

We may assume that x_0 is an inner point of V; otherwise we use a point close to x_0 and then pass to the limit.

Apply Claim I to the univariate polynomial $p|_{R \cap V}$, where R is a ray from x_0, to obtain

$$\max_V |p| = |p(x_0)| \leq t_k\left(\frac{2|V \cap R|}{|S \cap R|} - 1\right) \max_{S \cap R} |p|.$$

Taking ess inf over R and sup over S, and using monotonicity of t_k on $[1, +\infty)$ and (G.3) we derive that

$$\max_V |p| \leq t_k(2\gamma(\ell) - 1) \max_S |p| = t_k\left(\frac{1 + \sqrt[n]{1-\ell}}{1 - \sqrt[n]{1-\ell}}\right) \max_S |p|,$$

as required.

Finally, let us show that inequality (G.1) is sharp for the class of compact convex bodies. To this end we define V to be a circular cone of height 1, say,

$$V := \left\{x \in \mathbb{R}^n\,;\, x_1^2 \leq \sum_{i=2}^n x_i^2,\ 0 \leq x_1 \leq 1\right\}.$$

Let $\ell \in (0, 1)$ and V_h be a subcone of V of height h determined by the condition $|V \setminus V_h| = \ell|V|$; then $h = \sqrt[n]{1-\ell}$. Set now $S := V \setminus V_h$ and let

$$p(x) := t_k\left(\frac{2x_1 - 1 - h}{1 - h}\right)$$

be the Chebyshev polynomial associated to interval $[h, 1]$. Then $|S| = \ell|V|$ and

$$\max_V |p| = t_k\left(\frac{1+h}{1-h}\right) = t_k\left(\frac{1 + \sqrt[n]{1-\ell}}{1 - \sqrt[n]{1-\ell}}\right) \max_S |p|,$$

i.e., (G.1) becomes equality.

The result has been proved. $\square$

Appendix G. Local inequalities for polynomials

In applications the following consequence of the theorem is of common use.

Corollary G.2. *Under the assumption of Theorem G.1 it is true that*

$$\max_V |p| \leq \frac{1}{2}\left(\frac{4n}{\ell}\right)^k \max_S |p|. \tag{G.8}$$

Proof. The function $\ell \mapsto 1 - \sqrt[n]{1-\ell}$ is convex on $(0,1]$ and therefore

$$\frac{1 + \sqrt[n]{1-\ell}}{1 - \sqrt[n]{1-\ell}} \leq \frac{2n}{\ell} - 1.$$

This, the definition of t_k and its monotonicity on $[1,\infty)$ imply the result. $\square$

In fact, inequality (G.8) may be generalized to integral norms as follows.

Corollary G.3. *Let $0 < r \leq q \leq \infty$ and let S be a subset of V of relative measure $\ell \in (0,1]$. Then for every polynomial P of degree k the inequality*

$$\left\{\frac{1}{|V|}\int_{|V|} |P|^q dx\right\}^{\frac{1}{q}} \leq (rk+1)^{\frac{1}{r}}\gamma(k,n)\ell^{-k}\left\{\frac{1}{|S|}\int_S |P|^r dx\right\}^{\frac{1}{r}} \tag{G.9}$$

holds for $\gamma(k,n) := \frac{1}{2}(4n)^k$.

Proof. It suffices to consider the case $q = \infty$. Due to the homogeneity of (G.9) we may assume that

$$\max_V |P| = 1.$$

Further, for $t \in (0,1]$, we define the sublevel set of P by

$$L_t := \{x \in V\,;\, |P(x)| \leq t\}.$$

Applying to this subset inequality (G.8) we get

$$1 = \max_V |P| \leq \gamma(k,n)\left(\frac{|V|}{|L_t|}\right)^k \cdot t$$

which implies

$$|L_t| \leq |V|\bigl(\gamma(k,n)t\bigr)^{\frac{1}{k}}. \tag{G.10}$$

To proceed, we need the notion of *rearrangement*, see, e.g., [Zi-1989, sec. 1.8] for details.

Let (Σ, μ) be a measure space and $f : \Sigma \to \mathbb{R}$ be μ-measurable. The nonincreasing function $m(f) : (0, +\infty) \to \mathbb{R}_+ \cup \{+\infty\}$ is then given by

$$m(f\,;\,t) := \mu\{\sigma \in \Sigma\,;\, |f(\sigma)| > t\},$$

while the rearrangement $f^* : (0, \mu(\Sigma)] \to \mathbb{R}_+ \cup \{+\infty\}$ is defined by

$$f^*(s) := \inf\{t\,;\, m(f\,;\,t) \leq s\}.$$

The functions $|f|$ and f^* are *equimeasurable*; therefore, for $0 < r < \infty$,

$$\int_0^{\mu(\Sigma)} f^*(s)^r ds = \int_\Sigma |f|^r d\mu.$$

Using these definitions we relate $|L_t|$ to the rearrangement of the trace $P|_V$. Specifically,

$$|L_t| = |V| - m(P|_V\,;\,t)$$

and therefore the converse to the function $t \mapsto |L_t|$ is equal to the function $t \mapsto (P|_V)^*(|V| - t)$.

The latter is estimated by (G.10) to give

$$(P|_V)^*(|V| - t) \geq \frac{1}{\gamma(k,n)} \left(\frac{t}{|V|}\right)^k.$$

It remains to note that for $S \subset V$ and $0 \leq t \leq |S|$,

$$(P|_S)^*(t) \geq (P|_V)^*(t)$$

and therefore

$$\int_0^{|S|} \left[\frac{1}{\gamma(k,n)}\left(\frac{t}{|V|}\right)^k\right]^r dt \leq \int_0^{|S|} \left[(P|_S)^*(|S| - t)\right]^r dt$$

$$= \int_0^{|S|} (P|_S)^*(t)^r dt = \int_S |P|^r dx.$$

Integrating and raising to the power $\frac{1}{r}$ we get the inequality

$$\frac{1}{(rk+1)^{\frac{1}{r}}} \frac{1}{\gamma(k,n)} \left(\frac{|S|}{|V|}\right)^k \leq \left(\frac{1}{|S|} \int_S |P|^r dx\right)^{\frac{1}{r}}$$

which is equivalent to (G.8) with $q = \infty$. □

Remark G.4.

(a) Let $S = V$; then (G.9) yields the *inverse Hölder inequality* for polynomials. The constant obtained is, up to a numerical factor, optimal for $r \leq 1$ and $q = \infty$ but may be essentially improved for other values of r and q, see the paper [CW-2001] by Carbery and Wright and the references therein.

(b) Fix a compact convex body V and denote by $\rho_k(\ell, V)$ the optimal constant in inequality (G.2). By $\rho_k^{\text{conv}}(\ell, V)$ we denote the similar constant, where S now runs over *convex* subsets of V of relative measure ℓ. It is conjectured by Yu. Brudnyi and Ganzburg in [BrG-1973] that

$$\rho_k^{\text{conv}}(\ell, V) = \rho_k(\ell, V). \qquad (\text{G}.11)$$

If this were true, we would obtain the following asymptotic for $\rho_k(\ell, V)$ in the case of V being the Euclidean unit ball B^n:

$$\rho_k(\ell, B^n) = \frac{1}{2}\left(\frac{c_n}{\ell}\right)^k + o(\ell^{-k}),$$

where $c_1 := 8$ and

$$c_n := 4|B^n|\left(1 + \frac{1}{n}\right)^{\frac{n+1}{2}}\left(1 - \frac{1}{n}\right)^{\frac{n-1}{2}}, \quad n \geq 2,$$

see [BrG-1973, Corollary 1].

Finally, we present a Markov type inequality for derivatives of a polynomial. The classical Markov inequality asserts that

$$\max_{|t|\leq 1}|P'(t)| \leq k^2 \max_{|t|\leq 1}|P(t)|, \qquad (\text{G}.12)$$

provided that P is a univariate polynomial of degree k.

Now let $Q \subset \mathbb{R}^n$ be a cube of side length $2r$ and consider polynomial $P \in \mathcal{P}_{k,n}$. Then (G.12) immediately implies that

$$\max_Q |\partial^\alpha P| \leq \left(\frac{k^2}{r}\right)^{|\alpha|} \max_Q |P|.$$

Combining this and inequality (G.9) we get, for this setting and S being a subset of Q of relative measure $\ell \in (0,1)$, the following.

Corollary G.5. *For $0 < r \leq q \leq \infty$ the inequality*

$$\left\{\frac{1}{|Q|}\int_Q |D^\alpha P|^q dx\right\}^{\frac{1}{q}} \leq c\ell^{-k}\left(\frac{k^2}{r}\right)^{|\alpha|}\left\{\frac{1}{|S|}\int_S |P|^r dx\right\}^{\frac{1}{q}} \qquad (\text{G}.13)$$

holds with $c := (rk+1)^{\frac{1}{r}}\gamma(k,n)$.

The first inequality of this kind was due to Di Giorgi and was published in the paper [Cam-1964] by Campanato. It states that there exists an unspecified constant $c(k,n,\ell) < \infty$ such that (G.13) holds with the factor $\frac{c(k,n,\ell)}{r^{|\alpha|}}$ in the right-hand side. The proof is based on a compactness argument.

Part II

Topics in Geometry of and Analysis on Metric Spaces

If catastrophe theory is not predictive, it is precisely because this theory does not take into account a metric or a measure.

R. Thom

Chapter 3

Topics in Metric Space Theory

The chapter presents basic concepts and facts of the theory with an emphasis on the material required for the study of Lipschitz extension problems. As in the case of continuous extensions where the corresponding material covers an essential part of topology, the results presented in this and the next two chapters form a specifically oriented course of metric space theory. For the sake of completeness we include in the exposition several classical results that have played an important role in the development of the theory. The results will be accompanied by proofs, provided that their derivations are based on or related to the important results of this book. In other cases we refer to the original papers or the corresponding books.

3.1 Principal concepts and related facts

Metric spaces are among the most fundamental structures of mathematics, and there is no need for a detailed exposition of their background; the interested reader may find all of the required prerequisites in the textbook [STh-1967] by Singer and Thorpe. The main goal of this subsection is only to recall the basic definitions and to fix notation.

3.1.1 Pseudometrics, metrics and quasimetrics

Definition 3.1. A *pseudometric* on a set $\mathcal{M}$ is a function $d : \mathcal{M} \times \mathcal{M} \to \mathbb{R}_+ \cup \{+\infty\}$ satisfying the following conditions for all $m, m', m'' \in \mathcal{M}$:

- *Positivity*: $d(m, m') \geq 0$ and $d(m, m) = 0$.
- *Symmetry*: $d(m, m') = d(m', m)$.
- *Triangle inequality*: $d(m, m'') \leq d(m, m') + d(m', m'')$.

If a pseudometric d is *finite* and *positive definite*, i.e.,

$$0 < d(m, m') < \infty \quad \text{if} \quad m \neq m',$$

then it is called a *metric*.

Definition 3.2. A symmetric and positive definite function $d : \mathcal{M} \times \mathcal{M} \to \mathbb{R}_+$ is said to be a *quasimetric*, if for all $m, m', m'' \in \mathcal{M}$ and some constant $C > 1$,

$$d(m, m'') \leq C\{d(m, m') + d(m'm'')\} \tag{3.1}$$

and moreover $d(m, m) = 0$ for all m.

The next result by Frink [Fr-1937] clarifies a relation between the last two notions.

Proposition 3.3. *Let d be a quasimetric on $\mathcal{M}$. Then there exists a constant $0 < p < 1$ and a metric $\widetilde{d}$ such that*

$$\widetilde{d} \leq d^p \leq 4\widetilde{d}. \tag{3.2}$$

Proof. We define

$$p := \frac{\log 2}{\log(2C)},$$

where C is the constant in (3.1) for d. Then for any $m, m', m'' \in \mathcal{M}$,

$$d(m, m'')^p \leq 2\max\{d(m, m')^p, d(m', m'')^p\}. \tag{3.3}$$

Lemma 3.4. *If $m, m_1, \ldots, m_n, m'$ are any $n + 2$ points of $\mathcal{M}$, then*

$$d(m, m')^p \leq 2d(m, m_1)^p + 4 \cdot \sum_{j=1}^{n-1} d(m_j, m_{j+1})^p + 2d(m_n, m')^p. \tag{3.4}$$

Proof. Suppose the lemma false. Then there is some value of n for which (3.4) does not hold. Let N be the smallest such integer. Then

$$d(m, m')^p > 2d(m, m_1)^p + 4 \cdot \sum_{j=1}^{N-1} d(m_j, m_{j+1})^p + 2d(m_N, m')^p, \tag{3.5}$$

while (3.4) holds for $n < N$. Now $N > 1$, for with $n = 1$, the relation (3.4) is a consequence of (3.3). It follows from (3.3) that for every m_r either

$$d(m, m')^p \leq 2d(m, m_r)^p, \tag{3.6}$$

or

$$d(m, m')^p \leq 2d(m_r, m')^p. \tag{3.7}$$

3.1. Principal concepts and related facts

If $r = 1$, (3.6) does not hold because of (3.5), hence (3.7) does. Likewise (3.7) does not hold for $r = N$. Let k be the largest value of r for which (3.7) holds. Then $k < N$, and

$$d(m, m')^p \leq 2d(m_k, m')^p. \tag{3.8}$$

From the definition of k,

$$d(m, m')^p \leq 2d(m, m_{k+1})^p. \tag{3.9}$$

Since (3.4) holds for $n < N$,

$$d(m_k, m')^p \leq 2d(m_k, m_{k+1})^p + 4 \cdot \sum_{j=k+1}^{N-1} d(m_j, m_{j+1})^p + 2d(m_N, m')^p,$$

and

$$d(m, m_{k+1})^p \leq 2d(m, m_1)^p + 4 \cdot \sum_{j=1}^{k-1} d(m_j, m_{j+1})^p + 2d(m_k, m_{k+1})^p.$$

Adding the last two inequalities and combining (3.8) and (3.9) gives

$$d(m, m')^p \leq 2d(m, m_1)^p + 4 \cdot \sum_{j=1}^{N-1} d(m_j, m_{j+1})^p + 2d(m_N, m')^p, \tag{3.10}$$

which contradicts (3.5). □

Given two points $m, m' \in \mathcal{M}$, let $m_1, \ldots, m_n$ be any finite number of points of $\mathcal{M}$, not necessarily distinct from each other or from m and m'. We define

$$\widetilde{d}(m, m') := \inf \left\{ d(m, m_1)^p + \sum_{j=1}^{n-1} d(m_j, m_{j+1})^p + d(m_n, m')^p \right\} \tag{3.11}$$

where the infimum is taken over all possible selections of $m_1, \ldots, m_n \in \mathcal{M}$. Clearly $\widetilde{d}$ is a metric on $\mathcal{M}$ satisfying, due to Lemma 3.4, the required inequality

$$\widetilde{d} \leq d^p \leq 4\widetilde{d}.$$

The proof of the proposition is complete. □

3.1.2 Metric and quasimetric spaces

In the sequel we adopt metric space terminology and notation also for the quasimetric case. So, a (quasi-)*metric space* is a pair $(\mathcal{M}, d)$ where d is a metric or quasimetric. As before, we sometimes suppress the d referring to $\mathcal{M}$ as a metric or quasimetric space. Associated to this d, one has the *metric topology* with the basis formed by the *open balls*

$$B_r(m) := \{m' \in \mathcal{M} \ ; \ d(m, m') < r\}.$$

Using this metric topology one introduces all the related notions:

- *open* and *closed* subsets;
- *closure*, *interior* and *boundary*;
- *connectedness* and *arcwise connectedness*;
- *compactness* and *local compactness*;
- *continuous* and *uniformly continuous maps*,

and so on, see, e.g., [STh-1967]. In particular, the *closed ball*

$$\overline{B}_r(m) := \{m' \in \mathcal{M} \ ; \ d(m, m') \leq r\}$$

is closed if d is a quasimetric, because d is (uniformly) continuous in each variable. Note that this notation is a bit misleading as $\overline{B}_r(m)$ may be strictly larger than the closure of $B_r(m)$.

A metric topology is *normal*, i.e., every pair of disjoint closed subsets in $\mathcal{M}$ can be separated by open disjoint neighborhoods. For instance, if F_1, F_2 are closed in a *metric space* $\mathcal{M}$ and the distance

$$d(F_1, F_2) := \inf\{d(m_1, m_2) \ ; \ m_i \in F_i\} \tag{3.12}$$

between them is nonzero, say $r > 0$, then

$$U_i := \left\{m \in \mathcal{M} \ ; \ d(m, F_i) < \frac{r}{2}\right\}, \quad i = 1, 2,$$

are the required neighborhoods.

The same is true for a quasimetric space $(\mathcal{M}, d)$, because, by Proposition 3.3, its metric topology coincides with that of the metric space $(\mathcal{M}, \tilde{d})$.

A (quasi-)metric space $\mathcal{M}$ is *complete*, if every Cauchy sequence in it converges. If $(\mathcal{M}, d)$ is incomplete, it can be completed. In other words, there exists an essentially unique (quasi-)metric space $(\widetilde{\mathcal{M}}, \tilde{d})$, the *completion* of $(\mathcal{M}, d)$, such that $\mathcal{M}$ is a *dense subset*[1] of $\widetilde{\mathcal{M}}$ and the metric induced by $\tilde{d}$ on $\mathcal{M}$ (i.e., $\tilde{d}|_{\mathcal{M} \times \mathcal{M}}$) coincides with d.

The best known result of this relatively basic part of the theory is the *Baire Category Theorem* asserting that *a G_δ-set in a complete metric space is nonempty if it is a countable intersection of dense open sets*.

Recall that a G_δ-set is a countable intersection of open sets, and an F_σ-set is a countable union of closed sets.

---
[1] i.e., its closure in $\widetilde{\mathcal{M}}$ coincides with $\widetilde{\mathcal{M}}$.

3.1. Principal concepts and related facts

Another useful but rather unknown result is due to Sierpiński [Sie-1928]. It asserts that an incomplete metric space is a G_δ-set in its completion. From here, in particular, it follows that an incomplete metric space can be equipped with a topologically equivalent metric [2] under which the space is complete. Equivalently, every sequence converging in one of these metrics converges to the same limit in the other.

The original (and equivalent) version of the *Baire Category theorem* asserts that a complete metric space is a *union of no countable number of nowhere dense subsets*.

Recall that a subset of $\mathcal{M}$ is *nowhere dense*, if its closure does not contain a nonempty subset that is open in $\mathcal{M}$.

In order to animate and demonstrate in practice the basic definitions and concepts introduced, we now present an example of a *universal homogeneous Polish space*, a masterpiece due to Urysohn [Ur-1927]. To clarify the meaning of the italicized words, we formulate several definitions.

A *Polish space* is a complete separable metric space, and the Urysohn theorem in particular asserts that there is a Polish space such that every Polish space is isometric to a subspace of its universal space.

Recall that a *(metric) subspace* of $(\mathcal{M}, d)$ is a pair $\left(S, d|_{S\times S}\right)$ where S is an arbitrary subset of $\mathcal{M}$. In the sequel $d|_{S\times S}$, the restriction of d to $S \times S$, is called the *induced metric* on S.

Moreover, metric spaces $(\mathcal{M}_1, d_1)$ and $(\mathcal{M}_2, d_2)$ are *isometric*, if there is an *isometry* $\varphi : \mathcal{M}_1 \to \mathcal{M}_2$, i.e., a bijection preserving distance.

Notation. Iso$(\mathcal{M}, d)$ or simply Iso$(\mathcal{M})$ is the group of isometries of $\mathcal{M}$ onto itself.

Now we explain the notion of homogeneity used in the formulation of Urysohn's result.

Definition 3.5. (a) A metric space $\mathcal{M}$ is said to be *transitive*, if for every pair of its points there is an isometry of $\mathcal{M}$ that sends one of them to the other.

(b) A metric space $\mathcal{M}$ is said to be (metrically) *homogeneous*, if every pair of its finite isometric subspaces can be combined by an isometry of $\mathcal{M}$.

Theorem 3.6 (Urysohn). *There exists a universal homogeneous Polish space $\mathcal{U}$ such that every Polish space is isometric to a subspace of $\mathcal{U}$.*

The space with these properties is unique up to isometry.

Proof. We begin with the following criterion.

Claim I. A Polish space $(\mathcal{M}, d)$ is homogeneous and universal if and only if it satisfies the following property:

---

[2] Two metrics d_1, d_2 on $\mathcal{M}$ are topologically equivalent if they generate the same metric topology.

(∇) *For every pair of finite sets* $\{m_i\}_{i=1}^n \subset \mathcal{M}$ *and* $\{\lambda_i\}_{i=1}^n \subset \mathbb{R}_+$ *satisfying the inequalities*

$$|\lambda_i - \lambda_j| \leq d(m_i, m_j) \leq \lambda_i + \lambda_j, \quad 1 \leq i, j \leq n, \qquad (3.13)$$

there exists a point $m \in \mathcal{M}$ *such that*

$$d(m, m_i) = \lambda_i, \quad 1 \leq i \leq n. \qquad (3.14)$$

Proof. (Necessity) Let $(\mathcal{M}, d)$ be a universal Polish space. We show that it possesses the property (∇). To this end, define on the set $\mathbb{N}_{n+1} \times \mathbb{N}_{n+1}$, where $\mathbb{N}_n := \{1, \ldots, n\}$, a function $\tilde{d}$ by setting

$$\tilde{d}(i,j) := d(m_i, m_j) \quad \text{for} \quad 1 \leq i, j \leq n,$$

and then extending $\tilde{d}$ to the remaining pairs by

$$\tilde{d}(n+1, i) = \tilde{d}(i, n+1) := \begin{cases} \lambda_i, & \text{if } i \leq n, \\ 0, & \text{if } i = n+1. \end{cases}$$

Here $\{m_i\}$ and $\{\lambda_i\}$ are the n-sets satisfying (3.13). This $\tilde{d}$ is a metric on $\mathbb{N}_{n+1}$ because it is clearly symmetric and positive definite, while the triangle inequality for $\tilde{d}$ follows from its definition and (3.13). By the universality of $(\mathcal{M}, d)$, there is a subspace $\{\hat{m}_1, \ldots, \hat{m}_{n+1}\}$ of $\mathcal{M}$ which is isometric to $(\mathbb{N}_{n+1}, \tilde{d})$ under some isometry φ. Then the point $m := \varphi(n+1)$ satisfies (3.14):

$$d(m, m_i) = d(\varphi(n+1), \varphi(i)) = \tilde{d}(n+1, i) = \lambda_i \quad \text{for} \quad 1 \leq i \leq n.$$

We did not employ the homogeneity of $\mathcal{M}$ in this proof. Hence, every universal Polish space is homogeneous.

(Sufficiency) Assume that $(\mathcal{M}, d)$ is a Polish space satisfying the property (∇). We show first that it is universal. For an arbitrary Polish space $(\widetilde{\mathcal{M}}, \tilde{d})$, let $\{\tilde{m}_i\}_{i=1}^\infty$ be a countable dense subspace. By induction on n, let us construct a sequence of points $\{m_i\}_{i=1}^\infty$ in $\mathcal{M}$ such that

$$\tilde{d}(\tilde{m}_i, \tilde{m}_j) = d(m_i, m_j) \quad \text{for all} \quad 1 \leq i, j \leq n. \qquad (3.15)$$

We arbitrarily choose $m_1 \in \mathcal{M}$ and assume that (3.15) has been proved for $i, j \leq n$. Now we set

$$\lambda_i := \tilde{d}(\tilde{m}_i, \tilde{m}_{i+1}) \quad \text{for} \quad 1 \leq i < n.$$

Then the pair of sets $\{m_1, \ldots, m_n\} \subset \mathcal{M}$ and $\{\lambda_1, \ldots, \lambda_n\} \subset \mathbb{R}_+$ satisfies the inequalities in (3.13) (by (3.15) and the triangle inequality for $\tilde{d}$). Therefore there is a point, say m_{n+1}, in $\mathcal{M}$ such that for $1 \leq i \leq n$,

$$d(m_i, m_{n+1}) = \lambda_i = \tilde{d}(\tilde{m}_i, \tilde{m}_{n+1}).$$

3.1. Principal concepts and related facts

This proves (3.15) for $1 \leq i, j \leq n+1$ and hence for all i, j.

In this way we have obtained a countable subspace $\{m_i\}_{1 \leq i < \infty}$ of $\mathcal{M}$ isometric to a dense subspace of $\widetilde{\mathcal{M}}$. Therefore the closure of $\{m_i\}_{1 \leq i < \infty}$ in $\mathcal{M}$ is a subspace isometric to $\widetilde{\mathcal{M}}$.

We show now that $(\mathcal{M}, d)$ possessing the property (Δ) is homogeneous. Let $F_n^1 := \{m_i^1\}_{i=1}^n$ and $F_n^2 := \{m_i^2\}_{i=1}^n$ be an arbitrary pair of finite isometric subspaces in $\mathcal{M}$. We must find $\varphi \in \mathrm{Iso}(\mathcal{M})$ such that

$$\varphi(F_n^1) = F_n^2.$$

To this end we construct two countable subsets $F^1 := \{m_i^1\}_{i=1}^\infty$ and $F^2 := \{m_i^2\}_{i=1}^\infty$ containing F_n^1 and F_n^2 that satisfy the following conditions:

(a) $d(m_i^1, m_j^1) = d(m_i^2, m_j^2)$ for all i, j;

(b) F^1 and F^2 are dense in $\mathcal{M}$.

When this is done, we will define the required isometry φ as follows.

Given $m \in \mathcal{M}$, choose a subsequence $\{m_i^1\}_{i \in I}$ converging to m. By (a), the set $\{m_i^2\}_{i \in I}$ is a Cauchy sequence. Let $\hat{m}$ be its limit which is clearly independent of the choice of $\{m_i^1\}_{i \in I}$. The map $\varphi : m \mapsto \hat{m}$ is a bijection preserving distances. Moreover, by definition, $\varphi(m_i^1) = m_i^2$ for all i. Hence φ is the required isometry of $\mathcal{M}$ sending F_n^1 to F_n^2.

So it remains to construct the sequences F^1 and F^2. To this end we fix a dense countable set D in $\mathcal{M}$. Using this D and the property (Δ) we construct by induction on N subsets $F_{N+n}^k := \{m_i^k\}_{i=1}^{N+n} \supset F_n^k$, $k = 1, 2$, for which equalities (a) hold with $i, j \leq N + n$. For $N = 0$, these subsets coincide with isometric subspaces F_n^1 and F_n^2 and the result is clear. We now assume that, for $N \geq 0$, these sets have been constructed and define them for $N + 1$. First, let N be even and define m_{n+N+1}^1 to be equal to the element of $D \setminus F_{N+n}^1$ of the smallest index. By the triangle inequality and the induction hypothesis, the pair of sets F_{N+n}^2 and $\{\lambda_i\}_{i=1}^{N+1}$, where $\lambda_i := d(m_i^1, m_{N+n+1}^1)$, satisfies inequalities (3.13). Hence, there is a point, say, $m_{N+n+1}^2 \in \mathcal{M}$, such that for $1 \leq i \leq N + n$,

$$d(m_i^1, m_{N+n+1}^1) = \lambda_i = d(m_i^2, m_{N+n+1}^2).$$

In this case, $F_{N+n+1}^k := F_{N+n}^k \cup \{m_{N+n+1}^k\}$, $k = 1, 2$, are the sets for which the required equalities hold with $1 \leq i, j \leq N + n + 1$.

If, in turn, N is odd we simply repeat the above construction starting now with F_{N+n}^2.

Hence, we have defined two countable sets F^1, F^2 for which (a) holds. Since, by their definition, both of them contain the dense set D, they are dense as well.

This completes the proof of Claim I. $\square$

Corollary 3.7. *A universal homogeneous Polish space is unique up to isometry.*

Proof. Let $(\mathcal{M}_1, d_1)$ and $(\mathcal{M}_2, d_2)$ be two such spaces. Let D_i be a countable dense subspace of $\mathcal{M}_i$ for $i = 1, 2$. Using the previous construction, define sequences $F_i \subset \mathcal{M}_i$, $i = 1, 2$, such that F_1 and F_2 are isometric and $D_i \subset F_i$, $i = 1, 2$. By the choice of D_i, the sets F_i are dense; therefore $\mathcal{M}_1$ and $\mathcal{M}_2$ are isometric as well. □

To complete the proof of Urysohn's theorem, it remains to find a metric space satisfying the property (Δ). This will be done in two steps, defining first an appropriate metric on the set $\mathbb{N}$ of natural numbers.

Claim II. There is a metric d_0 on $\mathbb{N}$ assuming its values in the set of nonnegative rational numbers $\mathbb{Q}_+$ and satisfying the property (Δ) for positive rational numbers λ_i.

Proof. Let $[\mathbb{Q}]_{\text{fin}}$ be the class of all *finite* subsets of $\mathbb{Q}_+$. Since this class is countable, one enumerates it in such a way that the number $n(R)$ for $R \in [\mathbb{Q}]_{\text{fin}}$ satisfies the inequality

$$\operatorname{card} R < n(R). \tag{3.16}$$

Let us define the required metric d_0, setting $d_0(i, i) = 0$ for all $i \in \mathbb{N}$ and then proceeding inductively. Namely, we assume that d_0^n is a metric on $\mathbb{N}_n := \{1, \ldots, n\}$ and extend it to $\mathbb{N}_{n+1}$ as follows.

Let $R = \{\lambda_1, \ldots, \lambda_p\}$ be the element of $[\mathbb{Q}]_{\text{fin}}$ numerated by n, i.e., $n = n(R)$; then $p < n$ by (3.16). To define the required extension d_0^{n+1} of the metric d_0^n, consider two cases.

1. At least one of the inequalities

$$|\lambda_i - \lambda_j| \le d_0^n(i, j) \le \lambda_i + \lambda_j \tag{3.17}$$

   with $i, j \le p$ does not hold.

2. All these inequalities are true for $i, j \le p$.

   We then set for $1 \le j \le n$

$$d_0^{n+1}(n+1, j) = d_0^{n+1}(j, n+1) := \max_{i, k \le n} d_0^n(i, k) \tag{3.18}$$

in the first case and

$$d_0^{n+1}(n+1, j) = d_0^{n+1}(j, n+1) := \min_{k \le p}\{d_0^n(j, k) + \lambda_k\} \tag{3.19}$$

in the second case.

To show that these formulas define a metric on $\mathbb{N}_{n+1}$ which agrees with d_0^n on $\mathbb{N}_n \times \mathbb{N}_n$, we must only establish the triangle inequality for d_0^{n+1}. It suffices to prove the next inequalities for $i, j < n+1$:

$$d_0^{n+1}(i, n+1) + d_0^{n+1}(n+1, j) \ge d_0^n(i, j), \tag{3.20}$$

3.1. Principal concepts and related facts

$$d_0^{n+1}(n+1,i) + d_0^{n+1}(i,j) \geq d_0^{n+1}(n+1,j). \quad (3.21)$$

Both of these inequalities are evident for the case of definition (3.18). If, on the other hand, d_0^{n+1} is defined by (3.19), we denote by k and k' the numbers which give the minimum in (3.19) for $(n+1,i)$ and $(n+1,j)$, respectively. Hence, the left-hand side in (3.20) equals $\bigl(d_0^n(i,k)+d_0^n(j,k')\bigr)+(\lambda_k+\lambda_{k'})$. Since $k, k' \leq p$, inequality (3.17) holds with $i = k$ and $j = k'$, and therefore $\lambda_k + \lambda_{k'} \geq d_0^n(k,k')$. Using these and the triangle inequality for d_0^n, we prove (3.20).

The proof of (3.21) is similar.

We now define a metric d_0 on $\mathbb{N}$ by letting

$$d_0 := \lim_{n \to \infty} d_0^n$$

and show that d_0 satisfies the property (Δ) with rational numbers λ_i. Assume that the inequalities in (3.13) hold for a pair of increasing sequences $\{i_1, \ldots, i_k\} \subset \mathbb{N}$ and $\{\lambda_{i_1}, \ldots, \lambda_{i_k}\} \subset \mathbb{Q}_+$ with $d = d_0$. We extend these sets for indices $i < i_k$ with $i \neq i_s$, $1 \leq s \leq k$, by using the rational numbers

$$\lambda_i := \min_{s \leq k}\{d_0(i,i_s) + \lambda_{i_s}\}, \quad i \neq i_s.$$

It is easy to check that the inequalities in (3.13) hold for the pair $\{1, 2, \ldots, i_k\}$ and $\{\lambda_1, \lambda_2, \ldots, \lambda_k\}$ if they are true for the initial pair. This means that it suffices to prove the property (Δ) for pairs $\{1, 2, \ldots, p\} \subset \mathbb{N}$ and $\{\lambda_1, \ldots, \lambda_p\} \subset \mathbb{Q}_+$ satisfying condition (3.13) (which coincides in this case with (3.17)). To prove the property (Δ) for this case let us denote by n the number of the set $\{\lambda_1, \ldots, \lambda_p\}$ in the enumeration of the class $[\mathbb{Q}]_{\text{fin}}$. According to (3.16), $p < n$. Now let us show that $n+1$ is the required point, i.e., see (3.14),

$$d_0(n+1,i) = \lambda_i \quad \text{for} \quad 1 \leq i \leq p. \quad (3.22)$$

In fact, (3.17) holds for these λ_i and d_0^n and therefore

$$d_0(n+1,i) = \min_{k \leq p}\{d_0(i,k) + \lambda_k\}, \quad (3.23)$$

by (3.19). But the inequalities in (3.17) imply that

$$\lambda_i + d_0(i,i) \leq \lambda_k + d_0(i,k);$$

hence, the minimum in (3.23) equals λ_i, and (3.22) is true for all $1 \leq i \leq p$.

Thus the property (Δ) is satisfied for $(\mathbb{N}, d_0)$ and rational λ_i. □

Let us now define the required universal Polish space $(\mathcal{U}, d)$ as the completion of the space $(\mathbb{N}, d_0)$. In the sequel, $(\mathbb{N}, d_0)$ will be regarded as a dense subspace of $(\mathcal{U}, d)$. In particular, d_0 is the metric induced by d on $\mathbb{N} \subset \mathcal{U}$.

Claim III. $(\mathcal{U}, d)$ is a universal homogeneous Polish space.

According to Claim I it suffices to show that this space satisfies the property (Δ). In other words, given a pair $\{m_1, \ldots, m_n\} \subset \mathcal{U}$ and $\{\lambda_1, \ldots, \lambda_n\} \subset \mathbb{R}_+$ satisfying, for $1 \leq i, j \leq n$, the inequalities

$$|\lambda_i - \lambda_j| \leq d(m_i, m_j) \leq \lambda_i + \lambda_j, \tag{3.24}$$

we must find a point $m \in \mathcal{U}$ so that

$$d(m, m_i) = \lambda_i, \quad 1 \leq i \leq n. \tag{3.25}$$

Using the density of $\mathbb{N}$ in $\mathcal{U}$ and $\mathbb{Q}_+$ in $\mathbb{R}_+$, choose for a given *rational* $\varepsilon > 0$ corresponding ε-approximants $\{m_1^\varepsilon, \ldots, m_n^\varepsilon\} \subset \mathbb{N}$ for $\{m_1, \ldots, m_n\}$ and $\{\lambda_1^\varepsilon, \ldots, \lambda_n^\varepsilon\} \subset \mathbb{Q}_+$ for $\{\lambda_1, \ldots, \lambda_n\}$. For an appropriate ε, we can derive from (3.24) similar inequalities for the pair of ε-approximants. Applying Claim II to these approximants, we find a point $m_\varepsilon \in \mathbb{N}$ such that

$$d(m_\varepsilon, m_j^\varepsilon) = \lambda_j^\varepsilon \quad \text{for} \quad 1 \leq j \leq n. \tag{3.26}$$

Consider now a new pair consisting of the subsets $\{m_1^\varepsilon, \ldots, m_n^\varepsilon\} \cup \{m_\varepsilon\} \subset \mathbb{N}$ and $\{\lambda_1^\varepsilon, \ldots, \lambda_n^\varepsilon\} \cup \{\frac{\varepsilon}{2}\} \subset \mathbb{Q}_+$. Because of (3.25) this pair satisfies, for sufficiently small ε, the inequalities of the property (Δ) (for rational numbers). By Claim II there is a point $m_{\frac{\varepsilon}{2}} \in \mathbb{N}$ so that

$$d(m_{\frac{\varepsilon}{2}}, m_j^\varepsilon) = \lambda_j^\varepsilon \quad \text{for} \quad 1 \leq j \leq n, \quad \text{and} \quad d(m_{\frac{\varepsilon}{2}}, m_\varepsilon) = \frac{\varepsilon}{2}. \tag{3.27}$$

Replacing here ε by $\eta := 2^{-j}\varepsilon$ and setting $m_j := m_\eta$, we derive from the second equality in (3.27) that

$$d(m_i, m_j) \leq \sum_{i \leq k < j} d(m_k, m_{k+1}) \to 0 \quad \text{as} \quad i, j \to \infty.$$

Hence, $\{m_i\}$ is a Cauchy sequence. If $m \in \mathcal{U}$ is the limit of $\{m_i\}$, then passing to the limit in the equalities in (3.27) as $\varepsilon = 2^{-j} \to 0$, we prove that the limit point m satisfies equality (3.25), as required.

This completes the proof of Theorem 3.6. $\square$

Now we briefly discuss the bizarre geometry of the Urysohn space (U, d) using this topic to introduce (or recall) several important notions of the general theory of metric spaces.

Definition 3.8. A metric space $(\mathcal{M}, d)$ is said to be *convex* if for every pair of points $m, m' \in \mathcal{M}$ and numbers $a, b \geq 0$ related by $d(m, m') = a + b$ there exists a point $m'' \in \mathcal{M}$ satisfying

$$d(m, m'') = a \quad \text{and} \quad d(m'', m') = b.$$

3.1. Principal concepts and related facts

We will see that the space $(\mathcal{U}, d)$ is convex. To show this we use an equivalent definition of convexity based on the notion of a *geodesic segment*. Recall that a *curve* (or *path*) in a metric space $(\mathcal{M}, d)$ is a continuous map γ from an interval $[t_0, t_1]$ of the real line into $\mathcal{M}$. This curve *joins points* m and m' if $\gamma(t_0) = m$ and $\gamma(t_1) = m'$. If, in addition, γ is an *isometric embedding*, then it is called a *geodesic* joining m and m', and its image $\gamma([t_0, t_1])$ is the *(geodesic) segment with endpoints* m and m'.

It is readily seen that the following criterion holds.

Proposition 3.9. *A complete metric space $(\mathcal{M}, d)$ is convex if and only if every pair of its points can be joined by a segment.* □

Applying this to a subset S of $\mathcal{M}$ regarded as a metric subspace (with the induced metric) we will say that S is *convex* if every pair $m, m' \in S$ can be joined by a segment in $\mathcal{M}$ and *all* such segments are contained in S.

In particular, a segment may be not convex (e.g., in the case of the space (U, d) where *every* segment is not convex).

Remark 3.10. (a) A name commonly accepted now for the spaces of Definition 3.8 is a *geodesic (metric) space*. We will discuss the basic properties of such spaces in subsection 3.1.7.

(b) Geodesic curves in the sense of Differential Geometry need not be geodesics in the sense of our book; in general, they are *local geodesics*. Let us recall that a curve $\gamma : [0, \ell] \to \mathcal{M}$ is a *local geodesic* if for every point $t \in [0, \ell]$ there is an open interval $I \subset [0, \ell]$ such that $\gamma|_I$ is an isometric embedding.

For example, a locally isometric embedding of $[0, \ell]$ with $\ell < 2\pi$ into a great circle of $\mathbb{S}^2$ is a geodesic if and only if $\ell \le \pi$.

Corollary 3.11. *The space (U, d) is convex but every pair of distinct points can be joined by uncountably many segments.*

Proof. Let $d(m, m') = a > 0$. Consider the sphere

$$\mathbb{S}_a^n := \left\{ x \in \mathbb{R}^{n+1} \; ; \; \sum_{i=1}^{n+1} x_i^2 = \left(\frac{a}{\pi}\right)^2 \right\}$$

equipped with the *spherical distance*. In other words, the distance between points x and y of the sphere is equal to the length of the shorter arc of a great circle of $\mathbb{S}_a^n$ joining these points. Note that this shorter part is a geodesic segment with endpoints x and y, see Remark 1.39. This segment is unique if $x \ne -y$ while there are uncountably many such segments if x, y are antipodal points. In the latter case, the distance between x and $y \, (= -x)$ is precisely $a \, (= d(x, y))$. Using now the universality and homogeneity of the space (U, d) we find an isometric embedding φ of $\mathbb{S}_a^n$ into U such that $\varphi(x) = m$ and $\varphi(-x) = m'$ for some $x \in \mathbb{S}_a^n$. Then the image of every geometric segment connecting x and $-x$ is a segment in $\mathcal{U}$ joining m and m', and there are uncountably many such images. □

The convexity of $\mathcal{U}$, in particular, implies that this space is *arcwise connected* (and hence connected). Recall that a metric space $(\mathcal{M}, d)$ is arcwise connected if any two its points can be joined by a curve. In fact, the Urysohn space enjoys a much stronger connectedness property, *contractibility*, since it is homeomorphic to a Hilbert space, see the Uspenskij paper [Us-2004].

To formulate the final consequence, we set

$$S_r(m_0) := \{m \in \mathcal{U} \; ; \; d(m, m_0) = r\}.$$

This sphere is regarded as a metric space with the induced metric. We will see that the diameter of the sphere $(:= \sup\{d(m, m') \; ; \; m, m' \in S_r(m_0)\})$ equals $2r$.

Corollary 3.12. (a) *Every Polish space $(\mathcal{M}, d)$ of diameter at most $2r$ is isometric to a subspace of $S_r(m_0)$.*

(b) *$S_r(m_0)$ is homogeneous.*

(c) *Every universal and homogeneous Polish space of diameter $2r$ is isometric to $S_r(m_0)$.*

Proof. We prove only claim (a) leaving the proof of the other assertions to the reader (which can be done by using the argument of Theorem 3.6). Let $(\mathcal{M}, d)$ be a Polish space of diameter at most $2r$. Add to $\mathcal{M}$ a new point p and introduce a metric space $(\widetilde{\mathcal{M}}, \widetilde{d})$ with $\widetilde{\mathcal{M}} = \mathcal{M} \cup \{p\}$ and $\widetilde{d}|_{\mathcal{M} \times \mathcal{M}} := d$ and $\widetilde{d}(p, m) = r$ for $m \in \mathcal{M}$.

Using the universality and homogeneity of $\mathcal{U}$, we find an isometric embedding φ of $\widetilde{\mathcal{M}}$ into $\mathcal{U}$ such that $\varphi(p) = m_0$. Then $\varphi(\mathcal{M})$ is clearly a subset of $S_r(m_0)$ and $\varphi|_{\mathcal{M}}$ is an isometry of $\mathcal{M}$ onto $\varphi(\mathcal{M})$. □

Remark 3.13. The argument of the proof of Corollary 3.11 can be applied to show that every pair of points in $S_r(m_0)$ can be joined by uncountably many segments contained in this sphere. But the convexity of the sphere is an open question. In other words, is it true that *every* segment joining two points of $S_r(m_0)$ is contained in this sphere?

In conclusion, let us compare the Urysohn space with two familiar Polish spaces: the Hilbert space of square summable sequences and the space $C[0, 1]$ of continuous functions (with the uniform norm). The former is homogeneous but not universal, while the latter is universal (the Banach–Mazur theorem) but not homogeneous. In fact, even two-point isometric subsets may be not combined by an isometry of this space. A detailed discussion of the aforementioned facts is presented in the classical book [Ban-1932, Ch. XI] by Banach.

3.1.3 Paracompactness and continuous partitions of unity

In many problems of Real Analysis and Topology one can easily find local solutions, but adjusting the various pieces into a global solution is a rather intricate

3.1. Principal concepts and related facts

process. In such cases appropriate partitions of unity yield a simple and powerful tool for resolving the problem. We present here several basic definitions and facts concerning continuous partitions of unity. Though we require these facts only for metric spaces, the following exposition concerns the more general situation of *paracompact topological spaces*.

The introduction of the last notion requires several definitions of point-set topology which are important in their own right.

A collection $\{U_\alpha\}_{\alpha \in A}$ of subsets in a set S is said to be a *cover*, if

$$S = \bigcup_\alpha U_\alpha.$$

A cover $\{V_\beta\}_{\beta \in B}$ of S is a *refinement* of $\{U_\alpha\}_{\alpha \in A}$, if for every $\beta \in B$ there exists $\alpha \in A$ such that

$$V_\beta \subset U_\alpha.$$

Now let S be a topological space. Then a cover $\{U_\alpha\}$ is said to be *open (closed)*, if every U_α is an open (closed) subset. Finally, a cover $\{U_\alpha\}_{\alpha \in A}$ is *locally finite*, if for every point $s \in S$ there exists an open neighborhood U such that the set

$$\{\alpha \in A \ ; \ U_\alpha \cap U \neq \emptyset\}$$

is finite.

Using these concepts we now introduce the required

Definition 3.14. A Hausdorff topological space $\mathcal{S}$ is said to be *paracompact*, if each of its open covers admits a locally finite open refinement.

It is the matter of the definitions to check that a compact metric space is paracompact and that a paracompact space is normal. Rather more profound is the next fact.

Theorem 3.15 (A. H. Stone [Sto-1948]). *Every metric space is paracompact.*

Proof. (M. E. Rudin [MRu-1968]) Assume that $(\mathcal{M}, d)$ is a metric space and $\{U_\alpha\}_{\alpha \in A}$ is an open cover. By the classic Zermelo theorem, there exists an order "$\leq$" on A that converts this set into a *completely ordered* one. In other words, every subset A_0 of A contains its infimum (denoted by $\inf \alpha$), i.e., a (unique) element α_0 of A_0 such that $\alpha_0 \leq \alpha$ for every $\alpha \in A_0$.

We now introduce the required refinement of $\{U_\alpha\}_{\alpha \in A}$ as follows.

Given $\alpha \in A$, define for every integer $n \geq 1$ an open set $V_{\alpha n}$ (by induction on n) to be the union of the balls $B_{2^{-n}}(m)$ with centers m satisfying the following conditions

(a) $\alpha = \inf\{\tilde{\alpha} \in A \ ; \ m \in U_{\tilde{\alpha}}\}$;

(b) m does not belong to any $V_{\beta j}$ with $j < n$ and arbitrary $\beta \in A$;

(c) $B_{3 \cdot 2^{-n}}(m) \subset U_\alpha$.

We show that $\{V_{\alpha n}\}$ covers $\mathcal{M}$. Given $m \in \mathcal{M}$, take $\tilde{\alpha}$ to be the infimum of the set $\{\alpha \in A\,;\, m \in U_\alpha\}$ and then choose n so large that (c) holds. Then, by (b), m either lies in $V_{\tilde{\alpha} n}$ or belongs to $V_{\beta j}$ with some β and $j < n$. Hence, $\{V_\alpha\}_{\alpha \in A}$ is a cover. Since, by definition, $V_{\tilde{\alpha} n}$ is an open subset of $U_{\tilde{\alpha}}$, this cover is an open refinement of $\{U_\alpha\}$.

To prove that $\{V_{\alpha n}\}$ is locally finite, pick $m \in \mathcal{M}$ and let α be the smallest index so that m belongs to $V_{\alpha n}$ for some n; then choose j so that

$$B_{2^{-j}}(m) \subset V_{\alpha n}. \tag{3.28}$$

We show that $B_{2^{-n-j}}(m)$ intersects at most $n + j$ subsets of the refinement. This clearly means that the refinement is locally finite, and hence, S is paracompact.

We derive the required result from the following facts:

(a) If $i \geq n + j$, then $B_{2^{-n-j}}(m)$ does not intersect any $V_{\beta i}$ with $\beta \in A$.

(b) If, on the contrary, $i < n + j$, then this ball intersects at most one $V_{\beta i}$.

To prove the first, choose the ball $B_{2^{-i}}(\widetilde{m})$ used in the definition of $V_{\beta i}$. Since $i > n$, its center $\widetilde{m}$ lies outside $V_{\alpha n}$, see (b). Together with (3.28) this implies that $d(m, \widetilde{m}) > 2^{-j}$.

But $i \geq j + 1$ and $n + j \geq j + 1$, so

$$B_{2^{-n-j}}(m) \cap B_{2^{-i}}(\widetilde{m}) = \emptyset$$

and m does not lie in $V_{\beta i} := \bigcup_{\widetilde{m}} B_{2^{-i}}(\widetilde{m})$.

To prove the second assertion, we will show that for arbitrary $\beta' \neq \beta''$ and i satisfying $i < n + j$,

$$d(V_{\beta' i}, V_{\beta'' i}) := \inf\left\{d(m', m'')\,;\, m' \in V_{\beta' i}, m'' \in V_{\beta'' i}\right\} > 2^{-n-j+1}. \tag{3.29}$$

This clearly implies that $B_{2^{-n-j}}(m)$ intersects only one of the $V_{\beta i}$.

Let $m' \in V_{\beta' i}$ and $m'' \in V_{\beta'' i}$ and $i < n + j$. Assume, for the sake of definiteness, that $\beta' < \beta''$. By the definition of a refinement, there exist points $\widehat{m}', \widehat{m}''$ such that

$$m' \in B_{2^{-i}}(\widehat{m}') \subset V_{\beta' i}, \quad m'' \in B_{2^{-i}}(\widehat{m}'') \subset V_{\beta'' i}.$$

The former inclusion and (c) imply that $B_{3 \cdot 2^{-i}}(\widehat{m}') \subset U_{\beta'}$. On the other hand, $\beta'' > \beta'$ and therefore the latter inclusion and (a) imply that $\widehat{m}'' \notin U_{\beta'}$. Hence $d(\widehat{m}', \widehat{m}'') > 3 \cdot 2^{-i}$ and therefore

$$d(m', m'') > 2^{-i} \geq 2^{-n-j-1}.$$

This proves (3.29) and the theorem. $\square$

We now discuss the second basic fact regarding paracompact spaces, which was used in the proofs of Theorems 1.3 and 1.8. To this goal we need

3.1. Principal concepts and related facts

Definition 3.16. Let S be a topological space, and $C_b(S)$ be the space of bounded continuous functions on S.

A family $\{p_\alpha\}_{\alpha \in A} \subset C_b(S)$ is said to be a *continuous partition of unity*, if it satisfies the conditions

(i) $p_\alpha \geq 0$ for all $\alpha \in A$;

(ii) for every $s \in S$ there exists a neighborhood U of s and a finite set $A_s \subset A$ such that
$$p_\alpha|_U = 0 \quad \text{for all} \quad \alpha \in A \setminus A_s;$$

(iii) $\sum_{\alpha \in A} p_\alpha = 1$;

(iv) $\sup_S p_\alpha \neq 0$ for every $\alpha \in A$.

The following result relates this notion to paracompactness.

Proposition 3.17. *Let $\{U_\alpha\}_{\alpha \in A}$ be an open cover of a paracompact space S. Then there exists a continuous partition of unity $\{p_\beta\}_{\beta \in B}$ satisfying the property:*

for every $\beta \in B$ there is $\alpha \in A$ for which
$$\operatorname{supp} p_\beta := \overline{\{s \in S \,;\, p_\beta(s) \neq 0\}} \subset U_\alpha.$$

For S being a metric space, say $(\mathcal{M}, d)$, the proof is as follows.
Pick an open locally finite refinement $\{V_\beta\}_{\beta \in B}$ of $\{U_\alpha\}_{\alpha \in A}$ and set
$$p_\beta(m) := \frac{d(m, V_\beta^c)}{\sum_{\tilde\beta \in B} d(m, V_{\tilde\beta}^c)}, \quad m \in \mathcal{M}.$$

The denominator is finite and continuous, since $\{V_{\tilde\beta}\}$ is locally finite, and the numerator is zero outside the closure $\overline{V}_\beta$ and strictly positive in its interior. Therefore $\{p_\beta\}$ is the required partition of unity.

Remark 3.18. The constructed partition of unity satisfies the condition
$$\operatorname{supp} p_\beta \subset V_\beta, \quad \beta \in B.$$

In such cases, $\{p_\beta\}$ is said to be *subordinate to the cover* $\{V_\beta\}$.

3.1.4 Compact and precompact metric spaces

For the record, a Hausdorff topological space $\mathcal{T}$ is *compact* if every open cover $\mathcal{V}$ of $\mathcal{T}$ has a finite subcover, that is, there exists a finite number of sets, say $\{V_1, \ldots, V_n\} \subset \mathcal{V}$, such that $\mathcal{T} = \bigcup_{i=1}^{n} V_i$.

If $\mathcal{T}$ is *metrizable*, i.e., its topology is generated by some metric d on $\mathcal{T}$, the classical criterion of compactness is the Bolzano–Weierstrass property:

(BW) *Every sequence in $(\mathcal{T}, d)$ has a convergent subsequence.*

Compact metric spaces are therefore complete. The analog of this notion for noncomplete metric spaces is *precompactness*. Let us recall that a metric space $\mathcal{M}$ is *precompact if every sequence of $\mathcal{M}$ has a Cauchy subsequence.*

The classical Hausdorff criterion of precompactness is more convenient to formulate using the *covering number* of a metric space $\mathcal{M}$. This is a function of $\varepsilon > 0$ given by

$$\mathrm{Cov}(\mathcal{M}\,;\,\varepsilon) := \inf\left[\mathrm{card}\bigl\{\{m_i\} \subset \mathcal{M}\,;\, \mathcal{M} = \bigcup_i B_\varepsilon(m_i)\bigr\}\right]. \qquad (3.30)$$

Equivalently, one can use for this purpose the *ε-capacity* of $\mathcal{M}$ defined by

$$\mathrm{Cap}(\mathcal{M}\,;\,\varepsilon) := \sup\left[\mathrm{card}\{\{m_i\} \subset \mathcal{M}\,;\, d(m_i, m_j) \geq \varepsilon \text{ for } i \neq j\}\right]. \qquad (3.31)$$

Applying the above definitions to a subset S regarded as a metric subspace of $\mathcal{M}$, one obtains the functions Cov and Cap of the arguments $S \in 2^{\mathcal{M}}$ and $\varepsilon > 0$. For example,

$$\mathrm{Cov}(S\,;\,\varepsilon) := \inf\left[\mathrm{card}\bigl\{\{m_i\} \subset S\,;\, S \subset \bigcup_i B_\varepsilon(m_i)\bigr\}\right].$$

The properties of these functions are summarized by

Proposition 3.19. (a) *Cov and Cap are countably subadditive functions in S and nonincreasing and continuous from the right functions in ε. In particular, for every $\varepsilon > 0$ and $\{S_i\}_{i=1}^\infty \subset \mathcal{M}$,*

$$\mathrm{Cov}\left(\bigcup_i S_i\,;\,\varepsilon\right) \leq \sum_i \mathrm{Cov}(S_i\,;\,\varepsilon).$$

(b) *For every $S \subset \mathcal{M}$ and $\varepsilon > 0$,*

$$\mathrm{Cap}\{S\,;\,2\varepsilon\} \leq \mathrm{Cov}(S\,;\,\varepsilon) \leq \mathrm{Cap}(S\,;\,\varepsilon).$$

The proof is the matter of definitions. □

We now formulate the aforementioned Hausdorff criterion of precompactness (see its proof after Theorem 3.59 below).

(H) *A metric space $\mathcal{M}$ is precompact if and only if $\mathrm{Cov}(\mathcal{M}\,;\,\varepsilon)$ is finite for every $\varepsilon > 0$.*

The Hausdorff criterion also contains a quantitative characteristic of the *massiveness* of $\mathcal{M}$. Clearly there are compact metric spaces of an arbitrary growth of $\mathrm{Cov}(\mathcal{M}\,;\,\varepsilon)$ as $\varepsilon \to 0$. However, $\dim \mathcal{M}$ imposes a restriction on this growth as the following Pontriagin and Schnirelman's result of [PSch-1932] demonstrates.

3.1. Principal concepts and related facts

Theorem 3.20. *Suppose that $\mathcal{M}$ is a compact metric space of dimension n. Then, for some constant $c > 0$ and every $\varepsilon > 0$,*

$$c\varepsilon^{-n} \leq \mathrm{Cov}(\mathcal{M} \, ; \, \varepsilon).$$

Proof. Define the *Urysohn k-width* $u_k(\mathcal{M})$ as the infimum of all $\delta > 0$ satisfying the following property:

There is a finite cover of $\mathcal{M}$ of order at most $k+1$ consisting of open sets of diameter at most δ.

Since $\mathcal{M}$ is compact and $\dim \mathcal{M} = n$, Definition A.1 of dimension and the Lebesgue Lemma C.6 imply that for some $\varepsilon_0 > 0$, the inequality

$$\varepsilon_0 < \frac{1}{4} u_{n-1}(\mathcal{M}) \tag{3.32}$$

is true.

Now we find an optimal ε_0-cover by open balls, say $\mathcal{U} := \{B_{\varepsilon_0}(m_i)\}_{i=1}^{\ell}$, that is, $N = \mathrm{Cov}(\mathcal{M}, \varepsilon_0)$.

For this cover we define the canonical $\mathcal{U}$-map $\varphi_{\mathcal{U}} : \mathcal{M} \to [N_{\mathcal{U}}]$ where the target space is the underlying set of the *nerve* for the cover $\mathcal{U}$, see Appendix B. Recall that $\varphi_{\mathcal{U}}$ is defined (in this case) as follows.

Let $\{\psi_i\}_{i=1}^{N}$ be a partition of unity subordinate to $\mathcal{U}$. Associate to element $B_i = B_{\varepsilon_0}(m_i)$ of this cover the vector b_i of the standard basis of the Euclidean space $\mathbb{R}^N (= \ell_2^N)$, $1 \leq i \leq N$. Then the map $\varphi_{\mathcal{U}}$ is given by

$$\varphi_{\mathcal{U}}(m) := \sum_{i=1}^{N} \psi_i(m) b_i, \quad m \in \mathcal{M}.$$

The right-hand side is a point of the simplex $\mathrm{conv}\{b_j : \psi_j(m) > 0\}$ and all these simplices form the complex $N_{\mathcal{U}}$ while their union is the underlying set $[N_{\mathcal{U}}]$ of this complex. The nerve $N_{\mathcal{U}}$ is clearly a subcomplex of the complex Σ_N formed by all subsimplices of the simplex $\left\{x \in \mathbb{R}_+^N \, ; \, \sum_{i=1}^{N} x_i = 1\right\}$.

Let now $\mathrm{St}(b_i)$ be the *star* of vertex b_i, i.e., the union of all the interiors of simplices containing this vertex. For the following it is of use to note that the preimage of $\mathrm{St}(b_i)$ under $\varphi_{\mathcal{U}}$ is given by

$$\varphi_{\mathcal{U}}^{-1}(\mathrm{St}(b_i)) = \bigcup_j \{B_j \, ; \, B_i \cap B_j \neq \emptyset\}.$$

Specifically, this implies that

$$\mathrm{diam} \, \varphi_{\mathcal{U}}^{-1}(\mathrm{St}(b_i)) < 3\varepsilon_0. \tag{3.33}$$

Finally, $\varphi_\mathcal{U}$ is K-Lipschitz for some $K > 1$, if the partition of unity is defined for $m \in \mathcal{M}$ by

$$\psi_i(m) := \frac{d(m, B_i^c)}{\sum\limits_{j=1}^{N} d(m, B_j^c)},$$

see Proposition 3.17.

Now we derive from these facts and (3.32) that

$$\dim \varphi_\mathcal{U}(\mathcal{M}) = n. \tag{3.34}$$

Since a Lipschitz map does not increase dimension, $\dim \varphi_\mathcal{U}(\mathcal{M}) \leq \dim \mathcal{M} = n$. Suppose now that this dimension is less than n. Cover $\varphi_\mathcal{U}(\mathcal{M})$ by the stars $\text{St}(b_i)$, $1 \leq j \leq N$. Due to the definition of dimension, there is a refinement $\mathcal{V} := \{V_\alpha\}_{\alpha \in A}$ of this cover of order at most n. Since each $V_\alpha \subset \text{St}(b_i)$ for some i, inequality (3.33) implies that

$$\operatorname{diam} \varphi_\mathcal{U}^{-1}(V_\alpha) < 3\varepsilon_0.$$

Moreover, the preimage $\varphi_\mathcal{U}^{-1}(\mathcal{V}) := \{\varphi_\mathcal{U}^{-1}(V_\alpha)\}_{\alpha \in A}$ is a cover of $\mathcal{M}$ whose order ord $\mathcal{V}$ is at most n.

Thus we have found the cover of $\mathcal{M}$ of order $\leq n$ and diameter $< 3\varepsilon_0$. By the definition of Urysohn's width, this means that $u_{n-1}(\mathcal{M}) < 3\varepsilon_0$, in contradiction with (3.32). This completes the proof of (3.34).

Now, arguing as in the derivation (3.32), we conclude that the compact n-dimensional set $\varphi_\mathcal{U}(\mathcal{M})$ satisfies

$$u_{n-1}(\varphi_\mathcal{U}(\mathcal{M})) > 0. \tag{3.35}$$

Now let Π_n be the collection of n-dimensional coordinate subspaces in $\mathbb{R}^N$, and $\varphi_\pi(\mathcal{M})$ denote the orthogonal projection of $\varphi_\mathcal{U}(\mathcal{M})$ onto the subspace $\pi \in \Pi_n$. We show that at least one of these projections has dimension n.

Assume, on the contrary, that all of $\varphi_\pi(\mathcal{M})$ have dimension less than n. Then for every $\pi \in \Pi_n$, there is a cover $\mathcal{V}_\pi$ of $\varphi_\pi(\mathcal{M})$ of order $\leq n$ by arbitrarily small open n-cubes (compare with the argument in the proof of Theorem C.5).

Let Q be a cube of $\mathcal{V}_\pi$ and let C_Q be the cylinder over Q orthogonal to π. Choose a cube $Q \in \mathcal{V}_\pi$ for each $\pi \in \Pi_n$ and consider the intersection of C_Q generated by the chosen cubes Q. Then the nonempty intersections form an open cover of $\varphi_\mathcal{U}(\mathcal{M})$ such that every point of $\varphi_\mathcal{U}(\mathcal{M})$ is contained in at most n subsets of the cover. Moreover, the diameters of the intersection can be made arbitrarily small. Therefore the Urysohn $(n-1)$-width of $\varphi_\mathcal{U}(\mathcal{M})$ is zero in contradiction with (3.35).

Let now some $\varphi_\pi(\mathcal{M})$ be of dimension n. Since $\varphi_\pi(\mathcal{M})$ is an n-dimensional subset of the n-dimensional Euclidean space π, the interior of this subset is not empty, see, e.g., [HW-1941, Theorem IV.3]. Therefore there exists an open ball,

3.1. Principal concepts and related facts

say B, of π ($= \mathbb{R}^n$) contained in $\varphi_\pi(\mathcal{M})$. Since the composition of the K-Lipschitz map $\varphi_\mathcal{U}$ and the orthogonal projection of $\mathbb{R}^N$ onto π is K-Lipschitz as well, we have

$$\operatorname{Cov}(B \subset \mathbb{R}^n \; ; \; K\varepsilon) \leq \operatorname{Cov}(\mathcal{M} \; ; \; \varepsilon).$$

To complete the proof, it remains to use the classic Blichfeldt estimate:

$$cR^n \varepsilon^{-n} \leq \operatorname{Cov}(B \subset \mathbb{R}^n \; ; \; \varepsilon), \quad \varepsilon > 0,$$

where R is the radius of the ball B and $c > 0$ depends only on n, see, e.g., [KT-1959].

Together with the previous inequality this completes the proof. □

Remark 3.21. The second part of the Pontriagin–Schnirelman theorem also states that for a compact separable Hausdorff topological space $\mathcal{T}$ of dimension n, there is a metric d_0 on $\mathcal{T}$ generating the same topology and such that for some $c > 0$ and every $\varepsilon > 0$,

$$\operatorname{Cov}((\mathcal{T}, d_0) \; ; \; \varepsilon) \leq c\varepsilon^{-n}.$$

Together with Theorem 3.20 this implies a definition of dimension in metric terms:

$$\dim \mathcal{T} = \inf_d \left\{ \lim_{\varepsilon \to 0} \frac{\log \operatorname{Cov}((\mathcal{T}, d) \; ; \; \varepsilon)}{\log \frac{1}{\varepsilon}} \right\},$$

where d runs over all metrics on $\mathcal{T}$ which are compatible with the topology of the (compact Hausdorff) space $\mathcal{T}$.

3.1.5 Proper metric spaces

Definition 3.22. A metric space is said to be *proper* if each of its closed balls is compact.

It is clear that a proper metric space is complete and *locally compact*; the latter means that *for every point there exists an open set of compact closure containing this point*.

The following simple example shows that a complete and locally compact metric space may be not proper.

Example 3.23. Let $(\mathcal{M}, d)$ be a countable metric space such that for some constant $c > 0$ and all of elements $m \neq m'$,

$$d(m, m') \geq c.$$

The metric topology of $\mathcal{M}$ is discrete, that is, every point is open and closed simultaneously. Hence, $\mathcal{M}$ is locally compact. On the other hand, every Cauchy sequence in $\mathcal{M}$ consists of only a finite number of distinct points and therefore converges. Thus, $\mathcal{M}$ is also complete.

But every closed ball of $\mathcal{M}$ containing infinitely many points is non-compact by the Hausdorff criterion (H).

Now we present a condition under which a compete locally compact metric space is proper. For its formulation we use the metric space $\mathcal{B}_{\mathcal{H}}(\mathcal{M})$ defined in subsection 1.9.1. Let us recall that its elements are bounded closed subsets of $(\mathcal{M}, d)$ and the distance between $S_0, S_1 \in \mathcal{B}_{\mathcal{H}}(\mathcal{M})$ is measured by the *Hausdorff metric* $d_{\mathcal{H}}$, i.e.,

$$d_{\mathcal{H}}(S_0, S_1) := \sup\left\{\max_{i=0,1} d(m_i, S_{1-i}) \; ; \; m_i \in S_i\right\}.$$

Now we formulate the aforementioned condition.

Property (γ)

For every $m \in \mathcal{M}$, a set-valued function from $\mathbb{R}_+$ into $\mathcal{B}_{\mathcal{H}}(\mathcal{M})$ given by $R \mapsto \overline{B}_R(m)$ is continuous.

Theorem 3.24. *Assume that $\mathcal{M}$ is a complete locally compact metric space satisfying the property (γ). Then $\mathcal{M}$ is proper.*

Proof. Given $m \in \mathcal{M}$, define a set $\Lambda \subset \mathbb{R}_+$ by

$$\Lambda := \{r > 0 \; ; \; \overline{B}_r(m) \text{ is compact}\}.$$

Since a closed subset of a compact set is compact, Λ is a nonempty interval. If we show that Λ is simultaneously open and closed, then Λ must coincide with $\mathbb{R}_+$ and we will be done. We establish the required property of Λ in two steps.

Claim I. Λ is open.

Let $r \in \Lambda$; we must show that $[r, \tilde{r}) \subset \Lambda$ for some $\tilde{r} > r$. By the choice of r, the sphere
$$S_r(m) := \{m' \in \mathcal{M} \; ; \; d(m, m') = r\}$$
is compact. Because of the local compactness of $\mathcal{M}$, for every point of this sphere there is an open ball of compact closure centered at this point. As the sphere $S_r(m)$ is compact, there exists a finite number of these balls, say, $B_1, \ldots, B_n$ covering $S_r(m)$. Their closures $\operatorname{clos} B_i \subset \overline{B}_i$ are compact and therefore the set

$$F := \bigcup_i \operatorname{clos} B_i$$

is also compact. By the Lebesgue Lemma C.6, there is $\varepsilon > 0$ such that every subset of diameter $< \varepsilon$ having the nonempty intersection with $S_r(m)$ is contained in one of the balls of this cover. Hence, the ε-neighborhood of $S_r(m)$ is a subset of $F \cup \overline{B}_r(m)$. The same assertion is clearly true for the ε-neighborhood of $\overline{B}_r(m)$. Using now the property (γ) we obtain for a sufficiently small $\varepsilon > 0$, that every

3.1. Principal concepts and related facts

closed ball $\overline{B}_{\tilde{r}}(m)$ with $r \leq \tilde{r} < r + \varepsilon$ is contained in the ε-neighborhood of $\overline{B}_r(m)$, and therefore $\overline{B}_{\tilde{r}}(m)$ is a closed subset of the compact set $F \cup \overline{B}_r(m)$. Hence, $\overline{B}_{\tilde{r}}(m)$ is compact and $[r, r + \varepsilon) \subset \Lambda$.

Claim II. Λ is closed.

Let $r > 0$ be a limit point of Λ. We must show that $r \in \Lambda$, that is, $\overline{B}_r(m)$ is compact. It suffices to prove that every sequence $\{m_i\} \subset \overline{B}_r(m)$ contains a convergent subsequence.

Using the property (γ), let us find, for a given $\varepsilon > 0$, a number $r_\varepsilon \in \Lambda$ sufficiently close to r so that

$$d(m', \overline{B}_{r_\varepsilon}(m)) < \varepsilon \quad \text{for all} \quad m' \in \overline{B}_r(m). \tag{3.36}$$

If $\overline{B}_{r_\varepsilon}(m)$ contains infinitely many points of the sequence $\{m_i\}$, then it contains a convergent subsequence (by the compactness of $\overline{B}_{r_\varepsilon}(m)$). Hence, without loss of generality, we can assume that

$$\sigma_1 := \{m_i\} \subset \overline{B}_r(m) \setminus \overline{B}_{r_\varepsilon}(m).$$

Let $\tilde{m}_i$ be a point of $\overline{B}_{r_\varepsilon}(m)$ closest to m_i. Since

$$\tilde{\sigma}_1 := \{\tilde{m}_i\} \subset \overline{B}_{r_\varepsilon}(m),$$

there is an infinite sequence $\tilde{\sigma}_2 \subset \tilde{\sigma}_1$ such that

$$\operatorname{diam} \tilde{\sigma}_2 < \varepsilon.$$

Let σ_2 be the subsequence of σ_1 corresponding to $\tilde{\sigma}_2$. Then the previous inequality and (3.36) yield

$$\operatorname{diam} \sigma_2 < 3\varepsilon.$$

Now replace in this argument σ_1 with σ_2 and ε with $\frac{\varepsilon}{2}$. Then we first find $\tilde{\sigma}_3 \subset \tilde{\sigma}_2 \subset \overline{B}_{r_{\frac{\varepsilon}{2}}}(m)$ such that

$$\operatorname{diam} \tilde{\sigma}_3 < \frac{3\varepsilon}{2}$$

and then obtain, for the subsequence $\sigma_3 \subset \sigma_2$ corresponding to $\tilde{\sigma}_3$, the inequality

$$\operatorname{diam} \sigma_3 < \frac{3\varepsilon}{4}.$$

Note that in this derivation one may assume that $\sigma_1 \subset \overline{B}_{r_{\frac{\varepsilon}{2}}}(m)$ (throwing away, if necessary, a finite number of elements m_i).

Taking now $\varepsilon = 2^{-i}$, $i = 1, 2 \ldots$, and applying the result established above, we obtain a decreasing family $\{\sigma_i\}_{i=1}^\infty$ of infinite subsequences of $\sigma_1 = \{m_i\}$ such that

$$\operatorname{diam} \sigma_i < \frac{3}{2^{i-1}}, \quad i = 1, 2, \ldots. \tag{3.37}$$

Using the Cantor diagonal process, that is, forming the subsequence $\{\widehat{m}_i\}_{i=1}^\infty$, where $\widehat{m}_i$ is the first element of σ_i, we get from (3.37)

$$d(\widehat{m}_i, \widehat{m}_j) < 6\max(2^{-i}, 2^{-j}).$$

Hence $\{\widehat{m}_i\} \subset \sigma$ is a Cauchy sequence and, by completeness of $\mathcal{M}$, it converges to some $m \in \overline{B}_r(m)$.

This proves Claim II and the theorem. □

Proper metric spaces, even satisfying the property (γ), may be of a highly complicated geometric structure. The situation is somewhat simpler if a proper metric space is transitive, see Definition 3.5. In this case open balls of equal radii are isometric and therefore their covering numbers coincide.

The following property of $\mathcal{M}$ may be regarded as a substitution for transitivity.

Definition 3.25. A complete metric space $\mathcal{M}$ is said to be *uniformly proper* if its *doubling function*

$$\delta_{\mathcal{M}}(R) := \sup_{m \in \mathcal{M}} \mathrm{Cov}\left(B_R(m) \; ; \; \frac{R}{2}\right) \tag{3.38}$$

is finite for all $R > 0$.

Covering $B_R(m)$ by at most $\delta(R)(:= \delta_{\mathcal{M}}(R))$ balls of radii $\frac{R}{2}$, then every ball of the cover by at most $\delta\left(\frac{R}{2}\right)$ balls of radii $\frac{R}{4}$ and so on, we obtain the following estimate for the ε-covering number:

$$\mathrm{Cov}(B_R(m) \; ; \; \varepsilon) \leq \prod_{j \leq j_\varepsilon} \delta\left(\frac{R}{2^j}\right), \tag{3.39}$$

where $j_\varepsilon := \left[\log_2 \frac{R}{\varepsilon}\right] + 1$.

Hence, a uniformly proper space is proper, but the converse is clearly not true.

3.1.6 Doubling metric spaces

Definition 3.26. A metric space $\mathcal{M}$ is said to be *doubling* if its *doubling constant*

$$\delta_{\mathcal{M}} := \sup_{R > 0} \delta_{\mathcal{M}}(R) \tag{3.40}$$

is finite.

As in the proof of (3.39) one derives from the finiteness of $\delta_{\mathcal{M}}$ the inequality

$$\mathrm{Cov}(B_R(m) \; ; \; \varepsilon) \leq 2\left(\frac{R}{\varepsilon}\right)^s, \tag{3.41}$$

3.1. Principal concepts and related facts

where $s := \log_2 \delta_{\mathcal{M}}$.

Hence, an open ball of a doubling metric space is precompact, see criterion (H), and this space is proper if it is complete. Since every subspace of a doubling space is also doubling, any unclosed subspace is an example of a doubling but not of a proper metric space (e.g., $\mathbb{Q}^n$ as a metric subspace of $\mathbb{R}^n$).

Comparing estimate (3.41) with that of Theorem 3.20 we immediately conclude that every ball of a doubling space is finite-dimensional. To make this statement more precise we need

Definition 3.27 (Minkowski dimension[3]). The upper Minkowski dimension of a precompact subspace S of a metric space $\mathcal{M}$ is defined by

$$\overline{\dim}_M S := \varlimsup_{\varepsilon \to 0} \frac{\log \mathrm{Cov}(S\,;\,\varepsilon)}{\log \frac{1}{\varepsilon}}.$$

Replacing $\varlimsup$ with $\varliminf$ one defines the lower Minkowski dimension $\underline{\dim}_M S$. If both of these limits are equal, their common value is denoted by $\dim_M S$.

Using (3.41) we obtain the inequality

$$\overline{\dim}_M B_R(m) \leq \log_2 \delta_{\mathcal{M}};$$

using this we prove

Proposition 3.28. *If $\mathcal{M}$ is doubling, then its (topological) dimension*

$$\dim \mathcal{M} \leq \log_2 \delta_{\mathcal{M}}.$$

Proof. By Theorem 3.20 and Definition 3.27

$$\dim \overline{B} \leq \underline{\dim}_M \overline{B}$$

for every closed ball $\overline{B}$ of $\mathcal{M}$. Together with the previous inequality this bounds $\dim \overline{B}$ by $\log_2 \delta_{\mathcal{M}}$. But $\mathcal{M}$ is the union of a countable family of closed balls, and therefore $\dim \mathcal{M}$ is less than or equal to the supremum of the dimensions of these balls, see, e.g., [HW-1941, Theorem III 2]. □

By the Blichfeldt inequality, see the end of the proof of Theorem 3.20,

$$\dim_M \mathcal{U} = n$$

for every open bounded subset of $\mathbb{R}^n$. Since an n-dimensional metric subspace of $\mathbb{R}^n$ contains an open ball, see, e.g., [HW-1941, Thm. IV 3], the same is true for n-dimensional metric subspaces of $\mathbb{R}^n$. But the Minkowski dimensions of a finite-dimensional doubling metric space $\mathcal{M}$ may be arbitrarily large. In fact, replace the metric d of this space by d^s with an arbitrary $0 < s < 1$. By definition,

$$\dim_M(\mathcal{M}, d^s) = \frac{1}{s} \dim_M(\mathcal{M}, d),$$

---
[3] also known, for Euclidean spaces, as box dimension denoted by $\dim_B$.

and this tends to infinity as $s \to 0$.

Since a compact doubling metric space has finite dimension, it can be homeomorphically embedded into some Euclidean space (the Menger–Nöbeling theorem, see, e.g., [HW-1941, Theorem V.5]). However, this embedding may completely destroy the metric structure of the space. To preserve the structure, such an embedding must be bi-Lipschitz. The following simple result imposes some restriction on the existence of such an embedding into an $\mathbb{R}^n$.

Proposition 3.29. (a) *A subspace of a doubling metric space $\mathcal{M}$ is doubling and its doubling constant is bounded by $\delta_{\mathcal{M}}$.*

(b) *A bi-Lipschitz homeomorphism preserves the doubling property and the Minkowski and topological dimensions.*

Hence, to be a bi-Lipschitz copy of a closed subset of some $\mathbb{R}^n$, a metric space should be doubling and have metric and topological dimensions bounded by n.

An example presented below shows that these conditions are insufficient. However, after a small perturbation of its metric a doubling metric space admits a bi-Lipschitz embedding into a suitable $\mathbb{R}^n$ (see the Assoud theorem in Chapter 5).

The Heisenberg groups

As a set, the Heisenberg group H_n is $\mathbb{C}^n \times \mathbb{R}$, and its group operation is defined by

$$(z,t) \cdot (z',t') := (z+z',\, t+t'+2\operatorname{Im}(z \cdot z')). \tag{3.42}$$

Here $z \cdot z'$ is the standard scalar product on $\mathbb{C}^n$:

$$z \cdot z' := \sum_{j=1}^{n} z_j \bar{z}'_j$$

for $z = (z_1, \ldots, z_n)$, $z' = (z'_1, \ldots, z'_n)$. In particular,

$$|z|^2 := z \cdot z = \sum_{j=1}^{n} |z_j|^2.$$

It is readily seen that H_n is a noncommutative group with $(0,0)$ as the unit and $(-z, -t)$ as the inverse to (z, t). The Heisenberg group has a rich algebraic and analytic structure. It is important for our goal that there is a natural metric on H_n consistent with these structures. The metric, say d, is defined by means of a quasinorm $\|\cdot\| : \mathbb{C}^n \times \mathbb{R} \to \mathbb{R}_+$ given by

$$\|(z,t)\| := (|z|^4 + t^2)^{\frac{1}{4}}.$$

Then $d : H_n \times H_n \to \mathbb{R}_+$ is defined by

$$d(g_1, g_2) := \|g_1^{-1} \cdot g_2\|. \tag{3.43}$$

3.1. Principal concepts and related facts

This function is clearly nonnegative symmetric and positive definite. Surprisingly, it also satisfies the triangle inequality, see [Cy-1981].

The metric introduced is clearly left-invariant, i.e., for every $g, g_1, g_2 \in H_n$,

$$d(g \cdot g_1,\ g \cdot g_2) = d(g_1, g_2).$$

Moreover, d is equivariant with respect to the family of *dilations* $\delta_r : H_n \to H_n$, $r > 0$, the group automorphisms defined by

$$\delta_r(z, t) := (rz, r^2 t).$$

Namely, a simple calculation shows that

$$d(\delta_r g_1,\ \delta_r g_2) = r d(g_1, g_2). \tag{3.44}$$

Finally, H_n has a naturally defined measure μ which equals the Lebesgue measure on $\mathbb{C}^n \times \mathbb{R}$ (identified with $\mathbb{R}^{2n+1}$). In view of (3.42), a left translation $\lambda_g : H_n \to H_n$ defined by $\lambda_g(g') := g \cdot g'$ (and regarded as a map from $\mathbb{R}^{2n+1}$ into itself) is an affine transform. Moreover, the linear part of λ_g has determinant equal to 1. Hence, for a Borel subset $S \subset H_n$ we get

$$\mu(\lambda_g S) = \mu(S).$$

Similarly, the dilation δ_r is a linear transform of $\mathbb{R}^{2n+1}$ of determinant r^{2n+2} and therefore

$$\mu(\delta_r(S)) = r^{2n+2} \mu(S)$$

for Borel subsets $S \subset H_n$. These facts and the invariance of the metric with respect to left translations and dilations imply that

$$\mu(\overline{B}_r(g) \subset H_n) = c_n r^{2n+2},$$

where the constant equals the measure of the unit ball $\overline{B}_1(0)$.

The last equality allows us, as in the case of $\mathbb{R}^n$, to estimate the covering number of a closed ball in H_n as follows:

$$C_1 \left(\frac{r}{\varepsilon}\right)^{2n+2} \leq \operatorname{Cov}(\overline{B}_r(g)\ ;\ \varepsilon) \leq C_2 \left(\frac{r}{\varepsilon}\right)^{2n+2},$$

where the constants $0 < C_1 < C_2$ depend only on n.

This immediately implies that H_n is a doubling metric space. Moreover, its Minkowski dimension $\dim_M H_n = 2n + 2$ and is greater than $\dim H_n = 2n + 1$. In other words, (H^n, d) is a *fractal set*.

For the reader not familiar with this notion, let us only mention that the first fractal set was constructed by G. Cantor at the end of the 19th century. The classical Cantor set $C \subset [0, 1]$ has Minkowski dimension $\log 2 / \log 3 > \dim C = 0$, see, e.g., [Fal-1999] and subsection 4.2.2 for more information on fractals. Nevertheless, the geometric structure of (H_n, d) is much more complicated than that of the Cantor set. In particular, it will be proved in Section 5.1 that (H_n, d) does not admit a bi-Lipschitz embedding in *any* Euclidean space.

Many important metric spaces of Geometry and Analysis satisfy only a local form of the doubling condition. A quantitative version of the corresponding notion presents

Definition 3.30. A metric space $\mathcal{M}$ is *locally doubling* if there exist a constant $R > 0$ and and integer $N \geq 1$ such that the doubling function of $\mathcal{M}$ satisfies

$$\delta_{\mathcal{M}}(r) \leq N \quad \text{for all} \quad r \leq R.$$

The class of these $\mathcal{M}$'s is denoted by $\mathcal{D}(R, N)$.

As an example of a locally doubling space we mention the hyperbolic space $\mathbb{H}^n$ equipped with the geodesic metric d_g, see the text in Chapter 1 preceding (1.63). It is well known that the n-volume of every ball of radius R in $(\mathbb{H}^n, d_g)$ grows asymptotically at infinity as $\exp((n-1)R)$. Therefore, $\delta_{\mathbb{H}^n}(R)$ is finite but has an exponential growth as $R \to \infty$. Hence $(\mathbb{H}^n, d_g)$ is locally doubling but is not doubling.

However, $(\mathbb{H}^n, d_g)$ is uniformly proper (i.e., $\delta_{\mathbb{H}^n}$ is finite). It is worth noting that this fact reflects a general effect related to the so-called geodesic spaces: if such a space belongs to $\mathcal{D}(R_0, N_0)$, then it belongs to $\mathcal{D}(R, N)$ for every $R > 0$ and some $N = N(R)$, see Theorem 3.99 below.

3.1.7 Metric length structure

The concepts of curve and length have sporadically appeared in the preceding text. Now we discuss them more systematically and introduce the associated notions of *length* and *geodesic spaces*.

Let us recall that a *curve* in a metric space $\mathcal{M}$ is a continuous map from a closed interval of the real line, say I, in $\mathcal{M}$. The set of such curves is denoted by $\mathcal{C}(I, \mathcal{M})$. We equip this set with the (uniform) metric letting, for $\gamma_1, \gamma_2 \in \mathcal{C}(I, \mathcal{M})$,

$$d_\ell(\gamma_1, \gamma_2) := \max_{t \in I} d(\gamma_1(t), \gamma_2(t)).$$

Hence, $\mathcal{C}(I, \mathcal{M})$ becomes a metric space which is complete if $\mathcal{M}$ is. If, for instance, $\mathcal{M}$ is discrete $(d(m, m') = 1$ for all $m \neq m')$, then $\mathcal{C}(I, \mathcal{M})$ consists only of constant functions. Such pathological examples are of no interest for Metric Geometry. Below we introduce a class of metric spaces with geometric structure comparable to some extent with the Euclidean structure. A metric space of this class contains a lot of curves which can be effectively used to study its geometry. In the most important cases the set of these curves contains *geodesics*. Let us recall that $\gamma : [a, b] \to \mathcal{M}$ is a *geodesic* (with endpoints $\gamma(a)$ and $\gamma(b)$) if γ is an isometric embedding. Then the image $\gamma([a, b])$ is called a (geodesic) *segment*.

Segments are special cases of *rectifiable curves* introduced by

3.1. Principal concepts and related facts

Definition 3.31. The length of a curve $\gamma : I \to \mathcal{M}$ is

$$\ell(\gamma) := \sup\left\{\sum_i d\big(\gamma(t_i), \gamma(t_{i+1})\big)\right\},$$

where the supremum is taken over all finite increasing sequences $\{t_i\} \subset I$.

The curve γ is said to be *rectifiable* if $\ell(\gamma) < \infty$.

In a discrete metric space, the only rectifiable curves are constant functions.

A more informative example is the graph W of the Weierstrass nowhere differentiable function $w : [0, 2\pi) \to \mathbb{R}$ regarded as a metric subspace of $\mathbb{R}^2$. In this metric space every nontrivial curve is nonrectifiable. Otherwise $w\big|_{[a,b]}$ would be of bounded variation for some nontrivial interval $[a, b]$. But this would imply the differentiability of w almost everywhere in $[a, b]$ (the Lebesgue Differentiability Theorem), a contradiction.

The following result summarizes the main properties of length.

Proposition 3.32. *Let $\gamma : [a, b] \to \mathcal{M}$ be a curve. Then the following is true:*

(a) $\ell(\gamma) \geq \operatorname{diam} \gamma([a, b]) \geq d(\gamma(a), \gamma(b))$ *and* $\ell(\gamma) = 0$ *if and only if γ is constant.*

(b) *Change of parameters: If φ is a homeomorphism from $[a', b']$ onto $[a, b]$, then*

$$\ell(\gamma \circ \varphi) = \ell(\gamma).$$

(c) *Additivity: if $a = t_0 \leq t_1 \leq \cdots \leq t_n = b$ is a nondecreasing sequence, then*

$$\ell(\gamma) = \sum_{i=0}^{n-1} \ell\big(\gamma\big|_{[t_i, t_{i+1}]}\big).$$

(d) *Continuity: If γ is rectifiable of length ℓ, then a function $\lambda : [a, b] \to [0, \ell]$ defined by*

$$\lambda(t) := \ell\big(\gamma\big|_{[a,t]}\big)$$

is continuous.

(e) *Parametrization by arc length: In the previous setting there exists a unique curve $\tilde{\gamma} : [0, \ell] \to \mathcal{M}$ such that*

$$\gamma = \tilde{\gamma} \circ \lambda \quad \text{and} \quad \ell\big(\tilde{\gamma}\big|_{[0,t]}\big) = t.$$

(f) *Lower semicontinuity: If $\gamma_n : I \to \mathcal{M}$, $n \in \mathbb{N}$, converges in $\mathcal{C}(I, \mathcal{M})$ to a rectifiable curve γ, then*

$$\ell(\gamma) \leq \varliminf \ell(\gamma_n).$$

Proof. Assertions (a)–(c) are self-evident while (e) follows from (b) and (d). To prove (d) we use the uniform continuity of γ to find, for a given $\varepsilon > 0$, such $\delta > 0$ that $d(\gamma(t), \gamma(t')) < \frac{\varepsilon}{2}$ for all $t, t' \in [a, b]$ with $|t - t'| < \delta$. Because of the finiteness of $\ell(\gamma)$ one can also find a sequence $a = t_0 < t_1 < \cdots < t_n = b$ such that

$$\sum_{i=0}^{n-1} d(\gamma(t_i), \gamma(t_{i+1})) \geq \ell(\gamma) - \frac{\varepsilon}{2}.$$

Adding, if necessary, new points to $\{t_i\}$ we may assume that $\max_i(t_{i+1} - t_i) < \delta$. Finally, using the additivity of length one rewrites the above inequality as

$$0 \leq \sum_{i=0}^{n-1} \left(\ell\big(\gamma|_{[t_i, t_{i+1}]}\big) - d(\gamma(t_i), \gamma(t_{i+1})) \right) \leq \frac{\varepsilon}{2}.$$

Since each summand here is nonnegative, see (a), the inequality yields

$$\ell\big(\gamma|_{[t_i, t_{i+1}]}\big) \leq d(\gamma(t_i), \gamma(t_{i+1})) + \frac{\varepsilon}{2} < \varepsilon.$$

This proves the continuity of $\lambda(t) := \ell\big(\gamma|_{[a,t]}\big)$.

To prove (f), choose $\varepsilon > 0$ and $\{t_i\}$ as above. Then, for arbitrary N,

$$\ell(\gamma) \leq \frac{\varepsilon}{2} + \sum_{i=0}^{n-1} d(\gamma(t_i), \gamma(t_{i+1})) \leq \frac{\varepsilon}{2} + 2nd(\gamma, \gamma_N) + \ell(\gamma_N).$$

Passing to the limit as N tends to infinity through a suitable sequence of integers, one proves (f). □

Remark 3.33. We also have established within the proof of (d) that

$$\ell(\gamma) = \lim_{|\Delta| \to 0} \sum_{t_i \in \Delta} d(\gamma(t_i), \gamma(t_{i+1})),$$

where the limit is taken over an arbitrary sequence of the partitions $\Delta = \{t_i\}$ of $[a, b]$ such that $|\Delta| := \max_i(t_{i+1} - t_i) \to 0$.

If a curve $\gamma : I \to \mathcal{M}$ is Lipschitz, its length can be computed in a simpler way. Namely, define the *local Lipschitz constant* of $f : (\mathcal{M}, d) \to (\mathcal{M}', d')$ by

$$L(f, m) := \lim_{\varepsilon \to 0} L\big(f|_{B_\varepsilon(m)}\big), \quad m \in \mathcal{M}, \qquad (3.45)$$

cf. with the definition of the global Lipschitz constant in Section 1.7. In particular, $t \mapsto L(\gamma, t)$ is a nonnegative measurable function on I as it is the limit of nonincreasing continuous functions (in $\varepsilon > 0$).

3.1. Principal concepts and related facts

Proposition 3.34. (a) *If $\gamma : I \to \mathcal{M}$ is Lipschitz, then it is rectifiable and*

$$\ell(\gamma) = \int_a^b L(\gamma, t)dt;$$

(b) *if $f : \mathcal{M} \to \mathcal{M}'$ is a Lipschitz map with $L(f) \leq C$, then for a curve $\gamma : I \to \mathcal{M}$,*

$$\ell(f \circ \gamma) \leq C\ell(\gamma).$$

Assertion (b) is self-evident, while assertion (a) requires a (standard) derivation which informally can be presented as

$$\ell(\gamma) \approx \sum d(\gamma(t_i), \gamma(t_{i+1})) \approx \sum L(\gamma, t_i)\Delta t_i \approx \int_a^b L(\gamma, t)dt,$$

where $a = t_0 < t_1 < \cdots < t_n = b$ is a sequence with sufficiently small $\max \Delta t_i := \max(t_{i+1} - t_i)$. We leave the details of the proof to the reader; for a more general statement see [Ri-1961, p. 108].

The family of metric spaces $\mathcal{C}(I, \mathcal{M})$, $I \subset \mathbb{R}$, and the map $\ell : \bigcup_I \mathcal{C}(I) \to \mathbb{R}_+ \cup \{+\infty\}$ defined by length form the *metric length structure* of a metric space $\mathcal{M}$.

Using this, one introduces a pseudometric d_{in} called the *inner metric* of $\mathcal{M}$ by setting

$$d_{in}(m, m') := \inf_\gamma \ell(\gamma) \quad \text{for} \quad m, m' \in \mathcal{M}; \tag{3.46}$$

the infimum is taken over all curves $\gamma : [a,b] \to \mathcal{M}$ joining m and m' (i.e., $\gamma(a) = m$ and $\gamma(b) = m'$).

By Proposition 3.32 (a)

$$d \leq d_{in} \tag{3.47}$$

but $d_{in}(m, m')$ is ∞ if either there are no curves joining m and m' (i.e., $(\mathcal{M}, d)$ is not *arcwise connected*) or all such curves have infinite length.

By (3.47) the d_{in}-topology is stronger than the initial d-topology of $\mathcal{M}$. Their coincidence implies the *locally arcwise connectedness* of $(\mathcal{M}, d)$.

As examples presented below will show, arcwise connectedness is not sufficient for d_{in} to be a metric, and local arcwise connectedness does not imply the topological equivalence of d and d_{in}. For the example of the metric space $W \subset \mathbb{R}^2$ given above (the graph of the Weierstrass function), its inner metric $d_{in}(m, m') = \infty$ for all $m \neq m'$. Hence, d_{in} generates the discrete topology on W.

A slight modification of this example yields a metric space $\mathcal{M}$ for which d_{in} is a metric generating a stronger topology than the initial topology. Namely, let $\mathcal{M} \subset \mathbb{R}^3$ be a straight cone over W of height 1. Let γ be a curve joining points

$m \neq m' \in W$ such that its projection from the apex to $\mathbb{R}^2$ contains a connected component $K \subset W$ distinct from a single point. Then γ satisfies the inequality $\ell(\gamma) \geq C_K \ell(K)$ with some $C_K > 0$, see Proposition 3.34 (a). But $\ell(K) = \infty$ and therefore the only curve of finite length joining m and m' is the union of the straight lines joining the apex with these points.

Hence, d_{in} is a metric in this case, but it induces the discrete topology on W and therefore is stronger than the d-topology.

Definition 3.35. A metric space $(\mathcal{M}, d)$ is called a *length (metric) space*, if
$$d_{in} = d.$$

The following result is a criterion for a complete metric space to be a length space.

Proposition 3.36. *Let $(\mathcal{M}, d)$ be a length space. For every pair $m_0, m_1 \in \mathcal{M}$ and $\varepsilon > 0$ there is a point $m_{\frac{1}{2}}$ so that*
$$\max_{i=0,1} d(m_i, m_{\frac{1}{2}}) \leq \frac{1}{2} \left(d(m_0, m_1) + \varepsilon \right). \tag{3.48}$$

The converse is also true if $(\mathcal{M}, d)$ is complete.

Proof. Let $\gamma : [0,1] \to \mathcal{M}$ be an ε-optimal curve joining the points m_0 and m_1; that is (see Definition 3.35),
$$\ell(\gamma) \leq d(m_0, m_1) + \varepsilon.$$

Since $t \mapsto \ell(\gamma|_{[0,t]})$ is continuous, there is a point $t_0 \in (0,1)$ such that
$$\ell\left(\gamma|_{[0,t_0]}\right) = \ell\left(\gamma|_{[t_0,1]}\right) = \frac{1}{2} \ell(\gamma).$$

Taking $m_{\frac{1}{2}} := \gamma(t_0)$ we prove (3.48).

Now let $(\mathcal{M}, d)$ be complete and suppose that condition (3.48) holds. Given $\lambda > 0$, set $\varepsilon_n := \frac{\lambda}{n^2}$, $n \in \mathbb{N}$. Using induction on n we will find for every pair of dyadic rationals $\delta = \frac{p}{2^n}$ and $\delta_+ := \frac{p+1}{2^n}$ points m_δ and m_{δ_+} such that
$$d(m_\delta, m_{\delta_+}) \leq \frac{d(m_0, m_1)}{2^n} \prod_{k=1}^{n} (1 + \varepsilon_k). \tag{3.49}$$

Assume for the moment that this has been accomplished.

The function defined on the dyadic rationals by $\delta \mapsto m_\delta$ is uniformly continuous by (3.49). Since $\mathcal{M}$ is complete, this function admits a continuous extension, say $\gamma : [0,1] \to \mathcal{M}$. By (3.49), Remark 3.33 and the choice of ε_n,
$$\ell(\gamma) = \lim_{n \to \infty} \sum_{p=0}^{2^n - 1} d\left(\gamma\left(\frac{p}{2^n}\right), \gamma\left(\frac{p+1}{2^n}\right)\right) \leq d(m_0, m_1) \prod_{n=1}^{\infty} \left(1 + \frac{\lambda}{n^2}\right).$$

3.1. Principal concepts and related facts

Choosing here sufficiently small λ we make the right-hand side arbitrarily close to $d(m_0, m_1)$. This clearly implies the desired inequality

$$d_{in}(m_0, m_1) \le d(m_0, m_1).$$

It remains to find the family $\{m_\delta\}$. To this end, set $d := d(m_0, m_1)$ and apply (3.48) with $\varepsilon := \varepsilon_1 d$. Then we find a point $m_{\frac{1}{2}}$ such that

$$\max_{i=0,1} d(m_i, m_{\frac{1}{2}}) \le \frac{d}{2}(1 + \varepsilon_1).$$

Now for the pairs $m_i, m_{\frac{1}{2}}$, $i = 0, 1$, and $\varepsilon := \frac{d}{2}(1+\varepsilon_1)\varepsilon_2$ find points $m_{\frac{1}{4}}$ and $m_{\frac{3}{4}}$, respectively, such that all the distances $d(m_{\frac{p}{4}}, m_{p+\frac{1}{4}})$ with $p = 0, 1, 2, 3$, are at most

$$\frac{1}{2}\left(\frac{d}{2}(1+\varepsilon_1) + \varepsilon\right) = \frac{d}{4}(1+\varepsilon_1)(1+\varepsilon_2).$$

Proceeding this way we obtain the required family $\{m_\delta\}$ satisfying (3.49). □

A large class of examples of length spaces is given by

Proposition 3.37. *Let $(\mathcal{M}, d)$ be a metric space such that d_{in} is a metric on $\mathcal{M}$ (i.e., d_{in} is finite). Then $(\mathcal{M}, d_{in})$ is a length space.*

Proof. Let $\gamma : [a, b] \to (\mathcal{M}, d)$ be a curve. Denote its length in $(\mathcal{M}, d_{in})$ by $\ell_{in}(\gamma)$. We will show that

$$\ell_{in}(\gamma) = \ell(\gamma). \tag{3.50}$$

This clearly proves the proposition.

Since $t \mapsto \ell(\gamma|_{[a,t]})$ is continuous and therefore uniformly continuous on $[a, b]$, we find for each $\varepsilon > 0$ and $\eta > 0$ such that

$$\ell(\gamma|_{[t,t']}) < \varepsilon \quad \text{for} \quad |t - t'| < \eta.$$

Now choose a partition $\{t_i\}_{i=0}^n$ of $[a, b]$ such that $\max(t_{i+1} - t_i) < \eta$. Hence,

$$d_{in}(\gamma(t_i), \gamma(t)) \le \ell(\gamma|_{[t,t']}) < \varepsilon \quad \text{for} \quad t_i \le t \le t_{i+1}. \tag{3.51}$$

Now let $\gamma_i : [t_i, t_{i+1}] \to \mathcal{M}$ be a curve joining $\gamma(t_i)$ and $\gamma(t_{i+1})$, and such that

$$\ell(\gamma_i) \le d_{in}(\gamma(t_i), \gamma(t_{i+1})) + \frac{\varepsilon}{n}.$$

Denote by $\gamma_\varepsilon : [a, b] \to \mathcal{M}$ a curve defined on $[t_i, t_{i+1}]$ by $\gamma_\varepsilon|_{[t_i, t_{i+1}]} := \gamma_i$. Then, by additivity of length,

$$\ell(\gamma_\varepsilon) = \sum_{i=0}^{n-1} \ell(\gamma_i) \le \sum_{i=0}^{n-1} d_{in}(\gamma(t_i), \gamma(t_{i+1})) + \varepsilon \le \ell_{in}(\gamma) + \varepsilon.$$

Now let d_ℓ be the uniform metric in the space $C(I, \mathcal{M})$ of curves $\gamma : I \to \mathcal{M}$. By (3.51) and the triangle inequality we get

$$d_\ell(\gamma_\varepsilon, \gamma) = \max_i d_\ell\left(\gamma_i, \gamma\big|_{[t_i, t_{i+1}]}\right)$$
$$\leq \max_i \max_{t_i \leq t \leq t_{i+1}} \left\{ d(\gamma_i(t), \gamma(t_i)) + d(\gamma(t_i), \gamma(t)) \right\} \leq 2\varepsilon + \frac{\varepsilon}{n}.$$

Using the previous inequalities and the lower semicontinuity of length we then get

$$\ell(\gamma) \leq \lim_{\varepsilon \to 0} \ell(\gamma_\varepsilon) \leq \ell_{in}(\gamma).$$

Since the opposite inequality is evident, this proves (3.50) and the proposition. $\square$

As an example let us consider a metric subspace S of a real normed space B. It is easily seen that S is a length space (i.e., its inner metric equals to the metric induced from B) if and only if S is convex. This motivates the following

Definition 3.38. A metric subspace S of a metric space $\mathcal{M}$ is said to be *C-quasiconvex* ($C \geq 1$) if for every $m, m' \in S$ and $\varepsilon > 0$ there is a curve γ joining m and m' and contained in S so that

$$\ell(\gamma) \leq C d(m, m') + \varepsilon.$$

Let us note that a 1-quasiconvex subspace may be not a convex space in the sense of Definition 3.8 even for the stronger condition of 1-quasiconvexity with $\varepsilon = 0$. Nevertheless, such a subspace will be called convex, except for cases where ambiguity may appear.

Example 3.39. (a) Let $D \subset \mathbb{R}^n$ be a bounded domain with Lipschitz boundary ∂D. This means that for every $x \in \partial D$ there is a neighborhood U_x such that $U_x \cap \partial D$ is a bi-Lipschitz image of the open unit ball of $\mathbb{R}^{n-1}$. It is a simple exercise to show that D is C-quasiconvex for some $C > 1$.

(b) Let K_s be a cone obtained by rotating about the x_1-axis in $\mathbb{R}^2$ the graph of the function $x_1 \mapsto x_1^s$, $0 \leq x_1 \leq 1$. Then the domain $\mathbb{R}^2 \setminus K_s$ is not C-quasiconvex for any $C > 1$ if $0 < s < 1$, see the example after Definition 2.63 for details.

The basic property of length spaces is given by

Theorem 3.40 (Hopf–Rinow). *Let $\mathcal{M}$ be a length space. If $\mathcal{M}$ is complete and locally compact, then*

(a) *$\mathcal{M}$ is proper; equivalently, every bounded closed subset is compact.*

(b) *$\mathcal{M}$ is a geodesic space, i.e., every pair of its points can be joined by a geodesic.*

3.1. Principal concepts and related facts

Proof. (a) Every point of the boundary of a closed ball $\overline{B}_R(m)$ can be joined to its center by an approximate radius, a curve of length approximately R. Hence, the map $R \mapsto \overline{B}_R(m)$ is continuous in the Hausdorff metric, and the space $\mathcal{M}$ satisfies the conditions of Theorem 3.24 (of properness).

(b) Let $\gamma_n : [0,1] \to \mathcal{M}$ be a curve parametrized proportionally to arc length, joining m and m' and such that

$$\ell(\gamma_n) \leq \left(1 + \frac{1}{n}\right) d(m, m'), \quad n \in \mathbb{N}.$$

Then the image of each γ_n lies in the ball $\overline{B}_R(m)$ where $R := 2d(m, m')$ and for $0 \leq t, t' \leq 1$,

$$d(\gamma_n(t), \gamma_n(t')) \leq \ell\left(\gamma_n\big|_{[t,t']}\right) = \ell(\gamma_n)|t - t'| \leq R|t - t'|, \qquad (3.52)$$

see Proposition 3.32 (a).

Hence, the family of functions $t \mapsto \gamma_n(t)$ is equicontinuous and maps $[0,1]$ into a compact set. By the Arcelà–Ascoli theorem we can assume $\{\gamma_n\}$ to be uniformly convergent to a curve $\gamma : [0,1] \to \mathcal{M}$ joining m and m'. By the lower semicontinuity of length,

$$\ell(\gamma) \leq \lim_{n \to \infty} \ell(\gamma_n) \leq d(m, m').$$

Since the length of every curve joining m and m' is at least $d(m, m')$, the curve γ is of minimal length.

Using this γ we define the curve $\tilde{\gamma} : [0, d] \to \mathcal{M}$ where $d := d(m_1, m_2)$ by $\tilde{\gamma}(t) := \gamma(t/d)$. Passing to the limit in (3.52) as $n \to \infty$ we get for $0 \leq t < t' \leq d$,

$$d(\tilde{\gamma}(t), \tilde{\gamma}(t')) \leq |t - t'|. \qquad (3.53)$$

This implies that

$$d \leq d(m, \tilde{\gamma}(t)) + d(\tilde{\gamma}(t), \tilde{\gamma}(t')) + d(\tilde{\gamma}(t'), m') \leq t + |t - t'| + (d - t') = d$$

and therefore (3.53) is actually an equality. Hence, $\tilde{\gamma}$ is an isometric embedding, i.e., the required geodesic joining m and m'. $\square$

To formulate a corollary we need

Definition 3.41. Let a length space be *geodesic*, i.e., every pair of its points can be joined by a geodesic. This space is said to be *uniquely geodesic* if this geodesic is unique.

Corollary 3.42. *A closed 1-quasiconvex subspace of a complete locally compact length space is geodesic.*

Example 3.43. (a) The hyperbolic space $\mathbb{H}^n$ is uniquely geodesic under the (Riemannian) metric introduced in Section 1.10, see the text preceding Theorem 1.38.

(b) The sphere $\mathbb{S}^n$ equipped with the spherical metric, see Remark 1.39, is geodesic but not uniquely geodesic.

(c) The Urysohn universal space is geodesic but not uniquely geodesic, see Corollary 3.11.

(d) A Banach space B is clearly geodesic; B is uniquely geodesic if and only if its closed unit ball U_B is *strictly convex*, i.e., the middle point of every closed interval with endpoints in U_B lies in the interior of U_B; in particular, $\ell_\infty(\Gamma)$ is geodesic but not uniquely geodesic, while $\ell_p(\Gamma)$ with $1 < p < \infty$ is, see, e.g., [BL-2000].

Recall that $\ell_p(\Gamma)$ is defined by the norm

$$\|c\|_p := \left\{ \sum_{\gamma \in \Gamma} |c(\gamma)|^p \right\}^{\frac{1}{p}}.$$

The next result characterizes complete geodesic spaces.

Proposition 3.44 (Existence of a middle point). *A complete metric space $(\mathcal{M}, d)$ is (uniquely) geodesic if and only if for every pair $m_0, m_1 \in \mathcal{M}$, there exists a (unique) point $m_{\frac{1}{2}}$ such that*

$$d(m_i, m_{\frac{1}{2}}) = \frac{1}{2} d(m_0, m_1), \quad i = 0, 1. \tag{3.54}$$

Proof. Since a geodesic is isometric to a segment of the real line, (3.54) is clearly true for a (uniquely) geodesic space.

To prove the converse assertion we, as in the proof of Proposition 3.36, use (3.54) to construct a map f from the set of dyadic rationals of $[0, 1]$ into $\mathcal{M}$ such that $f(0) = m_0$, $f(1) = m_1$ and

$$d\left(f\left(\frac{p}{2^n}\right), f\left(\frac{p+1}{2^n}\right)\right) = \frac{d(m_0, m_1)}{2^n}$$

for $p = 0, 1, \ldots, 2^n - 1$. This clearly implies for $p < q$, the inequality

$$d\left(f\left(\frac{p}{2^n}\right), f\left(\frac{q}{2^n}\right)\right) \leq \frac{q-p}{2^n} d(m_0, m_1). \tag{3.55}$$

But (3.55) is an equality for $p = 0$ and $q = 2^n$ and therefore is an equality for all $p < q$, see the proof of Proposition 3.36. Hence, f admits a continuous extension $\tilde{f} : [0, 1] \to \mathcal{M}$ satisfying

$$d(\tilde{f}(t'), \tilde{f}(t'')) = |t' - t''| d(m_0, m_1) \quad \text{for} \quad 0 \leq t', t'' \leq 1.$$

Let $\ell := d(m_0, m_1)$ and $\gamma : [0, \ell] \to \mathcal{M}$ be given by $\gamma(t) := \tilde{f}(\frac{t}{\ell})$, $0 \leq t \leq \ell$. Then γ is clearly the required geodesic joining m_0 and m_1.

3.1. Principal concepts and related facts

Assume now that there is a unique middle point $m_{\frac{1}{2}}$ for every pair m_0, m_1. Then the geodesic constructed above, $\gamma : [0, \ell] \to \mathcal{M}$, is uniquely determined by its endpoints m_0 and m_1.

Let $\tilde{\gamma} : [0, \ell] \to \mathcal{M}$ be a geodesic joining m_0 and m_1. Then

$$d\left(m_0, \tilde{\gamma}\left(\frac{\ell}{2}\right)\right) = \frac{\ell}{2} = \frac{d(m_0, m_1)}{2},$$

i.e., $\tilde{\gamma}(\frac{\ell}{2})$ is a middle point and therefore $\tilde{\gamma}(\frac{\ell}{2}) = \gamma(\frac{\ell}{2})\left(= f(\frac{1}{2})\right)$. Repeating this argument we then obtain

$$\tilde{\gamma}(q\ell) = \gamma(q\ell)$$

for every dyadic rational of $[0,1]$ and therefore γ and $\tilde{\gamma}$ coincide.

Hence, $\mathcal{M}$ is a uniquely geodesic space. □

Remark 3.45. (a) An earlier version of the result (and the notion of a geodesic (convex) space) was due to Menger [Men-1928]. His result is essentially stronger than the middle point criterion; it states:

A complete metric space $(\mathcal{M}, d)$ is geodesic, if for every pair $m_0 \neq m_1$ from $\mathcal{M}$ there is an intermediate point, that is, a point m distinct from m_0 and m_1 and such that

$$d(m_0, m_1) = d(m_0, m) + d(m, m_1).$$

Menger's original proof is very long and complicated. Simplified versions were given by Aronszajn [Aro-1935] and Kirk [Kir-1976].

(b) The following problem is of considerable interest. Let $\mathcal{T}$ be a Hausdorff topological space. Under what conditions does there exist a geodesic metric on $\mathcal{T}$ whose metric topology coincides with that of $\mathcal{T}$?

A partial answer is given by the following theorem of Bing [Bin-1953].

If $\mathcal{T}$ is metrizable, compact, connected and locally connected, then the required geodesic metric exists. Moreover, there is a dense subset in $\mathcal{T}$ such that every pair of its points can be joined by a unique geodesic.

To formulate the final result of this subsection we need

Definition 3.46. A subset Γ of a metric space $\mathcal{M}$ is said to be an ε-*lattice with parameter* $c \in (0, \frac{1}{2}]$ if the open balls $B_\varepsilon(\gamma)$, $\gamma \in \Gamma$, cover $\mathcal{M}$ while the open balls $B_{c\varepsilon}(\gamma)$, $\gamma \in \Gamma$, are mutually disjoint.

Example 3.47. A subset $C \subset \mathcal{M}$ is ε-*separated* if

$$d(c, c') \geq \varepsilon \quad \text{for distinct} \quad c, c' \in C. \tag{3.56}$$

A maximal ε-separated set is called an ε-chain.

An ε-chain C is an ε-lattice with parameter $\frac{1}{2}$. In fact, the family $\{B_{\frac{1}{2}\varepsilon}(c)\}_{c\in C}$ is disjoint by (3.56) while $\{B_\varepsilon(c)\}_{c\in C}$ covers $\mathcal{M}$ because of the maximality of C.

Remark 3.48. (a) The dual notion to the one just introduced is that of an ε-*net*. This is an ε-*dense* subset N of $\mathcal{M}$ which is *minimal*. In other words, for every point $m \in \mathcal{M}$,

$$d(m, N) < \varepsilon \quad (\varepsilon\text{-density}) \tag{3.57}$$

and for every $n' \in N$,

$$\mathcal{M} \neq \bigcup_{n \in N\setminus\{n'\}} B_\varepsilon(n) \quad (minimality).$$

(b) The existence of ε-chains and ε-nets in a metric space follows from Zorn's lemma.

The next result shows that the points of a lattice in a locally doubling geodesic space are distributed in a rather regular way.

Proposition 3.49. *Let C be an ε-lattice in a locally doubling geodesic space $\mathcal{M}$. Then for every $R > 0$ the function*

$$\delta_C(R) := \sup_{m \in \mathcal{M}} \operatorname{card}(C \cap B_R(m)) \tag{3.58}$$

is finite.

Proof. We prove the result only for the case of ε-chains leaving the proof of the general result to the reader. Since $\mathcal{M}$ is locally doubling, there are constants $R_0 > 0$ and $N_0 > 1$ such that, for all $m \in \mathcal{M}$,

$$\operatorname{Cov}\left(B_{R_0}(m), \frac{R_0}{2}\right) \leq N_0. \tag{3.59}$$

We first show that

$$\delta_C(R_0) < \infty. \tag{3.60}$$

For $\varepsilon \geq R_0$, we consider an optimal cover of $B_{R_0}(m)$ by balls $B_{\frac{R_0}{2}}(m_i)$. Due to (3.59) the number of these balls is at most N_0. Every $B_{\frac{R_0}{2}}(m_i)$ can contain at most one point of the ε-chain C (as $\varepsilon \geq R_0$). Hence $\delta_C(R_0) \leq N_0$ and (3.60) is true.

Now let $\varepsilon < R_0$. Iterating inequality (3.59) and applying Proposition 3.19 (b) we get, for the ε-capacity of $B_{R_0}(m)$, the inequality

$$\operatorname{Cap}(B_{R_0}(m), 2\varepsilon) \leq N_0^\lambda,$$

3.1. Principal concepts and related facts

where $\lambda := \left[\log_2 \frac{R_0}{\varepsilon}\right] + 1$. Recalling the definition of ε-capacity, see (3.31), we derive from this

$$\text{card}(C \cap B_{R_0}(m)) \leq \text{Cap}(B_{R_0}(m), 2\varepsilon) \leq N_0^\lambda$$

and (3.60) holds in this case as well.

Now assume that $R > R_0$. Adding, if necessary, new points to C we may assume without loss of generality that $\varepsilon < R_0 (< R)$.

In this case we choose a point $c \in C \cap B_R(m)$ and denote by $\tilde{c}$ a middle point for the pair m, c, see Proposition 3.44. Let $\hat{c}$ be a point of the ε-chain C closest to $\tilde{c}$. Then $c \in B_{\frac{R}{2}+\varepsilon}(\hat{c})$ and $\hat{c} \in B_{\frac{R}{2}+\varepsilon}(m)$ whence

$$C \cap B_R(m) \subset \bigcup_{\hat{c} \in B_{\frac{R}{2}+\varepsilon}(m)} (C \cap B_{\frac{R}{2}+\varepsilon}(\hat{c})).$$

This embedding clearly implies that

$$\delta_C(R) \leq \left(\delta_C\left(\frac{R}{2}+\varepsilon\right)\right)^2.$$

Iterating this inequality and using (3.60) we get $\delta_C(R) < \infty$ for all $R > R_0$. $\square$

3.1.8 Basic metric constructions

Direct sums

We begin with

Definition 3.50. The p-sum, $1 \leq p \leq \infty$, of a family of metric spaces $(\mathcal{M}_i, d_i)$, $1 \leq i \leq N$, is a metric space with underlying set $\prod_{i=1}^{N} \mathcal{M}_i$ and a metric given for $m, m' \in \prod_{i=1}^{N} \mathcal{M}_i$ by

$$d_p(m, m') := \left(\sum_{i=1}^{N} d_i(m_i, m'_i)^p\right)^{\frac{1}{p}}. \tag{3.61}$$

We denote this space by $\bigoplus^{(p)} \{\mathcal{M}_i\}_{1 \leq i \leq N}$.

Proposition 3.51. Let $\mathcal{M}$ be the p-sum of the family $(\mathcal{M}_i, d_i)$, $1 \leq i \leq N$.

(a) $\mathcal{M}$ is complete if and only if all of $\mathcal{M}_i$ are.

(b) $\mathcal{M}$ is compact if and only if all of $\mathcal{M}_i$ are.

(c) $\mathcal{M}$ is connected if and only if all of $\mathcal{M}_i$ are.

(d) $\mathcal{M}$ is a length space if and only if all of $\mathcal{M}_i$ are.

(e) $\mathcal{M}$ is a complete geodesic space if and only if all of $\mathcal{M}_i$ are.

Proof. Assertions (a)–(c) are well known, see, e.g., [STh-1967]. To prove (d) and (e) it suffices to consider the case of two spaces $\mathcal{M}_1$ and $\mathcal{M}_2$ and then proceed by induction.

Assuming that $\mathcal{M}$ is a length space, we show that, say, $\mathcal{M}_1$ also is. Given points $m_1, m_1' \in \mathcal{M}_1$ and $\varepsilon > 0$, we must find a curve in $\mathcal{M}_1$ of length at most $d_1(m_1, m_1') + \varepsilon$ joining m_1 with m_1'. Let $\gamma : [0,1] \to \mathcal{M}$ be a curve joining $m := (m_1, m_2)$ and $m' = (m_1', m_2')$ with $m_2 = m_2'$ so that
$$\ell(\gamma) \leq d_p(m, m') + \varepsilon = d_1(m_1, m_1') + \varepsilon.$$
Let $p_1 : \mathcal{M} \to \mathcal{M}_1$ be the natural projection. Then the curve $\tilde{\gamma} := p_1 \circ \gamma$ joins m_1, m_1' and its length is bounded by $\ell(\gamma)$, since p_1 is 1-Lipschitz. Hence, $\tilde{\gamma}$ is the required curve.

Conversely, given $m = (m_1, m_2)$, $m' = (m_1', m_2')$ and $\varepsilon > 0$, we must find a curve $\gamma : [0,1] \to \mathcal{M}$ joining m and m' so that $\ell(\gamma) \leq d_p(m, m') + \varepsilon$. To this end, we choose curves $\gamma_i : [0,1] \to \mathcal{M}_i$ parametrized by arc length (see Proposition 3.32 (e)) and joining m_i to m_i', so that
$$\ell(\gamma_i) \leq d_i(m_i, m_i') + \frac{\varepsilon}{2}, \quad i = 1, 2. \tag{3.62}$$

Then the curve $\gamma := (\gamma_1, \gamma_2)$ joins m with m'. Since γ_i is parametrized by arc length, we have for every $n \in \mathbb{N}$ and $i = 1, 2$,
$$d_i\left(\gamma_i\left(\frac{k}{n}\right), \gamma_i\left(\frac{k+1}{n}\right)\right) \leq \ell\left(\gamma_i\big|_{\left[\frac{k}{n}, \frac{k+1}{n}\right]}\right) = \frac{1}{n}\ell(\gamma_i).$$
This, (3.62) and the Minkowski inequality then imply that
$$n d_p\left(\gamma\left(\frac{k}{n}\right), \gamma\left(\frac{k+1}{n}\right)\right) \leq \left(\ell(\gamma_1)^p + \ell(\gamma_2)^p\right)^{\frac{1}{p}} \leq d_p(m, m') + \varepsilon.$$
From this and from Remark 3.33 it follows that
$$\ell(\gamma) = \lim_{n \to \infty} \sum_{k=0}^{n-1} d_p\left(\gamma\left(\frac{k}{n}\right), \gamma\left(\frac{k+1}{n}\right)\right) \leq d_p(m, m') + \varepsilon,$$
as required.

To prove (e), we simply apply the criterion of Proposition 3.44. □

Remark 3.52. In the sequel we will encounter situations which require a special choice of p in (3.61) to define a case "naturally" associated to the direct sum. For instance, the p-direct sum of a family of Euclidean spaces is also an Euclidean space only for $p = 2$. Other examples are the classes of Riemannian manifolds (regarded as metric spaces), finitely generated groups with the word metric and Kobayshi hyperbolic spaces. In these cases, the "natural" choice of p is 2, 1, and ∞, respectively. We will discuss these examples in detail in Section 2.3.

3.1. Principal concepts and related facts

Quotient spaces

Given an equivalence relation $\sim$ on the underlying set of a metric space $(\mathcal{M}, d)$, we let $\widetilde{\mathcal{M}} := \mathcal{M}/\sim$; the latter is the set of equivalence classes. We first define on $\widetilde{\mathcal{M}}$ a pseudometric given by the distance function

$$\delta(\widetilde{m}, \widetilde{m}') := \inf\{d(m, m'); \, m \in \widetilde{m}, \, m' \in \widetilde{m}'\}. \tag{3.63}$$

The metric topology generated by δ is Hausdorff if and only if δ is a metric, i.e., if $\delta(\widetilde{m}, \widetilde{m}') \neq 0$ for all $\widetilde{m} \neq \widetilde{m}'$. The natural projection $p : \mathcal{M} \to \widetilde{\mathcal{M}}$ is continuous in this topology, since

$$\delta(p(m), p(m')) \leq d(m, m').$$

We then define a "length" pseudometric on $\widetilde{\mathcal{M}}$, denoted by $\widetilde{d}$, in the following way.

Let $\widetilde{m} \neq \widetilde{m}'$ be elements of $\widetilde{\mathcal{M}}$. A k-chain joining $\widetilde{m}$ to $\widetilde{m}'$ is a sequence $C := (m_1, m'_1, \ldots, m_k, m'_k)$ of points in $\mathcal{M}$ such that

$$m_1 \in \widetilde{m}, \; m'_k \in \widetilde{m}' \quad \text{and} \quad m_{i+1} \sim m'_i \quad \text{for} \quad 1 \leq i < k.$$

The *length* of C is given by

$$\ell(C) := \sum_{i=1}^{k} d(m_i, m'_i).$$

The pseudometric $\widetilde{d}$ is then defined by

$$\widetilde{d}(\widetilde{m}, \widetilde{m}') := \inf \ell(C), \tag{3.64}$$

where C runs over all chains joining $\widetilde{m}$ to $\widetilde{m}'$.

This clearly satisfies the axioms of a pseudometric, see Definition 3.1, and is finite, since for all $\widetilde{m}, \widetilde{m}'$,

$$\widetilde{d}(\widetilde{m}, \widetilde{m}') \leq \delta(\widetilde{m}, \widetilde{m}'). \tag{3.65}$$

However, it may be zero for $\widetilde{m} \neq \widetilde{m}'$, see examples below.

One of the reasons for introducing $\widetilde{d}$ is the following fact.

Proposition 3.53. *Let $(\mathcal{M}, d)$ be a length space with an equivalence relation. If the length pseudometric δ associated with this relation is a metric, then $(\widetilde{\mathcal{M}}, \delta)$ is a length space.*

Proof. Given $\widetilde{m}, \widetilde{m}' \in \widetilde{\mathcal{M}}$ and $\varepsilon > 0$, we find a k-chain $C := \{(m_i, m'_i)\}_{1 \leq i \leq k}$ joining $\widetilde{m}$ to $\widetilde{m}'$ so that

$$\sum_{i=1}^{k} d(m_i, m'_i) < \widetilde{d}(\widetilde{m}, \widetilde{m}') + \varepsilon. \tag{3.66}$$

Since d is a length metric, there exists a curve $\gamma_i : [0,1] \to \mathcal{M}$ of length smaller than $d(m_i, m_i') + \frac{\varepsilon}{k}$ joining m_i to m_i', $1 \leq i \leq k$. Define a map $\tilde{\gamma} : [0,1] \to \widetilde{\mathcal{M}}$ by

$$\tilde{\gamma}(t) := (p \circ \gamma_i)(kt - i) \quad \text{for} \quad \frac{i}{k} \leq t < \frac{i+1}{k}, \quad i = 0, 1, \ldots, k-1.$$

Further, we set $\tilde{\gamma}(1) := (p \circ \gamma_k)(1)$.

Since the natural projection p is 1-Lipschitz by (3.65), this map is continuous, i.e., $\tilde{\gamma}$ is a curve in $\widetilde{\mathcal{M}}$ which, by definition, joins $\tilde{m}$ with $\tilde{m}'$. Any 1-Lipschitz map is also length nonincreasing and therefore the length of $\tilde{\gamma}$ is not greater than $\sum_{i=1}^{k} \ell(\gamma_i)$ which, in turn, is less than $\sum_{i=1}^{k} d(m_i, m_i') + \varepsilon$. Together with (3.66) this proves the result. $\square$

Example 3.54. (a) Let L be a proper linear subspace of a Banach space $(X, \|\cdot\|)$. An equivalence relation is given by

$$x \sim y \iff x - y \in L.$$

Then the quotient space $\widetilde{X}$ is now the factor-space X/L, and the pseudonorms d and $\tilde{d}$ coincide and are determined by the factor-norm

$$\|\tilde{x}\|_{X/L} := \inf\{\|x - \ell\| \,;\, \ell \in L\}.$$

Note that X/L is incomplete if L is not closed, and the factor-norm is zero identically if L is dense.

(b) More generally, let $G \subset \mathrm{Iso}(\mathcal{M}, d)$ be a group acting on $(\mathcal{M}, d)$ by isometries. An equivalence relation is given by

$$m \sim m' \iff m = g(m') \quad \text{for some} \quad g \in G.$$

Then the quotient space $\widetilde{\mathcal{M}}$ is the set of G-orbits $G(m) := \{g(m)\,;\, g \in G\}$, $m \in \mathcal{M}$, equipped with the pseudonorm $\tilde{d}$. In this case,

$$\tilde{d}(\tilde{m}, \tilde{m}') = d(\tilde{m}, \tilde{m}'),$$

i.e., the infimum in (3.64) is attained on 1-chains.

In fact, if $C := \{(m_i, m_i')\}_{1 \leq i \leq k}$ is a k-chain with $k > 1$ and $m_1' = g(m_2)$ for some isometry $g \in G$, then for the $(k-1)$-chain

$$C' := \big(m_1, g(m_2'), g(m_3), g(m_3'), \ldots, g(m_k), g(m_k')\big)$$

we have $\ell(C') \leq \ell(C)$.

The conventional notation of the quotient space of this example is $G\backslash\mathcal{M}$ (the *space of G-orbits*).

3.1. Principal concepts and related facts

(c) Under the following condition the space of G-orbits is a metric space.

Assume that for each $m \in \mathcal{M}$ there is a number $r > 0$ such that

$$\operatorname{card}\{g \in G\,;\, g(B_r(m)) \cap B_r(m) \neq \emptyset\} < \infty. \tag{3.67}$$

In this case the group G is said to be *proper*[4].

We claim that if G is proper, then $\tilde{d}(=d)$ is a metric on $G\backslash\mathcal{M}$. In fact, let $d(G(m'), G(m)) = 0$. Then for some sequence $\{g_n\} \subset G$

$$\lim_n d(m', g_n(m)) = 0. \tag{3.68}$$

We derive from here that $m' = g(m)$ for some $g \in G$ and therefore $G(m) = G(m')$; that is, d is a metric on $G\backslash\mathcal{M}$. To see this, given $\varepsilon > 0$, we choose an integer n_0 so that $g_n(m) \in B_\varepsilon(m')$ for all $n \geq n_0$. Then for all such n,

$$g_{n_0}(B_{2\varepsilon}(m)) \cap g_n(B_{2\varepsilon}(m)) \neq \emptyset. \tag{3.69}$$

We claim that if ε is sufficiently small, then (3.69) implies that $g_n = g_{n_0}$ for $n \geq n_0$. Together with (3.68) this yields $m' = g_{n_0}(m)$, as required.

Indeed, by properness, the ball $B_r(m)$ in (3.67) meets only a finite number of points from $G(m)$. Hence, for sufficiently small $\varepsilon > 0$ we have $B_{4\varepsilon}(m) \cap G(m) = \{m\}$. If $B_{2\varepsilon}(m) \cap g(B_{2\varepsilon}(m)) \neq \emptyset$ for this ε, then $g(m) = m$. Applying this to (3.69) with $g := g_{n_0}^{-1} g_n$ we obtain the desired result.

Let, in particular, G be a subgroup of $\mathrm{Iso}(\mathbb{R}^n)$ generated by n linearly independent shifts $x \mapsto x + e_i$. Here $\{e_i\}_{i=1}^n$ is the standard basis of the Euclidean space $\mathbb{R}^n$. An equivalence relation $x \sim y$ is defined by $x - y \in \mathbb{Z}^n$ and $G\backslash\mathbb{R}^n$ is therefore the factor-group $\mathbb{R}^n/\mathbb{Z}^n$ endowed with the quotient metric. Note that $\mathbb{R}^n/\mathbb{Z}^n$ is the direct 2-sum of n circles $\mathbb{T} := \mathbb{R}/\mathbb{Z}$, i.e., $\mathbb{R}^n/\mathbb{Z}^n$ is a (flat) n-torus $\mathbb{T}^n$. The natural projection $p : \mathbb{R}^n \to \mathbb{T}^n$ is a *local isometry*, i.e., for each point $x \in \mathbb{R}^n$ there is its open neighborhood U such that $p|_U$ is an isometric embedding. In general, the natural projection $p : \mathcal{M} \to G\backslash\mathcal{M}$ is a local isometry for every proper action G (an exercise for the reader).

(d) (*Glueing along isometric subspaces.*) Let $(\mathcal{M}_i, d_i)$ be a metric space and S_i be a closed subspace isometric to a (fixed) metric space S, $i = 1, 2$. Let $\varphi_i : S \to S_i$ be the corresponding isometries. We define a quotient space whose underlying set is the disjoint union $\mathcal{M}_1 \sqcup \mathcal{M}_2$ transformed by a "glueing" relation. Namely, we say that $m_1 \in S_1$ is equivalent to $m_2 \in S_2$ whenever $\varphi_1^{-1}(m_1) = \varphi_2^{-1}(m_2)$. In the remaining case of $m \in (\mathcal{M}_1 \backslash S_1) \sqcup (\mathcal{M}_2 \backslash S_2)$, the point m is equivalent to m' whenever $m = m'$.

We denote the corresponding quotient space by $\mathcal{M}_1 \sqcup_S \mathcal{M}_2$. This space is called the *glueing* (or *amalgamation*) of the $\mathcal{M}_i$'s along S. In this notation, each space $\mathcal{M}_i$ is, in fact, identified with its image under the natural

---
[4] One also says that G acts on $\mathcal{M}$ *properly*.

projection $\mathcal{M}_1 \sqcup \mathcal{M}_2 \to (\mathcal{M}_1 \sqcup \mathcal{M}_2)/\sim$. Using this identification we present a formula giving the quotient metric (3.63) on the glueing. Using the triangle inequality, it is easily seen that the distance $\delta(m_1, m_2)$ in (3.63) equals $d(m_1, m_2)$ whenever these points belong to $\mathcal{M}_i$ for some $i \in \{0, 1\}$ (in this case $\widetilde{m}_i = m_i$, $i = 1, 2$). In other words, the infimum in (3.64) is attained on 1-chains. Moreover,

$$\widetilde{d}(m_1, m_2) = \inf\{d_1(m_1, s) + d_2(m_2, s) \, ; \, s \in S\} \qquad (3.70)$$

provided that $m_1 \in \mathcal{M}_1$ and $m_2 \in \mathcal{M}_2$, because any k-chain C in (3.64) can be replaced by a 2-chain C' with $\ell(C') \le \ell(C)$.

The equality $\widetilde{d}(\widetilde{m}_1, \widetilde{m}_2) = 0$ implies that $\widetilde{m}_1 = \widetilde{m}_2$ due to the fact that S is closed; hence $\widetilde{d}$ is a metric. It can be easily proved that *if the d_i are length metrics, then $\widetilde{d}$ also is. Moreover, $\widetilde{d}$ is the unique length metric on $\mathcal{M}_1 \sqcup_S \mathcal{M}_2$ such that the induced metric on each $\mathcal{M}_i \subset \mathcal{M}_1 \sqcup_S \mathcal{M}_2$ coincides with d_i*.

If, in addition, S is proper and each $\mathcal{M}_i$ is geodesic, then $\mathcal{M}_1 \sqcup_S \mathcal{M}_i$ is also geodesic.

In fact, by a compactness argument, the infimum in (3.70) is attained at some $s \in S$. This observation immediately leads to the existence of a geodesic joining every two points of $\mathcal{M}_1 \sqcup_S \mathcal{M}_2$.

Limits of metric spaces

Let $d_\mathcal{H}(\mathcal{M}_1, \mathcal{M}_2)$ be the Hausdorff distance between two subsets of a metric space $(\mathcal{M}, d)$, see (1.31) for its definition. The Hausdorff distance is a pseudometric on the set of all subsets of $\mathcal{M}$, since it may assign $+\infty$ and since $d_\mathcal{H}(\mathcal{M}_1, \mathcal{M}_2)$ may be zero for some $\mathcal{M}_1 \ne \mathcal{M}_2$ (e.g., if $\mathcal{M}_1$ is dense in $\mathcal{M}_2$). However, $d_\mathcal{H}$ is a metric on the set $\mathcal{B}(\mathcal{M})$ of all closed bounded subsets of $\mathcal{M}$. The classical Hausdorff theorem asserts that $(\mathcal{B}(\mathcal{M}), d_\mathcal{H})$ is compact if $\mathcal{M}$ is.

Gromov [Gr-1981] proposed a generalization of the Hausdorff construction allowing one to define limits for sequences of metric spaces which are not contained in an ambient metric space. The main point of his approach is the *Gromov–Hausdorff distance* whose definition we present now.

Let $(\mathcal{M}_i, d_i)$ be metric spaces, $i = 1, 2$. Let d be a metric on the disjoint union whose restriction to $\mathcal{M}_i$ coincides with d_i, $i = 1, 2$. Then each $\mathcal{M}_i$ is a subspace of the space $(\mathcal{M}_1 \sqcup \mathcal{M}_2, d)$, and the Hausdorff distance $d_\mathcal{H}(\mathcal{M}_1, \mathcal{M}_2)$ is well defined. Now we set

$$d_{\mathcal{GH}}(\mathcal{M}_1, \mathcal{M}_2) := \inf_d d_\mathcal{H}(\mathcal{M}_1, \mathcal{M}_2), \qquad (3.71)$$

where d runs over all such metrics on $\mathcal{M}_1 \sqcup \mathcal{M}_2$.

3.1. Principal concepts and related facts

Since the Hausdorff distance is a pseudometric, $d_{\mathcal{GH}}$ is not a metric, in general. Even the restriction of $d_{\mathcal{GH}}$ to the class of all complete bounded metric spaces is not a metric (though $\mathcal{M}_1, \mathcal{M}_2$ belong to $\mathcal{B}(\mathcal{M}_1 \sqcup \mathcal{M}_2, d)$ in this case).

To clarify the properties of the notion introduced we first establish its relation to the Hausdorff distance. We note that if $\mathcal{M}_1, \mathcal{M}_2$ are isometric, then $d_{\mathcal{GH}}(\mathcal{M}_1, \mathcal{M}_2) = 0$.

In fact, let $f : \mathcal{M}_1 \to \mathcal{M}_2$ be the corresponding isometry. We define on $\mathcal{M}_1 \sqcup \mathcal{M}_2$ a metric d which extends the metrics d_i and is given, for $m_i \in \mathcal{M}_i$, $i = 1, 2$, by

$$d(m_1, m_2) := d_2(f(m_1), m_2).$$

Then $\mathcal{M}_1$ is contained in every ε-neighborhood of $\mathcal{M}_2$ and vice versa, and therefore $d_{\mathcal{GH}}(\mathcal{M}_1, \mathcal{M}_2) = 0$.

We will show that the converse is not true even for complete metric spaces (see Example 3.57). However, it is true for compact metric spaces. These facts will be derived using the following

Proposition 3.55. *Let $(\mathcal{M}_i, d_i)$ be metric spaces, $i = 1, 2$. Then the inequality*

$$d_{\mathcal{GH}}(\mathcal{M}_1, \mathcal{M}_2) < \varepsilon \tag{3.72}$$

holds if and only if there is a metric space $(\mathcal{M}, d)$ and isometric embeddings $e_i : \mathcal{M}_i \to \mathcal{M}$ such that

$$d_{\mathcal{H}}\big(e_1(\mathcal{M}_1), e_2(\mathcal{M}_2)\big) < \varepsilon. \tag{3.73}$$

Proof. If (3.72) is true, then $d_{\mathcal{H}}(\mathcal{M}_1, \mathcal{M}_2) < \varepsilon$ for some metric d on $\mathcal{M}_1 \sqcup \mathcal{M}_2$ that extends the metrics d_i. Then (3.73) clearly holds for $\mathcal{M} := \mathcal{M}_1 \sqcup \mathcal{M}_2$ and e_i being the natural embeddings $\mathcal{M}_i \to \mathcal{M}_1 \sqcup \mathcal{M}_2$.

Conversely, suppose that (3.73) is true for some $(\mathcal{M}, d)$ and isometric embeddings $e_i : \mathcal{M}_i \to \mathcal{M}$. We replace $(\mathcal{M}, d)$ by the direct 1-sum $\widetilde{\mathcal{M}} := \mathcal{M} \oplus^{(1)} [0, \eta]$ with a given $\eta > 0$. The metric $\widetilde{d}$ of the direct 1-sum $\widetilde{\mathcal{M}}$ satisfies

$$d \le \widetilde{d} \le d + \eta. \tag{3.74}$$

Now we define maps $\widetilde{e}_i : \mathcal{M}_i \to \widetilde{\mathcal{M}}$ by $\widetilde{e}_1(m) := (e_1(m), 0)$ and $\widetilde{e}_2(m) := (e_2(m), \eta)$. Then $\widetilde{e}_i$ are isometric embeddings and their images are disjoint. We then define a metric $\overline{d}$ on $\mathcal{M}_1 \sqcup \mathcal{M}_2$ which extends the metrics d_i and is given for $m_i \in \mathcal{M}_i$, $i = 1, 2$, by

$$\overline{d}(m_1, m_2) := \widetilde{d}\big(\widetilde{e}_1(m_1), \widetilde{e}_2(m_2)\big).$$

By (3.73) and (3.74), we get

$$\overline{d}_{\mathcal{H}}(\mathcal{M}_1, \mathcal{M}_2) = \widetilde{d}_{\mathcal{H}}\big(\widetilde{e}_1(\mathcal{M}_1), \widetilde{e}_2(\mathcal{M}_2)\big) \le d_{\mathcal{H}}\big(e_1(\mathcal{M}_1), e_2(\mathcal{M}_2)\big) + \eta < \varepsilon + \eta.$$

This and definition (3.71) imply that

$$d_{\mathcal{GH}}(\mathcal{M}_1, \mathcal{M}_2) \leq \overline{d}_{\mathcal{H}}(\mathcal{M}_1, \mathcal{M}_2) < \varepsilon + \eta.$$

Choosing η to be sufficiently small we get (3.72). □

Corollary 3.56. *Let $(\mathcal{M}, d)$ and $(\mathcal{M}', d')$ be compact metric spaces. Then*

$$d_{\mathcal{GH}}(\mathcal{M}, \mathcal{M}') = 0 \qquad (3.75)$$

if and only if $\mathcal{M}$ and $\mathcal{M}'$ are isometric.

Proof. Since (3.75) holds if $\mathcal{M}$ and $\mathcal{M}'$ are isometric, we must only prove the converse.

Let $\{m_i\}$ be a dense countable subset of $\mathcal{M}$. By (3.75), given integers i, $k \geq 1$, we can find a point $m'_{k,i} \in \mathcal{M}'$ so that

$$\widetilde{d}(m_i, m'_{k,i}) < \frac{1}{k} \qquad (3.76)$$

for some metric $\widetilde{d}$ on $\mathcal{M} \sqcup \mathcal{M}$ extending the metrics d and d'. Since $\mathcal{M}$ is compact, we may assume, passing to a subsequence, that there is $m'_1 \in \mathcal{M}'$ such that $d'(m'_k, m'_1) \to 0$ as $k \to \infty$. By passing to a further subsequence we may assume that $d'(m'_k, m'_2) \to 0$ as $k \to \infty$ for some $m'_2 \in \mathcal{M}$ and so on. Since d' is uniformly continuous in its arguments, (3.76) implies that for all i, j and infinitely many k, ℓ,

$$|d(m_i, m_j) - d'(m'_{k,i}, m'_{\ell,j})| < \frac{1}{k} + \frac{1}{\ell}.$$

Passing to the limit as $k, \ell \to \infty$, we get $d(m_i, m_j) = d'(m'_i, m'_j)$, and the required isometry is a unique continuous extension of the map $m_i \mapsto m'_i$. □

Example 3.57. The assertion of Corollary 3.56 is not true for complete (even proper) metric spaces. To show this, we define a family of proper spaces $\mathcal{M}_\alpha$, where $0 \leq \alpha < 1$ and is rational, by

$$\mathcal{M}_\alpha := \bigsqcup_{k \in \mathbb{Z}} [k, k + |\sin(k + \alpha)|];$$

each $\mathcal{M}_\alpha$ is endowed with the metric d induced from $\mathbb{R}$. We prove that

$$d_{\mathcal{GH}}(\mathcal{M}_\alpha, \mathcal{M}_\beta) = 0. \qquad (3.77)$$

Since $\mathcal{M}_\alpha$ and $\mathcal{M}_\beta$ are not isometric for $\alpha \neq \beta$, this gives the desired counterexample.

It suffices to prove (3.77) for $\beta = 0$ and $0 < \alpha < 1$; the general case follows then from the triangle inequality for $d_{\mathcal{GH}}$.

3.1. Principal concepts and related facts

To establish this, given $\varepsilon > 0$ and $0 < \alpha < 1$, we first find integers $t = t(\alpha)$ and $s = s(\alpha)$, so that
$$\left|2\pi s - (t+\alpha)\right| < \varepsilon.$$
Existence of these integers follows from the classical Jacobi theorem (see, e.g., [La-1966]) which, in particular, implies the density of the set $\{2\pi s\}$, $s \in \mathbb{N}$, in $[0,1]$; here $\{x\}$ stands for the fractional part of $x \in \mathbb{R}$.

Using this we define maps $e_\alpha : \mathcal{M}_\alpha \to \mathbb{R}$, $0 \le \alpha < 1$, where e_0 is the identity map and e_α with $0 < \alpha < 1$ is given by $e_\alpha(x) := x + t(\alpha)$. These maps are clearly isometric embeddings. Further, by the choice of $s = s(\alpha)$, $t = t(\alpha)$, we have the estimate for the Hausdorff distance between $e_0(\mathcal{M}_0)$ and $e_\alpha(\mathcal{M}_\alpha)$:

$$d_\mathcal{H}\bigl(e_0(\mathcal{M}_0), e_\alpha(\mathcal{M}_\alpha)\bigr) \le \sup_k \left|\sin k - \sin(k + t + \alpha)\right|$$
$$= \sup_k \left|\sin(k + 2\pi s) - \sin(k + t + \alpha)\right| \le |2\pi s - t - \alpha| < \varepsilon.$$

Due to Proposition 3.55 this implies the desired equality (3.77).

Now we define the *limit of a sequence* $\{(\mathcal{M}_i, d_i)\}_{i \ge 1}$ *in the Gromov–Hausdorff metric* to be a metric space $(\mathcal{M}, d)$ such that

$$d_{\mathcal{GH}}(\mathcal{M}_i, \mathcal{M}) \to 0 \quad \text{as} \quad i \to \infty.$$

Example 3.57 shows that a sequence may have an uncountable set of mutually nonisometric limits. Nevertheless, these limits inherit some basic properties from the spaces in the sequence. The proof of the corresponding result is based on the Gromov–Hausdorff compactness criterion [Gr-1981] presented now.

Theorem 3.58 (Gromov). *Assume that a sequence of compact metric spaces* $\{(\mathcal{M}_i, d_i)\}_{i \ge 1}$ *satisfies the following condition:*

The diameters and covering numbers $\mathrm{Cov}(\mathcal{M}_i, \varepsilon)$ *of the spaces* $\mathcal{M}_i$ *are uniformly bounded for each* $\varepsilon > 0$.

Then there exists a subsequence of $\{\mathcal{M}_i\}$ *that converges in the Gromov–Hausdorff metric to a compact metric space.*

Proof. We first show that $\mathcal{M}_i$ can be realized as closed subspaces of a fixed compact metric space. To this end, we define the nondecreasing function

$$K(\varepsilon) := \sup_i \mathrm{Cov}(\mathcal{M}_i, \varepsilon), \quad \varepsilon \ge 0.$$

Due to our assumptions, $K(\varepsilon)$ is finite for all $\varepsilon \ge 0$; here we set

$$\mathrm{Cov}(\mathcal{M}_i, \varepsilon) := \mathrm{diam}\, \mathcal{M} \quad \text{for} \quad \varepsilon = 0.$$

Then we set $K_i = K(2^{-i})$ and $\varepsilon_i := 2^{-i}$. Hence each space $\mathcal{M}_i$ can be covered by K_i balls of radius ε_i. Using this notation we define a "parallelotope" Π_i as the set

$$\Pi_i := \{n = (n_1, \ldots, n_i) \in \mathbb{Z}^i\,;\, 1 \le n_\ell \le K_\ell,\quad \ell = 1, \ldots, i\}.$$

Note that Π_i is a "face" of Π_{i+1}. Hence the natural projection $p_i : \Pi_{i+1} \to \Pi_i$ is well-determined.

Now we need

Lemma 3.59. *For every pair $i, j \in \mathbb{N}$ there is a map $\varphi_j^i : \Pi_i \to \mathcal{M}_j$ such that*

(a) *the image $\varphi_j^i(\Pi_i)$ is ε_i-dense in $\mathcal{M}_j$;*

(b) *for every $n \in \Pi_{i+1}$ the point $\varphi_j^{i+1}(n)$ is contained in the ball of radius $2\varepsilon_i$ centered at $\varphi_j^i(p_i(n))$.*

Proof. Cover $\mathcal{M}_j$ by K_1 balls of radius ε_1 and take any bijection from the set $\{1, 2, \ldots, K_1\} =: \Pi_1$ to the set of the centers of these balls. Denote this bijection by φ_j^1. Then cover each of the balls of radius ε_1 by K_2 balls of radius ε_2 and map Π_2 onto the set of the centers of these new balls so that a point (n_1, n_2) of Π_2 goes to the center of a ball which has been used to cover the ball $B_{\varepsilon_1}(\varphi_j^1(n_1))$. This defines the map φ_j^2. Then cover each of K_2 balls of radius ε_2 by K_3 balls of radius ε_3 and map Π_3 onto the set of centers of these ε_3-balls so that a point (n_1, n_2, n_3) of Π_3 goes to the center of an ε_3-ball from the cover of the ball $B_{\varepsilon_2}(\varphi_j^2(n_1, n_2))$. Continuing in this way we define φ_j^i for all i (and j). According to this construction, for every $n \in \Pi_{i+1}$,

$$d_j(\varphi_j^i(p_i(n)), \varphi_j^{i+1}(n)) \leq \varepsilon_{i+1} + \varepsilon_i < 2\varepsilon_i. \tag{3.78}$$

Moreover, the definition of φ_j^i immediately implies the validity of property (a) of the lemma. □

Let $\Pi := \bigsqcup_{i=1}^{\infty} \Pi_i$. The maps φ_j^i, $i \geq 1$, define maps $\varphi_j : \Pi \to \mathcal{M}_j$, $j = 1, 2, \ldots$.

We now introduce the desired compact metric space $\mathcal{M}$ which will be an ambient space for all $\mathcal{M}_i$. It will be defined as a closed subset of the Banach space $\ell_\infty(\Pi)$; recall that the norm of this space $\|f\|_\infty := \sup_{n \in \Pi} |f(n)|$. Namely, $\mathcal{M}$ is defined as the closed subset of $\ell_\infty(\Pi)$ consisting of functions satisfying

$$\begin{cases} 0 \leq f(n) \leq K(0) := \sup_j \operatorname{diam} \mathcal{M}_j, & \text{if } n \in \Pi_1 \subset \Pi, \\ |f(n) - f(p_{i-1}(n))| \leq 2\varepsilon_{i-1}, & \text{if } n \in \Pi_i \subset \Pi, \ i > 1. \end{cases} \tag{3.79}$$

We show that $\mathcal{M}$ regarded as a metric subspace of $\ell_\infty(\Pi)$ is compact. In fact, $\mathcal{M}$ is closed and bounded. Further, let us define a map $p^i : \Pi \to \bigsqcup_{j=1}^{i} \Pi_j$ by the formula

$$p^i(n) := \begin{cases} n, & \text{if } n \in \Pi_j, \ 1 \leq j \leq i \\ (p_i \circ \cdots \circ p_{k-1})(n), & \text{if } n \in \Pi_k, \ k > i. \end{cases}$$

3.1. Principal concepts and related facts

Then for $f \in \mathcal{M}$ and all $n \in \Pi$ the second condition in (3.79) yields

$$\left|f(n) - f(p^i(n))\right| \leq 2\sum_{k \geq i} \varepsilon_k = 2^{-i+1}.$$

Therefore the distance in $\ell_\infty(\Pi)$ between $\mathcal{M}$ and the set $S_i := \{f \circ p^i \,;\, f \in \mathcal{M}\}$ is at most $\varepsilon_{i-1} = 2^{-i+1}$. But S_i is a closed bounded subset of a finite-dimensional vector subspace of $\ell_\infty(\Pi)$, hence, is compact. Then there exists a finite ε_{i-1}-dense subset of S_i which, in turn, is a $2\varepsilon_{i-1}$-dense subset of $\mathcal{M}$. By virtue of Hausdorff's compactness criterion, $\mathcal{M}$ is compact.

Let us define now a map $e_i : \mathcal{M}_i \to \mathcal{M}$ by

$$e_i(m)(n) := d_i(m, \varphi_i(n)), \quad m \in \mathcal{M}_i, \ n \in \Pi.$$

This is the required isometric embedding. In fact, $\varphi_i(\Pi)$ is dense in $\mathcal{M}_i$ by assertion (a) of Lemma 3.59 and therefore

$$\left\|e_i(m) - e_i(m')\right\|_\infty = \sup_{n \in \Pi} \left|d_i(m, \varphi_i(n)) - d_i(m', \varphi_i(n))\right| = d_i(m, m').$$

We now identify $\mathcal{M}_i$ with its image $e_i(\mathcal{M}_i)$ in the compact space $(\mathcal{M}, d)$; here d is the metric induced from $\ell_\infty(\Pi)$. The assumptions of the theorem remain true for this setting. Since the metric space $(\mathcal{B}(\mathcal{M}), d_\mathcal{H})$ is compact by the Hausdorff theorem, we can find a subsequence $\{\mathcal{M}_{i_k}\}_{k \geq 1}$ converging to a compact space $\mathcal{M}_\infty$ with respect to the Hausdorff metric. But

$$d_{\mathcal{GH}}(\mathcal{M}_i, \mathcal{M}_\infty) \leq d_\mathcal{H}(\mathcal{M}_i, \mathcal{M}_\infty)$$

by the definition of $d_{\mathcal{GH}}$, see (3.71). Hence, $\{\mathcal{M}_{i_k}\}_{k \geq 1}$ converges to $\mathcal{M}_\infty$ in the Gromov–Hausdorff metric.

The proof is complete. $\square$

For the sake of completeness we also outline the proof of the aforementioned Hausdorff theorem.

Let $\{\mathcal{M}_i\}_{i \geq 1}$ be a sequence of closed (hence compact) subspaces of a compact metric space $\mathcal{M}$. Assume that for all $\varepsilon \geq 0$,

$$K(\varepsilon) := \sup \mathrm{Cov}(\mathcal{M}_i, \varepsilon) < \infty. \tag{3.80}$$

Then there is a subsequence of $\{\mathcal{M}_i\}$ which converges to some closed subset of $\mathcal{M}$.

We construct the desired subsequence as follows. Cover each $\mathcal{M}_i$ by $K(\frac{\varepsilon}{3})$ balls, $B_{\frac{\varepsilon}{3}}(m_\ell^i)$, $1 \leq \ell \leq K(\frac{\varepsilon}{3})$, see (3.80). Since $K(\frac{\varepsilon}{3}) < \infty$, there is an infinite subsequence $I(\varepsilon) \subset \mathbb{N}$ and an integer $1 \leq p \leq K(\frac{\varepsilon}{3})$ such that each $\mathcal{M}_i$ with $i \in I(\varepsilon)$ is covered by exactly p balls of radius $\frac{\varepsilon}{3}$ centered at some m_ℓ^i. For $i, j \geq 1$, consider the p^2-vector $v^{i,j} = \left(d(m_\ell^i, m_{\ell'}^j)\right)_{1 \leq \ell,\, \ell' \leq p}$. Since each coordinate of this vector is bounded by diam $\mathcal{M}$, the sequence $\{v^{i,j}\}_{i,j\in I(\varepsilon)}$ lies in a p^2-dimensional cube of side length diam $\mathcal{M}$. Divide this cube into a finite number of subcubes of side length $\frac{\varepsilon}{3}$. Using the Cantor diagonal process we find an infinite subsequence $J(\varepsilon) \subset I(\varepsilon)$ such that all $v^{i,j}$ with $i,j \in J(\varepsilon)$ lie in one of these subcubes. Therefore, for $i,j \in J(\varepsilon)$, we get

$$|d(m_\ell^i, m_{\ell'}^i) - d(m_\ell^j, m_{\ell'}^j)| \leq \frac{2\varepsilon}{3} \quad \text{for all} \quad 1 \leq \ell, \ell' \leq p.$$

Setting $S_i := \{m_\ell^i\}_{1\leq \ell \leq p}$ we derive from here that

$$d_\mathcal{H}(S_i, S_j) \leq \frac{2\varepsilon}{3} \quad \text{for} \quad i,j \in J(\varepsilon).$$

Moreover, the family $\{B_{\frac{\varepsilon}{3}}(m_\ell^i)\}_{1\leq \ell \leq p}$ covers $\mathcal{M}_i$ and therefore

$$d_\mathcal{H}(\mathcal{M}_i, S_i) \leq \frac{2\varepsilon}{3}.$$

Combining these inequalities we then get

$$d_\mathcal{H}(\mathcal{M}_i, \mathcal{M}_j) \leq 2\varepsilon \quad \text{for} \quad i,j \in J(\varepsilon). \tag{3.81}$$

Applying this with $\varepsilon = \frac{1}{2}$ we obtain a subsequence $\{\mathcal{M}_i^1\}_{i\geq 1} \subset \{\mathcal{M}_i\}_{i\in I(1)}$ such that

$$d_\mathcal{H}(\mathcal{M}_i^1, \mathcal{M}_j^1) < 1 \quad \text{for all} \quad i,j.$$

Applying (3.81) to this subsequence with $\varepsilon = \frac{1}{4}$ we obtain a subsequence $\{\mathcal{M}_i^2\} \subset \{\mathcal{M}_i^1\}$ satisfying

$$d_\mathcal{H}(\mathcal{M}_i^2, \mathcal{M}_j^2) < \frac{1}{2} \quad \text{for all} \quad i,j.$$

Continuing this procedure with $\varepsilon = 2^{-\ell-1}$, $\ell = 1, 2 \ldots$, we obtain, for the diagonal sequence $\{\mathcal{M}_\ell^\ell\}$,

$$d_\mathcal{H}(\mathcal{M}_\ell^\ell, \mathcal{M}_{\ell'}^{\ell'}) < \frac{1}{2^\ell} \quad \text{for all} \quad \ell, \ell' \quad \text{with} \quad \ell < \ell'. \tag{3.82}$$

Consider sequences of points $m_\ell \in \mathcal{M}_\ell^\ell$, $\ell = 1, 2, \ldots$, satisfying $d(m_\ell, m_{\ell'}) \leq \frac{1}{2^\ell}$ for all $\ell \leq \ell'$. Let $\mathcal{M}_\infty$ be the closure of the limits of all these sequences. Then $\mathcal{M}_\infty$ is compact and

$$\lim_{\ell \to \infty} d_\mathcal{H}(\mathcal{M}_\ell^\ell, \mathcal{M}_\infty) = 0,$$

as required.

Example 3.60. (a) The *Chebyshev radius* of a metric space $\mathcal{M}$ is defined by

$$R(\mathcal{M}) := \inf\{R\,;\, B_R(m) = \mathcal{M}\}.$$

3.1. Principal concepts and related facts

It is a matter of definition to show that

$$d_{\mathcal{GH}}(\{m_0\}, \mathcal{M}) = R(\mathcal{M}).$$

In particular, a sequence of metric spaces $\{\mathcal{M}_i\}$ converges in the Gromov–Hausdorff metric to a space with only one point, provided that $R(\mathcal{M}_i) \to 0$ as $i \to \infty$.

(b) Example 3.57 shows that a sequence of metric spaces may have an uncountable number of mutually nonisometric limits.

(c) We leave to the reader the proof of the converse to Theorem 3.58. That is to say, it should be proved that if a sequence $\{\mathcal{M}_i\}$ of compact metric spaces converges in the Gromov–Hausdorff metric to a space $\mathcal{M}$ then, for all $\varepsilon > 0$,

$$\sup_i \operatorname{Cov}(\mathcal{M}_i, \varepsilon) < \infty.$$

Moreover, the completion of $\mathcal{M}$ is compact.

(Hint: Show that $\operatorname{Cov}(\mathcal{M}, \varepsilon) < \infty$ for all $\varepsilon > 0$ and then apply the Hausdorff compactness criterion.)

(d) Let d denote the inner (geodesic) metric of the hyperbolic space $\mathbb{H}^n$, and $\mathcal{M}_k := (B_k, \frac{1}{k} d)$ where B_k is the ball of radius k in $\mathbb{H}^n$. Then no subsequence of $\{\mathcal{M}_k\}$ converges in the Gromov–Hausdorff metric. In fact, if we define a number $K(\varepsilon)$ as $\sup_{k \in J} \operatorname{Cov}(\mathcal{M}_k, \varepsilon) < \infty$ for a subsequence $J \subset \mathbb{N}$ and all $\varepsilon > 0$, then every ball B_k with $k \in J$ can be covered by $K(\varepsilon)$ balls of radius ε from $\mathcal{M}_k$, and therefore

$$\operatorname{vol}_n(B_k) \leq K(\varepsilon) \operatorname{vol}_n(B_{k\varepsilon}), \quad k \in J.$$

But the n-volume of B_a grows like C^a as $a \to \infty$ for some $C > 1$, and this inequality results in a contradiction as $k \to \infty$.

The next result demonstrates that the limit of a sequence $\{\mathcal{M}_i\}$ inherits some basic properties of the metric spaces $\mathcal{M}_i$.

Proposition 3.61. *Let a complete metric space $\mathcal{M}$ be the limit of a sequence $\{\mathcal{M}_i\}$ in the Gromov–Hausdorff metric. Then the following is true:*

(a) *if every $\mathcal{M}_i$ is proper, then $\mathcal{M}$ is proper;*

(b) *if every $\mathcal{M}_i$ is a length space, then $\mathcal{M}$ is a length space;*

(c) *if every $\mathcal{M}_i$ is proper and geodesic, then $\mathcal{M}$ is proper and geodesic.*

Proof. (a) First, let all of $\mathcal{M}_i$ be closed subspaces of a fixed complete metric space converging to a closed subspace $\mathcal{M}$ in the Hausdorff metric. In virtue of the definition of this metric, every closed ball of $\mathcal{M}$ is the limit of closed balls of spaces $\mathcal{M}_i$. By Proposition 3.55, the same is true for convergence in the Gromov–Hausdorff metric. Hence, there is a sequence $\{\overline{B}_i \subset \mathcal{M}_i\}$ of closed balls converging to a given closed ball $\overline{B} \subset \mathcal{M}$ in the Gromov–Hausdorff metric. Since $\mathcal{M}_i$ is proper, all $\overline{B}_i$ are compact and therefore their limit $\overline{B}$ is a complete metric space. Hence, $\overline{B}$ is compact as well (see Example 3.60 (c)).

(b) Since $\mathcal{M}$ is complete, it suffices to verify that the approximate middle point condition holds for $\mathcal{M}$, see Proposition 3.36.

Suppose $m, m' \in \mathcal{M}$ and $\varepsilon > 0$ are given. We must find an ε-approximate middle point $\widetilde{m}$, i.e., $\widetilde{m}$ should satisfy

$$\max\{d(\widetilde{m}, m), d(\widetilde{m}, m')\} \leq \frac{1}{2} d(m, m') + \varepsilon. \tag{3.83}$$

To this end we choose an integer i and isometric embeddings $e_i : \mathcal{M}_i \to \widetilde{\mathcal{M}}$ and $e : \mathcal{M} \to \widetilde{\mathcal{M}}$ into some metric space $(\widetilde{\mathcal{M}}, \widetilde{d})$ so that

$$\widetilde{d}_{\mathcal{H}}\bigl(e_i(\mathcal{M}_i), e(\mathcal{M})\bigr) < \varepsilon,$$

see Proposition 3.55. Then there exist points m_i, m'_i from $\mathcal{M}_i$ such that

$$\widetilde{d}\bigl(e_i(m_i), e(m)\bigr) < \frac{\varepsilon}{3} \quad \text{and} \quad \widetilde{d}\bigl(e_i(m'_i), e(m')\bigr) < \frac{\varepsilon}{3}.$$

Since $\mathcal{M}_i$ is a length space, there is an ε-approximate middle point $\widetilde{m}_i$ for the pair m_i, m'_i, see Proposition 3.36. Hence, (3.83) holds for the points $e_i(\widetilde{m}_i), e_i(m_i), e_i(m'_i)$ and the metric $\widetilde{d}$. Further, we choose $\widetilde{m} \in \mathcal{M}$ so that

$$\widetilde{d}\bigl(e_i(\widetilde{m}_i), e(\widetilde{m})\bigr) < \frac{\varepsilon}{3}.$$

Combining all the inequalities above and applying the triangle inequality and Proposition 3.55 we establish (3.83) for the images of the points $\widetilde{m}, m, m'$ under the isometric embedding e.

(c) The statement follows straightforwardly from (a), (b) and the Hopf–Rinow theorem 3.40.

This completes the proof. $\square$

Remark 3.62. (a) In many naturally arising situations, the Gromov–Hausdorff convergence cannot be used directly to define limiting procedures of geometry. For example, it is natural to think of $\mathbb{R}^n$ as a limit of the Euclidean n-spheres of a fixed center and with radii tending to infinity. But this sequence diverges in the Gromov–Hausdorff metric, see Example 3.60 (a).

At least for the basic case of proper metric spaces this disadvantage can be overcome by using the following variant of the Gromov–Hausdorff convergence.

Let $\{\mathcal{M}_i\}$ be a sequence of *pointed* metric spaces with basepoints $m_i^* \in \mathcal{M}_i$. This sequence is said to converge to a pointed metric space $\mathcal{M}$ with

3.1. Principal concepts and related facts

a *basepoint* $m^* \in \mathcal{M}$ if, for every $R > 0$, the sequence of closed balls $\overline{B}_R(m_i^*)$ (with the metric induced from $\mathcal{M}_i$) converges to $\overline{B}_R(m^*) \subseteq \mathcal{M}$ in the Gromov–Hausdorff metric.

The reader can easily check that the limit of a sequence of proper pointed spaces is also proper, provided that this limit is complete, and that the limit of doubling pointed spaces is doubling.

A version of the Gromov–Hausdorff compactness theorem also holds for this case, see, e.g., [BH-1999, p. 77].

(b) Using the introduced variant of the Gromov–Hausdorff convergence one defines an analog of the concept of tangent space for Riemannian manifolds. Namely, for a proper and doubling metric space (M, d) and a point m, let us consider the family $\{M, \varepsilon^{-1}d, m\}_{\varepsilon > 0}$ of proper and doubling spaces. Exploiting the Gromov compactness theorem, one finds a sequence $(M, \varepsilon_n^{-1}d, m)$, $1, 2, \ldots$, in this family converging to a pointed metric space (M_∞, m_∞) as $\varepsilon_n \to 0$; clearly, M_∞ is doubling and proper. This limit space is said to be the *tangent space* of M at m. In general, the set of tangent spaces at a point is infinite, but for M being a Riemannian manifold all these spaces are isometric to some Euclidean spaces.

Another interesting example is the Heisenberg group H_n where each tangent space is the same Heisenberg group (more general results of this kind see in Example 3.141 and Theorem 3.142).

(c) The following generalization of the above definition is also in use.

Let $\{(\mathcal{M}_i, m_i^*)\}_{i \geq 1}$ be a sequence of pointed metric spaces. Denote by $\mathcal{M}_\infty$ the set of sequences $\{m_i \in \mathcal{M}_i\}_{i \geq 1}$ such that

$$\sup_i d_i(m_i, m_i^*) < \infty.$$

Consider an equivalence relation on $\mathcal{M}_\infty$ given by

$$(m_i) \sim (m_i') \iff \operatorname*{LIM}_{i \to \infty} d_i(m_i, m_i') = 0.$$

Here LIM denotes the *Banach limit*, a linear bounded functional defined on the space $\ell_\infty(\mathbb{N})$ of all bounded sequences of real numbers, see, e.g., [DS-1958, Sec. II.4.22]. Let us recall that $\operatorname{LIM} a_i = \lim a_i$ if the latter limit exists. Then we denote by $\mathcal{M}_\omega$ the set $\mathcal{M}_\infty/\sim$ of equivalence classes and endow $\mathcal{M}_\omega$ with the metric

$$d_\omega\big((m_i), (m_i')\big) = \operatorname*{LIM}_{i \to \infty} d_i(m_i, m_i').$$

The metric space $(\mathcal{M}_\omega, d_\omega)$ is said to be the ω-*limit of a sequence* $\{(\mathcal{M}_i, m_i^*)\}_{i \geq 1}$.

Consider, in particular, a fixed pointed metric space $(\mathcal{M}, d, m^*)$ and the sequence of the pointed metric spaces $(\mathcal{M}, \frac{1}{i} d)$ with the basepoint m^*. Then the ω-limit of this sequence is called an *asymptotic cone* of $\mathcal{M}$ and is denoted by $\mathrm{Cone}_\omega \mathcal{M}$.

3.2 Measures on metric spaces

3.2.1 Measure theory

We will freely use the basic concepts and results of Measure Theory. In order to fix terminology and notations we present some of them in this subsection. The reader may also consult the classical book of Halmos [Hal-1950] for the results in measure theory and the encyclopedic treatise [Fe-1969] by Federer and the modern book [Ma-1995] by Mattila for the results in geometric measure theory.

A function $\mu : 2^\mathcal{M} \to \mathbb{R}_+$ is called a *measure on a metric space* $(\mathcal{M}, d)$ if it satisfies the conditions

$$\mu(\emptyset) = 0 \quad \text{and} \quad \mu(S) \leq \sum_{n=1}^{\infty} \mu(S_n)$$

whenever $S \subset \bigcup_{n=1}^{\infty} S_n$.

In the standard way, $\mathcal{M}$ defines the σ-algebra of (μ-measurable) subsets of $2^\mathcal{M}$ denoted by $\Sigma(\mathcal{M}, \mu)$.

This function is *countably additive* on $\Sigma(\mathcal{M}, \mu)$, i.e.,

$$\mu\left(\bigsqcup_{n=1}^{\infty} S_n\right) = \sum_{n=1}^{\infty} S_n,$$

provided that the sets of the disjoint union belong to the σ-algebra $\Sigma(\mathcal{M}, \mu)$.

The smallest σ-algebra containing all open (or, equivalently, closed) subsets of $(\mathcal{M}, d)$ is said to be the σ-*algebra of Borel sets* (written $\mathcal{BS}(\mathcal{M})$).

Definition 3.63. A measure μ on $\mathcal{M}$ is said to be a Borel measure if

$$\mathcal{BS}(\mathcal{M}) \subset \Sigma(\mathcal{M}, \mu).$$

A Borel measure is said to be regular if for every $S \subset \mathcal{M}$ there is a Borel set $\widetilde{S}$ such that

$$S \subset \widetilde{S} \quad \text{and} \quad \mu(S) = \mu(\widetilde{S}).$$

Convention. In what follows "measure" means a *regular Borel measure*, if another meaning is not stated explicitly.

3.2. Measures on metric spaces

The linear space of μ-*measurable functions* on $\mathcal{M}$ is denoted by $L_0(\mathcal{M},\mu)$. This space is equipped with pseudometric d_0 defined by

$$d_0(f,g) := \inf\left[\varepsilon\,;\,\mu\Big(\{m\in\mathcal{M}\}\,;\,d\big(f(m),g(m)\big)\Big)\leq\varepsilon\right]. \tag{3.84}$$

This clearly may assign value $+\infty$ and be zero for $f\neq g$. In the latter case, $f(m)=g(m)$ for all m except for a subset of μ-measure zero (briefly, $f=g$ μ-*almost everywhere*). The space $L_0(\mathcal{M},\mu)$ is factorized with respect to this equivalence relation (preserving the same notation for the factorized linear space and factorized pseudometric). Convergence in the pseudometric d_0 is referred to as *convergence in measure*.

The basic properties of $L_0(\Sigma,\mu)$ are described in

Theorem 3.64. (a) *The space* $\big(L_0(\mathcal{M},\mu),d_0\big)$ *is complete. Moreover, every Cauchy sequence contains a subsequence converging both in measure and μ-almost everywhere.*

(b) *This space is separable if and only if μ is σ-finite.*

Let us recall that μ is σ-*finite* if its support is a countable union of sets of finite μ-measure.

The *support of a measure* μ (written $\operatorname{supp}\mu$) is defined by

$$\operatorname{supp}\mu := \mathcal{M}\setminus\mathcal{N}(\mu), \tag{3.85}$$

where $\mathcal{N}(\mu)$ is the union of all open subsets in $\Sigma(\mathcal{M},\mu)$ of μ-measure zero [5].

The next classical result is

Theorem 3.65 (Luzin's C-property). *If $\mu(\mathcal{M})<\infty$, then the space $C_b(\mathcal{M})$ of bounded continuous functions of $\mathcal{M}$ is dense in the space $L_0(\mathcal{M},\mu)$ equipped with the metric*

$$d_L(f,g) := \mu\{m\in\mathcal{M}\,;\,f(m)\neq g(m)\}.$$

In other words, in this case, given $\varepsilon>0$ and $f\in L_0(\mathcal{M},\mu)$, one can find a bounded continuous function $f_\varepsilon:\mathcal{M}\to\mathbb{R}$ and a subset (in fact, closed) $S_\varepsilon\subset\mathcal{M}$ so that $\mu(\mathcal{M}\setminus S_\varepsilon)<\varepsilon$ and

$$f_\varepsilon\big|_{S_\varepsilon} = f\big|_{S_\varepsilon}.$$

Hence, f can be changed on a subset of an arbitrarily small μ-measure to turn into a continuous function.

To formulate a useful consequence of this result we need

Definition 3.66. A function $f:\mathcal{M}\to\mathbb{R}$ is said to be *Borel measurable* if for every subset S from $\mathcal{BS}(\mathbb{R})$ the set $f^{-1}(S)$ belongs to $\mathcal{BS}(\mathcal{M})$. In other words, $f^{-1}(S)$ is Borel if S is.

The set of all Borel functions on $\mathcal{M}$ is denoted by $\mathcal{BF}(\mathcal{M})$.

---

[5] In [Fe-1969], the notation $\operatorname{spt}\mu$ is used for 'support'.

Corollary 3.67. *If μ is σ-finite, then for every function $f \in L_0(\mathcal{M}, \mu)$ there is a function $g \in \mathcal{BF}(\mathcal{M})$ such that $f = g$ μ-almost everywhere.*

Now let $F : (\mathcal{M}, d) \to (\mathcal{M}', d')$ be a μ-*measurable map* meaning that $F^{-1}(S')$ is μ-measurable for every open subset $S' \subset \mathcal{M}'$.

The following result emphasizes the role of the Borel measurability.

Theorem 3.68. *Assume that F is a μ-measurable map between metric spaces $\mathcal{M}_1$ and $\mathcal{M}_2$ and $g : \mathcal{M}_2 \to \mathbb{R}$ is Borel measurable. Then composition $g \circ f$ is μ-measurable.*

The result is not true, in general, for g being only μ-measurable.

Now let $\mathcal{M}_i$ be a metric space carrying a measure μ_i, $1 \leq i \leq N$. The tensor product of these measures, denoted by $\bigotimes_{i=1}^{N} \mu_i$, is defined first on measurable "polytopes" $\prod_{i=1}^{N} S_i$ with $S_i \in \Sigma(\mathcal{M}_i, \mu_i)$, $1 \leq i \leq N$, by

$$\left(\bigotimes_{i=1}^{N} \mu_i\right)\left(\prod_{i=1}^{N} S_i\right) := \prod_{i=1}^{N} \mu_i(S_i).$$

Then this tensor product is extended to the σ-algebra generated by all measurable polytopes using the classical Lebesgue–Carathéodory extension procedure.

For the case of a Polish (= complete and countable) metric space measure–theoretic considerations can be essentially simplified and reduced to the classical Lebesgue theory. Namely, the following is true.

Theorem 3.69 (von Neumann–Rochlin). *Let μ be a (Borel) measure on a Polish space $\mathcal{M}$. Then there exists a measure–theoretic isomorphism of $(\mathcal{M}, \mu)$ (regarded as a measure space) onto a real segment (bounded or unbounded) equipped with the Lebesgue measure attached with a finite or countable collection of atoms, i.e., isolated points p_i of positive measure (mass) m_i.*

3.2.2 Integration

Let μ be a measure on a metric space $\mathcal{M}$ and X be a Banach space. A μ-measurable map $f : \mathcal{M} \to X$ is *Bochner integrable* if $\int_{\mathcal{M}} \|f\|_X d\mu < \infty$. In this case, $\int_{\mathcal{M}} f d\mu$ is defined in a standard way starting with μ-measurable step functions $f := \sum_{i=1}^{n} x_i \chi_{S_i}$ where $\{x_i\} \subset X$. Hereafter, χ_S denotes the *characteristic function (indicator)* of a subset S. Then $\int_{\mathcal{M}} f d\mu$ is defined as the limit of integrals of the corresponding step functions.

The basic properties of the Bochner integral are similar to those of the Lebesgue integral, see, e.g., [DS-1958, Ch. 3]. The linear space of all Bochner integrable maps [6] $f : \mathcal{M} \to X$ is denoted by $L_1(\mathcal{M}, \mu, X)$. More generally, $L_p(\mathcal{M}, \mu, X)$ is defined to be the linear space of all μ-measurable maps $f : \mathcal{M} \to X$ such that

$$\|f\|_p := \left\{ \int_{\mathcal{M}} \|f\|^p d\mu \right\}^{\frac{1}{p}} < \infty.$$

This defines a Banach space if $1 \leq p \leq \infty$, and a quasi-Banach space if $0 < p < 1$.

Let us recall only two basic results concerning integration.

Proposition 3.70. *Let T be a linear bounded operator acting between Banach spaces X and Y. Then*

$$T\left(\int_{\mathcal{M}} f d\mu \right) = \int_{\mathcal{M}} T f d\mu$$

for every $f \in L_1(\mathcal{M}, \mu, X)$.

Theorem 3.71 (Fubini). *Let μ_i be a σ-finite measure on a metric space $\mathcal{M}_i$, $i = 0, 1$. Assume that $f : \mathcal{M}_0 \times \mathcal{M}_1 \to X$ is a Bochner integrable function with respect to the measure $\mu_0 \otimes \mu_1$. Then the following is true.*

(a) *For every $m_i \in \mathcal{M}_i$ except for a set of μ_i-measure zero, the function $m_{1-i} \mapsto f(m_0, m_1)$ is μ_{1-i}-measurable, $i = 0, 1$.*

(b) *The iterated integrals exist and*

$$\int_{\mathcal{M}_0 \times \mathcal{M}_1} f d(\mu_0 \otimes \mu_1) = \int_{\mathcal{M}_1} d\mu_1 \int_{\mathcal{M}_0} f(\cdot, m_1) d\mu_0 = \int_{\mathcal{M}_0} d\mu_0 \int_{\mathcal{M}_1} f(m_0, \cdot) d\mu_1.$$

3.2.3 Measurable selections

Let F be a map from a metric space $\mathcal{M}_1$ into the family of subsets of a metric space $\mathcal{M}_2$. Recall that a *selection* of F is a map $f : \mathcal{M}_1 \to \mathcal{M}_2$ such that for all $m \in \mathcal{M}_1$,

$$f(m) \in F(m).$$

In the sequel, we shall encounter with multivariate maps into the set of all closed subsets of $\mathbb{R}^n$ or, more generally, of a Banach space. The problem of existence of selections with prescribed characteristics is basic in many fields of Analysis. The simplest case concerns μ-measurable selections. We present only two results, negative and positive ones, in this direction. The first one answers the following question:

Does the Implicit Function theorem hold for the Borel measurable functions?

In the two-dimensional situation, given a Borel measurable function $F : \mathbb{R}^2 \to \mathbb{R}$, we are looking for a Borel measurable function $f : \mathbb{R} \to \mathbb{R}$ such

---

[6] more precisely, the equivalence classes modulo μ-almost everywhere equality.

that $F(x, f(x)) = 0$ for all x. In other words, we ask for a Borel measurable selection for the map

$$x \longmapsto \{y \in \mathbb{R}\,;\, f(x,y) = 0\}, \quad x \in \mathbb{R}. \tag{3.86}$$

The following negative result was due to P. S. Novikov [No-1931].

Theorem 3.72. *There is a continuous function $f : \mathbb{R}^2 \to \mathbb{R}$ such that any selection of the map (3.86) is not Borel measurable.*

There is a relatively complicated and hardly verified definition of μ-measurability for multivariate maps in subsets of $\mathbb{R}^n$. The classical Luzin–Yankov theorem asserts that a map satisfying this definition admits a μ-measurable selection.

A much more constructive criterion yields the following

Theorem 3.73. *Let F be a multivariate map from $\mathcal{M}$ into closed subsets of $\mathbb{R}^n$. Assume that μ is a measure on $\mathcal{M}$ such that the Euclidean distance $m \mapsto d_{\mathbb{R}^n}(x, F(m))$ is μ-measurable for all $x \in \mathbb{R}^n$. Then F admits a μ-measurable selection.*

The proofs of the results formulated above as well as the corresponding references see, e.g., in the book [IT-1974] by A. Ioffe and V. Tichomirov.

3.2.4 Hausdorff measures

In order to give the corresponding definition we first define the *approximate Hausdorff p-measure* $\mathcal{H}_p^\delta$; here $0 \leq p < \infty$ and $0 < \delta \leq \infty$. Namely, we set for $S \subset \mathcal{M}$,

$$\mathcal{H}_p^\delta(S) := \inf\left\{ \frac{\omega(p)}{2^p} \sum_n (\operatorname{diam} S_n)^p \right\}, \tag{3.87}$$

where the infimum is taken over all countable covers $\{S_n\}$ of the set S with $\operatorname{diam} S_n \leq \delta$.

The factor $\omega(p) := \pi^{\frac{p}{2}}/\Gamma(\frac{p}{2}+1)$; in particular, $\omega(p)$ for $p \in \mathbb{N}$, is the volume of the Euclidean unit p-ball.

It is easily seen that $\mathcal{H}_p^\delta$ is a measure on $\mathcal{M}$. Simple examples show that not all open subsets of $\mathcal{M}$ are $\mathcal{H}_p^\delta$-measurable. Hence, $\mathcal{H}_p^\delta$ is *not a Borel measure*.

However, the following deep result of R. Davies [Dav-1970] shows that $\mathcal{H}_p^\delta$ has an important property similar to that of regular Borel measures.

Theorem 3.74. *If $\{S_n\}$ is a sequence of subsets in $\mathcal{M}$ increasingly tending to S, then*

$$\lim_{n \to \infty} \mathcal{H}_p^\delta(S_n) = \mathcal{H}_p^\delta(S).$$

The *Hausdorff p-measure* $\mathcal{H}_p$ is now given by

$$\mathcal{H}_p := \sup_{\delta > 0} \mathcal{H}_p^\delta. \tag{3.88}$$

3.2. Measures on metric spaces

This defines a regular Borel measure, see, e.g., [Fe-1969, §2.10]. Since $\mathcal{H}_p$ is not, in general, σ-finite (e.g., $\mathcal{H}_p$ on $\mathbb{R}^q$ for $q > p$), it is essential to be assured on existence of sets with finite positive measure. The positive answer was due to Besikovich [Bes-1952] for $\mathcal{M}$ being a closed subset of $\mathbb{R}^q$ of infinite $\mathcal{H}_p$ measure and to Davies [Dav-1970] for $\mathcal{M}$ being a Borel (even analytic) subset of a complete separable metric space of infinite $\mathcal{H}_p$ measure. The key point of Davies' proof is the above cited result.

For the subject matter of the present book, it is essential that the measure $\mathcal{H}_p$ is well behaved with respect to Lipschitz maps. A simple example of such a behavior is the inequality

$$\mathcal{H}_p(f(S)) \leq C\mathcal{H}_p(S) \tag{3.89}$$

which is true for every C-Lipschitz map $f : (\mathcal{M}_1, d_1) \to (\mathcal{M}_2, d_2)$.

More considerable illustration of this behavior is the following result, see [Fe-1969, Cor. 2.10.11].

Theorem 3.75. *Let $(\mathcal{M}, d)$ be a Polish space and $f : (\mathcal{M}, d) \to (\mathcal{M}_1, d_1)$ be a C-Lipschitz map. Then for every Borel measurable set $S \subset \mathcal{M}$,*

$$\int_{\mathcal{M}_1} N(f|_S) d\mathcal{H}_p \leq C^p \mathcal{H}_p(S).$$

Here $N(g)$ is the *Banach indicatrix* of the map $g : \mathcal{M} \to \mathcal{M}_1$, i.e., for every $m_1 \in \mathcal{M}_1$,

$$N(g)(m_1) := \operatorname{card} g^{-1}(m_1).$$

(If $\mathcal{H}_p(f(S)) = 0$, then we assume that the integral on the left-hand side is 0 even if the indicatrix is infinite.) In the special case of $\mathcal{M} := \mathbb{R}^p$ and $\mathcal{M}_1 := \mathbb{R}^q$ with $p \leq q$, the left-hand side can be expressed with the help of the *p-Jacobian*

$$J_p(f) := \left(\sum_I \left| \det\left(\frac{\partial f_I}{\partial x}\right) \right|^2 \right)^{\frac{1}{2}}.$$

Here I runs over all p-point subsets of the set $\{1, \ldots, q\}$ and the $p \times p$ matrix $\frac{\partial f_I}{\partial x}$ has entries $\frac{\partial f_i}{\partial x_k}$, $i \in I$, $1 \leq k \leq p$.

The p-Jacobian $J_p(f)$ exists almost everywhere on $\mathbb{R}^p$ due to the Rademacher theorem on the differentiability of Lipschitz functions on $\mathbb{R}^p$, see subsection 4.5.1 below.

The following result was due, in essence, to Kolmogorov [Ko-1932], see also [Fe-1969, Thm. 3.2.3].

Theorem 3.76. *Let $f : \mathbb{R}^p \to \mathbb{R}^q$ be C-Lipschitz and $p \leq q$. Then, for a Borel measurable subset $S \subset \mathbb{R}^p$,*

$$\int_{\mathbb{R}^q} N(f|_S; y) d\mathcal{H}_p(y) = \int_S J_p(f; x) d\mathcal{L}_p(x).$$

Here $\mathcal{L}_p$ is the Lebesgue measure on $\mathbb{R}^p$.

Using these facts one can prove some results relating $\mathcal{H}_p$ with integer p to the basic metric concepts (length, area, volume, etc.).

Note first that $\mathcal{H}_0$ is simply *the counting measure* on $\mathcal{M}$, that is,
$$\mathcal{H}_0(S) = \operatorname{card} S.$$

Further, the Hausdorff 1-measure may be regarded as a generalized length of subsets in a metric space. The following results of Kolmogorov [Ko-1932, Thm. 1] justifies this claim

Theorem 3.77. *Let $\gamma : [a,b] \to \mathcal{M}$ be a Jordan curve, i.e., γ is a homeomorphism of $[a,b]$ onto $|\gamma| := \gamma([a,b])$. Then the equality*
$$\mathcal{H}_1(|\gamma|) = \ell(\gamma) \tag{3.90}$$
is true.

Proof. We need

Lemma 3.78. *Let $S \subset \mathcal{M}$ be a connected set. Then*
$$\mathcal{H}_1(S) \geq \operatorname{diam}(S).$$

Proof. Since $\mathcal{H}_1$ is Borel regular, there is a Borel set $\widetilde{S}$ containing S and such that $\mathcal{H}_1(\widetilde{S}) = \mathcal{H}_1(S)$. Define a function $f : \mathcal{M} \to \mathbb{R}_+$ by $f(m) := d(m, m_0)$, where $m_0 \in S$ is fixed. Then Theorem 3.75 implies that
$$\mathcal{H}_1(S) = \mathcal{H}_1(\widetilde{S}) \geq \int N(f|_{\widetilde{S}}; y) d\mathcal{H}_1(y).$$

By connectedness of S, its image $f(S)$ is an interval and therefore $N(f|_{\widetilde{S}}; y) \geq 1$ whenever $y \in f(S)$. Thus, the right-hand side of this inequality is at least
$$\int_{f(S)} d\mathcal{H}_1(y) = \mathcal{H}_1(f(S)) = \operatorname{diam} f(S) \geq d(m, m_0)$$
for each $m \in S$ (since $f(m) \in f(S)$ and $f(m_0) = 0$). Taking the supremum over all $m, m_0 \in S$ we get the result. □

Given $\varepsilon > 0$, choose now an increasing sequence $\{t_i\}_{i=1}^n \subset [a,b]$ so that
$$\sum_{i=1}^{n-1} d\bigl(\gamma(t_i), \gamma(t_{i+1})\bigr) \geq \ell(\gamma) - \varepsilon.$$

Set $\gamma_i := \gamma\big|_{[t_i, t_{i+1}]}$. By Lemma 3.78,
$$\mathcal{H}_1(|\gamma_i|) \geq \operatorname{diam} |\gamma_i|([t_i, t_{i+1}]) \geq d\bigl(\gamma(t_i), \gamma(t_{i+1})\bigr).$$

3.2. Measures on metric spaces

Using the additivity of length we then have

$$\mathcal{H}_1(|\gamma|) = \sum_i \mathcal{H}_1(|\gamma_i|) \geq \sum_i d\big(\gamma(t_i), \gamma(t_{i+1})\big) \geq \ell(\gamma) - \varepsilon.$$

Hence, $\mathcal{H}_1(|\gamma|) \geq \ell(\gamma)$, and it remains to prove the converse inequality. Let $\ell(\gamma) < \infty$. Equip $|\gamma|$ with a new metric $\tilde{d}$ given by

$$\tilde{d}\big(\gamma(s), \gamma(t)\big) := \ell\big(\gamma|_{[s,t]}\big), \quad a \leq s < t \leq b.$$

Then the identity map $id : (|\gamma|, \tilde{d}\,) \to (|\gamma|, d|_{|\gamma|})$ is 1-Lipschitz, and the function γ can be represented as $\gamma = f \circ s$, where $s : [a,b] \to \mathbb{R}_+$ is given by $s(t) := \ell(\gamma|_{[a,t]})$ and $f : s([a,b]) \to (\gamma, d|_\gamma)$ is the corresponding 1-Lipschitz function. By Theorems 3.76 (with $p = 1$) and 3.75, we get

$$\ell(\gamma) = s(b) - s(a) = \int_a^b d\mathcal{H}_1 \geq \int N\big(f|_{s([a,b])}\big) d\mathcal{H}_1 = \int N\big(\gamma|_{[a,b]}\big) d\mathcal{H}_1.$$

The last equality holds, since $N\big(f|_{s([a,b])}; m\big) = N\big(\gamma|_{[a,b]}; m\big)$ whenever the set $[a,b] \cap \gamma^{-1}(m)$ does not contain intervals of positive measure, i.e., for all $m \in \mathcal{M}$ excluding a countable set. But $N\big(\gamma|_{[a,b]}; m\big) \geq 1$ for $m \in |\gamma|$ and therefore the last integral is at least $\int_{|\gamma|} d\mathcal{H}_1 = \mathcal{H}_1(|\gamma|)$.

The result is established. □

Now we consider the case of the Euclidean space $\mathbb{R}^p$. Since the Hausdorff p-measure $\mathcal{H}_p$ and the Lebesgue p-measure $\mathcal{L}_p$ on $\mathbb{R}^p$ are translation and rotation invariant, they coincide up to a constant factor. Using the so-called isodiametric inequality, see, e.g., [Fe-1969, Sec. 3.2.43], one can prove that this factor is 1, i.e.,

$$\mathcal{H}_p = \mathcal{L}_p \quad \text{for} \quad p \in \mathbb{N}. \tag{3.91}$$

Finally, we consider the p-measure $\mathcal{H}_p$ with integer p on the Euclidean space $\mathbb{R}^q$ with $q > p$. It is clearly invariant under translations and rotations. Moreover, Theorem 3.77 suggests that $\mathcal{H}_p$ can be regarded as a generalized p-area of subsets in $\mathbb{R}^q$.

The following result, due to M. Kneser [Kn-1955], see also [Fe-1969, Thm. 3.2.39], confirms this claim. For its formulation, we define the *Minkowski p-area measure* of a subset $S \subset \mathbb{R}^q$ by

$$\mathcal{M}_p(S) = \lim_{\varepsilon \to 0} \frac{\mathcal{L}_q(S_\varepsilon)}{\omega(q-p)(2\varepsilon)^{q-p}}.$$

Here $S_\varepsilon := \{x \in \mathbb{R}^q ; d_{\mathbb{R}^q}(x, S) < \varepsilon\}$ is the ε-neighborhood of S in $\mathbb{R}^q$ and the function ω is defined in (3.87). Note that $\mathcal{M}_p$ is not a measure on the family of subsets on which it is defined.

Theorem 3.79. *Let S be the image of a Lipschitz map from a bounded subset of $\mathbb{R}^p$ into $\mathbb{R}^q$. If F is a closed subset of S, then $\mathcal{M}_p(F)$ exists and*

$$\mathcal{H}_p(F) = \mathcal{M}_p(F).$$

Remark 3.80. A set S satisfying the condition of this theorem is said to be *p-rectifiable*.

Finalizing this subsection we discuss relations between Hausdorff measures on the direct ∞-sum $\mathcal{M}_1 \oplus^{(\infty)} \mathcal{M}_2$ and those on $\mathcal{M}_i$'s. Namely, given $p, q \geq 0$, one can consider the tensor product $\mathcal{H}_p \otimes \mathcal{H}_q$ and compare it with the Hausdorff $(p+q)$-measure on $\mathcal{M}_1 \oplus^{(\infty)} \mathcal{M}_2$. The corresponding result was due to J. D. Kelly [Ke-1973]. Since this distinguished fact has never appeared in book form, we outline here its proof.

Theorem 3.81. $\mathcal{H}_{p+q}(\cdot \; ; \mathcal{M}_1 \oplus^{(\infty)} \mathcal{M}_2) \geq \mathcal{H}_p(\cdot \; ; \mathcal{M}_1) \mathcal{H}_q(\cdot \; ; \mathcal{M}_2).$

Proof. Surprisingly, there is no direct derivation of this inequality based on the definition and properties of the Hausdorff measure. Instead, one uses the so-called weighted Hausdorff p-measure whose definition we present now.

Let $S \subset \mathcal{M}$ and $\mathcal{Y}_\delta(S)$ be the set of sequences $\{(c_n, S_n)\}_{n \in \mathbb{N}}$ with $c_n > 0$ and $S_n \subset \mathcal{M}$ satisfying

$$\sum_n c_n \chi_{S_n} \geq \chi_S \quad \text{and} \quad \sup_n \operatorname{diam} S_n \leq \delta. \tag{3.92}$$

An element of $\mathcal{Y}_\delta(S)$ is called a *weighted δ-cover of the set S*. An ordinary δ-cover $\{S\}_{n \in \mathbb{N}}$ of S is, clearly, a weighted cover with all $c_n = 1$ and $\delta := \sup_n \operatorname{diam} S_n$.

Using this, we define an *approximate p-measure* given for $S \subset \mathcal{M}$ by

$$h_p^\delta(S) := \inf \left\{ \sum_n c_n (\operatorname{diam} S_n)^p \right\},$$

where the infimum is taken over all weighted δ-covers $\{(c_n, S_n)\}_{n \in \mathbb{N}}$ from $\mathcal{Y}_\delta(S)$.
Further, we define $h_p(S)$ by

$$h_p(S) := \sup_{\delta > 0} h_p^\delta(S).$$

Similarly to the case of Hausdorff p-measure, one can prove that h_p is a (Borel regular) measure. Using his Theorem 3.74, R. Davies [Dav-1970] established that, for $\mathcal{H}_p$-measurable subsets $S \subset \mathcal{M}$,

$$h_p(S) = \mathcal{H}_p(S). \tag{3.93}$$

Since h_p is Borel regular, for an arbitrary $S \subset \mathcal{M}$ there is a Borel set $\widetilde{S} \supset S$ such that $h_p(S) = h_p(\widetilde{S})$. Then the previous equality gets

$$h_p(S) = h_p(\widetilde{S}) = \mathcal{H}_p(\widetilde{S}) \geq \mathcal{H}_p(S).$$

3.2. Measures on metric spaces

Since $h_p \leq \mathcal{H}_p$ by the definition, the equality (3.93) is now true for *every* S.

Hence, it suffices to prove the inequality

$$h_{p+q}(S_1 \times S_2) \geq h_p(S_1)h_q(S_2) \tag{3.94}$$

for $S_1 \subset \mathcal{M}_1$ and $S_2 \subset \mathcal{M}_2$. Here, clearly, h_{p+q} is the weighted $(p+q)$-measure on $\mathcal{M}_1 \oplus^{(\infty)} \mathcal{M}_2$.

To prove this we will exploit an auxiliary measure $h_{p,q}$ defined on $\mathcal{M}_1 \oplus^{(\infty)} \mathcal{M}_2$ in the same way as h_p and $\mathcal{H}_p$. Actually, let $P_i : \mathcal{M}_1 \oplus^{(\infty)} \mathcal{M}_2 \to \mathcal{M}_i$ be the natural projection, $i = 1, 2$. Define first an approximate measure $h_{p,q}^\delta$ given on subsets $U \subset \mathcal{M}_1 \oplus^{(\infty)} \mathcal{M}_2$ by

$$h_{p,q}^\delta(U) := \inf\left\{\sum_n c_n (\operatorname{diam} P_1(U_n))^p (\operatorname{diam} P_2(U_n))^q\right\}, \tag{3.95}$$

where the infimum is taken over all weighted covers $\{(c_n, U_n)\}_{n \in \mathbb{N}} \in \mathcal{Y}_\delta(U)$.

Then we set

$$h_{p,q} := \sup_{\delta > 0} h_{p,q}^\delta$$

and prove that

$$h_{p,q}(S_1 \times S_2) \leq h_{p+q}(S_1 \times S_2). \tag{3.96}$$

Without loss of generality we may assume that the right-hand side is finite. Let $\{(c_n, U_n)\} \in \mathcal{Y}_\delta(S_1 \times S_2)$. In virtue of the definition of the metric on $\mathcal{M}_1 \oplus^{(\infty)} \mathcal{M}_2$,

$$\operatorname{diam} U_n = \max\{\operatorname{diam} P_1(U_n), \operatorname{diam} P_2(U_n)\}.$$

This yields the estimate

$$\sum_n c_n (\operatorname{diam} U_n)^{p+q} \geq \sum_n c_n (\operatorname{diam} P_1(U_n))^p (\operatorname{diam} P_2(U_n))^q \geq h_{p,q}^\delta(S_1 \times S_2).$$

Taking the infimum over all $\{c_n, U_n\}_{n \in \mathbb{N}}$ from $\mathcal{Y}_\delta(S_1 \times S_2)$ we then get

$$h_{p+q}^\delta(S_1 \times S_2) \geq h_{p,q}^\delta(S_1 \times S_2)$$

which implies (3.96).

Our next step will be to prove that

$$h_{p,q}(S_1 \times S_2) \geq h_p(S_1)h_q(S_2). \tag{3.97}$$

Together with (3.96) this leads to (3.94) and completes the proof.

We may assume that for all sufficiently small $\delta > 0$,

$$h_{p,q}^\delta(S_1 \times S_2) < \infty \quad \text{and} \quad h_q^\delta(S_2) > 0,$$

since otherwise (3.97) is trivial.

Let λ be an arbitrary number satisfying
$$0 < \lambda < h_q^\delta(S_2).$$
It is important that in the definition of $h_{p,q}^\delta(S_1 \times S_2)$, see (3.95), we may take only weighted covers of the form $\{(c_n, S_n^1 \times S_n^2)\}_{n \in \mathbb{N}} \in \mathcal{Y}_\delta(S_1 \times S_2)$. In fact, the inclusion $\{(c_n, U_n)\}_{n \in \mathbb{N}} \in \mathcal{Y}_\delta(S_1 \times S_2)$ implies that $\{(c_n, S_n^1 \times S_n^2)\}_{n \in \mathbb{N}}$ with $S_n^i := P_i(U_n)$, $i = 1, 2$, also belongs to $\mathcal{Y}_\delta(S_1 \times S_2)$.

Given now $\{(c_n, S_n^1 \times S_n^2)\}_{n \in \mathbb{N}}$ from $\mathcal{Y}_\delta(S_1 \times S_2)$, fix a point $m_1 \in S_1$ and denote by $N(m_1)$ the set of those $n \in \mathbb{N}$ for which $m_1 \in S_n^1$. Then for every $m_2 \in S_2$,
$$\sum_{n \in N(m_1)} c_n \chi_{S_n^2}(m_2) = \sum_{n \in \mathbb{N}} c_n \chi_{S_n^1 \times S_n^2}(m_1, m_2) \geq \chi_{S_1 \times S_2}(m_1, m_2) = 1.$$
Hence, $\{(c_n, S_n^2)\}_{n \in N(m_1)}$ is a weighted δ-cover of S_2 and, by the definition of h_q^δ and the choice of λ,
$$\sum_{n \in N(m_1)} c_n (\text{diam } S_n^2)^q > \lambda.$$
Setting $\tilde{c}_n := c_n \lambda^{-1} (\text{diam } S_n^2)^q$ we derive
$$\sum_{n \in \mathbb{N}} \tilde{c}_n \chi_{S_n^1}(m_1) > 1 = \chi_{S_1}(m_1), \quad m_1 \in S_1.$$
Then $\{(\tilde{c}_n, S_n^1)\}_{n \in \mathbb{N}}$ is a weighted δ-cover of S_1 and, consequently,
$$\sum_{n \in \mathbb{N}} c_n (\text{diam } S_n^1)^p (\text{diam } S_n^2)^q = \lambda \sum_{n \in \mathbb{N}} \tilde{c}_n (\text{diam } S_n^1)^p \geq \lambda h_p^\delta(S_1).$$
Sending λ to $h_q^\delta(S_2)$ and taking the infimum over all $\{(c_n, S_n^1 \times S_n^2)\}_{n \in \mathbb{N}}$ from $\mathcal{Y}_\delta(S_1 \times S_2)$, we get
$$h_{p+q}^\delta(S_1 \times S_2) \geq h_p^\delta(S_1) h_q^\delta(S_2)$$
whence (3.97) follows. $\square$

Remark 3.82. The equality $h_p(S) = \mathcal{H}_p(S)$ used in the proof was first established in [Fe-1969] for S being a countable union of subsets of finite $\mathcal{H}_p$-measure (see there Theorem 2.10.24 for $f = \chi_S$; then the upper integral $\int^* \chi_S d\mathcal{H}_p$ in its formulation equals precisely $h_p(S)$). Since $\mathcal{H}_p$ is not, in general, σ-finite, the assumption of the Federer theorem is too restrictive. The general result was due to R. Davies [Dav-1970] as a consequence of his Increasing Sequence Lemma, see Theorem 3.74, and the Federer result. These authors as well as J. D. Kelly [Ke-1973] studied a more general situation concerning the family $\{\mathcal{H}_\varphi\}_{\varphi \in \Phi}$ of generalized Hausdorff measures. Here Φ is the set of all nondecreasing continuous from the right functions $\varphi : \mathbb{R}_+ \to \mathbb{R}_+$, and $\mathcal{H}_\varphi$ is defined by replacing $(\text{diam } S_n)^p$ in (3.87) by $\varphi(\text{diam } S_n)$.

3.2.5 Doubling measures

The class of these (Borel regular) measures is described by

Definition 3.83. A measure μ on a metric space $\mathcal{M}$ is said to be doubling at a point $m \in \mathcal{M}$ if the μ-measure of every nonempty open ball is finite and strictly positive and the doubling constant

$$D_m(\mu) := \sup_{R>0} \frac{\mu(B_{2R}(m))}{\mu(B_R(m))}$$

is finite.

If μ is doubling at every point and its doubling constant

$$D(\mu) := \sup_{m \in \mathcal{M}} D_m(\mu)$$

is finite, then μ is called a doubling measure.

If a space $\mathcal{M}$ carries a doubling measure it should satisfy rather restricted conditions. For instance, a countable metric space without isolated points cannot carry such a measure. This follows from the next result of Coifman and Weiss [CW-1971].

Proposition 3.84. Let $(\mathcal{M}, d)$ carry a doubling measure μ. Then the following is true:

(a) for every nonisolated point m_0

$$\mu(\{m_0\}) = 0;$$

(b) the metric space $\mathcal{M}$ is doubling and its doubling constant, see (3.40), satisfies

$$\log_2 \delta_{\mathcal{M}} \leq 4 \log_2 D(\mu). \tag{3.98}$$

Proof. (a) Assume, on the contrary, that $\mu(\{m_0\}) = c > 0$. Let $\{m_i\}_{i \geq 1}$ be a sequence converging to m_0. Passing to a subsequence we may assume that the balls $B_{R_i}(m_i)$ with $R_i := \frac{2}{3} d(m_i, m_0)$ are mutually disjoint. Since $B_{2R_i}(m_i)$ contains m_0, the doubling condition yields

$$c = \mu(\{m_0\}) \leq \mu(B_{2R_i}(m_i)) \leq D(\mu) \mu(B_{R_i}(m_i)).$$

Therefore, for a ball $B_R(m_0)$ containing all $B_{R_i}(m_i)$ we get

$$\mu(B_R(m_0)) \geq \sum_i \mu(B_{R_i}(m_i)) \geq \sum_i \frac{c}{D(\mu)} = \infty.$$

This contradicts Definition 3.83.

(b) Given a ball $B_{2R}(m_0) \subset \mathcal{M}$, we choose there a *maximal R-net*, that is a subset $\{m_i\} \subset B_{2R}(m_0)$ satisfying

(i) $d(m_i, m_j) \geq R$ for $i \neq j$;

(ii) the distance from every $m \in B_{2R}(m_0)$ to this subset is less than R.

The existence of such a net easily follows from Zorn's lemma. Because of (ii),
$$B_{2R}(m_0) \subset \bigcup_i B_R(m_i);$$
hence, to estimate $\delta_{\mathcal{M}}$, it suffices to bound card$\{m_i\}$. To do this, we first note that the balls $B_{\frac{R}{2}}(m_i)$ are mutually disjoint by (i), and are contained in the ball $B_{\frac{5}{2}R}(m_0)$. This and the doubling inequality imply that
$$\sum \mu\bigl(B_{\frac{R}{2}}(m_i)\bigr) \leq \mu\bigl(B_{\frac{5}{2}R}(m_0)\bigr) \leq D(\mu)\mu\bigl(B_{2R}(m_0)\bigr).$$
On the other hand, $B_{2R}(m_0) \subset B_{4R}(m_i)$ and therefore
$$\mu\bigl(B_{2R}(m_0)\bigr) \leq \mu\bigl(B_{4R}(m_i)\bigr) \leq D(\mu)^3 \mu\bigl(B_{\frac{R}{2}}(m_i)\bigr),$$
by repeating the application of the doubling inequality.

Together with the previous inequality this yields
$$\frac{\mu\bigl(B_{2R}(m_0)\bigr)}{D(\mu)^3} \,\text{card}\{m_i\} \leq \sum \mu\bigl(B_{\frac{R}{2}}(m_i)\bigr) \leq D(\mu)\mu\bigl(B_{2R}(m_0)\bigr).$$
Since $\mu\bigl(B_{2R}(m_0)\bigr) \neq 0$, this implies the required inequality
$$\delta_{\mathcal{M}} \leq \text{card}\{m_i\} \leq D(\mu)^4. \qquad \square$$

Remark 3.85. If $\mathcal{M}$ contains at least two points, its doubling constant $\delta_{\mathcal{M}} \geq 2$. Therefore (3.98) implies that $D(\mu)$ cannot be arbitrarily close to 1. Namely, for a nontrivial $\mathcal{M}$,
$$D(\mu) \geq \sqrt[4]{2}. \tag{3.99}$$
A more accurate derivation gives $\sqrt{2}$ as a lower bound.

Definition 3.86. A metric space $(\mathcal{M}, d)$ with a fixed doubling measure is said to be of *homogeneous type*.

It should be emphasized that the object of this definition is the triple $(\mathcal{M}, d, \mu)$, since a metric space may carry even an uncountable family of mutually singular doubling measures. For instance, this is the case of a segment of the real line, see the paper [BA-1956] by Ahlfors and Beurling, see subsection 4.3.5. However, most metric spaces appearing in Analysis and Geometry carry naturally arising doubling measures, such as the Lebesgue measure on $\mathbb{R}^n$ or the Haar measure on the Heisenberg group.

The class of spaces of homogeneous type is stable under the most of metric constructions introduced in subsection 3.1.8. We consider here only the direct p-sum of metric spaces of homogeneous type. In fact, we establish a more general fact using the following

3.2. Measures on metric spaces

Definition 3.87. Let μ be a measure on $\mathcal{M}$ doubling at a point m. The dilation function of μ is given for $\ell \geq 1$ by

$$D_m(\mu\ ;\ \ell) := \sup\left\{\frac{\mu(B_{\ell R}(m))}{\mu(B_R(m))}\ ;\ R > 0\right\}. \tag{3.100}$$

If this μ is doubling, its dilation function is defined for $\ell \geq 1$ by

$$D(\mu\ ;\ \ell) = \sup\{D_m(\mu\ ;\ \ell)\ ;\ m \in \mathcal{M}\}. \tag{3.101}$$

It is clear that these functions are finite and submultiplicative, i.e.,

$$D_m(\mu\ ;\ \ell_1\ell_2) \leq D_m(\mu\ ;\ \ell_1) D_m(\mu\ ;\ \ell_2),$$

and the same is true for $D(\mu\ ;\ \cdot)$. Note also that

$$D(\mu\ ;\ 2) = D(\mu)$$

and therefore

$$D(\mu\ ;\ \ell) \leq D(\mu)^{1+\log_2 \ell}. \tag{3.102}$$

Now the desired result is given by

Proposition 3.88. *Let μ_i be a measure on $\mathcal{M}_i$ doubling at a point m_i for $1 \leq i \leq N$. Then, for $m := (m_1, \ldots, m_n) \in \oplus^{(p)}\{\mathcal{M}_i, d_i\}_{1 \leq i \leq N}$ and all $\ell \geq 1$,*

$$D_m(\otimes_i \mu_i\ ;\ \ell) \leq \prod_i D_{m_i}(\mu_i\ ;\ \ell). \tag{3.103}$$

Proof. It suffices to consider $N = 2$ and then complete the proof by induction on N.

Let $B_R(m)$ be an open ball in $\mathcal{M}_1 \oplus^{(p)} \mathcal{M}_2$, and $B_R(m_i)$ be the corresponding open balls in $\mathcal{M}_i$, $i = 1, 2$. Using the Fubini theorem we present $(\mu_1 \otimes \mu_2)(B_{\ell R}(m))$ in the form

$$\int_{B_{\ell R}(m_1)} \mu_2(B_{\ell \rho_1}(m_2)) d\mu_1(m_1') \quad \text{where} \quad \rho_1 := \left(R^p - \ell^{-p} d_1(m_1, m_1')^p\right)^{\frac{1}{p}}.$$

Applying the doubling condition and changing the order of integration we estimate this integral by

$$D_{m_2}(\mu_2\ ;\ \ell) \int_{B_R(m_2)} \mu_1(B_{\ell \rho_2}(m_1)) d\mu_2(m_2') \quad \text{where} \quad \rho_2 := \left(R^p - d_2(m_2, m_2')^p\right)^{\frac{1}{p}}.$$

Applying again the doubling condition and then the Fubini theorem we obtain the desired result:

$$(\mu_1 \otimes \mu_2)(B_{\ell R}(m)) \leq D_1(\ell) D_2(\ell)(\mu_1 \otimes \mu_2)(B_R(m)). \qquad \square$$

Examples of direct 2-sums of Euclidean spaces equipped with Lebesgue measures show sharpness of (3.103).

3.2.6 Families of pointwise doubling measures

Let $\mathcal{F}$ be a *family of pointwise doubling measures* on a metric space on $(\mathcal{M}, d)$ enumerated by points $m \in \mathcal{M}$, i.e., $\mathcal{F} := \{\mu_m\}_{m \in \mathcal{M}}$ where $D_m(\mu_m) < \infty$ for all m, see Definition 3.83.

Definition 3.89. (a) $\mathcal{F}$ is said to be *uniformly doubling* if
$$D(\mathcal{F}) := \sup_{m \in \mathcal{M}} D_m(\mu_m) < \infty. \tag{3.104}$$

(b) $\mathcal{F}$ is said to be *K-uniform*, $K \geq 1$, if for all $m_1, m_2 \in \mathcal{M}$ and $R > 0$,
$$\mu_{m_1}(B_R(m_1)) \leq K \mu_{m_2}(B_R(m_2)). \tag{3.105}$$

In particular, $K = 1$ amounts to $\mu_m(B_R(m)) = \varphi(R)$ for all m, where $\varphi : \mathbb{R}_+ \to \mathbb{R}_+$ is a nondecreasing function. In this case
$$D(\mathcal{F}) = \sup_{R > 0} \frac{\varphi(2R)}{\varphi(R)}.$$

To introduce the next concept, we need the definition of a quasiball. The set $S \subset \mathcal{M}$ is called a *λ-quasiball centered at m* ($\lambda \geq 1$) if for some ball $B_R(m)$
$$B_R(m) \subset S \subset B_{\lambda R}(m).$$
We denote a λ-quasiball satisfying these implications by $Q_R(m)$.

Definition 3.90. A family $\mathcal{F} = \{\mu_m\}_{m \in \mathcal{M}}$ of pointwise doubling measures on $(\mathcal{M}, d)$ is said to be *λ-consistent with the metric d* if there are a family of λ-quasiballs $\{Q_R(m)\,; m \in \mathcal{M},\ R > 0\}$ and a constant $C \geq 1$ such that for all $m_1, m_2 \in \mathcal{M}$ and $R > 0$ satisfying $d(m_1, m_2) \leq R$ the following inequality
$$|\mu_{m_1} - \mu_{m_2}|(Q_R(m)) \leq \frac{C \mu_m(Q_R(m))}{R} d(m_1, m_2) \tag{3.106}$$
holds for $m = m_1$ and m_2.

The optimal C denoted by $C_\lambda(\mathcal{F})$ is called the *consistency constant* of $\mathcal{F}$.
If (3.106) holds under the condition
$$d(m_1, m_2) \leq tR$$
for some $t > 0$, then the corresponding optimal constant is denoted by $C_\lambda(\mathcal{F}; t)$ and $t \mapsto C_\lambda(\mathcal{F}; t)$ is called the *consistency function* of $\mathcal{F}$.

Clearly, $C_\lambda(\mathcal{F}; \cdot)$ is nondecreasing and $C_\lambda(\mathcal{F}) = C_\lambda(\mathcal{F}; 1)$.

Let us recall that here $|\nu|$ is the *total variation* of a *signed measure* ν (= the difference of two measures), that is,
$$|\nu|(S) := \sup \sum_j |\nu(S_j)|,$$
where the supremum is taken over all Borel partitions $\{S_k\}$ of a Borel set S.

A family $\mathcal{F}$ on $\mathcal{M}$ satisfying condition (3.104) and condition (3.106) for some $\lambda > 0$ is said to be *coherent*. A metric space $(\mathcal{M}, d)$ equipped with a fixed coherent family $\mathcal{F}$ is said to be of *pointwise homogeneous type*.

We denote the class of these triples $(\mathcal{M}, d, \mathcal{F})$ by $\mathcal{PHT}$, and the class of spaces of homogeneous type by $\mathcal{HT}$. It is clear that

$$\mathcal{HT} \subset \mathcal{PHT}; \tag{3.107}$$

more precisely, a doubling measure μ on $(\mathcal{M}, d)$ determines the constant coherent family $\mathcal{F} := \{\mu_m := \mu\}_{m \in \mathcal{M}}$ with

$$D(\mathcal{F}) = D(\mu) \quad \text{and} \quad C(\mathcal{F}) = 0.$$

On the other hand, $\mathcal{PHT}$ is essentially larger than $\mathcal{HT}$ and consists of many important geometric objects of a nonhomogeneous type. As an example we only mention the classical hyperbolic space $\mathbb{H}^n$. A highly nontrivial construction of a coherent family on $\mathbb{H}^n$ is presented in [BSh-1999, pp. 437–540]. This construction will be presented in Section 4.4 for a more general class of metric spaces.

Similarly to the notion of homogeneous type, that of pointwise homogeneous type is bi-Lipschitz invariant. This fact is presented by

Proposition 3.91. *Let metric spaces $\mathcal{M}$ and $\widetilde{\mathcal{M}}$ be bi-Lipschitz homeomorphic. Then $\widetilde{\mathcal{M}}$ belongs to $\mathcal{PHT}$ whenever $\mathcal{M}$ does.*

Proof. Let $\mathcal{F} := \{\mu_m\}_{m \in \mathcal{M}}$ be a λ-coherent family of measures on $\mathcal{M}$. By the assumption, there is a bijection $f : (\mathcal{M}, d) \to (\widetilde{\mathcal{M}}, \widetilde{d})$ such that for some $\gamma \geq 1$ and all $m, m' \in \mathcal{M}$,

$$\gamma^{-1} d(m, m') \leq \widetilde{d}(f(m), f(m')) \leq \gamma d(m, m').$$

We define a family of measures $\widetilde{\mathcal{F}} := \{\widetilde{\mu}_{\widetilde{m}}\}_{\widetilde{m} \in \widetilde{\mathcal{M}}}$ as the image of $\mathcal{F}$ under f. That is to say, for a Borel set $\widetilde{S} \subset \widetilde{\mathcal{M}}$ we set

$$\widetilde{\mu}_{\widetilde{m}}(\widetilde{S}) := \mu_m(f^{-1}(\widetilde{S})) \quad \text{where} \quad \widetilde{m} := f(m).$$

It is the matter of the definitions to check that

$$D(\widetilde{\mathcal{F}}) \leq \left(D(\mathcal{F}; \gamma)\right)^2 D(\mathcal{F}) \quad \text{and} \quad C_\mu(\widetilde{\mathcal{F}}) \leq C_\lambda(\mathcal{F}) \quad \text{where} \quad \mu := \gamma^2 \lambda;$$

here $D(\mathcal{F}; \cdot)$ is the dilation function of the family $\mathcal{F}$, see (3.100) and (3.110) below. Due to (3.104) this function is finite and therefore also $D(\widetilde{\mathcal{F}})$ and $C_\mu(\widetilde{\mathcal{F}})$ are.

The result is proved. $\square$

To avoid complication of notation, in the subsequent proofs we will consider only the case of 1-coherent families of measures (called simply coherent ones). In this case the quasiball $Q_R(m)$ is the open ball $B_R(m)$. Changes in the proofs for $\lambda \neq 1$ are trivial. Moreover, for $\lambda = 1$ we will write $C(\mathcal{F})$ instead of $C_1(\mathcal{F})$, etc.

Let us single out a hereditary property of the class $\mathcal{HT}$. Every *closed* subspace S of a metric space $(\mathcal{M}, d) \in \mathcal{HT}$ belongs to $\mathcal{HT}$. A doubling measure on $(S, d|_{S \times S})$ may be distinct from $\mu|_S$, since the latter is, in general, not doubling. The existence of the required doubling measures follows from the Vol'berg–Koniagin theorem [VK-1987] presented in Section 4.3. An analogous hereditary property for the class $\mathcal{PHF}$ is unknown and probably incorrect.

Now we discuss stability of the class $\mathcal{PHT}$ under the operation of direct p-sum. To formulate the corresponding problem we define the tensor product of families $\mathcal{F}_j := \{\mu^j_{m_j}\}$, $1 \leq j \leq N$, by

$$\mathcal{F} := \bigotimes_{j=1}^{N} \mathcal{F}_j := \left\{ \bigotimes_{j=1}^{N} \mu^j_{m_j} \right\}_m \tag{3.108}$$

where $m := (m_1, \ldots, m_N)$ runs over all points of the Cartesian product $\prod_{j=1}^{N} \mathcal{M}_j$.

Further, for $1 \leq p \leq \infty$, we set

$$(\mathcal{M}, d_p) := \oplus^{(p)} \{(\mathcal{M}_j, d_j)\}_{1 \leq j \leq N}. \tag{3.109}$$

Let $\mathcal{F}_j$ be coherent on $(\mathcal{M}_j, d_j)$ for $1 \leq j \leq N$.

It can be proved that $\mathcal{F}$ is a coherent family on $(\mathcal{M}, d_p)$. To avoid some technicalities we, however, restrict ourselves to a partial result which is sufficient to the subsequent applications.

In its formulation, $D(\mathcal{F} ; \cdot)$ is recalled to be the dilation function of the family $\mathcal{F} := \{\mu_m\}_{m \in \mathcal{M}}$ given for $\ell \geq 1$ by

$$D(\mathcal{F} ; \ell) := \sup \left\{ \frac{\mu_m(B_{\ell R}(m))}{\mu_m(B_R(m))} \right\}, \tag{3.110}$$

where the supremum is taken over all open balls $B_R(m)$ of the space $(\mathcal{M}, d_p)$.

Proposition 3.92. *Assume that $\mathcal{F}_j$ is a coherent and K_j-uniform family on $(\mathcal{M}_j, d_j)$, $1 \leq j \leq N$. Then the family $\mathcal{F} := \bigotimes_{j=1}^{N} \mathcal{F}_j$ is coherent and K-uniform on the space $(\mathcal{M}, d_p)$ with $K := \prod_{j=1}^{N} K_j$. Moreover, the basic characteristics of $\mathcal{F}$ are estimated as*

$$D(\mathcal{F}) \leq \prod_{j=1}^{N} D(\mathcal{F}_j);$$

$$C(\mathcal{F};t) \leq \gamma_p(t) \left\{ \sum_{j=1}^{N} [K^j C(\mathcal{F}_j;t)]^q \right\}^{\frac{1}{q}} \quad \text{for } 0 < t \leq 1; \tag{3.111}$$

3.2. Measures on metric spaces

here $\frac{1}{p} + \frac{1}{q} = 1$, $K^j := \prod_{i>j} K_i$ for $j < N$, $K^N := 1$, and

$$\gamma_p(t) := \inf_{a>0}\left[\left((1+a)^p - t^p\right)^{-\frac{1}{p}} D(\mathcal{F};\, 1+a)\right].$$

In particular, we get

$$C(\mathcal{F}) \leq \gamma_p(1) K^1 \left\{ \sum_{i=1}^N C(\mathcal{F}_i)^q \right\}^{\frac{1}{q}}.$$

Proof. We consider the case $1 \leq p < \infty$ and derive the result for $p = \infty$ by passing to the lower limit (note that $\underline{\lim}_{p\to\infty} \gamma_p(t) \leq 1$).

Fix points $m := (m_1, \ldots, m_N)$ and $m' := (m'_1, \ldots, m'_N)$ and a number $R > 0$. To simplify the computations, we use the following notation:

$$\mu_i := \mu^i_{m_i}, \quad \mu'_i := \mu^i_{m'_i}, \quad 1 \leq i \leq N,$$

$$\mu_I := \bigotimes_{i \in I} \mu_i, \quad \mu'_I := \bigotimes_{i \in I} \mu'_i, \quad I \subset \{1, \ldots, N\};$$

here we set $\mu_\phi := 1$.

Further, m_I denotes a point $(m_i)_{i \in I}$ of the space $\oplus^{(p)}\{\mathcal{M}_i, d_i\}_{i \in I}$; the distance between points m_I and m'_I in this space is denoted by $d_p(m_I, m'_I)$; hence, $B_R(m_I)$ stands for the ball of this space. In the special case $I := \{1, \ldots, N\}\setminus\{i\}$, we denote m_I by m^i, $1 \leq i \leq N$.

We first prove that $\mathcal{F}$ is K-uniform starting with the trivial case of $N = 1$ and then proceeding by induction on N. Suppose that the claim is true for the family $\bigotimes_{j \in J} \mathcal{F}_j$ where $J := \{1, \ldots, N-1\}$, $N \geq 2$. Using Fubini's theorem and the induction hypothesis we then obtain the following chain of inequalities with $R(x_N) := \left(R^p - d_N(m_N, x_N)^p\right)^{\frac{1}{p}}$ and $R(x_J) := \left(R^p - d_p(m_J, x_J)^p\right)^{\frac{1}{p}}$,

$$\mu_m(B_R(m)) = \int_{B_R(m_N)} \mu_J\bigl(B_{R(x_N)}(m_J)\bigr) d\mu_N$$

$$\leq K_1 \cdot \ldots \cdot K_{N-1} \int_{B_R(m_N)} \mu'_J\bigl(B_{R(x_N)}(m'_J)\bigr) d\mu_N$$

$$= K_1 \cdot \ldots \cdot K_{N-1} \int_{B_R(m'_J)} \mu_N\bigl(B_{R(x_J)}(m_N)\bigr) d\mu'_J$$

$$\leq K_1 \cdot \ldots \cdot K_N \, \mu_{m'}\bigl(B_R(m')\bigr),$$

as required.

Further, the required estimate of the doubling constant $D(\mathcal{F})$ immediately follows from (3.103) which, in fact, implies a more general result:

$$D(\mathcal{F}\ ;\ \ell) \leq \prod_{j=1}^{N} D(\mathcal{F}_j\ ;\ \ell), \quad \ell \geq 1. \tag{3.112}$$

It remains to estimate the consistency function of $\mathcal{F}$. That is to say, we must estimate $|\mu_m - \mu_{m'}|(B_R(\widehat{m}))$ for $\widehat{m} = m$ and m' under the condition $d_p(m, m') \leq tR$, $0 < t \leq 1$. Let, for definiteness, $\widehat{m} = m$. By Fubini's theorem and the identity

$$\mu_m - \mu_{m'} = \sum_{i=1}^{N} \nu_i \otimes (\mu_i - \mu_i') \otimes \nu_i'$$

with $\nu_i := \bigotimes_{j<i} \mu_j$ and $\nu_i' := \bigotimes_{j>i} \mu_j'$, we obtain

$$|\mu_m - \mu_{m'}|(B_R(m)) \leq \sum_{i=1}^{N} \int_{B_R(m^i)} |\mu_i - \mu_i'|(B_\rho(m_i)) d(\nu_i \otimes \nu_i')(x^i);$$

here

$$\rho := \sqrt[p]{R^p - d_p(x^i, m^i)^p}.$$

We define ρ_a for $a > 0$ by replacing here R by $(1+a)R$, and then replace ρ by ρ_a in the above inequality. Since $d_i(m_i, m_i') \leq d_p(m, m') \leq tR$, we may apply the consistency inequality for $\mathcal{F}_i$, see (3.106), to bound the integrand in the i-th integral by

$$C(\mathcal{F}_i; t) \frac{\mu_i(B_{\rho_a}(m_i))}{\rho_a} d_i(m_i, m_i').$$

Since also $d_p(x^i, m^i) \leq tR$, the denominator here is at least $((1+a)^p - t^p)^{\frac{1}{p}} R$. Therefore, the i-th term is bounded by

$$\frac{C(\mathcal{F}_i; t) d_i(m_i, m_i')}{((1+a)^p - t^p)^{\frac{1}{p}} R} \int_{B_R(m^i)} \mu_i(B_{\rho_a}(m_i)) d(\nu_i \otimes \nu_i').$$

Using, as above, K_j-uniformity to replace μ_j' by μ_j, $j > i$, we estimate the last integral by

$$\left(\prod_{j>i} K_j\right) \int_{B_R(m^i)} \mu_i(B_{\rho_a}(m_i)) d\left(\bigotimes_{j \neq i} \mu_j\right)$$
$$\leq \left(\prod_{j>i} K_j\right) \mu_m(B_{R(1+a)}(m)) := K^j \mu_m(B_{R(1+a)}(m))$$
$$\leq K^j D(\mathcal{F}\ ;\ 1+a) \mu_m(B_R(m)).$$

3.2. Measures on metric spaces

Combining all these estimates we obtain

$$|\mu_m - \mu_{m'}|(B_R(m)) \leq \frac{D(\mathcal{F}\,;\,1+a)}{((1+a)^p - t^p)^{\frac{1}{p}}R} \cdot \sum_{j=1}^{N} K^j C(\mathcal{F}_j\,;t) d_i(m_i, m'_i).$$

Applying now the Hölder inequality to the sum on the right-hand side we get

$$|m_m - \mu_{m'}|(B_R(m)) \leq \frac{D(\mathcal{F}\,;\,1+a)}{((1+a)^p - t^p)^{\frac{1}{p}}R} \left\{ \sum_{j=1}^{N} [K^j C(\mathcal{F}_j\,;t)]^q \right\}^{\frac{1}{q}} d_p(m, m').$$

Taking the infimum over all $a > 0$ we prove the second inequality of (3.111). □

Unlike the estimates of Proposition 3.92 for the doubling constant and the constant of uniformity which are sharp, the estimate of the consistency constant $C(\mathcal{F})$ is far from being sharp for $p < \infty$. In fact, the constants $D(\mathcal{F}_j)$ and $C(\mathcal{F}_j)$ are the only data at our disposal. Therefore, to estimate $\gamma_p(t)$ in (3.106), we have to use the trivial and inaccurate inequalities

$$D(\mathcal{F}_j, \ell) \leq D(\mathcal{F}_j)^{\log_2 \ell + 1}, \quad 1 \leq j \leq N.$$

The situation becomes much better if one of the families $\mathcal{F}_j$, say $\mathcal{F}_N$, is of *n-homogeneous type*, meaning that, for all $m \in \mathcal{M}_N$ and $R > 0$,

$$\mu_m^N(B_R(m)) = cR^n \tag{3.113}$$

with some constants $c, n > 0$. An example of the space of n-homogeneous type is $\mathbb{R}^n$ equipped with the Lebesgue measure and some norm; the constant c equals the volume of the unit ball in this norm. However, (3.113) does not imply independence of μ_m^N of m; hence, $C(\mathcal{F}_N)$ may be not zero (while $K = 1$). The corresponding example is the hyperbolic space $\mathbb{H}^n$ equipped with a suitable coherent family satisfying (3.113). This will be discussed in detail later.

Now we present a variant of Proposition 3.92 with μ_N of n-homogeneous type. To avoid some technicalities, we consider only the case of $p = 1$ leaving that of $1 < p < \infty$ to the reader.

Corollary 3.93. *Under the notations and assumptions of Proposition 3.92 (with $p = 1$) and condition (3.113), the inequality*

$$C(\mathcal{F}) \leq \frac{6}{5} e^4 nK \max_{1 \leq j \leq N} C(\mathcal{F}_j) \tag{3.114}$$

holds for n satisfying

$$n \geq \lfloor \log_2 D \rfloor + 5. \tag{3.115}$$

Here D is a constant satisfying $D\left(\bigotimes_{j=1}^{N-1} \mathcal{F} \right) \leq D.$

Proof. We begin with a preliminary result that will be also used later.

Let $\mathcal{H} = \{\mu_m\}_{m \in \mathcal{M}}$ be a family of measures on a metric space $(\mathcal{M}, d)$ whose doubling constant satisfies

$$D(\mathcal{H}) \leq D \tag{3.116}$$

for some $D > 0$.

Let $\mathcal{H}' = \{\mu'_m\}_{m \in \mathcal{M}'}$ be a coherent family on a metric space $(\mathcal{M}', d')$ satisfying (3.113). We equip the metric space

$$(\widetilde{\mathcal{M}}, \tilde{d}) := (\mathcal{M}, d) \oplus^{(1)} (\mathcal{M}', d')$$

with the family of measures

$$\widetilde{\mathcal{F}} := \{\mu_{\widetilde{m}}\}_{\widetilde{m} \in \widetilde{\mathcal{M}}} := \{\mu_m \otimes \mu_{m'}\}_{(m,m') \in \widetilde{\mathcal{M}}}.$$

Then the following is true:

Lemma 3.94. *Assume that*

$$n \geq \lfloor \log_2 D \rfloor + 5. \tag{3.117}$$

Then for every $\widetilde{m} \in \widetilde{\mathcal{M}}$ and $R > 0$,

$$\mu_{\widetilde{m}}(B_{R_n}(\widetilde{m})) \leq \frac{6}{5} e^4 \mu_{\widetilde{m}}(B_R(\widetilde{m})); \tag{3.118}$$

here $R_n := \left(1 + \frac{1}{n}\right) R$.

Proof. The application of Fubini's theorem and (3.113) yield

$$\mu_{\widetilde{m}}(B_R(\widetilde{m})) = c \int_{B_R(m)} (R - d(m, x))^n d\mu_m(x). \tag{3.119}$$

We estimate this integral with R replaced by $R_n = \left(1 + \frac{1}{n}\right) R$. To this end, we split the integral into one over $B_{3R/4}(m)$ and one over the remaining part $B_{R_n}(m) \setminus B_{3R/4}(m)$. Denote these integrals by I_1 and I_2. For I_2 we get

$$I_2 \leq c(R_n - 3R/4)^n \int_{B_{R_n}(m)} d\mu_m = c\left(\frac{1}{4} + \frac{1}{n}\right)^n R^n \mu_m(B_{R_n}(m)).$$

Using the estimate for the doubling constant of $\mathcal{H} = \{\mu_m\}$, see (3.116), we further have

$$\mu_m(B_{R_n}(m)) \leq D \mu_m(B_{R_n/2}(m)).$$

Moreover, by (3.117), $D < 2^{\lfloor \log_2 D \rfloor + 1} \leq \frac{1}{16} 2^n$. Combining all these inequalities we obtain

$$I_2 \leq c \frac{1}{16} 2^{-n} \left(1 + \frac{4}{n}\right)^n R^n \mu_m(B_{R_n/2}(m)). \tag{3.120}$$

3.2. Measures on metric spaces

To estimate I_1 we present its integrand (which is equal to that in (3.119) with R replaced by R_n) in the following way.

$$\left(1 + \frac{1}{n}\right)^n (R - d(m, m'))^n \left(1 + \frac{d(m, m')}{(n+1)(R - d(m, m'))}\right)^n.$$

Since $d(m, m') \leq 3R/4$ for $m' \in B_{3R/4}(m)$, the last factor is at most $\left(1 + \frac{3}{n+1}\right)^n$. Hence we have

$$I_1 \leq c\left(1 + \frac{1}{n}\right)^n \left(1 + \frac{3}{n+1}\right)^n \int_{B_{3R/4}(m)} (R - d(m, x))^n d\mu_m(x).$$

Using again (3.119) we finally obtain

$$I_1 \leq ce^4 \mu_{\widetilde{m}}(B_R(\widetilde{m})). \tag{3.121}$$

To estimate the constant in (3.118), it remains to bound the fractions

$$\widetilde{I}_k := \frac{I_k}{\mu_{\widetilde{m}}(B_R(\widetilde{m}))}, \quad k = 1, 2.$$

Since $R_n < 2R$, the denominator in $\widetilde{I}_2$ is bounded from below by

$$c \int_{B_{R_n/2}(m)} (R - d(m, x))^n d\mu_m(x) \geq c2^{-n}\left(1 - \frac{1}{n}\right)^n R^n \int_{B_{R_n/2}(m)} d\mu_m$$

$$= c2^{-n}\left(1 - \frac{1}{n}\right)^n R^n \mu_m(B_{R_n/2}(m)).$$

Combining this and (3.120) we get

$$\widetilde{I}_2 \leq \frac{1}{16}\left(1 - \frac{1}{n}\right)^{-n}\left(1 + \frac{4}{n}\right)^n.$$

Since $\left(1 - \frac{1}{n}\right)^{-n} \leq \left(1 - \frac{1}{5}\right)^{-5}$ as $n \geq 5$, we finally obtain

$$\widetilde{I}_2 \leq \frac{1}{5} e^4.$$

As for $\widetilde{I}_1$, inequality (3.121) immediately gives

$$\widetilde{I}_1 \leq e^4.$$

Hence, the constant in (3.118) is at most $\frac{6}{5} e^4$. □

Returning to the proof of the corollary we apply Lemma 3.94 with $(\mathcal{M}, d) := \oplus^{(1)}(\mathcal{M}_j, d_j)_{1 \le j \le N-1}$ and $(\mathcal{M}', d') := (\mathcal{M}_N, d_N)$ and the families $\mathcal{H} := \bigotimes_{j<N} \mathcal{F}_j$ and $\mathcal{H}' := \mathcal{F}_N$. This yields

$$\mu_m\big(B_{(1+\frac{1}{n})R}(m)\big) \le \frac{6}{5} e^4 \mu_m(B_R(m));$$

whence

$$D\Big(\mathcal{F}\,;\,1+\frac{1}{n}\Big) \le \frac{6}{5} e^4.$$

Combining this and the second inequality of (3.111) with $p = 1$, we obtain

$$C(\mathcal{F})) \le \Big[\Big(1+\frac{1}{n}\Big) - 1\Big]^{-1} D\Big(\mathcal{F}\,;\,1+\frac{1}{n}\Big) K \max_{1 \le j \le N} C(\mathcal{F}_j) \le \frac{6}{5} e^4 n K \max_{1 \le j \le N} C(\mathcal{F}_j),$$

as required. $\square$

3.3 Basic classes of metric spaces

In this part, we present several classes of metric spaces playing an important role in applications. Throughout this chapter proofs will be presented only if they have not appeared in book form.

3.3.1 Ultrametric spaces

This class of spaces is introduced by

Definition 3.95. A metric space $(\mathcal{M}, d)$ is said to be *ultrametric* (or *non-Archimedean*) if its metric satisfies the strong triangle inequality

$$d(m, m') \le \max\{d(m, m''), d(m'', m')\} \quad \text{for all} \quad m, m', m''. \tag{3.122}$$

The structure of such a space is far beyond our geometric intuition. In particular, both open and closed balls of this space are simultaneously open and closed sets. Moreover, every point of an open ball is its center, i.e.,

$$m' \in B_r(m) \Rightarrow B_r(m) = B_r(m').$$

In other words, the inequality $d(m, m') < r$ is an *equivalence relation* on $\mathcal{M}$. Nevertheless, ultrametric spaces provide a geometric framework for several important fields of mathematics including Number Theory (p-adic numbers and their generalizations), Set Theory (the Baire spaces), Theoretical Computer Science (coding and decoding). We present here only one example.

3.3. Basic classes of metric spaces

Example 3.96 (*p*-adic numbers). Let p be a prime number. Define a *p-adic norm* $\|\cdot\|_p : \mathbb{Q} \to \mathbb{R}_+$ given for a rational number $r := \frac{m}{n} p^k$ by

$$\|r\|_p := p^{-k};$$

here m, n are integers which are not divided by p. It is easily seen that this function meets the the following conditions:

$$\|r_1 r_2\|_p = \|r_1\|_p \cdot \|r_2\|_p,$$
$$\|r_1 + r_2\|_p \leq \max\{\|r_1\|_p, \|r_2\|_p\}.$$

So, the set of rationals $\mathbb{Q}$ endowed with the distance

$$d_p(r_1, r_2) := \|r_1 - r_2\|_p$$

is an ultrametric space. This space is incomplete (e.g., $d_p(p^n, p^m) \to 0$ as $n, m \to \infty$, i.e., $\{p^n\}_{1 \leq n < \infty}$ is a Cauchy sequence). Its completion is, by definition, the space of *p-adic numbers* $\mathbb{Q}_p$ which is a field, since all arithmetic operations on $\mathbb{Q}$ are extended by continuity to $\mathbb{Q}_p$. Every element of $\mathbb{Q}_p$ can be represented as a sum $\sum_{i=-k}^{+\infty} a_i p^i$, where $0 \leq a_i \leq p-1$ are integers (convergence in $\mathbb{Q}_p$). The closure of the ring of integers $\mathbb{Z}$ in $\mathbb{Q}_p$ is also a ring denoted by $\mathbb{Z}_p$ (*the set of p-adic integers*). Every element of $\mathbb{Z}_p$ has a form $\sum_{i=0}^{\infty} a_i p^i$, where $0 \leq a_i \leq p-1$ are integers. In other words, $\mathbb{Z}_p$ is the closed ball $\overline{B}_1(0\,;\,\mathbb{Q}_p) := \{r \in \mathbb{Q}_p\,;\,\|r\|_p \leq 1\}$.

Let us show that $\mathbb{Z}_p$ is homeomorphic to a Cantor set and therefore is compact. In fact, define the Cantor set C_p as follows.

Set $I_{0,1} := [0,1]$ and let $I_{1,j} := \left[\frac{2j}{2p-1}, \frac{2j+1}{2p-1}\right]$, $j = 0, 1 \ldots, p-1$, be the intervals determined by the subdivision of $I_{0,1}$ into $(2p-1)$ intervals of length $\frac{1}{2p-1}$. We continue this process of selecting p subintervals of each already chosen interval. Namely, given the intervals $I_{k,j}$, $j = 0, 1, \ldots, p^k - 1$, we define intervals $I_{k+1,j}$, $j = 0, 1, \ldots, p^{k+1} - 1$, by dividing $I_{k,j}$ into $\frac{1}{2p-1}$ equally spaced intervals and selecting p of them as above. It remains to set $C_p := \bigcap_{k=0}^{\infty} \bigcup_{j=0}^{p^k - 1} I_{k,j}$. Since each point $x \in C_p$ has a form $\sum_{i=1}^{\infty} \frac{2a_i}{(2p-1)^i}$ with integers $0 \leq a_i \leq p-1$, we define the required homeomorphism $h : \mathbb{Z}_p \to C_p$ by $\sum_{i=0}^{\infty} a_i p^i \mapsto \sum_{i=1}^{\infty} \frac{2a_{i+1}}{(2p-1)^i}$.

Unlike its bizarre geometry, an ultrametric space can be seen as a (metric) subspace of a very nice geometric object, a Hilbert space. This basic result was due to Ismagilov [Ism-1966], and will be presented now.

Theorem 3.97. *Let $(\mathcal{M}, d)$ be an ultrametric space. Then it admits an isometric embedding into a Hilbert space. If, in addition, $\mathcal{M}$ is finite, then this Hilbert space is ℓ_2^{n-1} where $n := \mathrm{card}\,\mathcal{M}$.*

Proof. As was mentioned above, the inequality $d(m,m') < t$ is an equivalence relation on $\mathcal{M}$. Therefore $\mathcal{M}$ is the disjoint union of equivalence classes of this relation (open balls of radius t). Choosing a point from each class and denoting this set by $\mathcal{C}_t$ we then have

$$\mathcal{M} = \coprod_{m \in \mathcal{C}_t} B_t(m). \tag{3.123}$$

Let $F(\mathcal{M})$ be the space of real measures of finite support on $\mathcal{M}$. Then the family of δ-measures defined by

$$\delta_m(\{m'\}) := \begin{cases} 1, & \text{if } m = m' \\ 0, & \text{if } m \neq m' \end{cases}$$

is an algebraic basis of $F(\mathcal{M})$, i.e., for every $\nu \in F(\mathcal{M})$,

$$\nu = \sum_{m \in \mathcal{M}} \hat{\nu}(m)\delta_m,$$

where $\hat{\nu}(m) := \nu(\{m\})$.

The key object of the proof is a family of functions $g_{\mu,\nu} : \mathbb{R}_+ \to \mathbb{R}$ given for a fixed pair $\nu, \mu \in F(\mathcal{M})$ by

$$g_{\mu,\nu}(t) := \sum_{m \in \mathcal{C}_t} \mu(B_t(m))\nu(B_t(m)).$$

Fixing t we obtain a symmetric bilinear form on $F(\mathcal{M})$. By its definition,

$$g_{\delta_m,\delta_{m'}}(t) = \begin{cases} 0, & \text{if } t \leq d(m,m') \\ 1, & \text{if } t > d(m,m'). \end{cases} \tag{3.124}$$

Since every measure from $F(\mathcal{M})$ is a finite linear combinations of the measures δ_m, this formula implies that every function $g_{\mu,\nu}$ assumes only a finite number of values and is continuous from the left. Note also that

$$\begin{aligned} g_{\mu\nu}(0+) &= \sum_{m \in \mathcal{M}} \hat{\mu}(m) \cdot \hat{\nu}(m), \\ g_{\mu\nu}(+\infty) &= \mu(\mathcal{M})\nu(\mathcal{M}). \end{aligned} \tag{3.125}$$

The Lebesgue–Stiltjes measure on $\mathbb{R}_+$ generated by the function g_{μ_1,μ_2} is finitely supported, all functions are integrable and integrals are simply finite sums.

Define now, for a fixed $f : \mathbb{R}_+ \to \mathbb{R}_+$, a bilinear form B_f on $F(\mathcal{M})$ by

$$B_f(\mu,\nu) := \int_0^\infty f(t) dg_{\mu\nu}(t).$$

3.3. Basic classes of metric spaces

Then (3.124) yields
$$B_f(\delta_m, \delta_{m'}) = f(d(m, m'))$$
and this leads to one more representation of this bilinear form:
$$\int_0^\infty f(t) dg_{\mu\nu}(t) = \sum_{m,m' \in \mathcal{M}} f(d(m,m'))\hat{\mu}(m)\hat{\nu}(m'). \tag{3.126}$$

Now introduce a linear subspace of $F(\mathcal{M})$ given by
$$F_0(\mathcal{M}) := \{\mu \in F(\mathcal{M}); \mu(\mathcal{M}) = 0\}.$$

Choose in (3.126) $\nu := \mu \in F_0(\mathcal{M})$ and $f(t) := -\frac{t^2}{2}$ and then integrate the left-hand side by parts. Taking into account (3.125) we then obtain
$$B_f(\mu, \mu) := \int_0^\infty t g_{\mu\mu}(t) dt = -\frac{1}{2} \sum_{m,m' \in \mathcal{M}} d(m,m')^2 \hat{\mu}(m)\hat{\mu}(m'). \tag{3.127}$$

Since $g_{\mu\mu} \geq 0$, the quadratic form $B_f(\mu, \mu)$ is nonnegative. Consider the null set
$$N := \{\mu \in F_0(\mathcal{M}); B_f(\mu, \mu) = 0\}$$
and show that N is a linear subspace. In fact, applying the Cauchy–Schwartz inequality to the bilinear form
$$B_f(\mu, \nu) := \int_0^\infty t g_{\mu\nu}(t) dt$$
we immediately conclude that
$$B_f(\mu, \mu) = 0 \iff B_f(\mu, \nu) = 0 \quad \text{for all} \quad \nu \in F_0(\mathcal{M}); \tag{3.128}$$
hence, N is a linear space. Consider now the factor space
$$H := F_0(\mathcal{M})/N$$
and set for elements (classes) $h_i := [\mu_i]$, $i = 1, 2$, of this factor space
$$\langle h_1, h_2 \rangle := B_f(\mu_1, \mu_2).$$

Due to (3.128) this definition is independent of the choice of $\mu_i \in h_i$, and therefore $\langle \cdot, \cdot \rangle$ is a nonnegative and nondegenerate bilinear form on H (scalar product). Thus, H is a pre-Hilbert space with norm $\|h\| := \langle h, h \rangle^{\frac{1}{2}}$.

Finally, we define an isometric embedding of $\mathcal{M}$ into H. To this end we fix a point $m^* \in \mathcal{M}$ and define a map $I : \mathcal{M} \to F(\mathcal{M})$ by
$$I(m) := -\delta_m + \delta_{m^*}.$$

Its image clearly lies in $F_0(\mathcal{M})$ and therefore the composition of I and the canonical surjection $\pi : F_0(\mathcal{M}) \to H := F_0(\mathcal{M})/N$ is well defined. The map $J := \pi I$ is the desired isometric embedding. In fact, by (3.127),

$$\begin{aligned}\|J(m) - J(m')\|^2 &= B(\delta_m - \delta_{m'}, \delta_m - \delta_{m'}) \\ &= -\frac{1}{2}(-d(m,m')^2 - d(m,m')^2) = d(m,m')^2.\end{aligned}$$

for all $m, m' \in \mathcal{M}$.

It remains to note that if $\mathcal{M}$ is of cardinality n, the Hilbert space H introduced in the proof is of dimension $\dim(F_0(\mathcal{M})/N) = n - 1$. $\square$

3.3.2 Spaces of bounded geometry

This class is introduced by

Definition 3.98. A metric space is said to be of bounded geometry with parameters $n \in \mathbb{N}$, $R, D > 0$ (written $\mathcal{M} \in \mathcal{G}_n(R, D)$) if each of its open balls of radius R admits a bi-Lipschitz embedding into $\mathbb{R}^n$ with distortion D.

Hereafter symbol $\mathbb{R}^n$ stands for the Euclidean space ℓ_2^n.

Many examples of such spaces will be presented throughout this section (metric graphs and groups, Riemannian manifolds etc.). Among the spaces considered before, $\mathbb{R}^n$, $\mathbb{H}^n$ and $\mathbb{S}^n$ with the geodesic metrics are clearly of bounded geometry with parameters n, R and $D = C(R)$, where $R > 0$ may be chosen arbitrarily (of course, $D = 1$ for $\mathbb{R}^n$). Simple examples show that if $\mathcal{M} \in \mathcal{G}_n(R, C)$ then it may not belong to this class with any $R_1 > R$ and $n = n(R_1)$ and $C = C(R_1)$. Nevertheless, the following fact is true.

Theorem 3.99. Let a geodesic space $\mathcal{M}$ belong to $\mathcal{G}_{n_0}(R_0, D_0)$. Then for every $R > 0$ there is an integer $n = n(n_0, D_0)$ and a constant $D = D(n_0, R_0, D_0)$ such that $\mathcal{M}$ belongs to $\mathcal{G}_n(R, D)$.

We postpone the proof of this result to Chapter 5 (see the final part of subsection 5.1.4), since it requires some additional facts that will be established later.

For now, we only present two properties of these spaces.

Proposition 3.100. Let $(\mathcal{M}_i, d_i) \in \mathcal{G}_{n_i}(R_i, D_i)$, $1 \leq i \leq N$. Then the direct p-sum of these spaces belongs to $\mathcal{G}_n(R, D)$ with

$$n := \sum_{i=1}^n n_i, \quad R := N^{\frac{1}{p}} \min_{1 \leq i \leq N} R_i \quad \text{and} \quad D := N^{\frac{1}{p}} \max_{1 \leq i \leq N} D_i.$$

Proof. For $p = \infty$ the result is the matter of definitions. For $p < \infty$ it suffices to note that the p-sum of these spaces is bi-Lipschitz homeomorphic to that for $p = \infty$ with distortion $\leq N^{\frac{1}{p}}$. $\square$

3.3. Basic classes of metric spaces

Theorem 3.101. *If $\mathcal{M} \in \mathcal{G}_n(R, D)$, then*

$$\dim \mathcal{M} \leq n.$$

Proof. For a separable $\mathcal{M}$, we can present this space as a countable union of open balls B_i of radius R, $i \in \mathbb{N}$. Since every B_i is bi-Lipschitz homeomorphic to a subset of $\mathbb{R}^n$ and dimension is a topological (hence, bi-Lipschitz) invariant,

$$\dim B_i \leq \dim \mathbb{R}^n = n.$$

Moreover, by Theorem III 2 of the book [HW-1941], $\dim \mathcal{M} \leq \sup_i \dim B_i$. Hence, the assertion is true for this case.

For nonseparable $\mathcal{M}$ the above used covering $\{B_i\}$ should be replaced by a more involved family presented in

Lemma 3.102. *Let $\mathcal{M} \in \mathcal{G}_n(R, D)$. Then, for some integer N, there are disjoint subsets A_j of $\mathcal{M}$, $0 \leq j \leq N$, such that*

(a) *for every distinct points m, m' from A_j, $0 \leq j \leq N$,*

$$d(m, m') \geq R;$$

(b) $\mathcal{M} = \bigcup_{j=0}^{N} \coprod_{m \in A_j} B_{\frac{R}{4}}(m),$

(c) $N \leq (8D+1)^n.$

Proof. We define inductively the sequence of the required sets A_j and the associated metric subspaces $\mathcal{M}_j \subset \mathcal{M}$.

Set $\mathcal{M}_0 := \mathcal{M}$ and define A_0 to be a maximal R-separated subset of $\mathcal{M}$ (called an *R-net*). Its existence is an easy consequence of Zorn's lemma. Then $d(m, m') \geq R$ for points $m \neq m'$ from A_0, and for every $m \in \mathcal{M}$ there is $m' \in A_0$ such that $d(m, m') < R$.

Then set $\mathcal{M}_1 := \mathcal{M} \setminus \bigcup_{m \in A_0} B_{\frac{R}{4}}(m)$ and define A_1 to be an R-net in $\mathcal{M}_1$.
Proceeding this way we obtain the family of the subsets

$$\mathcal{M}_0 \supset \mathcal{M}_1 \supset \cdots \supset \mathcal{M}_i \supset \ldots$$

and the family of the corresponding R-nets $A_0, A_1, \ldots, A_i, \ldots$.

Show that $\mathcal{M}_i = \emptyset$, if $i > (8D+1)^2$. In fact, choose $\mathcal{M}_i \neq \emptyset$ for some i and show that $i \leq (8D+1)^2$. To this end, pick a point $m \in \mathcal{M}_i$. For every $0 \leq j \leq i$, there is a point $m_j \in A_j$ such that

$$d(m_j, m) < R,$$

since $\bigcup_{m \in A_j} B_R(m) \supset \mathcal{M}_j \supset \mathcal{M}_i$ by our construction.

Now consider the ball $B_R(m) \subset \mathcal{M}$ centered at this m which contains the points m_j, $0 \leq j \leq i$. Since $\mathcal{M} \in \mathcal{G}_n(R, D)$, there is a map $\psi_m : B_R(m) \to \mathbb{R}^n$ such that $\psi_m(m) = 0$, and for all $m', m'' \in B_R(m)$,

$$d(m', m'') \leq \|\psi_m(m') - \psi_m(m'')\|_2 \leq D d(m', m'').$$

This implies that

$$\|\psi_m(m_j)\|_2 \leq DR,$$

and that, for $0 \leq j < j' \leq i$,

$$\|\psi_m(m_j) - \psi_m(m_{j'})\|_2 \geq d(m_j, m_{j'}).$$

By the definition of the subspaces $\mathcal{M}_j$, the right-hand side here is at least $\frac{R}{4}$. Hence the (open) Euclidean balls $B_{\frac{R}{8}}(\psi_m(m_j))$, $0 \leq j \leq i$, are disjoint and all of them are contained in the Euclidean ball centered at $0 \in \mathbb{R}^n$ and of radius $R(D + \frac{1}{8})$. Comparing the volumes of these balls with that of $B_{R(D+\frac{1}{8})}(0)$, we obtain the inequality

$$i \left(\frac{R}{8}\right)^n \leq \left(R\left(D + \frac{1}{8}\right)\right)^n,$$

which implies the desired estimate $i \leq (8D + 1)^n$.

This proves assertion (b); the remaining assertion follows from the definitions of the A_i and $\mathcal{M}_i$. □

Now, to prove that $\dim \mathcal{M} \leq n$, it suffices to establish this inequality for every subspace $\widetilde{\mathcal{M}}_j := \coprod_{m \in A_j} B_{\frac{R}{8}}(m)$ and then to apply the aforementioned Theorem III 2 of [HW-1941] to the decomposition $\mathcal{M} = \bigcup_{j=0}^{N} \widetilde{\mathcal{M}}_j$.

In fact, $\widetilde{\mathcal{M}}_j$ is the disjoint union of open balls of radius $\frac{R}{8}$. It follows from the definition of the class $\mathcal{G}_n(R, D)$ that every such ball has dimension at most n. Since every open cover $\{U_\alpha\}$ of $\widetilde{\mathcal{M}}_j$ can be presented as the *disjoint* union of covers $\{U_\alpha \cap B_{\frac{R}{8}}(m)\}$, $m \in A_j$, Definition A.1 in Appendix A implies that $\dim \widetilde{\mathcal{M}}_j \leq n$.

The result is proved. □

3.3.3 Riemannian manifolds as metric spaces

Riemannian Geometry is a considerable source of important in applications length and geodesic metric spaces. Here we restrict ourselves only to the results of this field concerning the subject matter of this book. The reader may consult any textbook in Riemannian Geometry, e.g., [Cha-1993], on concepts and results which will be briefly discussed in this section.

Let $\mathcal{M}$ be a C^k-manifold, $k \geq 1$, of dimension n. By $T_m\mathcal{M}$ one denotes the tangent space to $\mathcal{M}$ at a point $m \in \mathcal{M}$. A *Riemannian structure* on $\mathcal{M}$ is an

3.3. Basic classes of metric spaces

assignment of a scalar product to each tangent space $T_m\mathcal{M}$ which continuously depends on m. In the classical notation, if $x_1, \ldots, x_n$ are local coordinates in a chart $\mathcal{U} \subset \mathcal{M}$ about m, then the scalar product of vectors $u = \sum_{i=1}^{n} u_i(m)\frac{\partial}{\partial x_i}$ and $v = \sum_{i=1}^{n} v_i(m)\frac{\partial}{\partial x_i}$ from $T_m\mathcal{M}$ is given by the formula

$$\langle u, v \rangle_m := \sum_{1 \leq i,j \leq n} g_{ij}(m)u_i(m)v_j(m),$$

where g_{ij} are continuous functions on $\mathcal{U}$.

The tangent vector to a C^1-curve $\gamma : [a,b] \to \mathcal{M}$ at a point t is denoted by $\dot{\gamma}(t)$. In local coordinates about $\gamma(t)$ it can be written as

$$\dot{\gamma}(t) = \left(\frac{d}{ds}x_1(\gamma(s)), \ldots, \frac{d}{ds}x_n(\gamma(s))\right)\bigg|_{s=t}.$$

The Riemannian length of γ is then defined by

$$\ell_R(\gamma) := \int_a^b \|\dot{\gamma}(t)\|_{\gamma(t)} dt,$$

where $\|u\|_m^2 := \langle u, u \rangle_m$ defines the Euclidean norm on $T_m\mathcal{M}$.

Now a *geodesic (inner)* metric on a Riemannian manifold $\mathcal{M}$ is defined by

$$d_g(m, m') := \inf_\gamma \ell_R(\gamma), \qquad (3.129)$$

where γ runs over all C^1-smooth curves joining m and m'.

Proposition 3.103. (a) *The function d_g is a metric on $\mathcal{M}$.*

(b) *The metric topology generated by d_g coincides with the (Hausdorff) topology of the manifold $\mathcal{M}$.*

(c) *$(\mathcal{M}, d_g)$ is a length metric space.*

(d) *The length of a C^1-curve in the metric space $(\mathcal{M}, d_g)$ is equal to its Riemannian length.*

For the proof see, e.g., [BH-1999, pp. 39–41].

Since a finite-dimensional manifold is locally compact, assertion (c) and the Hopf–Rinow theorem, see Theorem 3.40, immediately imply

Corollary 3.104. *A complete finite-dimensional Riemannian manifold endowed with the geodesic metric is a proper geodesic space.*

Let now $\mathcal{N}$ be a submanifold of a Riemannian manifold $\mathcal{M}$. Then the restriction of the scalar product on $T_m\mathcal{M}$ to its linear subspace $T_m\mathcal{N}$ defines a scalar product on the tangent space to $\mathcal{N}$. Hence, $\mathcal{N}$ inherits a Riemannian structure from $\mathcal{M}$. Using this structure we define the metric on $\mathcal{N}$ similarly to (3.129). On the other hand, $\mathcal{N}$ is equipped with the metric $d_g|_{\mathcal{N}\times\mathcal{N}}$ induced from $(\mathcal{M}, d_g)$. These metrics, in general, do not coincide.

Remark 3.105. Let f be a *Riemannian isometric embedding* of a Riemannian manifold $\mathcal{M}$ into a Riemannian manifold $\mathcal{M}'$. By definition, this means:

(a) f maps $\mathcal{M}$ onto the image $f(\mathcal{M}) \subset \mathcal{M}'$ homeomorphically;

(b) for every $m \in \mathcal{M}$ the differential $(df)_m : T_m\mathcal{M} \to T_{f(m)}\mathcal{M}'$ is injective;

(c) for each $m \in \mathcal{M}$ and $u, v \in T_m\mathcal{M}$,

$$\langle (df)_m(u), (df)_m(v) \rangle_{f(m)} = \langle u, v \rangle_m.$$

A Riemannian isometry clearly gives rise to an isometry of the associated length spaces. For finite-dimensional Riemannian manifolds of class C^2 the converse is also true, see, e.g., [Hel-1978].

According to the Nash theorem every n-dimensional Riemannian manifold of sufficiently large smoothness can be realized as a submanifold of a Euclidean space (seen as a Riemannian manifold with the standard scalar product). Actually, the following is true, see [Na-1966] and references therein.

Theorem 3.106. *Let $\mathcal{M}$ be an n-dimensional Riemannian manifold, analytic or of class C^k, $k = 3, 4, \ldots, \infty$. Then there is a Riemannian embedding of the same smoothness into $\mathbb{R}^N$ with $N \geq n^2 + 5n + 3$.*

The theorem is true also for $k = 1$. In this case N can be taken to equal $n+1$ (Nash–Kuiper theorem).

Finally, let us recall the notion of the *Riemannian measure (n-volume)* vol_n on a Riemannian n-manifold $\mathcal{M}$. It is defined as the Hausdorff n-measure on the metric space $(\mathcal{M}, d_g)$ using covers by balls. Another (equivalent) way is to use the Lebesgue integration. Namely, in local coordinates of a chart $\mathcal{U} \subset \mathcal{M}$, a Borel subset $\mathcal{S} \subset \mathcal{U}$ has n-volume given by

$$\mathrm{vol}_n(S) := \int_{\varphi(S)} \sqrt{\det(g_{ij}(x))}\, dx_1 \ldots dx_n.$$

Here $\varphi : \mathcal{U} \to \mathbb{R}^n$ is the coordinate system (smooth embedding into $\mathbb{R}^n$) and $g_{ij}(x) := \langle \frac{\partial}{\partial x_i}, \frac{\partial}{\partial x_i} \rangle_m$, where $x = (x_1, \ldots, x_n)$ are the coordinates of $\varphi(m)$ in $\mathbb{R}^n$.

Example 3.107 (The model spaces $\mathcal{M}_\kappa^n$). The Riemannian manifolds $\mathcal{M}_\kappa^n$, $\kappa \in \mathbb{R}$, are perfect models to which one can compare more general geodesic spaces. An advantage of such a comparison will be demonstrated below. We describe only the associated geodesic spaces $(\mathcal{M}_\kappa^n, d_g)$ for $\kappa = 0$ and $\kappa = \pm 1$. The space $(\mathcal{M}_0^n, d_g)$ coincides with the Euclidean space $\mathbb{R}^n (= \ell_2^n)$, while the space $(\mathcal{M}_{\pm 1}^n, d_g)$ coincides with the unit sphere $\mathbb{S}^n$ for $\kappa = 1$ and the hyperbolic space $\mathbb{H}^n$ for $\kappa = -1$.

3.3. Basic classes of metric spaces

Let us recall, see Section 1.10, that the geodesic distance between points x and y in $\mathbb{S}^n \subset \mathbb{R}^{n+1}$ is defined as the smallest nonnegative angle φ such that

$$\cos\varphi = \langle x, y\rangle \left(:= \sum_{i=1}^{n+1} x_i y_i\right). \tag{3.130}$$

In other words, $d_g(x, y)$ is the length of the smallest arc of a *great circle* in $\mathbb{S}^n$ joining x and y. This arc is a (geodesic) segment with the endpoints x and y. The corresponding geodesic $\gamma(t) : [0, a] \to \mathbb{S}^n$ is given by $\gamma(t) := x\cos t + v\sin t$, where v is a unit vector in $\mathbb{R}^{n+1}$ orthogonal to x and the number a is determined by the equation $\gamma(a) = y$. In fact, the image of γ is clearly contained in the great circle subject to the intersection of $\mathbb{S}^n$ with the vector 2-subspace $\mathrm{span}(x, v)$. Moreover, $d_g(\gamma(t), \gamma(t')) = |t - t'|$ for $t, t' \in [0, a]$; hence γ is the required geodesic. Note that there is a unique geodesic joining x and y provided that $d_g(x, y) < \pi$.

Further, recall, see Section 1.10, that the geodesic distance between points x and y in $\mathbb{H}^n \subset \mathbb{R}^{n+1}$ is given by

$$\cosh d_g(x, y) = -\sum_{i=1}^{n} x_i y_i + x_{n+1} y_{n+1}. \tag{3.131}$$

Here $\cosh t := \frac{1}{2}(e^t + e^{-t})$ and $\mathbb{H}^n$ is

$$\left\{ x \in \mathbb{R}^{n+1} \,;\, -\sum_{i=1}^{n} x_i^2 + x_{n+1}^2 = 1 \quad \text{and} \quad x_n > 0 \right\}.$$

The geodesic segment with the endpoints x and y is determined by the *unique* geodesic curve passing through these points. Namely, let $v \in \mathbb{R}^{n+1}$ be a vector defined by the condition

$$\langle v, v\rangle_{n,1} = 1 \quad \text{and} \quad \langle v, x\rangle_{n,1} = 0.$$

Here $\langle x, y\rangle_{n,1}$ stands for the right-hand side of (3.131). Then the required geodesic curve $\gamma : [0, a] \to \mathbb{H}^n$ is defined by $\gamma(t) := x\cosh t + v\sinh t$, where $\sinh t := \frac{1}{2}(e^t - e^{-t})$ and the number a satisfies the equation $\gamma(a) = y$.

Finally, given a real number κ, we denote by $(\mathcal{M}_\kappa^n, d_g)$ the following metric spaces:

(a) if $\kappa = 0$, then $(\mathcal{M}_0^n, d_g)$ is the Euclidean space $\mathbb{R}^n (= \ell_2^n)$;

(b) if $\kappa > 0$, then $(\mathcal{M}_\kappa^n, d_g)$ is obtained from the sphere $\mathbb{S}^n$ by multiplying its metric, see (3.130), by $\frac{1}{\sqrt{\kappa}}$;

(c) if $\kappa < 0$, then $(\mathcal{M}_\kappa^n, d_g)$ is obtained from the hyperbolic space $\mathbb{H}^n$ by multiplying its metric, see (3.131), by $\frac{1}{\sqrt{-\kappa}}$.

It follows from the above discussion that if $\kappa \leq 0$, then $(\mathcal{M}_\kappa^n, d_g)$ are uniquely geodesic spaces and all balls in $(\mathcal{M}_\kappa^n, d_g)$ are convex (i.e., every pair of points in a ball can be joined by a geodesic segment lying in the ball). If $\kappa > 0$, then there is a unique geodesic joining $x, y \in \mathcal{M}_\kappa^n$ if and only if $d(x,y) < \frac{\pi}{\sqrt{\kappa}}$. Moreover, if $\kappa > 0$, closed balls in $(\mathcal{M}_\kappa^n, d_g)$ of radius $< \frac{\pi}{2\sqrt{\kappa}}$ are convex and

$$\operatorname{diam} \mathcal{M}_\kappa = \frac{\pi}{\sqrt{\kappa}}.$$

Remark 3.108. From the point of view of Riemannian Geometry, $\mathcal{M}_\kappa^n$ is a complete simply connected Riemannian manifold of constant sectional curvature κ.

Now we present a wide class of Riemannian manifolds[7] $(\mathcal{M}, d_g)$ of bounded geometry. Its description is based on the Rauch comparison theorem whose assumptions include a two-sided estimate of *sectional curvature*. In keeping with the spirit of this book, we prefer an equivalent description using only metric concepts.

First, we define a *geodesic triangle* in a geodesic space $(\mathcal{M}, d)$. This consists of three *vertices* m_1, m_2, m_3 and three geodesic segments $[m_1, m_2], [m_2, m_3]$ and $[m_3, m_1]$, named the *sides* of the triangle. Such a triangle is denoted by $\Delta\big([m_1, m_2], [m_2, m_3], [m_3, m_1]\big)$ or simply $\Delta(m_1, m_2, m_3)$ if the geodesic segments (sides) are uniquely determined by their endpoints (e.g., if $\mathcal{M}$ is a uniquely geodesic space).

A triangle $\widetilde{\Delta}(\widetilde{m}_1, \widetilde{m}_2, \widetilde{m}_3)$ in the 2-dimensional model space $\mathcal{M}_\kappa^2$ is called a *comparison triangle* for $\Delta := \Delta\big([m_1, m_2], [m_2, m_3], [m_3, m_1]\big) \subset \mathcal{M}$ if

$$d(m_i, m_j) = d_g(\widetilde{m}_i, \widetilde{m}_j), \quad 1 \leq i, j \leq 3.$$

For $\kappa > 0$, the triangle $\widetilde{\Delta}$ exists if the perimeter $d(m_1, m_2) + d(m_2, m_3) + d(m_3, m_1)$ of Δ less than $\frac{2\pi}{\sqrt{\kappa}}$, see Example 3.107.

A point $\widetilde{m} \in [\widetilde{m}_i, \widetilde{m}_j]$ is called a *comparison point* for $m \in [m_i, m_j]$ if $d(m_i, m) = d_g(\widetilde{m}_i, \widetilde{m})$.

Definition 3.109. (a) A geodesic metric space $(\mathcal{M}, d)$ is said to be of *curvature* $\leq \kappa$ if for every $m_0 \in \mathcal{M}$ there is a ball $B_r(m_0)$ with $r = r(m_0) > 0$ such that the following condition holds.

For every geodesic triangle[8] $\Delta\big([m_1, m_2], [m_2, m_3], [m_3, m_1]\big) \subset B_r(m_0)$ and for every point $m \in [m_2, m_3]$ its comparison point $\widetilde{m} \in [\widetilde{m}_2, \widetilde{m}_3] \subset \Delta(\widetilde{m}_1, \widetilde{m}_2, \widetilde{m}_3)$ satisfies the inequality
(A_κ^+) $\qquad d(m_1, m) \leq d_g(\widetilde{m}_1, \widetilde{m}).$

(b) A geodesic metric space $(\mathcal{M}, d)$ is said to be of *curvature* $\geq \kappa$ if the above condition holds with the reverse to (A_κ^+) inequality (denoted by A_κ^-).

---

[7] We slightly abuse language referring to the metric space $(\mathcal{M}, d_g)$ as a Riemannian manifold.
[8] For $\kappa > 0$, we assume that the perimeter of each geodesic triangle considered is less than $\frac{2\pi}{\sqrt{\kappa}}$.

3.3. Basic classes of metric spaces

It was proved by A. D. Aleksandrov [Ale-1951] that an n-dimensional Riemannian manifold of class C^3 is of curvature $\leq \kappa$ if and only if its sectional curvature is at most κ. On the other hand, this manifold is of sectional curvature $\geq \kappa$ if it is complete and satisfies condition (A_κ^-), see the survey [BGP-1992] by Yu. Burago, M. Gromov and G. Perel'man.

These results and the Rauch comparison theorem lead to the following theorem giving sufficient conditions for $(\mathcal{M}, d_g)$ to be of bounded geometry (for the proof see the book [CE-1975] by Cheeger and Ebin or Section 8.7 of the book [Gr-2000] by Gromov).

Theorem 3.110. *Let $(\mathcal{M}, d_g)$ be a finite-dimensional complete Riemannian manifold of class C^3. Assume that $\mathcal{M}$ is simultaneously of curvature $\leq \kappa_+$ and of curvature $\geq \kappa_-$ where $\kappa_- \leq \kappa_+$. Then $(\mathcal{M}, d_g)$ is a metric space of bounded geometry.*

Remark 3.111. According to Theorem 3.99, the space $(\mathcal{M}, d_g)$ satisfying the conditions of this theorem belongs to $\mathcal{G}_n(R, C)$ for every $R > 0$ and some n, C depending on R.

Now we describe a wide class of Riemannian manifolds $(\mathcal{M}, d_g)$ which are doubling metric spaces. The corresponding result is based on the R. Bishop comparison theorem, see, e.g., [Gr-2000, pp. 275–277] where this theorem is presented in an appropriate form.

Theorem 3.112. *Let $(\mathcal{M}, d_g)$ be a Riemannian n-manifold of class C^2. Assume that its Ricci curvature $\mathrm{Ric}(\mathcal{M})$ is nonnegative. Then the Riemannian volume vol_n of $\mathcal{M}$ is a doubling measure.*

Hence, $(\mathcal{M}, d_g, \mathrm{vol}_n)$ is a metric space of homogeneous type if $\mathrm{Ric}(\mathcal{M}) \geq 0$. In particular, $(\mathcal{M}, d_g)$ is a doubling metric space, see Proposition 3.84.

Finally, we present a family of Riemannian manifolds of unbounded sectional curvature which are not doubling. Nevertheless, these are spaces of pointwise homogeneous type, see Chapter 4 below.

Example 3.113 (Generalized hyperbolic spaces). Let $\omega : (0, +\infty) \to \mathbb{R}_+$ be a nondecreasing C^2-function. The *generalized hyperbolic space* $\mathbb{H}_\omega^n$ is the half-space

$$H_+^n := \{x \in \mathbb{R}^n \,;\, x_n > 0\}$$

equipped with the Riemannian metric

$$ds^2 := \omega(x_n)^2 \sum_{i=1}^n dx_i^2.$$

If $\omega(t) := \frac{1}{t}$, $t > 0$, the manifold $\mathbb{H}_\omega^n$ coincides with the Poincaré half-space model for the hyperbolic space $\mathbb{H}^n$. In this case the geodesic metric is given by

$$\cosh\bigl(d_g(x, y)\bigr) := 1 + \frac{\|x - y\|_2^2}{x_n y_n}, \quad x, y \in H_+^n,$$

and sectional curvature equals -1. Moreover, $(\mathbb{H}^n_{t^{-1}}, d_g)$ is not doubling, since $\mathrm{vol}_n(B_r(x))$ increases exponentially in r at infinity.

Further, for $\omega(t) := \mathrm{const} > 0$, the space $(\mathbb{H}^n_\omega, d_g)$ coincides with the metric subspace $\mathbb{R}^n_+$ of the Euclidean space $\mathbb{R}^n$. Hence, $\mathbb{H}^n_{\mathrm{const}}$ is incomplete but doubling. In general, the geometry of the space $(\mathbb{H}^n_\omega, d_g)$ is essentially more complicated. To simplify computations, we consider the case of the space $\mathbb{H}^2_\omega$ with $\omega(t)^2 := t^{-\alpha}$, $t > 0$ and $\alpha > 0$. Then the (Gaussian) curvature of this space is given by the classical formula

$$K(x) = -\frac{1}{\omega^2}\Delta(\log\omega),$$

where Δ is Laplacian. Therefore, $K(x) = -\frac{\alpha}{2}x_2^{\alpha-2}$ in the case considered, and the curvature of $\mathbb{H}^2_\omega$ is unbounded near the boundary $\partial\mathbb{R}^2_+$ if $0 < \alpha < 2$, and unbounded at infinity if $\alpha > 2$. Moreover, this space is incomplete if $0 \leq \alpha < 2$, e.g., $\{(0, \frac{1}{i})\}_{i \in \mathbb{N}}$ is a Cauchy sequence in its geodesic metric. Evaluating the 2-volume of a disc in $\mathbb{H}^2_\omega$ we conclude that vol_2 is not a doubling measure on this space if $\omega(t) = t^{-\alpha}$ with $\alpha > 0$. It may be easily derived from here that $\mathbb{H}^n_\omega$ with this ω is not a doubling metric space. Nevertheless, for any ω satisfying the condition $\int_1^\infty \omega(t)dt = \infty$, the metric space $(\mathbb{H}^n_\omega, d_g)$ carries a family of pointwise doubling and consistent measures, i.e., this space is of pointwise homogeneous type. This highly nontrivial fact was firstly proved in the paper [BSh-1999] by Yu. Brudnyi and P. Shvartsman for ω satisfying some additional restrictions.

The proof of the general result is similar and is presented in Section 4.4.

3.3.4 Gromov hyperbolic spaces

The metric spaces of this class introduced by M. Gromov in [Gr-1987] play a considerable role in modern study of Global Geometry and Geometric Group Theory. There is a vast literature devoted to this theory, see, e.g., [BH-1999], [Vai-2005] and [BuSch-2007] and references therein. Our interest in this area is explained by its relation to Lipschitz extension problems. It will be shown in Chapter 6 that subspaces of a (Gromov) hyperbolic space are, in a sense, universal with respect to such extensions. In the consequent brief account we discuss (without proofs) results and examples which will be used in our settings.

Let δ be a nonnegative number. A geodesic triangle is said to be δ-*slim* if each of its sides is contained in the δ-neighborhood of the union of the other two sides.

In particular, a 0-slim triangle is degenerate, i.e., the triangle inequality for its vertices becomes equality.

Definition 3.114. A geodesic metric space is δ-hyperbolic if each of its triangles is δ-slim.

If such a space is δ-hyperbolic for some $\delta \geq 0$, it is said to be Gromov hyperbolic.

3.3. Basic classes of metric spaces

There exists an equivalent definition of Gromov hyperbolic spaces which sometimes is more suitable for applications. For its introduction one defines the *Gromov product* of points m_1, m_2 from a metric space $(\mathcal{M}, d)$ with respect to a basepoint m^*. This is given by

$$(m_1|m_2) := \frac{1}{2}\left[d(m_1, m^*) + d(m_2, m^*) - d(m_1, m_2)\right]. \tag{3.132}$$

We illustrate this notion by two examples.

Let $(\mathcal{M}, d)$ be the Euclidean plane. Then $(m|m')_{m^*}$ equals the distance from m^* to either of the two points where the circle inscribed in the triangle $\Delta(m^*, m, m')$ meets the sides $[m^*, m]$ and $[m^*, m']$.

Now let $(\mathcal{M}, d)$ be the hyperbolic space $(\mathbb{H}^n, dg)$. If $\gamma : [0, 1] \to \mathbb{H}^n$ is the unique geodesic joining m, m', then $(m|m')_{m^*}$ is roughly equal to the distance from m^* to $\gamma[0, 1]$. More precisely, for some numerical constant $\delta > 0$,

$$d_g(m^*, \gamma([0, 1])) - \delta \leq (m|m')_{m^*} \leq d_g(m^*, \gamma([0, 1])).$$

Definition 3.115. A geodesic metric space is Gromov hyperbolic, if for some $\delta' \geq 0$ and all triples $m_i \in \mathcal{M}$, $1 \leq i \leq 3$,

$$(m_1|m_2) \geq \min\{(m_1|m_3), (m_2|m_3)\} - \delta'. \tag{3.133}$$

Remark 3.116. (a) If (3.133) holds for a basepoint m^*, then it also holds for any other basepoint with δ' replaced by $2\delta'$. Hence, the definition does not depend on the choice of a basepoint.

(b) Omitting the assumption of geodesity one defines more general Gromov hyperbolic spaces, see the monograph [Vai-2005] by Väsälä. This generalization is not of considerable interest, since every generalized δ'-hyperbolic space can be isometrically embedded into a complete *geodesic* δ'-hyperbolic space (in the sense of Definition 3.115), see Theorem 4.1 of the paper [BSch-2000] by Bonk and Schramm.

The basic result whose proof may be found, e.g., in [BH-1999, p. 399] asserts that $\mathcal{M}$ is δ-hyperbolic in the sense of Definition 3.114 if and only if condition (3.133) holds for some δ' (with $C^{-1}\delta' \leq \delta \leq C\delta$ for a numerical constant $C > 0$). Hence, both definitions of Gromov hyperbolicity are equivalent.

In the theory of Gromov hyperbolic spaces, the following class of "rough" maps plays an essential role.

Definition 3.117. A map $f : (\mathcal{M}_1, d_1) \to (\mathcal{M}_2, d_2)$ is said to be roughly (C, k)-similar if for some constants $C \geq 1$ and $k \geq 0$ and for all $m, m' \in \mathcal{M}_1$

$$\left|d_2(f(m), f(m')) - Cd_1(m, m')\right| \leq k, \tag{3.134}$$

and, moreover, the image $f(\mathcal{M}_1)$ is k-cobound in $\mathcal{M}_2$, i.e.,

$$\sup\{d_2(m', f(\mathcal{M}_1)) \,;\, m' \in \mathcal{M}_2\} \leq k. \tag{3.135}$$

We say that $\mathcal{M}_1$ is *roughly similar* to $\mathcal{M}_2$ if such a map exists. *Rough similarity* is an equivalence relation on the class of metric spaces, since if $f : \mathcal{M}_1 \to \mathcal{M}_2$ is (C,k)-roughly similar, then a map $g : \mathcal{M}_2 \to \mathcal{M}_1$ determined by the condition $g(m') \in f^{-1}(\{m'\})$, $m' \in \mathcal{M}_2$ is $(C^{-1}, C^{-1}k)$-roughly similar.

Roughly similar maps may essentially destroy metric structure. For example, every bounded metric space is *roughly similar* to a one-point metric space. Nevertheless, rough similarity preserves some essential features of a metric space as the following fact demonstrates.

A straightforward computation shows that if $\mathcal{M}_1$ is δ'-hyperbolic in the sense of Definition 3.115 and a geodesic space $\mathcal{M}_2$ is (C,k)-roughly similar to $\mathcal{M}_1$, then condition (3.133) holds for $\mathcal{M}_2$ with $\delta'' := C\delta' + 3k$. Hence, roughly similar transforms preserve Gromov's hyperbolicity.

Example 3.118. (a) Let $\mathcal{M}$ be a geodesic space of curvature $\leq \kappa$, see Definition 3.109. Then $\mathcal{M}$ is δ-hyperbolic for some $\delta = \delta(\kappa)$ provided that $\kappa < 0$. In fact, the comparison of Definitions 3.109 and 3.114 shows that this claim is a straightforward consequence of the following fact.

The model space $\mathcal{M}_{-1}^2 (= \mathbb{H}^2)$ is δ_0-hyperbolic for some $\delta_0 > 0$.

The existence of such δ_0 follows from the inequality $\mathrm{vol}_2(\Delta) < \pi$ which holds for every geodesic triangle Δ in $\mathcal{M}_{-1}^2$, see, e.g., Section 4.3 of the book [CDP-1991] by Coornaert, Delzant and Papadopoulos.

Then the required $\delta(\kappa)$ equals $\frac{\delta_0}{\sqrt{\kappa}}$.

In particular, a convex subset of the hyperbolic space $\mathbb{H}^n$ is δ_0-hyperbolic.

(b) Unlike the case $\kappa < 0$, a geodesic space of nonpositive curvature may be not hyperbolic. A typical example is the Euclidean space $\mathcal{M}_0^n (= \mathbb{R}^n)$. In fact, an equilateral triangle in $\mathbb{R}^2$ with length sides tending to infinity demonstrates that the condition of Definition 3.114 does not hold for any δ.

(c) In particular, a geodesic metric space containing a roughly similar copy of $\mathbb{R}^2$ cannot be Gromov hyperbolic.

(d) Example (b) also shows that the direct sum of Gromov hyperbolic spaces may be not Gromov hyperbolic ($\mathbb{R}^2 = \mathbb{R} \oplus^{(2)} \mathbb{R}$ and $\mathbb{R}$ is 0-hyperbolic).

In a sense, the situation described in example (c) is typical. Namely, the following result is true, see, e.g., [BH-1999, p. 400] for the proof.

Theorem 3.119. *A proper cocompact[9] geodesic space of curvature ≤ 0 is Gromov hyperbolic if and only if it does not contain a subspace isometric to $\mathbb{R}^2$.*

Finally we illustrate the concept of rough similarity presenting a version of an important result of Bonk and Schramm [BSch-2000, Thm. 1.1]; the proof of their theorem will be discussed in Section 5.2.

---

[9] A metric space $\mathcal{M}$ is *cocompact* if its group of isometrics $\mathrm{Iso}(\mathcal{M})$ acts cocompactly, i.e., $\mathcal{M} = \cup \{I(K)\,;\, I \in \mathrm{Iso}(\mathcal{M})\}$ for some compact set $K \subset \mathcal{M}$.

3.3. Basic classes of metric spaces

Theorem 3.120. *If $\mathcal{M}$ is a Gromov hyperbolic locally doubling metric space, then $\mathcal{M}$ is roughly similar to a (hyperbolically) convex subset of some $\mathbb{H}^n$.*

Example 3.121. (a) Let $(\mathcal{M}, d_g)$ be a simply connected finite-dimensional Riemannian manifold with sectional curvature κ satisfying $-b^2 \leq \kappa \leq -a^2 < 0$. Then $(\mathcal{M}, d_g)$ is of curvature < 0 and therefore hyperbolic. Moreover, $\mathcal{M}$ is of bounded geometry, see Theorem 3.110. Hence, $(\mathcal{M}, d_g)$ admits a rough similar embedding into some $\mathbb{H}^n$.

(b) Another important class of Gromov hyperbolic spaces is formed by *uniform domains* in $\mathbb{R}^n$ introduced in subsection 2.6.1; see also Chapter 9 for a more detailed discussion. At first sight, this class has nothing in common with hyperbolicity. To show that this, nevertheless, is the case, we need two equivalent definitions of uniform domains.

The first one is a trivial reformulation of the original definition based on the notion of a *uniform curve*.

Let $D \subsetneq \mathbb{R}^n$ be a nonempty domain (open connected set), and $C \geq 1$ and $\lambda > 0$ are fixed constants. A curve $\gamma : [a,b] \to D$ is said to be (C, λ)-*uniform* if it satisfies the conditions

(i) $\ell(\gamma) \leq C \|\gamma(b) - \gamma(a)\|_2$;

(ii) for every $t \in (a, b)$,

$$\min\left\{\ell\left(\gamma|_{[a,t]}\right), \ell\left(\gamma|_{[t,b]}\right)\right\} \leq \lambda d(\gamma(t) \,;\, D^c);$$

here $d(x, D^c)$ is the distance in $\mathbb{R}^n$ from x to $D^c := \mathbb{R}^n \setminus D$.

Now we may reformulate the definition considered as follows.

A nonempty domain $D \subsetneq \mathbb{R}^n$ is uniform if for some fixed $C > 1$ and $\lambda > 0$ every pair of its points can be joined in D by a (C, λ)-uniform curve.

The second definition is based on a deep theorem of Gehring and Osgood [GO-1979]. For its formulation we use the notion of a *quasihyperbolic metric* of an arcwise connected domain $D \subset \mathbb{R}^n$. This is given for $x, y \in D$ by

$$K_D(x, y) := \inf_\gamma \int_0^1 \frac{d\ell(t)}{d(\gamma(t), D^c)},$$

where the infimum is taken over all rectifiable curves $\gamma : [0, 1] \to D$ joining x and y.

Here $\ell(t) := \ell(\gamma|_{[0,t]})$, $0 \leq t \leq 1$; the Stiltjes integral exists, as the integrand is continuous and ℓ is of bounded variation.

A standard compactness argument, see, e.g., [GO-1979, p. 53], shows that (D, K_D) is a geodesic metric space. We denote a geodesic joining points x and y in this space by $\gamma_D(x, y)$.

The following result is a slight reformulation of the aforementioned theorem of Gehring and Osgood.

Theorem 3.122. *An arcwise connected domain $D \subsetneq \mathbb{R}^n$ is uniform if and only if for some fixed $C \geq 1$ and $\lambda > 0$ and every pair $x, y \in D$ the geodesic $\gamma_D(x, y)$ is a (C, λ)-uniform curve.*

Now note that for the special case $D := \{x \in \mathbb{R}^n \, ; \, x_n > 0\}$ the quasi-hyperbolic metric K_D coincides with the hyperbolic metric of the half-space model of the classical hyperbolic space $\mathbb{H}^n (= \mathcal{M}^n_{-1})$. This motivates the following result of Bonk, Heinonen and Koskela [BHK-2001] which describes uniform domains as Gromov hyperbolic spaces.

Theorem 3.123. *Let $D \subsetneq \mathbb{R}^n$ be a uniform domain. Then (D, K_D) is a Gromov hyperbolic space.*

(c) In our final example, we freely use some notions and results of Complex Analysis, see, e.g., the classical book [Kob-2005] by Kobayashi.

Let D be a domain in the complex n-space $\mathbb{C}^n$. By $\mathrm{Hol}(\mathbb{D}, D)$ we denote the set of holomorphic maps from the unit disk $\mathbb{D} := \{z \in \mathbb{C}\, ; \, |z| < 1\}$ into D. Equip $\mathbb{D}$ with the Poincaré metric

$$d(z_1, z_2) := (\tanh)^{-1}\left(\left|\frac{z_1 - z_2}{1 - z_1 \bar{z}_2}\right|\right),$$

where $\tanh(t) := \frac{e^t - e^{-t}}{e^t + e^t}$, $t \in \mathbb{R}$, and consider all pseudometrics $\tilde{d}$ on D such that, for every $f \in \mathrm{Hol}(\mathbb{D}, D)$,

$$\tilde{d}(f(z_1), f(z_2)) \leq d(z_1, z_2), \quad z_1, z_2 \in \mathbb{D}.$$

Let k_D be the supremum over all such pseudometrics. Then k_D is, in general, a pseudometric, e.g., $k_{\mathbb{C}^n} = 0$. If k_D is a *metric*, the domain D is said to be *Kobayashi hyperbolic*.

According to the Schwarz–Ahlfors–Pick lemma, k_D coincides with the Poincaré metric, if $D := \mathbb{D}$. Hence, the space $(\mathbb{D}, d)$ is Kobayshi hyperbolic. More generally, *every bounded strictly pseudoconvex domain $D \subset \mathbb{C}^n$* (see, e.g., its definition in [Kob-2005]) *is Kobayshi hyperbolic under the metric k_D.* In this case, the explicit calculation of k_D can be done only for a few types of domains (e.g., symmetric domains). Therefore the above statement is proved in a different way.

A relation between Kobayashi and Gromov hyperbolity describes the following result of Balogh and Bonk [BaBo-2000].

Theorem 3.124. *Let $D \subset \mathbb{C}^n$ be a bounded strictly pseudoconvex domain. The metric space (D, k_D) is complete Gromov hyperbolic.*

3.3.5 Sub-Riemannian manifolds

To introduce this concept we need several definitions (see the book [Mon-2002] for details).

Let $\mathcal{M}$ be a smooth finite-dimensional manifold. A *distribution* on $\mathcal{M}$ is a subbundle $V(\mathcal{M}) = \{V_m\}_{m \in \mathcal{M}}$ of the tangent bundle $T(\mathcal{M}) = \{T_m \mathcal{M}\}_{m \in \mathcal{M}}$. Hence, V_m is a vector subspace of the tangent space $T_m \mathcal{M}$ smoothly depending on m and having dimension independent of m.

Given a distribution $V(\mathcal{M})$, a vector field $v : \mathcal{M} \to T(\mathcal{M})$ is said to be *horizontal* if $v(m)$ belongs to V_m for any m.

A distribution $V(\mathcal{M})$ is called *completely non-integrable* if for any m the tangent space $T_m \mathcal{M}$ is the linear span of the family of all iterated commutators of horizontal vector fields.

Let V_m^j be the linear subspace of V_m generated by all the iterated commutators of order j. Then $V_m := V_m^0 \subset V_m^1 \subset \cdots$. This sequence is stabilized at some $r \le \dim \mathcal{M}$ depending on m. For completely non-integrable $V(\mathcal{M})$ we have $V_m^r = T_m$ for all m. Set

$$d_j(m) := \dim V_m^j, \quad 0 \le j \le r.$$

Using these notions we define a *sub-Riemannian manifold* as a triple $(\mathcal{M}, V, g)$, where $\mathcal{M}$ is a finite-dimensional smooth manifold, V is completely non-integrable on $\mathcal{M}$ and $g = \{g_m\}_{m \in \mathcal{M}}$ is a family of strictly positive definite quadratic forms on $V = \{V_m\}_{m \in \mathcal{M}}$ smoothly depending on m.

A sub-Riemannian manifold carries the natural *(Carnot–Carathéodory) metric* defined in local coordinates by

$$d_C(m, m') := \inf \int_0^1 \sqrt{g_{\gamma(t)}(\dot{\gamma}(t), \dot{\gamma}(t))} \, dt$$

where the infimum is taken over all *horizontal curves* $\gamma : [0, 1] \to \mathcal{M}$ connecting points m, m'; the curve γ is horizontal if $\dot{\gamma}(t) \in V_{\gamma(t)}$ for any t.

The Chow–Rashevski theorem asserts that, for a connected sub-Riemannian manifold, d_C is indeed a metric, see [Mon-2002, Thm. 1.17].

As every finite-dimensional smooth manifold, $\mathcal{M}$ may be equipped with a smooth Riemannian metric. Fix this metric and denote by Vol the corresponding Riemannian measure on $\mathcal{M}$. Then the volume of the Carnot–Carathéodory ball

$$B_r^C(m) := \{m' \in \mathcal{M} \,;\, d_C(m, m') < r\}$$

depends on the choice of a Riemannian metric. Nevertheless, in the "generic" case the asymptotic behavior of this volume, as $r \to 0$, does not, up to constant multipliers, depend on this choice. It is worth noting that in spite of this fact, $B_r^C(m)$ is a very intricate geometric object, see, e.g., [VG-1994, Sec. 3.1.7].

To present the precise result due to Mitchell [Mitch-1985] we use the following notion.

A sub-Riemannian manifold $(\mathcal{M}, V, g)$ is said to be *regular* if the dimensions $d_j(m)$ of the subspaces V_m^j, $1 \leq j \leq \dim \mathcal{M}$, are independent of m. Denote these constant dimensions by d_j and set

$$Q := \sum_j j(d_j - d_{j-1}).$$

Then the following is true.

The metric space $(\mathcal{M}, d_C)$ is locally doubling. Actually, there are constants $C > 1$ and $r_0 > 0$ such that for all $0 < r \leq r_0$,

$$C^{-1} r^Q \leq \mathrm{Vol}\big(B_r^C(m)\big) \leq C r^Q.$$

The well-known example of a regular sub-Riemannian manifold is the Heisenberg group H_n. This will be discussed within a more general context in subsection 3.3.7 (see the part discussing Carnot groups there).

3.3.6 Metric graphs

Graphs are the most basic and well-known mathematical structures. Mostly for the record, we recall several notions and results of Graph Theory which will be used in this book.

A combinatorial *graph* G is a pair (V, E), where V is the set of *vertices* and E is the set of *edges*, and each edge $e \in E$ "joins" two vertices $v, w \in V$ named the *endpoints* of e. If $v = w$, this edge is called a *loop*. We will only work with *simple graphs* which do not contain loops or *multiple edges*, i.e., edges having the same set of endpoints. In particular, an edge of a simple graph is uniquely determined by its endpoints. Therefore an edge with endpoints v, w is also denoted by $[v, w]$.

In the sequel, the term "graph" stands for a simple graph.

A graph $G' = (V', E')$ is a *subgraph* of a graph $G = (V, E)$ if $G' \subset G$, $V' \subset V$ and every edge $e' \in E'$ has the same endpoints in G' and in G.

A graph is *connected* if it cannot be split into two nonempty subgraphs with disjoint vertex sets. Equivalently, a graph $G = (V, E)$ is connected if every pair v, w of its vertices can be joined by a path. In other words, there is a finite sequence $\{v_i\}_{1 \leq i \leq n} \subset V$ such that $v = v_1$, $w = v_n$ and $[v_i, v_{i+1}] \in E$ for all $i < n$.

To introduce the final combinatorial notion, *coloring of a graph*, we recall that two vertices of a graph are *adjacent* if they are the endpoints of an edge. Now, a *k-coloring* of a graph $G = (V, E)$ is a function $f : V \to \{1, \ldots, k\}$ such that adjacent vertices are assigned different numbers (*colors*). The minimal k for which a graph G is k-colorable is called the *chromatic number* of G (denoted by $\chi(G)$). We will use an estimate of this number by the *degree of a graph* denoted

3.3. Basic classes of metric spaces

by $\deg(G)$. This is defined as the supremum of $\deg(v)$ where the *degree of a vertex* v is equal to the number of edges having v as an endpoint. The aforementioned estimate is given by the following inequality:

$$\chi(G) \leq \deg(G), \tag{3.136}$$

see, e.g., the survey of White [Whi-2000].

A metric on a connected graph is obtained by metricizing its edges as bounded intervals of the real line and defining the distance between two points to be the infimum of the length of paths joining them. The length of a path is measured using the chosen metrics on the edges. This intuitively clear description requires a formalization to avoid possible pathologies.

To give a precise definition of *metric graphs* we realize a combinatorial connected (simple) graph $G := (V, E)$ as a connected one-dimensional CW-complex in a real vector space. Namely, consider the set of vertices V as a discrete metric space, i.e., the distance between two distinct vertices is 1. Then one isometrically embeds this metric space into the Banach space $\ell_\infty(V)$, see Lemma 1.13, and identify V with its image. Each edge $e = [v, w] \in E$ gives rise to the interval $I_e := \{x \in \ell_\infty(V) \,;\, x = (1-t)v + tw \text{ for some } t \in [0,1]\}$. Note that $I_e \cap I_{e'} \neq \emptyset$ if and only if the edges e and e' have a common endpoint, say v. Under the above identification, this intersection equals $\{v\}$ or is empty. So, the sets V and $\{I_e\}_{e \in E}$ are, respectively, zero- and one-dimensional skeletons of the corresponding CW-complex.

Now let $b_e : [0,1] \to I_e$ be the barycentric coordinate in I_e, that is, $b_e\big((1-t)v + tw\big) := t$ provided that $e = [v,w]$. Set $X_G := \bigcup_{e \in E} I_e$ and define a *(piecewise linear) path* in X_G to be a map $\gamma : [0,1] \to X_G$ for which there is a partition $0 = t_0 \leq t_1 \leq \cdots \leq t_n = 1$ such that

$$\gamma\big|_{[t_i, t_{i+1}]} = b_{e_i} \circ c_i;$$

here $e_i \in E$ and c_i is an affine map from $[t_i, t_{i+1}]$ onto $[0,1]$. We say that γ *joins* x and y if $\gamma(0) = x$ and $\gamma(1) = y$. Since the graph G is connected, every two points in X_G are joined by a path.

Further, let $w : E \to (0, \infty)$ be a *weight* on E. Then we define the *length* of the above introduced path $\gamma : [0,1] \to X_G$ by

$$\ell(\gamma) := \sum_{i=0}^{n-1} w(e_i) \big| c_i(t_i) - c_{i+1}(t_{i+1}) \big|.$$

Finally, we define a *pseudometric* $d_w : X_G \times X_G \to [a, +\infty)$ by setting

$$d_w(x, y) := \inf_\gamma \ell(\gamma), \tag{3.137}$$

where γ runs over all paths joining x and y in X_G.

The space (X_G, d_w) is called a *metric graph*.

Show that d_w may be not a metric.

Example 3.125. Let $s_i \subset \mathbb{R}^2$ be a piecewise linear curve with vertices v_k^i, $1 \leq k \leq i+1$, connecting points $x \neq y$, $i \in \mathbb{N}$ (in particular, $s_1 = [x, y]$). Assume that for each i, $s_i \cap s_{i+1} = \{x, y\}$, and define a graph whose vertex set is $\{v_k^i\,;\, 1 \leq k \leq i+1,\ i \in \mathbb{N}\}$ and edge set is formed by all edges of the piecewise linear curves. Define a weight w by $w\big([v_k^i, v_{k+1}^i]\big) := \frac{1}{2^i}$, $1 \leq k \leq i$, $i \in \mathbb{N}$. Then $d_w(x, y) = \inf_{i \geq 1} \frac{i}{2^i} = 0$, but $x \neq y$.

It is easy to check the following facts, see, e.g., [BH-1999, pp. 6–7].

Proposition 3.126. (a) *If w satisfies*

$$\inf\{w(e)\,;\, e \in E\} > 0,$$

then (X_G, d_w) is a length space.

(b) *If, in addition,*

$$\sup\{w(e)\,;\, e \in E\} < \infty,$$

then (X_G, d_w) is a complete geodesic space.

(c) *This geodesic space is proper if the degree of every vertex of G is finite.*

(d) *If, in addition,*

$$\sup\{\deg v\,;\, v \in V\} < \infty,$$

then (X_G, d_w) is of bounded geometry.

The next example of a metric graph plays an important role in several extension problems. The graph structure is determined by the family of parametrized metric balls of a nontrivial metric space $(\mathcal{M}, d)$, i.e., $\operatorname{card} \mathcal{M} > 1$.

By $\mathcal{B}(\mathcal{M})$ we denote the *set of open balls*:

$$\mathcal{B}(\mathcal{M}) := \{B_r(m)\,;\, m \in \mathcal{M},\ 0 < r \leq 2\operatorname{diam}\mathcal{M}\}.$$

This set may be identified with the direct product $\mathcal{M} \times (0, 2\operatorname{diam}\mathcal{M}]$. So two balls may be equal as subsets of $\mathcal{M}$ but distinct as members of $\mathcal{B}(\mathcal{M})$, e.g., a finite metric space has a finite number of balls but its associated space of parametrized balls is infinite. Further, the set theoretic embedding relation gives rise to that for $\mathcal{B}(\mathcal{M})$.

We introduce a graph structure on $\mathcal{B}(\mathcal{M})$ regarding the elements of this set as vertices and joining balls $B \neq B'$ from $\mathcal{B}(\mathcal{M})$ by an edge if one of them is contained in the other. The graph so obtained is denoted by $G_\mathcal{M}$; it is readily seen that $G_\mathcal{M}$ is simple. Let us show that $G_\mathcal{M}$ is connected. Actually, if $B' \neq B$ are two arbitrary balls of $\mathcal{B}(\mathcal{M})$ centered at m', m, respectively, then $B_R(m)$ with $R = 2\operatorname{diam}\mathcal{M}$ contains m' along with a ball $B_r(m') \subset B'$ with sufficiently small

3.3. Basic classes of metric spaces

r. Then $[B, B_R(m)]$, $[B_R(m), B_r(m')]$, $[B_r(m'), B']$ forms a path joining B' and B.

In applications we also use a version of this construction with $\mathcal{B}(\mathcal{M})$ replaced with the set of *closed* parametrized balls of $\mathcal{M}$. The corresponding structure coincides with that generated by $\mathcal{B}(\mathcal{M})$, since both of them are dealing with the same set of pairs $(m, r) \in \mathcal{M} \times (0, 2 \operatorname{diam} \mathcal{M}]$.

Hereafter we denote by c_B and r_B the center and the radius of a ball $B \in \mathcal{B}(\mathcal{M})$. We also write λB where $\lambda > 0$ for the ball of radius λr_B centered at c_B.

Now let $\omega : \mathbb{R}_+ \to \mathbb{R}_+$ be a 2-majorant, i.e., nondecreasing homeomorphism of $\mathbb{R}_+$ such that the function $t \mapsto \omega(t)/t^2$ is nonincreasing for $t > 0$.

We define a weight $\widehat{\omega}$ on the edge set of the graph $G_\mathcal{M}$ by setting

$$\widehat{\omega}([B, B']) := \frac{\omega(r_{B'})}{r_B} \tag{3.138}$$

for open balls $B \subset B'$.

By Proposition 3.126 (a), the distance $d_{\widehat{\omega}}$, see (3.137), is a length metric in case $G_\mathcal{M}$ is connected.

In applications we use instead of $d_{\widehat{\omega}}$ a simpler function $\widetilde{\omega} : \mathcal{B}(\mathcal{M}) \times \mathcal{B}(\mathcal{M}) \to \mathbb{R}_+$ given by

$$\widetilde{\omega}(B, B') := \int_m^M \frac{\omega(t)}{t^2} \, dt, \tag{3.139}$$

where

$$m = m(B, B') := \min(r_B, r_{B'}),$$
$$M = M(B, B') := r_B + r_{B'} + d(c_B, c_{B'}).$$

This function is not a metric, e.g., $\widetilde{\omega}(B, B) \neq 0$. However, out of the diagonal of the space $\mathcal{B}(\mathcal{M}) \times \mathcal{B}(\mathcal{M})$ (where $B = B'$) the metric $d_{\widehat{\omega}}$ is uniformly equivalent to $\widetilde{\omega}$, see Proposition 3.127 below. We will also use an equivalent to $\widetilde{\omega}$ weight w given at an edge $B \subset B'$ by

$$w(B, B') := \frac{\omega(r_B)}{r_B} + \frac{\omega(r_{B'})}{r_{B'}} + \int_m^{m+\widehat{d}(B,B')} \frac{\omega(t)}{t^2} \, dt, \tag{3.140}$$

where $\widehat{d}$ is a metric on $\mathcal{B}(\mathcal{M})$ given by

$$\widehat{d}(B, B') := |r_B - r_{B'}| + d(c_B, c_{B'}).$$

The main reason for replacing the metric $d_{\widehat{\omega}}$ by $\widetilde{\omega}$ is to reveal the intimate relation of the metric graph $(\mathcal{B}(\mathcal{M}), d_{\widehat{\omega}})$ to the generalized hyperbolic space $\mathbb{H}_\varphi^{n+1}$ where $\varphi(t) := \frac{\omega(t)}{t^2}$, $t > 0$, introduced in Example 3.113. In fact, the last summand in (3.140) is a metric equivalent to a length metric. Moreover, the space $\mathcal{B}(\mathbb{R}^n)$

equipped with this metric is bi-Lipschitz homeomorphic to the aforementioned space $\mathbb{H}_\varphi^{n+1}$. All these results will be proved in Section 4.4.

In the next result we change $\widetilde{\omega}$ on the diagonal by setting $\widetilde{\omega}(B,B) = 0$. We also assume that the associated to $\mathcal{B}(\mathcal{M})$ graph $G_\mathcal{M}$ is connected.

Proposition 3.127. (a) *The equivalence relation*
$$d_{\widehat{\omega}} \approx \widetilde{\omega}$$
holds with some numerical constants.

(b) *If $B \subset B'$, then*
$$w(B,B') \approx \widetilde{\omega}(B,B')$$
with numerical constants of equivalence.

Proof. (a) Show that for some constant C,
$$d_{\widehat{\omega}} \leq C\widetilde{\omega}. \tag{3.141}$$

First, let open balls $B \subset B'$ have a common center. Define an integer s by
$$2^s r_B < r_{B'} \leq 2^{s+1} r_B$$
and let $\{B_i\}_{0 \leq i \leq s+1}$ be the family of balls such that $B_i := 2^i B$ for $i \leq s$ and $B_{s+1} := B'$.

By the definition of $d_{\widehat{\omega}}$,
$$d_{\widehat{\omega}}(B,B') \leq \sum_{i=1}^{s} \widehat{\omega}\big([B_i, B_{i+1}]\big) = \sum_{i=0}^{s-1} \frac{\omega(2^{i+1} r_B)}{2^i r_B} + \frac{\omega(r_{B'})}{2^s r_B}.$$

Since $\omega(t)/t^2$ is nonincreasing, we have
$$\frac{\omega(a+b)}{4b} \leq \int_a^{a+b} \frac{\omega(t)}{t^2}\, dt \quad \text{for} \quad 0 < a \leq b.$$

Applying this to the i-th term of the previous inequality with $a = b := 2^i r_B$ and summing on i, we get
$$d_{\widehat{\omega}}(B,B') \leq 4 \int_{r_B}^{2^s r_B} \frac{\omega(t)}{t^2}\, dt + 4 \int_{2^s r_B}^{r_{B'}} \frac{\omega(t)}{t^2}\, dt$$
$$= 4 \int_{r_B}^{r_{B'}} \frac{\omega(t)}{t^2}\, dt \leq 4\widetilde{\omega}(B,B'). \tag{3.142}$$

This proves (3.141) for balls with a common center.

Now let $B \neq B'$ be arbitrary balls in $\mathcal{B}(\mathcal{M})$. Consider the open balls
$$\widehat{B} := B_M(c_B) \quad \text{and} \quad \widehat{B'} := B_M(c_{B'});$$

3.3. Basic classes of metric spaces

recall that $M = M(B, B') := r_B + r_{B'} + d(c_B, c_{B'})$. By the triangle inequality

$$d_{\widehat{\omega}}(B, B') \leq d_{\widehat{\omega}}(B, \widehat{B}) + d_{\widehat{\omega}}(B', \widehat{B}') + d_{\widehat{\omega}}(\widehat{B}, 2\widehat{B}) + d_{\widehat{\omega}}(2\widehat{B}, \widehat{B}'). \quad (3.143)$$

The first two summands are estimated by (3.142) and their contribution into the final inequality (3.141) is

$$4\left(\int_{r_B}^{M+r_B} + \int_{r_{B'}}^{M+r_{B'}}\right) \frac{\omega(t)}{t^2}\, dt \leq 8 \int_{r_B}^{2M} \frac{\omega(t)}{t^2}\, dt;$$

in the last inequality we assume for definiteness that $r_B \leq r_{B'}$. Since $M \geq 2r_B$ and $\omega(t)/t^2$ is nonincreasing,

$$\int_{M}^{2M} \frac{\omega(t)}{t^2}\, dt \leq 2 \int_{\frac{M}{2}}^{M} \frac{\omega(t)}{t^2}\, dt,$$

the right-hand side of the previous inequality is at most

$$8 \left(\int_{r_B}^{M} + 2\int_{\frac{M}{2}}^{M}\right) \frac{\omega(t)}{t^2}\, dt \leq 24\, \widetilde{\omega}(B, \widehat{B}).$$

It remains to obtain a similar bound for the sum of the final two terms in (3.143). Due to the definitions of the balls involved and the metric $d_{\widehat{\omega}}$ this sum is at most

$$\widehat{\omega}([\widehat{B}, 2\widehat{B}]) + \widehat{\omega}([2\widehat{B}, \widehat{B}']) \leq 2\frac{\omega(2M)}{M};$$

here we use the embedding $\widehat{B}' \subset 2\widehat{B}$ following from the definition of $M = M(B, B')$.

Clearly, the right-hand side does not exceed, cf. (3.142),

$$4\frac{\omega(M)}{M} \leq 4\frac{\omega(M)}{M - r_B} \leq 16 \int_{r_B}^{M} \frac{\omega(t)}{t^2}\, dt \leq 16\, \widetilde{\omega}(B, B').$$

Inequality (3.141) is established with $C = 40$.

For the proof of the converse inequality, consider a sequence of open balls B_i, $1 \leq i \leq n$, connecting given balls $B \neq B'$ (for definiteness we assume that $r_B \leq r_{B'}$). Hence, $B_1 = B$, $B_n = B'$ and one of the balls B_i, B_{i+1} is embedded in the other.

Set $r_i := r_{B_i}$, $m_i := \min(r_i, r_{i+1})$, $M_i := \max(r_i, r_{i+1})$ and $Q_i := [m_i, 3M_i]$. Since $t \mapsto \frac{\omega(t)}{t^2}$ is nonincreasing,

$$\widehat{\omega}([B_i, B_{i+1}]) := \frac{\omega(M_i)}{m_i} \geq \frac{1}{9}\frac{\omega(3M_i)}{m_i} \geq \frac{1}{9}\int_{Q_i} \frac{\omega(t)}{t^2}\, dt,$$

whence
$$\sum_{i=1}^{n-1} \widehat{\omega}([B_i, B_{i+1}]) \geq \frac{1}{9} \sum_{i=1}^{n-1} \int_{Q_i} \frac{w(t)}{t^2} dt.$$

By the same reason for any $h_i \geq 0$,
$$\int_{Q_i} \frac{w(t)}{t^2} dt \geq \int_{Q_i + h_i} \frac{w(t)}{t^2} dt.$$

Choosing numbers h_i so that $h_1 = 0$ and the segments $Q_i + h_i$ and $Q_{i+1} + h_{i+1}$ have the only common point (such h_i's exist as $Q_i \cap Q_{i+1} \neq \emptyset$), we get from here and the previous inequality
$$\sum_{i=1}^{n-1} \widehat{\omega}([B_i, B_{i+1}]) \geq \frac{1}{9} \int_{r_1}^{r_1+R} \frac{w(t)}{t^2} dt, \tag{3.144}$$

where $R := \sum_{i=1}^{n-1} \text{length}(Q_i)$ and $r_1 := r_B$. Since B_i, B_{i+1} is an embedded pair of balls,
$$d(c_{B_i}, c_{B_{i+1}}) \leq \max\{r_i, r_{i+1}\} =: M_i.$$

Moreover, $\text{length}(Q_i) \geq 2M_i$ and $r_{B'} =: r_n \leq \frac{1}{2}\text{length}(Q_{n-1})$.
We then conclude that $M(B, B')$ is bounded by
$$r_B + r_{B'} + \sum_{i=1}^{n-1} d(c_{B_i}, c_{B_{i+1}}) \leq r_B + r_{B'} + \frac{1}{2}\sum_{i=1}^{n-1} \text{length}(Q_i) \leq r_B + R.$$

Together with (3.144) this implies that
$$d_{\widehat{\omega}}(B, B') := \inf_{\{B_i\}} \sum_i \widehat{\omega}([B_i, B_{i+1}])$$
$$\geq \frac{1}{9} \int_{r_B}^{M(B,B')} \frac{w(t)}{t^2} dt =: \frac{1}{9}\widetilde{\omega}(B, B').$$

(b) Using the properties of ω we obtain
$$\widetilde{\omega}(B, B') = \frac{1}{2}\int_m^M \frac{w(t)}{t^2} dt + \frac{1}{2}\int_m^M \frac{w(t)}{t^2} dt$$
$$\geq \frac{1}{2}\int_m^{m+\widehat{d}(B,B')} \frac{w(t)}{t^2} dt + \frac{1}{2}\left(\frac{1}{2}\int_m^{2m} \frac{w(t)}{t^2} dt + \frac{1}{2}\int_m^M \frac{w(t)}{t^2} dt\right)$$
$$\geq \frac{1}{2}\int_m^{m+\widehat{d}(B,B')} \frac{w(t)}{t^2} dt + \frac{1}{2}\left(\frac{1}{2}\cdot\frac{w(2m)}{4m} + \frac{1}{2}\cdot\frac{w(M)}{M^2}(M-m)\right)$$
$$\geq \frac{1}{2}\int_m^{m+\widehat{d}(B,B')} \frac{w(t)}{t^2} dt + \frac{1}{24}\left(\frac{w(r_B)}{r_B} + \frac{w(r_{B'})}{r_{B'}}\right) \geq \frac{1}{24}w(B, B').$$

3.3. Basic classes of metric spaces

Conversely,

$$\widetilde{\omega}(B, B') = \int_m^{m+\widehat{d}(B,B')} \frac{\omega(t)}{t^2} dt + \int_{m+\widehat{d}(B,B')}^M \frac{\omega(t)}{t^2} dt$$

$$\leq \int_m^{m+\widehat{d}(B,B')} \frac{\omega(t)}{t^2} dt + \frac{\omega(m+\widehat{d}(B,B'))}{(m+\widehat{d}(B,B'))^2} \cdot m$$

$$\leq \int_m^{m+\widehat{d}(B,B')} \frac{\omega(t)}{t^2} dt + \frac{\omega(m)}{m^2} \cdot m \leq w(B, B').$$

The result is proved. □

An important role in extension problems plays a subclass of metric graphs formed by *metric trees*. Let us recall that:

A combinatorial graph is called a tree if it has no cycles as subgraphs.

In turn, a *cycle* is a path of a form $\{v_1, \ldots, v_n\}$ with $v_1 = v_n$ and $v_i \neq v_{i+1}$ for $1 < i < n$.

The definition implies that a tree is a connected graph in which every pair of vertices is joined by a *unique* path (this may be taken as an equivalent definition). The last characterization immediately implies

Proposition 3.128. *Let $T = (V, E)$ be a tree. The metric space (X_T, d_w) with the metric associated to any weight $w : E \to (0, +\infty)$ is a geodesic space. Moreover, this space is 0-hyperbolic.*

Proof. The proof of the first assertion is the matter of definitions. If now $[m_1, m_2, m_3] \subset X_T$ is a geodesic triangle, then one of its vertices, say m_2 should be between the others, i.e., $d(m_1, m_3) = d(m_1, m_2) + d(m_2, m_3)$. Otherwise, these vertices form a cycle.

Hence, every geodesic triangle is degenerate and (X_T, d_w) is δ-hyperbolic with $\delta = 0$. □

In the consequent examples we deal with trees whose vertices are subsets of a fixed set. This fixed set is a *root* of the corresponding tree. Let us recall that a vertex $r \in V$ is the *root* of a tree $T = (V, E)$ if all vertices incident to r are implicitly directed away from r. Since for every vertex v there is a unique path joining v and r, the root of a tree is unique.

A *child* of a vertex v in a rooted tree is the immediate successor of v on a path from the root. A *parent* of a vertex v in this tree is the immediate predecessor of v on a path from v to the root.

We use this terminology in the consequent part of the book.

Example 3.129 (Partition tree). Given a fixed set S, we let $\mathcal{P}_i$ denote a partition of S, $i \in \mathbb{N}$, i.e., the elements of $\mathcal{P}_i$ are pairwise disjoint subsets of S whose union equals S. The sequence $\{\mathcal{P}_i\}_{i \in \mathbb{N}}$ is called a *partition tree* if for every $i \geq 1$ the partition $\mathcal{P}_{i+1}$ is a refinement of $\mathcal{P}_i$. A tree structure on the family of subsets $\mathcal{P} := \bigcup_{i \in \mathbb{N}} \mathcal{P}_i$ is defined by regarding these subsets as vertices; a pair $S', S'' \in \mathcal{P}$ determines an edge if one of these sets, say S', embeds in the other one and, in addition, $S' \in \mathcal{P}_{i+1}$ and $S'' \in \mathcal{P}_i$ for some $i \geq 1$.

Adding to $\mathcal{P}$ the element $\{S\}$ we define in this way a rooted tree with the root S.

Using the weight 1 assigned to each edge, we obtain the *metric partition tree* denoted by $T(\mathcal{P})$.

Example 3.130 (Spatially colored cover graph). We briefly discuss a generalization of a partition tree having important applications in nonlinear approximation by wavelets, see the paper [Br-2004] by Yu. Brudnyi.

Let S be given. Let $\{\mathcal{U}_i\}_{i \in \mathbb{N}}$ be a sequence of covers of S such that $\mathcal{U}_1 := \{S\}$. We introduce a graph structure on the set $\mathcal{U} := \bigcup_{i \in \mathbb{N}} \mathcal{U}_i$ of subsets of S regarding $\mathcal{U}$ as a vertex set and defining an edge to be a pair $S', S'' \in \mathcal{U}$ such that $S' \subset S''$ and $S' \in \mathcal{U}_{i+1}$, $S'' \in \mathcal{U}_i$ for some $i \geq 1$. Defining a weight to be 1 at each edge, we obtain a metric graph called a *cover graph* and denoted by $G(\mathcal{U})$.

In general, the combinatorial structure of $G(U)$ is rather complicated and far from tree structure. The situation is considerably simpler for a subclass of cover graphs introduced as follows.

A cover graph is *spatially k-colored* if there is a coloring of the graph vertices into k colors, such that equally colored vertices (subsets) are either disjoint or one of them embeds in the other.

Such a graph can be presented as the disjoint union of trees, see Proposition 4.1 of [Br-2004]. As an illustration we consider the following special case of this proposition. One indicates a sequence of covers $\{\mathcal{U}_j\}_{j=0}^{\infty}$ of the closed interval $[0,3]$, where $\mathcal{U}_0 := \{[0,3]\}$, and $\mathcal{U}_j$ with $j \geq 1$ consists of all intervals $[0, 3 \cdot 2^{-j}] + h$, where the shift h runs over all binary fractions of j digits (i.e., $h = \sum_{i=1}^{j} d_i 2^{-i}$ where $d_i \in \{0,1\}$). The corresponding graph $G(\mathcal{U})$ is 3-colorable, and the disjoint tree decomposition looks as follows.

Given a color $\gamma \in \{1,2,3\}$ we define a γ-root R_γ to be a vertex (subset) of $G(\mathcal{U})$ which is not contained in any other γ-colored vertex. Then define $T(R_\gamma)$ to be the set of all γ-colored vertices contained in R_γ. Define a graph with $T(R_\gamma)$ as a vertex set; an edge of this graph is a pair $(S', S'') \subset T(R_\gamma)$ where S'' is the firstly entered element of $T(R_\gamma) \setminus \{S'\}$ in the unique path directed from S' to the root $[0,3]$. It is easily seen that $T(R_\gamma)$ is a rooted tree and that

$$G(U) = \coprod_{\gamma=1}^{3} \coprod_{R(\gamma)} T(R_\gamma).$$

The set of γ-roots here is infinite, see the cited paper.

A metric tree is a special case of a more general notion called $\mathbb{R}$-*tree*. The latter is a uniquely geodesic space $(\mathcal{M}, d)$ with the following property:

If $\gamma : [a, b] \to \mathcal{M}$ is a curve such that $\gamma|_{[a,c]}$ and $\gamma|_{[c,b]}$ with $a < c < b$, are geodesics then γ is a geodesic.
Hence, a metric space is an $\mathbb{R}$-tree if and only if it is 0-hyperbolic.

A (combinatorial) metric tree is clearly an $\mathbb{R}$-tree but $\mathbb{R}$-trees are, in general, essentially more complicated geometric objects. One gives an example of an $\mathbb{R}$-tree which is not a combinatorial metric tree, see [BH-1999, pp. 167–168]. Equip the set $(0, \infty) \times (0, \infty)$ with the distance $d(x, y)$ between $x = (x_1, x_2)$ and (y_1, y_2) by $d(x, y) := x_1 + y_1 + |x_2 - y_2|$ if $x_2 \neq y_2$ and $d(x, y) := |x_1 - y_1|$ if $x_2 = y_2$. The reader can easily see that this is an $\mathbb{R}$-tree and that there is no tree structure on $(0, +\infty) \times (0, +\infty)$ which gives rise to the metric tree coinciding with the $\mathbb{R}$-tree constructed.

Another example of an $\mathbb{R}$-tree is the asymptotic cone

$$\mathrm{Con}_\omega(\mathbb{H}^n) = \lim_\omega \left\{ \left(\mathbb{H}^n, \frac{1}{i} d_{\mathbb{H}^n}\right)_{i \geq 1} \right\},$$

see Remark 3.62 of subsection 3.1.8. The complexity of this $\mathbb{R}$-tree is emphasized by the following surprising property: for every $m \in \mathrm{Con}_\omega(\mathbb{H}^n)$ its complement has infinitely many connected components.

The role of $\mathbb{R}$-trees and their natural generalization, $\mathbb{R}$-*buildings*, in low-dimensional topology and Geometric Group Theory is discussed in the surveys [Shal-1991] and [D-1998] and the collection of papers [KL-1997].

3.3.7 Metric groups

A group G endowed with a metric is called a *metric group* if the group operations $(g_1, g_2) \mapsto g_1 \cdot g_2$ and $g \mapsto g^{-1}$ are continuous in the metric topology. This metric may be replaced by a topologically equivalent metric [10] compatible with the group structure. According to the G. Birkhoff–Kakutani theorem, see, e.g., [HR-1963, Sec. 2.8] for any metric group (G, d), there exists a topologically equivalent *left-invariant* metric d'. Hence, for all $g, g_1, g_2 \in G$,

$$d'(g \cdot g_1, g \cdot g_2) = d'(g_1, g_2).$$

The following example shows that such a group may not admit an *invariant* (i.e., simultaneously left- and right-invariant) topologically equivalent metric.

Example 3.131. The *linear group* $GL_n(\mathbb{R})$ of real invertible $n \times n$ matrices may be naturally regarded as a metric subspace of the Euclidean space $\mathbb{R}^{n^2}$. Then $GL_n(\mathbb{R}) \subset \mathbb{R}^{n^2}$ equipped with the induced metric is a metric group. Show that

---
[10] i.e., both metrics give rise to the same metric topology.

there is no topologically equivalent invariant metric, say d, on this group. In fact, invariance implies that for all $g_1, g_2 \in GL_n(\mathbb{R})$,

$$d(g_1 g_2, e) = d(g_2 g_1, e); \tag{3.145}$$

here e is the unit matrix.

Consider a matrix group consisting of all matrices $\begin{pmatrix} \alpha & \beta \\ 0 & \alpha^{-1} \end{pmatrix}$ where $\alpha > 0$, $\beta \in \mathbb{R}$. This group is naturally isomorphic and homeomorphic to a subgroup of $GL_n(\mathbb{R})$. Therefore it admits an invariant metric, if $GL_n(\mathbb{R})$ does. Denote this metric by $\tilde{d}$ and apply (3.145) with $g_n := \begin{pmatrix} \frac{1}{n} & \frac{1}{n} \\ 0 & n \end{pmatrix}$ and $h_n := \begin{pmatrix} n & \frac{1}{n} \\ 0 & \frac{1}{n} \end{pmatrix}$ and $d := \tilde{d}$. Passing to the limit as $n \to \infty$ and noting that $g_n h_n \to \begin{pmatrix} 1 & 0 \\ 0 & 1 \end{pmatrix} = e$ while $h_n g_n \to \begin{pmatrix} 1 & 2 \\ 0 & 1 \end{pmatrix}$ we get the contradiction

$$0 = \lim_{n \to \infty} \tilde{d}(g_n h_n, e) = \lim_{n \to \infty} \tilde{d}(h_n g_n, e) \neq 0.$$

Replacing the metric of a metric group by a topologically equivalent one may lead to the loss of completeness as the following example exhibits.

Let $\mathrm{Hom}[0,1]$ be the group of order preserving homeomorphisms of the interval $[0,1]$ with composition as multiplication. Equip $\mathrm{Hom}[0,1]$ with right-invariant metric

$$d_\infty(\varphi_1, \varphi_2) := \max_{0 \le t \le 1} |\varphi_1(t) - \varphi_2(t)|.$$

The metric space $(\mathrm{Hom}[0,1], d_\infty)$ is incomplete and its completion consists of all nondecreasing continuous functions with fixed points 0 and 1.

Nevertheless, there exists a metric d on $\mathrm{Hom}[0,1]$ topologically equivalent to d_∞ such that $(\mathrm{Hom}[0,1], d)$ is complete. To construct this metric, define $\mathcal{U}_n$ to be the set of nondecreasing functions φ on $[0,1]$ with fixed points 0 and 1 such that

$$\sum_{x \in [0,1]} \text{length}\left(\varphi^{-1}(x)\right) \ge \frac{1}{n}, \quad n \in \mathbb{N}.$$

Then define the required metric on $\mathrm{Hom}[0,1]$ by

$$d(\varphi_1, \varphi_2) := d_\infty(\varphi_1, \varphi_2) + \sum_{n \in \mathbb{N}} \frac{1}{2^n} \frac{\delta_n(\varphi_1, \varphi_2)}{1 + \delta_n(\varphi_1, \varphi_2)},$$

where

$$\delta_n(\varphi_1, \varphi_2) := \left| \frac{1}{d_\infty(\varphi_1, \mathcal{U}_n)} - \frac{1}{d_\infty(\varphi_2, \mathcal{U}_n)} \right|.$$

Since functions from $\mathrm{Hom}[0,1]$ are strictly monotone, δ_n (and therefore d) are well defined. It may be checked that d is a metric which is topologically equivalent on $\mathrm{Hom}[0,1]$ to d_∞.

Completeness of the space $(\mathrm{Hom}[0,1], d)$ may be established as follows.

3.3. Basic classes of metric spaces

Let $\{f_n\}$ be a Cauchy sequence in $(\text{Hom}[0,1], d)$. Then $\{f_n\}$ converges uniformly to a nondecreasing function f_0 with fixed points 0 and 1. This f_0 must belong to $\text{Hom}[0,1]$, since for otherwise f_0 is not strictly increasing and belongs to some set U_{n_0}. This implies that $\delta_{n_0}(f_n, f_m) \to \infty$ as $n, m \to \infty$ and leads to a contradiction:

$$0 = \lim_{n,m\to\infty} d(f_n, f_m) \geq \lim_{n,m\to\infty} \frac{1}{2^{n_0}} \frac{\delta_{n_0}(f_n, f_m)}{1 + \delta_{n_0}(f_n, f_m)} > 0.$$

Hence, f_0 is the limit of $\{f_n\}$ in $(\text{Hom}[0,1], d)$.

It is essential in some applications that an *invariant metric preserves completeness* (the Klee theorem [Kl-1952]). In other words, if (G, d) is a complete metric group, and d' is an invariant metric on G topologically equivalent to d, then (G, d') is also complete.

Now we describe several classes of metric groups which will be used mostly as illustrations to extension results and conjectures. The reader who is not versed in the material presented below may omit the corresponding part of the text.

Finitely generated groups with the word metric

A group G is *finitely generated* if there is a finite subset $A \subset G$ named a *generating set*, such that every element of G can be written as

$$\prod_{j=1}^{k} a_j^{\varepsilon_j} \quad \text{for some} \quad a_j \in A \quad \text{and} \quad \varepsilon_j \in \{1, -1\};$$

the empty product (with $k = 0$) is defined to be the unit e of the group.

Given a group G with a generating set A, we define the *word metric* d_A associated to A as follows: the distance $d_A(g, h)$ between elements $g, h \in G$ is the minimal k in the *word representation* $g^{-1} \cdot h = \prod_{j=1}^{k} a_j^{\varepsilon_j}$, where all $a_j \in A$ and $\varepsilon_j \in \{-1, 1\}$.

The word metrics associated to different finite generating sets A and A' are equivalent, i.e., for some $C > 0$ and all $g, h \in G$,

$$C^{-1} d_A(g, h) \leq d_{A'}(g, h) \leq C d_A(g, h).$$

To see this it suffices to express the elements of A as words generating by A' and vise versa.

Hence, every finitely generated group is a metric space defined up to a bi-Lipschitz homeomorphism.

Finitely generated groups of isometries appear in the situation described by the so-called Efremovich–Švarc–Milnor lemma, see, e.g., [BH-1999, pp. 140–141]. For its formulation we recall

Definition 3.132. A subgroup G of the group $\mathrm{Iso}(\mathcal{M}, d)$ acts

(a) *properly*, if for every compact set $K \subset \mathcal{M}$, the set

$$\{g \in G\,;\, g(K) \cap K \neq \emptyset\}$$

is finite;

(b) *freely*, if

$$g(m) = m \quad \text{for all } m \text{ implies that} \quad g = e;$$

(c) *cocompactly*, if there is a compact set $K_0 \subset \mathcal{M}$ such that

$$\mathcal{M} = G(K_0) := \cup\{g(K_0)\,;\, g \in G\}.$$

The aforementioned lemma then asserts

Theorem 3.133. *Let $(\mathcal{M}, d)$ be a length space. Assume that a subgroup G of the group $\mathrm{Iso}(\mathcal{M})$ acts on $\mathcal{M}$ properly, freely and cocompactly. Then G is finitely generated and for every generating set A and fixed $m \in \mathcal{M}$ there is a constant $C > 0$ depending only on $\mathcal{M}, G$ and A such that for all $g, h \in G$*

$$C^{-1} d_A(g, h) \leq d\big(g(m), h(m)\big) \leq C d_A(g, h).$$

Now let G be a finitely generated group with the word metric d_A associated to a generating set $A \subset G$. We say that G is of *polynomial growth* if there are constants $C > 0$ and $p \geq 0$ such that for every $R > 0$,

$$\operatorname{card} B_R(e) \leq C R^p. \tag{3.146}$$

Here $B_R(g)$ is the open ball of radius R centered at g in the metric space (G, d_A). Since all word metrics on G are equivalent, the definition does not depend on the choice of a generating set.

According to the Gromov theorem [Gr-1981], every finitely generated group of polynomial growth has a nilpotent subgroup of a finite index.

Combining this with the result of Bass [Ba-1972], we obtain

Theorem 3.134. *Let G be a finitely generated group of polynomial growth and let A be a generating set. Then there are constants $C_1, C_2 > 0$ depending only on G and A, and $Q \geq 0$ depending only on G so that for every ball $B_R(g) \subset (\mathcal{M}, d_A)$,*

$$C_1 R^Q \leq |B_R(g)| \leq C_2 R^Q; \tag{3.147}$$

here $|\cdot|$ is the counting measure.

In particular, $(\mathcal{M}, d_A, |\cdot|)$ is a metric space of homogeneous type.

Number Q is the *degree* of the maximal torsion free nilpotent subgroup of G. To recall this notion we let Γ be a nilpotent group and $\{\Gamma_n\}_{0\leq i\leq n}$ be its *lower central series*. So, $\Gamma_0 = \Gamma$, $\Gamma_n = \{e\}$, and Γ_i is the subgroup generated by all i-fold commutators [11]. Then Γ_i/Γ_{i-1} is an abelian group and its *rank* (the greatest integer d such that Γ_i/Γ_{i-1} has a subgroup isomorphic to $\mathbb{Z}^d$) is finite if Γ is finitely generated. Then the degree of Γ is

$$\deg \Gamma := \sum_{i=0}^{n-1}(i+1)\mathrm{rank}(\Gamma_i/\Gamma_{i+1}).$$

The aforementioned Bass theorem states that for a finitely generated nilpotent group G equipped with the word metric d_A inequality (3.147) is true. If, moreover, G is torsion free, then constants C_1, C_2 in this inequality depend only on Q.

To illustrate these notions and results we introduce

Example 3.135 (Discrete Heisenberg group). Define $H_n(\mathbb{Z})$ to be the set of triples (k, ℓ, m) where $k, \ell \in \mathbb{Z}^n$ and $m \in \mathbb{Z}$. The reader easily checks that this set equipped with the multiplication induced from H_n is a subgroup of the Heisenberg group H_n, see (3.42) for the definition of group operation on H_n. This subgroup is finitely generated and its "natural" generating set can be introduced as follows.

Let $\{e_i\}_{1\leq i\leq n}$ be the standard basic in $\mathbb{Z}^n$. Then we set

$$a_j := (e_j, 0, 0), \quad b_j := (0, e_j, 0), \quad 1 \leq j \leq n \quad \text{and} \quad c := (0, 0, 1).$$

These $2n+1$ elements generate $H_n(\mathbb{Z})$. In fact, if $(k, \ell, t) \in H_n(\mathbb{Z})$, then

$$(k, \ell, t) = a_1^{k_1} \cdot \ldots \cdot a_n^{k_n} \cdot b_1^{\ell_1} \cdot \ldots \cdot b_n^{\ell_n} \cdot c^m.$$

There is a finite number of relations between the generators:

For $i, j = 1, \ldots, n$ we have

$$a_i a_j = a_j a_i, \quad b_i b_j = b_j b_i$$

and

$$a_i c = c a_i, \quad b_j c = c b_j.$$

Moreover, for $1 \leq i, j \leq n$ and $i \neq j$,

$$a_i b_j = b_j a_i c^4$$

and for $i = j$,

$$b_i a_i = a_i b_i c^4.$$

These relations describe the group $H_n(Z)$ completely; every element (k, ℓ, m) of the group can be represented in a unique way by the above written product of the powers of the generators.

According to its definition, see the next subsection, the Cayley graph of $H_n(\mathbb{Z})$ related to this generating set has cycles determined by the above written relations. Hence, this graph is not a tree.

---

[11] recall that the commutator of elements $g, h \in \Gamma$ is $[g, h] := g \cdot h \cdot g^{-1} \cdot h^{-1}$; a 2-fold commutator is $[[g, h], k]$ etc.

Hyperbolic groups

Let G be a group with a (maybe infinite) generating set A. The *Cayley graph* $\mathcal{C}_A(G)$ of the pair (G, A) has the vertex set G and the edge set $E := \{(g, ag)\,;\, g \in G,\, a \in A \cup A^{-1}\}$. Note that (g, ag) has no direction; so (e, a) and $(a, a^{-1}a)$ determine a unique edge with the endpoints e and a. This notion has been introduced and used by A. Cayley [Cay-1878] and his followers as a powerful tool for the study of finitely generated groups. Endow $\mathcal{C}_A(G)$ with a metric $\tilde{d}_A$ associated to the weight $w : E \to (0, \infty)$ to be the constant function 1. Then $(\mathcal{C}_A(G), \tilde{d}_A)$ is a metric graph containing the metric space G with the word metric d_A as its subspace. The Cayley graph is connected: for instance, e and $g \neq e$ are joined in the combinatorial graph (G, E) by the path $e, a_1, a_1 a_2, \ldots, g = a_1 a_2 \ldots a_n$ where a_i are suitable elements of $A \cup A^{-1}$. Therefore, the curve defined by the concatenation of unit segments $[e, a_1 \cdot e], [a_1, a_1 a_2] \ldots [a_1 \ldots a_{n-1}, g]$ joins e and g in the metric graph $(\mathcal{C}_A(G), \tilde{d}_A)$. Choosing the representation $g = a_1 \cdot \ldots \cdot a_n$ such that $d_A(e, g) = n$, we conclude that this curve is a geodesic.

Hence, $(\mathcal{C}_A(G), \tilde{d}_A)$ is a *geodesic space*.

Now let B be another generating set of the group G, and $\mathcal{C}_B(G)$ be the Cayley graph associated to (G, B). It is easy to verify that the following is true.

Proposition 3.136. *Let G be a finitely generated group and A, B be its finite generating sets. Then the metric graphs $\mathcal{C}_A(G)$ and $\mathcal{C}_B(G)$ are bi-Lipschitz homeomorphic.*

Now the class of groups under consideration is introduced by

Definition 3.137. *A finitely generated group is said to be (Gromov) hyperbolic if one of its Cayley graphs is δ-hyperbolic for some $\delta \geq 0$.*

In fact, the definition does not depend on the choice of a generating set $A \subset G$ (and, therefore, on $\mathcal{C}_A(G)$). This follows from Proposition 3.136 and the following result, see, e.g., [Vai-2005, Thm. 3.18].

Theorem 3.138. *If a δ-hyperbolic space $(\mathcal{M}, d)$ is bi-Lipschitz homeomorphic to a geodesic space $(\mathcal{M}', d')$, then $\mathcal{M}'$ is δ'-hyperbolic with $\delta' \geq 0$ depending only on δ and the constant of the bi-Lipschitz homeomorphism.*

In the next examples we will say that a group G is δ-hyperbolic if there is a generating set such that the metric graph $(\mathcal{C}_A(G), \tilde{d}_A)$ is δ-hyperbolic.

Example 3.139. (a) Finite and cyclic groups are hyperbolic (they are called *elementary hyperbolic groups*).

(b) If a group G acts properly and cocompactly by isometries on $\mathbb{H}^n$, then G is hyperbolic; in particular, the fundamental group of a compact Riemann surface of genus ≥ 2 is hyperbolic.

For $n > 2$, groups acting properly and cocompactly on $\mathbb{H}^n$ were constructed by A. Borel [Bo-1963] (arithmetic groups) and by Gromov and Piatetski–Shapiro [GP-1988] (non-arithmetic groups).

3.3. Basic classes of metric spaces

(c) The previous statement remains to be true if $\mathbb{H}^n$ is replaced by a geodesic metric space of curvature ≤ 0 (see Definition 3.109) which does not contain an isometric copy of $\mathbb{R}^2$, see, e.g., [BH-1999, p. 459] for the proof.

(d) It was noted by Gromov and proved by Ol'shanski [Ol'-1992] that in some statistical sense almost every finitely generated group is hyperbolic.

(e) If a finitely generated group contains $\mathbb{Z}^2$ as a subgroup, then G *is not hyperbolic*. In particular, a direct sum of infinite hyperbolic groups is not hyperbolic, since every infinite hyperbolic group contains an element of infinite order.

For the same reason, a *uniform lattice*, i.e., cocompact discrete subgroup of a semi-simple Lie group of rank > 1 is not hyperbolic. In fact, for otherwise this Lie group would be also hyperbolic which is impossible, since such a Lie group contains an isometric copy of the Euclidean plane as a geodesic submanifold, see, e.g., [Hel-1978].

(f) A hyperbolic group of polynomial growth is either finite or contains a cyclic group of finite index; see Theorem 3.2 in [BH-1999, p. 459].

Free groups

Given a set A, the elements of the set A^{-1} are, by definition, the symbols a^{-1} where $a \in A$. The disjoint union $A^{\pm 1} := A \coprod A^{-1}$ is called an *alphabet*, and a word in this alphabet is a finite sequence $a_1 a_2 \ldots a_n$ where all $a_i \in A^{\pm 1}$. We may insert or delete a subword aa^{-1} and two words are said to be *equivalent* if we can pass from one to the other by a sequence of such deletions and insertions. The word $a_1 \ldots a_n$ is said to be *reduced* if each $a_i \neq a_{i-1}^{-1}$.

Now we define the *free group* $\mathcal{F}(A)$ to be the set of equivalence classes of words over the alphabet $A^{\pm 1}$. The *group operation* on $\mathcal{F}(A)$ is given by concatenation of words and the *unit* of $\mathcal{F}(A)$ is the empty word.

Since every equivalence class contains a unique reduced word, we may regard the elements of $\mathcal{F}(A)$ as reduced words. If w_1 and w_2 are two reduced words then $w_1 w_2$ equals to a unique reduced word in the equivalence class $[w_1, w_2]$; in particular, $ww^{-1} = \emptyset$.

It is easily seen that the Cayley graph of $\mathcal{F}(A)$ associated to the generating set $A^{\pm 1}$ is a rooted tree with the root $\emptyset$. Metricizing $\mathcal{C}_{A^{\pm 1}}(\mathcal{F}(A))$ with the help of the constant weight 1, we turn $\mathcal{F}(A)$ into a metric group. As every metric tree, $\mathcal{F}(A)$ is a 0-hyperbolic space.

Carnot groups

The Heisenberg group H_n, see section 3.1.6, and the usual additive group $\mathbb{R}^n$, are simplest examples of *Carnot groups*, connected and simply connected nilpotent Lie groups $G := (\mathbb{R}^n, \cdot)$, whose Lie algebras admit a *stratification*. That is to say,

$\mathbb{R}^n$ is equipped with two maps $(x,y) \mapsto x \cdot y$ and $x \mapsto x^{-1}$ acting, respectively, from $\mathbb{R}^n \times \mathbb{R}^n$ and $\mathbb{R}^n$ into $\mathbb{R}^n$ that are smooth and so that they turn $\mathbb{R}^n$ into a group. The Lie algebra $\mathfrak{G}$ of G can be identified with the vector space (over $\mathbb{R}$) of left-invariant vector fields [12] $X := \sum_{i=1}^{n} u_i \frac{\partial}{\partial x_i}$ on G with C^∞-coefficients equipped with the *Lie bracket*

$$[X, Y] = X \circ Y - Y \circ X.$$

An easy computation shows that this commutator is a vector field whose coefficients $\{w_i\}$ are expressed via the coefficients $\{u_i\}$ and $\{v_i\}$ of X and Y by

$$w_j = \sum_{i=1}^{n} \left(u_i \frac{\partial v_j}{\partial x_i} - v_i \frac{\partial u_j}{\partial x_i} \right), \quad 1 \leq j \leq n.$$

Finally, the Lie algebra $\mathfrak{G}$ is assumed to be stratified, i.e., as a vector space it can be decomposed into the direct sum of nontrivial vector spaces

$$\mathfrak{G} = \bigoplus_{j=1}^{m} V_j \tag{3.148}$$

such that $\dim V_1 \geq 2$ and

$$[V_1, V_j] = V_{j+1}, \quad 1 \leq j \leq n-1, \quad [V_1, V_m] = \{0\}.$$

Here $[V_1, V_j]$ is the subspace of $\mathfrak{G}$ generated by elements $[X, Y]$ with $X \in V_1$ and $Y \in V_j$.

This finalizes the definition of an *m-step Carnot group*.

The following facts on Carnot groups can be found in the books [CG-1990] by Corwin and Greenleaf and [FS-1982] by Folland and Stein.

(a) The group multiplication on $G = (\mathbb{R}^n, \cdot)$ can be written in the form

$$x \cdot y = x + y + P(x, y), \quad x, y \in \mathbb{R}^n, \tag{3.149}$$

where $P : \mathbb{R}^n \times \mathbb{R}^n \to \mathbb{R}^n$ is a polynomial map with the first k components being zero where $k := \dim V_1$. In particular, 0 is the unit of the group and the inverse x^{-1} to $x \in \mathbb{R}^n$ is $(-x_1, \ldots, -x_n)$.

(b) Via the exponential map $\exp : \mathfrak{G} \to G$ one induces on G a one-parametric family of automorphisms $\delta_\lambda : \mathbb{R}^n \to \mathbb{R}$, $\lambda > 0$, such that

$$\delta_\lambda(x_1, \ldots, x_n) = (\lambda^{\alpha_1} x_1, \ldots, \lambda^{\alpha_n} x_n), \tag{3.150}$$

where $1 = \alpha_1 = \cdots = \alpha_k \leq \alpha_{k+1} \leq \cdots \leq \alpha_n$.

(c) The Lebesgue measure on $\mathbb{R}^n$ is both left- and right-invariant with respect to the group multiplication, i.e., it is the Haar measure. We denote this measure by μ_h.

---

[12] A vector field X is *left-invariant* if $X(f_y) = (Xf)_y$ for all $y \in \mathbb{R}^n$ where $f \in C^\infty(\mathbb{R}^n)$ and $f_y(x) := f(y \cdot x)$, $x \in \mathbb{R}^n$.

3.3. Basic classes of metric spaces

(d) There exists a *homogeneous norm* on G, i.e., a continuous function $N : G \to \mathbb{R}_+$, that is smooth away from 0 and such that

$$N(x) = 0 \Leftrightarrow x = 0, \quad N(\delta_\lambda(x)) = \lambda N(x) \quad \text{and} \quad N(x^{-1}) = N(x).$$

This N is also a quasimetric, i.e., for some $C \geq 1$ and all $x, y \in \mathbb{R}^n$,

$$N(x+y) \leq C(N(x) + N(y)).$$

(e) Let us recall that on every Lie group (regarded as a smooth manifold) there exists a unique left-invariant Riemannian metric. The *length* $\ell(\gamma)$ of a curve $\gamma : [a, b] \to G$ is defined in this metric. But there is another notion of length related to sub-Riemannian structure of a Carnot group. Namely, let $X_1, \ldots, X_k$ be the left-invariant vector fields uniquely defined by the basis vectors $v_1, \ldots, v_k$ of the space V_1. A curve $\gamma : [a, b] \to G$ is said to be *horizontal* if $\gamma'(t)$ is a linear combination of vectors $X_i(\gamma(t))$, $1 \leq i \leq k$, for any t. Then the *Carnot-Carathéodory metric* on G is defined by

$$d_C(x, y) := \inf_\gamma \ell(\gamma), \quad x, y \in \mathbb{R}^n,$$

where γ runs over all horizontal curves connecting x and y.

Generally speaking, $d_C(x, y)$ may be $+\infty$ but, in the case of Carnot groups, this defines a metric on G which is homogeneous with respect to the dilations:

$$d_C(\delta_\lambda(x), \delta_\lambda(y)) = \lambda d_C(x, y).$$

Moreover, $x \mapsto d_C(x, x)$ is a homogeneous norm.

(f) The family of (Carnot–Carathéodory) balls

$$B_r^C(x) := \{y \in G \,;\, d_C(x^{-1} \cdot y) < R\}, \quad x \in \mathbb{R}^n, \ n > 0,$$

is left-invariant and homogeneous:

$$y \cdot B_r^C(x) = B_r^C(y \cdot x), \quad \delta_\lambda(B_r^C(x)) = B_{\lambda r}^C(\delta_\lambda(x)).$$

The Haar measure of such a ball is proportional to the power of radius r. More precisely,

$$\mu_h(B_r^C(x)) = ar^Q, \tag{3.151}$$

where $a := \mu_h(B_1(0))$ and

$$Q := \sum_{i=1}^m i \dim V_i. \tag{3.152}$$

This Q is called the *fractal (or homogeneous) dimension* of G and is denoted by $\dim_{hom}(G)$.

Due to (3.151) the Hausdorff p-measure of a ball $B_r^C(x)$ is zero if $p > Q$, and infinity if $p < Q$, i.e., the *Hausdorff dimension* $\dim_H\left(B_r^C(x)\right)$ of such a ball equals Q, see subsection 3.2.4 for the definition and properties of $\dim_H$. If G is non-abelian, then

$$Q = \dim_{\text{hom}}(G) > \dim G (= n). \tag{3.153}$$

Therefore, in this case, $\dim_H(G) > \dim G$; metric spaces whose Hausdorff dimension is greater than topological one are called *fractals*. The middle third Cantor set C is one of the best known fractals which displays many typical fractal characteristics (in this case, $\dim_H C = \log 2/\log 3 > \dim C = 0$), see subsection 4.2.3 for more examples.

So, it is not surprising that the geometry of Carnot–Carathéodory balls for non-abelian Carnot groups is highly intricate, see, e.g., the papers [VG-1994, Sec. 3.1.7] by Vershik and Gershkovich and [Kar-1994] by Karidi where schematic pictures of such balls and their spheres are depicted for some other nilpotent Lie groups.

Notwithstanding this complexity, the metric space (G, d_C) has some important features in common with $\mathbb{R}^n$. The next result presents two of them.

Proposition 3.140. (a) *Let G be a Carnot group. Then the metric space $(\mathcal{M}, d_C)$ is of Q-homogeneous type where $Q := \dim_{\text{hom}} G$.*

(b) *An open ball of this space is a uniform domain.*

Assertion (a) is the immediate consequence of (3.151) and the definition of n-homogeneity, see (3.113). Assertion (b) is formulated in the paper [VGr-1996, Proposition 2] by Vodop'anov and Greshnov. The definition of a uniform domain in a metric space is obtained from that in $\mathbb{R}^n$ by replacing the Euclidean metric by d_C (see, e.g., the text preceding Theorem 3.123 for the definition).

Example 3.141. (a) Consider $\mathbb{R}^n$ with its usual group structure:

$$x \cdot y := x + y \quad \text{and} \quad x^{-1} := -x.$$

This group is abelian, its Lie group is generated by linear differential operators $X = \sum_{i=1}^n a_i \frac{\partial}{\partial x_i}$ with constant coefficients, and therefore $[X, Y] = 0$ for all X and Y. Decomposition (3.148) contains only one term V_1 isomorphic to $\mathbb{R}^n$. Therefore, the Carnot–Carathéodory metric coincides with the Euclidean one; the dilations δ_λ are defined by $\delta_\lambda(x) := \lambda x$, $x \in \mathbb{R}^n$, since all $\alpha_i = 1$ in this case.

(b) The Heisenberg group H_n is an example of a noncommutative 2-step Carnot group. In this case, $H_n = \mathbb{C}^n \times \mathbb{R} (= \mathbb{R}^{2n+1})$ and the group multiplication is given by (3.149) where $x, y \in \mathbb{R}^{2n+1}$ and the components of the polynomial map $P : \mathbb{R}^{2n+1} \times \mathbb{R}^{2n+1} \to \mathbb{R}^{2n+1}$ are

$$P_j(x, y) := 0 \quad \text{if} \quad j \leq 2n$$

3.3. Basic classes of metric spaces

and
$$P_{2n+1}(x,y) := 2\sum_{j=1}^{n}(x_{2j+1}y_{2j} - x_{2j}y_{2j+1}),$$

see section 3.1.6 where this multiplication is presented in complex form. The Lie algebra of H_n is decomposed in the direct sum $V_1 \oplus V_2$. Here the basis of the horizontal space V_1 generates the vector fields

$$X_j := \frac{\partial}{\partial x_{2j}} + 2x_{2j+1}\frac{\partial}{\partial x_{2n+1}},$$

$$Y_j := \frac{\partial}{\partial x_{2j+1}} - 2x_{2j}\frac{\partial}{\partial x_{2n+1}}, \quad 1 \leq j \leq n.$$

The commutator relations for the vector fields of the basis are:

$$[X_j, Y_j] = -4\frac{\partial}{\partial x_{2n+1}}, \quad 1 \leq j \leq n,$$

and other brackets for $\{X_j, Y_j\}_{1 \leq j \leq n}$ are zeros.

Hence, V_2 is one-dimensional and

$$\dim_{\text{hom}} H_n = \dim V_1 + 2\dim V_2 = 2n+2,$$

while $\dim H_n = 2n+1$.

Dilations (3.150) are given on H_n by

$$\delta_\lambda(x) = (\lambda x_1, \ldots, \lambda x_{2n}, \lambda^2 x_{2n+1}), \quad \lambda > 0,$$

and a homogeneous norm is defined by

$$N(x) := \left(\left(\sum_{j=1}^{2n} x_j^2\right)^2 + x_{2n+1}^4\right)^{\frac{1}{4}};$$

recall that $d(x,y) := N(x^{-1} \cdot y)$, $x, y \in H_n$, is a metric on this group.

Since d and the Carnot–Carathéodory metric are left-invariant, a standard compactness argument shows that d and d_C are equivalent (on H_n). Hence the balls $B_R(x)$ and $B_R^C(x)$ satisfy, for some numerical constant $\lambda > 1$,

$$B_{\lambda^{-1}r}(x) \subset B_r^C(x) \subset B_{\lambda r}(x).$$

The ball $B_r^C(x)$ is of complicated geometric structure (with the fractal boundary) while $B_r(x)$ resembles an ellipsoid whose axes are changed in length and direction together with x. Nevertheless, the Haar measures of these balls coincide up to a numerical factor.

Hence,

$$\mu_h(B_r(x)) = cr^{2n+2} \tag{3.154}$$

and the same, with another numerical constant, holds for $B_r^C(x)$.

(c) Let T_n be the group of all $n \times n$ real matrices (t_{ij}) such that $t_{ii} = 1$ for $1 \leq i \leq n$ and $t_{ij} = 0$ when $i > j$. The smooth manifold structure is defined by identifying T_n with $\mathbb{R}^N$ where $N := \dim T_n = \frac{n(n-1)}{2}$. The matrix multiplication and inversion are clearly smooth functions in coordinates t_{ij}. The Lie algebra of T_n may be identified with the vector space T_n^0 of all upper triangle $n \times n$ matrices (t_{ij}) with $t_{ii} = 0$ for $1 \leq i \leq n$. Decomposition (3.148) for T_n^0 is as follows:

$$T_n^0 = \bigoplus_{k=1}^{n-1} V_k$$

where V_k contains all $n \times n$ matrices (t_{ij}) with all elements but t_{ij}, with $j - i = k$ equal to zero. The homogeneous dimension of T_n is therefore given by

$$\dim_{\text{hom}} T_n = \sum_{k=1}^{n-1} k \dim V_k = \sum_{k=1}^{n-1} k(n-k) = \frac{(n^2-1)n}{6}.$$

Finally, the dilations on T_n are defined by

$$\delta_\lambda(t_{ij}) = (\lambda^{j-i} t_{ij}), \quad (t_{ij}) \in T_n,$$

and a homogeneous norm may be defined by

$$N(t_{ij}) := \max_{j \geq i} |t_{ij}|^{\frac{1}{j-i}}.$$

The following result describes the structure of tangent spaces to a Carnot group equipped with the left invariant metric d_g, see the definition of a tangent space in Remark 3.62 (b).

Theorem 3.142. *Every space tangent to a Carnot group (G, d_g) at any point is isometric to this group.*

The result follows from a more general Mitchell's theorem [Mitch-1985] asserting that under mild additional restrictions the tangent spaces of a Carnot–Carathéodory metric space (M, d_C) at every point is a Carnot group arising from the Lie bracket structure on horizontal vector fields of M.

Remark 3.143. For the Heisenberg group H_n there is another limit relation between its geodesic structure and Carnot–Carathéodory metric, see [VG-1994, Sec. 2.6].

Comments

The analog of Frink's Proposition 3.3 is the Aoki–Rolewicz renormalization theorem for quasinormed spaces, see, e.g., the Peetre and Sparr paper [PSp-1972] for an elegant elementary proof of this theorem.

Theorem 3.6 and its corollaries are taken from Urysohn's posthumous paper [Ur-1927]which has mainly remained an object of respective references. The situation may change after the publication of Gromov's [Gr-2000] and Vershik's [V-1998] results relating Urysohn's crucial idea to some modern problems of analysis (classification of measure metric spaces, random distances, Hausdorff moduli space, etc.). We briefly discuss only one of these results showing that the Urysohn universal space $\mathcal{U}$ is, in a sense, *typical* among all Polish metric spaces.

Let $(\mathcal{M}, d, \mu)$ be a *metric triple*, a Polish metric space $(\mathcal{M}, d)$ equipped with a Borel probability measure μ of full carrier (i.e., $\mu(\mathcal{M}) = 1$ and the measure of every nonempty open sets is positive). Two triples, $(\mathcal{M}, d, \mu)$ and $(\mathcal{M}', d', \mu')$, are *isomorphic* if there is an isometry I from $(\mathcal{M}, d)$ onto $(\mathcal{M}', d')$ such that the image of μ under I equals μ'. The class of triples isomorphic to a given triple $(\mathcal{M}, d, \mu)$ is denoted by $[(\mathcal{M}, d, \mu)]$ and the set of all these classes is denoted by $\mathcal{T}$. The Gromov classification theorem [Gr-2000] (see its reformulation in [V-1998, Thm. 1] which we use here) corresponds to each class $\nu := [(\mathcal{M}, d, \mu)]$ a parameter M_η, a unique Borel probability measure on the convex cone $\mathcal{D}$ of infinite *distance matrices* $(d_{ij})_{i,j \in \mathbb{N}}$ (i.e., the function $(i,j) \mapsto d_{ij}$ is a pseudometric on $\mathbb{N}$). Therefore, the standard topology on the space of measures on $\mathcal{D}$ gives rise to a topology on the set of classes $\mathcal{T}$.

Theorem (Vershik [V-1998]). *The set of classes $[(\mathcal{U}, d_\mathcal{U}, \mu)]$ generated by the Urysohn space is dense and G_δ in the topological space $\mathcal{T}$.*

Due to Urysohn's theorem the isometry group $\mathrm{Iso}(\mathcal{U})$ is very large (n-fold transitive for every $n \geq 1$). The study of its algebraic and topological structure may be of considerable interest.

The invariants of compact metric spaces Cov and Cap introduced in subsection 3.1.4 are closely related to Shannon's concept of ε-*entropy*; in the simplest case entropy of a compact metric space $\mathcal{M}$ is a function of $\varepsilon > 0$ given by

$$H_\varepsilon(\mathcal{M}) := \log_2 \mathrm{Cov}(\mathcal{M}; \varepsilon).$$

For many important metric spaces the asymptotics of $H_\varepsilon(\mathcal{M})$ as $\varepsilon \to 0$ are known, see, in particular, the papers [KT-1959] by Kolmogorov and Tihomirov and [Lo-1966] by Lorentz.

In the case of metric triples, there is a variation of Shannon's concept which, in the simplest case, looks as follows. Let $(\mathcal{M}, \mu)$ be a metric triple [13] and $\varepsilon > 0$, $\delta \geq 0$. Then (ε, δ)-*entropy* of the triple is defined by

$$H_{\varepsilon,\delta}(\mathcal{M}, \mu) := \log N_{\varepsilon,\delta}$$

where $N_{\varepsilon,\delta}$ is the minimal number of open ε-balls B_i such that

$$\mu(\cup B_i) \geq 1 - \delta.$$

---

[13] here one does not require that $\mathrm{supp}\,\mu$ is a set of full measure.

This concept is studied in several papers, see, in particular, the paper [PRR-1967] by Posner, Rodemich and Rumsey. Nevertheless, up to our knowledge, the asymptotics of $H_{\varepsilon,\delta}$ as $\varepsilon,\delta \to 0$ for many important metric triples are unknown. As an interesting example, one points out the metric triple consisting of the Banach space $C_0[0,1]$ of continuous on $[0,1]$ functions vanishing at 0 equipped with the Wiener measure w. It seems to be natural to assume that $H_{\varepsilon,0}(C_0[0,1],w) \approx \varepsilon^{-2}$.

The concept of entropy relates to the problem of efficient data compression firstly posed and investigated in the classical paper by Shannon [Shan-1948]. In particular, $H_{\varepsilon,\delta}$ is a quantitative characteristic of data compression when the outcome is known within ε and one is allowed to ignore a certain part of the outcome, at most δ in this case. However, in a special case, ε-entropy of a metric compact had appeared much earlier in the Pontriagin and Schnirelman paper [PSch-1932] devoted to a metric characterization of topological dimension, see Theorem 3.20.

An abstract version of the classical Hopf–Rinow theorem, Theorem 3.24 in subsection 3.1.5, is, apparently, new. The Hopf–Rinow result follows from this theorem as a simple consequence, see the proof of Theorem 3.40.

Most results of subsection 3.1.8 can be found in the Bridson and Haefliger book [BH-1999] but the proof of Gromov's compactness criterion, see Theorem 3.58, follows his paper [Gr-1981]. This result and Corollary 3.56 allow us to bring together all compact metric spaces forming a universal metric space called the *Hausdorff moduli space* and denoted by $\mathfrak{H}_c$. Namely, let $[\mathcal{M}]$ be the class of metric spaces isometric to a compact metric space $\mathcal{M}$. The set of all these classes equipped with the distance $d_{\mathcal{GH}}$ (which is a metric by Corollary 3.56) is the required space $\mathfrak{H}_c$. Its properties and the relation to another universal object, the Urysohn space $\mathcal{U}$, are described in Gromov's book [Gr-2000]. In particular, $\mathfrak{H}_c$ is a complete, separable and contractible metric space.

In fact, its contractibility is obvious, since any metric space $(\mathcal{M},d)$ connects with the one point space by the curve $t \mapsto [(\mathcal{M},td)]$, $0 \le t \le 1$. The separability follows from density in $\mathfrak{H}_c$ of isometry classes of finite metric spaces. At last, the completeness may be easily proved by employing the the Gromov compactness theorem.

As for the relation to the Urysohn space, the following is presented as an exercise in [Gr-2000, p. 83].

Let $\mathcal{U}_c$ be the set of all compact subsets of the Urysohn space $\mathcal{U}$ equipped with the Hausdorff metric. Then isometrically

$$\mathfrak{H}_c = \mathcal{U}_c / \operatorname{Iso}(\mathcal{U}).$$

The theory of functions and function spaces on complete metric spaces of homogeneous type, see Definition 3.86, were firstly studied in the Coifman and Weiss paper [CW-1971]; some modern results are presented in Semmes' Appendix B_+ of the book [Gr-2000].

Theorem 3.97 of Ismagilov has been rediscovered several times, see references in [LuuM-L-1994] to the papers by J.B. Kellly, A. Timan and Vestfrid and Lemin.

The elegant version of Ismagilov's proof presented here was due to E. Gorin (personal communication). It was proved by A. Timan [Tim-1975] that an ultrametric space $\mathcal{M}$ admits an isometric embedding into an n-dimensional Hilbert space if and only if $\operatorname{card} \mathcal{M} = n + 1$.

Families of pointwise doubling measures and metric spaces of pointwise homogeneous type are introduced in the A. and Yu. Brudnyis paper [BB-2007a]. The results of subsection 3.2.6 are taken from there.

The concept of a geodesic metric space of bounded curvature in the sense of Definition 3.109 was introduced and studied by A. D. Alexandrov [Ale-1951] and was named after him. Many classical results of Riemannian geometry concerning global geometry of Riemannian manifolds with sectional curvature bounds are generalized to the case of the Alexandrov spaces, see the survey of Yu. Burago, Gromov and G. Perel'man [BGP-1992] and the modern book [BBI-2001] by D. and Yu. Burago and S. Ivanov.

The role of the generalized hyperbolic space $\mathbb{H}_\omega^n$, see Example 3.113, and of the related space of metric balls $\mathcal{B}(\mathcal{M})$, see the example after Proposition 3.126, in the solution to the Simultaneous Extension Problem for the spaces $\Lambda^{2,\omega}(\mathbb{R}^n)$ and $C^{1,\omega}(\mathbb{R}^n)$ were discovered in the Yu. Brudnyi and Schvartzman paper [BSh-1985], see the proofs in [BSh-1999].

A detailed account of the geometry of Carnot–Carathéodory spaces and its applications to nonholonomic variational problems of Analysis and Mechanics are presented in the survey [VG-1994] by Vershik and Gershkovich and in the book [Mon-2002] by R. Montgomery.

Chapter 4

Selected Topics in Analysis on Metric Spaces

We will present several topics in the field which are related to the main problems studied in the book. Because of diversity of methods and results involved in this study the chapter may be seen as a rather satisfactory introduction to Analysis on Metric Spaces.

Like those in the previous chapters, the results employed in the study of the extension-trace problems will be presented with detailed proofs if they have not appeared in book form. Other results are mostly surveyed; they allow us to include the proved theorems into the general framework of the theory.

We begin, in Section 4.1, with an elegant result describing almost non-Archimedean structure of large subspaces of finite metric spaces. The probability approach employed in the proof has become, recently, a considerable tool in the study of extension problems for Lipschitz functions on metric spaces.

Section 4.2 presents definitions and basic properties of three concepts of dimension for metric spaces, *metric dimension, Hausdorff dimension and Nagata dimension*. The last two are most important for our study but we also present a detailed discussion of the metric dimension to motivate the more elaborate concept of the Nagata dimension.

Section 4.3 is devoted to a measure theoretic characterization of doubling metric spaces, namely, the existence of a doubling measure on such spaces. The proof of the theorem has already appeared in the recent book [Hei-2001] by Heinonen. However, we present a detailed proof of the result, since, in our opinion, some missing details in the proof in [Hei-2001] may cause difficulties for the reader who is not acquainted with the technique of Geometric Measure Theory.

In Section 4.4, we study geometric and measure theoretic properties of the space of metric balls $\mathcal{B}^\omega(\mathcal{M})$ which initially appeared as a special metric graph in Section 3.3. In particular, we show that $\mathcal{B}^\omega(\mathbb{R}^n)$, under a mild restriction on ω, is a metric space of pointwise homogeneous type, the fact playing a decisive role in a few simultaneous extension problems.

Section 4.5 describes the modern research in Global Differential Analysis including analogs of the Rademacher differentiability theorem and Sobolev's embeddings for functions on metric space. Since these interesting results are relatively far from our main theme, we restrict ourselves to a brief survey. The reader is referred to the excellent survey [Hei-2007] by Heinonen for detailed information.

Finally, Section 4.6 studies some Banach properties of Lipschitz spaces $\text{Lip}(\mathcal{M}, d)$. In particular, we present the special case of the central result of the real interpolation theory (K-divisibility) and describe a Kantorivich–Rubinstein type space of measure which is predual to $\text{Lip}(\mathcal{M}, d)$. Both results will be applied to extension problems of Chapters 5 and 6.

4.1 Dvoretsky type theorem for finite metric spaces

The structure theory of finite metric spaces has been intensively developed under the strong influence of graph theory, local Banach space theory and the so-called theoretical computer science studying combinatorial algorithms. Our interest in this theory is in particular explained by its applications to the *finiteness phenomenon* which is described in subsection 2.4.2 and will be at length considered in Chapter 7. This allows us to reduce the trace and linear extension problems for functions defined on infinite subspaces to those defined on finite subspaces of a fixed cardinality.

In this section, we describe only one fundamental result of the structure theory related to the so-called *metric Ramsey theory*. Let us first recall one of the problems of the *combinatorial* Ramsey theory.

Given a "large" set S, a family $\mathcal{F}$ of its subsets and a positive integer r, one must prove or disprove the following claim.

> *For every partition of S into r subsets, at least one of these subsets contains a subset of the family $\mathcal{F}$.*

The typical example is the classical Ramsey theorem from his landmark paper [Ram-1930] concerning a generalization of the pigeonhole principle. In this case S is the complete graph K_N, i.e., K_N contains N vertices and every two vertices are joined by a unique edge, and $\mathcal{F}$ is a family of its complete subgraphs containing n vertices. The partial case of the Ramsey theorem asserts that for sufficiently large N and for every subgraph $C \subset K_N$ either C or its complement contains a complete graph of n vertices.

It was pointed out in the paper [BFM-1986] by Bourgain, V. Milman and Figiel that the classical Dvoretzky theorem may be treated in the vein of Ramsey's philosophy (large systems must contain highly-organized subsystems). These authors suggested a metric space analog of Dvoretzky's theorem formulated in the following way.

4.1. Dvoretsky type theorem for finite metric spaces

Given a target distortion $D > 1$ and an integer n, find the largest k such that every n-point metric space $(\mathcal{M}, d)$ has a subset S of cardinality k which embeds into Hilbert space with distortion D.

That is to say, there is a map f of S into L_2 such that for every $m, m' \in S$,

$$d(m, m') \leq \|f(m) - f(m')\|_2 \leq Dd(m, m').$$

We present a recent result on the metric Ramsey problem which is asymptotically optimal as distortion D tends to infinity. Constructive proof of a slightly weaker version was due to Bartal, Linial, Mendel and Naor [BLMN-2005]. In their result, the upper bound of the number $k(D)$ differs (asymptotically) from the lower one by the factor $\log D$ as $D \to \infty$. A nonconstructive (probabilistic) method for the upper estimate closing this gap was proposed by Mendel and Naor [MN-2007] whose proof we will reproduce. The main point of their proof is a construction of well-behaved stochastic partitions of a given n-point metric space $\mathcal{M}$ (so-called *stochastic padded decomposition*). The family of these decompositions gives rise to an ultrametric d_ω on $\mathcal{M}$ depending on stochastic variable ω from a probability space (Ω, μ). Estimating from below the expectation of a specially chosen event in Ω one then proves existence (only!) of a large subspace $\mathcal{M}_0 \subset \mathcal{M}$ and of a stochastic variable $\omega_0 \in \Omega$ such that d_{ω_0} is equivalent on $\mathcal{M}_0$ to d with the required distortion. Finally, one applies Theorem 3.97 to embed isometrically $(\mathcal{M}_0, d_{\omega_0})$ into a Hilbert space.

The realization of this plan is presented in the proof of

Theorem 4.1. *Let $(\mathcal{M}, d)$ be an n-point metric space and $\varepsilon \in (0, 1)$. Then there exists a subspace $\mathcal{M}_\varepsilon \subset \mathcal{M}$ of cardinality at least $n^{1-\varepsilon}$ which admits a bi-Lipschitz embedding into Hilbert space with distortion at most $\frac{128}{\varepsilon}$.*

An example constructed in [BLMN-2005] shows that the lower estimate is asymptotically sharp as $n \to \infty$.

Proof. We will freely use some elementary probabilistic notions and results. In particular, the *random partition* of a metric space $(\mathcal{M}, d)$ is a triple $(\Omega, \mu, \mathcal{P}(\cdot))$, where (Ω, μ) is a probability space and $\mathcal{P}(\omega)$, for every $\omega \in \Omega$, is a partition of the metric space $\mathcal{M}$. Since we will use only finite Ω and $\mathcal{M}$, all problems concerning measurability of sets and functions are immaterial.

We specialize a probability space for the n-point space $\mathcal{M}$ taking

$$\Omega_n := \Pi_n \times I_\Delta,$$

where Π_n is the set of all permutations of the index set $\{1, \ldots, n\}$ and $I_\Delta := [\frac{\Delta}{4}, \frac{\Delta}{2}] \subset \mathbb{R}$, $\Delta > 0$. This will be endowed with the probabilistic measure

$$\mu_n := \gamma_n \times \lambda_\Delta,$$

where γ_n is the normalized counting measure on Π_n and λ_Δ is the normalized Lebesgue measure on I_Δ. In other words,

$$\gamma_n(S \subset \Pi_n) = \frac{\operatorname{card} S}{n!} \quad \text{and} \quad \lambda_\Delta(S \subset I_\Delta) = \frac{4|S|}{\Delta}.$$

In the sequel, we will also use the symbol Prob instead of μ_n (or γ_n).

A random partition of $\mathcal{M} := \{m_1, m_2, \ldots, m_n\}$ related to the probability space (Ω_n, μ_n) which we deal with in the consequent proof is defined as follows.

Given a random variable $\omega := (\pi, R) \in \Omega_n$, one defines the first element C_1 of the desired partition denoted by $\mathcal{P}_n(\omega)$ by setting

$$C_1 := \overline{B}_R(m_{\pi(1)}).$$

Then we proceed inductively defining the j-th element of $\mathcal{P}_n(\omega)$ with $2 \leq j \leq n$ by

$$C_j := \overline{B}_R(m_{\pi(j)}) \setminus \bigcup_{i<j} C_i. \tag{4.1}$$

Let us note that the partition $\mathcal{P}_n(\omega)$ is Δ-bounded, i.e., for all ω,

$$\operatorname{diam} \mathcal{P}_n(\omega) := \sup_j (\operatorname{diam} C_j) \leq \Delta. \tag{4.2}$$

The following lemma is the main point of the proof. For its formulation, we denote by $\mathcal{P}_n(\omega)[m]$ the unique subset of $\mathcal{P}_n(\omega)$ containing a point $m \in \mathcal{M}$, and then consider the event

$$\mathcal{E}(m,t) := \{\omega \in \Omega_n \,;\, \overline{B}_t(m) \in \mathcal{P}_n(\omega)[m]\}. \tag{4.3}$$

Lemma 4.2. *For every $t \in \left(0, \frac{\Delta}{8}\right)$ and every $m \in \mathcal{M}$,*

$$\operatorname{Prob} \mathcal{E}(m,t) \geq \left(\frac{\varphi\left(m, \frac{\Delta}{8}\right)}{\varphi(m, \Delta)}\right)^{\frac{16t}{\Delta}},$$

where $\varphi(m, s)$ stands for the cardinality of the closed ball $\overline{B}_s(m) \subset \mathcal{M}$.

Proof. We begin with the estimate of the (conditional) probability of the event

$$\widehat{\mathcal{E}}(m, r, t) := \{\omega = (\pi, R) \in \mathcal{E}(m, t) \,;\, R = r\}.$$

To this end, we introduce a linear order on the points of the closed ball $\overline{B}_{r+t}(m)$ writing $m_i \leq m_j$ if $\pi^{-1}(i) \leq \pi^{-1}(j)$, $\pi \in \Pi_n$.

Let $m_{\pi(j^*)}$ be the minimal element of the ball $\overline{B}_{r+t}(m)$ with respect to this order. Consider the event

$$\widetilde{\mathcal{E}}(m, r, t) := \{\pi \in \Pi_n \,;\, m_{j(\pi)} \in \overline{B}_{r-t}(m)\}$$

4.1. Dvoretsky type theorem for finite metric spaces

and show that
$$\widetilde{\mathcal{E}}(m,r,t) \subset \widehat{\mathcal{E}}(m,r,t). \tag{4.4}$$

In fact, let π belong to the left-hand side. Then $m_{\pi(j^*)} \in \overline{B}_{r-t}(m)$ and the triangle inequality yields
$$\overline{B}_t(m) \subset \overline{B}_r(m_{\pi(j^*)}).$$

On the other hand, if $m_{\pi(j)} < m_{\pi(j^*)}$, then $m_{\pi(j)}$ does not belong to $\overline{B}_{r+t}(m)$ and the triangle inequality implies that
$$\overline{B}_t(m) \subset \mathcal{M}\setminus\overline{B}_r(m_{\pi(j)}) \subset \mathcal{M}\setminus C_j.$$

Due to (4.1) these and the previous embeddings give
$$B_t(m) \subset \overline{B}_r(m_{\pi(j^*)}) \setminus \bigcup_{j<j^*} C_j = C_{j(\pi)}.$$

Since $m \in C_{j^*}$, this set coincides with $\mathcal{P}_{(\pi,r)}[m]$, the unique element of $\mathcal{P}_n(\pi,r)$ containing m. Hence we have proved that $\omega \in \widetilde{\mathcal{E}}(m,r,t)$ implies $\omega \in \widehat{\mathcal{E}}(m,r,t)$, and (4.4) is established.

Now we compute $\text{Prob}\,\widetilde{\mathcal{E}}(m,r,t)$. For this we set
$$k := \text{card}\,\overline{B}_{r+t}(m) \quad \text{and} \quad \ell := \text{card}\,\overline{B}_{r-t}(m)$$

and show that
$$\text{Prob}\,\widehat{\mathcal{E}}(m,r,t) = \frac{\ell}{k}.$$

To this end, we denote by the same symbol π the permutation of the set $\mathcal{M} := \{m_1,\ldots,m_n\}$ generated by the permutation π of its indices. There are $k!(n-k)!$ permutations π such that $\pi(\overline{B}_{r+t}(m)) = \overline{B}_{r+t}(m)$. Further, every π uniquely determines the above introduced linear order on $\overline{B}_{r+t}(m)$, and there are ℓ possibilities from which to choose one of the elements of $\overline{B}_{r-t}(m)$ to be the minimal element $m_{\pi(j^*)}$. Moreover, there are $(k-1)!(n-k)!$ permutations π preserving the chosen minimal element and such that $\pi(\overline{B}_{r+t}(m)) = \overline{B}_{r+t}(m)$. Hence, there are $\ell(k-1)!(n-k)!$ permutations π preserving $\overline{B}_{r+t}(m)$ and such that $m_{\pi(j^*)} \in \overline{B}_{r-t}(m)$.

We therefore conclude that
$$\text{Prob}\,\widetilde{\mathcal{E}}(m,r,t) = \frac{\ell}{k} =: \frac{\varphi(m,r-t)}{\varphi(m,r+t)}.$$

Together with (4.4), this gives the required estimate:
$$\text{Prob}\{(\pi,R) \in \mathcal{E}(m,t)\,;\, R=r\} \geq \frac{\varphi(m,r-t)}{\varphi(m,r+t)},$$

which, in turn, leads to the inequality

$$\operatorname{Prob}\mathcal{E}(m,t) \geq \frac{4}{\Delta}\int_{I_\Delta} \frac{\varphi(m,r-t)}{\varphi(m,r+t)}\,dr.$$

Since $0 < t < \frac{\Delta}{8}$, we can write

$$\frac{\Delta}{8t} := k + \beta,$$

where k is a positive integer and $\beta \in [0,1)$, and then present the integral over I_Δ as the sum

$$\sum_{j=0}^{k-1}\int_{\frac{\Delta}{4}+2jt}^{\frac{\Delta}{4}+(2j+1)t} + \int_{\frac{\Delta}{4}+2kt}^{\frac{\Delta}{2}} =: \sum_{j=0}^{k-1} A_j + B_k.$$

Changing variables in each A_j and using the arithmetic–geometric mean inequality, we have

$$\sum_{j=0}^{k-1} A_j = \int_0^{2t}\left[\sum_{j=0}^{k-1}\frac{\varphi(m, \frac{\Delta}{4} + (2j-1)t + s)}{\varphi(m, \frac{\Delta}{4} + (2j+1)t + s)}\right]ds$$

$$\geq k\int_0^{2t}\left[\frac{\varphi(m, \frac{\Delta}{4} + s - t)}{\varphi(m, \frac{\Delta}{4} + t(2k-1) + s)}\right]^{\frac{1}{k}} ds \geq 2kt\left[\frac{\varphi(m, \frac{\Delta}{4} - t)}{\varphi(m, \frac{\Delta}{4} + t(2k+1))}\right]^{\frac{1}{k}}.$$

For the term B_k we simply write

$$B_k \geq \left(\frac{\Delta}{4} - 2kt\right)\frac{\varphi(m, \frac{\Delta}{4} + (2k-1)t)}{\varphi(m, \frac{\Delta}{2} + t)}.$$

Combining the estimates obtained we get

$$\operatorname{Prob}\mathcal{E}(m,t) \geq \frac{8kt}{\Delta}\left(\frac{\varphi(m, \frac{\Delta}{4} - t)}{\varphi(m, \frac{\Delta}{4} + t(2k+1))}\right)^{\frac{1}{k}} + \left(1 - \frac{8kt}{\Delta}\right)\frac{\varphi(m, \frac{\Delta}{4} + (2k-1)t)}{\varphi(m, \frac{\Delta}{2} + t)}.$$

Apply to the right-hand side the elementary inequality $\theta a + (1-\theta)b \geq a^\theta b^{1-\theta}$ which holds for all $\theta \in [0,1]$ and $a, b \geq 0$. Choosing here $\theta := \frac{8kt}{\Delta}$ we estimate this side from below by

$$\left(\frac{\varphi(m, \frac{\Delta}{4} - t)}{\varphi(m, \frac{\Delta}{4} + t(2k+1))}\right)^{\frac{8t}{\Delta}}\left(\frac{\varphi(m, \frac{\Delta}{4} + (2k-1)t)}{\varphi(m, \frac{\Delta}{2} + t)}\right)^{1-\frac{8kt}{\Delta}}$$

$$= \left[\frac{\varphi(m, \frac{\Delta}{4} - t)}{\varphi(m, \frac{\Delta}{4} + (2k+1)t)} \cdot \frac{\varphi(m, \frac{\Delta}{4} + (2k-1)t)}{\varphi(m, \frac{\Delta}{2} + t)}\right]^{\frac{8t}{\Delta}}$$

$$\times \left[\frac{\varphi(m, \frac{\Delta}{4} + (2k-1)t)}{\varphi(m, \frac{\Delta}{2} + t)}\right]^{\frac{8t}{\Delta}(\frac{\Delta}{8t} - k - 1)}.$$

4.1. Dvoretsky type theorem for finite metric spaces

By the definition of $k \in \mathbb{N}$ and $\beta \in [0,1)$, we have

$$\frac{\Delta}{8t} - k - 1 := \beta - 1 < 0 \quad \text{and} \quad \frac{\Delta}{4} + (2k-1)t \leq \frac{\Delta}{2} + t;$$

therefore the last factor is at least 1. Moreover, by the same reason,

$$\frac{\Delta}{4} + (2k-1)t \geq \frac{\Delta}{4} - t \quad \text{and} \quad \frac{\Delta}{4} + (2k-1)t \leq \frac{\Delta}{2} + t.$$

Hence the product of the first two factors is bounded from below by

$$\left(\frac{\varphi(m, \frac{\Delta}{4} - t)}{\varphi(m, \frac{\Delta}{2} + t)}\right)^{\frac{16t}{\Delta}} \geq \left(\frac{\varphi(m, \frac{\Delta}{8})}{\varphi(m, \Delta)}\right)^{\frac{16t}{\Delta}};$$

recall that $0 < t < \frac{\Delta}{8}$. Combining all the obtained inequalities we get

$$\operatorname{Prob} \mathcal{E}(m, t) \geq \left(\frac{\varphi(m, \frac{\Delta}{8})}{\varphi(m, \Delta)}\right)^{\frac{16t}{\Delta}},$$

as required. $\square$

Let now $\mathcal{P}_j$, $j \in \mathbb{N}$, be a random partition of the space $\mathcal{M} := \{m_1, \ldots, m_n\}$ defined as in Lemma 4.2 but with the parameters

$$\Delta_j := 8^{-j} \operatorname{diam} \mathcal{M}, \quad t_j := \frac{\Delta_j}{\alpha}; \tag{4.5}$$

here $\alpha > 1$ will be indicated later. Hence $\mathcal{P}_j$ is governed by probability space (Ω_j, μ_j) where $\Omega_j := \Pi_n \times I_{\Delta_j}$ and $\mu_j := \gamma_n \times \lambda_{\Delta_j}$. Let us recall that γ_n and λ_Δ are, respectively, normalized measures on the set of n-permutations Π_n and the interval $I_\Delta := \left[\frac{\Delta}{4}, \frac{\Delta}{2}\right]$.

We assume that the random partitions $\mathcal{P}_1, \mathcal{P}_2, \ldots$ are chosen independently. In other words, for every subset of integers $I \subset \mathbb{N}$ the probability of the event $\{\mathcal{E}_j \subset \Omega_j; j \in I\}$ equals $\prod_{j \in I} \operatorname{Prob} \mathcal{E}_j := \prod_{j \in I} \mu_j(\mathcal{E}_j)$. Hence we now consider every $\mathcal{P}_j$ as a random partition associated to the product probability space $(\Omega_\infty, \mu_\infty) := \bigotimes_{j=1}^\infty (\Omega_j, \mu_j)$. In particular, $\mathcal{P}_j(\omega) = \mathcal{P}_j(\omega_j)$ for $\omega = (\omega_1, \omega_2, \ldots) \in \Omega_\infty$.

Using this sequence of partitions, we construct the so-called *random partition tree* described by

Lemma 4.3. *There is a sequence of random partitions $\mathcal{Q}_j$ of $\mathcal{M}$, $j \in \mathbb{Z}_+$, governed by $(\Omega_\infty, \mu_\infty)$ such that:*

(a) *for every $\omega \in \Omega_\infty$ and $j \in \mathbb{Z}_+$ the partition $\mathcal{Q}_{j+1}(\omega)$ is a refinement of $\mathcal{Q}_j(\omega)$;*

(b) *for every point $m \in \mathcal{M}$ the probability of the event*

$$\mathcal{E}(m) := \{\omega \in \Omega_\infty\,;\, \overline{B}_{t_j}(m) \subset \mathcal{Q}_j(\omega)[m] \quad \text{for all} \quad j\}$$

is estimated by

$$\operatorname{Prob} \mathcal{E}(m) \geq (\operatorname{card} \mathcal{M})^{-\frac{16}{\alpha}} \left(= n^{-\frac{16}{\alpha}}\right).$$

As above, $\mathcal{P}[m]$ denotes a unique subset of a partition $\mathcal{P}$ containing a point m.

Proof. Set $\mathcal{Q}_0 := \{\mathcal{M}\}$ and define $\mathcal{Q}_j$, $j \geq 1$, inductively, to be the common refinement of $\mathcal{Q}_{j-1}$ and $\mathcal{P}_j$. In other words,

$$\mathcal{Q}_j(\omega) := \{C \cap C'\,;\, C \in \mathcal{Q}_{j-1}(\omega) \quad \text{and} \quad C' \in \mathcal{P}_j(\omega)\}.$$

This definition implies that for all $\omega \in \Omega_\infty$, $m \in \mathcal{M}$ and $j \in \mathbb{Z}_+$,

$$\mathcal{Q}_{j+1}(\omega)[m] = \mathcal{Q}_j(\omega)[m] \cap \mathcal{P}_{j+1}(\omega)[m]. \tag{4.6}$$

We derive from here the implication

$$\overline{B}_{t_j}(m) \subset \mathcal{P}_j(\omega)[m] \Rightarrow \overline{B}_{t_j}(m) \subset \mathcal{Q}_j(\omega)[m] \tag{4.7}$$

starting from the evident equality

$$\mathcal{Q}_1(\omega)[m] = \mathcal{Q}_0(\omega)[m] \cap \mathcal{P}_1(\omega)[m] =: \mathcal{P}_1(\omega)[m]$$

and then proceed inductively as follows.

If $\overline{B}_{t_{j+1}}(m) \subset \mathcal{P}_{j+1}(\omega)[m]$, then the induction hypothesis and (4.6) yield

$$\overline{B}_{t_{j+1}}(m) \subset \overline{B}_{t_j}(m) \cap \mathcal{P}_{j+1}(\omega)[m] \subset \mathcal{Q}_j(\omega)[m] \cap \mathcal{P}_{j+1}(\omega)[m] = \mathcal{Q}_{j+1}(\omega)[m].$$

Further, (4.7) and Lemma 4.2 lead to the estimate

$$\operatorname{Prob} \mathcal{E}(\omega) := \operatorname{Prob}\{\omega\,;\, \overline{B}_{t_j}(m) \subset \mathcal{Q}_j(\omega)[m] \quad \text{for all} \quad j \in \mathbb{N}\}$$
$$\geq \operatorname{Prob}\{\omega\,;\, \overline{B}_{t_j}(m) \subset \mathcal{P}_j(\omega)[m] \quad \text{for all} \quad j\}$$
$$= \prod_{j \in \mathbb{N}} \operatorname{Prob}\{\omega \in \Omega_j\,;\, \overline{B}_{t_j}(m) \subset \mathcal{P}_j(\omega)[m]\}$$
$$\geq \prod_{j \in \mathbb{N}} \left(\frac{\varphi(m, \frac{1}{8}\Delta_j)}{\varphi(m, \Delta_j)}\right)^{\frac{16 t_j}{\Delta_j}}.$$

Since $\frac{t_j}{\Delta_j} = \frac{1}{\alpha}$ and $\frac{1}{8}\Delta_j = \Delta_{j+1}$, we finally conclude that

$$\operatorname{Prob} \mathcal{E}(m) \geq \varphi(m, \Delta_1)^{-\frac{16}{\alpha}} := \left(\operatorname{card} \overline{B}_{t_1}(m)\right)^{-\frac{16}{\alpha}} \geq (\operatorname{card} \mathcal{M})^{-\frac{16}{\alpha}}. \qquad \square$$

4.1. Dvoretsky type theorem for finite metric spaces

Now, for $\omega \in \Omega_\infty$, we define the required stochastic ultrametric d_ω by setting, for $m \neq m' \in \mathcal{M}$,
$$d_\omega(m, m') := 8^{-J} \operatorname{diam} \mathcal{M} (=: \Delta_J),$$
where $J = J(m, m')$ is the largest integer j for which
$$\mathcal{Q}_j(\omega)[m] = \mathcal{Q}_j(\omega)[m'].$$
In addition, we set $d_\omega(m, m') = 0$ for $m = m'$. It is straightforward to check that d_ω is an ultrametric.

Now one defines a random subspace $\mathcal{M}(\omega) \subset \mathcal{M}$ given for $\omega \in \Omega_\infty$ by
$$\mathcal{M}(\omega) := \{m \in \mathcal{M}; \overline{B}_{t_j}(m) \subset \mathcal{Q}_j(\omega)[m] \text{ for all } j\}.$$
The expectation of $\mathcal{M}(\omega)$ is
$$\mathbb{E}(\mathcal{M}(\cdot)) = \sum_{m \in \mathcal{M}} \operatorname{Prob}\{\omega ; \overline{B}_{t_j}(m) \subset \mathcal{Q}_j(\omega)[m] \text{ for all } j\},$$
and due to Lemma 4.3 the right-hand side is at least $n^{1-\frac{16}{\alpha}}$.

Choose now $\frac{16}{\alpha} = \varepsilon$; the lower estimate for the expectation allows us to assert that there is $\omega_0 \in \Omega_\infty$ such that $\mathcal{M}_0 := \mathcal{M}(\omega_0)$ meets the conditions
$$\operatorname{card} \mathcal{M}_0 \geq n^{1-\varepsilon} \text{ and } \overline{B}_{t_j}(m) \subset \mathcal{Q}_j(\omega_0)[m] \text{ for all } j.$$
Using these, we show that d_{ω_0} and d are equivalent on $\mathcal{M}_0$ with the required estimate of distortion. First, one notes that for $J = J(m, m')$,
$$d(m, m') \leq \operatorname{diam} \mathcal{Q}_J(\omega_0)[m] \leq 8^{-J} \operatorname{diam} \mathcal{M} = d_{\omega_0}(m, m'),$$
since $\mathcal{Q}_J(\omega_0)[m] = \mathcal{Q}_J(\omega_0)[m']$, see (4.2).

On the other hand, by the maximality of $J = J(m, m')$,
$$\mathcal{Q}_{J+1}(\omega_0)[m] \neq \mathcal{Q}_{J+1}(\omega_0)[m'];$$
this and the definition of $\mathcal{M}(\omega)$ yield
$$m \notin \overline{B}_{t_{J+1}}(m') \text{ for } m, m' \in \mathcal{M}_0.$$
This immediately implies that
$$d(m, m') > t_{J+1} := \frac{1}{\alpha} 8^{-J-1} \operatorname{diam} \mathcal{M} = \frac{\varepsilon}{16} \cdot \frac{1}{8} d_{\omega_0}(m, m').$$
Hence $d|_{\mathcal{M}_0} \approx d_{\omega_0}|_{\mathcal{M}_0}$ with distortion $\frac{\varepsilon}{128}$.

It remains to embed isometrically the metric subspaces $(\mathcal{M}_0, d_{\omega_0})$ into a Hilbert space, see Theorem 3.97.

The proof is complete. $\square$

Remark 4.4. The method of stochastic padding decompositions was discovered by Calinescu, Karloff and Rabani [CKR-2004] for analysis of approximation algorithms of Computer Science. They, in particular, introduced the distribution used in the proof of Lemma 4.2.

4.2 Covering metric invariants

There are several bi-Lipschitz invariants of metric spaces whose role for Lipschitz extensions is comparable with that of topological (covering) dimension for continuous extensions. We will present two of them, Hausdorff and Nagata dimensions, but will begin with a metric counterpart of topological dimension, as a preparatory material for a more involved Nagata dimension. Proofs of the results for which there are no corresponding book references will be included; other proofs will be discussed on different levels of detalization.

Convention. Throughout this section $\mathcal{M}\,(=(\mathcal{M},d))$ stands for a metric space, and indexed or not letter S stands for a subspace of $\mathcal{M}$.

4.2.1 Metric dimension

To formulate the basic notion, let us recall that the *diameter of a family* $\mathcal{F} := \{S_\alpha\}_{\alpha \in A}$ is given by

$$\operatorname{diam} \mathcal{F} := \sup_{\alpha \in A} (\operatorname{diam} S_\alpha), \tag{4.8}$$

and the *order* of this family is defined as

$$\operatorname{ord} \mathcal{F} := \sup_{m \in \mathcal{M}} \operatorname{card} \{S_\alpha \in \mathcal{F} \,;\, S_\alpha \ni m\}. \tag{4.9}$$

Definition 4.5. The metric dimension of $\mathcal{M}$, denoted by $\mu \dim \mathcal{M}$, is the least integer n such that for every $\varepsilon > 0$ there exists a cover of $\mathcal{M}$ of diameter ε and order $n+1$.

If such n does not exist, $\mu \dim \mathcal{M}$ is assumed to be ∞.

The following result lists the basic properties of metric dimension (see, e.g., Engelking's book [En-1978] for missing details).

Theorem 4.6. (a) $\mu \dim$ *is a uniformly continuous invariant.*

(b) $\mu \dim \mathcal{M} \leq \dim \mathcal{M}$, *where the inequality may be strict.*

(c) *For a compact space* $\mathcal{M}$,

$$\mu \dim \mathcal{M} = \dim \mathcal{M}.$$

(d) *For the direct sum* $\mathcal{M} := \oplus^{(p)} \{\mathcal{M}_i, d_i\}_{1 \leq i \leq N}$, *see Definition 3.50,*

$$\mu \dim \mathcal{M} \leq \sum_{i=1}^{N} \mu \dim \mathcal{M}_i.$$

4.2. Covering metric invariants

Proof (Outline). (a) Let $f : \mathcal{M} \to \mathcal{M}'$ be a bijection such that f and f^{-1} are uniformly continuous. Let $\{\mathcal{F}'_j\}_{j \in \mathbb{N}}$ be a sequence of open covers of $\mathcal{M}'$ with the properties

$$\operatorname{diam} \mathcal{F}'_j \to 0 \quad \text{as} \quad j \to \infty, \quad \text{and} \quad \operatorname{ord} \mathcal{F}_j \leq n+1 \quad \text{for all} \quad j.$$

Then $\mathcal{F}_j := f^{-1}(\mathcal{F}'_j)$, $j \in \mathbb{N}$, is the sequence of open covers of $\mathcal{M}$ with the same properties. By definition, $\mu \dim \mathcal{M} \leq \mu \dim \mathcal{M}'$.

The converse inequality is derived similarly.

(b) Drawing a comparison between the definition of $\mu \dim$ and topological dimension, see Definition A.1 of Chapter 1, we immediately obtain the required inequality. In order to show that this inequality may be strict, we describe Sitnikov's example of 1954 (see, in particular, [Sit-1955]) who constructs a subset Σ in $\mathbb{R}^3$ with

$$\dim \Sigma = 2 \quad \text{and} \quad \mu \dim \Sigma = 1.$$

Let $\mathcal{K}_i$ be the family of cubes $\left[\frac{n}{i}, \frac{n+1}{i}\right]^3 \subset \mathbb{R}^3$, $n \in \mathbb{Z}$, where $i \geq 1$ is an integer. The union of edges of all these cubes is denoted by E_i.

Set $B_1 := E_1$ and define B_i for $i \geq 2$ inductively as a set obtained by all possible translations of E_i which do not intersect the previous B_j (with $j < i$). Further, set $B := \bigcup_{i=1}^{\infty} B_i$ and

$$\Sigma := (0,1)^3 \setminus B.$$

The set B is dense in $\mathbb{R}^3$ and therefore Σ does not contain an open (in $\mathbb{R}^3$) subset. Then, $\dim \Sigma \leq 2$, since a subset of dimension 3 in $\mathbb{R}^3$ must contain such a subset, see, e.g., [HW-1941, Theorem IV3]. If, however, $\dim \Sigma = 1$, then the classical Menger–Urysohn separation theorem strengthened by Mazurkiewicz, see, e.g., [Kur-1968, pp. 466–467], implies the following.

For every pair of points $x, y \in (0,1)^3 \setminus \Sigma =: B \cap (0,1)^3$, there exists a *connected* compact subset $K \subset B \cap (0,1)^3$ containing these points.

We then conclude that $K = \bigcup_{i=1}^{\infty} (B_i \cap K)$, i.e., K is the disjoint union of closed subsets $B_i \cap K$. By the Sierpiński theorem, see, e.g., [Kur-1968, pp. 173–175], such representation of a connected compact metric space is impossible.

Hence, $\dim \Sigma = 2$ and it remains to explain why $\mu \dim \Sigma = 1$. To this end, we choose in a cube Q from the family $\mathcal{K}_i$, a point ϑ which is a lattice point of the E_{i+1}. Using the central projection from this point, we map $\Sigma \cap Q$ into the boundary ∂Q. For a suitable choice of these lattice points for different cubes of $\mathcal{K}_i$, we obtain a map f whose image $f(\Sigma)$ is contained in the two-dimensional faces of cubes $Q \in \mathcal{K}_i$. The next step is to use a central projection to map any set $f(\Sigma) \cap Q$ lying in a two-dimensional face of Q onto the one-dimensional skeleton of this face. In this way we obtain a continuous map $g : \Sigma \to E_i \cap [0,1]^3$ which clearly satisfies the condition

$$\|g(x) - x\|_2 \leq \frac{\sqrt{3}}{i}, \quad x \in \Sigma.$$

Since $E_i \cap [0,1]^3$ is a one-dimensional compact set, for every $\varepsilon > 0$ there is an open cover $\{U_j\}$ of this set of diameter $\frac{\varepsilon}{2}$ and of order 2, see assertion (c). Then $\{g^{-1}(U_j)\}$ is an open cover of Σ of order 2 whose diameter is bounded by $\frac{2\sqrt{3}}{i} + \frac{\varepsilon}{2}$ which is at most ε for sufficiently large i.

Hence, for every $\varepsilon > 0$ there is an open cover of Σ of diameter ε and of order 2, i.e., $\mu \dim \Sigma = 1$.

(c) Let the metric dimension of a compact metric space $\mathcal{M}$ be n. To show that $\dim \mathcal{M} \leq n$, we must prove that every open cover $\{U_\alpha\}$ of $\mathcal{M}$ admits an open refinement of order $n+1$, see Definition A.1 of $\dim \mathcal{M}$ from Appendix A. According to Lebesgue's Lemma C.6 of Appendix C, for this cover there is $\varepsilon_0 > 0$ such that every open cover of diameter at most ε_0 is a refinement of the cover $\{U_\alpha\}$. Since $\mu \dim \mathcal{M} = n$, there exists an open cover of diameter ε_0 and of order $n+1$ which should be a refinement of $\{U_\alpha\}$. Hence, $\dim \mathcal{M} \leq n = \mu \dim \mathcal{M}$ which, together with assertion (a), implies the desired equality.

(d) Let $\mathcal{U} := \{U_\alpha\}, \mathcal{V} := \{V_\beta\}$ be covers of metric spaces $\mathcal{M}_1, \mathcal{M}_2$, respectively. Then the order $\mathrm{ord}(\mathcal{U} \times \mathcal{V})$ of the cover $\mathcal{U} \times \mathcal{V} := \{U_\alpha \times V_\beta\}$ of the space $\mathcal{M}_1 \oplus^{(p)} \mathcal{M}_2$ does not exceed $(\mathrm{ord}\,\mathcal{U}) \cdot (\mathrm{ord}\,\mathcal{V})$. Since, in addition, $\mathrm{diam}(\mathcal{U} \times \mathcal{V}) \leq 2^{\frac{1}{p}} \max\{\mathrm{diam}\,U, \mathrm{diam}\,V\}$, this implies the assertion. □

Remark 4.7. Katětov [Kat-1958] proved that

$$\mu \dim \mathcal{M} \leq \dim \mathcal{M} \leq 2\mu \dim \mathcal{M}.$$

Sitnikov's example shows that the right inequality is sharp.

In view of Sitnikov's example it is natural to seek a metric characterization of topological dimension (for metric spaces). This was found by Dowker and Hurewicz [DH-1956], then P. Vopenka [Vop-1959] strengthened their result as follows:

Theorem 4.8. $\dim \mathcal{M} \leq n$ if and only if there exists a sequence of open covers $\{\mathcal{U}_k\}_{k \in \mathbb{N}}$ of $\mathcal{M}$ such that, for every k,

(a) $\mathcal{U}_{k+1}$ is a refinement of $\mathcal{U}_k$;

(b) the order of $\mathcal{U}_k$ is at most $n+1$;

(c) $\mathrm{diam}\,\mathcal{U}_k \leq \frac{1}{k}$.

For the proof see, e.g., Nagata's book [Nag-1965, pp. 126–132].

Example 4.9. We illustrate applications of Theorem 4.8 by computing metric dimensions of the one-third Cantor set and the Sierpiński gasket, see Figure 4.1

4.2. Covering metric invariants

Since $C = \bigcap_{j=0}^{\infty} C_j$ where C_j contains 2^j intervals of $[0,1]$ of length 3^{-j}, the family $\{C_j\}$ is a cover of C of order 1 and of diameter 3^{-j}. Hence

$$\mu \dim C = 0 \ (= \dim C).$$

The Sierpiński gasket G is obtained by repeatedly removing (inverted) triangles from the initial equilateral triangle of side length 1, see e.g., the book [Fal-1999] by Falkoner. In this case, $G = \bigcap_{j=0}^{\infty} G_j$ where G_j consists of 3^j closed equilateral triangles of lengthside 2^{-j}, and each pair of triangles has at most one point in common.

Figure 4.1: Sierpínski gasket.

Hence, G_j is a cover of G of order 2 and of diameter 2^{-j} and therefore

$$\mu \dim G = 1 \ (= \dim G).$$

4.2.2 Hausdorff dimension

The concept employs the notion of the Hausdorff measure – its basic characteristics used below see in subsection 3.2.4.

Definition 4.10. Hausdorff dimension of a metric space $\mathcal{M}$ denoted by $\dim_H \mathcal{M}$ is defined by
$$\dim_H \mathcal{M} := \sup\{p\,;\, \mathcal{H}_p(\mathcal{M}) = \infty\}.$$

Remark 4.11. (a) For a metric subspace S of $(\mathcal{M}, d)$ one may equivalently use, for the definition of $\dim_H S$, either the Hausdorff measure of the space $(S, d|_S)$ or the restriction to S of the Hausdorff measure for $(\mathcal{M}, d)$.

(b) At the critical value $\bar{p} := \dim_H \mathcal{M}$, $\dim_{\bar{p}} \mathcal{M}$ may be zero, or infinity, or $0 < \dim_H \mathcal{M} < \infty$, see examples below.

The following theorem lists the basic characteristics of $\dim_H$.

Theorem 4.12. (a) *Monotonicity:*

$$\dim_H S \leq \dim_H S' \quad \text{if} \quad S \subset S'.$$

(b) *Countably stability:*

$$\dim_H \left(\bigcup_{j \in \mathbb{N}} S_j \right) = \sup_j (\dim_H S_j).$$

(c) *If $f : (\mathcal{M}, d) \to (\mathcal{M}', d')$ is a bijection satisfying the Hölder condition of exponent $\alpha \in (0, 1]$, then*

$$\dim_H \mathcal{M}' \leq \frac{1}{\alpha} \dim_H \mathcal{M}.$$

In particular, $\dim_H$ is a bi-Lipschitz invariant.

(d) *For the direct sum $(\mathcal{M}, d_s) := \oplus^{(s)} (\mathcal{M}_j, d_j)_{1 \leq j \leq N}$,*

$$\dim_H \mathcal{M} \geq \sum_{j-1}^{N} \dim_H \mathcal{M}_j.$$

(e) *Topological and Minkowski dimensions are related to $\dim_H$ by the inequality*

$$\dim \leq \dim_H \leq \underline{\dim}_M. \tag{4.10}$$

Proof. Assertions (a) and (b) follow straightforwardly from the definition of $\dim_H$. To prove (c), we first note that

$$\dim_H S = \inf\{p;\ \mathcal{H}_p(S) = 0\}. \tag{4.11}$$

In fact, $\mathcal{H}_p(S) < \infty$ implies that $\mathcal{H}_q(S) = 0$ for all $q > p$.

This means that the definition of $\dim_H$ uses only the sets of Hausdorff measure zero.

Let now $f : \mathcal{M} \to \mathcal{M}'$ be a surjection satisfying the Hölder condition

$$d'(f(m_1), f(m_2)) \leq C d(m_1, m_2)^\alpha, \quad m_1, m_2 \in \mathcal{M},$$

4.2. Covering metric invariants

with some constants $C > 0$ and $0 < \alpha \leq 1$. Due to the definition of Hausdorff measures, see (3.87) and (3.88), the equality $\mathcal{H}_p(\mathcal{M}) = 0$ and the Hölder condition imply that $\mathcal{H}_{\frac{p}{\alpha}}(\mathcal{M}') = 0$. Together with (4.11) this proves assertion (c).

To prove (d), we first note that the spaces $(\mathcal{M}, d_s)$ with distinct s are bi-Lipschitz equivalent. Applying (c) we therefore have

$$\dim_H(\mathcal{M}, d_s) = \dim_H(\mathcal{M}, d_\infty).$$

Further, according to Theorem 3.81 the inequality

$$\mathcal{H}_{\sum_{j=1}^N p_j}(\mathcal{M}, d_\infty) \geq \prod_{j=1}^N \mathcal{H}_{p_j}(\mathcal{M}_j, d_j)$$

holds for $1 \leq p_j \leq \infty$. If, first, all $\dim_H \mathcal{M}_j > 0$, the right-hand side is infinity for all positive numbers $p_j < \dim_H \mathcal{M}_j$, then the left-hand side is also infinity and, by definition, $\dim_H \mathcal{M} \geq \sum_{j=1}^n p_j$. The result follows.

In the remaining case we replace all $\mathcal{M}_j$ of Hausdorff dimension zero by the one-point metric spaces and apply statement (a) and then Theorem 3.81 to the direct sum of the remaining $\mathcal{M}_j$.

The left inequality of assertion (e), one of the most deep facts of the theory, is due to Szpilrain [Szp-1937]. Its simplified proof given by Eilenberg may be found in Heinonen's book [Hei-2001, pp. 62–63].

The proof of the right inequality in (4.10) is essentially simpler. For its proof, we set $s := \underline{\dim}_M \mathcal{M}$ and assume, without loss of generality, that $s < \infty$. Due to Definition 3.27 of $\underline{\dim}_M$, for every $p > s$ there is a constant $C_p > 0$ such that for some sequence $\varepsilon_j \to 0$,

$$\mathrm{Cov}(\varepsilon_j\,;\,\mathcal{M}) \leq C_p \varepsilon_j^{-p} \quad \text{for all} \quad j.$$

Hence, there is a cover of $\mathcal{M}$ by ε_j-balls B_i in $\mathcal{M}$, $1 \leq i \leq N_j$, such that $N_j \leq C_p \varepsilon_j^{-p}$, $j \in \mathbb{N}$. By the definition of the approximate Hausdorff measure we then have

$$\mathcal{H}_p^{\varepsilon_j}(\mathcal{M}) \leq \sum_{i=1}^{N_j} (2 r_{B_i})^p \leq 2^p C_p.$$

Hence, for every $p > s$,

$$\mathcal{H}_p(\mathcal{M}) := \sup_{\varepsilon > 0} \mathcal{H}_p^\varepsilon(\mathcal{M}) < \infty$$

and therefore $\dim_H \mathcal{M} \leq s$. □

Hausdorff measures and dimensions are powerful tools for investigating the geometry of and analysis on "thin" subsets of $\mathbb{R}^n$ (of Lebesgue n-measure zero). Many problems in Analysis and Geometry are closely related to and depend on this investigation (PDE and variational problems with initial data on Lipschitz surfaces, complex plane dynamics, removable sets of holomorphic functions, stochastic processes, etc.). In the present book, we encounter thin subsets (Ahlfors regular sets) in Chapter 9, where we will study the trace and extension problems for Lipschitz functions of higher order on such subsets.

We briefly discuss below selected items of this vast field referring the reader to the aforementioned books by Falconer, Federer and Mattila for detailed information, proofs and references.

The structure of p-sets

A Borel subset $S \subset \mathbb{R}^n$ is said to be a *p-set* ($0 \leq p \leq n$) if $0 < \mathcal{H}_p(S) < \infty$. The Hausdorff dimension of such a set is clearly p.

The classification of p-sets initiated in the 1930s by Besicovich is based on the concept of density.

The *upper density of a p-set S at a point x* denoted by $\overline{D}_p(x\,;S)$ is given by

$$\overline{D}_p(x\,;S) := \varlimsup_{r \to 0} \frac{\mathcal{H}_p(S \cap B_r(x))}{(2r)^p}.$$

Here $B_r(x)$ is a Euclidean ball.

The *lower density* $\underline{D}_p(x\,;S)$ is defined similarly. If $\underline{D}_p(x\,;S) = \overline{D}_p(x\,;S)$, their common value is referred to as the *density of S at the point x* and is denoted by $D_p(x\,;S)$.

It is clear that $D_p(x\,;S) = 0$ for $x \in \mathbb{R}^n \setminus \overline{S}$, and it was proved by Besicovich that $2^{-p} \leq \overline{D}_p(x\,;S) \leq 1$ for $\mathcal{H}_p$ almost all points x in S. Further, due to the chosen normalization of the Hausdorff p-measure, see (3.87) and (3.91), $D_p(x\,;\mathbb{R}^p) = 1$ for all x. More generally, let a p-set S be a *p-rectifiable subset* of $\mathbb{R}^n$ meaning that $1 \leq p \leq n$ is an integer and $S = f(\mathbb{R}^p)$, where $f : \mathbb{R}^p \to \mathbb{R}^n$ is Lipschitz. Then for $\mathcal{H}_p$ almost all points $x \in S$,

$$D_p(x\,;S) = 1.$$

For $p = n$ this is true even for every S of positive Lebesgue (=Hausdorff) n-measure due to the classical Lebesgue density theorem. However, the behavior of densities for p-sets with $p < n$ essentially differs from that with $p = n$. For instance, an elementary evaluation gives for the one-third Cantor set $C \subset \mathbb{R}$,

$$\underline{D}_p(x\,;C) < \overline{D}_p(x\,;C) \quad \text{for all} \quad x \in C;$$

here $p := \log 2 / \log 3$ is the Hausdorff dimension of C.

The examples presented make the following definitions natural.

Let $S \subset \mathbb{R}^n$ be a p-set. A point $x \in S$ is said to be a *regular* point of S if $D_p(x\,;S)$ exists and equals 1; otherwise x is an *irregular* point. A p-set is called *regular* if $\mathcal{H}_p$ almost all its points are regular, and *irregular* if $\mathcal{H}_p$ almost all its points are irregular.

4.2. Covering metric invariants

The fundamental theorem presented below shows that there are, in a sense, only two types of p-sets – regular and irregular. The main steps in its proof were done by Besicovich, Mastrand and Mattila within about the last fifty years. In the formulation below, we call a p-set $S \subset \mathbb{R}^n$ *countably p-rectifiable* if there is a countable family of Lipschitz functions $f_i : \mathbb{R}^p \to \mathbb{R}^n$ such that

$$\mathcal{H}_p(S \setminus \cup f_i(\mathbb{R}^p)) = 0.$$

Theorem 4.13. *Let $S \subset \mathbb{R}^n$ be a p-set. Then the following holds:*

(a) *S is irregular unless p is an integer.*

(b) *Let p be an integer. Then the subset of regular points in S forms a regular set; the subset of irregular points in S forms an irregular p-set.*

(c) *S is regular if and only if p is an integer and S is countably p-rectifiable.*

Recently Preiss [Pr-1987] discovered a new powerful method for establishing this and much more general results. The main point of this approach is the fundamental concept of a tangent measure which is introduced and studied in his paper.

Ahlfors regular metric spaces

In Chapter 9, we will deal with a class of thin subsets of $\mathbb{R}^n$ which possess a property with the following general description.

Definition 4.14. A metric space $\mathcal{M}$ is called *Ahlfors p-regular* ($p \geq 0$) if it supports a Radon measure μ satisfying, for some constants $\lambda_0, \lambda_1, r_0 > 0$ and every closed ball $\overline{B}$ of radius $r \leq r_0$, the inequality

$$\lambda_0 r^p \leq \mu(\overline{B}) \leq \lambda_1 r^p. \tag{4.12}$$

Let us recall that a *Radon measure* is a (regular) Borel measure uniquely determined by its values at compact subsets. In particular, for every open set $O \subset \mathcal{M}$,

$$\mu(O) = \sup\{\mu(K) ;\ K \subset O \quad \text{and} \quad K \text{ is compact}\}. \tag{4.13}$$

By the definition of the Hausdorff dimension, for every closed ball $\overline{B}$ of an Ahlfors p-regular space $\mathcal{M}$,

$$\dim_H \overline{B} = p.$$

If, in addition, $\mathcal{M}$ is the union of a countable family $\{\overline{B}_i\}$ of closed balls, e.g., $\mathcal{M}$ is separable or bounded, then

$$\dim_H \mathcal{M} = \sup_i \dim_H \overline{B}_i = p,$$

by Theorem 4.12 (b).

The following result shows that the definition of an Ahlfors p-regular space does not depend on the choice of measure μ.

Proposition 4.15. (a) *Let μ_1, μ_2 be measures on $\mathcal{M}$ satisfying the conditions of Definition 4.14. Then they are equivalent, i.e.,*
$$a\mu_1 \leq \mu_2 \leq b\mu_2$$
for some constants $a, b > 0$.

(b) *Let $\mathcal{M}$ be a proper metric space, and let measure μ satisfy the condition of Definition 4.14. Then μ is equivalent to the Hausdorff measure $\mathcal{H}_p$.*

For the proof see, e.g., Chapter II of the book [JW-1984] by Jonsson and Wallin.

Corollary 4.16. *A subspace $\mathcal{M}$ of a Euclidean space is Ahlfors p-regular if and only if (4.12) holds for μ being the restriction of $\mathcal{H}_p$ to $\mathcal{M}$.*

An easy consideration leads to the next example.

Example 4.17. (a) The classical Cantor set C is p-regular with $p = \frac{\log 2}{\log 3}$.

(b) A p-rectifiable subset of $\mathbb{R}^n$ is p-regular.

(c) The Heisenberg group H_n, see (3.42) and the following text, is $(2n+1)$-regular.

4.2.3 Hausdorff dimension of doubling metric spaces

Let $(\mathcal{M}, d)$ be a doubling metric space. We show that its Hausdorff dimension is finite. We prove this by comparing $\dim_H \mathcal{M}$ with another metric invariant, the *Assouad dimension* $\dim_A \mathcal{M}$, introduced in 1977 under the name of metric dimension, see [As-1980]. The concept is defined as follows.

We say that a subset $S \subset \mathcal{M}$ is (a, b)-*separated* if for every pair of its points $m \neq m'$,
$$a \leq d(m, m') \leq b.$$

Then $\dim_A \mathcal{M}$ is the infimum of all $p \in [0, +\infty]$ such that for some $C \geq 0$ and all $a < b$ every (a, b)-separated set has cardinality at most $C\left(\frac{b}{a}\right)^p$.

This immediately implies that for every $p > \dim_A \mathcal{M}$ and every $0 < \varepsilon < R$ the cardinality of ε-net in a ball of radius R in $\mathcal{M}$ is bounded by $O\left(\left(\frac{R}{\varepsilon}\right)^p\right)$. Then due to Proposition 3.19, for some $C \geq 1$ and every pair ε, R with $0 < \varepsilon \leq R$ and every $m \in \mathcal{M}$,

$$\text{Cov}\left(\overline{B}_R(m)\,;\,\varepsilon\right) \leq C\left(\frac{R}{\varepsilon}\right)^p. \tag{4.14}$$

If $\delta_\mathcal{M}$ is the doubling constant of $\mathcal{M}$, then by (3.41),

$$\dim_A \mathcal{M} \leq \log_2 \delta_\mathcal{M},$$

4.2. Covering metric invariants

i.e., the Assouad dimension of a doubling metric space is finite.

Let us show that

$$\dim_H \mathcal{M} \leq \dim_A \mathcal{M}. \tag{4.15}$$

First let $\mathcal{M}$ be compact. It will be proved in the next section that there is a finite (doubling) measure μ on $\mathcal{M}$ such that, given p satisfying (4.14), for all closed balls $\overline{B}_r(x) \subset \mathcal{M}$ and all $\lambda \geq 1$,

$$\mu(\overline{B}_{\lambda r}(x)) \leq c\lambda^p \mu(\overline{B}_r(x))$$

with some constant $c > 0$ independent of r and x. This immediately implies (for $r \leq 1$) that

$$\mu(\overline{B}_r(x)) \geq c^{-1} r^p \mu(\overline{B}_1(x)) \geq \tilde{c} r^p, \tag{4.16}$$

where $\tilde{c} := c^{-1} \inf_{x \in \mathcal{M}} \mu(\overline{B}_1(x))$. The latter is strongly positive, since for otherwise $\mu(\overline{B}_{\frac{1}{2}}(x_0)) \leq \mu(\overline{B}_1(x_j)) \to 0$ as $j \to \infty$ for some sequence of points x_j converging to x_0 (use compactness). This yields, for $R > \frac{1}{2}$,

$$\mu(\overline{B}_R(x_0)) \leq c(2R)^p \mu(\overline{B}_{\frac{1}{2}}(x_0)) = 0,$$

which contradicts the definition of a doubling measure.

Using a standard covering argument, see, e.g., [Fal-1999, pp. 51–52], we derive from (4.16) that

$$\mathcal{H}_p(\mathcal{M}) \leq 2^p \mu(\mathcal{M})/\tilde{c} < \infty.$$

Since p is an arbitrary number greater than $\dim_A \mathcal{M}$, this implies (4.15) for compact doubling spaces $\mathcal{M}$.

Now, let $\mathcal{M}$ be an arbitrary doubling space and $\mathcal{M}^a$ be its completion. By the definition of $\dim_H$,

$$\dim_H \mathcal{M} = \dim_H \mathcal{M}^a$$

and we may assume that $\mathcal{M}$ is complete. Then the closed ball $\overline{B} \subset \mathcal{M}$ is compact and therefore

$$\dim_H \overline{B} \leq \dim_A \overline{B} \leq \dim_A \mathcal{M}.$$

Since $\mathcal{M}$ is a countable union of such balls, we conclude, see Theorem 4.12 (b), that

$$\dim_H \mathcal{M} \leq \dim_A \mathcal{M}.$$

Fractals

The class of these subsets of $\mathbb{R}^n$ has no generally accepted definition. Its typical representatives are the Cantor, von Koch and Sierpiński sets, graphs of nowhere differentiable functions, Julia and Fatou sets of complex plane dynamics and paths of Brownian motion. All of these sets have common features which make the concept of fractal intuitively clear. A formal definition proposed by Mandelbrot calls a set of $\mathbb{R}^n$ a fractal if its topological dimension is strictly less than that of Hausdorff's. Unfortunately, this definition excludes too many sets which undoubtedly should be regarded as fractals (e.g., the graph of the Van-der-Waerden nowhere differentiable function has both dimensions equal to 1). Nevertheless, sets of $\mathbb{R}^n$ satisfying Mandelbrot's criteria should be definitely regarded as fractals, and one may use his criterion as a robust test. It may evidently be extended to a metric space; from this point of view, the Heisenberg group is a fractal, since $\dim_H H_n = 2n + 2 > \dim H_n = 2n + 1$.

We begin with a remarkable class of fractals called *self-similar sets*. Roughly speaking, a set is self-similar if it can be split into parts which are geometrically similar to the whole set.

To give a precise definition, we recall that a map $S : \mathbb{R}^n \to \mathbb{R}^n$ is said to be a *similitude* if, for some $r \in (0, 1)$ and $x, y \in \mathbb{R}^n$,

$$\|S(x) - S(y)\| = r\|x - y\|;$$

here $\|\cdot\|$ is the standard Euclidean norm of $\mathbb{R}^n$.

Definition 4.18. A compact set $K \subset \mathbb{R}^n$ is said to be self-similar with respect to the collection $\mathcal{S} := \{S_1, \ldots, S_N\}$ of similitudes if

(a) K is $\mathcal{S}$-invariant, i.e.,

$$K = \bigcup_{j=1}^{N} S_j(K); \qquad (4.17)$$

(b) for some $p \in [0, n]$,

$$\dim_H K = p \text{ and } \mathcal{H}_p(S_i(K) \cap S_j(K)) = 0 \text{ for } i \neq j. \qquad (4.18)$$

The following result of Hutchinson [Hut-1981] establishes existence of $\mathcal{S}$-invariant sets. In its formulation, one sets for $U \subset \mathbb{R}^n$,

$$\mathcal{S}(U) := \bigcup_{j=1}^{N} S_j(U).$$

We consider $\mathcal{S}$ as a map of the set of subsets of $\mathbb{R}^n$ into itself.

Theorem 4.19. *For every finite collection of similitudes $\mathcal{S}$ there exists an $\mathcal{S}$-invariant compact set K.*

Proof. Since all S_j are contractions, there is a closed Euclidean ball $\overline{D} \subset \mathbb{R}^n$ such that for each of its closed subsets U, $\mathcal{S}(U)$ is also a subset of $\overline{D}$. In other words, $\mathcal{S}$

4.2. Covering metric invariants

maps into itself the metric space $(\mathcal{B}(\overline{D}), d_\mathcal{H})$ of all closed subsets of $\overline{D}$ equipped with the Hausdorff metric, see (1.31) and (1.32). The classical Hausdorff theorem asserts that $(\mathcal{B}(\overline{D}), d_\mathcal{H})$ is compact (cf. Theorem 3.58). An easy evaluation shows that $\mathcal{S}$ is a contraction of this space, i.e., for some $\lambda < 1$ and for every $U, U' \subset \overline{D}$,

$$d_\mathcal{H}(\mathcal{S}(U), \mathcal{S}(U')) \leq \lambda d_\mathcal{H}(U, U').$$

Applying the Banach–Cacciopoli fixed point theorem one obtains a unique compact set $K \subset \overline{D}$ satisfying $\mathcal{S}(K) = K$. □

Remark 4.20. (a) The proof also yields

$$K = \lim_{j \to \infty} \mathcal{S}^j(U) \tag{4.19}$$

where U is an arbitrary nonempty compact subset of $\overline{D}$ such that $\mathcal{S}_j(U) \subset U$ for each j and the limit is taken with respect to $d_\mathcal{H}$.

(b) The result remains true for S_j being contractions; (4.19) also holds for this case.

The following simple condition provides self-similarity of $\mathcal{S}$-invariant compact sets.

Separation condition

$\mathcal{S} := \{S_1, \ldots, S_N\}$ satisfies the *open set condition* if for some open set $O \subset \mathbb{R}^n$,

$$\mathcal{S}(O) \subset O \quad \text{and} \quad \mathcal{S}_i(O) \cap \mathcal{S}_j(O) = \emptyset \quad \text{for} \quad i \neq j. \tag{4.20}$$

The following theorem is essentially contained in Moran's paper [Mor-1946], see also Hutchinson [Hut-1981].

Theorem 4.21. *If a set $\mathcal{S} := \{S_1, \ldots, S_N\}$ of similitudes in $\mathbb{R}^n$ satisfies the open set condition, then the Hausdorff dimension p of an $\mathcal{S}$-invariant compact set K is determined by the equation*

$$\sum_{j=1}^{N} r_j^p = 1, \tag{4.21}$$

where $r_j := L(S_j)$, $1 \leq j \leq N$. Moreover, $0 < \mathcal{H}_p(K) < \infty$.

Example 4.22 (Classical fractals). For the Cantor set C, see, e.g., Example 4.9, the required collection $\mathcal{S}$ of similitudes of $\mathcal{R}$ consists of maps S_1, S_2 given by

$$S_1(x) = \frac{1}{3}x, \quad S_2(x) = \frac{1}{3}(x+2), \quad x \in \mathbb{R},$$

and the open set condition holds for $O := (0, 1)$. Then $\dim_H C$ is determined by (4.21) with $r_1 = r_2 = \frac{1}{3}$ and this yields $\dim_H C = \log 2/\log 3$ and $0 < \mathcal{H}_p(C) < \infty$ for $p = \log 2/\log 3$.

For the Sierpiński gasket G, see, e.g., Example 4.9, $\mathcal{S} := \{S_1, S_2, S_3\}$ consists of the r-homotheties with $r := \frac{1}{2}$ centered at the vertices of the equilateral triangle, and the open set O is an interior of this triangle. In this case, (4.21) yields $\dim_H G = \log 3 / \log 2$ and $0 < \mathcal{H}_p(G) < \infty$ for the critical p.

The reader may check in this way (or consult [Fal-1999]) that for the von Koch snowflake curve K, see Figure 4.2, Hausdorff dimension is $\log 4 / \log 3 > \dim K = 1$.

Our last example concerns the classical *Antoine Necklace* [Ant-1921] (denoted here by A_3), a zero-dimensional subset of $\mathbb{R}^3$ "wildly" embedded into $\mathbb{R}^3$. For its description we define a *solid torus* as a closed subset, say T, of $\mathbb{R}^3$ obtained by rotation of the closed two-dimensional Euclidean ball (disk) D about an axis ℓ lying in the plane of D and not intersecting D. The points of D closest to ℓ then form the *inner circle* of T; it bounds the disc called the *inner disk* of T. The radii of the inner circle and of the initial disk D uniquely determine this torus; we call them the *parameters* of T.

Two solid tori T_1 and T_2 are *linked* if $T_1 \cap T_2 = \emptyset$ and one of them intersects the inner disk of the other. Further, the family of solid tori $T_1, \ldots, T_N$ forms a *closed chain* if they are pairwise disjoint and T_i, T_{i+1} are linked for every $i = 1, \ldots, N$; here we set $T_{N+1} := T_1$.

Now the required Antoine's set A_3 is constructed as follows. Fix a solid torus T_1^1 and embed into the interior of T_1^1 the union of a closed chain of solid tori T_j^2, $1 \leq j \leq N$, which winds around the axis of T_1^1 a number of times. We construct the chain so that for every T_j^2 there is a similitude S_j of $\mathbb{R}^3$ mapping T_1^1 onto T_j^2 and such that there is a fixed Lipschitz constant $t \in (0,1)$ for all S_j, $1 \leq j \leq N$.

The Antoine Necklace A_3 is the $\mathcal{S}$-invariant set for $\mathcal{S} := \{S_1, \ldots, S_N\}$. Hence, we have

$$A_3 := \bigcap_{j=1}^{\infty} C_j$$

where $C_1 := T_1^1$, $C_2 := \bigcup_{j=1}^N T_j^2$, $C_3 := \bigcup_{i,j=1}^N T_{ij}^3$ with $T_{ij}^3 := S_j(T_i^2)$, and so on.

The parameters of every torus of level j tend to zero as $j \to 0$. Therefore the criterion of Theorem 4.8 and Theorem 4.6 (c) imply that

$$\dim A_3 = 0.$$

The Hausdorff dimension of A_3 depends on the choice of the (common) Lipschitz constant t of the similitudes $\mathcal{S}_j$ and the number N of tori T_j^2. Using a variant of Moran's theorem, Rushing [Rush-1992] proved that for every $s \in (1,3)$ there exists the Antoine Necklace A_3 such that

$$\dim_H A_3 = s,$$

4.2. Covering metric invariants

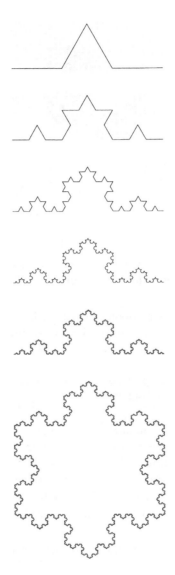

Figure 4.2: At each stage, the middle third of every interval is replaced by the other two sides of an equilateral triangle to obtain as limit the von Koch curve. Three von Koch curves fitted together form a snowflake curve.

where the parameters of A_3 satisfy the relation $Nt^s = 1$.

The restriction $1 < s < 3$ reflects some peculiarities of the construction. In particular, the chain $\{T_j^2\}_{1 \leq j \leq N}$ becomes too small and cannot wind around T_1^1

Figure 4.3: Antoine necklace.

if we try to accomplish the construction for $s < 1$, and N tends to infinity as s approaches 3 and, therefore, the number of turns of winding $\{T_j^2\}_{1 \leq j \leq N}$ around T_1^1 tends to infinity.

As for $s = 1$, there is a topological obstacle for A_3 to be of Hausdorff dimension 1. Actually, it was proved by Martio, Rickman and Väisälä [MRV-1971, Lemma 3.3] (see also Semmes [Sem-1996, Lemma 1.4]) that a zero-dimensional set S in $\mathbb{R}^3$ of Hausdorff dimension 1 has simply connected complement $\mathbb{R}^3 \setminus S$. But this contradicts the main property of the Antoine Necklace A_3 for which the fundamental group of $\mathbb{R}^3 \setminus A_3$ is nontrivial.

Remark 4.23. The construction of the Antoine Necklace is a variation of the original Antoine construction due to Blankenship [Blan-1951]. In this paper, the Antoine construction is generalized to $\mathbb{R}^n$, and for the Antoine set A_n obtained in this way, the fundamental group of $\mathbb{R}^n \setminus A_n$ is computed.

Rushing [Rush-1992] used the Blankenship construction to prove that for a judicious choice of parameters, $\dim_H A_n$ becomes a given s in the interval $(n-2, n)$, $n \geq 3$.

Once again, there is a topological obstacle for a zero-dimensional compact subset of $\mathbb{R}^n$, $n \geq 3$, to be of Hausdorff dimension less than $n - 2$, see the cited paper.

Finally, we present two important examples of fractal sets which have Hausdorff p-measure zero or infinity for the critical value of p.

Example 4.24 (Brownian motion). Brownian motion is the frequent, random fluctuation of particles suspended in some fluid. Its mathematical model is the *Wiener stochastic process* $\mathcal{W} := \{W_t\,;\, 0 \leq t \leq 1\}$ on the probability space $C_0\big([0,1]\big) := \{\omega \in C[0,1]\,;\, \omega(0) = 0\}$ with the *Wiener measure* w. The latter is uniquely determined by its values on cylinders:

4.2. Covering metric invariants

For all $0 \leq t_1 < \cdots < t_n \leq 1$ and $a_i < b_i$, $1 \leq i \leq n$,

$$w\bigl(\omega \in C_0([0,1])\,;\, a_i \leq \omega(t_i) \leq b_i,\ i = 1, \ldots, n\bigr)$$
$$:= \int_{a_1}^{b_1} \cdots \int_{a_n}^{b_n} \frac{1}{\sqrt{2\pi(t_i - t_{i-1})}} \exp\left(-\frac{(x_i - x_{i-1})^2}{t_i - t_{i-1}}\right) dx_1 \cdots dx_n,$$

where conventionally we set $x_0 := 0$.

Then the Wiener process is defined by the evolutions $W_t(\omega) := \omega(t)$.

It is known, see, e.g., [Fal-1999] for details and references, that for w almost all $\omega \in C_0([0,1])$ continuous plane curve $\gamma_\omega : t \mapsto (t, W_t(\omega))$, $0 \leq t \leq 1$, has Hausdorff dimension 2 (and topological dimension 1) but, nevertheless, $\mathcal{H}_2\bigl(\gamma_\omega([0,1])\bigr) = \infty$.

Example 4.25 (Fatou and Julia sets). Let $\mathcal{F}$ be a family of functions holomorphic in an open subset U of the Riemann sphere. $\mathcal{F}$ is said to be *normal* if every infinite sequence in $\mathcal{F}$ has a subsequence uniformly convergent on every compact subset of U.

Now let f be a holomorphic map of a domain $X \subset \mathbb{C}$ into itself. Let $\{f^{\langle j \rangle}\}_{j \in \mathbb{Z}_+}$ be the sequence of its iterates, i.e., $f^{\langle 0 \rangle} := f$ and $f^{\langle j+1 \rangle} := f \circ f^{\langle j \rangle}$. Then the *Fatou set* $\mathrm{Fat}(f)$ of f is the domain of normality of $\{f^{\langle j \rangle}\}$; in other words, $z_0 \in \mathrm{Fat}(f)$ if there is an open neighborhood U of z_0 such that $\{f^{\langle j \rangle}|_U\}$ is a normal family.

Further, the complement of $\mathrm{Fat}(f)$ in X is called the *Julia set* and is denoted by $\mathrm{Jul}(f)$.

Even for relatively simple holomorphic functions the Fatou and Julia sets have complicated fractal structures. For instance, the classical Fatou result asserts that for $f(z) = z^2 + c$, $z \in \mathbb{C}$, with real $c > \frac{1}{4}$, $\mathrm{Jul}(f)$ is a Cantor set.

Let now $f_\lambda(z) := \lambda e^z$, $z \in \mathbb{C}$. Then the following is true (see MacMillan [McM-1987]):

Julia set of f_λ for $0 < \lambda < \frac{1}{2}$ has Hausdorff dimension 2 but its Hausdorff 2-measure is zero.

The reader may find a rather complete account of topics in Complex Variable Dynamics and many nice pictures of fractal Fatou and Julia sets in Milnor's book [Mil-2000].

4.2.4 Nagata dimension

The invariant considered in this section was introduced under this name by P. Assouad [As-1982] who referred to J. Nagata's paper [Nag-1958] where this invariant was implicitly presented.

Definition 4.26. The Nagata (–Assouad) dimension of a metric space $\mathcal{M}$, denoted by $\dim_N \mathcal{M}$, is the least integer n with the following property.

For some constant $c > 0$ and every $t > 0$ there is a cover $\mathcal{U}$ of $\mathcal{M}$ such that $\mathrm{diam}\,\mathcal{U} \leq ct$ and every subset $S \subset \mathcal{M}$ of diameter at most t meets at most $n + 1$ subsets of $\mathcal{U}$.

In the sequel, we call a cover obeying the conditions of Definition 4.42 a (c, n, t)-*cover of* $\mathcal{M}$.

It may be easily seen that $\dim_N \mathcal{M}$ does not change if the covering sets are assumed to be either open or closed. The "test set" S may also be required to be an open or closed ball of radius $t/2$ (the constant c may increase in this case). It also follows from the definition that $\dim_N$ is a bi-Lipschitz invariant.

The basic properties of the Nagata dimension were established by Lang and Schlichenmaier whose paper [LSchl-2005] we will follow. Our first result relates the Nagata dimension to the Lipschitz maps of metric spaces to *metric polyhedral complexes*. As in the case of Euclidean polyhedral complexes, see Definition B.6 and the related text in Chapter 1, the basic block of this construction is a *metric polyhedral cell* (briefly, *cell*) meaning a subset of a metric space isometric to some polyhedral cell. Hence, we may speak about *faces, vertices* and *edges* of such a subset, and about its (relative) *interior, boundary* and *dimension*. If C is such a cell, then we use the symbols $C^\circ, \partial C$ and $\dim C$ for the last three notions.

Now the aforementioned basic concept is given by

Definition 4.27. A metric polyhedral complex (briefly, complex) is a pair $(\mathcal{M}, \mathcal{C})$ of a metric space $\mathcal{M}$ and its cover $\mathcal{C}$ such that:

(i) every $C \in \mathcal{C}$ is a cell;

(ii) for any $C \in \mathcal{C}$ each of its faces belongs to $\mathcal{C}$;

(iii) every two cells of $\mathcal{C}$ are either disjoint or intersect in a common face.

A *subcomplex* of a complex $(\mathcal{M}, \mathcal{C})$ is a complex $(\mathcal{M}_1, \mathcal{C}_1)$ such that $\mathcal{M}_1$ is a subspace of $\mathcal{M}$, and $\mathcal{C}_1$ is a subset of $\mathcal{C}$ which covers $\mathcal{M}_1$.

Now let $K := (\mathcal{M}, \mathcal{C})$ be a complex. We define its *barycentric subdivision* $\widehat{K} := (\mathcal{M}, \widehat{\mathcal{C}})$ beginning with the *barycentric subdivision of a cell* $C \subset \mathcal{M}$. Namely, let $i : C \to P$ be the isometry onto the associated polyhedron P. Then the barycentric subdivision $\widehat{P}$ of P, see the text after Definition B.6, gives rise to a complex $(C, \widehat{C})$, where the cover $\widehat{C}$ of C is given by

$$\widehat{C} := \{i^{-1}(S)\,;\, S \in \widehat{P}\}.$$

Consider now cells $C_1, C_2 \in \mathcal{C}$ with $C_1 \cap C_2 \neq \emptyset$. Let $i_k : C_k \to P_k$ be the corresponding isometry onto the associated polyhedron P_k, $k = 1, 2$. Due to Definition 4.27, $C_1 \cap C_2$ is a face, i.e., $F_k := i_k(C_1 \cap C_2)$ is a face of P_k, $k = 1, 2$, and F_1 and F_2 are isometric. Then the barycenters of these faces define a unique point of the face $C_1 \cap C_2$ called the *barycenter* of the face. Therefore the barycentric subdivision $\widehat{F}_1$ of F_1 determines a complex $(C_1 \cap C_2, \widehat{C_1 \cap C_2})$, where the cover $\widehat{C_1 \cap C_1}$ of $C_1 \cap C_2$ is $\{i_1^{-1}(S)\,;\, S \in \widehat{F}_1\} (= \{i_2^{-1}(S)\,;\, S \in \widehat{F}_2\})$. We conclude that this complex is a subcomplex both of $(C_1, \widehat{C}_1)$ and of $(C_2, \widehat{C}_2)$. Hence, the barycentric subdivisions of cells $C \in \mathcal{C}$ agree on intersections and determine the barycentric subdivision $\widehat{K} := (\mathcal{M}, \widehat{\mathcal{C}})$ of the complex $K := (\mathcal{M}, \mathcal{C})$. Note that all cells of $\widehat{K}$ are *metric simplices*, i.e., every cell $C \in \widehat{\mathcal{C}}$ is isometric to a Euclidean simplex. In the sequel such a complex is called *simplicial* and its cells are briefly called *simplices*.

Finally, we define the dimension of $K := (\mathcal{M}, \mathcal{C})$ by

$$\dim K := \sup\{\dim C \, ; \, C \in \mathcal{C}\},$$

and the *star of K at a point $m \in \mathcal{M}$* by

$$\operatorname{St}(m, K) := \bigcup_{C \ni C \ni m} C^\circ.$$

Now we are ready to formulate and prove the aforementioned characterization of the Nagata dimension.

Proposition 4.28. *Let $\mathcal{M}$ be a metric space. The following assertions are equivalent.*

1. *For some constant $c_1 > 0$ and every $t > 0$ there is a cover $\mathcal{U}$ of diameter at most $c_1 t$ such that every subset $S \subset \mathcal{M}$ of diameter at most t meets at most $n+1$ sets of $\mathcal{U}$.*

2. *For some constant $c_2 > 0$ and every $t > 0$ there exists a 1-Lipschitz map f of $\mathcal{M}$ into the metric space $\mathcal{M}_1$ of a complex $K_1 := (\mathcal{M}_1, \mathcal{C}_1)$ of dimension at most n such that:*

    (a) *every finite subcomplex of K_1 is isometric to a subcomplex of some regular Euclidean simplex[1] of edge length t;*

    (b) *for every vertex v of K_1,*

    $$\operatorname{diam} f^{-1}\bigl(\operatorname{St}(v, K_1)\bigr) \leq c_2 t.$$

3. *For some constant $c_3 > 0$ and all $t > 0$ there exists a map g of $\mathcal{M}$ into the metric space $\mathcal{M}_2$ of a complex $K_2 := (\mathcal{M}_2, \mathcal{C}_2)$ of dimension at most n such that:*

    (a) *every open ball of radius t in $\mathcal{M}_2$ is contained in a star $\operatorname{St}(v, \widehat{K}_2)$ of some vertex v from the barycentric subdivision $\widehat{K}_2$;*

    (b) *for every such vertex v,*

    $$\operatorname{diam} g^{-1}\bigl(\operatorname{St}(v, \widehat{K}_2)\bigr) \leq c_3 t.$$

4. *For some constant $c_4 > 0$ and all $t > 0$ the space $\mathcal{M}$ admits a cover $\mathcal{U}$ of diameter at most $c_4 t$ which is the union of $n+1$ families of subsets $\mathcal{F}_k$, $0 \leq k \leq n$, such that every $S \subset \mathcal{M}$ of diameter t meets at most one set from each of $\mathcal{F}_k$, $0 \leq k \leq n$.*

---
[1] i.e., a subcomplex formed by some faces of this simplex.

Proof. Clearly, (4) implies (1) with $c_1 = c_4$.

Show that (1) $\Rightarrow$ (2). Given $t > 0$, set

$$r := ct, \quad \text{where} \quad c := \sqrt{2}(4n+4)^{-2}.$$

Find a cover $\mathcal{U} := \{U_\alpha\}_{\alpha \in A}$ of the space $\mathcal{M}$ satisfying the conditions of (1) with t replaced by r. For every $\alpha \in A$, define a function $\varphi_\alpha : \mathcal{M} \to \mathbb{R}_+$ by

$$\varphi_\alpha(m) := \sup\left\{0, \frac{r}{2} - d(m, U_\alpha)\right\}.$$

By definition, the family $\{\varphi_\alpha\}_{\alpha \in A}$ satisfies:

(i) $\varphi_\alpha = \frac{r}{2}$ on U_α and equals zero outside the set

$$\widetilde{U}_\alpha := \left\{m \in \mathcal{M}; \, d(m, U_\alpha) < \frac{r}{2}\right\}.$$

Since $\operatorname{diam} \mathcal{U} \le c_1 r$, we have

$$\operatorname{diam} \widetilde{U}_\alpha \le (c_1 + 1)r.$$

(ii) Every φ_α is 1-Lipschitz.

(iii) Since every ball $B_{\frac{r}{2}}(m)$ meets at most $n+1$ sets U_α, for every $m \in \mathcal{M}$ one gets

$$\operatorname{card}\{\alpha \in A; \, \varphi_\alpha(m) > 0\} \le n+1.$$

In particular, the sum

$$\overline{\varphi}(m) := \sum_{\alpha \in A} \varphi_\alpha(m)$$

is finite for all m; moreover, $\overline{\varphi} \ge \frac{r}{2}$.

Now we consider a map from $\mathcal{M}$ into the Hilbert space $\ell_2(A)$ given by

$$f(m) := \left(\frac{r}{(4(n+1))^2} \varphi_\alpha(m)\right)_{\alpha \in A}.$$

Due to (ii) and (iii), f is a 1-Lipschitz function whose image $f(\mathcal{M})$ is contained in the set

$$\left\{y := (y_\alpha)_{\alpha \in A} \in \ell_2(A); \, \sum_\alpha y_\alpha = (4(n+1))^{-2} r \quad \text{and} \quad \operatorname{card}(\operatorname{supp} y) \le n+1\right\}.$$

In other words, $f(\mathcal{M})$ is contained in the n-skeleton Σ_n of the infinite-dimensional simplex $\Sigma \subset \ell_2(A)$ determined by the equation $\sum_{\alpha \in A} y_\alpha = (4(n+1))^{-2} r$.

Now define the required space $\mathcal{M}_1$ as a (metric) subspace of $\ell_2(A)$ given by

$$\mathcal{M}_1 := \cup S,$$

4.2. Covering metric invariants

where S runs over the set $\mathcal{C}_1$ of all simplices of Σ_n intersecting $f(\mathcal{M})$. In this way we obtain the simplicial complex $K_1 := (\mathcal{M}_1, \mathcal{C}_1)$ and the 1-Lipschitz map $f : \mathcal{M} \to \mathcal{M}_1$. By the definition of $\mathcal{C}_1$, each of its finite subcomplexes is isometric to a subset of $\ell_2(A)$ which is the union of the sets

$$\Sigma_J := \Big\{ y \in \ell_2(A) \,;\, \mathrm{supp}\, y = J \ \text{and}\ \sum_{\alpha \in J} y_\alpha = \big(4(n+1)\big)^{-2} r \Big\},$$

where J runs over all nonempty subsets of A of a fixed cardinality $m \leq n+1$.

Clearly, each Σ_J is isometric to the $m(= \mathrm{card}\, J)$-dimensional Euclidean simplex $\big\{ x \in \mathbb{R}^{m+1} \,;\, \sum_{i=1}^{m+1} x_i = \big(4(n+1)\big)^{-2} r \big\}$ of edge length $\sqrt{2}\big(4(n+1)\big)^{-2} r$. Hence, the complex K_1 possesses the required properties.

It remains to estimate the diameter of $f^{-1}\big(\mathrm{St}(v, K_1)\big)$ for every vertex v of K_1. By the definition of f, for every such vertex v, there is a unique $\alpha \in A$ such that $\varphi_\alpha(v) > 0$. Since $\mathrm{St}(v, K_1) := \bigcup_{v \in S} S^\circ$, every term of this union satisfies

$$\mathrm{diam}\, f^{-1}(S^\circ) \leq \mathrm{diam}(\mathrm{supp}\, \varphi_\alpha) \leq \mathrm{diam}\, \widetilde{U}_\alpha \leq (c_1+1)r = c_2 t,$$

where c_2 depends only on c_1 and n.

Implication $(1) \Rightarrow (2)$ is established.

To show that $(2) \Rightarrow (3)$ we need

Lemma 4.29. *Let $K := (\mathcal{M}, \mathcal{C})$ be a simplicial complex, and $\dim K \leq n$. Assume that for some constants $\delta, \vartheta > 0$ and every pair of simplices $S, S' \in \mathcal{C}$, the following holds:*

(A) $d(S, S') \geq \delta$ whenever $S \cap S' = \emptyset$;

(B) for otherwise, for all $m \in S$ and $m' \in S'$,
$$d(m, S \cap S') \leq \vartheta d(m, m').$$

Then there is a constant $c = c(\vartheta, n) > 0$ such that every ball $B_{c\delta}(m) \subset \mathcal{M}$ is contained in the star $\mathrm{St}(v, K)$ of some vertex v of K.

Proof. Given $m \in \mathcal{M}$, introduce the family

$$\mathcal{S}_m := \big\{ S \in \mathcal{C} \,;\, S^\circ \cap B_{c\delta}(m) \neq \emptyset \big\},$$

where the constant c will be determined later. We claim that $\bigcap_{S \in \mathcal{S}_m} S \neq \emptyset$; this implies the required assertion, since $B_{c\delta}(m)$ is clearly a subset of the star $\mathrm{St}(v, K)$ for every vertex v contained in this intersection.

Since $\dim K \leq n$, it suffices to prove that for every subfamily $\{S_i\}_{1 \leq i \leq n+2} \subset \mathcal{S}_m$ we will have

$$\bigcap_{i=0}^{n+1} S_i \neq \emptyset. \tag{4.22}$$

In fact, fix a simplex $S_0 \in \mathcal{S}_m$. Identifying S_0 with the associated Euclidean simplex of $\mathbb{R}^n$ we can treat the family $\mathcal{S}_m \cap S_0$ as a family of convex subsets of a simplex in $\mathbb{R}^n$. By (4.22) every $n+1$ subset of this family has a nonempty intersection. Therefore $\cap \mathcal{S}_m = \cap(\mathcal{S}_m \cap S_0) \neq \emptyset$ by the Helly criterion, see Theorem 1.22.

To prove (4.22) we pick, for every index i, a point $m_i \in S_i \cap \overline{B}_{cr}(m)$ and denote by $k \in \{0, 1, \ldots, n+2\}$ the maximal index such that $\bigcap_{i=0}^{k} S_i \neq \emptyset$. We claim that for every $j \leq k$ there is a point m^j in the simplex $T^j := \bigcap_{i=0}^{j} S_i$ such that

$$d(m, m^j) \leq \vartheta_j c\delta,$$

where $\vartheta_1 := 1$ and $\vartheta_j = 1 + \vartheta + \vartheta\vartheta_{j-1}$ for $2 \leq j \leq k$.

We define these points inductively putting $m^1 := m_1$; due to the choice of m_i,

$$d(m_1, m^1) \leq c\delta = \vartheta_1 c\delta.$$

Now let the required points $m^j \in T^j$ be defined for $2 \leq j \leq k-1$. Then condition (B) provides a point m^j in the (nonempty) simplex $S_j \cap T^{j-1} = T^j$ such that

$$d(m_j, m^j) \leq \vartheta d(m_j, m^{j-1}).$$

Together with the choice of the points m_i and the induction hypothesis this yields

$$d(m, m^j) \leq d(m, m_j) + d(m_j, m^j) \leq (1+\vartheta)d(m, m_j) + \vartheta d(m, m^{j-1}) \leq \vartheta_j c\delta.$$

The claim is proved.

If now $k < n+1$ then $S_{k+1} \cap T^k = \emptyset$, and therefore $d(S_{k+1}, T^k) \geq \delta$ by condition (A). On the other hand, $d(m, m^k) \leq \nu_k c\delta < \nu_n c\delta$ and therefore

$$d(S_{k+1}, T^k) \leq d(m_{k+1}, m^k) \leq d(m, m_{k+1}) + d(m, m^k) < c\delta + \nu_n c\delta.$$

Choosing the constant c from the equation $c(1+\nu_n) = 1$ we obtain a contradiction. Hence $k = n+1$ and (4.22) holds. □

Now we prove that (2) implies (3). Given $t > 0$, set $r := \lambda t$, where the constant $\lambda = \lambda(n) > 0$ will be determined later. Due to (2), with t replaced by r, there is a 1-Lipschitz map $g : \mathcal{M} \to \mathcal{M}_1$, where $\mathcal{M}_1$ is the underlying metric space of a simplicial complex $K_1 := (\mathcal{M}_1, \mathcal{C}_1)$. Moreover, $\dim K_1 \leq n$, every finite subcomplex of K_1 is isometric to a subcomplex of some regular Euclidean simplex of edge length r, and for every vertex v of K_1,

$$\operatorname{diam} g^{-1}\bigl(\operatorname{St}(v_1, K_1)\bigr) \leq c_2 r.$$

We show that the conditions of Lemma 4.29 hold for the simplices of the barycentric subdivision $\widehat{K}_1 = (\mathcal{M}_1, \widehat{\mathcal{C}}_1)$ of K_1 with some $\delta = \varepsilon r$ and ϑ, where ε and ϑ are some positive constants depending only on n.

4.2. Covering metric invariants

Let S, S' be such simplices. Due to condition (a) of assertion (2) from the proposition, there is a regular Euclidean simplex Σ of edge length r such that S and S' are simplices from the barycentric subdivisions of some subsimplices from Σ of dimension $\leq n$, say $\widetilde{S}$ and $\widetilde{S}'$. Since Σ is regular, $\widetilde{S}$ and $\widetilde{S}'$ are contained in the regular simplex isometric to

$$\Sigma_r^{2n+1} := \left\{ x \in \mathbb{R}^{2n+2} ; \sum_{i=1}^{2n+2} x_i = \frac{r}{\sqrt{2}} \right\}.$$

Hence, S and S' may be regarded as simplices of the barycentric subdivision $\widehat{\Sigma}_r^{2n+1}$ of the simplex Σ_r^{2n+1}.

There are only finitely many of such pairs S, S'. An easy elementary geometric consideration shows that there are positive constants $\varepsilon = \varepsilon(n)$ and $\vartheta = \vartheta(n)$ such that

$$d(S, S') \geq \varepsilon r \qquad (4.23)$$

whenever $S \cap S' = \emptyset$ and

$$d(m, S') \leq \vartheta d(m, m') \quad \text{for all} \quad m \in S \text{ and } m' \in S$$

whenever $S \cap S' \neq \emptyset$.

Then (4.23) and the latter assertion show that the conditions of Lemma 4.29 hold for simplices in $\mathcal{C}$ with $\delta := \varepsilon r$. Due to this lemma, every open ball in $\mathcal{M}$ of radius $c\varepsilon r$ is contained in the star $St(v_1, \widehat{K}_1)$ of some vertex v of $\widehat{K}_1$. Choose now λ (from the relation $t = \lambda r$) to be equal to $c\varepsilon$; since $c = c(\vartheta, n)$, where $\vartheta = \vartheta(n)$, and $\varepsilon = \varepsilon(n)$, λ depends only on n.

By this choice, every ball $B_t(m) \subset \mathcal{M}$ is contained in $St(v, \widehat{K}_1)$ for a suitable vertex v of $\widehat{K}_1$. Moreover, by the definition of a barycentric subdivision, $St(v, \widehat{K}_1)$ is contained in $St(v', K_1)$ for some vertex v' of K_1. Thus

$$\operatorname{diam} g^{-1}\bigl(St(v, \widehat{K}_1)\bigr) \leq \operatorname{diam} g^{-1}\bigl(St(v', K_1)\bigr) \leq c_2 r =: c_3 t,$$

where $c_3 := c_2(c\varepsilon)^{-1}$ depends only on c_2 and n.

Hence, $(2) \Rightarrow (3)$.

To complete the proof it remains to show that $(3) \Rightarrow (4)$. Let $t > 0$ and $g : \mathcal{M} \to \mathcal{M}_2$ be given as in (3), i.e., $\mathcal{M}_2$ is the underlying metric space of a complex $K_2 := (\mathcal{M}_2, \mathcal{C}_2)$, and K_2 and g have the properties described in (3). For every vertex v in the barycentric subdivision $\widehat{K}_2$ of K_2, we define the set

$$U_v := \bigl\{ m \in \mathcal{M} ; B_t(g(m)) \subset St(v, \widehat{K}_2) \bigr\}.$$

Since every open ball of radius t is contained in one of such stars, the sets U_v form a cover of $\mathcal{M}$ denoted by $\mathcal{U}$. For every pair of points $m, m' \in U_v$,

$$d(m, m') \leq \operatorname{diam} g^{-1}\bigl(St(v, \widehat{K}_2)\bigr) \leq c_3 t,$$

i.e., $\operatorname{diam} U_v \leq c_3 t$ and
$$\operatorname{diam} \mathcal{U} \leq c_3 t.$$

Each vertex v is a barycenter of the unique cell C_v of K_2. For $k = 0, 1, \ldots, n$ define $\mathcal{U}^k$ as the family of all U_v such that $\dim C_v = k$. If $U_v, U_{v'} \in \mathcal{U}^k$ for some k and $v \neq v'$, then the intersection $C_v^\circ \cap C_{v'}^\circ = \emptyset$ and therefore
$$\operatorname{St}(v, \widehat{K}_2) \cap \operatorname{St}(v', \widehat{K}_2) = \emptyset.$$

Further, let $m \in U_v$ and $m' \in U_{v'}$. Since $B_t(g(m)) \subset \operatorname{St}(v, \widehat{K}_2)$, and the same is true for m' and v', the previous relation implies that
$$d(m, m') \geq d(g(m), g(m')) > t.$$

Hence, every subset $S \subset \mathcal{M}$ of diameter t can intersect at most one set $U_v \in \mathcal{U}^k$.

Thus, we have found the cover $\mathcal{U}$ of $\mathcal{M}$ of diameter $c_3 t$ which can be presented as the union of the families of $\mathcal{U}^k$, $0 \leq k \leq n$, with the required property.

Hence, (3) implies (4) with $c_4 = c_3$ and the proof of Proposition 4.28 is complete. □

The proposition just established and Theorem 4.8 lead to a fairly simple derivation of the basic facts concerning the Nagata dimension. This is given by

Theorem 4.30. (a) *It is true that*
$$\dim_N \left(\oplus^{(p)} (\mathcal{M}_i, d_i)_{1 \leq i \leq k} \right) \leq \sum_{i=1}^{k} \dim_N (\mathcal{M}_i, d_i).$$

(b) *If a metric space is the union of its subspaces S_i, $1 \leq i \leq k$, then*
$$\dim_N \left(\bigcup_{i=1}^{k} S_i \right) = \max_{1 \leq i \leq k} \dim_N S_i.$$

(c) *The topological dimension of a metric space is bounded by its Nagata dimension.*

Proof. (a) It suffices to consider the case of two metric spaces with $n_i := \dim_N \mathcal{M}_i < \infty$, $i = 1, 2$. According to Proposition 4.28 (2) there is, for a given $r > 0$, a simplicial complex $K^i := (\mathcal{M}^i, \mathcal{C}^i)$ of dimension at most n_i and a 1-Lipschitz map $f_i : \mathcal{M}_i \to \mathcal{M}^i$ such that every finite subcomplex of K^i is isometric to a subcomplex of some regular Euclidean simplex of edge length r; moreover, for every vertex v of K^i,
$$\operatorname{diam} f_i^{-1}(\operatorname{St}(v, K^1)) \leq c_2^i r, \quad i = 1, 2.$$

Here c_2^i depends only on n_i and may be replaced by $c_2 := \max_{i=1,2} c_2^i$.

4.2. Covering metric invariants

Define the direct p-sum $K := K^1 \oplus^{(p)} K^2$ as a polyhedral complex with underlying metric space $\mathcal{M}^1 \oplus^{(p)} \mathcal{M}^2$ and the cover $\mathcal{C} := \mathcal{C}^1 \times \mathcal{C}^2 := \{S^1 \times S^2 \,;\, S^i \in \mathcal{C}^i,\ i=1,2\}$. Clearly, the dimension of this complex is at most $n_1 + n_2$ and each of its finite subcomplexes is isometric to a subcomplex of the polyhedron defined by the product of some regular Euclidean simplices of edge length r. Arguing as in the derivation of statement (3) from statement (2) in Proposition 4.28 we find constants $\varepsilon > 0$ and $\vartheta > 0$ depending only on n_1 and n_2 such that the conditions of Lemma 4.29 are satisfied for the barycentric subdivision $\widehat{K}$ of the direct p-sum K with $\delta := \varepsilon r$ and this ϑ. By the lemma, every open ball of radius $t := c \varepsilon r$ is contained in the star $\operatorname{St}(v, \widehat{K})$ of some vertex v of $\widehat{K}$, where the constant $c > 0$ depends only on $n_1 + n_2$. A product map $g : \mathcal{M}_1 \oplus^{(p)} \mathcal{M}_2 \to \mathcal{M}^1 \oplus^{(p)} \mathcal{M}^2$ given by $g(m_1, m_2) := \big(f_1(m_1), f_2(m_2)\big)$ is clearly 1-Lipschitz. Further, every vertex v of $\widehat{K}$ is of the form (v^1, v^2) for some vertices v^i of K^i. Therefore $\operatorname{St}(v, \widehat{K})$ is contained in the direct product of the stars $\operatorname{St}(v^i, \widehat{K}^i)$, $i = 1, 2$. Consequently,

$$\operatorname{diam} g^{-1}\big(\operatorname{St}(v, \widehat{K})\big) \le 2^{\frac{1}{p}} \max_{i=1,2} \operatorname{diam} f_i^{-1}\big(\operatorname{St}(v^i, \widehat{K}^i)\big) \le 2^{\frac{1}{p}} c_2 r.$$

Hence the space $\mathcal{M}_1 \oplus^{(p)} \mathcal{M}_2$ satisfies the conditions of Proposition 4.28 (3). By assertion (1) of this proposition,

$$\dim_N\big(\mathcal{M}_1 \oplus^{(p)} \mathcal{M}_2\big) \le n_1 + n_2 = \dim_N \mathcal{M}_1 + \dim_N \mathcal{M}_2.$$

(b) It suffices to consider the case of two subspaces S_i, $i = 1, 2$, of a metric space $\mathcal{M}$ such that $\mathcal{M} = S_1 \bigsqcup S_2$. It is clear that

$$\max \dim_N S_i \le \dim_N \mathcal{M}. \tag{4.24}$$

It suffices to prove the converse inequality only for the case of

$$n := \max_{i=1,2} \dim_N S_i < \infty.$$

In the subsequent derivation a cover $\mathcal{U}$ is recalled to be a (c, n, t)-cover, if $\mathcal{U}$ satisfies the conditions of Definition 4.27, i.e., $\operatorname{diam} \mathcal{U} \le ct$ and every subset S of this metric space of diameter t meets at most $n + 1$ sets of the cover $\mathcal{U}$.

Now, by (4.24) and the definition of the Nagata dimension, there exist a (c_2, n, t)-cover $\mathcal{U}_2 := \{U^2_\alpha\}_{\alpha \in A_2}$ of S_2 and a $\big(c_1, n, (3 + 2c_2)t\big)$-cover $\mathcal{U}_1 := \{U^1_\alpha\}_{\alpha \in A_1}$ of S_1. We may assume that $A_1 \cap A_2 = \emptyset$. Using these collections we construct, for some $c_3 = c_3(n)$ and every t, a (c_3, n, t)-cover of the space $S_1 \cup S_2$; this clearly yields $\dim_N(S_1 \cup S_2) \le n := \max_{i=1,2} \dim_N S_i$.

To this end, define the family $\widetilde{A}_2$ containing all indices $\alpha \in A_2$ such that there are no points m from the union of $\mathcal{U}^2$ and m_1 from U^1_α such that $d(m_1, m_2) \le t$. Due to this definition there exist maps $j : A_2 \setminus \widetilde{A}_2 \to A_1$ and $f_1, f_2 : A_2 \setminus \widetilde{A}_2 \to S_1 \bigsqcup S_2$ such that

$$f_1(\alpha) \in U^1_{j(\alpha)}, \quad f_2(\alpha) \in U^2_\alpha \quad \text{and} \quad d\big(f_1(\alpha), f_2(\alpha)\big) \le t.$$

Now we set

$$U_\alpha := \begin{cases} U_\alpha^1 \cup \left(\bigcup_{\beta \in j^{-1}(\alpha)} U_\beta^2 \right) & \text{if } \alpha \in A_1, \\ U_\alpha^2 & \text{if } \alpha \in \widetilde{A}_2. \end{cases} \quad (4.25)$$

Every U_α has diameter at most $c_1(3 + 2c_2)t + 2(1 + c_2)t =: c_3 t$. Moreover, the family $\mathcal{U} := \{U_\alpha\}_{\alpha \in A_1 \cup \widetilde{A}_2}$ is a cover of the space $S_1 \bigsqcup S_2$. To prove that $\mathcal{U}$ is a (c_3, n, t)-cover, it remains to check that every $S \subset S_1 \cup S_2$ of diameter t meets at most $n+1$ sets of $\mathcal{U}$. If such S is disjoint from the union of sets from $\mathcal{U}^1$, then S meets at most $n+1$ members of the (c_2, n, t)-cover $\mathcal{U}^2$ of S_2, each of which coincides with exactly one U_α from (4.25). On the other hand, if $S \cap U_\alpha^1 \neq \emptyset$ for some $\alpha \in A_1$, then every index $\alpha \in A_2$ satisfying $U_\alpha^2 \cap S \neq \emptyset$ must belong to the set $A_2 \setminus \widetilde{A}_2$. Then the set $S \cup \{f_1(\alpha) \,;\, U_\alpha^2 \cap S \neq \emptyset\}$ has diameter at most $(3+2c_2)t$ and meets no more than $n+1$ members of the $(c_1, n, (3 + 2c_2)t)$-cover $\mathcal{U}^1$.

This shows that $U = \{U_\alpha\}_{\alpha \in A_1 \cup \widetilde{A}_2}$ is a (c_3, n, t)-cover of the space $S_1 \cup S_2$, i.e., the Nagata dimension of this space is at most n.

(c) Suppose that a metric space $\mathcal{M}$ has the Nagata dimension $n < \infty$. Given $t > 0$, choose a (c, n, t)-cover $\mathcal{U} := \{U_\alpha\}_{\alpha \in A}$ of $\mathcal{M}$ and denote by $\widehat{U}_\alpha$ the open $\frac{t}{2}$-neighborhood of U_α. Then diam $\widehat{U}_\alpha \leq (c+1)t$; moreover, the open cover $\widehat{\mathcal{U}} := \{\widehat{U}_\alpha\}_{\alpha \in A}$ has order at most $n+1$. In fact, if m_0 is a common point of $n+2$ subsets $\widehat{U}_{\alpha_i}$, $i = 1, \ldots, n+2$, then the closed ball $\overline{B}_{\frac{t}{2}}(m_0)$ of diameter t intersects $n+2$ subsets of $\mathcal{U}$, a contradiction. Further, the cover $\{B_{\frac{t}{2}}(m)\}_{m \in \mathcal{M}}$ is a refinement of the cover $\widehat{\mathcal{U}}$ while the latter is a refinement of the cover $\{B_{(c+1)t}(m)\}_{m \in \mathcal{M}}$. Repeating this construction for $t_k := (2(c+1))^{-k}$, $k = 1, 2, \ldots$, we find a sequence of open covers $\mathcal{U}^k$, $k = 1, 2, \ldots$, satisfying the conditions

$$\operatorname{diam} \mathcal{U}^k \to 0 \quad \text{as} \quad k \to \infty \quad \text{and} \quad \operatorname{ord} \mathcal{U}^k \leq n + 1 \quad \text{for all} \quad k;$$

moreover, $\mathcal{U}^{k+1}$ is a refinement of $\mathcal{U}^k$. Hence this sequence satisfies the assumptions of Theorem 4.8 and therefore $\dim \mathcal{M} \leq n$. $\square$

An important class of metric spaces of the finite Nagata dimension describes the following

Theorem 4.31. *The Nagata dimension of a doubling metric space is finite.*

Proof. Let $\mathcal{M}$ be a doubling metric space with the doubling constant $\delta_\mathcal{M}$, see (3.40). So every closed ball of radius R can be covered by at most $\delta_\mathcal{M}$ balls of radius $\frac{R}{2}$. We show that

$$\dim_N \mathcal{M} \leq \delta_\mathcal{M}^2. \quad (4.26)$$

This immediately follows from assertion (4) of Proposition 4.28 and the following

4.2. Covering metric invariants

Lemma 4.32. *If every closed ball in a metric space $\mathcal{M}$ of radius $3t$ can be covered by $N+1$ sets of diameter t, then $\mathcal{M}$ admits a cover $\mathcal{U}$ by closed balls of radius t such that $\mathcal{U} = \bigsqcup_{k=0}^{N} \mathcal{U}^k$, where each family $\mathcal{U}^k$ satisfies the condition:*
Every subset $S \subset \mathcal{M}$ of diameter at most t meets at most one set of $\mathcal{U}^k$.

By the definition of the covering constant $\delta_{\mathcal{M}}$, every closed ball of radius $3t$ can be covered by $\delta_{\mathcal{M}}^2$ closed balls of radius t. Then the assertion of the lemma with N satisfying $N \leq \delta_{\mathcal{M}}^2 < N+1$ yields a cover $\mathcal{U}$ of diameter t such that every $S \subset \mathcal{M}$ of diameter t meets at most $N+1$ sets of $\mathcal{U} = \bigsqcup_{k=0}^{N} \mathcal{U}^k$.

This implies (4.26).

It remains to prove the lemma. Let Γ be a maximal t-net of $\mathcal{M}$, i.e., $d(m, m') > t$ for every $m \neq m'$ from Γ, and the family of closed balls $\{\overline{B}_{\frac{t}{2}}(m)\}_{m \in \Sigma}$ covers $\mathcal{M}$. Endow Γ with a graph structure regarding the points of Γ as vertices and joining m and m' from Γ by an edge if $0 < d(m, m') \leq 3t$. Show that the degree $\deg m$ of every vertex m of Γ is at most N. In fact, the closed ball $\overline{B}_{3t}(m)$ can be covered by $N+1$ sets of diameter at most t, each of which contains no more than one point of Γ. Thus, $\Gamma \cap \overline{B}_{3t}(m)$ has cardinality at most $N+1$ and therefore every vertex m of the graph Γ can be joined to at most N other vertices.

Thus, $\sup\{\deg m\,; m \in \Gamma\} \leq N$ and the Szekes–Wilf theorem [SzW-1968] imply that Γ can be colored in $N+1$ colors, i.e., there exists a function $\chi : \Gamma \to \{0, 1, \ldots, M\}$, where $M \leq N$, such that $\chi(m) \neq \chi(m')$ whenever m and m' are joined by an edge (i.e., $0 < d(m, m') \leq 3t$). For every $0 \leq k \leq N$, denote by $\mathcal{U}^k$ the family of all closed balls $\overline{B}_t(m)$ with $\chi(m) = k$. Then every subset $S \subset \mathcal{M}$ of diameter t can intersect no more than one of the balls in $\mathcal{U}^k$.

This completes the proof of the lemma and the theorem. □

Remark 4.33. (a) The inequality

$$\dim_N(\mathcal{M}_1 \oplus^{(p)} \mathcal{M}_2) \leq \dim_N \mathcal{M}_1 + \dim_N \mathcal{M}$$

may be strict, as the following example shows. Equip $\mathbb{Z}$ and $I := [0, 1]$ with the metric induced from $\mathbb{R}$. It is a matter of definition to check that

$$\dim_N \mathbb{Z} = \dim_N I = \dim_N(\mathbb{Z} \oplus^{(p)} I) = 1.$$

(b) Using the polyhedral complex generated by the partition of $\mathbb{R}^n$ into cubes of edge length t and applying Proposition 4.28 (3) we easily obtain

$$\dim_N \mathbb{R}^n \leq n.$$

Since for every subset S of $\mathbb{R}^n$ with nonempty interior $\dim S = n$ (see, e.g., [HW-1941, Thm. IV3]), the Nagata dimension of S equals n. In fact, by Theorem 4.30 (c),

$$n = \dim S \leq \dim_N S \leq \dim_N \mathbb{R}^n \leq n.$$

(c) Let $\mathcal{M}$ be a compact n-dimensional Riemannian manifold regarded as a metric space with the geodesic metric. Then $\mathcal{M}$ is a finite union of its charts which are bi-Lipschitz isomorphic to an open ball of $\mathbb{R}^n$. Therefore the previous result and Theorem 4.30 (b) imply that

$$\dim_N \mathcal{M} = n. \tag{4.27}$$

Now we consider another class of metric spaces of finite Nagata dimension which consists of Gromov hyperbolic spaces of bounded geometry, see Definitions 3.98 and 3.114. For some spaces of the class, e.g., for "pinched" Hadamard manifolds (see the definition below) one can even prove the coincidence of Nagata and topological dimensions.

Theorem 4.34. *Let $(\mathcal{M}, d)$ be a $\delta/4$-hyperbolic space of bounded geometry with $\delta > 0$. Then*

$$\dim_N \mathcal{M} < \infty.$$

Proof. We will use equivalent Definition 3.115 of Gromov hyperbolicity. Hence, given a basepoint $m^* \in \mathcal{M}$, there exists $\delta' > 0$ such that for every triple $m_i \in \mathcal{M}$, $i = 1, 2, 3$,

$$(m_1|m_2) \geq \min\{(m_1|m_3), (m_2|m_3)\} - \delta'. \tag{4.28}$$

Let us recall that

$$(m'|m'') := \frac{1}{2}\big[d(m', m^*) + d(m'', m^*) - d(m', m'')\big].$$

We need the following version of the so-called "tripod lemma".

Lemma 4.35. *Let $\widehat{m}_i$ be a point of the geodesic segment $[m_i, m^*]$, $i = 1, 2$. Assume that*

$$d(m^*, \widehat{m}_1) = d(m^*, \widehat{m}_2) \leq (m_1|m_2). \tag{4.29}$$

Then $d(\widehat{m}_1, \widehat{m}_2) \leq 4\delta'$.

Proof. Set $t := d(m^*, \widehat{m}_i)$. By the definition of the Gromov product and the choice of $\widehat{m}_i$,

$$\begin{aligned}(\widehat{m}_i|m_i) &= \frac{1}{2}\big[d(m_i, m^*) + d(\widehat{m}_i, m^*) - d(\widehat{m}_i, m_i)\big] \\ &= \frac{1}{2}\big[2d(m_i, m^*) + d(\widehat{m}_i, m_i) - d(\widehat{m}_i, m_i)\big] \\ &= d(m_i, m^*) =: t.\end{aligned}$$

This, (4.28) and (4.29) imply that

$$t - \frac{1}{2}d(\widehat{m}_1, \widehat{m}_2) := (\widehat{m}_1|\widehat{m}_2) \geq \min\{(\widehat{m}_1|m_1), (m_1, m_2), (m_2, \widehat{m}_2)\} - 2\delta' = t - 2\delta',$$

whence $d(\widehat{m}_1, \widehat{m}_2) \leq 4\delta'$. $\square$

4.2. Covering metric invariants

In the sequel we fix m^* and set

$$\delta := 4\delta'.$$

Further, let $\mathcal{M}$ belong to the class $\mathcal{G}_n(R, D)$ of spaces of bounded geometry, see Definition 3.98. Due to Theorem 3.99 $\mathcal{M}$ also belongs to every class $\mathcal{G}_{n'}(R', D')$ with given $R' > R$ and some n', D' depending only on n, R, D. Therefore, in the sequel, we may and will assume that $\delta = R$; for now, $\mathcal{M}$ is δ-hyperbolic and

$$\mathcal{M} \in \mathcal{G}_n(\delta, D).$$

To prove the theorem we, for every $t > 0$, must construct a (c_1, n_1, t)-cover of $\mathcal{M}$ with c_1, n_1 independent of t. For $t \leq \delta$ we immediately derive this result from Lemma 3.102. According to the lemma every space $\mathcal{M} \in \mathcal{G}_n(t, D)$ (contained in $\mathcal{G}_n(\delta, D)$ as $t \leq \delta$) can be presented in the form

$$\mathcal{M} = \bigcup_{j=0}^{N} \bigsqcup_{m \in A_j} B_{\frac{t}{2}}(m)$$

where $N \leq (8D+1)^n$ and every pair of distinct points m', m'' from A_j, $1 \leq j \leq N$, satisfies

$$d(m, m') \geq 2t.$$

Due to this inequality every subset S of diameter $\leq t$ meets at most one ball of the family $\{B_{\frac{t}{2}}(m)\}_{m \in A_j}$. Hence, the balls $B_{\frac{t}{2}}(m)$, $m \in A_j$, $1 \leq j \leq N$, form $(1, N-1, t)$-cover of $\mathcal{M}$ with $N \leq (8D+1)^n$.

We fix a special case of this result as

Lemma 4.36. *The family* $\{B_{\frac{\delta}{2}}(m)\}_{m \in A}$ *where* $A := \bigsqcup_{j=1}^{N} A_j$ *is a* $(1, N-1, \delta)$-*cover of* $\mathcal{M}$.

It remains to prove the following

Claim. *For some* $n_1 \in \mathbb{N}$ *and* $c_1 > 0$ *and every* $t > \delta$, *there exists a* (c_1, n_1, t)-*cover of* $\mathcal{M}$.

To prove the claim we first cover $\mathcal{M}$ by a countable family $\{U_k\}_{k \in \mathbb{Z}_+}$ of order 2. Using then the cover of Lemma 4.36 we refine every U_k, forming a cover of diameter $\leq c_1 t$ and of order $\leq N$. Collecting all these covers for $k = 0, 1, \ldots$ we obtain the required (c_1, n_1, t)-cover with $n_1 \leq 2N + 1 < 2(8D+1)^n + 1$.

To realize this plan, we need

Lemma 4.37. *Let* $\mathcal{M}$ *be a* $\delta/4$-*hyperbolic space and let* $m^* \in \mathcal{M}$ *be a basepoint. There exists a map* $F : \mathcal{M} \times \mathbb{R}_+ \to \mathcal{M}$ *such that*

(a) *for every* $(m,s) \in \mathcal{M} \times \mathbb{R}_+$,

$$d(m, m_s) \leq s \quad \text{where} \quad m_s := F(m,s); \tag{4.30}$$

(b) *if* $m, m' \in \mathcal{M}$ *and* $s \geq 0$ *satisfy*

$$d(m, m^*) = d(m', m^*) \quad \text{and} \quad s \geq \frac{1}{2} d(m, m'), \tag{4.31}$$

then

$$d(m_s, m'_s) \leq \delta. \tag{4.32}$$

Proof. For $(m,s) \in \mathcal{M} \times \mathbb{R}_+$ we define $F(m,s)$ to be the point in the geodesic segment $[m, m^*]$ at distance $\min\{s, d(m, m^*)\}$ from m.

By definition we get, for the nontrivial case of $s \leq d(m, m^*) \left(= d(m', m^*)\right)$,

$$d(m^*, m_s) = d(m^*, m) - d(m, m_s) = d(m^*, m) - s,$$

and the same holds for m'_s.

This and the equality in (4.31) yield

$$d(m^*, m_s) = d(m^*, m'_s).$$

Moreover, the inequality in (4.31) implies that

$$(m|m') := \frac{1}{2} \left[d(m^*, m) + d(m^*, m') - d(m, m') \right]$$

$$= d(m^*, m) - \frac{1}{2} d(m, m') > d(m^*, m) - s = d(m^*, m_s).$$

Therefore condition (4.29) of Lemma 4.35 holds for the points m, m', m_s, m'_s and this yields the required inequality $d(m_s, m'_s) \leq \delta$. $\square$

We now cover $\mathbb{R}_+$ by intervals I_k, $k \in \mathbb{Z}_+$, so that

$$\operatorname{ord}\{I_k\}_{k \in \mathbb{Z}_+} = 2 \quad \text{and} \quad |I_k| = t \quad \text{for all} \quad k. \tag{4.33}$$

Then we define a cover $\{U_k\}_{k \in \mathbb{Z}_+}$ of $\mathcal{M}$, where U_k is the preimage of I_k under the map $g : \mathcal{M} \to \mathbb{R}_+$, given by

$$g(m) := d(m^*, m).$$

Hence, $U_k := \{m \in \mathcal{M} \,;\, d(m^*, m) \in I_k\}$. Since g is 1-Lipschitz, we have

$$\operatorname{ord}\{U_k\}_{k \in \mathbb{Z}_+} = 2. \tag{4.34}$$

At the next stage we employ

4.2. Covering metric invariants

Lemma 4.38. *For every $k \in \mathbb{Z}_+$ there is a map $h_k : U_k \to \mathcal{M}$ such that for all $m \in U_k$,*

$$d(m^*, h_k(m)) = \inf I_k \quad \text{and} \quad d(m, h_k(m)) \leq |I_k| (= t). \tag{4.35}$$

Proof. Let $I_k := [a_k, b_k]$. For $n \in U_k$ we define $h_k(m)$ to be the point of the geodesic segment $[m^*, m]$ at distance a_k from m^*. Then for all m

$$d(m^*, h_k(m)) = a_k,$$

and, moreover,

$$b_k \geq d(m^*, m) = d(m^*, h_k(m)) + d(h_k(m), m) = a_k + d(h_k(m), m).$$

This implies (4.35). □

Now we are ready to prove the claim.

Let $\mathcal{B} := \{B_{\frac{\delta}{2}}(m)\}_{m \in A}$ be the cover of $\mathcal{M}$ from Lemma 4.36. For $k \in \mathbb{Z}_+$ and $m \in A$ we define the set

$$V_{km} := \{m' \in \mathcal{M} \,;\, F(h_k(m'), s) \in B_{\frac{\delta}{2}}(m)\}, \tag{4.36}$$

where $s := \frac{3}{2} t$.

If m', m'' are the points of V_{km}, then

$$d(m', m'') \leq d(m', h_k(m')) + d(h_k(m'), h_k(m''))$$
$$+ d(h_k(m''), m'') \leq 2t + d(h_k(m'), h_k(m'')),$$

see (4.35). Using then Lemma 4.37 and definition (4.36), we estimate the last term by

$$d(h_k(m'), F(h_k(m'), s)) + d(F(h_k(m'), s), F(h_k(m''), s))$$
$$+ d(F(h_k(m''), s), m'') \leq 2s + \delta < 4t.$$

Combining these estimates we conclude that diam $V_{k\alpha} < 6t$, i.e., for the cover $\mathcal{V} := \{V_{k\alpha} \,;\, k \in \mathbb{Z}_+,\, m \in A\}$ we have

$$\text{diam}\, \mathcal{V} < 6t. \tag{4.37}$$

To estimate the order of $\mathcal{V}$ we first will show that for every subset $S \subset U_k$ of diameter at most t the associated set $\hat{S} := \{F_k(m, s) \,;\, m \in S\}$ is of diameter at most δ. In fact, if $m', m'' \in S$, then

$$d(h_k(m'), h_k(m'')) \leq 2t + d(m', m'') \leq 2t + t := 2s.$$

Hence, the assumptions in (4.31) hold for the pair $h_k(m'), h_k(m'')$ and therefore Lemma 4.37 implies that

$$d\Big(F\big(h_k(m'), s\big),\ F\big(h_k(m''), s\big)\Big) \leq \delta.$$

Thus, diam $\widehat{S} \leq \delta$ and $\widehat{S}$ meets at most $N+1$ balls of the $(1, N, \delta)$-cover $\mathcal{B}$. It follows from here and (4.36) that the set $S \subset U_k$ intersects at most $N+1$ subsets $V_{k\alpha} \in \mathcal{V}$. Since $\{U_k\}_{k \in \mathbb{Z}_+}$ is a cover of $\mathcal{M}$ of order 2, every subset of diameter at most t meets at most $2N+2$ subsets from $\mathcal{V}$. Together with (4.37) this shows that $\mathcal{V}$ is the required $(6t, 2N+1, t)$-cover of $\mathcal{M}$.

The claim and the theorem are proved. □

In Lipschitz extension results presented in Chapter 5, Nagata dimension bounds from above the corresponding extension constants. The estimates of $\dim_N \mathcal{M}$ given by Theorems 4.31 and 4.34 are very rough (exponentially growing in topological dimension of $\mathcal{M}$). However, for some Gromov hyperbolic spaces one may obtain essentially better estimates. We present two such results whose proofs, in essence, follow the line of that for Theorem 4.34; we refer the reader to the paper [LSchl-2005] for the proof of the second result.

The first one concerns direct sums of $\mathbb{R}$-trees (0-hyperbolic spaces), see the end of subsection 3.3.6.

Theorem 4.39. *Let T_i, $1 \leq i \leq n$, be $\mathbb{R}$-trees each containing more than one point. Then*

$$\dim_N \bigoplus_{i=1}^{N} {}^{(p)}T_i = n. \tag{4.38}$$

Proof. A nontrivial tree contains a subset isomorphic to an open interval; therefore the left-hand side in (4.38) is at least $\dim_N (0,1)^n = n$, see Theorem 4.30 (c) and Remark 4.33 (c).

To prove the upper estimate, show that the Nagata dimension of an $\mathbb{R}$-tree, say T, is at most 1; then Theorem 4.30 (c) implies that the left-hand side of (4.38) is at most n.

To estimate $\dim_N T$, we will follow the argument of Theorem 4.34 preserving the same notation. First, one decomposes $\mathbb{R}_+$ into closed intervals I_k of length $2t$ such that

$$\mathrm{card}(I_k \cap I_{k+1}) = 1, \quad k \in \mathbb{Z}_+.$$

Then one defines $U_k := g^{-1}(I_k) \subset T$ using the 1-Lipschitz map $g : T \to \mathbb{R}_+$. Let $m \sim m'$ be an equivalence relation on U_k given by $F(h_k(m), s) = F(h_k(m'), s)$ where $s := \frac{3}{2}t$. Using Lemmas 4.38 and 4.37 (a) with $\delta = 0$ we conclude that every class of equivalence is of diameter $\leq 2(t+s) = 5t$. Moreover, every $S \subset U_k$ of diameter s meets at most one such class. In fact, if $m, m' \in S$, then $d\big(h_k(m), h_k(m')\big) \leq 2s+s = 3s =: 2t$ and therefore, by Lemma 4.37 (a), $d\big(F(h_k(m), s), F(h_k(m'), s)\big) \leq \delta = 0$, i.e., m and m' belong to the same class.

Collecting all the classes of equivalence for $k = 0, 1, 2, \ldots$, we then obtain, for every $t > 0$, a $(5, 1, t)$-cover.

Hence $\dim_N T \leq 1$, and the result is done. $\square$

The last result concerns *Hadamard manifolds*, complete simply connected Riemannian manifolds of nonpositive sectional curvature (or, what is the same, of nonpositive curvature in the Alexandrov sense, see Definition 3.109).

Theorem 4.40. *Let $\mathcal{M}$ be a "pinched" n-dimensional Hadamard manifold, i.e., the sectional curvature of $\mathcal{M}$ is bounded by constants $-\infty < c_1 < c_2 < 0$. Then*

$$\dim_N \mathcal{M} = n.$$

4.3 Existence of doubling measures

The goal of this section is to prove that every *complete* doubling metric space carries a doubling measure (the converse to Proposition 3.84 (b)); completeness here is essential. For instance, the metric space of rational numbers $\mathbb{Q}$ with the standard metric is countable and has no isolated points. Therefore, a doubling measure on $\mathbb{Q}$ should be identically zero by Proposition 3.84 (a), a contradiction to the definition of doubling measures.

To formulate the main result we need a more quantitative classification of doubling measures.

Definition 4.41. A measure μ on a metric space is called (C, s)-homogeneous ($C \geq 1, s \geq 0$) if μ-measure of every open ball is separated from 0 and $+\infty$ and the dilation of μ satisfies, for every $\ell \geq 1$, the inequality

$$D(\mu \,;\, \ell) \leq C\ell^s. \tag{4.39}$$

We say that μ is *s-homogeneous* if (4.39) holds for some C.

It is clear that a (C, s)-homogeneous measure μ is doubling with the doubling constant $D(\mu) \leq 2^s C$. Conversely, due to (3.102), a doubling measure μ is $(1, s)$-homogeneous with $s := \log_2 D(\mu) + 1$.

In the same vein, we classify doubling metric spaces employing

Definition 4.42. A metric space $\mathcal{M}$ is called (C, s)-homogeneous ($C \geq 1, s \geq 0$) if for each $\ell \geq 1$ and $R > 0$ every closed ball of radius ℓR can be covered by at most $C\ell^s$ closed balls of radius R.

A (C, s)-homogeneous metric space is clearly doubling with the doubling constant bounded by $2^s C$. Conversely, a doubling metric space $\mathcal{M}$ is $(1, s)$-homogeneous with $s := \log_2 \delta_{\mathcal{M}} + 1$.

The following main result was due to Vol'berg and Koniagin [VK-1987] for the basic case of compact doubling metric spaces.

Theorem 4.43. *Every (C, s)-homogeneous complete metric space carries, for each $\tilde{s} > s$, a $(\tilde{C}, \tilde{s})$-homogeneous measure; here $\tilde{C}$ depends only on C, s and $\tilde{s}$ and tends to infinity as $\tilde{s} - s$ tends to zero.*

We divide the proof into three parts beginning with finite metric spaces, then deriving from there the Vol'berg–Koniagin theorem and concluding with the general result.

4.3.1 Finite metric spaces

The main result of this subsection is

Proposition 4.44. *Theorem 4.43 is true for finite metric spaces.*

Proof. Let $(\mathcal{M}, d)$ be a finite (C, s)-homogeneous metric space. Assume without loss of generality that

$$\operatorname{diam} \mathcal{M} < 1. \tag{4.40}$$

Given $\tilde{s} > s$, we then define the constants

$$A := \max(C^{\frac{1}{\tilde{s}-s}}, 2^{\frac{1}{\tilde{s}-s}}, 21) \quad \text{and} \quad D := (2A)^{\tilde{s}}. \tag{4.41}$$

Now let S_0 be a maximal 1-net in $\mathcal{M}$. Due to (4.40) S_0 contains a single point, say $\widehat{m}$. Then choose a maximal A^{-1}-net containing S_0. Proceeding this way, we construct a family

$$S_0 \subset S_1 \subset \cdots \subset S_N = \mathcal{M},$$

where S_i is a maximal A^{-i}-net containing S_{i-1}. By definition, for all $m \neq m'$ from S_i,

$$d(m, m') \geq A^{-i} \tag{4.42}$$

and for every $m \in \mathcal{M}$,

$$d(m, S_i) < A^{-i}. \tag{4.43}$$

Now we define a map T_i from S_i into the set of subsets of S_{i+1} by

$$T_i(m) := \{m' \in S_{i+1} \,;\, d(m', m) = d(m', S_i)\}. \tag{4.44}$$

In other words, a point m' belongs to $T_i(m)$ if m' either equals m or is a point of S_{i+1} closest to m. In particular, $T_0(\widehat{m}) = S_1$ and $T_i(m)\setminus\{m\}$ are one-point sets if all distances between points S_i and $S_{i+1}\setminus S_i$ are pairwise disjoint. By the reason explained in Corollary 4.46, we will call the points of $T_i(m)\setminus\{m\}$ *children* of m.

The basic properties of the maps T_i are described by

Lemma 4.45. (a) *The family $\{T_i(m)\}_{m \in S_i}$ is a partition of S_{i+1};*

4.3. Existence of doubling measures

(b) $\operatorname{card} T_i(m) \leq D$.

Proof. (a) If $m \neq m'$ points of S_i and $T_i(m) \cap T_i(m') \neq \emptyset$, then by (4.43) and (4.44) for every $\widetilde{m} \in T_i(m) \cap T_i(m')$,

$$d(m, m') \leq d(m, \widetilde{m}) + d(\widetilde{m}, m') = d(m, S_{i+1}) + d(m', S_{i+1}) < 2A^{-i} < A,$$

a contradiction to (4.42).

It is also clear that $\bigcup_{m \in S_i} T_i(m) = S_{i+1}$.

(b) Due to (4.43) and (4.44), for every $m \in S_i$,

$$T_i(m) \subset B_{r_i}(m) \quad \text{where} \quad r_i := A^{-i}.$$

Because of (C, s)-homogeneity, the ball of radius A^{-i} contains at most

$$C\left(\frac{2A^{-i}}{A^{-i-1}}\right)^s = C2^s A^s$$

points with pairwise distances $\geq A^{-i-1}$. This implies that

$$\operatorname{card} T_i(m) \leq \operatorname{card}\bigl(S_{i+1} \cap B_{r_i}(m)\bigr) \leq C2^s A^s = (2A)^{\tilde{s}} =: D,$$

see (4.41). □

Now we define a graph structure on $\mathcal{M}$ joining $m \neq m'$ by a (directed) edge whenever $m' \in T_i(m)$ for some i. In particular, every $m \in S_1 \setminus S_0$ is joined with $\hat{m}$ by an edge directed to $\hat{m}$, see Figure 4.4 below.

Corollary 4.46. *The graph introduced is a rooted tree with the root $\hat{m}$.*

Proof. By definition, a graph is a tree if it is connected and every pair of its vertices is joined by a *unique* path. Let us recall, see subsection 3.3.6, that the sequence of vertices $v_1, \ldots, v_n$ of a graph forms a *path* if v_i and v_{i+1} are joined, for each i, by an edge.

By Lemma 4.45 (a), for every $m' \in S_{i+1} \setminus S_i$ there is a unique $m \in S_i$ such that $m' \in T_i(m)$. Hence, every pair of vertices (points) is joined by a unique path, and the graph is a tree. Since every m is joined by a path with $\hat{m}$, the latter is the root. □

In the sequel, m is called the *parent* of a *child* m', if $m' \in T_i(m) \setminus \{m\}$ for some i.

Now we define the required $\tilde{s}$-homogeneous measure by the process of transferring mass from points of S_i to points of $S_{i+1} \setminus S_i$ starting with the δ-measure $\delta_{\hat{m}}$. So we set

$$\mu_0 = \delta_{\hat{m}},$$

i.e., $\mu_0(S) = 1$, and define μ_1 by transferring the $\frac{1}{D}$-part of the mass at $\hat{m}$ to each its child. Hence, $\mu_1(\{m\}) := \frac{1}{D}$ for $m \in S_1 \setminus S_0$ and $\mu_1(\{m\}) := 1 - \frac{1}{D}(\operatorname{card} T_0(m) - 1)$ for $m \in S_0$ (i.e., for $m = \hat{m}$).

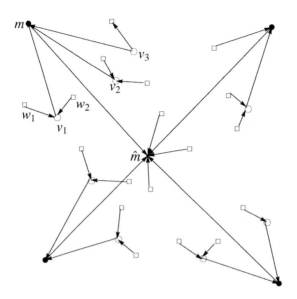

Figure 4.4: $S_0 := \{\widehat{m}\}$; $S_1 \setminus S_0$ is the set of black dots; $S_2 \setminus S_1$ is the set of circles; $S_3 \setminus S_2$ is the set of squares; $T_1(m) := \{m, v_1, v_2, v_3\}$; $T_2(v_1) := \{v_1, w_1, w_2\}$.

In general, given the measure μ_i defined on S_i, we define μ_{i+1} by transferring from every $m \in S_i$ the $\frac{1}{D}$-part of $\mu_i(\{m\})$ to each child of m. This procedure defines the sequence of measures $\mu_0, \ldots, \mu_N$ such that μ_i is carried by S_i and μ_i and μ_{i+1} are related by

$$\mu_{i+1}(\{m'\}) := w_{i+1}(m)\mu_i(\{m\}); \tag{4.45}$$

here $m' \in T_i(m) \, (\subset S_{i+1})$ and the weight $w_{i+1} : S_{i+1} \to \mathbb{R}_+$ is given by

$$w_{i+1}(m) := \begin{cases} \frac{1}{D} & \text{if } m \in S_{i+1} \setminus S_i, \\ 1 - \frac{1}{D}(\operatorname{card} T_i(m) - 1) & \text{if } m \in S_i. \end{cases}$$

The properties of the sequence $\{\mu_i\}$ are described by

Lemma 4.47. (a) $\mu_i(S_i) = 1$, $0 \leq i \leq N$.

(b) $\mu_{i+1}(T_i(m)) = \mu_i(\{m\})$ for all $m \in S_i$.

Proof. By the definition of the weights,

$$\sum_{m' \in T_i(m)} w_{i+1}(m') = 1$$

4.3. Existence of doubling measures

and therefore for every $m \in S_i$,

$$\sum_{m' \in T_i(m)} \mu_{i+1}(\{m'\}) = \mu_i(\{m\}). \tag{4.46}$$

This implies (b); then the equality

$$\mu_{i+1}(S_{i+1}) = \sum_{m \in S_i} \mu_{i+1}(\{T_i(m)\}) = \sum_{m \in S_i} \mu_i(\{m_i\}) = \mu_i(S_i)$$

which follows from Lemma 4.45 (a) together with the equality $\mu_0(S_0) = 1$ imply (a). □

Now we define the desired measure μ by

$$\mu := \mu_N.$$

We must prove that μ is $\tilde{s}$-homogeneous. This requires some preparatory work presented in the next subsection.

Technical lemma

In its formulation, we denote by $[S]_\varepsilon$ the closed ε-neighborhood of $S \subset \mathcal{M}$, i.e.,

$$[S]_\varepsilon := \{m \in \mathcal{M}\,;\, d(m, S) \le \varepsilon\}.$$

Lemma 4.48. (a) *If $m, m' \in S_i$ and $d(m, m') \le 8A^{-i}$, then*

$$\mu_i(\{m\}) \le D^2 \mu_i(\{m'\}). \tag{4.47}$$

(b) *Let $m \in S_{i+1}$ and let m' be a closest to m point of S_i. Then*

$$\mu_{i+1}(\{m\}) \le \mu_i(\{m'\}) \le D\mu_{i+1}(\{m\}). \tag{4.48}$$

(c) *For every $S \subset \mathcal{M}$,*

$$\mu_i(S) \le \mu_{i+1}([S]_\varepsilon) \quad \text{and} \quad \mu_{i+1}(S) \le \mu_i([S]_\varepsilon) \quad \text{where} \quad \varepsilon := A^{-i}. \tag{4.49}$$

Proof. (c) Since $T_i(m) \subset B_\varepsilon(m)$ with $\varepsilon := A^{-i}$ for every $m \in S_i$, the assertion follows from (4.46).

(b) First let $m \in S_{i+1} \setminus S_i$. Then by (4.46) and the definition of w_i, we have

$$\mu_i(\{m'\}) = \mu_{i+1}(\{m\})/w_{i+1}(m) = D\mu_{i+1}(\{m\}),$$

as required.

Now let $m \in S_i$. Then $m' = m$ and, moreover,

$$\mu_{i+1}(\{m\}) = \left(1 - \frac{1}{D}\left(\operatorname{card} T_i(m') - 1\right)\right)\mu_i(\{m'\}).$$

These and Lemma 4.45 (b) yield

$$\mu_i(\{m\}) = \mu_i(\{m'\}) = \left(1 - \frac{1}{D}\left(\operatorname{card} T_i(m) - 1\right)\right)^{-1} \mu_{i+1}(\{m\}) \leq D\mu_{i+1}(\{m\}).$$

(a) Without loss of generality we assume that $m \neq m'$.

Let now $m =: m_i, m_{i-1}, \ldots, m_0 := \widehat{m}$ be the unique path connecting m and the root $\widehat{m}$. That is to say, we define $m_{k-1} \in S_{k-1}$ to be the parent of m_k; if, however, m_k belongs also to S_{k-1} we set $m_{k-1} := m_k$. In particular, this path may stabilize after some m_k so that $m_k = m_{k-1} = \cdots = m_0 := \widehat{m}$.

Let $m' =: m'_i, m'_{i-1}, \ldots, m'_0 := \widehat{m}$ be a similar path from m'. The sequences have common points, e.g., $m_0 = m'_0$. Set now

$$J := \max\{j\,;\, m_j = m'_j\} \qquad (4.50)$$

and assume first that $i \leq J + 2$. Then we have, for some $k \in \{1, 2\}$,

$$m =: m_i \neq m'_i := m' \quad \text{but} \quad m_{i-k} = m'_{i-k}.$$

(Without loss of generality we may also assume that $m'_{i-2} \neq m'_{i-1}$.)

Let, for definiteness, $k = 2$, i.e.,

$$m := m_i \neq m'_i =: m' \quad \text{but} \quad m_{i-2} = m'_{i-2}.$$

Applying twice (4.48) we then get

$$\mu_i(\{m\}) \leq \mu_{i-2}(\{m_{i-2}\}) = \mu_{i-2}(\{m'_{i-2}\}) \leq D^2\mu_i(\{m'\}),$$

as required.

Alternatively, let

$$J + 2 < i. \qquad (4.51)$$

We will derive from here that

$$m_j \neq m_{j-1} \quad \text{and} \quad m'_j \neq m'_{j-1} \qquad (4.52)$$

for all j satisfying $J + 2 \leq j \leq i$. Assuming for a while that this is true we get, for $j = J+2, \ldots, i$,

$$m_j, m'_j \in S_j \setminus S_{j-1}.$$

Since for any $m \in S_j \setminus S_{j-1}$ and its child $\widetilde{m}$ we have $\mu_j(\{\widetilde{m}\}) = \frac{1}{D}\mu_{j-1}(\{m\})$, the following is true:

$$\mu_i(\{m\}) =: \mu_i(\{m_i\}) = D^{J-i+1}\mu_{J+1}(\{m_{J+1}\}).$$

The same equality is clearly true for the points with primes.

4.3. Existence of doubling measures

On the other hand, inequality (4.48) applied to m_J gives

$$\mu_{J+1}(\{m_{J+1}\}) \leq \mu_J(\{m_J\}) = \mu_J(\{m'_J\}).$$

Similarly, (4.48) implies

$$\mu_J(\{m'_J\}) \leq D\mu_{J+1}(\{m'_{J+1}\}) \leq D^{i-J}\mu_i(\{m'_i\}) := D^{i-J}\mu_i(\{m'\}).$$

Combining these inequalities with the previous equalities we get

$$\mu_i(\{m\}) = D^{J-i+1}\mu_{J+1}(\{m_{J+1}\})$$
$$\leq D^{J-i+1}\mu_J(\{m'_J\}) \leq D^{J-i+2}\mu_{J+1}(\{m'_{J+1}\})$$
$$\leq D\mu_i(\{m'\}),$$

as required.

It remains to establish (4.52). We prove that $m_j \neq m_{j-1}$; the case of points with primes is considered similarly.

Let, on the contrary,

$$m_j = m_{j-1} \quad \text{for some} \quad J+2 \leq j \leq i. \tag{4.53}$$

By the definition of the sequence $\{m_k\}$ and (4.43), $d(m_k, m_{k-1}) = d(m_k, S_{k-1}) < A^{-k+1}$ and therefore

$$d(m, m_j) := d(m_i, m_j) \leq \sum_{k=j}^{\infty} A^{-k} = \frac{A^{-j+1}}{A-1}. \tag{4.54}$$

On the other hand, $j - 1 \geq J + 1$, and by (4.53) and the maximality of J,

$$m_{j-1} \neq m'_{j-1} \quad (\text{while} \quad m_{j-1} = m_j).$$

Further, these points belong to S_{j-1} and therefore

$$d(m_{j-1}, m'_{j-1}) \geq A^{-j+1}. \tag{4.55}$$

Moreover, m'_{j-1} is the point of S_{j-1} closest to m'_j and so

$$d(m'_j, m'_{j-1}) \leq d(m'_j, m_{j-1}) = d(m'_j, m_j).$$

Using the triangle inequality we then have

$$d(m'_j, m'_{j-1}) \leq d(m_j, m) + d(m, m') + d(m', m'_j).$$

Now (4.54), a similar estimate for $d(m', m'_j)$ and the assertion $d(m, m') \leq 8A^{-i}$ imply that

$$d(m'_j, m'_{j-1}) \leq \frac{2A^{-j+1}}{A-1} + 8A^{-i}.$$

This and (4.53) lead to the inequality

$$d(m_{j-1}, m'_{j-1}) = d(m_j, m'_{j-1}) \le d(m_j, m'_j) + d(m'_j, m'_{j-1})$$
$$\le d(m_j, m'_j) + \frac{2A^{-j+1}}{A-1} + 8A^{-i}.$$

In turn, the first term on the right-hand side is bounded by

$$d(m, m_j) + d(m, m') + d(m'_j, m') \le \frac{2A^{-j+1}}{A-1} + 8A^{-i}$$

so that

$$d(m_{j-1}, m'_{j-1}) \le \frac{4A^{-j+1}}{A-1} + 16A^{-i}.$$

Together with (4.55) and the definition of A, see (4.41), this leads to the contradictory inequality

$$1 \le \frac{4}{A-1} + 16A^{j-i-1} \le \frac{4}{A-1} + \frac{16}{A} \le \frac{4}{21-1} + \frac{16}{21} < 1.$$

We have established (4.50) and proved the lemma. □

Upper bound for $D(\cdot\,;\mu)$

In order to achieve the required result we estimate the μ-measure of a ball $B_{\ell R}(m_0)$ with $\ell > 1$ from above and that of the ball $B_R(m_0)$ from below. Since $\operatorname{diam}\mathcal{M} < 1$, we consider only the case $\ell R < 1$ deriving from that the result for $R < 1 \le \ell R$ by setting $\ell := \frac{1}{R}$.

Let i, I be uniquely defined integers given by

$$\ell R \le A^{-i} < A\ell R \quad \text{and} \quad A^{-1}R \le A^{-I} < R. \tag{4.56}$$

As $\ell > 1$ and $\ell R < 1$ we get

$$0 \le i < I.$$

We will consider two cases:

(1) $I < N$;

(2) $I \ge N$.

In case (1) we define a sequence m_j, $i \le j \le I+1 \le N$, starting from a point m_{I+1} from S_{I+1} closest to the center of the ball $B_R(m_0)$, i.e., satisfying

$$d(m_0, S_{I+1}) = d(m_0, m_{I+1}).$$

Then there exists a unique path from m_{I+1} to a point from S_i. The successive points of this path $m_{I+1}, m_I, \ldots, m_i$ give the required sequence.

4.3. Existence of doubling measures

Further, we will prove that

$$\mu(B_{\ell R}(m_0)) \leq 8^s C D^{I-i+3} \mu_{I+1}(\{m_{I+1}\}) \tag{4.57}$$

and, moreover, that

$$\mu(B_R(m_0)) \geq \mu_{I+1}(\{m_{I+1}\}). \tag{4.58}$$

Since by (4.56) and (4.41), $A^{I-i} \leq A^2 \ell$ and $D := (2A)^{\tilde{s}}$, this implies the required inequality

$$D(\mu; \ell) \leq C_1 \ell^{s_1}$$

where $C_1 := 8^s C A^{5s_1}$ and $s_1 := (1 + \tilde{s} - s)\tilde{s}$.

We prove (4.57) in three steps. For the first one we set

$$R_k := \ell R + \sum_{j=i}^{k} A^{-j}, \quad k = i+1, \ldots, N-i.$$

By Lemma 4.48 (c),

$$\mu_{i+1}(B_{\ell R}(m_0)) \leq \mu_i([B_{\ell R}(m_0)]_{A^{-i}}) \leq \mu_i(B_{R_1}(m_0))$$

and, further,

$$\mu_{i+2}(B_{R_1}(m_0)) \leq \mu_{i+1}([B_{\ell R}(m_0)]_{A^{-i-1}}) \leq \mu_i(B_{R_2}(m_0)),$$

and so on until the index $i + (N - i) = N$. Since $\mu_N =: \mu$ and

$$R_N \leq \ell R + \frac{A^{-i+1}}{A-1} =: \widehat{R},$$

we finally get

$$\mu(B_{\ell R}(m_0)) \leq \mu_i(B_{\widehat{R}}(m_0)). \tag{4.59}$$

Now show that

$$S_i \cap B_{\widehat{R}}(m_0) \subset B_{4A^{-i}}(m_0). \tag{4.60}$$

In fact, if m belongs to the left-hand side, then (4.56), the definition of $\widehat{R}$ and the choice of m_i imply

$$d(m, m_i) \leq d(m, m_0) + d(m_0, m_i) \leq \widehat{R} + \frac{A^{-i+1}}{A-1}$$
$$= \ell R + \frac{2A^{-i+1}}{A-1} \leq A^{-i}\left(1 + \frac{2A}{A-1}\right) < 4A^{-i}$$

as $A \geq 21$, see (4.41).

In turn, it follows from (4.60) that for every $m \in S_i \cap B_{\widehat{R}}(m)$,

$$d(m, m_i) \leq 8A^{-i}$$

and we can employ Lemma 4.48 (a) to obtain for such m,

$$\mu_i(\{m\}) \leq D^2 \mu_i(\{m_i\}).$$

Summing over all $m \in S_i \cap B_{\widehat{R}}(m)$ and using (4.59), we then get

$$\mu(B_{\ell R}(m_0)) \leq D^2 \operatorname{card}(S_i \cap B_{\widehat{R}}(m_0)) \mu_i(\{m_i\}).$$

Next, we apply Lemma 4.48 (b) to get

$$\mu_i(\{m_i\}) \leq D^{I-i+1} \mu_{I+1}(\{m_{I+1}\}),$$

whence

$$\mu(B_{\ell R}(m_0)) \leq D^{I-i+3} \mu_{I+1}(\{m_{I+1}\}) \operatorname{card}(S_i \cap B_{\widehat{R}}(m_0)). \tag{4.61}$$

At the third step we estimate the last factor in (4.61). By (4.60), $S_i \cap B_{\widehat{R}}(m_0)$ is an A^{-i}-net in the ball $B_{4A^{-i}}(m_0)$. By (C, s)-homogeneity of $\mathcal{M}$, the number of points in this net is at most $C\left(\frac{8A^{-i}}{A^{-i}}\right)^s = 8^s C$. Inserting this estimate in (4.61) we prove inequality (4.57).

It remains to prove the lower estimate for $\mu_i(B_R(m_0))$ presented in (4.58). Subsequently applying (4.49), we have

$$\mu_{I+1}(\{m_{I+1}\}) \leq \mu_{I+2}(B_{A^{-I-1}}(m_{I+1}))$$
$$\leq \mu_{I+3}(B_{A^{-I-1}+A^{-I-2}}(m_{I+1}))$$
$$\leq \cdots \leq \mu_N(B_\rho(m_{I+1})) =: \mu(B_\rho(m_{I+1})),$$

where

$$\rho := A^{-I-1} + \cdots + A^{-N+1} \leq \frac{A^{-I}}{A-1}.$$

On the other hand, the definition of the path $\{m_j\}_{i \leq j \leq I+1}$, (4.43) and (4.56) imply

$$d(m_0, m_{I+1}) = d(m_0, S_{I+1}) < A^{-I-1} < A^{-1} \widehat{R}.$$

Moreover, by (4.56) and the inequality $A \geq 21$ we also have

$$\rho + A^{-1} R \leq \frac{A^{-I}}{A-1} + A^{-1} R < R,$$

whence

$$B_\rho(m_{I+1}) \subset B_{\rho + A^{-1} R}(m_0) \subset B_R(m_0).$$

4.3. Existence of doubling measures

Combining this with the first inequality, we obtain

$$\mu_{I+1}(\{m_{I+1}\}) \leq \mu(B_\rho(m_{I+1})) \leq \mu(B_R(m_0)).$$

This proves (4.58).

Let us consider now the case $I \geq N$. By the definition of the final net S_N we have

$$A^{-N} \leq \min\{d(m, m'); m \neq m' \in \mathcal{M}\},$$

see (4.42). In particular, we may assume that $\ell R \geq A^{-N}$, since otherwise $\mu(B_{\ell R}(m_0)) = \mu(B_R(m_0))$ by the definition of $\mu = \mu_N$. Then inequality (4.56) shows that $i \leq N$.

Now we define a sequence of points m_j, $i \leq j \leq N$, starting with $m_N := m_0 \in S_N(\mathcal{M})$ and joining m_0 by a unique path with some point $m_i \in S_i$; the successive points of this path $m_N, m_{N-1}, \ldots, m_i$ give the required sequence. Using this sequence and arguing as in case (1) we then obtain

$$\mu(B_{\ell R}(m_0)) \leq 8^s C D^{N-i+3} \mu_N(\{m_N\}) \leq 8^s C D^{I-i+3} \mu(\{m_0\}).$$

Together with the evident inequality $\mu(B_R(m_0)) \geq \mu(\{m_0\})$ this leads to the required estimate for case (2).

Proposition 4.44 is thus established. □

4.3.2 Compact metric spaces

Now we derive, from Proposition 4.44,

Proposition 4.49. *Let $(\mathcal{M}, d)$ be a (C, s)-homogeneous compact metric space. Then for every $\tilde{s} > s$, $\mathcal{M}$ carries a $(\widetilde{C}, \tilde{s})$-homogeneous measure with $\widetilde{C}$ depending only on C and $(\tilde{s} - s)^{-1}$.*

Proof. Let S_j be a maximal $\frac{1}{j}$-net in $\mathcal{M}$, $j \in \mathbb{N}$. Show that the metric subspace S_j is $(3^s C, s)$-homogeneous, i.e., for each $\ell \geq 1$, all of its closed balls $\overline{B}_{3\ell R}(m; S_j) = \overline{B}_{3\ell R}(m) \cap S_j$ can be covered by at most $3^s C \ell^s$ balls of radius $3R$.

First let $R \geq \frac{1}{2j}$. Cover the ball $\overline{B}_{3\ell R}(m)$ by a family of balls $\{B_R(m_j)\}_{1 \leq j \leq N}$ with $N \leq (3\ell)^s C$ and denote by $\widetilde{m}_j$ a point from S_j closest to m_j. Then $d(m_j, \widetilde{m}_j) < \frac{1}{j} \leq 2R$ and therefore $B_R(m_j) \subset B_{3R}(\widetilde{m}_k)$.

Hence $\overline{B}_{3\ell R}(m; S_j)$ is covered by the family of balls $B_{3R}(\widetilde{m}_k; S_j)$, where $1 \leq k \leq N \leq 3^s C \ell^s$, as required.

Now let $R < \frac{1}{2j}$. In this case every ball $B_R(m_k)$ contains at most one point of S_j, say $\widetilde{m}_k$, and the ball $\overline{B}_{3\ell R}(m; S_j)$ is covered by the balls $B_{2R}(\widetilde{m}_k; S_j)$, $1 \leq k \leq N$.

Now we apply Proposition 4.44 to the s-homogeneous metric space S_j in order to find a $(\widetilde{C}_1, \tilde{s})$-homogeneous measure μ_j on S_j which we then extend by zero to subsets of $\mathcal{M}\setminus S_j$; here $\widetilde{C}_1 := 3^s\widetilde{C}$ also depends only on C and $(\tilde{s}-s)^{-1}$. Preserving the same letter for the extension we have

$$\mu_j(\mathcal{M}) = 1 \quad \text{and} \quad \operatorname{supp}\mu_j = S_j. \tag{4.62}$$

Further, every sequence of probability measures on a compact metric space contains a subsequence which is $*$-weakly convergent to a (regular) Borel measure, see, e.g., [Ma-1995]. Passing to a subsequence we may therefore assume that $\{\mu_j\}$ itself is $*$-weak convergent to some measure μ. In other words, for every $f \in C(\mathcal{M})$,

$$\lim_{j\to\infty} \int_\mathcal{M} f\, d\mu_j = \int_\mathcal{M} f\, d\mu. \tag{4.63}$$

Then the A. Alexandrov theorem [Ale-1943] asserts that (4.63) is equivalent to one of the conditions:

(a) For every open subset $G \subset \mathcal{M}$,

$$\varliminf_{j\to\infty} \mu_j(G) \geq \mu(G). \tag{4.64}$$

(b) For every closed subset $F \subset \mathcal{M}$,

$$\varlimsup \mu_j(F) \leq \mu(F), \tag{4.65}$$

see, e.g., Billingsley [Bil-1968, Thm. 1.2] where the Alexandrov theorem is presented in this form.

We derive from here that μ is $(\widetilde{C}_1, \tilde{s})$-homogeneous.

Let $B_{\ell R}(m_0)$ be an arbitrary open ball where $\ell \geq 1$, and let m_j be a point of S_j such that $d(m_0, m_j) < \frac{1}{j}$. Given $0 < q < 1$, set $\tilde{\ell} := q^{-1}\ell$ and $R_j := qR + \frac{1}{j}$. Then $B_{\ell R}(m_0) \subset B_{\tilde{\ell} R_j}(m_j)$ and the applications of (4.64) and $(\widetilde{C}_1, \tilde{s})$-homogeneity of μ_j yield

$$\mu\big(B_{\ell R}(m_0)\big) \leq \varliminf_{j\to\infty} \mu_j\big(B_{\ell R}(m_0)\big) \leq \varliminf_{j\to\infty} \mu_j\big(B_{\tilde{\ell} R_j}(m_j)\big)$$
$$\leq q^{-\tilde{s}} C_1 \ell^{\tilde{s}} \varliminf_{j\to\infty} \mu_j\big(B_{R_j}(m_j)\big).$$

Further, for any $\delta > 0$ and sufficiently large j,

$$B_{R_j}(m_j) \subset \overline{B}_{qR+\delta}(m_0)$$

and the applications of (4.65) and the previous inequality give

$$\mu\big(B_{\ell R}(m_0)\big) \leq q^{-\tilde{s}} C_1 \ell^{\tilde{s}} \varlimsup_{j\to\infty} \mu_j\big(\overline{B}_{qR+\delta}(m_0)\big) \leq q^{-\tilde{s}} C_1 \ell^s \mu\big(\overline{B}_{qR+\delta}(m_0)\big).$$

4.3. Existence of doubling measures 369

Letting δ go to zero and noting that $\overline{B}_{qR}(m_0) \subset B_R(m_0)$ we get

$$\mu(B_{\ell R}(m_0)) \leq q^{-\tilde{s}} C_1 \ell^s \mu(B_R(m_0)).$$

Since $q < 1$ is arbitrary, we obtain the required inequality.

It remains to check that $\mu(B) > 0$ for every open ball in $\mathcal{M}$. This follows from the inequality

$$1 = \mu(\mathcal{M}) \leq \widetilde{C} \ell^{\tilde{s}} \mu(B)$$

where $\ell := \operatorname{diam} \mathcal{M} / \operatorname{diam} B$. $\square$

4.3.3 Complete metric spaces

Let $(\mathcal{M}, d)$ be a (C, s)-homogeneous *complete* metric space. We fix $m_0 \in \mathcal{M}$ and set $\mathcal{M}_j := \overline{B}_j(m_0)$, $j \in \mathbb{N}$. Since $\mathcal{M}$ is doubling and complete, $\mathcal{M}_j$ is a compact metric subspace of $\mathcal{M}$ which is clearly $(\widetilde{C}, \tilde{s})$-homogeneous. Using Proposition 4.49 we find a $(\widetilde{C}_1, \tilde{s})$-homogeneous measure μ_j on $\mathcal{M}$ supported on $\mathcal{M}_j$. Normalizing μ_j, if necessary, we may assume that $\mu_j(\mathcal{M}_1) = 1$. Then we have, for every $k \leq j$,

$$\mu_j(\mathcal{M}_k) \leq \widetilde{C}_1 k^{\tilde{s}} \mu_j(\mathcal{M}_1) = \widetilde{C}_1 k^{\tilde{s}}.$$

Applying the Cantor diagonal process to the family $\{\mu_j\}_{j \in \mathbb{N}}$ and employing, as in the proof of Proposition 4.49, the $*$-weak compactness argument, we construct a subfamily $\{\mu_j\}_{j \in J}$, $J \subset \mathbb{N}$, such that for some measure μ on $\mathcal{M}$ and every $k \in \mathbb{N}$ the sequence $\{\mu_j|_{\mathcal{M}_k}\}_{j \in J}$ is $*$-weakly convergent to $\mu|_{\mathcal{M}_k}$. Hence, (4.63) holds for every *compactly supported* continuous function f.

As in the proof of Proposition 4.49, we derive from this version of (4.63) that $D(\ell; \mu) \leq \widetilde{C}_1 \ell^{\tilde{s}}$ for all $\ell \geq 1$. Moreover, since $\mathcal{M}_1$ is contained in every open ball $B_{\ell R}(m_0)$ with sufficiently large ℓ, we get

$$1 = \mu(\mathcal{M}_1) \leq \mu(B_{\ell R(m_0)}) \leq \widetilde{C} \ell^{\tilde{s}} \mu(B_R(m_0)),$$

i.e., the μ-measure of every open ball in $\mathcal{M}$ is positive.

This completes the proof of the main result, Theorem 4.43.

4.3.4 Dyn'kin conjecture

The Vol'berg–Koniagin result was motivated by the papers [Dy-1983, Dy-1984] by Dyn'kin where the existence of a doubling measure was proved for "porous" subsets of $[0, 1]$, and a very attractive way of deriving the key part of the Vol'berg–Koniagin theorem concerning finite metric space (Proposition 4.44) was proposed. We present the Dyn'kin combinatorial conjecture as a challenge for the reader and show that it, in fact, is equivalent to Proposition 4.44.

Let $(\mathcal{M}, d)$ be a metric space and $\mathcal{F}$ be a family of its subsets. By $\mathrm{ord}(\mathcal{F}; m)$ we denote the number of subsets from $\mathcal{F}$ containing m, i.e.,

$$\mathrm{ord}(\mathcal{F}; m) := \sum_{S \in \mathcal{F}} \chi_S(m). \tag{4.66}$$

If then $\mathcal{F}$ is a family of balls B, then notation $2\mathcal{F}$ stands for the family of balls $2B$, i.e., the balls of the same centers as B's by twice bigger radii.

Conjecture 4.50. *Let $(\mathcal{M}, N)$ be a doubling finite metric space with the doubling constant $\delta_\mathcal{M}$. Then there exists a constant D depending only on $\delta_\mathcal{M}$ such that for every finite family of balls in M there is a point $\hat{m}$ such that*

$$\mathrm{ord}(2\mathcal{F}; \hat{m}) \leq D\,\mathrm{ord}(\mathcal{F}; \hat{m}). \tag{4.67}$$

It is important to note that $\mathcal{F}$ may contain several balls with the same center and radius.

We will show that the assertion of the conjecture is equivalent to that of Proposition 4.44; the direct proof of the conjecture is unknown.

Proposition 4.51. *If the conjecture is true for a finite metric space $\mathcal{M}$, then M carries a doubling measure with the doubling constant bounded by D.*

Proof. Let $\mathcal{M} := \{m_1, \ldots, m_N\}$ be a finite doubling metric space, and $\{k_j\}_{1 \leq j \leq N}$ be a family of positive integers. We consider a cover $\mathcal{F}$ of $\mathcal{M}$ consisting of k_i balls B_i of common center m_i and of the same radius for all $1 \leq i \leq N$. According to the conjecture a function $f : \mathcal{M} \to \mathbb{R}$ given by

$$f(m) := \mathrm{ord}(2\mathcal{F}; m) - D\,\mathrm{ord}(\mathcal{F}; m)$$

attains nonpositive values. By (4.66), this function can be presented as

$$f = \sum_{i=1}^{N} k_i (\chi_{2B_i} - D\chi_{B_i}).$$

Replacing k_i by positive rationals and passing to the limit we derive from here that every linear combination $\sum_{i=1}^{N} \alpha_i (\chi_{2B_i} - D\chi_{B_i})$ with nonnegative $\alpha_i \in \mathbb{R}$ is nonpositive on $\mathcal{M}$. Let us denote the cone of all these combinations by K and the cone of all strictly positive functions on $\mathcal{M}$ by P_+. Then we conclude that

$$P_+ \cap K = \phi.$$

By the Hahn–Banach theorem, there exists a linear functional on $\ell_\infty(\mathcal{M})$ (i.e., a discrete measure μ on $\mathcal{M}$) that separates these cones. Hence, for this μ,

$$\mu|_{P_+} > 0 \quad \text{and} \quad \mu|_{K} \leq 0,$$

which clearly means that, for every B_i,

$$0 < \mu(B_i) \quad \text{and} \quad \mu(2B_i) \leq D\mu(B_i).$$

In other words, μ is the required doubling measure. □

Proposition 4.52. *Conjecture* 4.50 *is true.*

Proof. Let $\mathcal{M}$ be a finite doubling metric space with the doubling constant $\delta_{\mathcal{M}}$. Let $\mathcal{F}$ be a finite family of balls in $\mathcal{M}$. By Proposition 4.44, there exists a measure μ on $\mathcal{M}$ such that for some constant $C(\delta_{\mathcal{M}})$ and every open ball $B_r(m)$,

$$\mu(B_{2r}(m)) \leq C\mu(B_r(m)). \tag{4.68}$$

As before, we consider the cone P_+ of all (strictly) positive functions on $\mathcal{M}$ and the cone K spanned by the functions $\chi_{2B} - C\chi_B$ where B runs over $\mathcal{F}$. Regarding μ as a linear functional on the space $\ell_\infty(\mathcal{M})$ we derive from (4.68) that

$$\mu\big|_{P_+} > 0 \quad \text{and} \quad \mu\big|_K \leq 0.$$

Hence $P_+ \cap K = \emptyset$ and therefore every function from K is nonpositive at some point $\widehat{m} \in \mathcal{M}$. This is clearly equivalent to (4.67) with $D = C(\delta_{\mathcal{M}})$. □

4.3.5 Concluding remarks

(a) In [VK-1987] the authors construct, for a given $0 < s < n$, an s-homogeneous compact metric subspace of $\mathbb{R}^n$ which does not carry any s-homogeneous measure. Hence, the condition $\tilde{s} > s$ in Theorem 4.43 cannot be removed.

However, *every n-homogeneous compact subspace of $\mathbb{R}^n$ does carry an n-homogeneous measure.*

The proof of this result in [VK-1987] should be corrected in one point concerning the choice of A^{-j}-nets S_j, $0 \leq j \leq N$, see Luukainen [Lu-1998, subsec. 6.13] for details.

(b) It is readily seen that an Ahlfors s-regular metric space $\mathcal{M}$ carries an s-homogeneous measure, e.g., the Hausdorff measure $\mathcal{H}_s$ on $\mathcal{M}$ is s-homogeneous.

If, in addition, $\mathcal{M}$ is a compact subspace of $\mathbb{R}^n$, then all s-measures on $\mathcal{M}$ are mutually absolutely continuous, see [Jons-1995].

In particular, every classical fractal and, more generally, every self-similar subset of $\mathbb{R}^n$ satisfying the condition of Theorem 4.21 carries an s-homogeneous measure where s is the Hausdorff dimension of the subset. Moreover, every such s-homogeneous measure and the Hausdorff measure $\mathcal{H}_s$ are mutually absolutely continuous.

However, even a compact interval of the real line carries an uncountable family of doubling measures which are mutually singular. This was pointed out in [VK-1987] by referring to the following construction of Beurling and Ahlfors [BA-1956].

Let $I \subset [0,1]$ be a compact interval of length $< \frac{2\pi}{3}$ and let k be the smallest integer such that $|I| > \frac{2\pi}{3^k}$. Then the required family is comprised by the measures

$$\mu := \prod_{j=1}^{k-1}(1 + a_j \cos 3^j x)\mu_k$$

where

$$\mu_k := \prod_{j=k}^{\infty}(1 + a_j \cos 3^j x)$$

and $a := (a_j)_{j \in \mathbb{N}}$ is a sequence of real numbers satisfying $\sup_j |a_j| < 1$.

In this definition, the infinite product which defines μ_k is regarded as the limit of finite products taken in the sense of distributions. Therefore, μ_k is a linear continuous functional on the space of 2π-periodic C^∞-functions, which, in this case, can be extended to a continuous linear functional on $C[0, 2\pi]$. Using the Riesz representation theorem we then conclude that every μ_k and, hence, μ can be regarded as a (Borel regular) measure.

A detailed discussion of this subject is presented in the Semmes survey [Sem-1999].

(c) The assumption of compactness in the Vol'berg–Koniagin theorem cited in (c) cannot be removed. In fact, in every $\mathbb{R}^n$ there exists an open connected metric subspace which does not carry any doubling measure, see Saksman [Sa-1999] and the discussion in Heinonen [Hei-2001, p. 107], where the result for $n = 1$ is derived from the work of Tukia [Tu-1989].

4.4 Space of balls

The space of balls $\mathcal{B}(\mathcal{M})$ is recalled to consist of nonempty open balls of a *length space* $(\mathcal{M}, d)$. In subsection 3.3.6, $\mathcal{B}(\mathcal{M})$ is considered as a metric graph whose metric is determined by a weight $\widehat{\omega}$ defined on each edge $B \subset B'$ by

$$\widehat{\omega}([B, B']) := \frac{\omega(r_{B'})}{r_B}.$$

Then it was proved in Proposition 3.127 that this graph metric is equivalent to another one defined by the weight

$$w([B, B']) := \frac{\omega(r)}{r} + \frac{\omega(r')}{r'} + \int_r^{r+\hat{d}(B,B')} \frac{\omega(t)}{t^2}\, dt, \qquad (4.69)$$

4.4. Space of balls

where $B \subset B'$ are balls (the endpoints of the edge $[B, B']$) of radii r, r' and of centers m, m', respectively; moreover, $\widehat{d}$ is a metric on $\mathcal{B}(\mathcal{M})$ given by

$$\widehat{d}(B_1, B_2) := |r_1 - r_2| + d(m_1, m_2) \tag{4.70}$$

for each pair $B_i := B_{r_i}(m_i)$, $i = 1, 2$ (not necessarily embedded).

Let us recall that the proposition was proved under the assumption that the function $t \mapsto \frac{\omega(t)}{t^2}$ is nonincreasing and ω is nondecreasing and $\omega(0+) = 0$.

Now we will prove a general result concerning weights similar to the integral part in (4.69) but extended to all pairs B, B', i.e., for a function $d_\omega : \mathcal{B}(\mathcal{M}) \times \mathcal{B}(\mathcal{M}) \to \mathbb{R}_+$ given by

$$d_\omega(B, B') := \int_{\min(r_B, r_{B'})}^{\min(r_B, r_{B'}) + \widehat{d}(B, B')} \omega(t) dt; \tag{4.71}$$

here $\widehat{d}$ is given by (4.70).

We show that under mild assumptions on $\omega : (0, +\infty) \to \mathbb{R}_+$ this is a metric equivalent to the length metric on $\mathcal{B}(\mathcal{M})$ associated to d_ω. Moreover, the corresponding space $\mathcal{B}(\mathbb{R}^n)$ of Euclidean balls is, as we will see, of pointwise homogeneous type, see subsection 3.2.6 for the definition. As a consequence we will show that the generalized hyperbolic space $\mathbb{H}_\omega^{n+1}$ of Example 3.113 is of pointwise homogeneous type for this choice of ω.

4.4.1 $\mathcal{B}(\mathcal{M})$ as a length space

A length space is recalled to be a metric space where the distance between any pair of points equals the infimum of lengths of curves joining these points. Hence, if we define the inner metric $\widetilde{d} : \mathcal{M} \to [0, +\infty]$ of a metric space $(\mathcal{M}, d)$ by

$$\widetilde{d}(m_1, m_2) := \inf_\gamma \ell(\gamma), \tag{4.72}$$

where γ's are curves joining m_1 and m_2, then $\mathcal{M}$ is a length space if and only if $d = \widetilde{d}$. Moreover, if $\widetilde{d} < +\infty$, then $(\mathcal{M}, \widetilde{d})$ is a length space, see Proposition 3.36.

The main result of this subsection is

Theorem 4.53. *Let $(\mathcal{M}, d)$ be a length space. Assume that $\omega : (0, +\infty) \to \mathbb{R}_+$ is continuous and nonincreasing. Then the following is true.*

(a) *The function d_ω given by (4.71) is a metric on $\mathcal{B}(\mathcal{M})$.*

(b) *If δ is a metric on $\mathcal{M}$ equivalent to d, then the associated metrics d_ω and δ_ω are equivalent.*

(c) *The inner metric $\widetilde{d}_\omega$ generated by d_ω satisfies*

$$d_\omega \leq \widetilde{d}_\omega \leq 3 d_\omega. \tag{4.73}$$

Proof. For simplicity we assume that diam $\mathcal{M} = \infty$ and use the bijection $(r, m) \mapsto B_r(m)$ to identify $\mathcal{B}(\mathcal{M})$ with $(0, +\infty) \times \mathcal{M}$. In the sequel, we refer to (r_B, c_B) as the coordinates of the open ball B.

(a) Since ω is nonincreasing, we have for $0 < a - h < b - h$ and $h > 0$,

$$\int_a^b \omega(t)dt \leq \int_{a-h}^{b-h} \omega(t)dt. \tag{4.74}$$

Using this we prove the triangle inequality

$$d_\omega(B, B') \leq d_\omega(B, B'') + d_\omega(B'', B'), \tag{4.75}$$

the only metric axiom to be checked for d_ω.

Without loss of generality we assume that $r \leq r'$; here $(m, r), (m', r'), (m'', r'')$ are the coordinates of the balls in (4.75). Hence, the left-hand side of (4.75) equals

$$\int_r^{r+\hat{d}(m,m')} \omega(t)dt$$

and the terms of the right-hand side are written similarly.

We have the following three possibilities:

(i) $r'' \leq r$;

(ii) $r \leq r'' \leq r'$;

(iii) $r' \leq r''$.

In case (i) we first apply (4.74) with $h := r - r''$ and the triangle inequality for $\hat{d}$, see (4.70), to get

$$d_\omega(B, B') \leq \int_{r''}^{r''+\hat{d}(m,m')} \omega(t)dt \leq \int_{r''}^{r''+\hat{d}(m,m'')+\hat{d}(m'',m')} \omega(t)dt$$

$$= d_\omega(B, B'') + \int_{r''+\hat{d}(m,m'')}^{r''+\hat{d}(m,m'')+\hat{d}(m'',m')} \omega(t)dt.$$

Applying to the second summand (4.74) with $h := \hat{d}(m, m'')$ we estimate it by $d_\omega(B', B'')$ and prove (4.75) for this case.

In case (ii), we write as before

$$d_\omega(B, B') \leq d_\omega(B, B'') + \int_{r+\hat{d}(m,m'')}^{r+\hat{d}(m,m'')+\hat{d}(m'',m')} \omega(t)dt \tag{4.76}$$

4.4. Space of balls

and estimate the second summand by (4.74) with $h := r + \widehat{d}(m, m'') - r''$ which is nonnegative as $\widehat{d}(m, m'') \geq |r - r''|$, see (4.70).

In the last case, we estimate the second summand in (4.76) by (4.74) with $h := r + \widehat{d}(m, m'') - r' \geq r + |r - r''| - r'' \geq 0$.

(b) It suffices to prove that if two metrics on $\mathcal{B}(\mathcal{M})$, say, d and δ, satisfy $d \leq N\delta$ for some integer $N \geq 1$, then associate metrics d_ω and δ_ω satisfy $d_\omega \leq N\delta_\omega$. For this we (assuming that $r \leq r'$) simply write

$$d_\omega(B, B') \leq \int_r^{r+N\hat{\delta}(m,m')} \omega(t)dt = \left(\sum_{j=0}^{N-1} \int_{r+j\hat{\delta}(m,m')}^{r+(j+1)\hat{\delta}(m,m')}\right) \omega(t)dt$$

and then apply (4.74) with $h := j\hat{\delta}(m, m')$ to the j-th integral of the sum for $j = 0, 1, \ldots, N-1$.

This proof shows that the assertion remains true if we replace $\widehat{d}$ in the definition of d_ω, see (4.71), by the maximum of the standard metric of $\mathbb{R}_+$ and a metric equivalent to d. We use below this variant of (b).

(c) Let $B, B' \subset \mathcal{B}(\mathcal{M})$ have coordinates (r, m) and (r', m'), respectively. Without loss of generality, we assume that $r \leq r'$.

We will prove, given $\varepsilon > 0$, that there is a curve $\gamma : [0, 1] \to \mathcal{B}(\mathcal{M})$ joining B and B' such that its length satisfies

$$\ell(\gamma) \leq (3 + \varepsilon)d_\omega(B, B').$$

We construct γ by juxtaposing two curves: γ_1 joining B with the ball B'' whose coordinates are (r', m) and γ_2 joining B'' and B'. It will be proved that for the first curve

$$\ell(\gamma_1) = d_\omega(B, B''), \qquad (4.77)$$

while for the second one

$$\ell(\gamma_2) \leq (3 + \varepsilon)(1 + \varepsilon')d_\omega(B'', B') \qquad (4.78)$$

with arbitrary $\varepsilon' > 0$. Because of the choice of B'',

$$d_\omega(B, B') = d_\omega(B, B'') + d_\omega(B'', B'),$$

see (4.71). Therefore the previous inequalities give

$$\widetilde{d}_\omega(B, B') \leq \ell(\gamma) \leq \ell(\gamma_1) + \ell(\gamma_2) \leq (3 + \varepsilon)(1 + \varepsilon')d_\omega(B, B').$$

Letting ε and ε' to 0 we obtain the required result.

We define the first curve by $\gamma_1(t) := B_t$ where the ball B_t has coordinates $(r + (r' - r)t, m)$, $0 \leq t \leq 1$. Clearly, γ_1 joins B and B''. Such a curve will be called in the sequel a "vertical" one.

Lemma 4.54. *Let $B_1, B_2 \in \mathcal{B}(\mathcal{M})$ have coordinate (r_1, m_1) and (r_2, m_2) and $m_1 = m_2 =: \widetilde{m}$. Then the vertical curve v joining B_1 and B_2 satisfies*

$$\ell(v) = d_\omega(B_1, B_2).$$

Proof. Let $\pi := \{B^0, B^1, \ldots, B^n\}$ be a partition of $v([0,1])$ and $0 =: t_0 < t_1 < \cdots < t_n := 1$ be the associated partition of $[0,1]$, i.e., B^i has coordinates $(r_1 + (r_2 - r_1)t_i, \widetilde{m})$, $i = 0, 1, \ldots, n$. By the definition of d_ω,

$$\ell(\pi) := \sum_{j=0}^{n-1} d_\omega(B^j, B^{j+1}) = d_\omega(B^0, B^k) = d_\omega(B_1, B_2)$$

and therefore

$$\ell(v) = \lim_{\max(t_{i+1} - t_i) \to 0} \ell(\pi) = d_\omega(B_1, B_2). \qquad \square$$

Hence, joining the balls B and B'' by the vertical curve (denoted by γ_1), we prove (4.77).

Now we will construct the second curve γ_2 joining B'' and B' by juxtaposing alternatively vertical and horizontal curves. The latter is defined as follows.

Let B_1, B_2 be open balls in $\mathcal{M}$ having equal radii, say $s > 0$, and centers at m_1 and m_2. Given $q > 1$, there is a curve $\gamma : [0,1] \to \mathcal{M}$ joining m_1 and m_2 such that

$$\ell(\gamma) < q \cdot d(m_1, m_2). \tag{4.79}$$

This γ exists because $\mathcal{M}$ is a length space.

We then define a *horizontal q-curve* $h : [0,1] \to \mathcal{B}(\mathcal{M})$ by

$$h(t) := B_s(\gamma(t)), \quad 0 \leq t \leq 1.$$

Clearly, h joins B_1 and B_2.

We need a result estimating the length of a horizontal curve. For its formulation one determines a number $q > 1$ by $q^3 = 1 + \varepsilon$. Employing then uniform continuity of $\log \omega$ on compact intervals we find a number $\delta(q) > 0$ such that

$$\omega(t_2) \leq \omega(t_1) < q\omega(t_2) \tag{4.80}$$

whenever $r \leq t_1 \leq t_2 \leq 2(d(m, m') + r)$; recall that r, r' and m, m' are the radii and the centers of the initial balls B and B' and $r \leq r'$.

Lemma 4.55. *Let $m_1, m_2 \in \mathcal{M}$ satisfy*

$$d(m_1, m_2) < \delta(q). \tag{4.81}$$

Then for every $s \in [r, 2(d(m, m') + r) - \delta(q)]$ there is a horizontal q-curve h_s joining $B_s(m_1)$ and $B_s(m_2)$ such that

$$\ell(h_s) \leq q^3 d_\omega(B_s(m_1), B_s(m_2)). \tag{4.82}$$

4.4. Space of balls

Proof. By definition, $h_s(t) = B_s(\gamma(t))$, $0 \le t \le 1$, for some curve γ in $\mathcal{M}$ joining m_1 and m_2 and obeying (4.79). Let $\pi := \{m^0, m^1, \ldots, m^n\}$ be a partition of $\gamma([0,1])$. We choose π to be so fine that

$$\text{mesh } \pi := \max_i d(m^i, m^{i+1}) < \delta(q) \tag{4.83}$$

and, moreover,

$$\ell(\pi) := \sum_{i=0}^{n-1} d(m_i, m_{i+1}) \le q\ell(\gamma). \tag{4.84}$$

Let $0 =: t_0 < t_1 < \cdots < t_n := 1$ be the partition of $[0,1]$ associated to π and let $\widehat{\pi} := \{B_s(m_0), \ldots, B_s(m_n)\}$ be the corresponding partition of $h_s([0,1])$. By the mean value theorem and the equality $\widehat{d}(B_s(m_i), B_s(m_{i+1})) = d(m_i, m_{i+1})$, see (4.70), we get

$$\ell(\widehat{\pi}) := \sum_{i=0}^{n-1} d_\omega(B_s(m_i), B_s(m_{i+1})) = \sum_{i=0}^{n-1} \int_s^{s+d(m_i, m_{i+1})} \omega(t) dt$$
$$= \sum_{i=0}^{n-1} \omega(\tau_i) d(m_i, m_{i+1}), \tag{4.85}$$

where $\tau_i \in (s, s + d(m_i, m_{i+1})) \subset (s, s + \delta(q))$, see (4.83).

By the mean value theorem, there is t satisfying

$$\omega(t) = \frac{1}{d(m_1, m_2)} \int_s^{s+d(m_1, m_2)} \omega(u) du.$$

Since the values of the integrand lie in the interval $\big(\omega(s + d(m_1, m_2)), \omega(s)\big)$, this t satisfies

$$s \le t \le s + d(m_1, m_2) < s + \delta(q).$$

Hence, $|t - \tau_i| < \delta(q)$ for any i, and by (4.80) and the conditions of the lemma, $\omega(\tau_i) < q\omega(t)$.

It follows from this and (4.85) that

$$\ell(\widehat{\pi}) < q\omega(t)\ell(\pi).$$

Using estimates (4.85), (4.79) and the choice of t we bound the right-hand side by

$$q^3 \cdot \int_s^{s+d(m_1, m_2)} \omega(t) dt = q^3 d_\omega(B_s(m_1), B_s(m_2)).$$

It remains to note that if mesh π tends to zero, then mesh $\widehat{\pi}$ does as well. Hence,
$$\ell(h_s) = \lim_{\text{mesh } \pi \to 0} \lim \ell(\widehat{\pi}) \leq q^3 d_\omega\big(B_s(m_1), B_s(m_2)\big),$$
as required. □

We now construct the required curve γ_2 joining $B := B_r(m)$ and $B' = B_{r'}(m')$.

Let γ be a curve in $\mathcal{M}$ joining m and m' and satisfying
$$\ell(\gamma) \leq 2\big(r' + d(m, m') - \delta(q)\big). \tag{4.86}$$

By $\pi := \{m_0, m_1, \ldots, m_n\}$ we denote a partition of $\gamma([0,1])$ such that $d(m_i, m_{i+1}) < \delta(q)$, $i = 0, 1, \ldots, n-1$. We further define a sequence of balls with centers m_i and of radii r_i, where $m_0 = \gamma(0) = m$, $m_n := \gamma(1) = m'$ and
$$r_i := r + \sum_{j=0}^{n-1} d(m_j, m_{j+1}), \quad 0 \leq i \leq n-1, \quad \text{and} \quad r_0 := r.$$

Actually, we define the sequence as
$$(B =) B_{r_0}(m_0), B_{r_0}(m_1), B_{r_1}(m_1), B_{r_1}(m_2),$$
$$\ldots, B_{r_{n-1}}(m_n), B_{r_n}(m_n), B'(= B_{r'}(m_n)).$$

Connect every pair $B_{r_i}(m_i), B_{r_i}(m_{i+1})$ by a horizontal q-curve h_i as in Lemma 4.55, and every pair $B_{r_i}(m_i), B_{r_{i+1}}(m_i)$ by the vertical curve v_i as in Lemma 4.54. So we have, for $i = 0, 1, \ldots, n-1$,
$$\ell(h_i) \leq q^3 d_\omega\big(B_{r_i}(m_i), B_{r_i}(m_{i+1})\big),$$
$$\ell(v_i) = d_\omega\big(B_{r_i}(m_{i+1}), B_{r_{i+1}}(m_{i+1})\big).$$

At the final step we join the final pair $B_{r_n}(m'), B' = B_{r'}(m')$ by the vertical curve v_n so that
$$\ell(v_n) = d_\omega\big(B_{r_n}(m'), B_{r'}(m')\big).$$

Summing all these inequalities and noting that
$$d_\omega\big(B_{r_i}(m_i), B_{r_i}(m_{i+1})\big) = \int_{r_i}^{r_i + d(m_i, m_{i+1})} \omega(t)\,dt = \int_{r_i}^{r_{i+1}} \omega(t)\,dt$$

and that the same is true for the pair $B_{r_i}(m_i), B_{r_{i+1}}(m_i)$, we get, for the curve $\widetilde{\gamma}_2$ formed by all h_i and v_i,
$$\ell(\widetilde{\gamma}_2) = \sum_{i=0}^{n-1} \big(\ell(h_i) + \ell(v_i)\big) + \ell(v_n) \leq (2 + q^3) \int_r^{r_n} \omega(t)\,dt.$$

4.4. Space of balls

Now note that
$$r_n := r + \sum_{i=0}^{n-1} d(m_i, m_{i+1}) =: r + \ell(\pi)$$

and that $\lim_{\text{mesh } \pi \to 0} \ell(\pi) = \ell(\gamma)$. So we may, given $\eta > 0$, choose the partition π so that
$$\ell(\tilde{\gamma}_2) \leq (2 + q^3) \int_r^{r+\eta \ell(\gamma)} \omega(t) dt.$$

Moreover, $\mathcal{M}$ is a length space and therefore $d(m, m') = \inf_\gamma \ell(\gamma)$, where γ runs over all curves joining m and m' and satisfying (4.86). So we may assume that the previous inequality holds with $\ell(\gamma)$ replaced by $d(m, m')$. This leads to the inequality
$$\ell(\tilde{\gamma}_2) < (2 + q^3) \left(\int_{r_1}^{r_1 + d(m_1, m_2)} \omega(t) dt + \varepsilon(\eta) \right) = (2 + q^3)\bigl(d_\omega(B', B) + \varepsilon(\eta)\bigr),$$

where $\varepsilon(\eta) \to 0$ as $\eta \to 1$.

Since $q^3 = 1 + \varepsilon$, this proves the desired inequality (4.78) and completes the proof of the theorem. $\square$

4.4.2 $\mathcal{B}(\mathbb{R}^n)$ as a space of pointwise homogeneous type

According to the definition, see subsection 3.2.6, we must find a family of measures on $\mathcal{B}(\mathbb{R}^n)$ which is uniform and consistent with the metric d_ω. By Theorem 4.53 (b), it suffices to find such a family for the space $\mathcal{B}(\ell_\infty^n)$. Since the balls of ℓ_∞^n are cubes, we will denote the points of $\mathcal{B}(\ell_\infty^n)$ by Q, Q', etc. Let us recall that an open cube of side length $2r$ and of center $c \in \mathbb{R}^n$ denoted by $Q_r^o(c)$ is given by
$$Q_r^o(c) := \{x \in \mathbb{R}^n \,;\, |x - c| < r\}$$

where $|y| := \max_{1 \leq i \leq n} |y_i|$.

It is reasonable to simply identify a point $Q_r^o(c) \in \mathcal{B}(\ell_\infty^n)$ with the point (c, r) of the upper half-space
$$H_+^{n+1} := \{(x, r) \,;\, x \in \mathbb{R}^n, \ r > 0\}$$

to make more transparent the subsequent calculations.

Now we construct the required family of measures $\{\mu_Q\}_{Q \in \mathcal{B}(\ell_\infty^n)}$ for the metric d_ω where the nonincreasing function ω satisfies for all $r > 0$ the condition
$$\int_r^\infty \omega(s) ds = \infty. \tag{4.87}$$

Let $S \subset H_+^{n+1}$ be a Borel set and S^+ be its "upper" one-half given by

$$S^+ := \{(c,r) \in S \,;\, r \geq M_S\}$$

where M_S is a "vertical" median of S given by

$$M_S := \frac{1}{2}\left[\sup_{(c,r)\in S} r + \inf_{(c,r)\in S} R\right].$$

Given $Q = Q_r^o(c)$, a measure μ_Q is then defined by

$$\mu_Q(S) := \int_{S^+} \left(\prod_{i=1}^n \omega(r + |x_i - c_i|)\right) \omega(r + |x_{n+1} - r|) dx_1 \ldots dx_{n+1}. \qquad (4.88)$$

Setting

$$\Omega_s(t) := \int_s^{s+t} \omega(u) du, \quad s, t > 0, \qquad (4.89)$$

we rewrite (4.88) for $Q := Q_{x_{n+1}}^o(x)$ in the form

$$\mu_Q(S) = \int_{S_+} \left(\prod_{i=1}^{n+1} \Omega'_{x_{n+1}}(|x_i - y_i|)\right) dy_1 \ldots dy_{n+1}. \qquad (4.90)$$

Since by the definition of d_ω, see (4.71), for $Q = Q_r^o(c)$, $Q' = Q_{r'}^o(c')$ the identity

$$d_\omega(Q, Q') = \Omega_{\min(r,r')}(|c - c'|) = \max_{1 \leq i \leq n} \Omega_{\min(r,r')}(|c_i - c'_i|)$$

is true, we may write the upper half-ball $B_R^+(Q) \subset \mathcal{B}(\ell_\infty^n)$, where $Q = Q_r^o(c)$, as follows:

$$B_R^+(Q) = \left\{(x,r) \in H_+^{n+1} \,;\, \Omega_r(|x_i - c_i|) \leq R, \ i = 1, \ldots, n, \ \Omega_r(|x_{n+1} - r|) \leq R\right\}.$$

The right-hand side is the direct product

$$B_R^+(Q) = \prod_{j=1}^{n+1} I_j \qquad (4.91)$$

of the sets given for $1 \leq j \leq n$ by

$$I_j := \{s \in \mathbb{R} \,;\, \Omega_r(|s - c_j|) \leq R\}$$

and for $j = n+1$ by

$$I_{n+1} := \{s \in \mathbb{R} \,;\, \Omega_r(|s - r|) \leq R \ \text{ and } \ s \geq r\}.$$

4.4. Space of balls 381

Setting for brevity $r := c_{n+1}$ we then rewrite (4.90) as

$$\mu_Q(B_R(Q)) = \prod_{j=1}^{n+1} \int_{I_j} \Omega'_r(|s-c_j|)\,ds. \qquad (4.92)$$

Since $t \mapsto \Omega_r(t)$ is strictly increasing from 0 to $+\infty$ on $[r,+\infty)$, see (4.87), each set I_j with $j \leq n$ is an interval determined by the inequality $|s-c_j| \leq \Omega_r^{-1}(R)$. Therefore the corresponding integral in (4.92) equals $2\Omega_r(\Omega_r^{-1}(R)) = 2R$. Similarly, I_{n+1} is determined by the inequality $0 \leq s-r \leq \Omega_r^{-1}(R)$ and the $(n+1)$-th integral equals R. Hence,

$$\mu_Q(B_R(Q)) = 2^n R^{n+1} \qquad (4.93)$$

and the family $\mathcal{F}$ is $(n+1)$-*homogeneous*. In particular, $\mathcal{F}$ is 1-*uniform* with $D(\mathcal{F}) = 2^{n+1}$, see Definition 3.89.

It remains to show that $\mathcal{F}$ is consistent with the metric d_ω. Having in mind consequent applications we will prove a slight generalization of the consistency inequality required by Definition 3.90. Actually, we will prove that for every pair $Q_1 = Q^o_{r_1}(c^1)$ and $Q_2 = Q^o_{r_2}(c^2)$ such that $d_\omega(Q_1, Q_2) \leq \lambda R$, where $0 < \lambda \leq 1$, we have

$$|\mu_{Q_1} - \mu_{Q_2}|(B_R(Q_2)) \leq \varphi_n(\lambda)\frac{B_R(Q_2)}{R} d_\omega(B_R(Q_1), B_R(Q_2)); \qquad (4.94)$$

here we set $\varphi_n(\lambda) := \frac{3(n+1)}{2} \cdot \frac{(1+\lambda)^{n+1}-1}{\lambda}$. Changing places of Q_1 and Q_2 we obtain the same inequality for $B_R(Q_1)$.

To prove (4.94) we set, for brevity, $B := B_R(Q_2)$ and

$$v_i(t) := \Omega_{r_1}(|t-c_i^1|), \quad w_i(t) := \Omega_{r_2}(|t-c_i^2|), \qquad (4.95)$$

where $1 \leq i \leq n+1$; we also convene that $c^1_{n+1} := r_1$, $c^2_{n+1} := r_2$. Further, by I_j we denote the corresponding intervals in (4.91) for the ball $B_{(1+\lambda)R}(Q_1)$ and by J_j the intervals for the ball $B_R(Q_2)$. Since the centers Q_1, Q_2 satisfy the inequality $d_\omega(Q_1, Q_2) \leq \lambda R$, the embedding $B := B_R(Q_2) \subset B_{(1+\lambda)R}(Q_1)$ is true and we conclude that $J_j \subset I_j$, $1 \leq j \leq n+1$.

Therefore equality (4.92) and the embeddings give

$$|\mu_{Q_1} - \mu_{Q_2}|(B) = \int_{B^+} \left|\prod_{j=1}^{n+1}(v'_j(x_j) - w'_j(x_j))\right| dx_1 \ldots dx_{n+1}$$

$$\leq \sum_{k=1}^{n+1} \left(\prod_{j=1}^{k-1} \int_{J_j} w'_j(x_j)\,dx_j \prod_{j=k+1}^{n+1} \int_{I_j} v'_j(x_j)\,dx_j\right) \int_{J_k} |v'_k(x_k) - w'_k(x_k)|\,dx_k.$$

The argument used for the proof of (4.93) gives, for the product in the brackets of the right-hand side, the bound

$$2^{k-1}R^k \cdot 2^{n-k}((1+\lambda)R)^{n-k+1} = 2^{n-2}(1+\lambda)^{n-k+1}R^n.$$

This leads to the inequality

$$|\mu_{Q_1} - \mu_{Q_2}|(B) \leq 2^{n-2} R^n \sum_{k=1}^{n+1} (1+\lambda)^{n-k+1}$$

$$\times \max_{1 \leq k \leq n+1} \int_{J_k} |v'_k(x_k) - w'_k(x_k)| dx_k = 2^{n-2} \cdot \frac{(1+\lambda)^{n+1} - 1}{\lambda} \quad (4.96)$$

$$\times \max_{1 \leq k \leq n+1} \int_{J_k} \left| \omega(r_1 + |s - c_k^1|) - \omega(r_2 + |s - c_k^2|) \right| ds.$$

We now show that the maximum is at most $6 d_\omega(Q_1, Q_2)$. Together with (4.93) this leads to the inequality

$$|\mu_{Q_1} - \mu_{Q_2}|(B) \leq 6 \cdot \frac{(1+\lambda)^{n+1} - 1}{\lambda} (n+1) 2^{n-2} R^n d_\omega(Q_1, Q_2)$$

$$= \varphi_n(\lambda) \frac{\mu_{Q_2}(B)}{R} d_\omega(Q_1, Q_2).$$

Hence, the required inequality (4.94) would be established together with the estimate $C(\mathcal{F}) \leq 3(n+1)2^n$ (for $\lambda = 1$).

To avoid problems related to integrability of ω over $\mathbb{R}_+$ we first estimate the k-th integral in (4.96) for ω replaced by an integrable approximate function ω_N defined by

$$\omega_N := [\omega]^N \chi_{[0,N]}$$

where $[\omega]^N$ is the cut-off of ω at the level $\omega(\frac{1}{N})$, $N \in \mathbb{N}$.

Assuming for definiteness that $r_1 \leq r_2$ we write

$$\int_{J_k} \left| \omega_N(r_1 + |s - c_k^1|) - \omega_N(r_2 + |s - c_k^2|) \right| ds \leq A_1 + A_2,$$

where we set

$$A_1 := \int_{\mathbb{R}} \left| \omega_N(r_1 + |t - c_k^1|) - \omega_N(r_2 + |t - c_k^1|) \right| dt,$$

$$A_2 := \int_{\mathbb{R}} \left| \omega_N(r_2 + |t - c_k^1|) dt - \omega_N(r_2 + |t - c_k^2|) \right| dt.$$

Since ω_N is nonincreasing, we have

$$A_1 = \int_{c_k^1}^{\infty} \left(\omega_N(r_1 + t - c_k^1) - \omega_N(r_2 + t - c_k^1) \right) dt$$

$$= 2 \int_{r_1}^{r_2} \omega_N(s) ds \leq 2 \cdot \int_{r_1}^{r_1 + |r_1 - r_2| + |c^1 - c^2|} \omega(s) =: d_\omega(Q_1, Q_2),$$

4.4. Space of balls

see (4.71) and (4.70).

To estimate A_2 we present the integral over $\mathbb{R}$ as

$$\int_{\mathbb{R}} = \int_{c_k^1}^{c_k^2} + \int_{-\infty}^{c_k^1} + \int_{c_k^2}^{\infty} =: I_1 + I_2 + I_2;$$

here we assume, for definiteness, that $c_k^1 \leq c_k^2$.

Using monotonicity of ω we find by the change of variable

$$I_1 = 2\left(\int_0^{\frac{\ell}{2}} \omega_N(r_2 + s)ds - \int_{\frac{\ell}{2}}^{\ell} \omega_N(r_2 + s)ds\right),$$

where $\ell := c_k^2 - c_k^1$.

By the same reason

$$I_k = \int_0^\ell \omega_N(r_2 + s)ds, \quad k = 2, 3,$$

and we have

$$A_2 = I_1 + I_2 + I_3 = 4 \cdot \int_{r_2}^{r_2 + \frac{|c_k^1 - c_k^2|}{2}} \omega_N(s) \leq 4 d_\omega(Q_1, Q_2).$$

Combining the estimates for A_1 and A_2 we get (4.96) for ω_N.

Since ω_N converges pointwise to ω on $(0, +\infty)$, we have by the Fatou lemma

$$\int_{J_k} \left|\omega(r_1 + |s - c_k^1|) - \omega(r_2 + |s - c_k^2|)\right| ds$$

$$\leq \lim_{N \to \infty} \int_{J_k} \left|\omega_N(r_1 + |s - c_j^1|) - \omega_N(r_2 + |s - c_k^2|)\right| ds \leq 6 d_\omega(Q_1, Q_2),$$

as required.

Now we collect the results proved in the following

Theorem 4.56. *Assume that $\omega : [0, +\infty) \to \mathbb{R}_+$ is a nonincreasing continuous function satisfying for all $r > 0$,*

$$\int_r^\infty \omega(s)ds = \infty. \tag{4.97}$$

Then the space of Euclidean balls $\mathcal{B}(\mathbb{R}^n)$ equipped with the metric d_ω is of pointwise homogeneous type.

Moreover, the corresponding family of measures $\mathcal{F}$ is $(n+1)$-homogeneous (so, $D(\mathcal{F}) = 2^{n+1}$) and $C(\mathcal{F}) \leq 3 \cdot 2^n (n+1)\sqrt{n}$.

In the latter estimate we use the two-sided inequality between the ℓ_2^n- and ℓ_∞^n-norms and the estimate of Theorem 4.53 (b).

Concluding Remark 4.57. If integrability condition (4.87) for ω does not hold, the function $\Omega_r(t) := \int_r^{r+t} \omega(s) ds$, $r, t > 0$, assigns its values from interval $[0, \Omega_r(+\infty)]$ and therefore the length of the interval I_j, $1 \leq j \leq n+1$, in (4.91) stabilizes for $R > \Omega_r(+\infty)$. This leads to homogeneity condition (4.93) and consistency condition (4.95) only under restrictions $R \leq \Omega_r(+\infty)$ and $0 < r < R$.

It is worth noting that in Chapter 7 we will use the family of measures $\{\mu_Q\}$ to construct linear Lipschitz extensions from subsets of a metric space equipped with such a family. If condition (4.87) does not hold, the corresponding linear operator extends Lipschitz functions only to a δ-neighborhood of the initial domain for some $\delta = \delta(\omega) > 0$. However, ε-perturbation of ω allows us to reduce this case to the above studied one and to obtain the desired extension by letting ε to zero.

4.4.3 Generalized hyperbolic spaces $\mathbb{H}_\omega^{n+1}$

The Riemannian manifold $\mathbb{H}_\omega^{n+1}$ was introduced in Example 3.113; now we assume that $\omega : (0, +\infty) \to \mathbb{R}_+$ is only continuous and nondecreasing. Its underlying C^∞-manifold is recalled to be the open half-space $H_+^{n+1} := \{x \in \mathbb{R}^{n+1}; x_{n+1} > 0\}$, and the Riemannian length of a C^1-curve $\gamma : [0, 1] \to \mathbb{H}_\omega^{n+1}$ is given by

$$\ell_R(\gamma) := \int_0^1 \omega(\gamma_{n+1}(t)) \left(\sum_{j=1}^{n+1} \dot\gamma_j(t)^2 \right)^{\frac{1}{2}} dt,$$

see subsection 3.3.3 for more details.

We regard $\mathbb{H}_\omega^{n+1}$ as a metric space equipped with the geodesic (inner) metric d_g given by

$$d_g(x, y) := \inf \ell_R(\gamma),$$

where γ runs over all C^1-curves in $\mathbb{H}_\omega^{n+1}$ joining x and y.

One knows that $(\mathbb{H}_\omega^{n+1}, d_g)$ is a length space, see Proposition 3.103.

Our main result is

Theorem 4.58. *Assume that ω is a nonincreasing continuous function satisfying condition (4.97). Then $(\mathbb{H}_\omega^{n+1}, d_g)$ is a space of pointwise homogeneous type. Moreover, the corresponding family of measures $\mathcal{F}$ satisfies, for some $c_1, c_2 > 0$ depending only on n, the inequalities*

$$D(\mathcal{F}) \leq c_1 \quad \text{and} \quad C(\mathcal{F}) \leq c_2.$$

Proof. We identify H_+^{n+1} and $\mathcal{B}(\mathbb{R}^n)$ using the bijection $I : x \mapsto B_{x_{n+1}}(x')$ where $x' := (x_1, \ldots, x_n)$. Using the identification we transfer the metric d_ω from $\mathcal{B}(\mathcal{M})$ to H_+^{n+1} and show that

$$d_\omega \leq d_g \leq 3 d_\omega. \tag{4.98}$$

4.4. Space of balls

Let $\gamma : [0,1] \to \mathbb{H}_\omega^{n+1}$ be a curve joining points x and y. Then
$$d_g(x,y) \leq \ell_R(\gamma).$$

Using uniform continuity of $\log \omega$ we, given $\varepsilon > 0$, find $\delta > 0$ so that
$$(\omega(s) \leq) \; \omega(t) \leq (1+\varepsilon)\omega(s) \tag{4.99}$$
for every pair s,t from the interval with endpoints x_{n+1}, y_{n+1} satisfying $t \leq s \leq t + \delta$.

Let $\pi := \{\gamma(t_i)\}_{i=0}^N$ be a partition of $\gamma([0,1])$ such that $|\gamma_{n+1}(t_{i+1}) - \gamma_{n+1}(t_i)| < \delta$. Then
$$\ell_R(\gamma) = \sum_{j=0}^{N-1} \int_{t_j}^{t_{j+1}} \omega(\gamma_{n+1}(s)) \left(\sum_{j=1}^{n+1} \dot\gamma_i(s)^2 \right)^{\frac{1}{2}} ds$$
$$= \sum_{j=1}^{N-1} \sum_{i=1}^{n+1} \omega(\xi_j) \|\gamma(t_{j+1}) - \gamma(t_j)\|_2,$$
where ξ_j is a point of the interval with endpoints $\gamma_{n+1}(t_j), \gamma_{n+1}(t_{j+1})$.

By the choice of π and (4.99), this implies
$$\ell_R(\gamma) \leq (1+\varepsilon) \sum_{j=0}^{N-1} \int_{M_j}^{M_j + \|\gamma(t_{j+1}) - \gamma(t_j)\|_2} \omega(s)\,ds,$$
where $M_j := \min\{\gamma_{n+1}(t_j), \gamma_{n+1}(t_{j+1})\}$. By the definition of d_ω, see (4.71), the right-hand side equals
$$(1+\varepsilon) \sum_{j=0}^{N-1} d_\omega(\gamma(t_i), \gamma(t_{i+1})) =: (1+\varepsilon)\ell(\pi).$$

If $\varepsilon \to 0$, then $\operatorname{mesh} \pi := \max_j \|\gamma(t_{j+1}) - \gamma(t_i)\|_2 \to 0$ and therefore $\ell(\pi)$ tends to the d_ω-length of γ denoted by $\ell_\omega(\gamma)$.

Hence we conclude that
$$d_g(x,y) \leq \ell_R(\gamma) \leq \ell_\omega(\gamma).$$

Taking the infimum over all γ joining x and y we obtain in the right-hand side the distance $\tilde{d}_\omega(x,y)$, where $\tilde{d}_\omega$ is the inner metric of the space $(\mathcal{B}(\mathbb{R}^n), d_\omega)$. By Theorem 4.53 (c), $\tilde{d}_\omega \leq 3 d_\omega$ and the right-side inequality of (4.98) is established.

The proof of the left-side inequality goes on by the very same argument.

Now we transfer the family of measures $\mathcal{F} := \{\mu_B\}_{B \in \mathcal{B}(\mathbb{R}^n)}$ constructed in the proof of Theorem 4.56 to the space $(\mathbb{H}_\omega^{n+1}, d_g)$ using the identification map I.

As it was proved, I is a bi-Lipschitz homeomorphism and therefore the image $\widetilde{\mathcal{F}} := I(\mathcal{F}) := \{\mu_{I(x)}\}_{x \in H_+^{n+1}}$ is a uniformly doubling family consistent with the metric d_ω (hence, d_g) and $D(\widetilde{\mathcal{F}})$ and $C(\widetilde{\mathcal{F}})$ are bounded by constants depending only on n.

Theorem 4.58 is proved. $\square$

Corollary 4.59. *Assume that ω_i are nonincreasing continuous functions satisfying condition* (4.97). *Then the metric space* $\bigoplus_{i=1}^{N}{}^{(p)}(\mathbb{H}_{\omega_i}^{n_i}, d_g)$, $1 \leq p \leq \infty$, *is of pointwise homogeneous type.*

Proof. The result is an immediate consequence of Theorem 4.58 and Proposition 3.92. $\square$

Remark 4.60. The family of pointwise doubling measures (4.90) is introduced in the Yu. Brudnyi and Shvartsman paper [BSh-1999].

4.5 Differentiability of Lipschitz functions

In this section, we briefly discuss several differentiability results for Lipschitz functions whose domains and target spaces have a suitable geometric structure. A model case is the classical Rademacher theorem [Ra-1919] which will be presented together with its generalizations.

Results of another kind concern Lipschitz functions on doubling metric spaces with sufficiently rich length structure. This clearly requires a new concept of differentiability as the spaces considered may not have an infinitesimal affine structure which is the key point of the classical concept.

4.5.1 Lipschitz functions on $\mathbb{R}^n$

The classical Lebesgue differentiability theorem implies that a real Lipschitz function on $\mathbb{R}$ is differentiable almost everywhere (a.e.). The result is clearly true for Lipschitz maps from $\mathbb{R}$ to Euclidean spaces but is not true for the case of maps to Banach spaces, as the Aronszajn's example shows.

Let $: \mathbb{R}_+ \to L_1(\mathbb{R}_+)$ be the "moving" characteristic function given by the formula $f(t) := \chi_{[0,t]}$, $t \geq 0$. Then

$$\|f(t+h) - f(t)\|_1 = \int_t^{t+h} ds = h \quad \text{for all} \quad h \geq 0,$$

i.e., $g \in \text{Lip}(\mathbb{R}_+, L_1(\mathbb{R}_+))$.

4.5. Differentiability of Lipschitz functions

However, f is nowhere differentiable. In fact, if $h^{-1}\big(f(t_0 + h) - f(t_0)\big)$ converges in $L_1(\mathbb{R}_+)$ as $h \to 0^+$ to some function $g \in L_1(\mathbb{R}_+)$, then for every compactly supported continuous function $\varphi : \mathbb{R}_+ \to \mathbb{R}$ we get

$$\frac{1}{h} \int_{t_0}^{t_0+h} \varphi(s)ds = \int_{\mathbb{R}_+} \left(\frac{f(t_0+h) - f(t_0)}{h}\right)(s)\varphi(s)ds$$

$$\longrightarrow \int_{\mathbb{R}_+} g(t_0)(s)\varphi(s)ds \quad \text{as} \quad h \to 0^+.$$

Hence, $\varphi(t_0) = \int_{\mathbb{R}_+} g(t_0)(s)\varphi(s)ds$, i.e., $g(t_0)$ should be the Dirac measure δ_{t_0} which is impossible.

However, Lipschitz functions from $\mathbb{R}$ to $L_p(\mathbb{R})$ are differentiable a.e. if $1 < p < \infty$; more generally, the result holds for Lipschitz functions from $\mathbb{R}$ to reflexive or separable dual spaces, a consequence of Gelfand's theorem [Gel-1938].

A Banach space X for which every Lipschitz map $f : \mathbb{R} \to X$ is differentiable a.e. is said to have the *Radon–Nikodym property* (RNP). Hence, reflexive and dual separable Banach spaces have RNP. An extensive discussion of the results and conjectures related to RNP may be found in Chapters 5 and 6 of the Benyamini and Lindenstrauss book [BL-2000].

Passing to the differentiability results concerning Lipschitz functions of arbitrary order on domains of $\mathbb{R}^n$, we begin with the classical

Theorem 4.61 (Rademacher). *A Lipschitz map $f : \mathbb{R}^n \to \mathbb{R}^m$ is differentiable a.e. on $\mathbb{R}^n$.*

Proof. We follow the derivation of L. Simon [Sim-1983] which may be easily generalized to the case of Lipschitz maps from $\mathbb{R}^n$ to a Banach space with RNP.

Passing to the coordinate functions of f, we may and will assume that $m = 1$. In this case, we must show that for almost all $x \in \mathbb{R}^n$ there is a Taylor polynomial $T_x f(y) := f(x) + \sum_{i=1}^{n} f_i(x)(y_i - x_i)$ such that

$$f(x) - T_x f(y) = o(|x - y|) \quad \text{as} \quad y \to x; \tag{4.100}$$

as before $|x| := \max_{1 \leq i \leq n} |x_i|$ for $x \in \mathbb{R}^n$.

Let $D_e f(x) := \lim_{t \to 0} \frac{f(x+te) - f(x)}{t}$ be the directional derivative along a vector $e \in \mathbb{S}^{n-1}$. By the Lebesgue theorem $D_e f(x)$ exists on every line in direction e except a set of Lebesgue 1-measure zero. In turn, the Fubini theorem shows that the union of these exceptional sets for all lines in $\mathbb{R}^n$ in direction e is measurable and of Lebesgue n-measure zero. In particular, the partial derivatives $D_i f$, $1 \leq i \leq n$, exist a.e. on $\mathbb{R}^n$. Regarding $D_e f$ and $D_i f$ as distributional derivatives (linear functionals on $C_0^\infty(\mathbb{R}^n)$), we may write

$$D_e f = \sum_{i=1}^{n} e_i D_i f.$$

But these derivatives are L_∞ functions as f is Lipschitz and therefore this equality holds a.e. in $\mathbb{R}^n$.

By the definition of D_e, this implies that for all $x \in \mathbb{R}^n \setminus S_e$ where $|S_e| = 0$ it is true that

$$f(x+y) = f(x) + \sum_{i=1}^{n} D_i f(x) y_i + o(|y-x|), \qquad (4.101)$$

provided that $y := te \to 0$.

Moreover, for all $x \in \mathbb{R}^n \setminus S_e$,

$$|D_i f(x)| \le L(f), \quad 1 \le i \le n. \qquad (4.102)$$

It remains to show that (4.101) holds for all $e \in \mathbb{S}^{n-1}$ and all $x \in \mathbb{R}^n \setminus E$, where E is a subset of n-measure zero. To this end, we choose a dense family $\{e^j\}_{j \in \mathbb{N}} \subset \mathbb{S}^{n-1}$ containing the standard basis $\{e^1, \ldots, e^n\}$ of $\mathbb{R}^n$. We define the required exceptional set by $E := \bigcup_{j=1}^{\infty} S_{e^j}$. Then $|E| = 0$ and (4.101) and (4.102) hold for all $x \in \mathbb{R}^n \setminus E$.

Let $R_x(f; y) := f(x+y) - f(x) - \sum_{i=1}^{n} D_i f(x) \cdot y_i$. By (4.102), we get for $x \in \mathbb{R}^n \setminus E$,

$$|R_x(f; y) - R_x(f; y')| \le (n+1) L(f) |y - y'|. \qquad (4.103)$$

Given $\varepsilon > 0$, we let $\{e^j\}_{1 \le j \le N}$ be an ε-net in $\mathbb{S}^{n-1}$. By the definition of E, for every $x \in \mathbb{R}^n \setminus E$, there is $\delta = \delta_x > 0$ such that for all $1 \le j \le N$, and all $0 < t < \delta$,

$$|R_x(f; te^j)| < \varepsilon.$$

Now, given $e \in \mathbb{S}^{n-1}$, we choose e^j, $1 \le j \le N$, so that $|e - e^j| < \varepsilon$, to obtain

$$|R_x(f; te)| \le |R_x(f; te) - R_x(f; te^j)| + |R_x(f; te^j)| \le (n+1) L(f) \varepsilon + \varepsilon.$$

This clearly implies (4.101) for every $x \in \mathbb{R}^n \setminus E$ and every $y := te$. □

Using the definition of Banach spaces with RNP and the very same argument we easily extend the Rademacher theorem to Lipschitz maps from $\mathbb{R}^n$ to such a space.

A localized version of Rademacher's theorem was due to V. Stepanov [Step-1925] who also removed superfluous assumptions in Rademacher's theorem.

Theorem 4.62 (Stepanov). *A measurable function $\mathbb{R}^n \to \mathbb{R}$ is differentiable a.e. on a measurable set S if and only if*

$$\varlimsup_{y \to x} \frac{|f(y) - f(x)|}{|y - x|} < \infty$$

for almost all $x \in S$.

4.5. Differentiability of Lipschitz functions

Finally, we formulate the generalization of the Rademacher–Stepanov theorem to higher derivatives, see [Br-1994] for the proof.

For the formulation of this result we need

Definition 4.63. A function $f : \mathbb{R}^n \to \mathbb{R}$ is k-differentiable at a point x if there is a polynomial p_x of degree k such that
$$f(y) - p_x(y) = o(|y-x|^k) \quad \text{as} \quad y \to x.$$
The (Taylor) polynomial p_x is clearly unique.

Theorem 4.64. (a) *A measurable function $f : \mathbb{R}^n \to \mathbb{R}$ is k-differentiable a.e. on a measurable set S if and only if for almost all $x \in S$,*
$$\varlimsup_{h \to 0} \frac{|\Delta_h^k f(x)|}{|h|^k} < \infty.$$

(b) *If S is a bounded subset of positive measure and (4.103) holds, then, for given $\varepsilon > 0$, there is a C^k-function $f_\varepsilon : \mathbb{R}^n \to \mathbb{R}$ such that*
$$\operatorname{mes}_n(\{x \in S \,;\, f(x) \neq f_\varepsilon(x)\}) < \varepsilon.$$

The results remain to be true for maps from $\mathbb{R}^n$ to Banach spaces with RNP.

Remark 4.65. For univariate functions assertion (a) of the theorem was due to Marcinkiewicz and Zygmund [MZ-1936].

4.5.2 Lipschitz functions on metric spaces

Even in the presence of the basic ingredients involved in Rademacher's theorem Lipschitz functions on metric spaces locally homeomorphic to $\mathbb{R}^n$ may be nowhere differentiable. A simple example is the Weierstrass nowhere differentiable function $\sum_{j=1}^{\infty} 2^{-j\alpha} \cos 2^j x$, $0 < \alpha < 1$, which is Lipschitz on the metric space $(\mathbb{R}, d_\alpha)$ with $d_\alpha(x, y) := |x - y|^\alpha$.

Nevertheless, there are few classes of metric spaces for which some analogs of Rademacher's theorem are true.

The simplest is the class of *Lipschitz manifolds*, metric spaces equipped with Lipschitz structure. A metric space $(\mathcal{M}, d)$ belongs to this class if there is a *Lipschitz atlas* $\mathcal{A} := \{U_\alpha, \varphi_\alpha\}_{\alpha \in \mathcal{A}}$, where $\{U_\alpha\}$ is an open cover of $\mathcal{M}$, and each φ_α is a bi-Lipschitz embedding of U_α into a domain of $\mathbb{R}^n$. By virtue of Rademacher's theorem, every Lipschitz map f from $\mathcal{M}$ into a Euclidean space is *differentiable a.e. with respect to the atlas $\mathcal{A}$*, i.e., every composite map $f \circ \varphi_\alpha^{-1}$ has this property.

Remark 4.66. Lipschitz manifolds are introduced by Whitney [Wh-1957] to develop the de Rham theory for differential forms with L_∞ coefficients. From the topological point of view, this class is rather rich. It is known that every topological manifold of dimension $n \neq 4$ admits a Lipschitz manifold structure, see Sullivan [Sul-1979].

More interesting is the class of Carnot groups regarded as metric spaces, see the end of subsection 3.3.7 for definitions and examples. To formulate the corresponding result, a special case of the Pansu theorem [Pa-1989], we denote by (G, d_G) a Carnot group with the Carnot–Carathéodory metric. Let μ_G be the Haar measure on G and let e be the unit of G.

Theorem 4.67. *Let f be a Lipschitz map from (G, d_G) into the Euclidean space $\mathbb{R}^n$. Then for μ_G almost all $g \in G$, there is a group homomorphism $D_g : G \to \mathbb{R}^n$ such that*

$$f(hg) = f(g) + D_g h + o\bigl(d_G(h, e)\bigr) \quad \text{as} \quad h \to e. \tag{4.104}$$

Since $\mathbb{R}^n$ is a Carnot group, the Rademacher theorem is a consequence of this result. As $\mathbb{R}^n$ is abelian, every homomorphism from G to $\mathbb{R}^n$ vanishes on the *commutant* of G (a normal subgroup formed by commutators $ghg^{-1}h^{-1}$ of G). This implies

Corollary 4.68. *Under the assumptions of Theorem 4.67, for μ_G almost all $g \in G$,*

$$\lim_{h \to e} \frac{\|f(hg) - f(g)\|_{\mathbb{R}^n}}{d_G(h, e)} = 0$$

whenever h tends to e along the commutant of G.

Remark 4.69. In general, Pansu's theorem asserts that a Lipschitz map f from (G, d_G) to another Carnot group (H, d_H) is differentiable μ_G a.e. In this case, (4.104) is replaced by

$$d_H\bigl(f(hg), f(g) D_g h\bigr) = o\bigl(d_G(h, e)\bigr) \quad \text{as} \quad h \to e.$$

In the special case of the Heisenberg group H_n, see Example 3.141, the target space in Theorem 4.67 can be replaced by a Hilbert space and even by a Banach space with RNP, see the recent papers [ChK-2006a] and [ChK-2006b] by Cheeger and Kleiner.

Even more surprising is that a version of the result is true for the target space $L_1(\mathbb{R}^n)$ (cf. the Aronszajn counterexample presented above).

The proofs of these results employ a differentiation theory for functions on metric spaces developed by Cheeger [Ch-1999] which we briefly discuss.

The theory concerns metric spaces with relatively rich length structure. One of the possibilities to characterize such spaces is proposed by Heinonen and Koskela [HK-1996]. Their approach is based on the following definitions.

Let $(\mathcal{M}, d)$ be a metric space. We say that a function $f : \mathcal{M} \to \mathbb{R}$ has an *upper gradient* $g : \mathcal{M} \to \mathbb{R} \cup \{+\infty\}$ if for every rectifiable curve $\gamma : [0, \ell] \to \mathcal{M}$ parametrized by arclength s,

$$\bigl|f(\gamma(\ell)) - f(\gamma(0))\bigr| \leq \int_0^\ell g(\gamma(s)) ds. \tag{4.105}$$

4.5. Differentiability of Lipschitz functions

Evidently, $g = \infty$ is an upper gradient for every f, and $L(f)$ is the upper gradient for a Lipschitz function f.

A more interesting example is the functional

$$\|\nabla f\| := \left(\sum_{i=1}^{n} (D_i f)^2 \right)^{\frac{1}{2}}$$

defined on smooth functions on $\mathbb{R}^n$ (or its analog for a Riemannian manifold). In this case, the Newton–Leibnitz rule and the Cauchy–Schwartz inequality give, for the left-hand side of (4.105), the bound

$$\int_0^\ell \|\nabla f(\gamma(s))\| \cdot \left\| \frac{d}{ds} \gamma(s) \right\| ds = \int_0^\ell \|\nabla f(\gamma(s))\| ds.$$

Further, we set for a ball $B \subset \mathcal{M}$,

$$f_B := \frac{1}{\mu(B)} \int_B f d\mu,$$

and denote by λB the ball centered at c_B and of radius λr_B.

Definition 4.70. A metric measure space $(\mathcal{M}, d, \mu)$ is said to be a PI space if

(a) $\mathcal{M}$ is proper;

(b) μ is locally doubling;

(c) there exist constants $c, r_0 > 0$ and $\lambda \geq 1$ such that the Poincaré inequality

$$\frac{1}{\mu(B)} \int_B |f - f_B| d\mu \leq c \frac{\operatorname{diam} B}{\mu(\lambda B)} \int_{\lambda B} g d\mu \qquad (4.106)$$

holds for every ball B in $\mathcal{M}$ of radius at most r_0 and every upper gradient g of f.

The simplest example of a PI space is $\mathbb{R}^n$. In this case, the classical Sobolev–Poincaré inequality, see, e.g., [Ste-1970], implies inequality (4.106) with $g := \|\nabla f\|$ and with the Lebesgue n-measure.

However, the class of PI spaces also contains even fractal spaces as the following two examples show.

Example 4.71. (a) *Carnot–Carathéodory spaces.* Let $\mathcal{M}$ be a sub-Riemannian manifold equipped with the Carnot–Carathéodory metric d_C and the canonical measure Vol, see subsection 3.3.5 for the definitions and facts used now. If $(\mathcal{M}, d_C)$ is complete, then this space is proper. In fact, by the Mitchell theorem, Vol is locally doubling and therefore every closed ball is compact. Hence, $(\mathcal{M}, d_C, \text{Vol})$ would be a PI space if it supported the Poincaré inequality. The latter fact can be found, e.g., in Section 11 of the memoir [HaK-2000] by Hajlacz and Koskela. In particular, Carnot groups are PI spaces.

(b) *Laakso spaces.* It was shown by Laakso [Laa-2000] that for every $p > 1$ there is an Ahlfors p-regular metric space (Λ_p, d_p) such that $(\Lambda_p, d_p, \mathcal{H}_p)$ is a PI-space. We describe a special case of the Laakso construction for $\widehat{p} := 1 + \log 2 / \log 3$ and then briefly discuss the general case.

Let C be the classical Cantor set. The Laakso space $\Lambda_{\widehat{p}}$ is obtained from the space $[0,1] \times C$ of Hausdorff's dimension $1 + \dim_{\mathcal{H}} C = 1 + \log 2 / \log 3 =: \widehat{p}$ by an ingenious identification procedure started with a nested family $\{C_a;\, a \in \{0,1\}^{\mathbb{N}}\}$ of subsets of C. Namely, C clearly satisfies the relation

$$C = \frac{1}{3} C \cup \left(\frac{1}{3} C + \frac{2}{3} \right). \quad (4.107)$$

Set $C_0 := \frac{1}{3} C$, $C_1 := \frac{1}{3} C + \frac{2}{3}$, and define C_{00}, C_{01} and C_{10}, C_{11} as the similar parts of C_0 and C_1, respectively (e.g., $C_{00} = \frac{1}{9} C$ and $C_{11} = \frac{1}{3} C_1 + \frac{2}{3} = \frac{1}{9} C + \frac{2}{9} + \frac{2}{3}$). Proceeding this way, one defines for every infinite binary word $a = a_1, a_2, \ldots$ from $\{0,1\}^{\mathbb{N}}$ the nested sequence

$$C_{a_1} \supset C_{a_1 a_2} \supset C_{a_1 a_2 a_3} \supset \cdots.$$

Now we label some points of $[0,1]$ using a special sequence of integers $b_1, b_2, b_3, \ldots \in \{3, 4\}$ defined later. Namely, for every set of integers $m_1, \ldots, m_k$ satisfying $0 \leq m_i < b_i$ for $i \leq k-1$ and $0 < m_i < b_i$ for $i = k$, we put

$$w(m_1, \ldots, m_k) := \sum_{i=1}^{k} m_i \left(\prod_{j=1}^{i} b_j \right)^{-1}. \quad (4.108)$$

It is easily seen that these points belong to $[0,1]$ and are pairwise distinct because of the condition $m_k < b_k$.

Further, we define the underlying set $\Lambda_{\widehat{p}}$ of the Laakso space by the identification of the point $(x_1, x_2) \in w(m_1, \ldots, m_k) \times C_{a0} \subset [0,1] \times C$ where $|a| = k - 1$ with the point $\left(x_1, x_2 + \frac{2}{3^k}\right) \in w(m_1, \ldots, m_k) \times C_{a1}$.

The resulting set of this identification we denote by $\Lambda_{\widehat{p}}$. Let π be the natural projection from $[0,1] \times C$ onto $\Lambda_{\widehat{p}}$.

Using the projection we define in this set a metric setting for $m, m' \in \Lambda_{\widehat{p}}$

$$d_{\widehat{p}}(m, m') := \inf \mathcal{H}_1(S),$$

where the infimum is taken over all subset $S \subset [0,1] \times C$ such that $\pi(S)$ is a curve in $\Lambda_{\widehat{p}}$ joining m and m'.

It is proved in [Laa-2000] that $(\Lambda_{\widehat{p}}, d_{\widehat{p}})$ is a complete unbounded Ahlfors $\widehat{p}$-regular metric space; in particular, its Hausdorff $\widehat{p}$-measure is doubling. Moreover, $(\Lambda_{\widehat{p}}, d_{\widehat{p}}, \mathcal{H}_{\widehat{p}})$ is a PI space, since it carries the Poincaré inequality (4.106).

4.5. Differentiability of Lipschitz functions

It remains to define the sequence $\{b_i\} \subset \{3,4\}$ involved in the labeling procedure in (4.108). We define it using for every $k \in \mathbb{N}$ the inequality

$$\frac{3}{4}\left(\prod_{j=1}^{k} b_j\right)^{-1} \leq \frac{1}{3^k} < \frac{4}{3}\left(\prod_{j=1}^{k} b_j\right)^{-1} \tag{4.109}$$

and the conditions $b_j \in \{3,4\}$.

The general case differs from the presented one only by the choice of a Cantor set. Namely, instead of the classical Cantor set C we use a Cantor set $K = K(t)$ satisfying

$$K = tK \cup (tK + 1 - t)$$

for $t \in (0,1)$ defined by

$$p = 1 + \log 2 / \log 1/t;$$

here $p \in (1,2)$ is a fixed number.

Then the Laakso construction with $[0,1] \times K$ goes in a very similar way with the sequence $b_1, b_2, \ldots$ defined as follows.

Let n be an integer such that $\frac{1}{n+1} \leq t < \frac{1}{n}$. Then the integers b_i are defined by the inequalities

$$\left(\prod_{j=1}^{k} b_j\right)^{-1} \leq t^k < \frac{n+1}{n}\left(\prod_{j=1}^{k} b_j\right)^{-1}$$

and the conditions $b_j \in \{n, n+1\}$.

This gives the Laakso space $(\Lambda_p, d_p, \mathcal{H}_p)$ of Hausdorff dimension p possessing properties similar to those for the case $p = \widehat{p}$

To construct the Laakso space of Hausdorff dimension ≥ 2 it suffices to consider the direct product $[0,1] \times K(t)^M$ equipped with similar identifications of sets as those in each $K(t)$. In this way 2^M points are identified with each other and the final Laakso space will be p-regular with $p := 1 + M \log 2 / \log 1/t$.

Now we return to Cheeger's differentiation theory. Its main result states, roughly speaking, that a PI space may be equipped with a "measurable" atlas similar to that for Lipschitz manifolds. A *measurable atlas* $\mathcal{A}$ on a metric measure space $(\mathcal{M}, d, \mu)$ is a collection of triples $(U_\alpha, \varphi_\alpha, \mathbb{R}^{N_\alpha})_{\alpha \in A}$ such that:

(a) $\{U_\alpha\}$ is a μ *measurable partition* of $\mathcal{M}$, i.e., every U_α is μ measurable, the distinct sets of this family are disjoint and $\bigcup_{\alpha \in A} U_\alpha$ is a set of full μ-measure.

(b) Every φ_α is a Lipschitz map from U_α in the Euclidean space $\mathbb{R}^{N_\alpha}$.

The Cheeger result is given by

Theorem 4.72. *Let $(\mathcal{M}, d, \mu)$ be a PI space. There is a countable measurable atlas $\mathcal{A} := \{U_j, \varphi_j, \mathbb{R}^{N_j}\}_{j \in \mathbb{N}}$ on $\mathcal{M}$ and a collection of maps a_j from $\mathrm{Lip}(\mathcal{M})$ to the space $B(U_j)$ of bounded functions on U_j, $j \in \mathbb{N}$, such that for every $f \in \mathrm{Lip}(\mathcal{M})$,*

$$f(m) = f(m') + \langle a_j(f)(m), \varphi_j(m) - \varphi_j(m')\rangle_{\mathbb{R}^{N_j}} + o\bigl(d(m, m')\bigr) \quad \text{as} \quad m \to m'$$

for μ almost every $m \in \mathcal{M}$.

In particular, Carnot–Carathéodory spaces (e.g., Carnot groups) and Laakso fractals are covered by Cheeger's result. However, the Pansu theorem is not a consequence of Theorem 4.72 as it employs a different concept of differentiability.

4.6 Lipschitz spaces

In this section, we discuss several results concerning Lipschitz functions on metric spaces and on subsets of $\mathbb{R}^n$. In the latter case, we overview results on Lipschitz spaces of higher order similar to those on metric spaces.

We begin with a discussion of properties, old and relatively new, of modulus of continuity including its relation to K-functionals of Lipschitz couples. This naturally leads to interpolation theorems for operators acting in Lipschitz spaces presented in the next subsection.

The subsection is completed by duality theorems for spaces $\mathrm{Lip}(M)$ and its generalization to Lipschitz spaces of higher order.

As before, proofs mostly accompany the results related to the main theme of the present book.

4.6.1 Modulus of continuity

Modulus of continuity of a function $f : (\mathcal{M}, d) \to \mathbb{R}$ is recalled to be a function $\omega(f) : \mathbb{R}_+ \to \mathbb{R}_+$ given by

$$\omega(t; f) := \sup\{f(m) - f(m') \,;\, d(m, m') \leq t\}.$$

This clearly is a nondecreasing function satisfying $\omega(0+; f) = 0$ if and only if f is uniformly continuous (in particular, if f is continuous on a compact metric space). In general, these properties of $\omega(f)$ characterize modulus of continuity up to equivalence with numerical factors, see Theorem 4.76 below, but this functional possesses some additional characteristics for the case of *convex* metric spaces, see Definition 3.8. To formulate the result we use

4.6. Lipschitz spaces

Definition 4.73. Let $\varphi : \mathbb{R}_+ \to \mathbb{R}_+$ be a measurable function. Its convexification denoted by $\widehat{\varphi}$ is a function given for $t > 0$ by

$$\widehat{\varphi}(t) := \inf\{\omega(t)\,;\, \varphi \le \omega\},$$

where the infimum is taken over all concave functions[2] $\omega : \mathbb{R}_+ \to \mathbb{R}_+$ majorating φ.

It is easily seen that $\widehat{\varphi}$ is a concave function.

Proposition 4.74. *Let f be a real function on a convex metric space $(\mathcal{M}, d)$. Then its modulus of continuity is subadditive. In particular, its convexification satisfies*

$$\omega(f) \le \widehat{\omega}(f) \le 2\omega(f). \tag{4.110}$$

Proof. Since every pair of points in $\mathcal{M}$ can be joined by a curve isometric to a real segment, the result reduces to that for functions on the real line. In this case, the result is known, see, e.g., [Tim-1963]. □

Condition (4.110) imposes a restriction on behavior of $\omega(f)$ near 0 for nonconstant f.

Corollary 4.75. *A function f on a convex metric space is a constant if and only if*

$$\lim_{t \to 0} \frac{\omega(t\,;\, f)}{t} = 0. \tag{4.111}$$

In fact, if a nondecreasing nonnegative concave function on $\mathbb{R}_+$ satisfies this condition, then it is zero identically.

In general, any rate of convergence of $\omega(f)$ to zero at 0 cannot provide constancy of f as the following result by Besicovich and Schoenberg [BeSch-1961] demonstrates.

Theorem 4.76. *Given an arbitrary nondecreasing continuous function $\omega : \mathbb{R}_+ \to \mathbb{R}_+$ such that $\omega(0) = 0$ there is a compact connected metric space $(\mathcal{M}, d)$ and a nonconstant function $f_0 : \mathcal{M} \to \mathbb{R}$ satisfying*

$$\sup_{t > 0} \frac{\omega(t\,;\, f_0)}{\omega(t)} < \infty. \tag{4.112}$$

In the proof presented in [BeSch-1961], $\mathcal{M}$ is the image of a *screw* closed curve $\gamma : \mathbb{T} \to L_2(\mathbb{T})$ (the Wiener spiral), where $\mathbb{T} := \mathbb{R}/2\pi\mathbb{Z}$ is the unit circle. This means that for some $F \in L_2(\mathbb{T})$ the curve satisfies

$$F(t - t') = \|\gamma(t) - \gamma(t')\|_2$$

for every pair $t, t' \in \mathbb{T}$. Existence and properties of screw curves are discussed, e.g., in the von Neumann and Schoenberg paper [NSch-1941]. This F relates in a

---

[2] i.e., ω satisfies: $\omega\big((1-\lambda)t_1 + \lambda t_2\big) \ge (1-\lambda)\omega(t_1) + \lambda\omega(t_2)$ for all $t_1, t_2 \in \mathbb{R}_+$ and $0 < \lambda < 1$.

simple way to the initial function ω so that a function $f_0 : \gamma(\mathbb{T}) \to \mathbb{R}$ given by $f_0(\gamma(t)) := t$ satisfies (4.112).

The example considered shows that modulus of continuity may have a highly irregular behavior even for the case of metric spaces with good topological characteristics. However, restrictions imposed on Hausdorff or Minkowski dimensions of a metric space (see Definitions 4.10 and 3.27) essentially improve this behavior.

Theorem 4.77. *Let $\mathcal{M}$ be a compact connected metric space and let f be a real function on $\mathcal{M}$. Then f is a constant if one of the following conditions holds:*

(a) *Upper Minkowski dimension $\overline{\dim}_M \mathcal{M} \leq n$ and*

$$\omega(t\,;f) = o(t^n) \quad as \quad t \to 0. \tag{4.113}$$

(b) *$\mathcal{M}$ is a metric subspace of an n-dimensional Euclidean space (hence, $\overline{\dim}_M \mathcal{M} \leq n$) and*

$$\omega(t\,;f) = O\!\left(t^n |\log t|^m\right) \quad as \quad t \to 0,$$

where m is an arbitrary number less than $n - 1$.

The result is not true if $m > n$.

(c) *The space $\mathcal{M}$ of part (b) has Hausdorff dimension p where $0 < p < n$ and*

$$\omega(t\,;f) = o(t^p) \quad as \quad t \to 0.$$

Proof. We derive only assertion (a) referring the reader to the M. Brodskii paper [Bro-1967] for a sharper result implying (b), and to the aforementioned paper [BeSch-1961] for assertion (c).

Assume that $\overline{\dim}_M \mathcal{M} \leq n$ and that (4.113) holds for a nonconstant function $f_0 : \mathcal{M} \to \mathbb{R}$. Since $\mathcal{M}$ is connected and compact, $f_0(\mathcal{M})$ is a nontrivial closed interval which (after scaling) may be assumed to be $[0,1]$. Given $N \in \mathbb{N}$, we denote by m_j an arbitrary point from $f^{-1}(\frac{j}{N})$, $j = 0, 1, \ldots, N$. By a compactness argument, (4.113) implies that for all $j \neq j'$ and some $\varepsilon_N \to 0$ as $N \to \infty$,

$$|f(m_j) - f(m_{j'})| \leq d(m_j, m_{j'})^n \varepsilon_N.$$

Hence, $\{m_j\}_{0 \leq j \leq N}$ satisfies

$$d(m_j, m_{j'}) \geq \left(\frac{1}{N\varepsilon_N}\right)^{\frac{1}{n}}, \quad 0 \leq j \neq j' \leq N,$$

i.e., this family is an ε-chain in $\mathcal{M}$ with $\varepsilon := \left(\frac{1}{N\varepsilon_N}\right)^{\frac{1}{n}}$.

On the other hand, the definition of the Minkowski dimension and Proposition 3.19 imply that the cardinality of any ε-chain in $\mathcal{M}$ is at most $C\!\left(\frac{1}{\varepsilon}\right)^n$ where $C > 0$ is a constant. We therefore conclude that

$$N + 1 = \operatorname{card}\{m_j\} \leq C\, N\varepsilon_N = o(N) \quad as \quad N \to \infty,$$

a contradiction. $\square$

4.6. Lipschitz spaces

We complete this part with an important property of modulus of continuity which was discovered a century after this concept had appeared. The result looks to be completely elementary but the only known proof is based on some deep results of Interpolation Space Theory. We will briefly discuss some aspects of this theory in the next subsection and then derive a theorem as a consequence of the facts presented.

Let $\{\varphi_j\}_{j\in\mathbb{N}}$ be a sequence of nonnegative concave functions on $\mathbb{R}_+$ satisfying the condition

$$\sum_j \varphi_j(1) < \infty; \qquad (4.114)$$

in particular, $\sum \varphi_j$ converges uniformly on compact subsets of $(0,+\infty)$.

Then the following is true.

Theorem 4.78. *Assume that a real function f on a metric space $\mathcal{M}$ satisfies*

$$\omega(f) \leq \sum_j \varphi_j.$$

Then there is a sequence of functions $\{f_j\}_{j\in\mathbb{N}} \subset \ell_\infty(\mathcal{M})$ such that

$$f = \sum_j f_j \quad \text{(uniform convergence on } \mathcal{M}\text{),}$$

and for every $j \in \mathbb{N}$ and $t > 0$,

$$\omega(t; f_j) \leq 6\varphi_j(t).$$

4.6.2 Real interpolation of Lipschitz spaces

The real interpolation method studies Banach spaces defined by the following construction.

Let $\vec{X} := (X_0, X_1)$ be a *Banach couple*, i.e., a pair of Banach spaces continuously embedded into a topological vector space. The *K-functional* of $\vec{X}$ is recalled (see Definition 2.76) to be a function $K(\vec{X}) : \mathbb{R}_+ \times (X_0 + X_1) \to \mathbb{R}_+$ given by

$$K(t; x; \vec{X}) := \inf_{x=x_0+x_1} \{\|x_0\|_{X_0} + t\|x_1\|_{X_1}\}.$$

Now let Φ be a Banach function space on $\mathbb{R}_+$. Then a Banach space $K_\Phi(\vec{X})$ is given by the norm

$$\|x\|_{K_\Phi(\vec{X})} := \|K(\cdot; x; \vec{X})\|_\Phi.$$

It may be easily checked that $K_\Phi(\vec{X})$ is an *interpolation space* of $\vec{X}$. This means that the norm of every linear operator $T: X_0 + X_1 \to X_0 + X_1$ mapping X_i into itself, $i = 0, 1$, satisfies

$$\|T|_{K_\Phi(\vec{X})}\| \leq \max_{i=0,1}\{\|T|_{X_i}\|\}.$$

In particular, $T(K_\Phi(\vec{X})) \subset K_\Phi(\vec{X})$, i.e., this space is stable under actions of such operators.

In general, a Banach space X satisfying
$$X_0 \cap X_1 \subset X \subset X_0 + X_1$$
is said to be an interpolation space of $\vec{X}$ if $T(X) \subset X$ for all linear operators described above.

One of the basic problems of Interpolation Space Theory is to find methods to construct interpolation spaces of Banach couples. For several couples the real method allows us to find all of them. Such couples are called *Calderón* (or *C*-) *couples* after A. Calderón who discovered that the classical couple (L_1, L_∞) is of this type. Below we describe couples of Lipschitz spaces which are Calderón. The proof of this and many other results of the real interpolation is based on the following fundamental fact (the so-called *K-divisibility theorem*) proved by Yu. Brudnyi and Krugljak in 1981, see [BK-1991, Chapter 3] for details and references.

Theorem 4.79. *Let $\{\varphi_j\}_{j\in\mathbb{N}}$ be a family of nondecreasing concave functions on $\mathbb{R}_+$ satisfying (4.114). Then for every $x \in X_0 + X_1$ satisfying*
$$K(x; \vec{X}) \leq \sum_{j\in\mathbb{N}} \varphi_j$$
there exists a decomposition
$$x = \sum_{j\in\mathbb{N}} x_j \quad \text{(convergence in } X_0 + X_1\text{)}$$
such that for some numerical constant $C > 1$ and all j,
$$K(x_j; \vec{X}) \leq C\varphi_j.$$

Remark 4.80. Let us define the K-divisibility constant
$$\kappa := \inf C.$$
In the initial proof this was estimated by $\kappa \leq 14$, see [BK-1981a] and the manuscript [BK-1981b].

The method presented in this manuscript was improved by Cwikel [Cw-1984] and then by Cwikel, Jawerth and M. Milman [CwJM-1990] to yield the estimate $\kappa \leq (\sqrt{2}+1)^2$.

It was conjectured in [BK-1991, p. 492] that $\kappa \leq 4$. For the subclass of couples of Banach functional spaces on $\mathbb{R}_+$, this was recently proved by Cwikel [Cw-2003].

Now we derive Theorem 4.78 from the K-divisibility theorem. This will immediately follow from the next fact relating the K-functional of the couple $(\ell_\infty(\mathcal{M}), \text{Lip}(\mathcal{M}))$ to convexification of modulus of continuity.

4.6. Lipschitz spaces

Theorem 4.81. *For every $f \in \ell_\infty(\mathcal{M})$ and $t > 0$,*

$$K\bigl(t\,;\,f\,;\,(\ell_\infty, \mathrm{Lip})(\mathcal{M})\bigr) = \frac{1}{2}\,\widehat{\omega}(f\,;\,2t).$$

The result was firstly proved by Peetre [Peet-1979] for functions on the real line. An approach giving the result for the general case was presented in the book [BK-1991, Prop. 3.1.19].

We complete this subsection by two interpolation results for Lipschitz couples. The first of them, a remote consequence of the K-divisibility theorem, gives a criterion for these couples to be Calderón, while the second generalizes Theorem 4.81 to arbitrary couples.

In order to formulate the first theorem due to Yu. Brudnyi and Shteinberg [BSht-1996], we denote by LC the set of pairs $(\log \omega_0, \log \omega_1)$, where ω_i are nondecreasing concave functions on $\mathbb{R}_+$. It is easily seen that LC is a convex cone in the linear space of pairs of measurable functions on $(0, +\infty)$.

A point of a subset of a linear space is recalled (see, e.g., [Bo-1953]) to be *c-interior* if it is the middle point of an open interval lying in this set.

Let now $\mathrm{Lip}^\omega(\mathcal{M})$ be a space of functions on $(\mathcal{M}, d)$ defined by the seminorm

$$|f|_{\mathrm{Lip}^\omega(\mathcal{M})} := \sup\left\{ \frac{|f(m) - f(m')|}{\omega(d(m, m'))}\,;\; m, m' \in \mathcal{M}\right\};$$

here and below ω is a nondecreasing concave function on $\mathbb{R}_+$.

Theorem 4.82. $\bigl(\mathrm{Lip}^{\omega_0}(\mathcal{M}), \mathrm{Lip}^{\omega_1}(\mathcal{M})\bigr)$ *is a Calderón couple if and only if $(\log \omega_0, \log \omega_1)$ is a c-interior point of the cone LC.*

In other words, the couple

$$\mathrm{Lip}^{\vec{\omega}}(\mathcal{M}) := \bigl(\mathrm{Lip}^{\omega_0}(\mathcal{M}), \mathrm{Lip}^{\omega_1}(\mathcal{M})\bigr) \qquad (4.115)$$

is Calderón if and only if there are two pairs of concave nondecreasing functions (φ_0, φ_1) and (ψ_0, ψ_1) such that

$$\omega_0 = \sqrt{\varphi_0 \varphi_1} \quad \text{and} \quad \omega_1 = \sqrt{\psi_0 \psi_1}.$$

In particular, $\mathrm{Lip}^{\vec{\omega}}(\mathcal{M})$ is such if $\omega_i(t) = t^{\theta_i}$, $0 < \theta_i < 1$, $i = 0, 1$, while the couple $\bigl(\ell_\infty(\mathcal{M}), \mathrm{Lip}(\mathcal{M})\bigr)$ of Theorem 4.72 straightforwardly related to the case of $\theta_0 = 0$ and $\theta_1 = 1$ is not.

Now we present a result which computes interpolation spaces $K_\Phi(\vec{X})$ for the couple in (4.115). Clearly, it suffices to evaluate the K-functional of this pair, to find an analog of Theorem 4.81 for this case. In fact, we prove a sharp result for K_∞-*functional* given by

$$K_\infty(t\,;\,f\,;\,\vec{X}) := \inf_{f = f_0 + f_1}\bigl\{\max\bigl[\|f_0\|_{X_0}, t\|f_1\|_{X_1}\bigr]\bigr\}. \qquad (4.116)$$

This is related to the K-functional by the formula

$$K(t;f;\vec{X}) = \sup_{s>0} \frac{s}{s+t} K_\infty(s;f;\vec{X}),$$

see, e.g., [BLo-1976, Sec. 3.13], which then allows us to find the required formula for the K-functional. Since

$$K_\infty(\vec{X}) \leq K(\vec{X}) \leq 2K_\infty(\vec{X}),$$

we can use, in the definition of $K_\Phi(\vec{X})$ the functional $K_\infty(\vec{X})$ instead of $K(X)$. It is worth noting that, in fact, the interpolation inequality

$$\|T|_{K_\phi(\vec{X})}\| \leq \max_{i=0,1}\{\|T|_{X_i}\|\}$$

remains to be true for this definition.

We formulate the result due in an equivalent form by Shvartsman [Shv-1985] and present its proof, since the cited paper is now hardly available.

Let $\Delta(\mathcal{M})$ denote the class of pairs $S, S' \in \mathcal{M}$ such that

$$\operatorname{card} S = \operatorname{card} S' < \infty.$$

We extend the metric of $\mathcal{M}$ to $\Delta(\mathcal{M})$ by setting

$$d_\Delta(S, S') := \inf_\varphi \sum_{m \in S} d(m, \varphi(m)), \tag{4.117}$$

where φ runs over all bijections of S onto S'.

Next, for $f : \mathcal{M} \to \mathbb{R}$ and $S \subset \mathcal{M}$ we set

$$\tilde{f}(S) := \sum_{m \in S} f(m)$$

and define a functional Ω by

$$\Omega(f) := \sup\left\{ \frac{\tilde{f}(S) - \tilde{f}(S')}{d_\Delta(S, S')} \,;\, (S, S') \in \Delta(\mathcal{M}) \right\}. \tag{4.118}$$

Further, given $t > 0$, we associate to the couple $\operatorname{Lip}^{\vec{\omega}}(\mathcal{M})$ a new metric $d_t^{\vec{\omega}}$ on $\mathcal{M}$ given by

$$d_t^{\vec{\omega}} := (\omega_0 + t^{-1}\omega_1)(d) \tag{4.119}$$

and denote by Ω_t functional (4.118) with respect to the metric $(d_t^{\vec{\omega}})_\Delta$.

Theorem 4.83. *For every* $f \in \operatorname{Lip}^{\omega_0}(\mathcal{M}) + \operatorname{Lip}^{\omega_1}(\mathcal{M})$ *and* $t > 0$,

$$K_\infty(t;f;\operatorname{Lip}^{\vec{\omega}}(\mathcal{M})) = \Omega_t(f). \tag{4.120}$$

4.6. Lipschitz spaces

Proof. By scaling, we reduce the proof of (4.120) to the case of

$$\Omega_t(f) = 1 \quad \text{and} \quad t = 1. \tag{4.121}$$

We denote for simplicity the left-hand side of (4.120), for $t = 1$, by $K_\infty(f)$ and prove first that

$$1 \leq K_\infty(f).$$

Given $\varepsilon > 0$, there is a decomposition $f = f_0 + f_1$ such that

$$|f_i|_{\text{Lip}^{\omega_i}(\mathcal{M})} \leq K_\infty(f) + \varepsilon, \quad i = 0, 1,$$

see (4.116). Then, for every $(S, S') \in \Delta(\mathcal{M})$ and every pair of bijections (φ_0, φ_1) of S onto S',

$$|\tilde{f}(S) - \tilde{f}(S')| \leq \sum_{i=0,1} \sum_{m \in S} |f_i(m) - f_i(\varphi_i(m))|$$

$$\leq \left(K_\infty(f) + \varepsilon\right) \sum_{i=0,1} \sum_{m \in S} (\omega_i \circ d)(m, \varphi_i(m)).$$

Passing to infimum over all pairs (φ_0, φ_1) and using (4.117) and (4.119), we then have

$$|\tilde{f}(S) - \tilde{f}(S')| \leq \left(K_\infty(f) + \varepsilon\right) \left(d_1^{\vec{\omega}}\right)_\Delta (S, S')$$

which, due to (4.118), implies, for $\varepsilon \to 0$,

$$1 = \Omega_1(f) \leq K_\infty(f),$$

as required.

To convert this inequality, we first obtain from (4.118) and (4.121) the following:

For every $(S, S') \in \Delta(\mathcal{M})$ and every pair (φ_0, φ_1) of bijections of S onto S',

$$|\tilde{f}(S) - \tilde{f}(S')| \leq \sum_{i=0,1} \sum_{m \in S} (\omega_i \circ d)(m, \varphi_i(m)). \tag{4.122}$$

Using this, we will find a decomposition $f = f_0 + f_1$ satisfying

$$|f_i|_{\text{Lip}^{\omega_i}(\mathcal{M})} \leq \Omega_1(f) = 1, \quad i = 0, 1, \tag{4.123}$$

which clearly implies the desired converse inequality.

To achieve this aim, we define a *complete* oriented graph $\Gamma(\mathcal{M})$ whose vertices are points of m and directed edges are all pairs $(m, m') \in \mathcal{M} \times \mathcal{M}$ denoted by $m \to m'$. On the set of edges we define a weight w given, for $m \to m'$, by

$$w(m, m') := \min\{(\omega_0 \circ d)(m, m'), (\omega_1 \circ d)(m, m') + f(m) - f(m')\}. \tag{4.124}$$

Now we fix a point $m_0 \in \mathcal{M}$ and define a function $g : \mathcal{M} \to \mathbb{R} \cup \{-\infty\}$ by

$$g(m) = \inf \sum w(m_i, m_{i+1}), \qquad (4.125)$$

where the infimum is taken over all directed paths $m_0 \to m_1 \to \cdots \to m_k := m$ joining m_0 and m.

We will show that $g > -\infty$ and that

$$f = -g + (f + g)$$

is a decomposition satisfying (4.123) for $f_0 := -g$ and $f_1 := f + g$.

The main point is

Lemma 4.84. *Let π be a closed path $m_0 \to m_1 \to \cdots \to m_{k+1} := m_0$ with the initial (and final) point m_0. Then*

$$w(\pi) := \sum_{i=0}^{k} w(m_i, m_{i+1}) \geq 0. \qquad (4.126)$$

Proof. One says that $m \to m'$ is an edge of the *first kind* if

$$(\omega_0 \circ d)(m, m') \leq (\omega_1 \circ d)(m, m') + f(m) - f(m'),$$

i.e., $w(m, m') = \omega_0(d)(m, m')$, see (4.124). Otherwise $m \to m'$ is an edge of the *second kind* and then $w(m, m') = \omega_1(d)(m, m') + f(m) - f(m')$.

Let us show that it suffices to prove (4.126) for a path π with alternating edges of the first and the second kinds. In fact, let $m_i \to m_{i+1}$ and $m_{i+1} \to m_{i+2}$ be adjacent edges of the same kind. Let π' be the path with discarded edges $m_i \to m_{i+1}$ and $m_{i+1} \to m_{i+2}$ replaced by the edge $m_i \to m_{i+2}$. Considering separately the case of the edges of the first and of the second type we obtain the inequality

$$w(\pi') \leq w(\pi).$$

In fact, let, e.g., both of the edges be of the first kind. Then

$$w(m_i, m_{i+2}) \leq (\omega_0 \circ d)(m_i, m_{i+2})$$
$$\leq (\omega_0 \circ d)(m_i, m_{i+1}) + (\omega_0 \circ d)(m_{i+1}, m_{i+2})$$
$$= w(m_i, m_{i+1}) + w(m_{i+1}, m_{i+2})$$

and the inequality follows.

Further, we may assume that π is non self-intersecting. Otherwise, one can divide π into a finite number of non self-intersecting closed paths and prove (4.126) for each of them to derive the required result for π.

4.6. Lipschitz spaces

Hence, it suffices to prove (4.126) for a path π of an even number of vertices $m_0 \to m_1 \to \cdots \to m_{2k} := m_0$ such that, say, every $m_{2i} \to m_{2i+1}$ is of the first kind and every $m_{2i+1} \to m_{2i+2}$ is of the second one. Then we have

$$w(\pi) = \sum_{i=0}^{k-1} f(m_{2i+1}) - \sum_{i=0}^{k-1} f(m_{2i+2})$$
$$+ \sum_{i=0}^{k-1} \omega_0(d)(m_{2i}, m_{2i+1}) + \sum_{i=0}^{k-1} \omega_1(d)(m_{2i+1}, m_{2i+2}).$$

Now set $S_0 := \{m_0, m_2, \ldots, m_{2k-2}\}$, $S_1 := \{m_1, m_3, \ldots, m_{2k-1}\}$ and define bijections $\varphi_i : S_0 \to S_1$ by

$$\varphi_0(m_{2j}) = m_{2j+1}, \quad \varphi_1(m_{2j+1}) := m_{2j+2}, \quad j = 0, 1, \ldots, k-1.$$

Then the previous equality can be rewritten as

$$w(\pi) = \tilde{f}(S_1) - \tilde{f}(S_0) + \sum_{i=0,1} \sum_{m \in S_i} \omega_i(d)(m, \varphi_i(m))$$

and due to (4.122), the right-hand side is nonnegative, as required. $\square$

Now let π be a path $m_0 \to m_1 \to \cdots \to m_k := m$. Closing this path by the edge $m_k \to m_0$ and denoting the new path by π' we get, according to the lemma,

$$g(m) = \inf_\pi w(\pi) = \inf\bigl(w(\pi') - w(m, m_0)\bigr) \geq -w(m, m_0) > -\infty.$$

Hence $f = -g + (f - g)$ is a decomposition of f by finite functions and it remains to prove (4.123) for its terms.

By the definition of g and w, see (4.125), for all m, m',

$$g(m') \leq g(m) + w(m, m')$$
$$= g(m) + \min\{\omega_0(d)(m, m'), \omega_1(d)(m, m') + f(m) - f(m')\} \quad (4.127)$$

whence

$$g(m') \leq g(m) + \omega_0(d)(m, m').$$

This clearly implies for $f_0 := -g$ the inequality

$$\sup_{m \neq m'} \frac{|f_0(m) - f_0(m')|}{\omega_0(d)(m, m')} \leq 1,$$

which is equivalent to the desired estimate

$$|f_0|_{\mathrm{Lip}^{\omega_0}(\mathcal{M})} \leq 1.$$

Further, for $f_1 := f + g$ we get from (4.127)

$$f_1(m') = f(m) + g(m) + f(m') - f(m)$$
$$+ \min\{\omega_0(d)(m, m'),\ \omega_1(d)(m, m') + f(m) - f(m')\}$$
$$\leq f(m) + g(m) + \omega_1(d)(m, m') := f_1(m) + \omega_1(d)(m, m').$$

This is clearly equivalent to the desired inequality

$$|f_1|_{\text{Lip}^{\omega_1}(\mathcal{M})} \leq 1.$$

Inequality (4.123) and the theorem are proved. □

Finally we briefly discuss the results concerning Lipschitz spaces of higher order. For simplicity, we restrict ourselves to the case of bounded functions on $\mathbb{R}^n$. The corresponding results for p-integrable functions on a Lipschitz domain $G \subset \mathbb{R}^n$ are also true up to dependence of the constants involved on G and p.

Let $f \in \ell_\infty(\mathbb{R}^n)$ and

$$\omega_k(t;\ f) := \sup_{|h| \leq t} \sup_x |\Delta_h^k f(x)|$$

be the k-modulus of continuity for f. Theorem 2.77 relates the K-functional of the couple $(\ell_\infty, C^k)(\mathbb{R}^n)$ to the k-modulus of continuity. In fact, the same argument gives the result for the couple $(\ell_\infty, \dot{\Lambda}^{k,\omega})(\mathbb{R}^n)$ where $\omega(t) := t^k$, $t \in \mathbb{R}_+$.

Let us recall, see Definition 2.9, that the Lipschitz space $\dot{\Lambda}^{k,\omega}(\mathbb{R}^n)$ is defined by the seminorm

$$|f|_{\Lambda^{k,\omega}(\mathbb{R}^n)} := \sup_{t > 0} \frac{\omega_k(t;\ f)}{t^k}, \qquad (4.128)$$

where $\omega : \mathbb{R}_+ \to \mathbb{R}_+$ is a k-majorant.

Since $\dot{\Lambda}^{k,\omega}(\mathbb{R}^n)$ with $\omega(t) := t^k$ is isomorphic (in fact, isometric) to $\dot{C}^{k-1,1}(\mathbb{R}^n)$ (see Theorem 2.10), Theorem 2.77 may be rewritten as follows.

For some constants of equivalence depending only on k and n and all $f \in \ell_\infty(\mathbb{R}^n)$ and $t > 0$,

$$\omega_k(t;\ f) \approx K\big(t^k;\ f;\ (\ell_\infty, C^{k-1,1})(\mathbb{R}^n)\big). \qquad (4.129)$$

Together with the K-divisibility theorem this leads to

Theorem 4.85. *Let $\{\omega_j\}_{j \in \mathbb{N}}$ be a sequence of k-majorants such that $\sum_{j \in \mathbb{N}} \omega_j(1) < \infty$. Assume that a function $f \in \ell_\infty(\mathbb{R}^n)$ satisfies*

$$\omega_k(f) \leq \sum_{j \in \mathbb{N}} \omega_j. \qquad (4.130)$$

4.6. Lipschitz spaces

Then there is a decomposition

$$f = \sum_{j \in \mathbb{N}} f_j \quad \text{(uniform convergence on } \mathbb{R}^n)$$

such that for every j,

$$\omega_k(f_j) \leq C\omega_j;$$

here $C > 1$ is a constant depending only on k and n.

In fact, if ω is a k-majorant, then $t \mapsto \omega(\sqrt[k]{t})$ is a 1-majorant, see Definition 2.8, and (4.130) may be rewritten as

$$\omega_k(f) \leq 2 \sum_{j \in \mathbb{N}} \widehat{\omega}_j,$$

where $\widehat{\omega}$ is convexification of the function $t \mapsto \omega(\sqrt[k]{t})$. This, (4.129) and the K-divisibility theorem prove the result.

As a remote consequence of K-divisibility we single out one more property of a k-modulus of continuity, a plain consequence of Krugljak's inversion theorem for a K-functional, see [BK-1991, Thm. 4.5.7].

Theorem 4.86. *For every k-majorant ω there is a function $f \in \ell_\infty(\mathbb{R}^n)$ such that*

$$\omega_k(f) \approx \omega,$$

where the constants of equivalence depend only on k and n.

Finally, we describe Calderón couples among those of the Lipschitz spaces of higher order. The result due to Y. Brudnyi and Shteinberg [BSht-1996] shows that the condition of Theorem 4.82 is sufficient in this case. So, the following is true.

Theorem 4.87. *The couple $(\dot{\Lambda}^{k,\omega_0}, \dot{\Lambda}^{k,\omega_1})(\mathbb{R}^n)$ is a Calderón couple if the pair $(\log \widehat{\omega}_0, \log \widehat{\omega}_1)$, where $\widehat{\omega}_i$ is convexification of the function $t \mapsto \omega_i(\sqrt[k]{t})$, $i = 0, 1$, is a c-interior point of the cone of $\log$-concave functions LC.*

In particular, couples of Besov spaces are Calderón. On the other hand, the couple $(\ell_\infty, \dot{\Lambda}^{k,\omega})(\mathbb{R}^n)$ with $\omega(t) := t^k$, $t \in \mathbb{R}^n$, is not, see the Yu. Brudnyi and Shteinberg paper [BSht-1997] for $k = 1$; the argument there works for $k > 1$ as well. This makes the claim of the necessity of the condition in Theorem 4.87 highly plausible.

4.6.3 Duality theorem

In the study of simultaneous Lipschitz extensions, a considerable role is played by the result describing a predual space to the space of Lipschitz functions on a pointed metric space. The first result of this kind was due to Kantorovich and Rubinshtein [KR-1958] whose theorem we briefly discuss before proving the required result.

Let $(\mathcal{M}, m^*, d)$ be a pointed metric space and let $\mathrm{Lip}_0(\mathcal{M})$ be the space of Lipschitz functions on $\mathcal{M}$ vanishing at m^*. Then the Lipschitz constant

$$L(f) := \sup_{m \neq m'} \frac{|f(m) - f(m')|}{d(m, m')}$$

is clearly a (Banach) norm on $\mathrm{Lip}_0(\mathcal{M})$.

Let $K(\mathcal{M})$ be the linear space of bounded (regular Borel) measures μ on $\mathcal{M}$ satisfying

$$\mu(\mathcal{M}) = 0.$$

For every $\mu \in K(\mathcal{M})$ one defines the set ψ_μ consisting of bounded measures ν on $\mathcal{M} \times \mathcal{M}$ satisfying the so-called *Monge–Kantorovich balance condition*:

$$\nu(S \times \mathcal{M}) - \nu(\mathcal{M} \times S) = \mu(S)$$

for every Borel set $S \subset \mathcal{M}$.

Using this, we define a norm on $K(\mathcal{M})$ by

$$\|\mu\|_{KR} := \sup \left\{ \int_{\mathcal{M} \times \mathcal{M}} d(m, m') d\nu(m, m') \, ; \, \nu \in \psi_\mu \right\}.$$

Theorem 4.88. *Let $\mathcal{M}$ be a compact metric space. Then $K(\mathcal{M})$ equipped with the norm $\|\cdot\|_{KR}$ is a Banach space and its dual is linearly isometric to $\mathrm{Lip}_0(\mathcal{M})$.*

Dudley [Dud-1989] extended the Kantorovich–Rubinshtein theorem to a separable metric space; the result for the nonseparable case is unknown.

Now we introduce the canonical predual space to $\mathrm{Lip}_0(\mathcal{M})$, the so-called *Lipschitz-free space*, denoted by $\mathcal{F}(\mathcal{M})$ which is defined as the closed (linear) hull of the point evaluations

$$\delta_\mathcal{M}(m)(f) := f(m), \quad f \in \mathrm{Lip}_0(\mathcal{M}) \text{ and } m \in \mathcal{M},$$

in the dual space $\mathrm{Lip}_0(\mathcal{M})^*$. The definition is correct as

$$|\delta_\mathcal{M}(m)(f)| = |f(m) - f(m^*)| \leq L(f) d(m, m^*),$$

i.e., the linear functional $\delta_\mathcal{M}(m) : \mathrm{Lip}_0(\mathcal{M}) \to \mathbb{R}$ has the norm in $\mathrm{Lip}_0(\mathcal{M})^*$ bounded by $d(m, m^*)$.

By the definition of the dual norm, the space $\mathrm{Lip}_0(\mathcal{M})^*$ is a closed linear subspace of the space $\ell_\infty(B_\mathcal{M})$, where $B_\mathcal{M} := \{f \in \mathrm{Lip}_0(\mathcal{M}) \, ; \, L(f) \leq 1\}$ is the closed unit ball of $\mathrm{Lip}_0(\mathcal{M})^*$.

Therefore we can write

$$\mathcal{F}(\mathcal{M}) = \overline{\mathrm{hull}\, \delta_\mathcal{M}(\mathcal{M})} \quad \text{(closure in } \ell_\infty(B_\mathcal{M})\text{)}. \tag{4.131}$$

Theorem 4.89. *The dual space $\mathcal{F}(\mathcal{M})^*$ is linearly isometric to $\mathrm{Lip}_0(\mathcal{M})$.*

4.6. Lipschitz spaces

Proof. The required linear isometry I is defined on functionals $\ell \in \mathcal{F}(\mathcal{M})^*$ by

$$I(\ell) \longmapsto \ell(\delta|_B),$$

where we set for brevity, $\delta := \delta_\mathcal{M}$. In other words, $I(\ell)$ is a function on $\mathcal{M}$ given for $m \in \mathcal{M}$ by $I(\ell)(m) := \ell(\delta(m))$.

Claim I. It is true that

$$\operatorname{Lip}_0(\mathcal{M}) \subset I(\mathcal{F}(\mathcal{M})^*) \tag{4.132}$$

and, moreover,

$$\|I\| \geq 1. \tag{4.133}$$

To prove (4.132), we must show that

$$I(\ell)(m^*) = 0$$

and that for every $f \in \operatorname{Lip}_0(\mathcal{M})$ there is a functional $\ell_f \in \mathcal{F}(\mathcal{M})^*$ such that

$$f = I(\ell_f).$$

The first is the matter of definitions:

$$I(\ell)(m^*) = \ell\big(\delta(m^*)\big|_B\big) = 0$$

as $\delta(m^*)(f) = f(m^*) = 0$ for all $f \in \operatorname{Lip}_0(\mathcal{M})$.

As for the second, one defines ℓ_f by employing the canonical projections $\pi_b : \ell_\infty(B) \to \mathbb{R}$, $b \in B_\mathcal{M}$, given by

$$\pi_b(F) := F(b).$$

Then $\pi_b\big|_{\mathcal{F}(\mathcal{M})}$ belongs to $\mathcal{F}(\mathcal{M})^*$, see (4.131), and for every $m \in \mathcal{M}$,

$$I\big(\pi_b\big|_{\mathcal{F}(\mathcal{M})}\big)(m) := \pi_b\big(\delta(m)\big) = b(m); \tag{4.134}$$

here $\delta(m)$ is regarded as a function on $\ell_\infty(B_\mathcal{M})$. Since b is an arbitrary function from $\operatorname{Lip}_0(\mathcal{M})$ satisfying $L(b) \leq 1$, this, by homogeneity, implies (4.132). Equality (4.134) also implies (4.133):

$$\|I\| \geq \sup\Big\{L\big(I\big(\pi_b\big|_{\mathcal{F}(\mathcal{M})}\big)\big);\ b \in B\Big\} = 1.$$

Claim II. It is true that

$$I(\mathcal{F}(\mathcal{M})^*) \subset \operatorname{Lip}_0(\mathcal{M}) \quad \text{and} \quad \|I\| \leq 1. \tag{4.135}$$

Let $\ell \in \mathcal{F}(\mathcal{M})^*$ and let $m_1, m_2 \in \mathcal{M}$. To prove (4.135), it suffices to show that

$$|I(\ell)(m_1) - I(\ell)(m_2)| \le \|\ell\| d(m_1, m_2). \tag{4.136}$$

Since $I(\ell)(m^*) = 0$, this clearly proves the claim.

To establish this, we extend ℓ to the linear functional $\Lambda \in \ell_\infty(B)^*$ by the Hahn–Banach theorem; hereafter B stands for $B_\mathcal{M}$. Hence,

$$\Lambda|_{\mathcal{F}(\mathcal{M})} = \ell \quad \text{and} \quad \|\Lambda\| = \|\ell\|.$$

Then we equip the ball B with the discrete topology denoting the space obtained by B^{dsc}. Every bounded function on B is continuous in this topology, i.e., isometrically

$$\ell_\infty(B) = C_b(B^{dsc}).$$

By βB we denote the Stone–Čech compactification of B^{dsc}. Let us recall that βB is a compact topological space containing B^{dsc} as a dense subspace. Moreover, every $f \in C_b(B^{dsc})$ is uniquely extended to a function $\hat{f}$ continuous on βB and such that

$$(\|f\|_{\ell_\infty(B)} =) \|f\|_{C_b(B^{dsc})} = \|\hat{f}\|_{C(\beta B)},$$

see, e.g., [HR-1963].

The linear continuous functional Λ on $\ell_\infty(B)$ uniquely defines that on $C(\beta B)$, say $\widetilde{\Lambda}$, so that

$$\|\Lambda\| = \|\widetilde{\Lambda}\|.$$

By the F. Riesz representation theorem, see, e.g., [DS-1958], there is a bounded Borel regular measure μ_Λ on βB such that, for every $F \in C(\beta B)$,

$$\widetilde{\Lambda}(F) = \int_{\beta B} F \, d\mu_\Lambda \quad \text{and} \quad \|\widetilde{\Lambda}\| = \operatorname{var} \mu_\Lambda.$$

In particular, for every function $f \in \mathcal{F}(\mathcal{M})(\subset \ell_\infty(B))$, the extended function $\hat{f}$ satisfies

$$\Lambda(f) = \int_{\beta B} \hat{f} \, d\mu_L \quad \text{and} \quad \|\Lambda\| = \|\ell\| = \operatorname{var} \mu_\Lambda.$$

Now, given such f and $\varepsilon > 0$, we choose a finite open cover $\{U_i\}$ of βB such that the oscillation of $\hat{f}$ over every U_i is at most ε. Since B^{dsc} is dense in βB, every U_i contains a point $b_i \in B$, and therefore for every $\omega \in U_i$,

$$|\hat{f}(\omega) - f(b_i)| = |\hat{f}(\omega) - \hat{f}(b_i)| < \varepsilon.$$

This leads to the inequality

$$\left| \int_{\beta B} \hat{f} \, d\mu_\Lambda - \sum_i f(b_i) \mu_\Lambda(U_i) \right| < \varepsilon \operatorname{var} \mu_\Lambda = \varepsilon \|\ell\|.$$

4.6. Lipschitz spaces

The sum here can be written as $\ell_\varepsilon(f)$, where $\ell_\varepsilon \in \mathcal{F}(\mathcal{M})$ is given by

$$\ell_\varepsilon := \sum_i \left(\pi_{b_i}\big|_{\mathcal{F}(\mathcal{M})}\right) \mu_\Lambda(U_i).$$

We apply the above inequality to the function $f := \bigl(\delta(m_1) - \delta(m_2)\bigr)\big|_B$. Since

$$\int_{\beta B} \hat{f} d\mu_\Lambda = \Lambda(f) = \ell\left(\bigl(\delta(m_1) - \delta(m_2)\bigr)\big|_B\right) = I(\ell)(m_1) - I(\ell)(m_2),$$

this yields

$$\left|\bigl(I(\ell)(m_1) - I(\ell)(m_2)\bigr) - \bigl(I(\ell_\varepsilon)(m_1) - I(\ell_\varepsilon)(m_2)\bigr)\right| < \varepsilon \|\ell\|.$$

Moreover, $I(\ell_\varepsilon)(m) := \sum_i b_i(m) \mu_\Lambda(U_i)$, $m \in \mathcal{M}$, i.e., $I(\ell_\varepsilon)$ belongs to $\mathrm{Lip}_0(\mathcal{M})$ and its Lipschitz constant satisfies

$$L\bigl(I(\ell_\varepsilon)\bigr) \le \bigl(\max_i L(b_i)\bigr) \sum_i |\mu_\Lambda(U_i)| \le \operatorname{var} \mu_\Lambda = \|\ell\|.$$

Combining this with the previous inequality we get

$$\bigl|I(\ell)(m_1) - I(\ell)(m_2)\bigr| \le L\bigl(I(\ell_\varepsilon)\bigr) d(m_0, m_1) + \varepsilon \|\ell\| \le \bigl(d(m_1, m_2) + \varepsilon\bigr) \|\ell\|.$$

Letting ε to 0 we prove (4.136).

It remains to show that I is an injection. But if $I(\ell) = 0$ for some $\ell \in \mathcal{F}(\mathcal{M})^*$, then $\ell\big|_{\delta(\mathcal{M})} = 0$ and $\ell = 0$ as the set span $\delta(\mathcal{M})$ is dense in $\mathcal{F}(\mathcal{M})$, see (4.131).

Thus, the linear map I is a bijection of $\mathcal{F}(\mathcal{M})^*$ onto $\mathrm{Lip}_0(\mathcal{M})$ and $\|I\| = 1$. The result is proved. □

Remark 4.90. It is readily seen that $\mathcal{F}(\mathcal{M})$ can be regarded as the completion of the set of all finitely supported measures μ on $\mathcal{M}$ under the norm

$$\|\mu\| := \sup\left\{\int_\mathcal{M} f d\mu \,;\, L(f) \le 1\right\}.$$

Finally we present a property of the Lipschitz-free space construction which will be used for extension results for Banach-valued Lipschitz functions.

Let X be a Banach space, and let $\mathrm{Lip}_0(\mathcal{M}\,;\,X)$ be a Banach space of functions $f : \mathcal{M} \to X$ on pointed metric space $(\mathcal{M}, m^*, d)$ such that $f(m^*) = 0$ that is defined by the norm

$$L(f\,;\,X) := \sup_{m \neq m'} \frac{\|f(m) - f(m')\|_X}{d(m, m')}.$$

Theorem 4.91. For every $F \in \mathrm{Lip}_0(\mathcal{M}; X)$ there is a unique linear map $T_F : \mathcal{F}(\mathcal{M}) \to X$ such that
$$T_F \delta_\mathcal{M} = F \quad \text{and} \quad \|T_F\| \leq L(F; X). \tag{4.137}$$

Proof. We begin with

Lemma 4.92. (a) $\delta_\mathcal{M}$ is a (nonlinear) isometric embedding of $\mathcal{M}$ in $\mathcal{F}(\mathcal{M})$;

(b) $\{\delta_\mathcal{M}(m)\}_{m \in \mathcal{M}}$ is an algebraic (Hamel) basis of the linear span of $\delta_\mathcal{M}(\mathcal{M})$.

Proof. (a) By the definition of $\mathcal{F}(\mathcal{M})$, see (4.131),
$$\|\delta_\mathcal{M}(m) - \delta_\mathcal{M}(m')\|_{\mathcal{F}(\mathcal{M})} = \sup_{b \in B_\mathcal{M}} |b(m) - b(m')|$$
$$\leq \left(\sup_{b \in B_\mathcal{M}} L(b) \right) d(m, m') = d(m, m').$$

To prove the converse inequality, one defines a function g on the set $\{m, m', m^*\}$ by
$$g(n) := d(n, m') - d(m^*, m').$$

Then $g(m^*) = 0$ and the Lipschitz constant $L(g) = 1$. By the McShane extension Theorem 1.27, there exists a function $\hat{g} \in \mathrm{Lip}_0(\mathcal{M})$ which coincides with g on $\{m, m', m^*\}$, and such that $L(\hat{g}) = 1$. For this $\hat{g}$ we get
$$\|\delta_\mathcal{M}(m) - \delta_\mathcal{M}(m')\|_{\mathcal{F}(\mathcal{M})} \geq |\hat{g}(m) - \hat{g}(m')| = |g(m) - g(m')| = d(m, m').$$

Hence, $\delta_\mathcal{M} : \mathcal{M} \to \mathcal{F}(\mathcal{M})$ is an isometric embedding.

(b) We must show that, for every finite proper set $\{m_1, \ldots, m_2\} \subset \mathcal{M}$ of pairwise distinct points, the associated set $\{\delta_\mathcal{M}(m_1), \ldots, \delta_\mathcal{M}(m_n)\} \subset \mathcal{F}(\mathcal{M})$ is linearly independent.

Let for some constants λ_i,
$$\sum_i \lambda_i \delta_\mathcal{M}(m_i) = 0. \tag{4.138}$$

Due to the McShane extension theorem, there exists a function $f \in \mathrm{Lip}_0(\mathcal{M})$ which equals 0 at all m_i distinct from a given m_j and equals 1 at m_j. Evaluating the left-hand side of (4.138) at this f, we get
$$\lambda_j = \sum_i \lambda_i f(m_i) = 0,$$

i.e., all λ_j are zeros. $\square$

4.6. Lipschitz spaces

Given $F \in \mathrm{Lip}_0(\mathcal{M}; X)$, we now define the required linear operator T_F on all elements $\sum_{i=1}^{n} \lambda_i \delta_{\mathcal{M}}(m_i)$ of the linear span of $\delta_{\mathcal{M}}(\mathcal{M})$ by setting

$$T_F\left(\sum_{i=1}^{n} \lambda_i \delta_{\mathcal{M}}(m_i)\right) := \sum_{i=1}^{n} \lambda_i F(m_i).$$

One shows that T_F is continuous on the span and that its norm is bounded by $L(F; X)$. To this end, one chooses an arbitrary linear continuous functional ℓ on X of the norm $\|\ell\|_{X^*} = 1$ and considers the function $\ell \circ F : \mathcal{M} \to \mathbb{R}$. This function is clearly Lipschitz and of norm

$$L(\ell \circ F) \le \|\ell\|_{X^*} L(F; X) = L(F; X).$$

As every function from $\mathrm{Lip}_0(\mathcal{M})$, the function $\ell \circ F$ can be regarded as a linear continuous functional on the predual space $\mathcal{F}(\mathcal{M})$. We denote this functional by $\langle \ell \circ F, \cdot \rangle$ and note that

$$\langle \ell \circ F, \cdot \rangle \big|_{\delta_{\mathcal{M}}(\mathcal{M})} = \ell \circ F = (\ell \circ T_F) \big|_{\delta_{\mathcal{M}}(\mathcal{M})} \tag{4.139}$$

by definition.

But $\delta_{\mathcal{M}}(\mathcal{M}) = \{\delta_{\mathcal{M}}(m)\}_{m \in \mathcal{M}}$ is an algebraic basis of $\mathrm{hull}\,\delta_{\mathcal{M}}(\mathcal{M})$, and therefore every linear functional has a unique linear extension from $\delta_{\mathcal{M}}(\mathcal{M})$ to the hull. Hence, (4.139) implies that

$$\langle \ell \circ F, \cdot \rangle \big|_{\mathrm{hull}\,\delta_{\mathcal{M}}(\mathcal{M})} = \ell \circ T_F.$$

This, in turn, implies for every $\mu \in \mathrm{hull}\,\delta_{\mathcal{M}}(\mathcal{M})$,

$$|(\ell \circ T_F)(\mu)| = |\langle \ell \circ F, \mu \rangle| \le \|\ell\|_{X^*} L(F; X) \|\mu\|_{\mathcal{F}(\mathcal{M})} = L(F; X)\|\mu\|_{\mathcal{F}(\mathcal{M})}.$$

Taking supremum over all such ℓ, we get

$$\|T_F \mu\| \le L(F; X)\|\mu\|_{\mathcal{F}(\mathcal{M})}.$$

Hence, T_F is continuously extended to the closure $\overline{\mathrm{hull}\,\delta_{\mathcal{M}}(\mathcal{M})} = \mathcal{F}(\mathcal{M})$, see (4.131), and its norm is bounded by $L(F; X)$. Let us show that this norm equals $L(F; X)$. In fact, we have, by Lemma 4.92 (a) and the definition of T_F,

$$\|F(m) - F(m')\|_X := \|T_F[\delta(m) - \delta(m')]\|_X$$
$$\le \|T_F\| \|\delta(m) - \delta(m')\|_{\mathcal{F}(\mathcal{M})} = \|T_F\| d(m, m').$$

Dividing by $d(m, m')$ and taking supremum over all $m \ne m'$, we bound $\|T_F\|$ from below by $L(F; X)$.

Hence, $\|T_F\| = L(F; X)$; together with the equality $T_F \delta = F$ following from the definition, this proves (4.137). $\square$

It remains to discuss a similar duality theorem for Lipschitz functions of higher order. For simplicity, we consider the case of functions on $\mathbb{R}^n$. So, given a k-majorant $\omega : \mathbb{R}_+ \to \mathbb{R}_+$, one defines the space $\dot{\Lambda}^{k,\omega}(\mathbb{R}^n)$ by the seminorm given in (4.128). The subspace $\Lambda_0^{k,\omega}(\mathbb{R}^n)$ of this space is defined by factorizing by the space $\mathcal{P}_{k-1,n}$ of polynomials of degree $k-1$. More precisely, let $J_{k,n}$ be a finite interpolating subset of $\mathbb{R}^n$ for these polynomials. Hence, for every numerical data $\Delta : J_{k,n} \to \mathbb{R}$, there is a *unique* polynomial $P_\Delta \in \mathcal{P}_{k-1,n}$ such that

$$P_\Delta\big|_{J_{k,n}} = \Delta.$$

We now set

$$\Lambda_0^{k,\omega}(\mathbb{R}^n) := \{f \in \dot{\Lambda}^{k,\omega}(\mathbb{R}^n);\ f\big|_{J_{k,n}} = 0\}. \qquad (4.140)$$

It is readily seen that this is a Banach space under the norm defined by (4.128).

We define a predual to this space using, as before, the point evaluations $\delta_x(f) := f(x)$, $x \in \mathbb{R}^n$. Let us explain why δ_x may be regarded as a linear continuous functional on the space (4.140).

To this end, given $f \in \Lambda_0^{k,\omega}(\mathbb{R}^n)$ and $x \in \mathbb{R}^n$, one denotes by $P(f)$ a polynomial of degree $k-1$ such that

$$E_k(f\,;S) := \max_S |f - P(f)|,$$

where $S = S(x) := \mathrm{conv}(\{x\} \cup J_{k,n})$. Due to Theorem 2.37,

$$E_k(f\,;S) \leq c(S)\omega_k(\mathrm{diam}\,S\,;f) \leq c(S)\omega(\mathrm{diam}\,S)|f|_{\Lambda^{k,\omega}(\mathbb{R}^n)}.$$

This, in turn, implies that

$$|\delta_x(f)| = |f(x)| \leq |P(f)(x)| + c(S)\omega(\mathrm{diam}\,S)|f|_{\Lambda^{k,\omega}(\mathbb{R}^n)}.$$

Let now $\{P_y\}_{y \in J_{k,n}} \subset \mathcal{P}_{k-1,n}$ be the Lagrange basis of $\mathcal{P}_{k-1,n}$ associated to the interpolating set $J_{k,n}$. Then $P(f)$ can be represented as

$$P(f) = \sum_{y \in J_{k,n}} P(f)(y) P_y,$$

and therefore

$$|P(f)(x)| = \left| \sum_{y \in J_{k,n}} [P(f) - f](y) P_y(x) \right| \leq E_k(f\,;S) \sum_{y \in J_{k,n}} |P_y(x)|.$$

Combining this with the previous two inequalities we get

$$|\delta_x(f)| \leq c|f|_{\Lambda^{k,\omega}(\mathbb{R}^n)},$$

where c is independent of f; hence,

$$\delta_x \in \Lambda_0^{k,\omega}(\mathbb{R}^n)^* \subset \ell_\infty(B),$$

where B stands for the closed unit ball of the space $\Lambda_0^{k,\omega}(\mathbb{R}^n)$.

Now we define the required predual space denoted by $\mathcal{F}^{k,\omega}(\mathbb{R}^n)$ by

$$\mathcal{F}^{k,\omega}(\mathbb{R}^n) := \overline{\text{hull}\{\delta_x\}_{x\in\mathbb{R}^n}} \quad (\text{closure in } \ell_\infty(B)) \tag{4.141}$$

We leave it to the reader to repeat, with evident changes, the proof of Theorem 4.89 to establish the following

Theorem 4.93. *The dual space $\mathcal{F}^{k,\omega}(\mathbb{R}^n)^*$ is linearly isomorphic to the space $\Lambda_0^{k,\omega}(\mathbb{R}^n)$.*

Using this and the argument in the proof of Theorem 4.91 we also easily obtain

Theorem 4.94. *For every Banach-valued function $F : \mathbb{R}^n \to X$ belonging to the space $\Lambda_0^{k,\omega}(\mathbb{R}^n; X)$ there exists a uniquely defined linear operator $T_F : \mathcal{F}^{k,\omega}(\mathbb{R}^n) \to X$ whose norm is equivalent to $|F|_{\Lambda^{k,\omega}(\mathbb{R}^n;X)}$ such that $T_F \delta_\mathcal{M} = F$.*

Here the vector-valued analog of the space $\Lambda_0^{k,\omega}(\mathbb{R}^n)$ is defined by the k-modulus of continuity

$$\omega_k(t; f; X) := \sup_{|h|\leq t} \sup_x \left\|\Delta_h^k f(x)\right\|_X.$$

Remark 4.95. In some cases, it would be more appropriate to work with the Banach space $\Lambda^{k,\omega}(\mathbb{R}^n)$ whose norm differs from (4.128) by the summand $\sup_{\mathbb{R}^n}|f|$. To within the equivalence of the norms, this space coincides with the direct sum of $\Lambda_0^{k,\omega}(\mathbb{R}^n)$ and $\mathcal{P}_{k-1,n}$. Therefore its predual can be naturally identified with $\mathcal{F}^{k,\omega}(\mathbb{R}^n) \oplus \mathcal{P}_{k-1,n}^*$.

Comments

Along with the metric invariants discussed in Section 4.2, the following two deserve mentioning as well. The first of them, the *Assouad dimension*, $\dim_A$ is defined in subsection 4.2.3 where we show that it is bounded by the Hausdorff dimension $\dim_H$. However, $\dim_A$, in a sense, is dual to $\dim_H$; sharing with the latter properties (a)–(c) and (e) of Theorem 4.12, $\dim_A$ satisfies the converse inequality for direct sums:

$$\dim_A\bigl(\mathcal{M}_1 \oplus^{(p)} \mathcal{M}_2\bigr) \leq \dim_A \mathcal{M}_1 + \dim_A \mathcal{M}_2,$$

see the survey in the Luukainen paper [Lu-1998].

The second concept, the *packing dimension* was introduced by Tricot in 1984, see his book [Tr-1995]. Its definition is similar to that of Hausdorff with the approximate Hausdorff measure $\mathcal{H}_s^\delta$ substituted for a set function $\overline{\mathcal{H}}_s^\delta$ defined for a subset S of $\mathcal{M}$ by

$$\overline{\mathcal{H}}_s^\delta(S) := \sup_\pi \left\{ \sum_{\overline{B} \in \pi} (2r_B)^s \right\},$$

where π runs over all *disjoint* families of closed balls $\overline{B}$ centered at S.

Unlike $\mathcal{H}_s^\delta$ this is not an (outer) measure; in the standard way, it gives rise to a (regular) Borel measure denoted by $\mathcal{P}_s$ and called a *packing measure*. Using $\mathcal{P}_s$ instead of the Hausdorff s-measure we define the *packing dimension* $\dim_{\mathcal{P}}(\mathcal{M})$ similarly to that of Hausdorff, see (4.11).

Packing dimension shares with $\dim_H$ properties (a)–(c) and (e) of Theorem 4.12 but, like Assouad dimension, satisfies the opposite inequality for direct sums.

As proved in subsection 4.2.3, $\dim_H \leq \dim_A$, where the inequality may be strict; e.g., unlike the Hausdorff ones, the Assouad dimensions of a separable metric space and its dense countable metric subspace are the same.

Since $\mathcal{H}_s \leq \mathcal{P}_s$, the associated dimensions satisfy $\dim_{\mathcal{H}} \leq \dim_{\mathcal{P}}$, where the inequality may be strict. A spectacular example of a Cantor type subset in $\mathbb{R}^2$ whose Hausdorff dimension is strictly less than its packing dimension, is presented in the Peres paper [Per-1994].

It was noted by Coifman and Weiss [CW-1971] that a metric space of homogeneous type is doubling. Assouad [As-1980] and Dyn'kin [Dy-1984] independently conjectured that the converse is true for complete doubling metric spaces. Dyn'kin [Dy-1983] proved this for subsets S of the real line satisfying the condition equivalent to $\dim_A S < 1$ and then formulated, in [Dy-1984], the combinatorial conjecture presented in subsection 4.3.4. Motivated by Dyn'kin's papers, Vol'berg and Kongagin [VK-1987] proved the Assouad–Dyn'kin conjecture for the compact case. A simplified version of their proof discussed in subsection 4.3.1 is due to Wu [Wu-1998]. Using her approach Wu also proved the following result which is highly nontrivial even for subsets of the real line.

Given a compact doubling metric space $\mathcal{M}$ and a number $s > 0$, there exists a doubling measure μ on $\mathcal{M}$ such that $\dim_H(\operatorname{supp} \mu) < s$.

Finally, we mention two fields of research where doubling measures appear as the basic objects. The first of them is presented in the Garnett and Marshall book [GM-2005] on *harmonic measures*, the fundamental concept of Harmonic Analysis and PDF. The second one, studying the connections between deformation of metric spaces and related doubling measures, is presented in the David and Semmes book [DaSe-1997] and the recent books [Hei-2001] by Heinonen and [Sem-2007] by Semmes.

There is a way, different from that in Remark 4.69, to define differentiability of functions on Carnot groups. Actually, the structure of a stratified Lie group allows us to define polynomials on a Carnot group G which, in an appropriate coordinate system $(x_1, \ldots, x_q)$, can be written in the standard form $\sum_\alpha c_\alpha x^\alpha$. The *homogeneous degree* of such a polynomial is defined as $\max\{d(\alpha)\,;\, c_\alpha \neq 0\}$, where $d(\alpha) := \sum_{k=1}^{q} d_k \alpha_k$ and the numbers d_k are the dimensions of the linear spaces V_k used in decomposition (3.148) of the Lie algebra for G, see Chapter 1.C of the Folland and Stein book [FS-1982] for details.

Having this one defines the k-differential of a function $f : G \to \mathbb{R}$ similarly to that in Definition 4.63. In this setting, the Rademacher type theorem was due to Magnani [Mag-2005]. Apparently, the analog of Theorem 4.58 is also true for this case.

The Brodskii theorem [Bro-1967] which was used in assertion (b) of Theorem 4.62 estimates a functional dual to the modulus of continuity of a map $f : [0,1]^n \to [0,1]^m$ given by

$$\Omega(t\,;\,f) := \inf\{\|f(x) - f(y)\|\,;\,\|x - y\| \geq t\}.$$

These estimates also include the factor $\log|\log t|$ in a certain power.

The K-divisibility theorem presented in subsection 4.6.2 admits a linearization. Namely, if (X_0, X_1) is a K-linearized Banach couple, see Remark 2.78 for the definition, then there exist linear operators $T_j : X_0 + X_1 \to X_0 \cap X_1$, $j \in \mathbb{Z}$, such that the decomposition $x = \sum_j x_j$ in Theorem 4.79 is given by $x = \left(\sum_j T_j\right)(x)$. The proof of this theorem follows, with simple changes, the line presented in the book [BK-1991, Sec. 3.3.2].

This version of K-divisibility is used in the study of simultaneous Lipschitz extensions.

The predual space $\mathcal{F}(\mathcal{M})$ is defined in Chapter 2 of the Weaver book [W-1999] for bounded metric spaces; the proof of Theorem 4.89 is taken from A. and Yu. Brudnyi's paper [BB-2007b]. The name of a Lipschitz-free space for $\mathcal{F}(\mathcal{M})$ is coined by Godefroy and Kalton [GK-2003] whose paper contains several deep results of Banach space theory.

The map T_F linearizing a Lipschitz map F, see Theorem 4.82, was firstly used by Bachir [Bach-2001] to prove a nonconvex analog of the classical Fenchel duality theorem.

An extension of the Kantorovich–Rubinshtein duality theorem to the space $\Lambda_0^{k,\omega}(\mathbb{R}^n)$ of Lipschitz functions of higher order was due to Hanin [Han-1997]. He exploits the space of bounded Borel measures μ on $\mathbb{R}^n$ denoted by $\mathfrak{M}_0^k$ which satisfy

$$\int_{\mathbb{R}^n} x^\alpha d\mu = 0 \quad \text{for all} \quad |\alpha| \leq k - 1$$

and defines the analog of the set ψ_μ of Theorem 4.79 to be the set of bounded

Borel measures ν on $\mathbb{R}^n \times \mathbb{R}^n$ satisfying

$$\int_{\mathbb{R}^n} f d\mu = \int_{\mathbb{R}^{2n}} \Delta_h^k f(x) d\nu(x,h).$$

An analog of the Kantorovich–Rubinshtein norm $\|\cdot\|_{KR}$ is then given by

$$\|\mu\| := \inf_{\psi \in \psi_\mu} \int_{\mathbb{R}^n} \omega(|h|) d|\nu|(x,h)$$

and an argument similar to that in [KR-1958] shows that the dual to the Banach space $(\mathfrak{M}_0^k, \|\cdot\|)$ is linearly isomorphic to $\Lambda_0^{k,\omega}(\mathbb{R}^n)$.

Chapter 5

Lipschitz Embedding and Selections

In this chapter, we discuss the results of two areas intimately related to the main theme of the book. The first one studies (bi-)Lipschitz embeddings of metric spaces into the space forms of nonpositive curvature, Euclidean and hyperbolic spaces, while the main objective of the second area is Lipschitz selections for maps from metric spaces to the space of convex subsets in $\mathbb{R}^n$.

The results directly used for the proofs of extension and trace theorems are concentrated in subsections 5.1.2, 5.1.4, 5.3.2 and Section 5.4. The first contains a theorem asserting that every infinite tree with uniformly bounded degrees of vertices admits a bi-Lipschitz embedding into the direct product of the hyperbolic plane and a Euclidean space with distortion bounded by a numerical constant. In contrast, we prove in subsection 5.1.1 the Bourgain theorem showing that any n-point metric space admits a bi-Lipschitz embedding into a Euclidean space with distortion of sharp order $\log n$.

In subsection 5.1.4 we prove that every Gromov hyperbolic space of bounded geometry admits a bi-Lipschitz embedding into the direct product of some Euclidean and hyperbolic spaces. An essential ingredient of the proof is the Bonk–Schramm theorem on rough similarity of such a space to a convex subset of some hyperbolic space. This important result is proved in Section 5.2. In turn, its proof is based on the Assouad embedding theorem for doubling metric spaces which we present in subsection 5.1.3.

The last two sections, 5.3 and 5.4, contain Lipschitz selection theorems playing an essential role in study of the trace and extension problems for smooth functions, see Chapter 10. The main result of Section 5.3 is the Shvartsman selection theorem for set-valued maps from a metric space $\mathcal{M}$ into the set of affine subspaces $\mathrm{Aff}(\mathbb{R}^n)$.

The final section is devoted to the Yu. Brudnyi and Shvartsman simultaneous selection theorem dealing with a family of set-valued maps from $\mathcal{M}$ into $\mathrm{Aff}(\mathbb{R}^n)$ linearly parameterized by functions from $\mathrm{Lip}(\mathcal{M},\mathbb{R}^n)$. We ask for existence of a simultaneous Lipschitz selection of this family linearly depending on the parameter.

5.1 Embedding of metric spaces into the space forms of nonpositive curvature

5.1.1 Finite metric spaces

Every N-point metric space allows an isometric embedding into an N-dimensional subspace of the space $\ell_\infty(\mathcal{M})$. Composing this embedding and a linear operator of minimal norm mapping the subspace onto an N-dimensional Euclidean space $\mathbb{R}^N$, we obtain a bi-Lipschitz embedding of $\mathcal{M}$ into $\mathbb{R}^N$ with distortion at most $\sqrt{N}$.

This result may be essentially improved as the following theorem of Bourgain [Bou-1986] shows.

Theorem 5.1. *Every N-point metric space $(\mathcal{M},d)$ admits a bi-Lipschitz embedding into an N-dimensional Euclidean space with distortion bounded by $c\log(N+1)$, where $c>1$ is a numerical constant.*

Proof. As above, we construct the required embedding as a composition of two maps, the first of which, denoted by F, sends a point $m \in \mathcal{M}$ to the bounded function[1] $m \mapsto d(m,S)$ defined on the set $2^\mathcal{M}$ of all nonempty subsets of $\mathcal{M}$. Since
$$|d(m,S) - d(m',S)| \leq d(m,m'),$$
the Lipschitz constant of F satisfies
$$L(F) \leq 1. \tag{5.1}$$

To define the second map, we first consider a linear operator T acting from $\ell_\infty(2^\mathcal{M})$ into $\ell_2(2^\mathcal{M})$ as follows.

Let $\{\delta_S\}_{S \in 2^\mathcal{M}}$ be the canonical basis of $\ell_\infty(2^\mathcal{M})$, i.e., $\delta_S(S') := 1$ if $S' = S$ and $\delta_S(S') := 0$, otherwise. Then T is defined on the basis by

$$T(\delta_S) := T_S \delta_S, \quad \text{where} \quad T_S := \frac{1}{\ell}\binom{N}{\ell}^{-1} \quad \text{and} \quad \ell := \mathrm{card}\, S.$$

The norm of $T: \ell_\infty(2^\mathcal{M}) \to \ell_1(2^\mathcal{M})$ is by definition

$$\max\left\{ \left| \sum_{S \in 2^\mathcal{M}} T_S \varphi(S) \right| ; |\varphi(S)| \leq 1 \text{ for all } S \right\} = \sum_{\ell=1}^{N} \sum_{S \in \mathcal{M}\langle \ell \rangle} T_S,$$

---
[1] Recall that $d(m,S) := \inf\limits_{m' \in S} d(m,m')$.

5.1. Embedding of metric spaces into the space forms

where hereafter we set

$$\mathcal{M}^{(\ell)} := \{S \in 2^{\mathcal{M}} ; \operatorname{card} S = \ell\}.$$

As the number of elements in this set is $\binom{N}{\ell}$, we get

$$\|T\|_{\ell_\infty \to \ell_1} = \sum_{\ell=1}^{N} \frac{1}{\ell} =: H_N \leq \log(N+1). \tag{5.2}$$

Now we factorize the (diagonal) operator T through $\ell_2(2^{\mathcal{M}})$ by setting $T = T''T'$, where

$$T'(\delta_S) = T''(\delta_S) := \sqrt{T_S}\, \delta_S, \quad S \in 2^{\mathcal{M}}.$$

A simple evaluation gives

$$\|T'\|_{\ell_\infty \to \ell_2} = \sqrt{H_N}, \quad \|T''\|_{\ell_2 \to \ell_1} \leq \sqrt{H_N}.$$

The required embedding of $\mathcal{M}$ is now given by

$$I := T' \circ F.$$

This is clearly a bi-Lipschitz embedding of $\mathcal{M}$ into an N-dimensional (Euclidean) subspace of $\ell_2(2^{\mathcal{M}})$ and its Lipschitz constant satisfies

$$L(I) \leq L(F)\|T'\|_{\ell_\infty \to \ell_2} \leq \sqrt{H_N}. \tag{5.3}$$

To estimate the distortion $D(I)$ of I, we must evaluate the Lipschitz constant for the inverse $I^{-1} : I(\mathcal{M}) \to \mathcal{M}$. In other words, we must bound from above the quantity

$$L(I^{-1}) := \sup_{m \neq m'} \frac{d(m, m')}{\|(T' \circ F)(m) - (T' \circ F)(m')\|_2}.$$

In turn, the denominator is bounded from below by

$$\|T''\|_{\ell_2 \to \ell_1}^{-1} \|(T \circ F)(m) - (T \circ F)(m')\|_1$$

and therefore

$$L(I^{-1}) \leq \sqrt{H_N} \sup_{m \neq m'} \frac{d(m, m')}{\sum_{\ell=1}^{N} \frac{1}{\ell} J_\ell(m, m')}, \tag{5.4}$$

where we set

$$J_\ell(m, m') := \frac{1}{\binom{N}{\ell}} \sum_{S \in \mathcal{M}^{(\ell)}} |d(m, S) - d(m', S)|. \tag{5.5}$$

Finally, we will use

Proposition 5.2. *There exists a numerical constant $c > 0$ such that for all m, m',*

$$\sum_{\ell=1}^{N} \frac{1}{\ell} J_\ell(m, m') \geq c d(m, m'). \tag{5.6}$$

Proof. The left-hand side of (5.6) is bounded from below by

$$\frac{1}{2} \sum_{j=1}^{j^*} \min\left\{ J_\ell(m, m') \, ; \, \frac{1}{8} 2^{-j} N < \ell \leq \frac{1}{4} 2^{-j} N \right\},$$

where the integer $j^* \leq \log_2 N$ will be chosen later.

Hence, we can write

$$\sum_{\ell=1}^{N} \frac{1}{\ell} J_\ell(m, m') \geq \frac{1}{2} \sum_{j=1}^{j^*} J_{\ell_j}(m, m'), \tag{5.7}$$

where ℓ_j are integers satisfying

$$\frac{1}{8} 2^{-j} N < \ell_j \leq \frac{1}{4} 2^{-j} N. \tag{5.8}$$

To estimate the terms on the right-hand side of (5.7) we define an increasing sequence r_j, $j = 1, \ldots, j^*$, by the conditions:

For every $r > r_j$,

$$\min\{\text{card } B_r(m), \text{card } B_r(m')\} \geq 2^j,$$

while for every $0 < r < r_j$,

$$\max\{\text{card } B_r(m), \text{card } B_r(m')\} < 2^j.$$

We also set $r_0 := 0$ and define j^* to be the minimal j satisfying

$$\frac{d(m, m')}{4} < r_j \leq \frac{d(m, m')}{2}. \tag{5.9}$$

The desired estimate of the terms in (5.7) is given by

Lemma 5.3. *For every $1 \leq j \leq j^*$ and $N \geq 6$,*

$$J_{\ell_j}(m, m') \geq \frac{1}{16} (r_j - r_{j-1}). \tag{5.10}$$

5.1. Embedding of metric spaces into the space forms

Proof. Fix numbers r', r satisfying

$$r_{j-1} < r' < r < r_j.$$

By the definition of r_j,

$$2^{j-1} \leq \operatorname{card} B_r(m), \ \operatorname{card} B_{r'}(m') \leq 2^j. \tag{5.11}$$

Since $j \leq j^*$, we also have, see (5.9),

$$B_r(m) \cap B_{r'}(m') = \emptyset.$$

Therefore, for every subset $S \subset \mathcal{M}$ satisfying

$$S \cap B_{r'}(m') \neq \emptyset \quad \text{and} \quad S \cap B_r(m) = \emptyset, \tag{5.12}$$

the following is true:

$$\bigl|d(m, S) - d(m', S)\bigr| \geq r - r'.$$

This implies

$$J_{\ell_j}(m, m') \geq \frac{r - r'}{\binom{N}{\ell_j}} \operatorname{card}\{S \in \mathcal{M}^{\langle \ell_j \rangle}; \ S \text{ satisfies (5.12)}\}. \tag{5.13}$$

The cardinal in (5.13) may be easily found. In fact,

$$\operatorname{card}\{S \in \mathcal{M}^{\langle \ell_j \rangle}; \ S \text{ satisfies (5.12)}\} = M' \binom{N - M - 1}{\ell_j - 1},$$

where we set

$$M := \operatorname{card} B_r(m), \quad M' := \operatorname{card} B_{r'}(m').$$

We first estimate this for $\ell_j = 1$; actually we get, by (5.8) and (5.11),

$$M' \binom{N - M - 1}{0} \geq 2^{j-1} \geq \frac{N}{16} = \frac{1}{16} \binom{N}{1}$$

and therefore for $\ell_j = 1$,

$$J_{\ell_j}(m, m') \geq \frac{1}{16} \cdot (r - r').$$

Let now $\ell_j \geq 2$. Write

$$M' \binom{N - M - 1}{\ell_j - 1} = M' \binom{N}{\ell_j} \frac{\ell_j}{N} \prod_{i=1}^{\ell_j - 1} \left(1 - \frac{M}{N - i}\right).$$

Using, as above, (5.8) and (5.11) we get

$$M'\frac{\ell_j}{N} \geq 2^{j-1} \cdot \frac{1}{8} \cdot 2^{-j} = \frac{1}{16}.$$

Further, the product $\prod_{i=1}^{\ell_j-1}\left(1 - \frac{M}{N-i}\right)$ is bounded from below by

$$1 - M\sum_{i=1}^{\ell_j-1}\frac{1}{N-i} \geq 1 - M\log\frac{N-2}{N-\ell_j} \geq 1 - M\frac{\ell_j+2}{N-2}.$$

Since $\ell_j \geq 2$ and $N \geq 6$, we get $\frac{\ell_j+2}{N-2} \leq 2\frac{\ell_j}{N}$, and therefore,

$$\prod_{i=1}^{\ell_j-1}\left(1 - \frac{M}{N-i}\right) \geq 1 - 2M\frac{\ell_j}{N} \geq 1 - 2^{j+1}\cdot\frac{1}{4}\,2^{-j} = \frac{1}{2}.$$

Combining all these estimates we obtain that the cardinal in (5.13) is bounded from below by $\binom{N}{\ell_j}$.

The result is thus established. □

To complete the proof of the proposition, it remains to sum (5.10) over $j = 1, \ldots, j^*$ to obtain

$$\sum_{j=1}^{j^*} J_{\ell_j}(m,m') \geq \frac{1}{16}(r_{j^*} - r_0) = \frac{1}{16}r_{j^*} \geq \frac{1}{64}d(m,m').$$

The result is proved. □

Now, by (5.7) and (5.4), we have for $N \geq 6$,

$$L(I^{-1}) \leq \sqrt{H_N}\sup_{m\neq m'}\frac{d(m,m')}{\frac{1}{128}d(m,m')} = 128\sqrt{H_N},$$

which, together with (5.3), gives the required estimate of the distortion of I:

$$D(I) := L(I)L(I^{-1}) \leq 128 H_N \leq 128\log(N+1).$$

Since for $N \leq 6$ this distortion is bounded by $\sqrt{N} \leq \sqrt{6}$, the theorem is proved. □

The Bourgain estimate $O(\log N)$ is optimal up to a numerical factor, as was shown in the paper [LLR-1995] by Linial, London and Rabinovich. Additional information on the geometric structure of a finite metric space allows us to essentially improve Bourgain's estimate for a wide class of metric spaces, see the paper [KLMN-2005] by Krauthgamer, Lee, Mendel and Naor. One of their results asserts that an N-point doubling metric space $(\mathcal{M}, d)$ allows a bi-Lipschitz embedding in the Euclidean space $\mathbb{R}^N$ with distortion $O(\sqrt{\log \delta_\mathcal{M} \log N})$.

Since the doubling constant $\delta_\mathcal{M}$ is clearly at most N, this implies the Bourgain result. The proof of this result is much more involved and exploits the technique of "padded decompositions" used for a similar purpose in Section 4.1.

5.1.2 Infinite metric trees

The simplest infinite metric spaces are (combinatorial) metric trees. In view of Bourgain's Theorem 5.1, it seems probable that such infinite objects do not allow embeddings of finite distortion in Euclidean spaces. Strikingly, even the following negative result of Bourgain [Bou-1986] is true.

> *The dyadic metric tree does not admit a bi-Lipschitz embedding of finite distortion into a Hilbert space.*

However, Proposition 5.5 below implies that the dyadic, and even essentially more general metric trees, admit bi-Lipschitz embeddings into the *hyperbolic plane*.

To present this and related results of this part we recall several notions concerning metric trees, see subsection 3.3.6.

Let $T = (V, E)$ be a tree (i.e., a graph without cycles). An edge $e \in E$ joining vertices $v, v' \in V$ is denoted by $[v, v']$. The *degree* $\deg v$ of a vertex $v \in V$ equals 1 plus the number of immediate ancestors of v called its *children* (and v is called the (unique) *parent* of each of its children).

Given a *weight* $w : E \to (0, +\infty)$, the *length metric* d_w associated to w is defined on vertices $v \neq v'$ of T by

$$d_w(v, v') = \sum_{i=1}^{p} w([v_i, v_{i+1}]),$$

where $v =: v_1, v_2, \ldots, v_{p+1} := v'$ is a (unique) *path* joining v and v' (i.e., every pair v_i, v_{i+1} is joined by a (unique) vertex denoted by $[v_i, v_{i+1}]$).

This definition is naturally extended to all pairs of points situated in edges, since every edge e is (isometrically) identified with the segment of the real line of length $w(e)$.

For instance, in the dyadic tree (denoted below by $\mathcal{T}_1$) every vertex has two children and every edge is identified with the unit segment, i.e., the weight equals 1 in this case.

Finally, a *metric tree* associated to T and d_w is denoted by (T, d_w), deviating from the notation used for this object in subsection 3.3.6.

Our main result, Theorem 5.4, asserts that under mild restrictions posed on (T, d_w) this space admits a bi-Lipschitz embedding into the hyperbolic plane $\mathbb{H}^2$. It is essential for applications that the restriction of this embedding to the set of vertices of T has distortion bounded by a *numerical constant*. Assumption (a) of the theorem is necessary. Actually, $\mathbb{H}^2$ is of bounded geometry and therefore (T, d_w) should have this characteristic meaning, for graphs, boundedness of $\deg T := \sup\{\deg v \, ; \, v \in V\}$. It is unclear (but highly probable) whether assumption (b) is necessary.

Theorem 5.4. *Let (T, d_w) be an infinite rooted metric tree satisfying the conditions:*

(a)
$$\deg T < \infty;$$

(b)
$$\inf w > 0.$$

Then (T, d_w) admits a bi-Lipschitz embedding into $\mathbb{H}^2$ whose restriction to the vertex set of T has distortion bounded by 257.

Proof. We begin with the following result proved by A. and Yu. Brudnyi in [BB-2007b].

Let $\mathcal{T}_k := (\mathcal{V}_k, \mathcal{E}_k)$ be an infinite rooted tree with vertices of degree $k+2$, i.e., each vertex v has $k+1$ children. By d we denote the length metric of $\mathcal{T}_k$ generated by the constant weight equals 1 at every edge.

Proposition 5.5. *There is a Lipschitz map of $(\mathcal{T}_k, d)$, $k \geq 2$, into $\mathbb{H}^2$ whose restriction to the set of vertices $\mathcal{V}_k$ of $\mathcal{T}_k$ is a bi-Lipschitz embedding with distortion at most 256.*

Proof. We will work with the Poincaré model of $\mathbb{H}^2$ where the underlying set is the open half-plane $\mathbb{H}^2_+ := \{x = (x_1, x_2) \in \mathbb{R}^2 \,;\, x_2 > 0\}$ and the geodesic metric ρ is given by

$$\cosh \rho(x, y) := 1 + \frac{\|x - y\|^2}{2x_2 y_2}, \qquad (5.14)$$

see (5.72).

Hereafter $\|x\|$ stands for the Euclidean norm of $x \in \mathbb{R}^2$, and log is the natural logarithm.

In the proof, we also use an auxiliary metric ρ_0 given for $x, y \in \mathbb{H}^2_+$ by

$$\rho_0(x, y) := \max_{i=1,2} \log \left(1 + \frac{|x_i - y_i|}{\min(x_2, y_2)} \right). \qquad (5.15)$$

We leave it to the reader to check that ρ_0 satisfies the triangle inequality.

The relation between these two metrics is described by

Lemma 5.6. (a) $\rho \leq 4\rho_0$.

(b) *If $\|x - y\| \geq \frac{1}{2} \min(x_2, y_2)$, then*

$$\rho(x, y) \geq \frac{1}{8} \rho_0(x, y).$$

Proof. An easy computation allows us to rewrite formula (5.14) as

$$\rho(x, y) = \log \frac{(\|x - y^+\| + \|x - y\|)^2}{\|x - y^+\|^2 - \|x - y\|^2}$$

where $y^+ := (y_1, -y_2)$ is the reflection of y in the x_1-axis. The denominator here equals $4x_2 y_2$ and

$$\|x - y^+\| + \|x - y\| \leq 2\|x - y\| + \|y^+ - y\| = 2(\|x - y\| + y_2).$$

5.1. Embedding of metric spaces into the space forms

Assuming for definiteness that

$$\min(x_2, y_2) = y_2, \tag{5.16}$$

we then conclude that

$$\rho(x, y) \leq 2\log\left(\frac{y_2 + \|x - y\|}{y_2}\right) \leq 4\log\left(1 + \frac{\max_{i=1,2} |x_i - y_i|}{y_2}\right).$$

Comparing this with (5.15) we prove assertion (a).

In case (b), we use (5.14) and the inequality $\cosh t \leq e^t$, $t \geq 0$, to have

$$\rho(x, y) \geq \log\left(1 + \frac{\|x - y\|^2}{2x_2 y_2}\right). \tag{5.17}$$

Consider two cases:

(i) $y_2 \leq x_2 \leq 2y_2$;

(ii) $2y_2 < x_2$.

In case (i), we use (5.16) and the assumption in (b) to derive from (5.17)

$$\rho(x, y) \geq \log\left(1 + \left(\frac{\|x - y\|}{2y_2}\right)^2\right) \geq \frac{1}{8}\log\left(1 + \frac{\|x - y\|}{y_2}\right)$$

$$\geq \frac{1}{8}\log\left(1 + \frac{\max_{i=1,2} |x_i - y_i|}{y_2}\right) =: \frac{1}{8}\rho_0(x, y),$$

as required.

In case (ii), $\|x - y\|$ is at least $|x_2 - y_2| \geq \frac{1}{2} x_2$. Inserting this in (5.17) we obtain

$$\rho(x, y) \geq \log\left(1 + \frac{\|x - y\|}{4y_2}\right) \geq \frac{1}{4}\log\left(1 + \frac{\max_{i=1,2} |x_i - y_i|}{y_2}\right) =: \frac{1}{4}\rho_0(x, y).$$

The result is established. □

To construct the required embedding $I: \mathcal{T}_k \to \mathbb{H}^2$, we assign coordinates to vertices $v \in V_k$ of the tree $\mathcal{T}_k$. Actually, the coordinates (j_v, ℓ_v) of a vertex v are defined as follows.

The number ℓ_v is the *level* of v, the length of the unique path joining v and the root of $\mathcal{T}_k$. To define j_v we visualize the tree $\mathcal{T}_k$ using its natural embedding into $\mathbb{R}^2$. Then j_v is the number of v in ordering of the ℓ_v-th level from the left to the right. We use numbers $0, 1, 2, \ldots$ and therefore

$$0 \leq j_v < (k+1)^{\ell_v},$$

since the number of children for v equals $k+1$.

Finally, we assign coordinates $(0,0)$ to the root R of $\mathcal{T}_k$.

Using the $(k+1)$-ary digital system, the first coordinate of v can be written as
$$j_v = \sum_{s=1}^{\ell_v} \delta_s(v)(k+1)^{s-1}$$
where $\delta_s(v) \in \{0, 1, \ldots, k\}$ are the uniquely determined *digits* of v.

In particular, the coordinates of v and its parent denoted by v^+ are linked by
$$\ell_{v^+} = \ell_v - 1, \quad \delta_s(v^+) = \delta_{s+1}(v), \tag{5.18}$$
where $s = 1, \ldots, \ell_{v^+}$.

Now we compute distance between vertices v and w using their coordinates. For this aim, we define their *common ancestor*, denoted by $a(v, w)$, to be the vertex of highest level in the intersection of the paths joining the root with v and w, respectively. In particular, if $\ell_v = \ell_w$ then $a(v, w)$ is the root R whenever $j_v = 0$, $j_w = (k+1)^{\ell_v} - 1$.

Lemma 5.7. $d(v, w) = \ell_v + \ell_w - 2\ell_{a(v,w)}$.

Proof. There are uniquely defined paths $v =: v_1, \ldots, v_{p+1} := a(v, w)$ and $w =: w_1, \ldots, w_{q+1} := q(v, w)$ such that $v_i^+ = v_{i-1}$, $w_i^+ = w_{i-1}$ and $v_p \ne w_q$. Since the length of edges $[v_i, v_{i+1}]$ and $[w_i, w_{i+1}]$ is 1, we have
$$d(v,w) = p + q = \bigl(\ell_v - \ell_{a(v,w)}\bigr) + \bigl(\ell_w - \ell_{a(v,w)}\bigr),$$
as required. □

In order to define the desired embedding of $\mathcal{T}_k$ in $\mathbb{H}^2$ we now use the following *assignment* of squares in H_+^2 indexed by vertices of $\mathcal{V}_k$.

First we associate to the root $R \in \mathcal{V}_k$ a square $Q(R)$ with the sides parallel to the coordinate axes whose center $c(R)$ and length side $\mu(R)$ are given by
$$c(R) := (0, 1), \quad \mu(R) := \frac{2(n-1)}{n+1}; \tag{5.19}$$
here and below
$$n := k^2 + 1.$$

Let now v be a child of R. In this case, $\ell_v = 1$ and $j_v = \delta_1(v)$. Then we define $Q(v)$ as the square homothetic to $Q(R)$ whose center $c(v) = \bigl(c_1(v), c_2(v)\bigr)$ and length side $\mu(v)$ are given by
$$c_1(v) := \frac{1}{2}\mu(v)\bigl(2\delta_1(v)k - k^2\bigr), \quad c_2(v) := \frac{1}{n}, \quad \mu(v) := \frac{2(n-1)}{n+1} \cdot \frac{1}{n}.$$

5.1. Embedding of metric spaces into the space forms

Hence, the set $\{Q(v) \subset H^2_+ \,;\, \ell_v = 1\}$ is defined by the following geometric construction.

Divide the bottom side of $Q(R)$ into n equal segments and draw n congruent squares (with sides parallel to the coordinate axes) so that the segments become the bisectors of these squares. Apply this construction to each square $Q(v)$ with $\ell_v = 1$ to obtain $(k+1)^2$ squares associated to the vertices of level $\ell_v = 2$, and so on, see Figure 5.1 below.

Straightforward computation gives the following formulas relating centers of the squares associated to v and to its parent v^+:

$$c_1(v) = c_1(v^+) + \frac{1}{2}\mu(v)(2\delta_1(v)k - k^2), \quad c_2(v) = \frac{1}{n}c_2(v^+); \quad (5.20)$$

the corresponding formula for their length side is:

$$\mu(v) = \frac{1}{n}\mu(v^+). \quad (5.21)$$

In particular, we have

$$c_2(v) = \frac{1}{n^{\ell_v}}, \quad \mu(v) = \frac{2(n-1)}{n+1} \cdot \frac{1}{n^{\ell_v}}. \quad (5.22)$$

Hence, $\sum_{\ell_v=1}^{\infty} \frac{1}{2}\mu(v) < 1$ and therefore all squares $Q(v)$ lie in the open upper half-plane H^2_+.

We now compare the metrics ρ and ρ_0 restricted to the set of centers $\{c(v) \in H^2_+ \,;\, v \in V_k\}$.

Let $v \neq w$ be fixed vertices. Assuming without loss of generality that $\ell_v \geq \ell_w$, we consider two cases:

(i) $\ell_v = \ell_w$;

(i) $\ell_v > \ell_w$.

In case (i), let $\hat{v}$ be a neighbor of v in the ℓ_v-th level, i.e., $\ell_{\hat{v}} = \ell_v (= \ell_w)$ and $|j_{\hat{v}} - j_v| = 1$. Since w belongs to the same level,

$$|c_1(v) - c_1(w)| \geq |c_1(v) - c_1(\hat{v})|. \quad (5.23)$$

As v and $\hat{v}$ have a common parent, (5.20) and (5.22) show that the right-hand side equals $\frac{2(n-1)k}{(n+1)n^{\ell_v}}$. On the other hand, by (5.22), we get that

$$\min\{c_2(v), c_2(w)\} = \frac{1}{n^{\ell_v}}. \quad (5.24)$$

These inequalities imply that

$$\frac{\|c(v) - c(w)\|}{\min\{c_2(v), c_2(w)\}} \geq \frac{n-1}{n} \geq \frac{1}{2}. \quad (5.25)$$

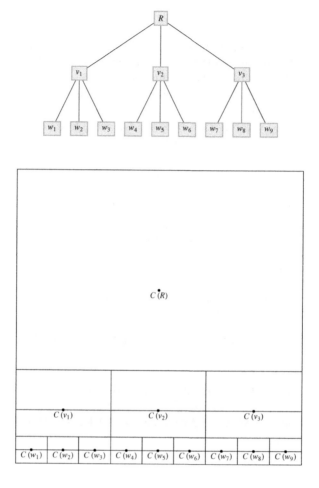

Figure 5.1: Tree associated to assignment of squares.

In case (ii), we choose $\hat{v}$ to be the parent of v. Then $\ell_{\hat{v}} = \ell_v - 1 \geq \ell_w$, and due to our geometric construction (5.23) holds for this $\hat{v}$. Exploiting (5.20) and (5.22) again, we show that the right-hand side of (5.23) in this case is equal to $\frac{n-1}{n \cdot n^{\ell_v}}$. Since (5.24) is also true, we obtain for the left-hand side of (5.25) the lower bound $\frac{n-1}{n} \geq \frac{1}{2}$.

Hence, (5.25) is true. This means that the assumption of Lemma 5.6 (b) holds for $x := c(v)$ and $y := c(w)$. Applying this lemma we get the required result:

$$\frac{1}{8}\rho_0\big(c(v), c(w)\big) \leq \rho\big(c(v), c(w)\big) \leq 4\rho_0\big(c(v), c(w)\big). \qquad (5.26)$$

The inequality obtained allows us to work in the sequel with the metric ρ_0.

5.1. Embedding of metric spaces into the space forms

Lemma 5.8. *For every* $v \in \mathcal{V}_k$,

$$\rho_0\big(c(v), c(v^+)\big) = \log n. \tag{5.27}$$

Proof. By (5.20),

$$\log\left(1 + \frac{|c_2(v) - c_2(v^+)|}{\min\{c_2(v), c_2(v^+)\}}\right) = \log n.$$

On the other hand, the similar expression with c_2 replaced by c_1 in the numerator equals, by (5.20), $\log\big(1 + \frac{n-1}{n+1} \cdot |2\delta_1(v)k - k^2|\big)$. Since $0 \le \delta_1(v) \le k$, this quantity is at most $\log n$. By the definition of ρ_0, see (5.15), this yields (5.27). □

Now we are in a position to define the desired embedding $I : \mathcal{T}_k \to \mathbb{H}^2$ on the set of vertices $\mathcal{V}_k$ regarded as a subset of $\mathcal{T}_k$. Actually, we define I by setting, for $v \in \mathcal{V}_k$,

$$I(v) := c(v). \tag{5.28}$$

Let us estimate the distortion $D(I|_{\mathcal{V}_k})$. Given $v, w \in \mathcal{V}_k$, by $v =: v_1, v_2, \ldots, v_p := a(v,w)$ and $w =: w_1, w_2, \ldots, w_q := a(v,w)$, we denote the paths of minimal length joining v and w with their common ancestor $a(v,w)$. Thus, $v_{i+1} = v_i^+$, $w_{i+1} = w_i^+$, and

$$p = \ell_v - \ell_{a(v,w)} + 1, \quad q = \ell_w - \ell_{a(v,w)} + 1.$$

By the triangle inequality, (5.27) and Lemma 5.7, we then have

$$\rho_0\big(c(v), c(w)\big) \le \sum_{i=1}^{p-1} \rho_0\big(c(v_i), c(v_{i+1})\big) + \sum_{i=1}^{q-1} \rho_0\big(c(w_i), c(w_{i+1})\big)$$
$$\le \big(\ell_v - \ell_{a(v,w)}\big) \log n + \big(\ell_w - \ell_{a(v,w)}\big) \log n = \log n \cdot d(v, w).$$

Hence, we have shown that for all $v, w \in \mathcal{V}_k$,

$$\rho_0\big(I(v), I(w)\big) \le \log n \cdot d(v, w). \tag{5.29}$$

To estimate $D(I|_{\mathcal{V}_k})$, we must prove the converse inequality. Given $v, w \in G$, we assume without loss of generality that $\ell_v \ge \ell_w$ and consider two cases:

(i) w lies in the unique path joining v with the root R, i.e.,

$$w = a(v, w);$$

(ii) $w \ne a(v, w)$.

In case (i), (5.22), (5.24) and Lemma 5.7 yield

$$\rho_0\bigl(c(v),c(w)\bigr) \geq \log\left(1 + \frac{|c_2(v)-c_2(w)|}{\min\{c_2(v),c_2(w)\}}\right) = \log(n^{\ell_v-\ell_w}) = \log n \cdot d(v,w).$$

In case (ii), we prove a similar estimate with the factor $\frac{1}{8}$. To this end, we use the inequality

$$\rho_0\bigl(c(v),c(w)\bigr) \geq \log\left(1 + \frac{|c_1(v)-c_1(w)|}{\min\{c_2(v),c_2(w)\}}\right) \qquad (5.30)$$
$$\geq \ell_v \log n + \log|c_1(v)-c_1(w)|.$$

The second summand here satisfies

$$\log|c_1(v)-c_1(w)| \geq -\left(\ell_{a(v,w)} + \frac{1}{2}\right)\log n. \qquad (5.31)$$

To prove this, we note that the orthogonal projection of the square $Q(v)$ onto the straight line passing through the bottom side of the square $Q(v^+)$ lies inside this side. Applying this fact to squares $Q(v_i)$ and $Q(w_i)$, where $\{v_i\}_{1\leq i\leq p}$ and $\{w_i\}_{1\leq i\leq q}$ are the vertices of the paths joining v and w with their common ancestor $a(v,w)$, and taking into account that $v_{i+1} = v_i^+$ and $w_{i+1} = w_i^+$, we conclude that the orthogonal projections of $c(v) =: c(v_1)$ and $c(w) =: c(w_1)$ onto the straight line determined by the bottom side of the square $Q(a(v,w)) = Q(v_{p-1}^+) = Q(w_{q-1}^+)$ lie, respectively, inside the bisectors of the squares $Q(v_{p-1})$ and $Q(v_{q-1})$ adjacent to $Q(a(v,w))$. This implies that

$$|c_1(v) - c_1(w)| \geq \operatorname{dist}\bigl(Q(v_{p-1}), Q(w_{q-1})\bigr).$$

By definition, the distance between the squares involved is at least $\frac{k\mu(a(v,w))}{n}$. Since $n := k^2 + 1 \geq 5$, this and (5.22) imply that

$$|c_1(v) - c_1(w)| \geq \log\left(\frac{2(n-1)^{\frac{3}{2}}}{n(n+1)} \cdot \frac{1}{n^{\ell_{a(v,w)}}}\right) \geq -\left(\ell_{a(v,w)} + \frac{1}{2}\right)\log n,$$

and (5.31) is established.

Further, (5.30) and (5.31) yield

$$\rho_0\bigl(c(v),c(w)\bigr) \geq \left(\ell_v - \ell_{a(v,w)} - \frac{1}{2}\right)\log n.$$

Then the inequalities $\ell_v \geq \ell_w$ and $\ell_v - \ell_{a(v,w)} \geq 1$ and Lemma 5.7 imply that

$$\ell_v - \ell_{a(v,w)} - \frac{1}{2} \geq \frac{1}{8}(\ell_v + \ell_w - 2\ell_{a(v,w)}) = \frac{1}{8}d(v,w).$$

5.1. Embedding of metric spaces into the space forms

Hence, for all pairs v, w,

$$\rho_0\bigl(c(v), c(w)\bigr) \geq \frac{1}{8}\log n \cdot d(v, w);$$

combining this with (5.29) we then obtain, for all $v, w \in \mathcal{V}_k$,

$$\frac{1}{8}\log n \cdot d(v, w) \leq \rho_0\bigl(I(v), I(w)\bigr) \leq \log n \cdot d(v, w). \tag{5.32}$$

In particular, for the distortion of $I|_{\mathcal{V}_k}$ we get

$$D\bigl(I|_{\mathcal{V}_k}\bigr) \leq 8.$$

It remains to extend the map I to all the edges of the metric space $(\mathcal{T}_k, d)$. According to the definition of a metric graph, see subsection 3.3.6, every edge $[v, v^+]$ is identified with the unit segment of the real line. Therefore there is a curve $j_v : [0, 1] \to [v, v^+]$ so that for $0 \leq t, t' \leq 1$,

$$d\bigl(\gamma_v(t), \gamma_v(t')\bigr) = |t - t'|. \tag{5.33}$$

To extend the map $I|_{\mathcal{V}_k}$, we join $c(v)$ and $c(v^+)$ by a unique geodesic in $\mathbb{H}^2$ (the subarc of an Euclidean circle or a straight line intersecting the x_1-axis orthogonally). Denote this geodesic by $[c(v), c(v^+)]$. By virtue of our geometric construction, every pair of geodesics either coincides or has no common interior points. Therefore, the union of all geodesics $[c(v), c(v^+)]$, $v \in \mathcal{V}_k \setminus \{R\}$, forms a metric tree such that every edge $[c(v), c(v^+)]$ is isometric to a segment of the real line of length $\rho_v := \rho\bigl(c(v), c(v^+)\bigr)$.

Now let $\widetilde{\gamma}_v : [0, \rho_v] \to [c(v), c(v^+)]$ be the arc length parametrization of the geodesic $[c(v), c(v^+)]$ so that for $0 \leq t, t' \leq 1$,

$$\rho\bigl(\widetilde{\gamma}_v(t), \widetilde{\gamma}_v(t')\bigr) = |t - t'|.$$

We extend the map $I|_{\mathcal{V}_k}$ to a point $m = \gamma_v(t) \in [v, v'] \subset \mathcal{T}_k$, $0 \leq t \leq 1$, by setting

$$I(m) := \widetilde{\gamma}(\rho_v t).$$

Then, for $m, m' \in [v, v^+]$, we have

$$\rho\bigl(I(m), I(m')\bigr) = \rho_v |t - t'| = \rho_v d(m, m'),$$

see (5.33).

By Lemma 5.8 and (5.26),

$$\frac{1}{8}\log n \leq \rho_v \leq 4 \log n$$

and therefore for these pairs

$$\frac{1}{8}\log n \cdot d(m, m') \leq \rho\bigl(I(m), I(m')\bigr) \leq 4 \log n \cdot d(m, m').$$

Let now $m \in [v, v^+]$ and $m' \in [w, w^+]$, where $v \neq w$. Since d is a length metric, the last inequality, together with (5.32) and (5.26), yields

$$\rho(I(m), I(m')) \leq 4\log n \cdot d(m, m'), \quad m, m' \in \mathcal{T}_k, \quad \text{and}$$

$$\frac{1}{64}\log n \cdot d(m, m') \leq \rho(I(m), I(m')) \leq 4\log n \cdot d(m, m'), \quad m, m' \in \mathcal{V}_k.$$

Hence, $I|_{\mathcal{V}_k}$ is a bi-Lipschitz embedding of distortion at most $4 \cdot 64 = 256$.

We leave it to the reader to check that for pairs m, m' out of $\mathcal{V}_k$,

$$\rho(I(m), I(m')) \geq c_0 n^{-1} d(m, m'),$$

where $c_0 > 0$ is a numerical constant, and the estimate is sharp. So, unlike $I|_{\mathcal{V}_k}$ distortion of the map I tends to infinity along with $\deg \mathcal{T}_k (= 2 + \sqrt{n-1})$.

Proposition 5.5 has been proved. □

The above arguments show that I is a bi-Lipschitz embedding of $\mathcal{T}_k$ into $\mathbb{H}^2$. It may be interesting to compare this result with the situation for finite metric graphs. Due to the *Kuratowski planarity theorem*, see, e.g., [Har-1969], no graph containing as a subgraph the complete graph K_5 or the complete bipartite graph $K_{2,2}$ allows even a homeomorphic embedding into $\mathbb{H}^2$ (or, what is the same, into $\mathbb{R}^2$).

Let us recall the notions of Graph Theory used in this assertion. The *complete graph* K_n has n vertices and $\binom{n}{2}$ edges (so every pair of vertices is joined by an edge). In turn, the complete bipartite graph $K_{n,m}$ has $n+m$ vertices and nm edges, and the set of vertices of $K_{n,m}$ is the disjoint union of subgraphs without edges having, respectively, n and m vertices, and every vertex of one of the subgraphs is joined with every edge of the other.

Now we derive Theorem 5.4 from the proposition. Let $T = (V, E)$ be an infinite rooted tree equipped with a weight $w : E \to (0, +\infty)$ and (T, d_w) be the associated metric tree. By the assumption of the theorem,

$$c := \deg T < \infty \quad \text{and} \quad \delta := \inf w > 0$$

and we must find a bi-Lipschitz embedding $i : (T, d_w) \to \mathbb{H}^2$ such that distortion $D(i|_V) \leq 257$.

The result is true for $c = 2$ because T is a one-line tree in this case. Hence, (T, d_w) is isometrically embedded into the real line which, in turn, is isometrically embedded into $\mathbb{H}^2$ via the map $x \mapsto e^x \cdot (0, 1), \, x \in \mathbb{R}$.

So, now we assume that $c \geq 3$. Then we replace w by a new weight $\widetilde{w}$ given, for $e \in E$, by

$$\widetilde{w}(e) := \left\lfloor \frac{m}{\delta} w(l) \right\rfloor$$

where $\lfloor x \rfloor$ denotes the largest integer bounded from above by x and $m \in \mathbb{N}$ is fixed. Then the length of each edge of the new metric tree $(T, d_{\widetilde{w}})$ is a natural number. For every $e \in E$ we insert into this edge $\widetilde{w}(e) - 1$ equally distributed new vertices (recall that e as a subset of $(T, d_{\widetilde{w}})$ is identified with the segment of the real line of length $\widetilde{w}(e)$). In this way we obtain a new rooted tree denoted by $\widetilde{T}$, a triangulation of T, which we equip with the constant weight equal to 1. Let d_1 be the metric on $\widetilde{T}$ associated with the weight. Then the length of each edge in $(\widetilde{T}, d_1)$ equals 1 and $\deg \widetilde{T} = \deg T = c$. Since $c \geq 3$, the space $(\widetilde{T}, d_1)$ admits an isometric embedding into the space $(\mathcal{T}_k, d)$ with $k := c - 2$ which, in turn, determines an isometric embedding $j_m : (T, d_{\widetilde{w}}) \to (\mathcal{T}_k, d)$.

By Proposition 5.5 there exists a bi-Lipschitz embedding $I : (\mathcal{T}_k, d) \to \mathbb{H}^2$ such that its restriction to the vertex set $\mathcal{V}_k$ satisfies $D(I|_{\mathcal{V}_k}) \leq 256$. Then $I \circ j_m$ gives a bi-Lipschitz embedding of $(T, d_{\widetilde{w}})$ into $\mathbb{H}^2$ and

$$D((I \circ j_m)|_V) \leq 256.$$

Finally, by the definition of $\widetilde{w}$,

$$\frac{m-1}{\delta} d_w \leq d_{\widetilde{w}} \leq \frac{m}{\delta} d_w$$

and therefore the identity map $i_m : (T, d_w) \to (T, d_{\widetilde{w}})$ has distortion bounded by $\frac{m}{m-1}$. Hence, $I \circ j_m \circ i_m$ is a bi-Lipschitz embedding of (T, d_w) into $\mathbb{H}^2$ and its restriction to the vertex set V has distortion bounded by $\frac{m}{m-1} 256 < 257$ for a sufficiently large m.

This completes the proof of Theorem 5.4. □

Now we present a variant of Theorem 5.4 where a tree (T, d_w) is embedded with low distortion into $\mathbb{H}^2 \oplus \mathbb{R}^n$ with $n = O(\log \deg T)$.

Theorem 5.9. *Let (T, d_w) be an infinite rooted metric tree of uniformly bounded degree and with $\inf w > 0$. Then it admits a bi-Lipschitz embedding into the direct sum of $\mathbb{H}^2$ and the Euclidean space $\mathbb{R}^n$ with $n \leq \log_2(\deg T) + 3$ and distortion bounded by a numerical constant.*

Proof. As in Theorem 5.4 we prove the result for the tree $(\mathcal{T}_k, d)$ of Proposition 5.5 where $(k+2 =) \deg \mathcal{T}_k = \deg T \geq 3$. We then obtain the result for (T, d_w) by the trick used in the final part of the proof of Theorem 5.4.

The main point of the subsequent derivation is the presentation of $\mathcal{T}_k$ as the covering space of a rather simple finite metric graph which, in turn, admits an embedding of low distortion into $\mathbb{R}^n$ with $n = O(\log(k+1))$. Combining the resulting map with that constructed in Proposition 5.5 we obtain the required embedding into $\mathbb{H}^2 \oplus \mathbb{R}^n$.

To realize this program we introduce a topological graph $\mathcal{C}_s$ consisting of a single vertex denoted by o and of s loops (cycles) L_i, $1 \leq i \leq s$, with endpoints at o. Each L_i is homeomorphic to the unit circle. Hence, $\mathcal{C}_s$ is homeomorphic to a topological space obtained by gluing together a collection of s circles along a single point. This object is known in topology as a *rose* (or a *bouquet of circles*).

The *universal covering* of $\mathcal{C}_s$ is a pair $(\widetilde{\mathcal{C}}_s, p)$ where $\widetilde{\mathcal{C}}_s$ is a (topological) infinite rooted tree with degree of each vertex equal to $2s$ and $p : \widetilde{\mathcal{C}}_s \to \mathcal{C}_s$ a continuous surjection such that every $p^{-1}(L_i)$ is a disjoint union of one-line subtrees of the tree emanating from the points of $p^{-1}(o)$. So, one may think of each connected component of $p^{-1}(L_i)$ as an infinite spiral over the circle $L_i \subset \mathbb{R}^2$ which is projected onto L_i, see, e.g., the book by Munkres [Mun-2000] for details.

We need a metric version of this construction with each L_i being isometric to the boundary of the unit square

$$\partial Q_0 := \left\{ (x,y) \in \mathbb{R}^2 \ ; \ \max(|x|,|y|) = \frac{1}{2} \right\},$$

where the common point o is identified with $(-\frac{1}{2}, -\frac{1}{2})$. As above we denote this combinatorial one-vertex graph by $\mathcal{C}_s$ and equip it with a metric generated by a constant weight w_s equal to 4 on every L_i. Here we relate s and k by

$$s := \left\lfloor \frac{k+3}{2} \right\rfloor \tag{5.34}$$

and denote the resulting metric graph by $(\mathcal{C}_s, d_s)$.

Further, $(\widetilde{\mathcal{C}}_s, \widetilde{d}_s)$ now is an infinite rooted metric tree where $\widetilde{d}_s$ is a length metric such that every edge of $\widetilde{\mathcal{C}}_s$ is isometric to the segment of length 4.

Finally, the map $p : \widetilde{\mathcal{C}}_s \to \mathcal{C}_s$ is a locally isometric surjection such that the set of interior points of every edge of $\widetilde{\mathcal{C}}_s$ is mapped isometrically onto some $L_i \setminus \{o\}$. Specifically, if $m, m' \in \widetilde{\mathcal{C}}_s$ and $\widetilde{d}_s(m, m') < 4$, then

$$d_s(p(m), p(m')) = \widetilde{d}_s(m, m') \tag{5.35}$$

and, otherwise,

$$d_s(p(m), p(m')) \leq \widetilde{d}_s(m, m'). \tag{5.36}$$

Let us introduce a new rooted tree $\mathcal{T}_k(l) = (\mathcal{V}_k(l), \mathcal{E}_k(l))$, $l \in \mathbb{N}$, whose vertices and edges are defined as follows.

Insert $l - 1$ equally distributed new vertices in each edge $e \in \mathcal{E}_k$ of $\mathcal{T}_k$. In this way we obtain the required new rooted tree $\mathcal{T}_k(l)$, the triangulation of $\mathcal{T}_k$, which we endow with the metric $d_l := l \cdot d$ where d is the metric of $\mathcal{T}_k$. Then every edge of $\mathcal{T}_k(l)$ is of length 1 with respect to d_l. Moreover, every vertex $v \in \mathcal{V}_k(l)$ has degree $\leq k + 2$. The corresponding metric space $(\mathcal{T}_k(l), d_l)$ is therefore isometric to a metric subspace of the metric space $(\mathcal{T}_k, d)$. This isometry determines a bi-Lipschitz embedding $I_l : (\mathcal{T}_k, d) \to (\mathcal{T}_k, d)$ such that

$$d(I_l(m), I_l(m')) = l \cdot d(m, m'), \quad m, m' \in \mathcal{T}_k.$$

Further, since the degree of each vertex of $\mathcal{T}_k$ is $k + 2$ and therefore $\leq 2s$, there exists an embedding J of $\mathcal{T}_k$ into $\widetilde{\mathcal{C}}_s$ which maps vertices of $\mathcal{T}_k$ to some

5.1. Embedding of metric spaces into the space forms

vertices of $\widetilde{\mathcal{C}}_s$ and edges of $\mathcal{T}_k$ onto some edges of $\widetilde{\mathcal{C}}_s$. Since every edge of $(\mathcal{T}_k, d)$ is of length 1, we have for the composite map $J_l := J \circ I_l : (\mathcal{T}_k, d) \to (\widetilde{\mathcal{C}}_s, \widetilde{d}_s)$ and every $m, m' \in (\mathcal{T}_k, d)$

$$\widetilde{d}_s(J_l(m), J_l(m')) = 4l \cdot d(m, m'). \tag{5.37}$$

(Observe that by the definition $J_1 := J$.)

Next, we embed $(\mathcal{C}_s, d_s)$ bi-Lipschitzly into $\mathbb{R}^{q+2}$ with $q := \lfloor \log_2 s \rfloor + 1$. For this aim we consider a $\frac{1}{2}$-chain in the unit Euclidean ball $B^q \subset \mathbb{R}^q$, that is, a maximal subset of the ball such that pairwise distances between its points are $\geq \frac{1}{2}$ and Euclidean balls of radius $\frac{1}{2}$ centered at these points cover B^q (hence Euclidean balls of radius $\frac{1}{4}$ centered at these points are mutually disjoint). Comparing volumes of the corresponding covers by balls we obtain that the number N_q of points in the chain satisfies

$$2^q \leq N_q \leq 5^q.$$

Let $M \subset (\mathcal{C}_s, d_s)$ be a subset consisting of the common point o and points $m_i \in L_i$ which correspond in the identification of L_i with ∂Q_0 to the point $(\frac{1}{2}, \frac{1}{2})$. The above inequality for N_q implies that $M \setminus \{o\}$ admits a bi-Lipschitz embedding into the $\frac{1}{2}$-chain in B^q. Moreover, since for distinct points x, y of the chain $\frac{1}{2} \leq \|x - y\| \leq 2$ and $d_s(m_i, m_j) = 4$ for all $i \neq j$, distortion of this embedding is ≤ 4. (Here and below by $\|\cdot\|$ we denote the Euclidean norm.)

Now present $\mathbb{R}^{q+1}$ as $\mathbb{R}^q \times \mathbb{R}$ and extend this bi-Lipschitz embedding to the point o by sending o to the point $\widetilde{o} := (0, 1) \in \mathbb{R}^q \times \mathbb{R}$. The resulting embedding of M into $\mathbb{R}^{q+1}$ will be denoted by E. Since for each $m \in E(M) \setminus \{\widetilde{o}\}$,

$$1 \leq \|\widetilde{o} - m\| \leq \sqrt{2}$$

and distances in $(\mathcal{C}_s, d_s)$ between o and points of $M \setminus \{o\}$ are 2, distortion of E is at most 4.

Further, for each point $m \in E(M) \setminus \{\widetilde{o}\}$ consider a two-dimensional affine subspace in $\mathbb{R}^{q+2} := \mathbb{R}^{q+1} \times \mathbb{R}$ containing the interval $[\widetilde{o}, m]$ and orthogonal to the subspace $\mathbb{R}^{q+1} \subset \mathbb{R}^{q+2}$. Let S_m denote the square in this plane with vertices at $\widetilde{o}$ and m and the *center* at the midpoint of the segment $[\widetilde{o}, m]$. The union of all boundaries ∂S_m determines a rose $\mathcal{R}$ in $\mathbb{R}^{q+2}$ homeomorphic to $\mathcal{C}_s$.

We determine a bi-Lipschitz isomorphism $T : (\mathcal{C}_s, d_s) \to (\mathcal{R}, \|\cdot\|)$ by mapping each $L_j \subset (\mathcal{C}_s, d_s)$ (identified with ∂Q_0) linearly onto $\partial S_{E(m_j)}$ so that $T(m_j) = E(m_j)$ and $T(o) = \widetilde{o}$. By this definition, for all $m, m' \in L_j$,

$$\frac{\|\widetilde{o} - E(m_j)\|}{2\sqrt{2}} d_s(m, m') \leq \|T(m) - T(m')\|$$
$$\leq \frac{\|\widetilde{o} - E(m_j)\|}{\sqrt{2}} d_s(m, m'). \tag{5.38}$$

Let us prove similar estimates for points m, m' belonging to distinct L_j.

Let $\pi: \mathbb{R}^{q+1} \times \mathbb{R} \to \mathbb{R}^{q+1}$ be the natural projection. Then for $x, y \in \mathcal{R}$ we have
$$||x - y|| \geq ||\pi(x) - \pi(y)||.$$
Assuming that x, y belong to distinct ∂S_m we obtain that $\pi(x)$ and $\pi(y)$ belong to distinct sides of a triangle where one side ends at $\tilde{o}$ and two other sides, say, v_1, v_2, at some points of the $\frac{1}{2}$-chain in B^q. We estimate the angle θ of this triangle with vertex at $\tilde{o}$.

According to the cosine theorem,
$$\cos\theta = \frac{||\tilde{o} - v_1||^2 + ||\tilde{o} - v_2||^2 - ||v_1 - v_2||^2}{2||\tilde{o} - v_1|| \cdot ||\tilde{o} - v_2||}.$$

Also, by our construction we have
$$1 \leq ||\tilde{o} - v_1|| \leq \sqrt{2}, \quad 1 \leq ||\tilde{o} - v_2|| \leq \sqrt{2}, \quad ||v_1 - v_2|| \geq \frac{1}{2}.$$

Solving the corresponding extremal problem we obtain
$$\cos\theta \leq \frac{(\sqrt{2})^2 + (\sqrt{2})^2 - \left(\frac{1}{2}\right)^2}{2\sqrt{2} \cdot \sqrt{2}} = \frac{15}{16}.$$

Applying then to the triangle with vertices $\pi(x), \pi(y)$ and $\tilde{o}$ the cosine theorem and the inequality $4ab \leq (a+b)^2$, $a, b \geq 0$, we obtain
$$||\pi(x) - \pi(y)|| \geq (||\tilde{o} - \pi(x)|| + ||\tilde{o} - \pi(y)||)\sqrt{\frac{1-\cos\theta}{2}}.$$

Thus
$$||\pi(x) - \pi(y)|| \geq \frac{1}{4\sqrt{2}}(||\tilde{o} - \pi(x)|| + ||\tilde{o} - \pi(y)||).$$

Together with the inequality
$$||\tilde{o} - \pi(x)|| \geq \frac{1}{\sqrt{2}}||\tilde{o} - x|| \geq \frac{||\tilde{o} - v_1||}{4}d_s(o, T^{-1}(x)) \geq \frac{1}{4}d_s(o, T^{-1}(x)),$$

see (5.38), and the similar inequality for $||\tilde{o} - \pi(y)||$, the above inequalities imply that
$$||x - y|| \geq \frac{1}{16\sqrt{2}}(d_s(o, T^{-1}(x)) + d_s(o, T^{-1}(y))) \geq \frac{1}{16\sqrt{2}}d_s(T^{-1}(x), T^{-1}(y)).$$

From this and (5.38) we get
$$\frac{1}{16\sqrt{2}}d_s(m, m') \leq ||T(m) - T(m')|| \leq d_s(m, m'), \quad m, m' \in C_s. \tag{5.39}$$

5.1. Embedding of metric spaces into the space forms

Further, let us consider the map $\widetilde{J}_l := I \circ J_l : (\mathcal{T}_k, d) \to \mathbb{H}^2$ where $I : (\widetilde{\mathcal{C}}_s, \widetilde{d}_s) \to \mathbb{H}^2$ is the Lipschitz map of Proposition 5.5. Then by (5.37) and the estimates of this proposition we have

$$\rho(\widetilde{J}_l(m), \widetilde{J}_l(m')) \leq 4l \log(s^2 + 1) \cdot d(m, m'), \quad m, m' \in \mathcal{T}_k,$$

and for points $m, m' \in \mathcal{V}_k(l) \subset \mathcal{T}_k$,

$$\frac{l}{64} \log(s^2 + 1) \cdot d(m, m') \leq \rho(\widetilde{J}_l(m), \widetilde{J}_l(m')) \leq 4l \log(s^2 + 1) \cdot d(m, m').$$

Finally, we construct a bi-Lipschitz embedding $R_l : (\mathcal{T}_k, d) \to \mathbb{H}^2 \times \mathbb{R}^n$ where $n := q + 2$ by the formula

$$R_l(m) := (\widetilde{J}_l(m), \log(s^2 + 1) \cdot (T \circ p \circ J)(m)), \quad m \in \mathcal{T}_k.$$

Let us recall that $p : (\widetilde{\mathcal{C}}_s, \widetilde{d}_s) \to (\mathcal{C}_s, d_s)$ is the covering map and $J : (\mathcal{T}_k, d) \to (\widetilde{\mathcal{C}}_s, \widetilde{d}_s)$ is the isometric embedding defined prior to formula (5.37).

Then according to the inequalities for $\widetilde{J}_l$ and (5.36), (5.37), (5.39), for the direct 2-sum metric on $\mathbb{H}^2 \times \mathbb{R}^n$ denoted by d_* we have

$$d_*(R_l(m), R_l(m'))$$
$$:= \sqrt{(\rho(\widetilde{J}_l(m), \widetilde{J}_l(m')))^2 + (\log(s^2+1))^2 \|(T \circ p \circ J)(m) - (T \circ p \circ J)(m')\|^2}$$
$$\leq \sqrt{(4l)^2 + 4^2} \cdot \log(s^2 + 1) \cdot d(m, m').$$

Also, for $d(m, m') \leq \frac{1}{2}$ from (5.35), (5.39) we have

$$d_*(R_l(m), R_l(m')) \geq \log(s^2 + 1) \|(T \circ p \circ J)(m) - (T \circ p \circ J)(m')\|$$
$$\geq \frac{\log(s^2+1)}{16\sqrt{2}} d_s(p(J(m)), p(J(m')))$$
$$= \frac{\log(s^2+1)}{16\sqrt{2}} \widetilde{d}_s(J(m), J(m'))$$
$$= \frac{\log(s^2+1)}{4\sqrt{2}} d(m, m').$$

If, on the other hand, $d(m, m') > \frac{1}{2}$, then we choose vertices $\widetilde{m}, \widetilde{m}' \in \mathcal{V}_k(l)$ such that $d(m, \widetilde{m}) \leq \frac{1}{l}$ and $d(m', \widetilde{m}') \leq \frac{1}{l}$. It follows by the triangle inequality and the above estimates for $\widetilde{J}_l$ that

$$d_*(R_l(m), R_l(m')) \geq \rho(\widetilde{J}_l(\widetilde{m}), \widetilde{J}_l(\widetilde{m}')) - \rho(\widetilde{J}_l(m), \widetilde{J}_l(\widetilde{m})) - \rho(\widetilde{J}_l(m'), \widetilde{J}_l(\widetilde{m}'))$$
$$\geq \frac{l}{64} \log(s^2+1) \cdot d(\widetilde{m}, \widetilde{m}') - 8l \log(s^2+1) \cdot \frac{1}{l}$$
$$\geq \frac{l}{64} \log(s^2+1) \cdot \left(d(m,m') - \frac{2}{l}\right) - 8 \log(s^2+1).$$

Choosing here $l := 1092$, from the condition $d(m,m') > \frac{1}{2}$ we get for $R := R_l$,

$$d_*(R(m), R(m')) \geq \log(s^2 + 1) \cdot d(m,m').$$

Together with the previous inequalities we obtain that $R : (\mathcal{T}_k, d) \to (\mathbb{H}^2 \times \mathbb{R}^n, d_*)$ is a bi-Lipschitz embedding with distortion

$$D \leq 16\sqrt{2} \cdot \sqrt{1092^2 + 1} < 24710.$$

The proof of Theorem 5.9 is complete. □

5.1.3 Doubling metric spaces

Since a doubling metric space has finite topological dimension, it can be homeomorphically embedded in a finite-dimensional Euclidean space (the Menger–Nöbeling theorem). However, this embedding may completely destroy the geometric structure of the space. On the other hand, bi-Lipschitz embeddings, which clearly preserve the main features of this structure, may not exist as the examples in the final part of this subsection show.

Strikingly, such an embedding does exist after a small "snowflake" perturbation of the metric. This important result of Assouad [As-1983] is now presented.

Theorem 5.10. *Let $(\mathcal{M}, d)$ be a doubling metric space. Given $0 < s < 1$, the metric space $(\mathcal{M}, d^s)$ allows a bi-Lipschitz embedding into some Euclidean space $\mathbb{R}^n$ with distortion at most $C \geq 1$. Here n and C depend only on s and the doubling constant $\delta_\mathcal{M}$.*

Proof. The main technical tool is the next lemma, whose proof will be presented later.

Fix $m_0 \in \mathcal{M}$ and let N be the smallest integer satisfying

$$N \geq \delta_\mathcal{M}^3 + 1. \tag{5.40}$$

Lemma 5.11. *For every $j \in \mathbb{Z}$ there is a map $\varphi_j : \mathcal{M} \to \mathbb{R}^N$ such that*

(a) *φ_j is zero at m_0;*

(b) *if points $m_1, m_2 \in \mathcal{M}$ satisfy*

$$1 < 2^j d(m_1, m_2) \leq 2,$$

then

$$\|\varphi_j(m_1) - \varphi_j(m_2)\| \geq 1; \tag{5.41}$$

(c) *for an arbitrary pair $m_1, m_2 \in \mathcal{M}$,*

$$\|\varphi_j(m_1) - \varphi_j(m_2)\| \leq 4N \min\{2, 2^j d(m_1, m_2)\}. \tag{5.42}$$

5.1. Embedding of metric spaces into the space forms

Hereafter $\|x\|$ stands for the standard Euclidean norm of the vector $x \in \mathbb{R}^N$.
Now we define the required bi-Lipschitz embedding as follows.

Let $\{e_i\}_{i=1}^N$ be the standard basis of $\mathbb{R}^N$ and $\ell = \ell(N,s)$ be an integer that will be chosen later. Define a periodic sequence of vectors v_j from $\mathbb{R}^{2\ell}$, $j \in \mathbb{Z}$, by

$$v_j := e_j, \quad 1 \le j \le 2\ell, \quad v_{j+2\ell} = v_j \quad \text{for} \quad j \in \mathbb{Z}. \tag{5.43}$$

Let $\mathbb{R}^N \otimes \mathbb{R}^{2\ell}$ be the tensor product of the Euclidean spaces. We regard $x \otimes y$ (the rank-one matrix $(x_i y_j)_{1 \le i \le N, 1 \le j \le 2\ell}$) as a vector in $\mathbb{R}^{2\ell N}$; the Euclidean structure of $\mathbb{R}^N \otimes \mathbb{R}^{2\ell}$ is defined by the scalar product

$$\langle x \otimes y, x' \otimes y' \rangle := \langle x, x' \rangle_{\mathbb{R}^N} \cdot \langle y, y' \rangle_{\mathbb{R}^{2\ell}}.$$

Using the functions φ_j of Lemma 5.11 we define the required bi-Lipschitz map $\varphi : \mathcal{M} \to \mathbb{R}^{2\ell N}$ by setting

$$\varphi(m) := \sum_{j \in \mathbb{Z}} 2^{-js} \varphi_j(m) \otimes v_j, \quad m \in \mathcal{M}. \tag{5.44}$$

Let us show that the series in (5.44) is absolutely convergent. Since $\varphi_j(m_0) = 0$ for all j, it suffices to prove that for all $m, m' \in \mathcal{M}$,

$$\sum_{j \in \mathbb{Z}} \|f_j(m) - f_j(m')\| < \infty$$

where one sets

$$f_j := 2^{-js} \varphi_j \otimes v_j.$$

To this end, define an integer $p = p(m, m')$ by the condition

$$2^{-p} \le d(m, m') < 2^{-p+1}. \tag{5.45}$$

Then the sum under consideration is bounded by

$$\left(\sum_{j \ge p+1} + \sum_{j \le p} \right) \left(2^{-js} \|\varphi_j(m) - \varphi_j(m')\| \right).$$

Applying (5.42) to the first sum and (5.42) and (5.45) to the second sum we obtain

$$\|\varphi(m) - \varphi(m')\| \le 4N \left[\frac{2^{-(p+1)s}}{1 - 2^{-s}} + \frac{2^{p(1-s)}}{1 - 2^{s-1}} d(m, m') \right] \le 4NC(s) d^s(m, m').$$

Hence, φ is well defined and satisfies the Lipschitz condition with respect to the metric d^s with a constant depending only on s and $\delta_\mathcal{M}$.

It remains to show that under a suitable choice of $\ell = \ell(N, s)$ we have, for every pair m, m',

$$\|\varphi(m) - \varphi(m')\| \ge \frac{1}{2} 8^{-s} d^s(m, m'). \tag{5.46}$$

This clearly implies that φ is a bi-Lipschitz embedding of $(\mathcal{M}, d^s)$ into $\mathbb{R}^n$, where $n := 2\ell N$ with distortion at most $8^{s+1} NC(s) =: C(n,s)$.

Using the same notation as above, we write
$$\|\varphi(m) - \varphi(m')\| \geq I_1 - I_2 - I_3,$$
where
$$I_1 := \Big\| \sum_{-\ell < j-p \leq \ell} f_j(m) - f_j(m') \Big\|,$$
$$I_2 + I_3 := \Big(\sum_{j-p > \ell} + \sum_{j-p \leq -\ell} \Big) \|f_j(m) - f_j(m')\|.$$

To evaluate I_1, we note that the vectors
$$w_j := f_j(m) - f_j(m') := 2^{-js}\big(\varphi_j(m) \otimes v_j - \varphi_j(m') \otimes v_j\big)$$
are mutually orthogonal whenever
$$-\ell < j - p \leq \ell.$$

In fact, for these j's the set of vectors v_j coincides with the standard basis $\{e_j\}_{j=1}^{2\ell}$ of $\mathbb{R}^{2\ell}$, see (5.43). By the orthogonality of w_j,
$$I_1 = \Big\| \sum_{-\ell<j-p\leq\ell} w_j \Big\| = \Big(\sum_{-\ell<j-p\leq\ell} \|w_j\|^2 \Big)^{\frac{1}{2}} \geq \|w_p\| = 2^{-sp}\|\varphi_p(m) - \varphi_p(m')\|.$$

Applying (5.45) and (5.41) to the right-hand side, we then have
$$I_1 \geq 2^{-sp} \geq 8^{-s} d^s(m, m').$$

In turn, (5.45) and (5.41) imply
$$I_2 := \sum_{j-p>\ell} 2^{-js}\|\varphi_j(m) - \varphi_j(m')\| \leq 4N \sum_{j-p>\ell} 2^{-js} \leq c_1(s) N \cdot 2^{-\ell s} d^s(m, m').$$

Similarly, for I_3 we obtain
$$I_3 \leq c_2(s) N \cdot 2^{-\ell(1-s)} d^s(m, m').$$

Combining all these inequalities we finally have
$$\|\varphi(m) - \varphi(m')\| \geq \Big[8^{-s} - \big(c_1(s) 2^{-\ell s} + c_2(s) 2^{-\ell(1-s)}\big) N \Big] \cdot d^s(m, m').$$

Choose now $\ell = \ell(N, s)$ such that the term in the square brackets will be more than $\frac{1}{2} 8^{-s}$. This gives (5.46) and proves the theorem. $\square$

5.1. Embedding of metric spaces into the space forms

Proof of Lemma 5.11. Since the scaling $d \mapsto cd$ with any $c > 0$ does not change the doubling constant, we may assume that

$$\operatorname{diam} \mathcal{M} > 4.$$

For the same reason (with d replaced by $2^j d$), it suffices to prove the lemma for $j = 0$.

Let Γ be a maximal 1-net in $\mathcal{M}$, that is, $d(\gamma, \gamma') \geq 1$ for every pair $\gamma \neq \gamma'$ from Γ and $d(m, \Gamma) < 1$ for every $m \in \mathcal{M}$. Endow Γ with a graph structure connecting $\gamma \neq \gamma'$ by a (unique) edge (written $\gamma \leftrightarrow \gamma'$) if $d(\gamma, \gamma') \leq 4$. Let us show that $\deg \gamma \leq N - 1$ for every point (vertex) $\gamma \in \Gamma$. Actually, the set of vertices adjoint to $\gamma \in \Gamma$ is contained in the closed ball of radius 4 which can be covered by at most $(\delta_\mathcal{M})^3$ open balls of radius $\frac{1}{2}$. Every one of these balls clearly contains at most one element of Γ. Hence, $\deg \gamma \leq (\delta_\mathcal{M})^3 \leq N - 1$, see (5.40).

The Szekeres–Wilf theorem, see, e.g., [Har-1969], applied to the graph Γ asserts that Γ may be colored by at most N colors. That is to say, there exists a function $c : \Gamma \to \{1, 2, \ldots, N\}$ such that

$$c(\gamma) \neq c(\gamma'), \quad \text{if} \quad \gamma \leftrightarrow \gamma'.$$

Define now a function $\psi : \mathcal{M} \to \mathbb{R}^N$ by

$$\psi(m) := \sum_{\gamma \in \Gamma} \big(2 - d(\gamma, m)\big)_+ e_{c(\gamma)};$$

here $\{e_j\}_{j=1}^N$ is the standard orthogonal basis of $\mathbb{R}^N$ and $t_+ := \max(t, 0)$.

For $m \in \mathcal{M}$, set

$$\Gamma(m) := \Gamma \cap \overline{B}_1(m).$$

By the definition, $\Gamma(m_1) \cap \Gamma(m_2) = \emptyset$, if $d(m_1, m_2) > 2$. Moreover, if $d(m_1, m_2) \leq 2$ and points $\gamma_1 \neq \gamma_2$ belong to $\Gamma(m_1) \cup \Gamma(m_2)$, then $d(\gamma_1, \gamma_2) \leq 4$ (and therefore $\gamma_1 \leftrightarrow \gamma_2$). Hence, for m_1, m_2 satisfying

$$1 < d(m_1, m_2) \leq 2,$$

the restriction of the coloring function c to the set $\Gamma(m_1) \cup \Gamma(m_2)$ is a bijection. This and orthogonality of vectors e_j imply that

$$\|\psi(m_1) - \psi(m_2)\|^2 = \sum_{\gamma \in \Gamma(m_1)} \big(2 - d(\gamma, m_1)\big)^2 \\ + \sum_{\gamma \in \Gamma(m_2)} \big(2 - d(\gamma, m_2)\big)^2. \tag{5.47}$$

Since Γ is a maximal 1-net, there is an element γ_1 in $\Gamma(m_1)$ such that $d(\gamma_1, m_1) < 1$, and the same is true for m_2 (the corresponding γ_2 may coincide with γ_1). Hence, the right-hand side in (5.47) is at least 1, and (5.41) holds for $\|\psi(m_1) - \psi(m_2)\|$, as required.

Now we prove that (5.42) also holds. By the definition of ψ,

$$\|\psi(m_1) - \psi(m_2)\| \leq \sum_{\gamma \in \Gamma} \Big|(2 - d(\gamma, m_1))_+ - (2 - d(\gamma, m_2))_+\Big|$$

$$\leq \operatorname{card}\Big[\Gamma \cap \big(\overline{B}_2(m_1) \cup \overline{B}_2(m_2)\big)\Big] \cdot \min\big(2, d(m_1, m_2)\big).$$

Further, every two points $\gamma_1 \neq \gamma_2$ from a ball of radius 2 are joined by an edge and, moreover, Γ is N-colored. This clearly implies that

$$\operatorname{card}\big(\Gamma \cap \overline{B}_2(m_i)\big) \leq N \quad \text{for} \quad i = 1, 2,$$

whence

$$\|\psi(m_1) - \psi(m_2)\| \leq 2N \min\big(2, d(m_1, m_2)\big).$$

Now we define the required function φ_j for $j = 0$ by setting

$$\varphi_0 := \psi - \psi(m_0).$$

Then $\varphi_0(m_0) = 0$ and

$$\|\varphi_0(m_1) - \varphi_0(m_2)\| \leq 4N \min\big(2, d(m_1, m_2)\big),$$

provided that $1 \leq d(m_1, m_2) \leq 2$.

This proves (5.42) for φ_0 and (by scaling $d \to 2^j d$) for every φ_j, $j \in \mathbb{Z}$. Since (5.41) is also true, Lemma 5.11 and the theorem are proved. □

Remark 5.12. (a) It will be interesting to have an asymptotically sharp estimate of the dimension $n = n(s, \delta_{\mathcal{M}})$ of the ambient Euclidean space $\mathbb{R}^n$. The proof presented gives apparently a very rough estimate

$$n \leq \left(\frac{a}{s(1-s)}\right)^2 (\delta_{\mathcal{M}})^{\frac{b}{s}},$$

where $a, b > 0$ are numerical constants.

(b) The operation $(\mathcal{M}, d) \to (\mathcal{M}, d^s)$, $0 < s < 1$, is called the *snowflake functor*. The following example explains the name of this operation.

The metric space $([0, 1], d^s)$ where $d(x, y) := |x - y|$, $0 \leq x, y \leq 1$, under a special choice of $0 < s < 1$ is bi-Lipschitz equivalent to the classical von Koch snowflake curve, see, e.g., [Sem-1999].

The snowflake functor does not change topological dimension, but essentially destroys the geometric structure of a metric space. In particular,

5.1. Embedding of metric spaces into the space forms

$(\mathcal{M}, d^s)$ is a fractal if its Hausdorff dimension is distinct from 0 and $+\infty$. In fact, by Theorem 4.12 and the definition of topological dimension,

$$\dim_H(\mathcal{M}, d^s) = \frac{1}{s} \dim_H(\mathcal{M}, d) > \dim(\mathcal{M}, d) = \dim(\mathcal{M}, d^s),$$

that is, $(\mathcal{M}, d^s)$ satisfies the Mandelbrot definition of a fractal.

Now we present the aforementioned examples of doubling metric spaces which do not allow bi-Lipschitz embeddings in Euclidean spaces. The first of them, due to Assouad and Semmes, see, e.g., [Sem-1999, p. 469], is the classical Heisenberg group $H_n = \mathbb{C}^n \times \mathbb{R}$ equipped with the group operation described in subsection 3.3.6 and the equivariant metric d given by formula (3.44) of this subsection.

Assume, on the contrary, that there is a map F of H_n in some Euclidean space $\mathbb{R}^N$ so that for some constant $c \geq 1$ and all $g, h \in H_n$,

$$C^{-1} d(g, h) \leq \|F(g) - F(h)\| \leq C d(g, h). \tag{5.48}$$

Due to Theorem 4.67, the Lipschitz map F has a derivative at some point $g_0 \in H_n$, i.e., there is a homomorphism $D_{g_0} : H_n \to \mathbb{R}^N$ satisfying, for all $h \in H_n$,

$$\|F(g_0 h) - F(g_0) - D_{g_0} h\| = o(d(h, e)) \tag{5.49}$$

as h tends to the identity element $e \in H_n$.

Since H_n is noncommutative, its *commutant* C_n (the subgroup generated by commutators, elements of the form $g_1 g_2 g_1^{-1} g_2^{-1}$) is nontrivial. But every homomorphism of a noncommutative group into an abelian group sends each commutator to zero; hence $D_{g_0}(C_n) = 0$.

Now choose in (5.49) h from C_n. Then (5.49) and (5.48) imply for this h

$$C^{-1} d(g_0 h, g_0) \leq \|F(g_0 h) - F(g_0)\| = o(d(h, e)) \quad \text{as} \quad h \to e.$$

Since d is left-invariant,

$$d(g_0 h, g_0) = d(h, e)$$

and the previous relation is contradictory.

Using the Cheeger differentiation theory it is possible to prove that H_n does not allow embeddings even into some *infinite-dimensional* Banach spaces, in particular, into $L_2[0, 1]$ and $L_1[0, 1]$, see the papers [ChK-2006a, ChK-2006b] by Cheeger and Kleiner.

The above result is not very surprising, since, like every Carnot group, H_n has a very intricate geometry. However, there exist even planar metric graphs which are nonembeddable into any Hilbert space. We present below an example of this kind due to Lang and Plaut [LaPl-2001]. Their construction is a variation of examples from the Laakso paper [Laa-2000] and therefore the space presented is said to be the *Laakso graph* and is denoted by $\mathcal{L}$. This is a metric graph defined as the Gromov–Hausdorff limit of a sequence of compact metric graphs $\{\mathcal{L}_i\}_{i \in \mathbb{Z}_+}$ introduced as follows.

$\mathcal{L}_0$ is the interval $[0,1]$ regarded as a metric graph with one edge and two vertices (the endpoints). Then $\mathcal{L}_1$ is constructed by using six copies of $\mathcal{L}_0$ re-scaled by the factor $\frac{1}{4}$. Namely, one glues four of them cyclically by identifying pairs of the endpoints and then attaching to the opposite gluing points the remaining two copies:

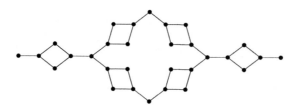

Figure 5.2: A fragment of the Laakso graph.

In general, given $\mathcal{L}_i$, one constructs $\mathcal{L}_{i+1}$ by applying the above operation to six copies of $\mathcal{L}_i$ re-scaled by the factor $\frac{1}{4}$.

In this way, we obtain the metric graph $\mathcal{L}_i$ with 6^i edges, of which all but two endpoints are of degree 3. The length metric of $\mathcal{L}_i$ denoted by d_i is determined by the constant weight equals $\frac{1}{4}$ at each edge. Hence, $(\mathcal{L}_i, d_i)$ is a compact geodesic space of diameter 1.

Now let h_i be a map of $\mathcal{L}_{i-1}$ in $\mathcal{L}_i$ sending every vertex of $\mathcal{L}_{i-1}$ into a vertex of $\mathcal{L}_i$ and endpoints to endpoints. It is easily seen that h_i is an isometric embedding. Regarding every $\mathcal{L}_i$ as a subset of $\mathbb{R}^2$ symmetric with respect to the axes, we choose h_i such that it maps $\mathcal{L}_{i-1}$ into the parts of $\mathcal{L}_i$ corresponding to the upper half of $\mathcal{L}_{i-1}$ (contained in $\mathbb{R}^2_+$). Due to the construction of $\mathcal{L}_i$ we have for this choice of h_i:

$$d_{\mathcal{H}}\bigl(h_i(\mathcal{L}_{i-1}), \mathcal{L}_i\bigr) := \sup\{d(m, h_i(\mathcal{L}_{i-1})) \, ; \, m \in \mathcal{L}_i\} = \frac{1}{4^i}.$$

By the definition of the Gromov–Hausdorff metric, see (3.72), this implies that

$$d_{\mathcal{GH}}(\mathcal{L}_{i-1}, \mathcal{L}_i) \leq \frac{1}{4^i},$$

i.e., $\{\mathcal{L}_i\}_{i \in \mathbb{Z}_+}$ is a Cauchy sequence in the space of the isometric classes of compact metric spaces of diameter at most 1 equipped with the metric $d_{\mathcal{GH}}$. According to the Gromov–Hausdorff compactness criterion, see Theorem 3.58, this space is compact, hence, complete. The limit of this Cauchy sequence denoted by $(\mathcal{L}, d)$ is a compact geodesic metric space of diameter at most 1, see Proposition 3.61. By the properties of the Gromov–Hausdorff limit, see Proposition 3.55, for every i there exist isometric embeddings of $\mathcal{L}_i$ into $\mathcal{L}$.

Now the Lang–Plaut theorem, in particular, asserts the following.

5.1. Embedding of metric spaces into the space forms

Theorem 5.13. (a) *The metric space $\mathcal{L}$ is doubling.*

(b) *There is no bi-Lipschitz embedding of $\mathcal{L}$ into a Hilbert space.*

Proof. (a) We first show that every open ball $B_{2r}^i(m)$ in $\mathcal{L}_i$ with $\frac{r}{2} \leq \frac{1}{4^i} < 2r$ can be covered by at most six closed balls of radius r. To this end, one defines a finite set $Z_i \subset \mathcal{L}_i$ by

$$Z_i := \left(\overline{B}_r^i(m) \setminus B_r^i(m)\right) \cup \left(B_r^i(m) \cap \{p, q\}\right),$$

where p, q are endpoints of $\mathcal{L}_i$.

Since $r \leq 2 \cdot \frac{1}{4^i}$, it follows that card $Z_i \leq 6$ and the closed balls of radius r centered at Z_i cover $B_{2r}^i(m)$.

Let now $B_{2r}(m)$ be an open ball in $\mathcal{L}$. To show that it can be covered by six closed balls of radius r, choose an index i and an isometry $h : \mathcal{L}_i \to \mathcal{L}$ so that

$$\frac{r}{2} \leq \frac{1}{4^i} < 2r \quad \text{and} \quad h(m') = m \quad \text{for some} \quad m' \in \mathcal{L}_i.$$

Let us show that six closed balls of radius r centered at $h(Z_i)$ cover $B_{2r}(m)$. In fact, the space $\mathcal{L}$, compared with $\mathcal{L}_i$, has only new edge cycles added after the i-th step of the construction of $\mathcal{L}$. But the lengths of these cycles are at most $4(\frac{1}{4})^{i+1} < 2r$, and therefore every open ball of radius r centered at such a cycle should contain the whole cycle. Together with the covering property of $\mathcal{L}_i$ this implies that $B_{2r}(m)$ is covered by at most six closed balls of radius r centered at $h(Z_i)$.

Hence, the doubling constant $\delta_\mathcal{L} \leq 6$, and the result is proved.

(b) Let $f : \mathcal{L}_i \to H$ be a map into a Hilbert space satisfying

$$\|f(m) - f(m')\|_H \geq d_i(m, m') \tag{5.50}$$

for all $m, m' \in \mathcal{L}_i$.

One proves by induction on i that there exist adjacent vertices $m, m' \in \mathcal{L}_i$ such that

$$\|f(m) - f(m')\|_H^2 \geq \left(1 + \frac{i}{4}\right) d_i(m, m')^2. \tag{5.51}$$

This is trivially true for $i = 0$. Let now $i \geq 1$ and $f : \mathcal{L}_i \to H$ satisfy (5.50). Since $\mathcal{L}$ contains an isometric copy of $\mathcal{L}_{i-1}$, there exist, by the induction hypothesis, vertices $m_0, m_2 \in \mathcal{L}_i$ corresponding to two adjacent vertices of $\mathcal{L}_{i-1}$, and such that

$$\|f(m_0) - f(m_2)\|_H^2 \geq \left(1 + \frac{i-1}{4}\right) d_i(m_0, m_2)^2.$$

Since m_0 and m_2 correspond to adjacent vertices of $\mathcal{L}_{i-1}$, there are, by the construction, precisely four vertices between them forming a diamond. Let m_1, m_3 be those of the four which are not adjacent to either m_0 or m_2. In particular, $d(m_k, m_{k+1}) = \frac{1}{2} d(m_0, m_2)$ for $k \in \{0, 1, 2\}$ and between every pair m_k, m_{k+1}

with this k there is a unique vertex denoted by $\widehat{m}_k$. By the parallelogram identity, we have

$$\sum_{k=0}^{3}\|f(m_k)-f(m_{k+1})\|_H^2 = \|f(m_0)-f(m_2)\|_H^2 + \|f(m_1)-f(m_3)\|_H^2$$

and, moreover,

$$\|f(m_1)-f(m_3)\|_H \geq d_i(m_1,m_3) = \frac{1}{2}d_i(m_0,m_2).$$

Combining these inequalities we obtain

$$\sum_{k=0}^{3}\|f(m_k)-f(m_{k+1})\|^2 \geq \left(1+\frac{i-1}{4}+\frac{1}{4}\right)d_i(m_0,m_2)^2.$$

Hence, for some $k \in \{0,1,2\}$,

$$\|f(m_k)-f(m_{k+1})\|_H^2 \geq \frac{1}{4}\left(1+\frac{i}{4}\right)d(m_0,m_2)^2 = \left(1+\frac{i}{4}\right)d(m_k,m_{k+1})^2.$$

This and the triangle inequality imply, for the middle vertex $\widehat{m}_k$,

$$\left(\|f(m_k)-f(\widehat{m}_k)\|_H + \|f(\widehat{m}_k)-f(m_{k+1})\|_H\right)^2 \geq \left(1+\frac{i}{4}\right)d(m_k,m_{k+1})^2,$$

whence for some $m \in \{m_k, m_{k+1}\}$,

$$\|f(m)-f(\widehat{m}_k)\|^2 \geq \frac{1}{4}\left(1+\frac{i}{4}\right)d(m_k,m_{k+1})^2 = \left(1+\frac{i}{4}\right)d(\widehat{m}_k,m)^2.$$

Since m_k, m are adjacent vertices, (5.51) has been proved.

Now let $f : \mathcal{L} \to H$ be a bi-Lipschitz embedding into a Hilbert space with bounded distortion, say $D \geq 1$. Without loss of generality we may assume that for all $m, m' \in \mathcal{L}$,

$$d(m,m') \leq \|f(m)-f(m')\|_H \leq Dd(m,m'). \tag{5.52}$$

Since $\mathcal{L}$ for all $i \geq 0$ contains an isometric copy of $\mathcal{L}_i$, for m, m' from this copy we have, from (5.51),

$$\|f(m)-f(m')\|_H \geq \left(1+\frac{i}{4}\right)d(m,m')^2.$$

For sufficiently large i, this contradicts (5.52).

The result is established. $\square$

5.1.4 Gromov hyperbolic spaces

Since the classical space forms, in particular hyperbolic and Euclidean spaces, are of bounded geometry, a Gromov hyperbolic space admitting embeddings in these spaces should possess this property. We will show that this property is also sufficient for an embedding (this result was proved in [BB-2007c]).

Theorem 5.14. *A Gromov hyperbolic space of bounded geometry admits a bi-Lipschitz embedding into a direct sum $\mathbb{H}^{n_1} \oplus \mathbb{R}^{n_2}$ for some $n_1, n_2 \in \mathbb{N}$, depending only on the bounded geometry parameters of this space.*

Proof. Let $(\mathcal{M}, d)$ be a δ-hyperbolic space of bounded geometry, see Definition 3.98. Due to Definition 3.114 of Gromov hyperbolicity, $\mathcal{M}$ is a geodesic space. Then Theorem 3.99, whose proof will be presented below, reads as follows.

Claim. For every $R > 0$ there exists an integer $n \geq 1$ and a constant $D > 1$ such that $\mathcal{M}$ belongs to the class $\mathcal{G}_n(R, D)$.

Further, we use the version of the Bonk–Schramm theorem formulated as Theorem 3.120 in subsection 3.3.4 (the proof of the theorem will be discussed in the next section). It asserts that under the conditions of Theorem 5.14 there is a map ϕ of $(\mathcal{M}, d)$ into a hyperbolic space $\mathbb{H}^n$ such that for some constants $C \geq 1$ and $k \geq 0$ and for all $m, m' \in \mathcal{M}$,

$$Cd(m, m') - k \leq d_g\bigl(\phi(m), \phi(m')\bigr) \leq Cd(m, m') + k; \tag{5.53}$$

here d_g is the geodesic metric of $\mathbb{H}^n$.

If $k = 0$, the map ϕ gives the required embedding (even C-isometry) of $\mathcal{M}$ into $\mathbb{H}^n$.

Now let $k > 0$. Set

$$\varepsilon := \frac{2k}{C}$$

and denote by N_ε a maximal ε-net in $\mathcal{M}$. Hence, for all $m, m' \in N_\varepsilon$,

$$d(m, m') \geq \varepsilon, \tag{5.54}$$

and for every $m \in \mathcal{M}$,

$$d(m, N_\varepsilon) := \inf_{m' \in N_\varepsilon} d(m, m') < \varepsilon. \tag{5.55}$$

Due to (5.53)–(5.55) and the choice of ε, we have, for all $m, m' \in N_\varepsilon$,

$$\frac{C}{2} d(m, m') \leq d_g\bigl(\phi(m), \phi(m')\bigr) \leq \frac{3C}{2} d_g(m, m').$$

Hence, the restriction $\phi\big|_{N_\varepsilon}$ is a $\frac{3C}{2}$-Lipschitz map of N_ε into $\mathbb{H}^n$. By the Lang–Pavlović–Schroeder theorem [LPSch-2000], which will be presented in Chapter 6, see Theorem 6.42 of Volume II, this restriction allows an extension, say ψ, to the whole of $\mathcal{M}$ with the Lipschitz constant satisfying

$$L(\psi) \leq \frac{3}{2} c(n) C \tag{5.56}$$

for some $c(n) > 1$. Clearly, ψ also satisfies at all points m, m' from N_ε, the condition

$$d_g\big(\psi(m), \psi(m')\big) = d_g\big(\phi(m), \phi(m')\big) \geq \frac{C}{2} d(m, m'). \tag{5.57}$$

Using these properties of ψ and choosing a suitable R in the Claim we will reconstruct this map to obtain the desired bi-Lipschitz embedding of $\mathcal{M}$ into a direct sum $\mathbb{H}^n \oplus \mathbb{R}^N$. The main tool for this procedure is the following auxiliary result of Naor, Y. Peres, Schramm and Sheffield [NPSS-2006, Cor. 6.2].

Proposition 5.15. *Let $(\mathcal{M}, d)$ be a metric space of the class $\mathcal{G}_n(R, D)$ and let $F : (\mathcal{M}, d) \to (\mathcal{M}_1, d_1)$ be a map between metric spaces satisfying the following conditions:*

(i) *F is a Lipschitz map whose Lipschitz constant satisfies*

$$L(F) \leq A \tag{5.58}$$

for some $A \geq 1$.

(ii) *There exists an ε-net N_ε and a constant $\mu \in (0, 1]$ such that for all $m, m' \in N_\varepsilon$,*

$$d_1\big(F(m), F(m')\big) \geq \mu d(m, m'). \tag{5.59}$$

Assume that the parameters are linked by

$$\mu R = 64\varepsilon. \tag{5.60}$$

Then there exists an integer $N = N(n, D) \geq 1$ and a constant $K = K(n, D, R, \mu, A)$ such that $\mathcal{M}$ admits a bi-Lipschitz embedding into the direct sum $\mathcal{M}_1 \oplus \mathbb{R}^N$ with distortion at most K.

Before proving the proposition we explain how to derive from it the theorem. Choose R in the Claim such that

$$R = \frac{384 k c(n)}{C}$$

and apply Proposition 5.15 with

$$\mathcal{M}_1 := \mathbb{H}^n, \quad \varepsilon := \frac{2k}{C} \quad \text{and} \quad A := \frac{3}{2} c(n) C.$$

Due to (5.56) and (5.57), conditions (5.58)–(5.60) hold under this choice of parameters. Hence, there is the desired bi-Lipschitz embedding of $\mathcal{M}$ into $\mathbb{H}^n \oplus \mathbb{R}^N$ for a suitable N. $\square$

5.1. Embedding of metric spaces into the space forms

Proof of Proposition 5.15. We use Lemma 3.102, according to which the implication $(\mathcal{M}, d) \in \mathcal{G}_n(R, D)$ implies existence of pairwise disjoint subsets A_j in $\mathcal{M}$, $1 \leq j \leq N$, satisfying the conditions:

(a) For all $m \neq m'$ from A_j,
$$d(m, m') > R;$$

(b) $\mathcal{M} = \bigcup_{j=1}^{N} \coprod_{m \in A_j} B_{\frac{R}{4}}(m);$

(c) $N \leq N_0 := (8D+1)^n$.

Using this, we prove

Lemma 5.16. *Let $(\mathcal{M}, d) \in \mathcal{G}_{n_0}(R_0, D_0)$. Then there exist an integer $N = N(n_0, D_0)$, a constant $C = C(n_0, R_0, D_0)$ and a Lipschitz map G from $\mathcal{M}$ into the Euclidean space $\mathbb{R}^{nN}$ such that $L(G) \leq C$, and for every pair $m, m' \in \mathcal{M}$ satisfying $d(m, m') \leq \frac{R_0}{8}$ we have*

$$d(m, m') \geq \|G(m) - G(m')\|.$$

Proof. The fact that $\mathcal{M}$ belongs to $\mathcal{G}_{n_0}(R_0, D_0)$ means that for every ball $B_{R_0}(m)$ there is a map $\psi_m : B_{R_0}(m) \to \mathbb{R}^n$ satisfying, for $m', m'' \in B_{R_0}(m)$, the inequality

$$d(m', m'') \leq \|\psi_m(m') - \psi_m(m'')\| \leq D_0 d(m' m'')$$

and such that $\psi_m(m) = 0$.

In particular, for a (unique) point $\widetilde{m}$ from $A_j \cap B_{R_0}(m)$, we have

$$\|\psi_m(\widetilde{m})\| = \|\psi_m(\widetilde{m}) - \psi_m(m)\| \leq D_0 d(\widetilde{m}, m) < D_0 R_0.$$

Using these facts we now define maps $f_j : \mathcal{M} \to \mathbb{R}^n$, $1 \leq j \leq N$, which will be used to define the required map G.

Let $\ell : \mathbb{R}_+ \to [0, 1]$ be a continuous piecewise linear function which equals 1 on $\left[0, \frac{3}{8} R_0\right]$, 0 outside $\left[0, \frac{R_0}{2}\right]$, and is linear in-between. Clearly,

$$|\ell(t) - \ell(t')| \leq \frac{8}{R_0} |t - t'|, \quad t, t' \in \mathbb{R}.$$

We first define f_j to be zero outside the disjoint union $\coprod_{\widetilde{m} \in A_j} B_{\frac{R_0}{2}}(\widetilde{m})$ and then define f_j on every ball $B_{\frac{R_0}{2}}(\widetilde{m})$ with $\widetilde{m} \in A_j$ by setting

$$f_j(m) := \ell(d(m, \widetilde{m})) \psi_{\widetilde{m}}(m), \quad m \in B_{\frac{R_0}{2}}(\widetilde{m}).$$

Check that $L(f_j) \leq 8D_0$. Let m, m' lie in one of the balls $B_{\frac{R_0}{2}}(\widetilde{m})$ with $\widetilde{m} \in A_j$. Then

$$\|f_j(m) - f_j(m')\| \leq |\ell(d(m, \widetilde{m})) - \ell(d(m', \widetilde{m}))| \cdot \|\psi_{\widetilde{m}}(m)\| + \|\psi_{\widetilde{m}}(m) - \psi_{\widetilde{m}}(m')\|$$

$$\leq \frac{8}{R_0} d(m, m') \cdot D_0 R_0 + D_0 d(m, m') = 9 D_0 d(m, m').$$

The cases $m \in B_{\frac{R_0}{2}}(\widetilde{m})$ and $m' \notin B_{\frac{R_0}{2}}(\widetilde{m})$ and $m, m' \notin B_{\frac{R_0}{2}}(\widetilde{m})$ for any $\widetilde{m} \in \bigcup_{j=1}^{N} A_j$, are considered similarly.

Now we define the desired map $G : \mathcal{M} \to \mathbb{R}^{nN}$ by setting

$$G(m) := (f_1(m), \ldots, f_N(m)).$$

Then $L(G) \leq \sqrt{N} \cdot 9D_0$. Further, let $m, m' \in \mathcal{M}$ satisfy the condition $d(m, m') \leq \frac{R_0}{8}$. By assertion (b), there exist an index $1 \leq j \leq N$ and a point $\widetilde{m} \in A_j$ such that $d(m, \widetilde{m}) \leq \frac{R_0}{4}$; therefore

$$d(m', \widetilde{m}) \leq d(m, m') + d(m, \widetilde{m}) \leq \frac{3R_0}{8}.$$

Hence, $m, m' \in \overline{B}_{\frac{3R_0}{8}}(\widetilde{m})$, $\widetilde{m} \in A_j$, and by the definition of f_j,

$$\|G(m) - G(m')\| \geq \|f_j(m) - f_j(m')\| = \|\psi_{\widetilde{m}}(m) - \psi_{\widetilde{m}}(m')\| \geq d(m, m'). \quad \square$$

Now we complete the proof of Proposition 5.15. We relate the constant R_0 of Lemma 5.16 to the basic parameters of the proposition by setting

$$R_0 := \frac{\mu + A}{2} R. \tag{5.61}$$

By the Claim, the space $\mathcal{M}$ (from $\mathcal{G}_n(R, D)$) also belongs to $\mathcal{G}_{n_0}(R_0, D_0)$ with n_0, D_0 depending on R_0 (hence depending only on μ, A, R, n and D). Applying Lemma 5.16 we find a C-Lipschitz map $F : \mathcal{M} \to \mathbb{R}^{nN}$ with $C > 1$ and $N \in \mathbb{N}$ depending only on the basic parameters of the proposition. Then we use the given map $F : \mathcal{M} \to \mathcal{M}_1$ to define the desired map $H : \mathcal{M} \to \mathcal{M}_1 \oplus \mathbb{R}^{nN}$ by setting $H := F \oplus G$. To show that H is a bi-Lipschitz embedding, we first estimate its Lipschitz constant. By the definition of the metric of the direct sum,

$$\widetilde{d}(H(m), H(m')) = \sqrt{d_1(F(m), F(m'))^2 + \|G(m) - G(m')\|^2}$$

and therefore $L(H) \leq \sqrt{2} \max(A, C)$, as required.

In the opposite direction, first let $m, m' \in \mathcal{M}$ satisfy $d(m, m') \leq \frac{R_0}{8}$. By Lemma 5.16, we then have

$$\widetilde{d}(H(m), H(m')) \geq \|G(m) - G(m')\| \geq d(m, m').$$

So, if $d(m, m') > \frac{R_0}{8}$, we choose points $\widetilde{m}, \widetilde{m}'$ from the ε-net N_ε so that

$$\max\{d(m, \widetilde{m}), d(m', \widetilde{m}')\} < \varepsilon.$$

Using the definition of H and the triangle inequality we then have

$$\widetilde{d}(H(m), H(m')) \geq d_1(F(\widetilde{m}), F(\widetilde{m}')) - d_1(F(m), F(\widetilde{m})) - d_1(F(m'), F(\widetilde{m}')).$$

5.1. Embedding of metric spaces into the space forms

By (5.58) and (5.59) and the triangle inequality, we estimate the right-hand side from below by

$$\mu[d(m,m') - d(m,\tilde{m}) - d(m',\tilde{m}')] - Ad(m,m') - Ad(m',\tilde{m}')$$
$$> \mu d(m,m') - 2(\mu + A)\varepsilon.$$

By (5.60) and (5.61) and the condition on $d(m,m')$, we can now finally derive from here

$$\tilde{d}(H(m), H(m')) \geq \mu d(m,m') - 2(\mu + A)\frac{\mu R}{64}$$
$$= \mu d(m,m') - R_0 > \mu d(m,m') - \frac{\mu}{2} d(m,m')$$
$$= \frac{\mu}{2} d(m,m').$$

Hence, H is the desired bi-Lipschitz embedding. □

To complete the proof of Theorem 5.14, it remains to prove the Claim (i.e., Theorem 3.99).

Proof of Theorem 3.99. Given a metric space $(\mathcal{M}, d)$ of bounded geometry, say $\mathcal{M} \in \mathcal{G}_{n_0}(R_0, D_0)$, and a constant $R > 0$, we must prove that every ball $B_R(m) \subset \mathcal{M}$ allows a bi-Lipschitz embedding into some $\mathbb{R}^N$ with distortion at most $C \geq 1$ where constants N and C are independent of the center m. To this end, we choose a maximal ε-net $N_\varepsilon \subset \mathcal{M}$, where ε will be determined later. By the definition of N_ε,

$$\mathcal{M} = \bigcup_{m \in N_\varepsilon} B_\varepsilon(m),$$

while the family $\{B_{\frac{\varepsilon}{2}}(m)\}_{m \in N_\varepsilon}$ is disjoint.
Set for brevity
$$B_m := B_\varepsilon(m), \quad \tilde{B}_m := B_{\frac{\varepsilon}{2}}(m).$$

Lemma 5.17. (a) *If $0 < \varepsilon \leq \frac{R_0}{2}$, then the order of the cover $\{B_m\}_{m \in N_\varepsilon}$ is at most $(2D_0)^{n_0}$.*

(b) *For the same ε and $m \in N_\varepsilon$ there exists a linear extension operator $E_m : \mathrm{Lip}(B_m \cap N_\varepsilon) \to \mathrm{Lip}(B_m)$ whose norm is bounded by $c(n_0)D_0$.*

Proof. (a) By the definition of $\mathcal{G}_{n_0}(R_0, D_0)$, every ball $B_R(m)$ with $R \leq R_0$ admits an embedding $\psi_m : B_R(m) \to \mathbb{R}^{n_0}$, such that for all $m', m'' \in B_R(m)$,

$$d(m', m'') \leq \|\psi_m(m') - \psi_m(m'')\| \leq D_0 d(m', m''). \tag{5.62}$$

Using this, we estimate the required order

$$\mathrm{ord}\, N_\varepsilon := \sup_{m \in \mathcal{M}} \mathrm{card}\{m' \in N_\varepsilon ; B_{m'} \ni m\}$$

in the following way.

The union $\bigcup\{B_{m'}\,;\,B_{m'}\ni m\}$ is contained in the ball $B_{2\varepsilon}(m)$. Since $2\varepsilon \leq R_0$, the image $\psi_m(B_{2\varepsilon}(m))$ is contained, by (5.62), in the Euclidean ball $B_\rho(\psi_m(m))$ of radius $\rho := 2D_0\varepsilon$. On the other hand, the left inequality in (5.62) implies that the image $\{\psi_m(\widetilde{B}_{m'})\,;\,m' \in N_\varepsilon\}$ of the disjoint family $\{\widetilde{B}_{m'}\,;\,m' \in N_\varepsilon\}$ is also disjoint. Comparing the n_0-Lebesgue measures of the sets $\bigcup\{\psi_m(\widetilde{B}_{m'})\,;\,B_{m'}\ni m\}$ and $B_\rho(\psi_m(m))$ we conclude that

$$\varepsilon^{n_0}\,\mathrm{card}\{m' \in N_\varepsilon\,;\,B_{m'}\ni m\} \leq (2D\varepsilon)^{n_0}.$$

This gives the required result.

(b) Fix $m \in \mathcal{M}$. By the Whitney extension theorem, see Theorem 2.19 for $k = 0$, there exists a linear extension operator T from the space of Lipschitz functions defined on $\psi_m(N_\varepsilon \cap B_m) \subset \mathbb{R}^{n_0}$ into that of Lipschitz functions defined on $\mathbb{R}^{n_0}$ whose norm is bounded by a constant $c(n_0) > 1$. Setting now $E_m := \psi_m^{-1} \circ T \circ \psi_m$, we obtain the required operator from $\mathrm{Lip}(N_\varepsilon \cap B_m)$ into $\mathrm{Lip}(B_m)$ with norm bounded by $Dc(n_0)$. $\square$

Using the lemma we find a linear extension operator denoted by E from $\mathrm{Lip}(N_\varepsilon \cap B_R(m))$ into $\mathrm{Lip}(B_R(m))$ with norm independent of m, and then use the predual to this operator as a key point in the construction of the required bi-Lipschitz embedding of the ball $B_R(m)$.

To find E, we exploit a suitable Lipschitz partition of unity subordinate to the cover $\{B_{m'}\,;\,m' \in N_\varepsilon \cap B_R(m)\}$ of the ball $B_R(m)$ to glue together the operators $E_{m'}$, $m' \in N_\varepsilon \cap B_R(m)$, of Lemma 5.17. Denoting this partition by $\{p_{m'}\,;\,m' \in N_\varepsilon \cap B_R(m)\}$, we set

$$E := \sum p_{m'} E_{m'},$$

where m' runs over all points of $N_\varepsilon \cap B_R(m)$. In this way, we obtain the required extension operator from $\mathrm{Lip}(N_\varepsilon \cap B_R(m))$ into $\mathrm{Lip}(B_R(m))$ whose norm is bounded by a constant $K > 1$ depending only on the order of the cover $\{B_m\}_{m \in N_\varepsilon}$ and on $\sup_{m'}\|E_{m'}\|$, see Lemma 7.9 of Chapter 7 for missing details. Due to Lemma 5.17, the constant K depends only on n_0 and D_0.

Fix a point $m^* \in N_\varepsilon \cap B_R(m)$ and denote by $\mathrm{Lip}_0(N_\varepsilon \cap B_R(m))$ a subspace of $\mathrm{Lip}(N_\varepsilon \cap B_R(m))$ determined by the condition $f(m^*) = 0$; the subspace $\mathrm{Lip}_0(B_R(m))$ is defined similarly. The restriction of E to the former subspace denoted by E_0 is a linear extension operator mapping $\mathrm{Lip}_0(N_\varepsilon \cap B_R(m))$ into $\mathrm{Lip}_0(B_R(m))$ whose norm is bounded by $K(n_0, D_0)$.

Now we use (duality) Theorem 4.89. According to this result, for the metric space $(\mathcal{M}, d)$, the associated Lipschitz-free space $\mathcal{F}(\mathcal{M})$ (see (4.131) for its definition) is predual to $\mathrm{Lip}_0(\mathcal{M})$. We apply this to the subspace $N_\varepsilon \cap B_R(m)$ of $\mathcal{M}$. Since $\mathrm{Lip}_0(N_\varepsilon \cap B_R(m))$ is finite-dimensional and $E : \mathrm{Lip}_0(N_\varepsilon \cap B_R(m)) \to \mathrm{Lip}_0(B_R(m))$ is an extension operator, there exists an operator $P : \mathcal{F}(B_R(m)) \to$

5.1. Embedding of metric spaces into the space forms

$\mathcal{F}(N_\varepsilon \cap B_R(m))$ such that $P^* = E$ and P is a *linear projection* onto the target space, in particular, $\|P\| = \|E\| \leq K(n_0, D_0)$.

Now let $\delta_{B_R(m)}$ be the isometric embedding of $B_R(m)$ into $\mathcal{F}(B_R(m))$, see Lemma 4.92. Setting $I := P \circ \delta_{B_R(m)}$ we define a $K(n_0, D_0)$-Lipschitz map from $B_R(m)$ into $\mathcal{F}(N_\varepsilon \cap B_R(m))$ which, in addition, satisfies, for all m', m'' belonging to $N_\varepsilon \cap B_R(m)$, the condition

$$\|I(m') - I(m'')\|_{\mathcal{F}(N_\varepsilon \cap B_R(m))} = \|\delta_{B_R(m)}(m') - \delta_{B_R(m)}(m'')\|_{\mathcal{F}(B_R(m))} = d(m', m'').$$

Now we impose the assumptions of Proposition 5.15 with $\mathcal{M} := B_R(m)$, $\mathcal{M}_1 := \mathcal{F}(N_\varepsilon \cap B_R(m))$, $A := K(n_0, D_0)$, $F := I$ and $\mu = 1$. Choosing here $\varepsilon := \frac{R_0}{64}$, we derive from this proposition the following statement.

There exists an integer N and a constant D depending only on n_0, D_0 and R_0, such that $B_R(m)$ allows a bi-Lipschitz embedding into the direct sum $\mathcal{F}(N_{\varepsilon_0} \cap B_R(m)) \oplus \mathbb{R}^N$.

Further, we claim that for some constant N_1 independent of the center m,

$$\dim \mathcal{F}(N_{\varepsilon_0} \cap B_R(m)) = \mathrm{card}(N_{\varepsilon_0} \cap B_R(m)) \leq N_1. \tag{5.63}$$

This result, which will be proved later, implies that the space $\mathcal{F}(N_{\varepsilon_0} \cap B_R(m))$ admits a bi-Lipschitz embedding into the Euclidean space $\mathbb{R}^{N_1}$ with distortion at most N_1.

In fact, let $\{e_i\}_{1 \leq i \leq N_1}$ be an Auerbach basis of $\mathcal{F}(N_\varepsilon \cap B_R(m))$, i.e., for every linear combination $\sum_{i=1}^{N_1} \lambda_i e_i$,

$$\max_{1 \leq i \leq N_1} |\lambda_i| \leq \left\| \sum_{i=1}^{N_1} \lambda_i e_i \right\| \leq \sum_{i=1}^{N_1} |\lambda_i|,$$

see, e.g., [LT-1977, Prop. 1.c.3].

Define a Euclidean norm on $\mathcal{F}(N_\varepsilon \cap B_R(m))$ by

$$\left\| \sum_{i=1}^{N_1} \lambda_i e_i \right\| := \left\{ \sum_{i=1}^{N_1} \lambda_i^2 \right\}^{\frac{1}{2}}.$$

Then the identity map from $\mathcal{F}(N_\varepsilon \cap B_R(m))$ into itself equipped with the Euclidean norm has distortion $\sup\left\{ \dfrac{\sum_{i=1}^{N_1} |\lambda_i|}{\max_{1 \leq i \leq N_1} |\lambda_i|} \right\} \leq N_1$.

We therefore conclude that $B_R(m)$ admits a bi-Lipschitz embedding into the Euclidean space $\mathbb{R}^{N_1} \oplus \mathbb{R}^N = \mathbb{R}^{N_1+N}$ with distortion at most DN_1, as required.

It remains to explain the validity of inequality (5.63). Since $\mathcal{M} \in \mathcal{G}_{n_0}(R_0, D_0)$, its *doubling function* $\delta_{\mathcal{M}}(R)$, the minimal number of closed balls of radius $\frac{R}{2}$

covering an open ball of radius R, is finite at the point R_0. That is to say, $\mathcal{M}$ is a *locally doubling* metric space, see Definition 3.30. Then Proposition 3.49 applied to this $\mathcal{M}$ and ε-net N_ε gives precisely inequality (5.63).

This completes the proof of the Claim. □

5.2 Roughly similar embeddings of Gromov hyperbolic spaces

The main goal of this section is to prove the Bonk–Schramm theorem whose version, formulated as Theorem 3.120, was used in subsection 5.1.4. The proof exploits concepts and results of the so-called *Coarse Geometry* which studies large scale invariants of Gromov hyperbolic spaces. One of them, *boundary at infinity* (or *Gromov boundary*) and its characteristics, is a key point of the forthcoming consideration. The reader may find a detailed exposition of the results surveyed below in the books [GH-1990] and [BH-1999] for proper hyperbolic spaces, and in the minimonograph [Vai-2005] and the paper [BSch-2000] for generalized hyperbolic spaces. The necessity of such a generalization is explained by its considerable role within the proof of the Bonk–Schramm theorem (which, however, concerns the classical Gromov hyperbolic spaces subject to each of the equivalent Definitions 3.114 and 3.115).

5.2.1 Coarse Geometry, a survey

We begin with the definitions of several basic objects. To this aim, we need "rough" substitutions for geodesic segments and geodesic rays [2].

Definition 5.18. A curve $\gamma : [0,1] \to (\mathcal{M}, d)$ is called a *k-roughly geodesic*, $k \geq 0$, if for all $0 \leq t, t' \leq 1$,

$$-k + |t - t'| \leq d(\gamma(t), \gamma(t')) \leq |t - t'| + k. \tag{5.64}$$

A map $\gamma : \mathbb{R}_+ \to (\mathcal{M}, d)$ is called a *k-roughly geodesic ray*, $k \geq 0$, if (5.64) holds for all $0 \leq t, t' < \infty$. For $k = 0$ these objects clearly coincide with a geodesic segment and a geodesic ray, respectively.

Now we introduce the aforementioned generalization of the Gromov hyperbolicity by discarding or relaxing the condition for a metric space in Definition 3.115 to be geodesic.

Definition 5.19. (a) A metric space $(\mathcal{M}, d)$ is said to be *δ-hyperbolic*, $\delta \geq 0$, if for some point $m^* \in \mathcal{M}$ and all $m, m', m'' \in \mathcal{M}$,

$$(m|m'')_{m^*} \geq \min\{(m|m')_{m^*}, (m'|m'')_{m^*}\} - \delta. \tag{5.65}$$

---

[2] The adjective "rough" refers to the change of equalities and inequalities defining the notions by inequalities containing additional additive constants.

5.2. Roughly similar embeddings of Gromov hyperbolic spaces

(b) A metric space $(\mathcal{M}, d)$ is said to be (δ, k)-hyperbolic, $k, \delta \geq 0$, if it satisfies (5.65), and every two points in $\mathcal{M}$ can be joined by a k-roughly geodesic.

Let us recall that the Gromov product used in (5.65) is defined by

$$(m|m')_{m^*} := \frac{1}{2}\left[d(m', m^*) + d(m'', m^*) - d(m', m'')\right].$$

Clearly, a $(\delta, 0)$-hyperbolic space is the object of Gromov's Definition 3.115. As in the latter case, Definition 5.19 does not depend on the choice of a basepoint (apart from the replacement of δ by 2δ).

In the sequel, the space of Definition 5.19 (a) is called *hyperbolic* if the requirement of Definition 5.19 (b) holds for some $\delta \geq 0$ and $k = 0$.

Remark 5.20. Inequality (5.65) is equivalent to the more symmetric inequality

$$\begin{aligned}&d(m_1, m_3) + d(m_2, m_4) \\ &\leq \max\{d(m_1, m_2) + d(m_3, m_4), d(m_2, m_3) + d(m_1, m_4)\} + 2\delta,\end{aligned} \quad (5.66)$$

which will be used below.

The key concept of the theory is the *boundary at infinity* of a metric space $\mathcal{M}$, denoted by $\partial_\infty \mathcal{M}$. To introduce it, we need some preliminary definitions.

We say that a sequence $\{m_j\} \subset (\mathcal{M}, d)$ *converges at infinity*, if

$$\lim_{i,j \to \infty} (m_i|m_j)_{m^*} = \infty.$$

Further, two such sequences $\{m_j\}$ and $\{m'_j\}$ are *equivalent* if

$$\lim_{j \to \infty} (m_j|m'_j)_{m^*} = \infty.$$

Since, by definition,

$$\left|(m|m')_{m^*} - (m|m')_{m^{**}}\right| \leq d(m^*, m^{**}), \quad (5.67)$$

neither of these notions depend on the choice of the basepoint.

The term "equivalent" is misleading, since it is not an equivalence relation (in general, transitivity does not hold). But for the class of hyperbolic spaces this is, indeed, an equivalence relation, as the inequality

$$(m_j|m''_j)_{m^*} \geq \min\{(m_j|m'_j)_{m^*}, (m'_j|m''_j)_{m^*}\} - \delta$$

demonstrates.

Definition 5.21. The boundary at infinity $\partial_\infty \mathcal{M}$ of a hyperbolic space $(\mathcal{M}, d)$ is the set of all equivalence classes of sequences in $\mathcal{M}$ convergent at infinity.

Note that $\partial_\infty \mathcal{M} = \phi$, if $\mathcal{M}$ is bounded. To exclude this trivial case, we assume in the sequel that

$$\operatorname{diam} \mathcal{M} = \infty. \tag{5.68}$$

In the case of a proper (unbounded) Gromov hyperbolic space $\mathcal{M}$, every two points $m \in \mathcal{M}$ and $\widehat{m} \in \partial_\infty \mathcal{M}$ can be, in a sense, joined by a geodesic ray. This fact is not true in general but its weaker version is. In its formulation, we will say that a ray $\gamma : \mathbb{R}_+ \to \mathcal{M}$ emanating from a point m *joins m with a point $\widehat{m} \in \partial_\infty \mathcal{M}$* if, for every unbounded sequence $\{t_j\} \subset \mathbb{R}_+$, the associated sequence $\{\gamma(t_j)\}$ belongs to the point (equivalence class) $\widehat{m}$. We will call $m, \widehat{m}$ the *endpoints* of γ.

In the sequel, two rays $\gamma_1, \gamma_2 : \mathbb{R}_+ \to \mathcal{M}$ are said to be *equivalent* if they have a common endpoint at $\partial_\infty \mathcal{M}$.

Proposition 5.22. *Let $(\mathcal{M}, d)$ be a (δ, k)-hyperbolic space. There exists a constant $\tilde{k} = \tilde{k}(\delta, k) \geq 0$ such that every two points $m \in \mathcal{M}$ and $\widehat{m} \in \partial_\infty \mathcal{M}$ can be joined by a $\tilde{k}$-roughly geodesic ray.*

This result allows us, for the case of (δ, k)-hyperbolic spaces, to define the boundary at infinity using equivalence classes of $\tilde{k}$-roughly geodesic rays.

Finally, we single out relations between the class of δ-hyperbolic spaces and its subclass consisting of (δ, k)-hyperbolic spaces. To this end, we need

Definition 5.23. A metric space $(\mathcal{M}, d)$ is said to be *k-visual* with respect to some point $m^* \in \mathcal{M}$ if every point lies in the image of a k-roughly geodesic ray emanating from m^*.

Using this, we formulate the required result.

Proposition 5.24. *Let $(\mathcal{M}, d)$ be a k-visual δ-hyperbolic space. Then $\mathcal{M}$ is $(\delta, \tilde{k})$-hyperbolic for some $\tilde{k} = \tilde{k}(\delta, k) \geq 0$.*

Visuality of hyperbolic spaces may be established by exploiting the following criterion.

Proposition 5.25. *Let $(\mathcal{M}, d, m^*)$ be a pointed δ-hyperbolic space. Assume that for some $k, c \geq 0$ the union of all k-roughly geodesic rays in $\mathcal{M}$ emanating from m^* is c-co-bounded*[3]. *Then $\mathcal{M}$ is $\tilde{k}$-visual with respect to m^* for some $\tilde{k} = \tilde{k}(\delta, k, c) \geq 0$.*

We now metricize $\partial_\infty \mathcal{M}$ by introducing a family of so-called *visual metrics*. To this end, we extend the Gromov product from a hyperbolic space $(\mathcal{M}, d)$ to the set $\mathcal{M} \cup \partial_\infty \mathcal{M}$. To simplify notation, we fix the basepoint m^* and write $(m|m')$ instead of $(m|m')_{m^*}$.

First, for $\widehat{m}, \widehat{m}' \in \partial_\infty \mathcal{M}$ we set

$$(\widehat{m}|\widehat{m}') := \sup \left\{ \lim_{j \to \infty} (m_j|m'_j) \, ; \, \{m_j\} \in \widehat{m}, \{m'_j\} \in \widehat{m}' \right\}. \tag{5.69}$$

---

[3] Let us recall that a subset $S \subset \mathcal{M}$ is *c-co-bounded* if $\sup_{m \in \mathcal{M}} d(m, S) \leq c$.

5.2. Roughly similar embeddings of Gromov hyperbolic spaces

Then for $\widehat{m} \in \partial_\infty \mathcal{M}$ and $m' \in \mathcal{M}$ we define $(\widehat{m}|m)$ by using (5.67) with $\{m'_j\}$ given by $m'_j = m'$ for all j.

The following properties of the extended product will be used in the proof of the main result.

Proposition 5.26. *Let $(\mathcal{M}, d, m^*)$ be a pointed δ-hyperbolic space. Then the following is true:*

(a) *For all $\widehat{m}, \widehat{m}', \widehat{m}'' \in \partial_\infty \mathcal{M}$,*

$$(\widehat{m}|\widehat{m}'') \geq \min\{(\widehat{m}|\widehat{m}'), (\widehat{m}'|\widehat{m}'')\} - 2\delta.$$

(b) *Assume that points $m_i \in \mathcal{M}, \widehat{m}_i \in \partial_\infty \mathcal{M}$ satisfy*

$$(\widehat{m}_i|m_i) \geq d(m_i, m^*) - \lambda, \quad i = 1, 2,$$

for some $\lambda \geq 0$. Then there exists a constant $c = c(\lambda, \delta) > 0$ such that

$$d(m_1, m_2) = \sum_{i=1,2} d(m_i, m^*) - 2\min\{(\widehat{m}_1|\widehat{m}_2), d(m_1, m^*), d(m_2, m^*)\}$$

$$+ O_{\delta,\lambda}(1).$$

Convention. In the sequel, the notation $O_{\alpha,\beta,\gamma,\ldots}(1)$ stands for a function satisfying, for some constant $c = c(\alpha, \beta, \gamma, \ldots)$ and all values of the involved variables, the inequality $|O_{\alpha,\beta,\gamma,\ldots}(1)| \leq c(\alpha, \beta, \gamma, \ldots)$.

Corollary 5.27. *Let γ_1, γ_2 be k-roughly geodesic rays of a δ-hyperbolic space $(\mathcal{M}, d)$ with common endpoints. Then there is a constant $c = c(k, \delta) > 0$ such that*

$$\sup_{t>0} d(\gamma_1(t), \gamma_2(t)) \leq c(k, \delta).$$

Now we use the extended Gromov product to define the family of metrics $\{d_\varepsilon\}_{\varepsilon>0}$ of $\partial_\infty \mathcal{M}$ by setting, for $\widehat{m}, \widehat{m}' \in \partial_\infty \mathcal{M}$,

$$d_\varepsilon(\widehat{m}, \widehat{m}') := \inf \sum_j e^{-\varepsilon(\widehat{m}_j|\widehat{m}_{j+1})}, \qquad (5.70)$$

where the infimum is taken over all finite sequences $\{\widehat{m}_j\}_{1 \leq j \leq n} \subset \partial_\infty \mathcal{M}$ such that $\widehat{m}_1 = \widehat{m}$ and $\widehat{m}_n = \widehat{m}'$.

Because of their definition, these metrics, at least for small ε, closely relate to the Gromov product. Actually, the following is true.

Proposition 5.28. *If $(\mathcal{M}, d)$ is a δ-hyperbolic space, then for $\varepsilon > 0$ satisfying $\varepsilon\delta \leq \frac{1}{5}$ and all $\widehat{m}, \widehat{m}' \in \partial_\infty(\mathcal{M})$,*

$$\frac{1}{2} e^{-\varepsilon(\widehat{m}|\widehat{m}')} \leq d_\varepsilon(\widehat{m}, \widehat{m}') \leq e^{-\varepsilon(\widehat{m}|\widehat{m}')}.$$

In particular, for these ε,
$$|\varepsilon(\widehat{m}|\widehat{m}') + \log d_\varepsilon(\widehat{m}, \widehat{m}')| \leq \log 2. \tag{5.71}$$

In the sequel, we consider only metrics d_ε for which (5.65) holds. However, the following two important results are true for every $\varepsilon > 0$.

Proposition 5.29. *The metric space $(\partial_\infty \mathcal{M}, d_\varepsilon)$ is bounded and complete.*

Finalizing the survey, we present a result which is one of the basic ingredients in the forthcoming proof. For its formulation we need

Definition 5.30. A metric space is of *bounded growth in some scale* if there are constants $0 < r < R$ and $N \in \mathbb{N}$ such that every open ball of radius R can be covered by at most N open balls of radius r.

Clearly, locally doubling metric spaces, see Definition 3.27, and metric spaces of bounded geometry satisfy this definition.

Theorem 5.31. *Let $(\mathcal{M}, d)$ be a Gromov hyperbolic space of bounded growth in some scale. Then the metric space $(\partial_\infty \mathcal{M}, d_\varepsilon)$ is doubling.*

As a consequence of this result and Proposition 5.29 we have

Corollary 5.32. *Under the conditions of Theorem 5.31 the space $(\partial_\infty \mathcal{M}, d_\varepsilon)$ is compact.*

5.2.2 Coarse geometry of $\mathbb{H}^n$

We illustrate the concepts introduced above by three examples related to $\mathbb{H}^n$ and convex subsets of $\mathbb{H}^n$. The facts presented below will be used in the proof of the Bonk–Schramm theorem and therefore we accompany them by a brief demonstration.

Example 5.33. Let $\mathbb{H}^{n+1}$ be the Poincaré half-space model of the Lobachevski space. The points of the underlying subspace $H_+^{n+1} \subset \mathbb{R}^{n+1}$ will be denoted by $(x, r), (x', r')$, etc., where $x, x' \in \mathbb{R}^n$ and $r, r' > 0$. The geodesic distance between these points is determined by the formula

$$\cosh[d_g((x, r), (x', r'))] = \frac{rr' + (r - r')^2 + \|x - x'\|^2}{rr'} \tag{5.72}$$

where $\|\cdot\|$ is the standard Euclidean norm of $\mathbb{R}^n$.

Since $\cosh(t) = \frac{1}{2}e^t + O(1)$ for $t \geq 0$ and

$$\frac{1}{2}(\max\{r, r'\})^2 \leq rr' + (r - r')^2 \leq 2(\max\{r, r'\})^2,$$

we can rewrite (5.72) as

$$d_g((x, r), (x', r')) = 2\log \frac{\max\{r', r\} + \|x - x'\|}{\sqrt{rr'}} + O(1). \tag{5.73}$$

5.2. Roughly similar embeddings of Gromov hyperbolic spaces

Since $\mathbb{H}^{n+1}$ is a proper Gromov hyperbolic space, its boundary at infinity is determined by equivalent geodesic rays. These are straight half-lines orthogonal to $\mathbb{R}^n$ and half-circles orthogonal to and centered at $\mathbb{R}^n$. It is easily derived from (5.72) that all half-lines orthogonal to $\mathbb{R}^n$ are equivalent and therefore determine a single point of $\partial_\infty \mathbb{H}^{n+1}$, naturally denoted by ∞.

Let now $\gamma, \gamma' : \mathbb{R}_+ \to \mathbb{H}^{n+1}$ be two half-circles which meet $\mathbb{R}^n$ at points $(x, 0)$ and $(x', 0)$. We next show that they are equivalent if and only if $x = x'$, which implies
$$\partial_\infty \mathbb{H}^{n+1} = \mathbb{R}^n \cup \{\infty\}.$$

Given the points $w := (x, r), w' := (x', r') \in \mathbb{H}^{n+1}$, we fix a point $w^* := (x^*, r^*)$ satisfying the condition
$$r^* \geq \max\{r, r'\}. \tag{5.74}$$

Using (5.73), (5.74) and the definition of the Gromov product we then have
$$(w|w')_{w^*} = -\log(\|x - x'\| + \max\{r, r'\}) \\ + \log \frac{r^* + \|x^* - x\|}{\sqrt{r^*}} + \log \frac{r^* + \|x^* - x'\|}{\sqrt{r^*}} + O(1),$$

whence
$$(w|w')_{w^*} = -\log(\|x - x'\| + \max\{r, r'\}) + O_{r^*, D}(1), \tag{5.75}$$

where $D := \max\{\|x^* - x\|, \|x^* - x'\|\}$.

Applying this to the points $\gamma(t) := (x(t), r(t))$, $\gamma'(t) := (x'(t), r'(t))$ and noting that $\max\{r(t), r'(t)\} \to 0$ as $t \to \infty$, we conclude that $(\gamma(t)|\gamma'(t))_{w^*}$ does not tend to infinity if and only if
$$x = \lim_{t \to \infty} x(t) \neq x' = \lim_{t \to \infty} x'(t).$$

Hence γ and γ' are equivalent if and only if $x = x'$, as required.

The above argument also implies that the extended Gromov product of the points x, x' lying in a compact subset $S \subset \mathbb{R}^n$ ($\subset \partial_\infty \mathbb{H}^{n+1}$) satisfies
$$(x|x')_w = -\log \|x - x'\| + O_{w, D}(1), \tag{5.76}$$

where D is the Euclidean diameter of S and w is an arbitrary point of $\mathbb{H}^{n+1}$.

It is worth noting that $\partial_\infty \mathbb{H}^{n+1}$ can be metricized in a direct manner using the formula
$$d_w(\widehat{m}, \widehat{m}') := e^{-(\widehat{m}|\widehat{m}')_w},$$

which determines a metric equivalent to any of the metrics defined by (5.70). The space $\partial_\infty \mathbb{H}^{n+1}$ equipped with this metric is bi-Lipschitz equivalent to the unit n-sphere equipped with the standard spherical metric.

Example 5.34. Let S be a compact subset of $\mathbb{R}^n \subset \partial_\infty \mathbb{H}^{n+1}$ containing more than one point. Its *convex hull*, denoted by $\mathrm{hull}(S)$, is the intersection of all closed half-spaces of $\mathbb{H}^{n+1}$ whose images under the orthogonal projection $(x, r) \mapsto (x, 0)$ contain S.

Let us recall that a *half-space* is determined by a hyperplane of $\mathbb{R}^{n+1}$ orthogonal to $\mathbb{R}^n$ or by an n-sphere orthogonal to and centered at $\mathbb{R}^n$. Due to the definition and the condition card $S > 1$, the set $\mathrm{hull}(S) \neq \emptyset$.

Since $\mathrm{hull}(S)$ is hyperbolically convex, every geodesic joining two of its points is contained in this hull. Hence, the space $(\mathrm{hull}(S), d_g)$ is geodesic and therefore Gromov hyperbolic.

We establish the following facts for this space:

(a) $\partial_\infty \mathrm{hull}(S) = S$;

(b) $\mathrm{hull}(S)$ is visual.

(a) Since $\mathrm{hull}(S)$ is hyperbolically convex, every geodesic ray (half-circle) joining a point in $\mathrm{hull}(S)$ with a point of S lies in $\mathrm{hull}(S)$. Hence, $S \subset \partial_\infty \mathrm{hull}(S)$.

Let now $\tilde{x} \notin S$. Since S is compact, there is a closed ball centered at $\tilde{x}$ which does not intersect S. Then the half-space determined by this ball intersects every geodesic ray γ joining a point in $\mathrm{hull}(S)$ and $\tilde{x}$. Hence, γ is not a geodesic ray of $\mathrm{hull}(S)$; this clearly means that $\tilde{x} \notin \partial_\infty \mathrm{hull}(S)$.

(b) We will use the following simple fact. Given a point $w \in \mathrm{hull}(S)$, there exists a point $w' \in \mathrm{hull}(S)$ such that w' lies in the image of a geodesic ray joining w^* with some point of S $(= \partial_\infty \mathrm{hull}(S))$ and satisfies, for some constant $x = c(w^*, S)$,

$$d_g(w, w') \leq c. \tag{5.77}$$

The simplest way to prove this result is to exploit the Klein ball model of the hyperbolic $(n+1)$-space (isometric to $\mathbb{H}^{n+1}$). The underlying set now is the unit open ball $B^{n+1} := \{x \in \mathbb{R}^{n+1}\,;\, \|x\| < 1\}$, geodesics and geodesic rays are straight line intervals, and half-spaces are determined by spherical cups. Since the isometry group of this space (Möbius group) is transitive, we may and will assume that w is the center O of B^{n+1}. Joining w^* by line segments with all points of $S \subset \mathbb{S}^n (= \partial_\infty B^{n+1})$, we obtain the set of geodesic rays emanating from w^* to the points of S. The union of these (closed) segments is compact and therefore there exists a point w' in B^{n+1} closest to O in the Euclidean metric. The geodesic distance between $O =: w$ and w' in the Klein model is given by

$$d_g(w, w') := \frac{1}{2} \log \frac{\|w - w'_\infty\| \cdot \|w' - w_\infty\|}{\|w - w_\infty\| \cdot \|w' - w'_\infty\|},$$

where w_∞, w'_∞ are the endpoints of the diameters passing though w, w' which are closest, respectively, to w and w'. This distance is clearly finite and depends only on w^* and S.

Hence, (5.77) is true.

The result just proved means that the union of all geodesic rays emanating from w^* is c-co-bounded in $\text{hull}(S)$. Applying now Proposition 5.25, we derive from this that $\text{hull}(S)$ is visual.

Example 5.35. Generalizing the previous results we consider the hull of an arbitrary subset $S \subset \mathbb{H}^{n+1}$, the intersection of all closed subspaces in $\mathbb{H}^{n+1}$ containing S. Assume that S is co-bounded in the set consisting of the union of all geodesic segments with endpoints in S. Then one can prove that the inclusion map $i : S \hookrightarrow \text{hull}(S)$ is a rough similarity.

For instance, this is true if $\partial_\infty \text{hull}(S)$ is a compact subset in $(\mathbb{R}^n, \|\cdot\|)$ (in this case, exploiting an argument similar to that in Example 5.34, one can prove that the union of geodesic segments with endpoints in S is co-bounded in $\text{hull}(S)$).

5.2.3 The Bonk–Schramm theorem

After the preliminary work presented in the previous subsections we are ready to formulate and prove the Bonk–Schramm theorem [BSch-2000, Thm. 1.1].

Theorem 5.36. *Let $(\mathcal{M}, d)$ be a Gromov hyperbolic space of bounded growth in some scale. Then $\mathcal{M}$ is roughly similar to a convex subset of some $\mathbb{H}^n$.*

Remark 5.37. The converse of the theorem is also true. In fact, a geodesic space $\mathcal{M}$ that is roughly similar to some convex subset of $\mathbb{H}^n$ is Gromov hyperbolic, see the text after Definition 3.117. Moreover, it is the matter of definitions to check that $\mathcal{M}$ is also of bounded growth in some scale.

Proof. Due to Definition 3.117 (of rough similarity) we must prove that for some constants $C \geq 1, k \geq 0$ and $n \in \mathbb{N}$, there exists a map $F : \mathcal{M} \to \mathbb{H}^n$ such that

(i) $F(\mathcal{M})$ is k-co-bounded in some convex subset of $\mathbb{H}^n$;

(ii) for all $m, m' \in \mathcal{M}$,
$$-k + Cd(m, m') \leq d_g\big(F(m), F(m')\big) \leq Cd(m, m') + k.$$

Hereafter a map $F : (\mathcal{M}_1, d_1) \to (\mathcal{M}_2, d_2)$ satisfying similar conditions is called (C, k)-*roughly similar* (or simply *roughly similar* if such C and k exist but are not specified). Rough similarity of spaces $\mathcal{M}_1$ and $\mathcal{M}_2$ is recalled to be an equivalence relation denoted in the sequel by $\mathcal{M}_1 \simeq \mathcal{M}_2$. Hence, claims (i), (ii) may be equivalently reformulated as

$$(\mathcal{M}, d) \simeq (V, d_g),$$

where V is a convex subset of $\mathbb{H}^n$.

The proof of this result is divided into several steps. The key observation is that the operation $\mathcal{M} \mapsto \partial_\infty \mathcal{M}$ can be converted to within rough similarity in the class of visual hyperbolic spaces. The converse operation assigns to every bounded metric space $(\mathcal{M}, d)$ its space of open balls $\mathcal{B}(\mathcal{M})$ equipped with a special metric ρ_d. This operation will be denoted by $\mathcal{B}^*$.

At the first stage we study properties of $\mathcal{B}^*$ which will be used in the proof. In particular, we show that $\mathcal{B}^*(\mathcal{M}, d)$ is a visual hyperbolic space, and if $(\mathcal{M}', d')$ is bi-Lipschitz equivalent to $(\mathcal{M}, d)$, then

$$\mathcal{B}^*(\mathcal{M}, \sqrt{d}) \simeq \mathcal{B}^*(\mathcal{M}, d) \simeq \mathcal{B}^*(\mathcal{M}', d').$$

At the second stage we prove that for every visual hyperbolic space $\mathcal{M}$,

$$\mathcal{B}^*(\partial_\infty \mathcal{M}, d_\varepsilon) \simeq (\mathcal{M}, d).$$

Then, if $\mathcal{M}$ is of bounded growth in some scale, Theorem 5.31 and Assouad's embedding Theorem 5.10 imply that $(\partial_\infty \mathcal{M}, \sqrt{d_\varepsilon})$ is bi-Lipschitz equivalent to a metric subspace $(S, \|\cdot\|)$ of some Euclidean space $(\mathbb{R}^n, \|\cdot\|)$. Together with the above this yields

$$(\mathcal{M}, d) \simeq \mathcal{B}^*(\partial_\infty \mathcal{M}, \sqrt{d_\varepsilon}) \simeq \mathcal{B}^*(S, \|\cdot\|).$$

At the next step, we show that

$$\mathcal{B}^*(S, \|\cdot\|) \simeq \bigl(\mathrm{hull}(S), d_g\bigr) \subset \mathbb{H}^{n+1}$$

provided that $\partial_\infty \mathcal{M}$ contains more than one point.

This leads to the required relation

$$(\mathcal{M}, d) \simeq \bigl(\mathrm{hull}(S), d_g\bigr)$$

under the additional assumptions of visuality for $\mathcal{M}$ and that $\partial_\infty \mathcal{M}$ contains more than one point.

At the final step (only outlined) the additional restrictions will be dropped by using an isometric embedding of the space $(\mathcal{M}, d)$ into a visual Gromov hyperbolic space $(\widehat{\mathcal{M}}, \widehat{d})$ of bounded growth at some scale whose boundary at infinity contains more than one point.

We begin the realization of this plan with

Definition 5.38. The operation $\mathcal{B}^*$ on the class of bounded metric spaces is defined by

$$\mathcal{B}^*(\mathcal{M}, d) := \bigl(\mathcal{B}(\mathcal{M}), \rho_d\bigr),$$

where the function ρ_d is given for points (open balls in $\mathcal{M}$) $B_r(m), B_{r'}(m')$ by the formula [4]

$$\rho_d\bigl(B_r(m), B_{r'}(m')\bigr) := 2\log \frac{d(m, m') + \max\{r, r'\}}{\sqrt{rr'}}. \tag{5.78}$$

---

[4] Cf. the metric for the space of balls $\mathcal{B}^\omega(\mathcal{M})$ in Section 4.4 (see formula (4.71)) and that of Example 5.33 given by formula (5.67).

5.2. Roughly similar embeddings of Gromov hyperbolic spaces

Let us show that ρ_d is a metric on $\mathcal{B}(\mathcal{M})$. Using the triangle inequality for d and the inequality

$$r' \max\{r, r''\} \le \max\{r, r'\} \cdot \max\{r', r''\},$$

we obtain

$$r'\big(d(m, m'') + \max\{r, r''\}\big) \le \big(d(m, m') + \max\{r, r'\}\big) \cdot \big(d(m', m'') + \max\{r', r''\}\big),$$

whence

$$\frac{d(m, m'') + \max\{r, r''\}}{\sqrt{rr''}} \le \frac{d(m, m') + \max\{r, r'\}}{\sqrt{rr'}} \cdot \frac{d(m', m'') + \max\{r', r''\}}{\sqrt{r'r''}}.$$

This is clearly equivalent to the triangle inequality for ρ_d.

If now the right-hand side of (5.78) is zero, then

$$d(m, m') + \max\{r, r'\} = \sqrt{rr'} \le \max\{r, r'\}.$$

Hence, $d(m, m') = 0$ and $r = r'$, i.e., $B_r(m) = B_{r'}(m')$.

Now we describe some basic properties of the metric space $\mathcal{B}^*(\mathcal{M}, d)$.

Proposition 5.39. *Let $(\mathcal{M}, d)$ be a metric space of diameter $D > 0$. There exist numerical constants k and $\delta > 0$ such that $\mathcal{B}^*(\mathcal{M}, d)$ is a k-visual (δ, k)-hyperbolic space.*

Proof. We first show that $\mathcal{B}^*(\mathcal{M}, d)$ is δ-hyperbolic for $\delta = \log 2$. Due to (5.66) we must prove, for the metric $\rho := \rho_d$ and arbitrary four points (balls) $B_i := B_{r_i}(m_i)$, $i = 1, 2, 3, 4$, that

$$\rho_d(B_1, B_2) + \rho_d(B_3, B_4)$$
$$\le \max\{\rho_d(B_1, B_3) + \rho_d(B_2, B_4), \rho_d(B_1, B_4) + \rho_d(B_2, B_3)\} + 2\log 2.$$

To this end, we need

Lemma 5.40. *Let R be a metric on the index set $\{1, \ldots, 4\}$. Then*

$$\log R(1, 2) + \log R(3, 4) \qquad (5.79)$$
$$\le \max\{\log R(1, 3) + \log R(2, 4), \log R(1, 4) + \log R(2, 3)\} + 2\log 2.$$

Proof. Let, for definiteness, $R(1, 3)$ be the smallest of the numbers $R(i, j)$. Then, by the triangle inequality, $R(1, 2) \cdot R(3, 4) \le 2R(2, 3) \cdot 2R(1, 4)$ and the result follows by taking logarithms. □

Now use the lemma for $R(i, j) := d(m_i, m_j) + \max\{r_i, r_j\}$, $1 \le i, j \le 4$. This clearly defines a metric, and (5.79) becomes the required inequality for ρ_d.

Now fix $m^* \in \mathcal{M}$ and let $B^* := B_D(m^*) \in \mathcal{B}(\mathcal{M})$. Then, given an arbitrary point $B := B_r(m) \in \mathcal{B}(\mathcal{M})$, there is a k-roughly geodesic ray $\gamma : \mathbb{R}_+ \to \mathcal{B}^*(\mathcal{M}, d)$

with $k = 2\log 2$ which emanates from B^* and passes through B.[5] This ray can be defined by
$$\gamma(0) := B^* \quad \text{and} \quad \gamma(t) := B_{De^{-t}}(m), \quad t > 0.$$
Then $\gamma(t_0) = B$ for $t_0 := \log \frac{D}{r}$, and for $0 < t < t'$,
$$\rho(\gamma(t), \gamma(t')) := 2\log \frac{d(m,m) + D\max\{e^{-t}, e^{-t'}\}}{De^{-\frac{t+t'}{2}}} = t' - t.$$
For the remaining case of $t > 0$ and $t' = 0$ we have
$$\rho(\gamma(0), \gamma(t)) := 2\log \frac{d(m, m^*) + D}{De^{-\frac{t}{2}}} = t + 2\log\left(1 + \frac{d(m, m^*)}{D}\right),$$
whence
$$\left|(\rho(\gamma(0), \gamma(t)) - t\right| \le 2\log 2,$$
as required.

Finally, we apply Proposition 5.24 to the (δ, k)-hyperbolic space $\mathcal{B}^*(\mathcal{M}, d)$ with $\delta = \log 2$ and $k = 2\log 2$. Since the union of all k-roughly geodesic rays of $\mathcal{B}^*(\mathcal{M}, d)$ coincides with $\mathcal{B}(\mathcal{M})$, the proposition implies that $\mathcal{B}^*(\mathcal{M}, d)$ is $\tilde{k}$-visual for some numerical constant $\tilde{k} > 0$. □

Proposition 5.41. *If $(\mathcal{M}, d)$ is a (δ, k)-hyperbolic space k-visual with respect to a point $m^* \in \mathcal{M}$, then $(\mathcal{M}, d)$ is roughly similar to the space $\mathcal{B}^*(\partial_\infty \mathcal{M}, d_\varepsilon)$ for a sufficiently small $\epsilon > 0$.*

Proof. Due to Proposition 5.22, for every point $\widehat{m} \in \partial_\infty \mathcal{M}$ there is a $\tilde{k} = \tilde{k}(\delta, k)$-roughly geodesic ray $\gamma_{\widehat{m}} : \mathbb{R}_+ \to \mathcal{M}$ which joins $\widehat{m}$ and m^*. We define the desired roughly similar map $f : \mathcal{B}^*(\partial_\infty \mathcal{M}, d_\varepsilon) \to \mathcal{M}$ by setting
$$f(B_r(\widehat{m})) := \gamma_{\widehat{m}}\left(\frac{D}{r}\right), \tag{5.80}$$
where now $D := \text{diam}(\partial_\infty \mathcal{M}, d_\varepsilon)$.

Rough similarity of f means that for some constant $c \ge 1$ and all points $B_r(\widehat{m}), B_{r'}(\widehat{m}')$ in $\mathcal{B}(\partial_\infty \mathcal{M})$,
$$d(f(B_r(\widehat{m})), f(B_{r'}(\widehat{m}'))) = c\rho(B_r(\widehat{m}), B_{r'}(\widehat{m}')) + O_{k,\delta}(1), \tag{5.81}$$
where ρ stands for ρ_{d_ε}.

To prove this, we first show that
$$(\widehat{m}|f(B_r(\widehat{m})))_{m^*} = \varepsilon^{-1}\log\frac{D}{r} + O_{k,\delta}(1). \tag{5.82}$$

---
[5] i.e., the image $\gamma(\mathbb{R}_+)$ contains B.

5.2. Roughly similar embeddings of Gromov hyperbolic spaces

By the definition of the Gromov product on $\mathcal{M} \cup \partial_\infty \mathcal{M}$, see (5.69) and (5.80), the left-hand side equals

$$\sup_{\{t_j\} \to \infty} \lim_{j \to \infty} \left(\gamma_{\widehat{m}}(t_j), \gamma_{\widehat{m}}\left(\varepsilon^{-1} \log \frac{D}{r}\right) \right)_{m^*}$$

which, in turn, equals the limit of the expression

$$\frac{1}{2}\left[d(\gamma_{\widehat{m}}(t_j), \gamma_{\widehat{m}}(0)) + d\left(\gamma_{\widehat{m}}\left(\varepsilon^{-1}\log\frac{D}{r}\right), \gamma_{\widehat{m}}(0)\right) - d\left(\gamma_{\widehat{m}}(t_j), \gamma_{\widehat{m}}\left(\varepsilon^{-1}\log\frac{D}{r}\right)\right) \right]$$

as $t_j \to \infty$.

By the definition of a $\tilde{k}$-roughly geodesic, see (5.64), this expression may be written for large t_j as

$$\frac{1}{2}\left[t_j + \varepsilon^{-1}\log\frac{D}{r} - \left(t_j - \varepsilon^{-1}\log\frac{D}{r}\right)\right] + O_{\tilde{k}}(1) = \varepsilon^{-1}\log\frac{D}{r} + O_{k,\delta}(1),$$

as required.

Now let $m := f(B_r(\widehat{m}))$, $m' := f(B_{r'}(\widehat{m}))$ be arbitrary points of the image $f(\partial_\infty \mathcal{M})$. Then, as before,

$$d(m, m^*) := d\left(\gamma_{\widehat{m}}\left(\varepsilon^{-1}\log\frac{D}{r}\right), \gamma_{\widehat{m}}(0)\right)$$

$$\leq \varepsilon^{-1}\log\frac{D}{r} + \tilde{k} \leq (\widehat{m}|m)_{m^*} + O_{k,\delta}(1),$$

and the same holds for $d(m', m^*)$.

Hence, the assumptions of Proposition 5.26 (b) hold for the points m, m', m^* and $\widehat{m}, \widehat{m}'$. This yields

$$d\big(f(B_r(\widehat{m})), f(B_{r'}(\widehat{m}'))\big) = d(m, m')$$
$$= \varepsilon^{-1}\log\frac{D}{r} + \varepsilon^{-1}\log\frac{D}{r'} - 2\min\left\{ (\widehat{m}|\widehat{m}')_{m^*},\ \varepsilon^{-1}\log\frac{D}{r},\ \varepsilon^{-1}\log\frac{D}{r} \right\} + O_{k,\delta}(1).$$

Now, $(\widehat{m}|\widehat{m}')_{m^*} = -\varepsilon^{-1}\log d_\varepsilon(\widehat{m}, \widehat{m}') + O(1)$ for a sufficiently small ϵ, see (5.71), and therefore the right-hand side equals

$$\varepsilon^{-1}\left[-\log r - \log r' + 2\log\max\{d_\varepsilon(\widehat{m}, \widehat{m}'), r, r'\}\right] + O_{k,\delta}(1).$$

Finally, using the inequality $\frac{a+b}{2} \leq \max\{a,b\} \leq a+b$, we obtain

$$d\big(f(B_r(\widehat{m})), f(B_{r'}(\widehat{m}'))\big) = 2\varepsilon^{-1}\log\frac{d_\varepsilon(\widehat{m}, \widehat{m}') + \max\{r, r'\}}{\sqrt{rr'}} + O_{\delta,k}(1)$$
$$= \varepsilon^{-1}\rho_{d_\varepsilon}\big(B_r(\widehat{m}), B_{r'}(\widehat{m}')\big) + O_{k,\delta}(1).$$

This proves that f is $(c, \bar{k})$-similar for $c := \varepsilon^{-1}$ and $\bar{k} := O_{k,\delta}(1)$.

To obtain the desired result it remains to show that the image $f(\mathcal{B}^*(\partial_\infty \mathcal{M}))$ is co-bounded in $\mathcal{M}$. Since $\mathcal{M}$ is k-visual and δ-hyperbolic, given $m \in \mathcal{M}$, there exists a $\tilde{k} = \tilde{k}(\delta, k)$-roughly geodesic ray $\gamma : \mathbb{R}_+ \to \mathcal{M}$ emanating from the basepoint m^* and passing through m. Let $\widehat{m} \in \partial_\infty \mathcal{M}$ be the endpoint of γ and $[\widehat{m}]$ be the equivalence class of all $\tilde{k}$-roughly geodesic rays emanating from m^* with the endpoint $\widehat{m}$. Since γ and $\gamma_{\widehat{m}}$ from the construction of the map f, see (5.80), belong to $[\widehat{m}]$, Corollary 5.27 implies that $d(m, \gamma_{\widehat{m}}(\mathbb{R}_+)) \leq c(k, \delta)$. Hence the distance from m to $\gamma_{\widehat{m}}(\mathbb{R}_+) \subset f(\mathcal{B}^*(\partial_\infty \mathcal{M}))$ is bounded by $c(k, \delta)$ and the image of f is co-bounded.

The result is established. □

Now we prove a special case of Theorem 5.36 formulated as

Proposition 5.42. *Let $(\mathcal{M}, d)$ be a k-visual δ-hyperbolic space of bounded growth in some scale. Assume that $\partial_\infty \mathcal{M}$ contains more than one point. Then $\mathcal{M}$ is roughly similar to a convex subset of some $\mathbb{H}^n$.*

Proof. By Propositions 5.24 and 5.41 for a sufficiently small $\epsilon > 0$,

$$(\mathcal{M}, d) \simeq \mathcal{B}^*(\partial_\infty \mathcal{M}, d_\varepsilon). \tag{5.83}$$

Further, due to Theorem 5.31 and Corollary 5.32, $(\partial_\infty \mathcal{M}, d_\varepsilon)$ is doubling and compact; Assouad's Theorem 5.10 then asserts that $(\partial_\infty \mathcal{M}, \sqrt{d_\varepsilon})$ admits a bi-Lipschitz embedding into some Euclidean space $(\mathbb{R}^n, \|\cdot\|)$. Let $(S, \|\cdot\|)$ be a compact metric subspace of $\mathbb{R}^n$ which is bi-Lipschitz equivalent to $(\partial_\infty \mathcal{M}, \sqrt{d_\varepsilon})$. We prove that

$$\mathcal{B}^*(\partial_\infty \mathcal{M}, d_\varepsilon) \simeq \mathcal{B}^*(\partial_\infty \mathcal{M}, \sqrt{d_\varepsilon}) \simeq \mathcal{B}^*(S, \|\cdot\|). \tag{5.84}$$

The first equivalence is given by a map $g : \mathcal{B}(\partial_\infty \mathcal{M}) \to \mathcal{B}(\partial_\infty \mathcal{M})$ defined by

$$g(B_r(\widehat{m})) := B_{\sqrt{r}}(\widehat{m}).$$

Actually, denoting the metric of $\mathcal{B}^*(\partial_\infty \mathcal{M}, d_\varepsilon)$ by ρ and that of $\mathcal{B}^*(\partial_\infty \mathcal{M}, \sqrt{d_\varepsilon})$ by $\bar{\rho}$, see (5.78), and using the inequality $a + b \leq (\sqrt{a} + \sqrt{b})^2 \leq 2(a+b)$, we have

$$\bar{\rho}\big(g(B_r(\widehat{m})), g(B_{r'}(\widehat{m}'))\big) := 2 \log \frac{\sqrt{d_\varepsilon(\widehat{m}, \widehat{m}')} + \sqrt{\max(r, r')}}{\sqrt{\sqrt{r_1} \cdot \sqrt{r_2}}}$$

$$= \log \frac{d_\varepsilon(\widehat{m}, \widehat{m}') + \max(r, r')}{\sqrt{r_1 r_2}} + O(1) = \frac{1}{2}\rho(B_r(\widehat{m}), B_{r'}(\widehat{m}')) + O(1).$$

Since g is a bijection, the first equivalence in (5.84) is proved.

Let now $h : (\partial_\infty \mathcal{M}, \sqrt{d_\varepsilon}) \to (S, \|\cdot\|)$ be a bijection satisfying, for some $C \geq 1$ and all $\widehat{m}, \widehat{m}' \in \partial_\infty \mathcal{M}$, the condition

$$C^{-1}\sqrt{d_\varepsilon(\widehat{m}, \widehat{m}')} \leq \|h(\widehat{m}) - h(\widehat{m}')\| \leq C\sqrt{d_\varepsilon(\widehat{m}, \widehat{m}')}. \tag{5.85}$$

5.2. Roughly similar embeddings of Gromov hyperbolic spaces

Define a map $H : \mathcal{B}^*(\partial_\infty \mathcal{M}, \sqrt{d_\varepsilon}) \to \mathcal{B}^*(S, \|\cdot\|)$ by

$$H(B_r(\hat{m})) := B_r(h(\hat{m})).$$

Denoting the hyperbolic metric of $\mathcal{B}^*(S, \|\cdot\|)$, see (5.78), by ρ_S, we get from (5.85), for all balls $B := B_r(\hat{m}), B' := B_{r'}(\hat{m}')$ in $\mathcal{B}^*(\partial_\infty \mathcal{M}, \sqrt{d_\varepsilon})$:

$$-2\log C + \bar{\rho}(B, B') \leq \rho_S(H(B), H(B')) \leq \bar{\rho}(B, B') + 2\log C.$$

Since H is a bijection, this proves the second equivalence in (5.84).
Now we show that

$$\mathcal{B}^*(S, \|\cdot\|) \simeq (\mathrm{hull}(S), d_g) \subset \mathbb{H}^{n+1}. \tag{5.86}$$

Let us recall that the definition of a convex hull in $\mathbb{H}^{n+1}$ is given in Example 5.34 where it was proved that the space $(\mathrm{hull}(S), d_g)$ is a visual (Gromov hyperbolic) space if card $S > 1$. The latter inequality is true, since S is a bijection of $\partial_\infty \mathcal{M}$ which, by the assumption, contains more than one point. Therefore Proposition 5.41 implies that

$$(\mathrm{hull}(S), d_g) \simeq \mathcal{B}^*(\partial_\infty \mathrm{hull}(S), \overline{d}_\varepsilon),$$

where by $\overline{d}_\varepsilon$ we denote a visual metric associated to d_g, see (5.70). It was also proved in Example 5.34 that

$$\partial_\infty \mathrm{hull}(S) = S$$

if S is compact. The latter is true in our case.
So, it remains to show that for some $\varepsilon > 0$,

$$\mathcal{B}^*(S, \overline{d}_\varepsilon) \simeq \mathcal{B}^*(S, \|\cdot\|). \tag{5.87}$$

Let $\varepsilon > 0$ be chosen so small that inequality (5.71) holds for the visual metric $\overline{d}_\varepsilon$. In this special case, this inequality is as follows:

$$|\varepsilon(x|x')_w + \log \overline{d}_\varepsilon(x, x')| \leq \log 2,$$

where $w \in \mathbb{H}^{n+1}$ is fixed and $x, x' \in S$ are arbitrary.
Further, formula (5.76) of Example 5.33 asserts, for the points x, x',

$$(x|x')_w = -\log \|x - x'\| + O_{w,D}(1),$$

where D is the Euclidean diameter of S. Comparing this with the previous inequality we obtain, for some constant $C = C(D, w) > 1$ and all $x, x' \in S$, the inequality

$$C^{-1}\|x - x'\|^\varepsilon \leq \overline{d}_\varepsilon(x, x') \leq C\|x - x'\|^\varepsilon,$$

meaning that $(S, \overline{d}_\varepsilon)$ and $(S, \|\cdot\|^\varepsilon)$ are bi-Lipschitz equivalent. As in the proof of (5.85), we derive from this that

$$\mathcal{B}^*(S, \|\cdot\|^\varepsilon) \simeq \mathcal{B}^*(S, \overline{d}_\varepsilon).$$

Now we check that

$$\mathcal{B}^*(S, \|\cdot\|^\varepsilon) \simeq \mathcal{B}^*(S, \|\cdot\|). \tag{5.88}$$

Recall that a similar fact was proved for the first equivalence in (5.84) with $\varepsilon = \frac{1}{2}$. The very same argument works in this case too. Actually, the map g there should be now replaced by $g_\varepsilon : B_r(x) \mapsto B_{r^\varepsilon}(x)$, where $x \in S$; then the required relation between the metrics of the spaces in (5.88) denoted by $\bar{\rho}$ and ρ, respectively, is

$$\bar{\rho}\big(g_\varepsilon(B_r(x)), g_\varepsilon(B_{r'}(x'))\big) = \varepsilon \rho\big(B_r(x), B_{r'}(x')\big) + O_\varepsilon(1).$$

Its proof immediately follows from the numerical inequality

$$a + b \leq (a^\varepsilon + b^\varepsilon)^{\frac{1}{\varepsilon}} \leq 2^{\frac{1-\varepsilon}{\varepsilon}}(a+b).$$

This proves (5.88) which, in turn, implies (5.87) and (5.86). Finally, equivalence relations (5.83), (5.84) and (5.86) yield

$$(\mathcal{M}, d) \simeq \big(\mathrm{hull}(S), d_g\big),$$

as required. □

Remark 5.43. Instead of visuality in Proposition 5.42 one can use its weaker form assuming only that the union of all geodesic rays emanating from some point of $\mathcal{M}$ is co-bounded. In fact, visuality of $\mathcal{M}$ is used only for the proof of equivalence (5.83) which, in turn, is a consequence of Proposition 5.41. The proof of this proposition exploits visuality only to check that the image $f(\mathcal{B}^*(\partial_\infty \mathcal{M}))$ is co-bounded. However, this fact easily follows from the above formulated weaker form of visuality.

Now we explain how to derive Theorem 5.36 from its special case presented in Remark 5.43.

Let $(\mathcal{M}, d)$ be a Gromov hyperbolic space of bounded growth in some scale. Since a bounded metric space is roughly similar to a point, we assume that $\mathrm{diam}\,\mathcal{M} = \infty$.

Let $N_R \subset \mathcal{M}$ be a maximal $5R$-net. At each point $m_0 \in N_R$ we glue to $\mathcal{M}$ an isometric copy of the real-half line $[0, +\infty)$ by identifying m_0 with the initial point of the copy of $[0, +\infty)$. Since N_R is a proper subspace of $\mathcal{M}$ (every closed ball is compact), there is a unique geodesic metric on this new space denoted by $\widehat{\mathcal{M}}$, which agrees with d on $\mathcal{M}$ and with the Euclidean metric on each copy of $[0, +\infty)$, see Example 3.54 (d). We identify $\widehat{\mathcal{M}}$ with the set of pairs (m, t) where

5.2. Roughly similar embeddings of Gromov hyperbolic spaces

$m \in \mathcal{M}$ and $t = 0$ if $m \in N_R$ and $t \in \mathbb{R}_+$ if $m \in \mathcal{M} \setminus N_R$, and denote the geodesic metric on $\widehat{\mathcal{M}}$ by $\widehat{d}$.

To verify that the geodesic space $(\widehat{\mathcal{M}}, \widehat{d})$ is Gromov hyperbolic we use Rips' Definition 3.114 of this concept. According to this definition and the assumption on $\mathcal{M}$ every geodesic triangle in $\mathcal{M}$ is δ-slim, $\delta \geq 0$, i.e., the distance from every point of one of its sides to the union of two others is at most δ, and we must prove the same for geodesic triangles in $\widehat{\mathcal{M}}$. Let $\widehat{m}_i := (m_i, t_i)$ be the vertices of this geodesic triangle $\widehat{\Delta}$, $i = 1, 2, 3$. Then the side $[\widehat{m}_i \widehat{m}_j] \subset \widehat{\Delta}$, $i \neq j$, is the union of a geodesic segment $[(m_i, 0), (m_j, 0)]$ and two line segments $I_k := [(m_k, 0), (m_k, t_k)]$, $k \in \{i, j\}$. Since each of these line segments belongs simultaneously to two sides of $\widehat{\Delta}$ and the triangle with vertices $(m_i, 0)$, $i = 1, 2, 3$, is δ-slim, the triangle $\widehat{\Delta}$ is also δ-slim.

To show that the Gromov hyperbolic space $\widehat{\mathcal{M}}$ is of bounded growth at some scale, we denote by $0 < r < R$ and $N \in \mathbb{N}$ the parameters of this characteristic for $(\mathcal{M}, d)$. If an open ball $B_R((m, t)) \subset \widehat{\mathcal{M}}$ does not contain a point of N_R, then either $t = 0$ and $B_R((m, t))$ is isometric to $B_R(m)$ or $m \in N_R$ and $B_R((m, t))$ belongs to the ray in $\widehat{\mathcal{M}}$ attached to m. In the first case, by our assumption, the ball $B_R((m, 0)) \subset \mathcal{M}$ can be covered by at most N open balls of radius r lying in $\mathcal{M}$, while in the second one the ball $B_R((m, t))$ can be covered by $\lfloor \frac{R+1}{r} \rfloor$ open balls of radius r in $\widehat{\mathcal{M}}$. Thus $B_R((m, t))$ can be covered by at most $\max\{N, \lfloor \frac{R+1}{r} \rfloor\}$ open balls of radius r of $\widehat{\mathcal{M}}$.

Suppose now that an open ball $B_R((m, t)) \subset \widehat{\mathcal{M}}$ contains at most one point, say m_0, of the maximal $5R$-net N_R. Therefore the ball belongs to the union of the open ball $B_R((m_0, 0))$ isometric to $B_R(m_0) \subset \mathcal{M}$ and the line interval $I := \{(m_0, t) \,;\, 0 \leq t < 2R\}$. To cover $B_R((m, t))$, it suffices to add to N open balls of radius r covering $B_R((m, 0))$ new open balls of radius r covering I. The number of the new balls is clearly bounded by $\lfloor \frac{R+1}{r} \rfloor$. Hence $(\widehat{\mathcal{M}}, \widehat{d})$ is of bounded growth at some scale.

Further, each half-line $\{(m_0, t) \,;\, t \in \mathbb{R}_+\}$ glued at $m_0 \in N_R$ is a geodesic ray which determines a point in $\partial_\infty \widehat{\mathcal{M}}$ and therefore

$$\operatorname{card} \partial_\infty \widehat{\mathcal{M}} \geq \operatorname{card} N_R = \infty.$$

Finally, every point of $\widehat{\mathcal{M}}$ either lies in a geodesic ray $\{m_0\} \times \mathbb{R}_+$ where $m_0 \in N_R$ or is at distance at most $5R$ from a point $(m_0, 0)$, where $m_0 \in N_R$. Hence, the distance from every point of $\widehat{\mathcal{M}}$ to the union of geodesic rays is at most $5R$, i.e., this union is $5R$-co-bounded in $\widehat{\mathcal{M}}$.

Summarizing, we conclude that the Gromov hyperbolic space $(\widehat{\mathcal{M}}, \widehat{d})$ satisfies the assumptions of the weak visuality and nontriviality of $\partial_\infty \widehat{\mathcal{M}}$ formulated in Remark 5.43. Hence, there is a convex subset $\widehat{W}$ in some $(\mathbb{H}^n, d_g)$ such that

$$(\widehat{\mathcal{M}}, \widehat{d}) \simeq (\widehat{W}, d_g).$$

Let $f : \widehat{\mathcal{M}} \to \widehat{W}$ be the corresponding (c, k)-roughly similar map, $c \geq 1, k \geq 0$, and let $W := f(\mathcal{M}) \subset \widehat{W}$. Then

$$(\mathcal{M}, d) \simeq (W, d_g),$$

and it remains to show that W is roughly similar to $\operatorname{hull}(W)$.

Due to the proof, W is co-bounded in the set consisting of the union of all geodesic segments with endpoints at W. Then the result of Example 5.35 implies that $W \simeq \operatorname{hull}(W)$.

The proof of Theorem 5.36 is complete. $\square$

5.3 Lipschitz selections

We return to the problem of Section 1.9 concerning the existence of Lipschitz selections for maps from a metric space into the space $\mathcal{C}(X)$ of convex subsets of a Banach space X. The answer, in general, is negative for $\dim X = \infty$, see Theorem 1.20, but positive for $\dim X < \infty$ as we will see below. As in Section 1.9, we consider two versions of the problem, geometric and combinatorial.

5.3.1 Barycenter and Steiner selectors

Let F be a map from a metric space $(\mathcal{M}, d)$ into the space $\mathcal{C}_\mathcal{H}(X)$ of convex compact subsets of a Banach space X equipped with the Hausdorff metric $d_\mathcal{H}$. In this case, $d_\mathcal{H}$ may be defined by the formula

$$d_\mathcal{H}(S_0, S_1) := \inf\{r \geq 0 \,;\, S_i \subset [S_{1-i}]_r, \ i = 0, 1\},$$

where $[S]_r$ is a closed rounded neighborhood of S given by

$$[S]_r := S + \overline{B}_r(0). \tag{5.89}$$

For convex S, the arithmetic (Minkowski) sum of S and a closed ball centered at 0 is called a *parallel body of S*.

Problem. *Under what conditions on $F : \mathcal{M} \to \mathcal{C}_\mathcal{H}(X)$ does a Lipschitz selection exist? The latter is a function $s : \mathcal{M} \to X$ satisfying the conditions:*

(a) *s is a selection, i.e.,*

$$s(m) \in F(m) \quad \text{for every} \quad m \in \mathcal{M};$$

(b) *s is Lipschitz.*

In the sequel, we assume that $\dim X < \infty$, if the opposite is not stated explicitly. We identify X with $\mathbb{R}^n$ equipped with the norm denoted by $\|\cdot\|_X$.

5.3. Lipschitz selections

Assume for the moment that there exists a Lipschitz map $\varphi : \mathcal{C}_\mathcal{H}(X) \to X$, such that $\varphi(V) \in V$ for every $V \in \mathcal{C}_\mathcal{H}(X)$. If also $F : \mathcal{M} \to \mathcal{C}_\mathcal{H}(X)$ is Lipschitz, then the composition $\varphi \circ F$ is clearly the required Lipschitz selection for F. To find such φ, it would be natural to consider some barycenter type constructions.

Let us recall that the *barycenter* of a bounded *convex body* $C \subset \mathbb{R}^n$, i.e., a convex set with nonempty interior, is defined by

$$b(C) := \frac{1}{\lambda_n(C)} \int_C x d\lambda_n; \tag{5.90}$$

here λ_n is the Lebesgue n-measure of $\mathbb{R}^n$. If, however, the interior $C^\circ = \emptyset$, this formula should be modified by replacing λ_n by the Lebesgue measure on the *affine span* of C, the smallest affine subspace of $\mathbb{R}^n$ containing C.

The map $b : \mathcal{C}_\mathcal{H}(X) \to X$ is such that $b(C) \in C$ for each $C \in \mathcal{C}_\mathcal{H}(X)$. In fact, if $b(C) \notin C$, then, (because C is compact) by the Hahn–Banach separation theorem, there exists an affine function $a : X \to \mathbb{R}$ such that $\sup_C a < a(b(C))$. This and (5.90) then lead to a contradiction, since

$$a(b(C)) = \frac{1}{\lambda_n(C)} \int_C a(x) d\lambda_n \le \sup_C a.$$

Since the limit of a sequence of convex bodies may be of empty interior, the barycentric selector $b : \mathcal{C}_\mathcal{H}(X) \to X$ is not continuous. For example, the sequence of triangles in the Euclidean plane $\Delta_n := \operatorname{conv}\{(0,0), (1, \frac{1}{n}), (1, -\frac{1}{n})\}$, $n \in \mathbb{N}$, converges to the segment $\Delta_\infty := \operatorname{conv}\{(0,0), (1,0)\}$, while $b(\Delta_n) = (\frac{2}{3}, 0)$ for all n and $b(\Delta_\infty) = (\frac{1}{2}, 0)$.

Being restricted to the subspace of $\mathcal{C}_\mathcal{H}(X)$ consisting of all convex *bodies*, the barycentric selector is continuous but, however, is not Lipschitz (take, e.g., $\widetilde{\Delta}_n := \operatorname{conv}\{(0, \frac{1}{n}), (0, -\frac{1}{n}), (1, 0)\}$ to obtain $d_\mathcal{H}(\widetilde{\Delta}_n, \Delta_n) = \frac{1}{n}$ and $\|b(\widetilde{\Delta}_n) - b(\Delta_n)\| = \frac{1}{3}$).

Nevertheless, a generalized barycenter selector $b_\mu : \mathcal{C}_\mathcal{H}(X) \to X$ given by

$$b_\mu(C) := \int x d\mu(x),$$

where μ is a suitably chosen probabilistic measure supported on C, may be Lipschitz. Of course, μ should be strongly related to the geometric structure of C. One of the possible choices of μ is the surface measure of C, in which case the corresponding barycenter is denoted by $b(\partial C)$. Hence,

$$b(\partial C) := \frac{1}{\sigma(C)} \int_{\partial C} x d\sigma,$$

where σ is the restriction of the Lebesgue n-measure λ_n to ∂C. This selector is not yet Lipschitz but is continuous on a wider subspace of $\mathcal{C}_\mathcal{H}(X)$ than that for $b(C)$.

To motivate the choice of μ giving the required Lipschitz selection b_μ, we consider a formula for the parallel body of a convex polygon, see (5.89), given by Steiner [S-1840].

Let $P \subset \mathbb{R}^2$ be a polygon with vertices v_i, $1 \leq i \leq k$, area $|P|$ and length of its boundary (perimeter) $\ell(P)$. Then

$$b([P]_r) = \frac{|P|b(P) + 2\ell(P)rb(\partial P) + \pi r^2 s(P)}{|P| + 2\ell(P)r + \pi r^2}$$

where $s(P)$, called the *Steiner point* of P, is given by

$$s(P) = \sum_{i=1}^k \frac{\pi - \alpha_i}{2\pi} v_i,$$

and α_i is the angle at v_i, $1 \leq i \leq k$.

Note that $\sum_{i=1}^k \frac{\pi - \alpha_i}{2\pi} = 1$, i.e., $s(P)$ is a generalized barycenter of P; therefore, $s(P) \in P$.

It is easy to derive a similar result for a convex polytope P in the Euclidean n-space with vertices v_i, $1 \leq i \leq k$. In this case, the corresponding Steiner point $s_n(P)$ is given by

$$s_n(P) = \sum_{i=1}^k \frac{\omega_i}{\sigma_n} v_i,$$

where σ_n is the area of the sphere $\mathbb{S}^{n-1}$ and ω_i is the spherical angle of the cone formed by the normals to the facets of P incident to v_i. It is easily seen that $\sum_{i=1}^k \omega_i = \sigma_n$. So, $s_n(P)$ is a generalized barycenter and $s_n(P) \in P$. Grünbaum [G-1963] discovered a presentation of $s_n(P)$ via the *support function* of P which is more suitable for our aim. Let us recall that the support function $h_S : E_n \to \mathbb{R} \cup \{+\infty\}$ of a subset S of an n-dimensional Euclidean space E_n is given by

$$h_S(x) := \sup\{x \cdot y\,;\, y \in S\} \qquad (5.91)$$

where $x \cdot y$ is a fixed scalar product of E_n. The Grünbaum formula was then extended to convex subsets of E_n by Sheppard [Sh-1966], and now is used as the definition of the Steiner point.

Definition 5.44. Let C be a convex compact subset of E_n. The Steiner point of C denoted by $s_n(C)$ is given by

$$s_n(C) := \frac{1}{|B|} \int_{\partial B} x h_C(x) d\sigma, \qquad (5.92)$$

where $B := \{x \in E_n\,;\, x \cdot x \leq 1\}$ is the unit ball and $|B|$ stands for its volume.

5.3. Lipschitz selections

To derive the required properties of the map $s_n : \mathcal{C}_\mathcal{H}(E_n) \to E_n$, we will use some properties of the support function, see, e.g., the Rockafellar book [Rock-1970] for proofs.

Proposition 5.45. (Additivity) *Let S, S' be subsets of E_n and λ, λ' be real numbers. It is true that*

$$h_{\lambda S + \lambda' S'} = \lambda h_S + \lambda' h_{S'}. \tag{5.93}$$

(Invariance) *Let T be a linear operator acting in E_n, and let T^* be its conjugate (i.e., $Tx \cdot y = x \cdot T^* y$ for all $x, y \in E_n$). Then the equality*

$$h_{T(S)}(x) = h_S(T^* x) \tag{5.94}$$

holds for all $x \in E_n$.

(Differentiability) *Let C be a convex subset of E_n. Then h_C is differentiable almost everywhere and its gradient ∇h_C belongs to C at every existence point.*

Note that (5.93) and (5.94) follow straightforwardly from the definition of h_S, see (5.91). It also follows from here that h_S is Lipschitz for bounded S. Hence, its almost everywhere differentiability is a consequence of Rademacher's theorem, see Theorem 4.61. So, the only nontrivial (but simple) fact is the inclusion $\nabla h_C \in C$.

Now we formulate and prove the required result, see Aubin and Frankowska [AF-1990, Thm. 4.6.1] and Przeslawski and Yost [PY-1989, Thm. 6.1] for other variants of the proof.

Theorem 5.46. *The map $s_n : \mathcal{C}_\mathcal{H}(E_n) \to E_n$ is a Lipschitz selector whose Lipschitz constant is bounded by n.*

Proof. First show that $s_n(C) \in C$. Since h_C is Lipschitz, its gradient ∇h_C is an L_∞-function. Regarding every point of the unit sphere ∂B as an outward normal of ∂B, we apply the Stokes formula to the right-hand side of (5.92) to obtain

$$s_n(C) := \frac{1}{|B|} \int_B \nabla h_C \, d\lambda_n.$$

As ∇h_C belongs to C for almost all points of B, the integral in the right-hand side also belongs to C. Hence, s_n is a selector.

To estimate the Lipschitz constant

$$L(s_n) := \sup \left\{ \frac{\|s_n(C) - s_n(C')\|}{d_\mathcal{H}(C, C')} \,;\, C, C' \in \mathcal{C}_\mathcal{H}(E_n) \right\},$$

we need the inequality

$$\sup_{x \in B} \left| h_C(x) - h_{C'}(x) \right| \leq d_\mathcal{H}(C, C'). \tag{5.95}$$

Let $r \geq 0$ be such that
$$[C]_r \supset C' \quad \text{and} \quad [C']_r \supset C. \tag{5.96}$$

Due to the additivity of h_C we then have
$$0 \leq h_{[C]_r}(x) - h_{C'}(x) = h_C(x) - h_{C'}(x) + rh_B(x) = h_C(x) - h_{C'}(x) + r\|x\|.$$

Then, for $\|x\| \leq 1$, we get
$$-r \leq h_C(x) - h_{C'}(x).$$

Changing the positions of C and C' we then bound this difference by r from above. Hence,
$$\sup_{x \in B} |h_C(x) - h_{C'}(x)| \leq r,$$

and taking the infimum over all r satisfying (5.96) we prove (5.95).

Now, applying (5.95), we have
$$|s_n(C) - s_n(C')| \leq \frac{1}{|B|} \int_{\partial B} \sup_B |h_C(x) - h_{C'}(x)| \cdot \|x\| d\sigma$$
$$\leq d_{\mathcal{H}}(C, C') \frac{1}{|B|} \int_{\partial B} d\sigma = n d_{\mathcal{H}}(C, C').$$

The result is established. $\square$

The inequality $L(s_n) \leq n$ may be essentially improved. The following result was proved by Posicelski [Pos-1971] and rediscovered by Vitale [Vit-1985], see, e.g., [BL-2000, Ch. 3] for the proof.

Theorem 5.47. (a) *It is true that*
$$L(s_n) = \gamma_n := \frac{2}{\sqrt{\pi}} \Gamma\left(\frac{n}{2} + 1\right) \Big/ \Gamma\left(\frac{n+1}{2}\right).$$

(b) *For every Lipschitz selector* $s : \mathcal{C}_{\mathcal{H}}(E_n) \to E_n$ *its Lipschitz constant satisfies*
$$L(s) \geq \gamma_n.$$

Applying the Stirling formula we obtain $\gamma_n \sim \sqrt{\frac{2n}{\pi}}$ as $n \to \infty$. Hence this result considerably improves the estimate $L(s_n) \leq n$.

As a consequence we have

Corollary 5.48. *Let F be a Lipschitz map from a metric space $(\mathcal{M}, d)$ into the space $\mathcal{C}_{\mathcal{H}}(X)$ of convex compact subsets on an n-dimensional normed space X. Then there is a selection $s : \mathcal{M} \to X$ of the map F such that*
$$L(s) \leq \gamma_n \sqrt{n} L(F). \tag{5.97}$$

5.3. Lipschitz selections

Proof. Show that $s := s_n \circ F$ is the required selection. To this end, we use the analytic form of the John theorem [Jo-1948, Thm. 5.6(i)] which asserts the existence of a Euclidean norm, say $\|\cdot\|$, on X such that for all $x \in X$,

$$\|x\| \leq \|x\|_X \leq \sqrt{n}\|x\|. \tag{5.98}$$

Using the second inequality and Theorem 5.47 we then obtain

$$L(s) := \sup_{m \neq m'} \frac{\|s_n(F(m)) - s_n(F(m'))\|_X}{d(m,m')} \leq \sqrt{n}\,\gamma_n \sup_{m \neq m'} \frac{\widetilde{d}_{\mathcal{H}}(F(m), F(m'))}{d(m,m')},$$

where $\widetilde{d}_{\mathcal{H}}$ is the Hausdorff metric on $(X, \|\cdot\|)$. Since $\widetilde{d}_{\mathcal{H}} \leq d_{\mathcal{H}}$ by the first inequality of (5.98), the right-hand side is bounded by $\sqrt{n}\,\gamma_n L(F)$, as required. □

Now we briefly discuss a possible generalization to infinite-dimensional Banach spaces. First, let X be a *Hilbert space of infinite dimension*. The second part of Theorem 5.47, see also Theorem 1.20, shows that there are no Lipschitz selections for the set-valued maps from $\mathcal{C}_{\mathcal{H}}(X)$ into X. Therefore we restrict our discussion to the subspace $\mathcal{C}_{\mathcal{H}}^{\text{fin}}(X) \subset \mathcal{C}_{\mathcal{H}}(X)$ consisting only of *finite-dimensional* convex subsets.

Problem. *Does there exist a selector* $s : \mathcal{C}_{\mathcal{H}}^{\text{fin}}(X) \to X$ *satisfying, for all* $C, C' \in \mathcal{C}_{\mathcal{H}}^{\text{fin}}(X)$, *the inequality*

$$\|s(C) - s(C')\|_X \leq \gamma d_{\mathcal{H}}(C, C') \tag{5.99}$$

where γ depends only on the dimensions of C and C'?

The answer is positive and the required selector may be defined by a version of formula (5.92), with E_n replaced by the linear span of C and with n equal to $\dim C$. The basic point of the proof exploits the following uniqueness theorem proved independently by Meyer [Mey-1970] and Posicelski [Pos-1973]. The elegant Posicelski proof is presented in the book [BL-2000, Thm. 3.18].

Theorem 5.49. *Let* $s : \mathcal{C}_{\mathcal{H}}(E_n) \to E_n$ *be a continuous additive function which commutes with affine isometries*[6] *of E_n. Then $s = s_n$.*

We derive from here the following.
Let H be a Euclidean subspace of the Hilbert space X containing C. Set

$$s_H(C) := \frac{1}{|B|} \int_{\partial B} x h_C(x) dx,$$

where B is the unit ball of H.

Using Proposition 5.45 we conclude that the map $s_H : \mathcal{C}_{\mathcal{H}}(H) \to H$ is continuous, additive and invariant with respect to affine isometries of H. Since affine

---
[6] combinations of shifts, mirror reflections and rotations.

isometries of a subspace $H' \subset H$ naturally generate affine isometries of H, the restriction of s_H to $\mathcal{C}_\mathcal{H}(H')$ is also continuous, additive and invariant with respect to affine isometries of H'. Applying Theorem 5.49 we therefore obtain, for $C \in \mathcal{C}_\mathcal{H}(H)$,

$$s_H(C) = s_{\text{span }C}(C) \; (=: s_X(C)). \tag{5.100}$$

Now let C, C' be subsets from $\mathcal{C}_\mathcal{H}^{\text{fin}}(X)$, and $H := \text{span}(C \cup C')$. Applying (5.100) and Theorem 5.47 we then have

$$\left\| s_X(C) - s_X(C') \right\|_X = \left\| s_H(C) - s_H(C') \right\|_H$$
$$\leq \gamma_n d_\mathcal{H}(C, C'; H) = \gamma_n d_\mathcal{H}(C, C'),$$

where $n := \dim\bigl[\text{span}(C \cup C')\bigr] \leq \dim C + \dim C' + 2$.

This proves (5.99).

Passing to the Banach case we immediately note that equality (5.100) is not true (together with several other elements of the above proof). For example, the Steiner point of the triangle $\Delta := \text{conv}\{(0,0,0),(1,1,0),(0,0,1)\}$, regarded as a subset of ℓ_1^3, differs from that for Δ regarded as a subset of the affine subspace of ℓ_1^3 generated by Δ, see the corresponding computation in [PY-1989, p. 128]. Therefore formula (5.93) cannot be used to construct a Lipschitz selector from $\mathcal{C}_\mathcal{H}^{\text{fin}}(X)$ to X.

Shvartsman [Shv-2004] proposed a more involved construction for this case in order to define an analog of the Steiner point. His result asserts the following.

Theorem 5.50. *Let X be a Banach space of infinite dimension. There exists a map $s : \mathcal{C}_\mathcal{H}^{\text{fin}}(X) \to X$ satisfying the conditions:*

(a) *For all C, C',*
$$\|s(C) - s(C')\|_X \leq \gamma d_\mathcal{H}(C, C'),$$
where γ depends only on dimensions of C and C'.

(b) *s is invariant with respect to homotheties*
$$s(\lambda C + x) = \lambda s(C) + x, \quad \lambda \in \mathbb{R}, \; x \in X;$$
here $\lambda C := \{\lambda x \, ; \, x \in C\}$.

5.3.2 Helly type result: a conjecture

The conjecture appears to be the basic geometric ingredient for a putative solution of the Finiteness Problem for the traces of Lipschitz functions of higher order to arbitrary subsets of $\mathbb{R}^n$ (see Conjecture 2.60 in subsection 2.4.4). It is formulated as follows.

Let F map a metric space $(\mathcal{M}, d)$ into the set of convex compact subsets $\mathcal{C}(\mathbb{R}^n)$.

5.3. Lipschitz selections

Conjecture 5.51 (Yu. Brudnyi, 1985). *Assume that the trace $F|_S$, for every subset $S \subset \mathcal{M}$ of cardinality 2^n, admits a selection with the Lipschitz constant at most 1. Then F has a C-Lipschitz selection where the constant $C = C(n) \geq 1$ (i.e., depends only on n).*

The constant 2^n is unusual for Helly type results, it first appeared in an attempt to prove the conjecture using an inductive procedure doubling the finiteness constant at each step. Since it equals 2 for $n = 1$ (see Proposition 1.24 in subsection 1.9.2), this gives 2^n for $\mathbb{R}^n$. In spite of the shaky argument, the value 2^n appears to be the only one possible, as follows from a result due to Shvartsman [Shv-2002].

Theorem 5.52. (a) *Assume that the conjecture is true for the finiteness constant $N > 2^n$. Then it is also true for 2^n.*

(b) *There exists a metric space $(\mathcal{M}_0, d_0)$ and a map $F_0 : \mathcal{M}_0 \to \mathcal{C}(\mathbb{R}^n)$ such that the assumption of the conjecture is true for the finiteness constant $2^n - 1$, but F_0 has no Lipschitz selection.*

Proof. (a) We derive the required result from

Proposition 5.53. *Let $(\widetilde{\mathcal{M}}, \widetilde{d})$ be a finite metric space consisting of at least $2^n + 1$ points. Assume that a map $F : \widetilde{\mathcal{M}} \to \mathcal{C}(\mathbb{R}^n)$ satisfies the assumption of the conjecture with the finiteness constant $-1 + \operatorname{card} \widetilde{\mathcal{M}}$. Then F admits a C-Lipschitz selection with a constant $C > 1$ depending only on n and $\operatorname{card} \widetilde{\mathcal{M}}$.*

Before proving the proposition, we explain how to derive assertion (a) from it. Let S be an arbitrary subspace of $\mathcal{M}$ of cardinality $2^n + 1$. By assumption, the trace of F to every subspace of S consisting of 2^n points has a 1-Lipschitz selection. Applying the proposition we conclude that $F|_S$ has a c_1-Lipschitz selection with some $c_1 = c_1(n) > 1$.

Further, assume that $2^n + 2 \leq N$ and consider an arbitrary subspace $S \subset \mathcal{M}$ consisting of $2^n + 2$ points. Due to the previous step $F|_{S'}$, for every $S' \subset S$ of cardinality $2^n + 1$, admits a $c_1(n)$-Lipschitz selection. Then the proposition implies that for some constant $c_2 = c_2(n) \geq 1$, the trace $F|_S$ admits a c_3-Lipschitz selection where $c_3 := c_1(n)c_2(n)$. Proceeding this way we finally conclude that for some constant $c(n) > 1$ and every subspace $S \subset \mathcal{M}$ of cardinality N the trace $F|_S$ has a $c(n)$-Lipschitz selection. By the assumption of Theorem 5.52 (a), this implies that F itself has the required Lipschitz selection.

Proof of Proposition 5.53. As we will show, the desired Lipschitz selection can be relatively easily found by exploiting Helly's theorem, see Theorem 1.22, for $\widetilde{\mathcal{M}}$ being a metric tree which has a vertex of degree at least $n + 1$. Therefore we impose on $\widetilde{\mathcal{M}}$ the required tree structure using

Lemma 5.54. *There exists a metric tree (T, ρ) and a bi-Lipschitz map of $(\widetilde{\mathcal{M}}, \widetilde{d})$ onto this tree such that:*

(i) *Distortion of this map is bounded by a constant depending only on* card $\widetilde{\mathcal{M}}$.

(ii) *There exists a vertex* $v \in T$ *whose degree satisfies*

$$\deg v \geq \lceil \log_2(\operatorname{card} \widetilde{\mathcal{M}}) \rceil ; \tag{5.101}$$

here $\lceil x \rceil$ stands for the smallest integer which bounds x.

Proof. (*Induction on* card $\widetilde{\mathcal{M}}$) The result is trivial for metric spaces of cardinality 1 and 2. Assume that the lemma is true for metric spaces of cardinality $k \geq 2$ and prove it for $k+1$. Without loss of generality, we assume that

$$\operatorname{diam} \widetilde{\mathcal{M}} = 1. \tag{5.102}$$

Let m_0, m_0' be points of $\widetilde{\mathcal{M}}$ such that

$$d(m_0, m_0') = 1. \tag{5.103}$$

By S we denote a subspace of $\widetilde{\mathcal{M}}$ containing m_0 and all points m such that there exists a sequence $\{m_0, m_1, \ldots, m_j\} \subset \widetilde{\mathcal{M}}$ satisfying $m_j = m$ and

$$d(m_i, m_{i+1}) < \frac{1}{k}, \quad i = 0, \ldots, j-1. \tag{5.104}$$

It may be that S consists only of m_0. Otherwise $1 \leq j \leq k$, where the second inequality follows from (5.102) and (5.103). In particular, m_0' does not belong to S, and therefore

$$S' := \widetilde{\mathcal{M}} \setminus S \neq \emptyset.$$

Show that for all $m \in S$ and $m' \in S'$,

$$d(m, m') \geq \frac{1}{k}. \tag{5.105}$$

Let, on the contrary, the converse be true for some $m \in S$ and $m' \in S'$. Join m_0 with m by a sequence $\{m_0, \ldots, m_j = m\}$ satisfying (5.104). All points of the sequence belong to the set S and therefore differ from m'. Then the sequence $\{m_0, m_1, \ldots, m_j, m'\}$ satisfies (5.104). Hence, m' belongs to S, a contradiction.

Since card $S \leq k$, by the induction hypothesis, there exists a tree (T, ρ) whose vertex set we identify with S such that, for all $m, m' \in S$,

$$\rho(m, m') \leq d(m, m') \leq c(k)\rho(m, m') \tag{5.106}$$

and, moreover, for some vertex $\widehat{m} \in S$ its degree satisfies

$$\deg_T m \geq \lceil \log_2(\operatorname{card} S) \rceil . \tag{5.107}$$

Since card $S' \leq k$, the similar inequality is true for S'; the corresponding metric tree with the vertex set S' is denoted by (T', ρ').

5.3. Lipschitz selections

Assume, for definiteness, that

$$\operatorname{card} S \geq \operatorname{card} S'.$$

Then $k+1 = \operatorname{card} \widetilde{\mathcal{M}} \leq 2\operatorname{card} S$ and (5.107) implies that, for some vertex $\widehat{m} \in S$,

$$\deg_T \widehat{m} \geq \lceil \log_2(k+1) \rceil - 1. \tag{5.108}$$

Now we construct the required tree (denoted by $(\widetilde{T}, \tilde{\rho})$) for $(\widetilde{\mathcal{M}}, \tilde{d})$. To this end, fix a point $\widehat{m}' \in S'$ and define the vertex set of $\widetilde{T}$ to be $T \cup T'$, and the edge set to be the union of edges in T and T' and the new edge $[\widehat{m}, \widehat{m}']$. Further, the metric $\tilde{\rho}$ is determined by a weight which equals $\rho(m_1, m_2)$ or $\rho'(m_1, m_2)$ whenever $[m_1, m_2]$ is the edge of T or T', respectively. For the remaining edge we define the weight to be equal $d(\widehat{m}, \widehat{m}')$.

Clearly, $(\widetilde{T}, \tilde{\rho})$ is a metric tree and the degree of $\widehat{m}$ in $\widetilde{T}$ satisfies, by (5.108), the inequality

$$\deg_{\widetilde{T}} \widehat{m} = \deg_T \widehat{m} + 1 \geq \lceil \log_2(k+1) \rceil.$$

This proves part (ii) of the lemma, see (5.101).

Now show that the identity map of $\widetilde{\mathcal{M}}$ onto the vertex set of $\widetilde{T}$ is the required bi-Lipschitz equivalence. First, let $m \in S$ and $m' \in S'$. Then, by definition,

$$\tilde{\rho}(m, m') = \rho(m, \widehat{m}) + d(\widehat{m}, \widehat{m}') + \rho'(\widehat{m}', m'),$$

and the induction hypothesis implies that

$$\tilde{\rho}(m, m') \leq c(k)d(m, \widehat{m}) + d(\widehat{m}, \widehat{m}') + c(k)d(\widehat{m}', m')$$
$$\leq (2c(k) + 1) \operatorname{diam} \widetilde{\mathcal{M}} =: c_1(k).$$

This and (5.105) then yield

$$\tilde{\rho}(m, m') \leq c_1(k) k d(m, m').$$

On the other hand, by the induction hypothesis,

$$\tilde{\rho}(m, m') \geq d(m, \widehat{m}) + d(\widehat{m}, \widehat{m}') + d(\widehat{m}', m') \geq d(m, m').$$

Hence, for $m \in S$ and $m' \in S'$,

$$d(m, m') \leq \tilde{\rho}(m, m') \leq c_1(k) k d(m, m').$$

Now let both m and m' belong either to S or to S'. Then the above inequality follows from the induction hypothesis (with $c(k)$ instead of $c_1(k)k$). Hence, the distortion of the identity map is bounded by $kc_1(k)$.

This proves assertion (i) of the lemma. $\square$

Now we establish Proposition 5.53. By the lemma, we may and will assume that $(\widetilde{\mathcal{M}}, \widetilde{d})$ is a finite metric tree consisting of at least $2^n + 1$ points and having a vertex $\widehat{m}$ satisfying

$$\deg \widehat{m} \geq \lceil \log(\operatorname{card} \widetilde{\mathcal{M}}) \rceil \geq \lceil \log(2^n + 1) \rceil = n + 1.$$

Now, by $I(\widehat{m})$ we denote the family of vertices incident to $\widehat{m}$. Then

$$\operatorname{card} I(\widehat{m}) \geq n + 1. \qquad (5.109)$$

Further, we introduce a family of subtrees $\mathcal{T} := \{T_m \subset \widetilde{\mathcal{M}} ; m \in I(\widehat{m})\}$, where T_m consists of all vertices joined with m by a (unique) path which does not contain $\widehat{m}$. Since there exists a unique path in T_m joining m with each of its points, T_m is a tree rooted at m. Therefore, T_m is a metric subtree of $\widetilde{\mathcal{M}}$.

By definition, $\operatorname{card} T_m < \operatorname{card} \widetilde{\mathcal{M}}$ and, moreover, $\mathcal{T}$ is a *disjoint* family such that the disjoint union of subtrees in $\mathcal{T}$ satisfies

$$\bigsqcup \mathcal{T} = \widetilde{\mathcal{M}} \setminus \{\widehat{m}\}. \qquad (5.110)$$

Now let F be a set-valued map from $\widetilde{\mathcal{M}}$ into $\mathcal{C}(\mathbb{R}^n)$ satisfying the assumptions of the proposition. Then, for every $T \in \mathcal{T}$, the restriction $F|_T$ has a selection $f_T : \widetilde{\mathcal{M}} \to \mathbb{R}^n$ such that $L(f_T) \leq 1$.

Using the family $\{f_T ; T \in \mathcal{T}\}$ we first define the required Lipschitz selection of F, denoted by f, on the set $\widetilde{\mathcal{M}} \setminus \{\widehat{m}\} = \bigsqcup \mathcal{T}$. Actually, we set for $m \in I(\widehat{m})$,

$$f|_{T_m} := f_{T_m}. \qquad (5.111)$$

Then we define f at the remaining point $\widehat{m}$.

To this end, we use the set of all 1-Lipschitz selections of the trace $F|_T$, $T \in \mathcal{T}$, denoted by $\Sigma(T)$. This set is, clearly, a convex and nonempty (as $f_T \in \Sigma(T)$) subset of $\operatorname{Lip}(T, \mathbb{R}^n)$.

Further, for every point $m \neq \widehat{m}$, we denote by $\operatorname{Orb}(m)$ its orbit under the action of $\Sigma(T_m)$. In other words,

$$\operatorname{Orb}(m) := \{x \in \mathbb{R}^n ; x = f(m) \quad \text{for some} \quad \Sigma(T_m)\}. \qquad (5.112)$$

By definition, $\operatorname{Orb}(m)$ is a *nonempty convex subset* of $\mathbb{R}^n$ contained in the set $F(m)$.

Now we define the family $\mathcal{G} := \{G_m ; m \in \{\widehat{m}\} \cup I(\widehat{m})\}$ of nonempty convex subsets of $\mathbb{R}^n$ by setting

$$G_m := \operatorname{Orb}(m) + B_m \qquad (5.113)$$

for $m \in I(\widehat{m})$, and $G_{\widehat{m}} := F(\widehat{m})$ for the remaining point.

Here B_m stands for a closed Euclidean ball in $\mathbb{R}^n$ given by

$$B_m := \{x \in \mathbb{R}^n ; \|x\| \leq 2d(m, \widehat{m})\}.$$

5.3. Lipschitz selections

Prove that

$$\cap \mathcal{G} \neq \emptyset. \tag{5.114}$$

This will follow from the Helly theorem, see Theorem 1.22, if we show that

$$\cap \mathcal{G}' = \emptyset \tag{5.115}$$

for every subfamily $\mathcal{G}' \subset \mathcal{G}$ of cardinality at most $n+1$.

Since $\operatorname{card} \mathcal{G} = 1 + \operatorname{card} I(\widehat{m}) \geq n+2$ and $\operatorname{card} \mathcal{G}' \leq n+1$, there exists a point $\widetilde{m}$ in $\{\widehat{m}\} \cup I(\widehat{m})$ which does not belong to $\mathcal{G}'$. Due to the assumption, the restriction of F to $\widetilde{\mathcal{M}} \setminus \{\widetilde{m}\}$ has a 1-Lipschitz selection $\tilde{f} : \widetilde{\mathcal{M}} \setminus \{\widetilde{m}\} \to \mathbb{R}^n$.

Using this selection, we find a point in the intersection $\cap \mathcal{G}'$ as follows.

Let I' be the subset of vertices in $\{\widehat{m}\} \cup I(\widehat{m})$ associated to the family $\mathcal{G}'$, and let $\overline{m}$ be a point of I' closest to $\widehat{m}$ (it may be that $\overline{m} = \widehat{m}$). Show that $\tilde{f}(\overline{m}) \in \cap \mathcal{G}'$. Since $L(\tilde{f}) \leq 1$ and $d(\widehat{m}, \overline{m}) \leq d(\widehat{m}, m)$ for each $m \in I'$, we have

$$\|\tilde{f}(m) - \tilde{f}(\overline{m})\| \leq d(m, \overline{m}) + d(\widehat{m}, \overline{m}) \leq 2d(\widehat{m}, m).$$

Moreover, $\tilde{f}(m) \in \operatorname{Orb}(m)$ for every $m \in I'$ by the definition of orbits. From here and (5.113) it follows that $\tilde{f}(\overline{m})$ belongs to $\operatorname{Orb}(m) + B_m =: \mathcal{G}_m$ for every $m \in I'$.

This proves (5.115) and, hence, establishes (5.114).

Now we define the value of the required Lipschitz selection f at the point $\widehat{m}$. Actually, (5.114) implies that there are points $p_{\widehat{m}} \in F(\widehat{m})$ and

$$p_m \in \operatorname{Orb}(m) \subset F(m), \quad m \in I(\widehat{m}),$$

such that for all these m,

$$\|p_{\widehat{m}} - p_m\| \leq 2d(\widehat{m}, m). \tag{5.116}$$

By the definition of $\operatorname{Orb}(m)$, we may, without loss of generality, assume that the selections f_{T_m} used in (5.111) satisfy

$$f_{T_m}(m) = p_m, \quad m \in I(\widehat{m}).$$

Now we set

$$f(\widehat{m}) := p_{\widehat{m}}$$

and show that the selection of F defined by this and (5.111) has Lipschitz constant bounded by 2.

First, let m', m'' be points of $\widetilde{\mathcal{M}}$ joined by an edge. If they differ from $\widehat{m}$, then they belong to some subtree T_m and therefore

$$\|f(m') - f(m'')\| := \|f_{T_m}(m') - f_{T_m}(m'')\| \leq d(m', m'').$$

For otherwise, one of them, say m', equals $\widehat{m}$, and (5.116) implies that the previous inequality holds with $2d(m', m'') = 2d(\widehat{m}, m'')$ in the right-hand side.

Finally, let m', m'' be joined by a (unique) path $m' = m_1, \ldots, m_j =: m''$. Then the inequality just proved yields

$$\|f(m') - f(m'')\| \le 2 \sum_{i=1}^{j-1} d(m_i, m_{i+1}).$$

According to the definition of the metric in a graph, see (3.137), the right-hand side equals $2d(m', m'')$.

The proposition and hence assertion (a) of Theorem 5.52 are established. □

(b) We consider only the two-dimensional case, referring to [Shv-2002] for the general situation. So, we must find a metric space $(\mathcal{M}_0, d_0)$ and a map $F_0 : \mathcal{M} \to \mathcal{C}(\mathbb{R}^2)$ such that the restriction of F_0 to every 3-point subset of $\mathcal{M}_0$ admits a 1-Lipschitz selection, while F_0 itself does not admit any selection whose Lipschitz constant is bounded by a fixed real number.

Let $(\mathcal{M}_0, d_0)$ consist of points m_i, $1 \le i \le 4$, and let the metric d_0 be defined by

$$d_0(m_i, m_j) := \begin{cases} \varepsilon & \text{if } (i,j) = (1,2) \text{ or } (3,4), \\ 1 & \text{otherwise}; \end{cases}$$

here $\varepsilon > 0$ is a fixed small number.

Given $N \ge 1$, the points A, B, C, D are the vertices of a rectangle in the Euclidean space $\mathbb{R}^2$ defined by

$$A := (0,1), \ B := (N,1), \ C := (0,0), \ D := (N,0).$$

Now we define a map $F_0 : \mathcal{M} \to \mathcal{C}(\mathbb{R}^2)$ by setting

$$F_0(m_1) := [A,B], \ F_0(m_2) = F_0(m_3) := [A,D] \text{ and } F_0(m_4) := [C,D];$$

here $[x,y]$ stands for the closed interval with the endpoints $x, y \in \mathbb{R}^2$.

Let us show that the restriction of F_0 to every three-point subset of $\mathcal{M}$ has a 1-Lipschitz selection. Actually, we set, for $\mathcal{M}' := \{m_1, m_2, m_3\}$,

$$f_{\mathcal{M}'}(m_i) := A, \quad 1 \le i \le 3,$$

and for $\mathcal{M}' := \{m_2, m_3, m_4\}$,

$$f_{\mathcal{M}'}(m_i) := D, \quad 2 \le i \le 4.$$

In the remaining two cases, we set, for $\mathcal{M}' := \{m_1, m_2, m_4\}$,

$$f_{\mathcal{M}'}(m_1) = f_{\mathcal{M}'}(m_2) := A, \ f_{\mathcal{M}'}(m_4) := C$$

and for $\mathcal{M}' := \{m_1, m_3, m_4\}$,

$$f_{\mathcal{M}'}(m_1) := B, \ f_{\mathcal{M}'}(m_3) = f_{\mathcal{M}'}(m_4) = D.$$

5.3. Lipschitz selections

It is easily seen that all these functions are 1-Lipschitz selections of the corresponding traces. Now assume that F_0 has a Lipschitz selection $f : \mathcal{M} \to \mathbb{R}^2$ whose Lipschitz constant is bounded by some number $c \geq 1$ (independent of N and ε). Then

$$\|f(m_1) - f(m_2)\| \leq cd(m_1, m_2) := c\varepsilon,$$

while $f(m_1) \in [A, B]$ and $f(m_2) \in [A, D]$. Hence, these points lie in the disc $\overline{B}_r(A)$ of radius $r := c\varepsilon \cdot \sqrt{1 + N^2}$ centered at A. Similarly, $f(m_3)$ and $f(m_4)$ lie in the disc $\overline{B}_r(D)$.

This implies that

$$\sqrt{1 + N^2} = \|A - D\| \leq cd(m_2, m_3) + 2r = c + 2c\varepsilon \cdot \sqrt{1 + N^2}.$$

Letting N to ∞ and ε to 0, we see that the conjecture for $n = 2$ is false for the finiteness constant less than 2^2. $\square$

Remark 5.55. Shvartsman [Shv-2002] formulates the following more general conjecture.

By $\mathcal{C}_k(\mathbb{R}^n)$ we denote the set of all k-dimensional convex compact subsets of $\mathbb{R}^n$, $1 \leq k \leq n$.

Conjecture. *Let the restriction of a map $F : \mathcal{M} \to \mathcal{C}_k(\mathbb{R}^n)$ to every subset of cardinality $2^{\min(k+1,n)}$ admit a 1-Lipschitz selection. Then there exists a Lipschitz selection of F with a constant $c > 1$ depending only on n and k.*

Shvartsman proved this conjecture for $n = 2$, $k = 1$ and $n = 2$, $k = 2$, see [Shv-2002].

5.3.3 A Sylvester type selection result

The solution of the finiteness problem for Lipschitz functions of order 2 requires a simpler selection result dealing with set-valued functions into the set $\mathrm{Aff}_k(\mathbb{R}^n)$ of all k-dimensional affine subspaces of $\mathbb{R}^n$, $1 \leq k \leq n - 1$ (cf. the Sylvester–Gallai Theorem 1.23). The corresponding result was proved by Shvartsman [Shv-1986], see also [Shv-1992]. We present a slight generalization of this result which will be used in the aforementioned application. In this case, we deal with a pseudometric space whose distance function may assign values 0 and $+\infty$. In fact, we do not need this for the application, but such pseudometrics will play an essential role in the proof.

Let G be a connected graph with the vertex and edge sets V and E, respectively. We define a distance function on V using a weight $w : E \to [0, +\infty]$ which may be 0 or $+\infty$ at some edges. Hence, the distance between vertices v_1, v_2 given by

$$d_w(v_1, v_2) := \inf \sum w(e_i), \qquad (5.117)$$

where the infimum is taken over all finite paths $\{e_i\} \subset E$ joining v_1 and v_2, is a *pseudometric*.

In this setting, we define the Lipschitz constant of a map $f : V \to \mathbb{R}^n$ by

$$L(f) := \inf\{\lambda > 0 \,;\, \|f(v) - f(v')\| \leq \lambda d_w(v, v') \text{ for all } v, v' \in V\}.$$

In the formulation of the main result, we will use the following finiteness property for set-valued maps F from (V, d_w) into the set of all k-dimensional affine subspaces $\mathrm{Aff}_k(\mathbb{R}^n)$ of $\mathbb{R}^n$, $0 \leq k \leq n - 1$.

(Φ_k) *The restriction of F to every subset of V of cardinality 2^{k+1} without isolated points[7] admits a 1-Lipschitz selection.*

Theorem 5.56. *Assume that $F : V \to \mathrm{Aff}_k(\mathbb{R}^n)$ satisfies condition (Φ_k). Then F admits a γ-Lipschitz selection where $\gamma = \gamma(n, k) > 1$.*

Proof. It suffices to establish the result for the target space ℓ_∞^n. To simplify notation we denote its norm (only in this subsection) by $|\cdot|$. Hence,

$$|x| := \max_{1 \leq i \leq n} |x_i|, \quad x \in \mathbb{R}^n.$$

In particular, the closed balls of ℓ_∞^n are closed cubes denoted by $Q_r(x)$, etc.

We will use induction on $k = 0, 1, \ldots, n - 1$. For $k = 0$, F is a single-valued function and the condition (Φ_0) simply states that $L(F) \leq 1$. Assuming that the theorem is true for $k \geq 0$, we prove it for maps from (V, d_w) into $\mathrm{Aff}_{k+1}(\mathbb{R}^n)$ satisfying the condition (Φ_{k+1}). To this end, given such a map $F : V \to \mathrm{Aff}_{k+1}(\mathbb{R}^n)$, we find for every $v \in V$ a closed cube $Q(v)$ such that the set-valued map

$$G : v \mapsto F(v) \cap Q(v)$$

satisfies the Lipschitz condition in the Hausdorff metric of ℓ_∞^n, i.e., for every pair $v, v' \in V$,

$$d_\mathcal{H}(G(v), G(v')) \leq c(n, k+1) d_w(v, v'). \tag{5.118}$$

The desired selection of F will be then defined by

$$f := s_n \circ G$$

where s_n is the Steiner selection, see (5.92). Then the previous inequality and Theorem 5.46 will imply that $L(f)$ is bounded by a constant depending only on n, and $k + 1$, as required.

The construction of the family of cubes $\{Q(v)\}$ and the proof of (5.118) are divided into the next three steps.

A. For every pair $v_1, v_2 \in V$ joined by an edge we define a centrally symmetric polytope $P(v_1, v_2)$ and then find a certain point of $P(v_1, v_2)$ which will

---

[7] i.e., every point of the subset is joined by an edge with one of its points.

5.3. Lipschitz selections

be used as the "center" of the cube $Q(v)$. To define $P(v_1, v_2)$, we use the restriction $F|_{\{v_1,v_2\}}$. Since $2^{k+2} \geq 2$ and v_1, v_2 are joined by an edge, $F|_{\{v_1,v_2\}}$ has a 1-Lipschitz selection. Hence, there exist points $x^i = x^i(v_1, v_2) \in F(v_i)$, $i = 1, 2$, such that

$$|x^1 - x^2| \leq d_w(v_1, v_2). \tag{5.119}$$

Denoting the closed cube of radius $2d_w(v_1, v_2)$, centered at 0, by $Q(v_1, v_2)$, we then set

$$P(v_1, v_2) := F(v_1) \cap \{F(v_2) + Q(v_1, v_2) + x^1 - x^2\}. \tag{5.120}$$

In particular, it equals $F(v_1)$ if either $d_w(v_1, v_2) = \infty$ or $F(v_2)$ is parallel to $F(v_1)$. Due to (5.119) and its definition, $P(v_1, v_2)$ is, in other cases, a nonempty centrally symmetric convex polytope with a (unique) center at x^1. Therefore, $P(v_1, v_2)$ can be presented as the intersection of its facet [8]-determined half-spaces of the affine space $F(v_1)$, see, e.g., Schneider's book [Sch-1993, Cor. 2.4.4]. Because of central symmetry of $P(v_1, v_2)$ the set of these half-spaces is subdivided into pairs with parallel boundaries. The intersection of such a pair forms a strip and the intersection of all these strips gives $P(v_1, v_2)$. In turn, every strip can be presented in the form $F(v_1) \cap \{L + \overline{B}_r(0)\}$, where L is a k-dimensional affine subspace of $F(v_1)$ passing through x^1, and $\overline{B}_r(0)$ is a uniquely determined closed Euclidean ball in $\mathbb{R}^n$ with $0 < r < \infty$. Enumerating the set of strips for $P(v_1, v_2)$ by integers and denoting the corresponding interval of indices by $J(v_1, v_2)$, we then have

$$P(v_1, v_2) = \bigcap_{i \in J(v_1,v_2)} \left[F(v_1) \cap (L_i + \overline{B}_{r_i}(0))\right]. \tag{5.121}$$

Note that for the exceptional cases $(d_w(v_1, v_2) = \infty$ or $F(v_1) \| F(v_2))$ this representation is also true with $J(v_1, v_2) = \{1\}$ and $r_1 = \infty$ (where $\overline{B}_\infty(0) = \mathbb{R}^n$) and with an arbitrarily chosen k-dimensional subspace L_1 of $F(v_1)$ passing through x^1.

Now we introduce a pseudometric space $(\widetilde{V}, \widetilde{d})$ whose underlying set consists of all triples (v_1, v_2, i), where $[v_1, v_2]$ is an *edge* of the graph $G = (V, E, d_w)$ and $i \in J(v_1, v_2)$. We equip $\widetilde{V}$ with a metric $d_{w'}$, see (5.117), regarding $\widetilde{V}$ as the vertex set of a *complete graph* and defining a weight $\widetilde{w}$ on an edge with endpoints $\widetilde{v} = (v_1, v_2, i)$ and $\widetilde{v}' = (v_1', v_2', i')$ by

$$\widetilde{w}(\widetilde{v}, \widetilde{v}') := w(v_1, v_1') + r_i + r_{i'}. \tag{5.122}$$

In the sequel, we will write $\widetilde{d}$ instead of $d_{\widetilde{w}}$. Clearly, $\widetilde{d}$ may assign values 0 or $+\infty$.

Further, we associate to the set-valued map F a map $\widetilde{F} : \widetilde{V} \to \mathrm{Aff}_k(\mathbb{R}^n)$ by setting, for $\widetilde{v} = (v_1, v_2, i) \in \widetilde{V}$,

$$\widetilde{F}(\widetilde{v}) := L_i. \tag{5.123}$$

---

[8] Recall that a facet is a proper face of maximal dimension (equals k for $P(v_1, v_2) \neq F(v_1)$).

Unlike the map F, this one has $\mathrm{Aff}_k(\mathbb{R}^n)$ as a target space, and therefore the induction hypothesis may be applied to $\widetilde{F}$ to prove

Lemma 5.57. $\widetilde{F}$ has a selection $\tilde{f} : (\widetilde{V}, \tilde{d}) \to \ell_\infty^n$ satisfying

$$L(\tilde{f}) \leq \gamma(n, k). \tag{5.124}$$

Recall that $\gamma(n, k)$ is the constant from the formulation of the theorem.

Proof. Since $\widetilde{V}$ may be regarded as the vertex set of the complete graph, every subset of cardinality at least 2 has no isolated points. Therefore we must check that the restriction of $\widetilde{F}$ to every subset $\widetilde{S} \subset \widetilde{V}$ of cardinality 2^{k+1} admits a 1-Lipschitz selection. To this end, we associate to $\widetilde{S}$ a subset S of V given by

$$S := pr_1(\widetilde{S}) \cup pr_2(\widetilde{S}),$$

where the projections pr_k, $k = 1, 2, 3$, are defined for $\tilde{v} = (v_1, v_2, i) \in \widetilde{V}$ by

$$pr_k(\tilde{v}) := v_k, \ k = 1, 2, \ pr_3(\tilde{v}) := i. \tag{5.125}$$

We also will write below $v_k(\tilde{v})$, $k = 1, 2$, for the first two coordinates and $i(\tilde{v})$ for the third one.

By definition, $\mathrm{card}\, S = 2\,\mathrm{card}\,\widetilde{S} = 2^{k+2}$ and S does not contain isolated points, therefore the assumption of the theorem implies that $F|_S$ has a selection $f_S : V \to \mathbb{R}^n$ with $L(f_S) \leq 1$.

We now show that

$$f_S(v_1) \in P(v_1, v_2). \tag{5.126}$$

In fact, it belongs to $F(v_1)$. Moreover, due to (5.119),

$$|(f_S(v_1) - f_S(v_2)) - (x^1 - x^2)| \leq d_w(v_1, v_2) + d_w(v_1, v_2) = 2d_w(v_1, v_2).$$

Hence, $f_S(v_1)$ also belongs to the set $F(v_2) + Q(v_1, v_2) + x^1 - x^2$, see (5.120), and therefore (5.126) holds.

Now the required selection $\tilde{f}_{\widetilde{S}} : \widetilde{S} \to \mathbb{R}^n$ of the restriction $\widetilde{F}|_{\widetilde{S}}$ is given, for $\tilde{v} \in \widetilde{S}$, by

$$\tilde{f}_{\widetilde{S}}(\tilde{v}) := \text{ closest to } f_S(v_1(\tilde{v})) \text{ point of } \widetilde{F}(\tilde{v}).$$

Due to (5.126) and (5.121), for every $\tilde{v} \in \widetilde{S}$,

$$f_S(v_1(\tilde{v})) \in L_{i(\tilde{v})} + \overline{B}_{r_{i(\tilde{v})}}(0) \subset L_{i(\tilde{v})} + Q_{r_{i(\tilde{v})}}(0).$$

Further, $\tilde{f}_{\widetilde{S}}(\tilde{v})$ is the closest point to $f_S(v_1(\tilde{v}))$ of $\widetilde{F}(\tilde{v})$, and therefore the ℓ_∞-distance between them is bounded by $r_{i(\tilde{v})}$. Hence, for all $\tilde{v}, \tilde{v}' \in \widetilde{S}$,

$$|\tilde{f}_{\widetilde{S}}(\tilde{v}) - \tilde{f}_{\widetilde{S}}(\tilde{v}')| \leq r_{i(\tilde{v})} + r_{i(\tilde{v}')} + L(f_S)d_w(v_1(\tilde{v}), v_1(\tilde{v}')).$$

5.3. Lipschitz selections

By (5.122) and the definition of $f|_S$, the right-hand side is at most $\tilde{d}(\tilde{v},\tilde{v}')$ and therefore $L(\tilde{f}_{\tilde{S}}) \leq 1$.

Hence, the condition (Φ_k) holds for $\widetilde{F}$ and, by the induction hypothesis, $\widetilde{F}$ admits a selection $\tilde{f}$ satisfying (5.124). $\square$

B. Now we use the selection $\tilde{f}$ to find, for every $v \in V$, the above mentioned cubes $Q(v)$. The set of centers for these cubes forms the target set of a function $g : V \to \mathbb{R}^n$, which will be introduced within the proof of

Proposition 5.58. *There exists a map* $g : (V, d_w) \to \mathbb{R}^n$ *such that*

(a) $L(g) \leq \gamma(n,k)$;

(b) *for every* $\tilde{v} \in \widetilde{V}$,

$$\left| g(v_1(\tilde{v})) - \tilde{f}(\tilde{v}) \right| \leq \gamma(n,k) r_{i(\tilde{v})}. \tag{5.127}$$

Proof. We define the required map g to be a selection of an auxiliary map K from $\widetilde{V}$ into the set of closed cubes in $\mathbb{R}^n$ for $\tilde{v} \in \widetilde{V}$ defined by

$$K(\tilde{v}) := Q_{r(\tilde{v})}(\tilde{f}(\tilde{v})), \quad \text{where} \quad r(\tilde{v}) := \gamma(n,k) r_{i(\tilde{v})}. \tag{5.128}$$

We will show that K has a selection $\tilde{g} : \widetilde{V} \to \mathbb{R}^n$ satisfying, for $\tilde{v}, \tilde{v}' \in \widetilde{V}$, the Lipschitz condition:

$$\left| \tilde{g}(\tilde{v}) - \tilde{g}(\tilde{v}') \right| \leq \gamma(n,k) d_w\bigl(v_1(\tilde{v}), v_1(\tilde{v}')\bigr). \tag{5.129}$$

If the first coordinates of $\tilde{v}$ and $\tilde{v}'$ coincide, the right-hand side is zero. Hence, $\tilde{g}$ will define a function $g : (V, d_w) \to \mathbb{R}^n$ given, for $v \in V$, by

$$g(v) := \tilde{g}(\tilde{v}), \tag{5.130}$$

where $\tilde{v}$ is arbitrary satisfying $v_1(\tilde{v}) = v$.

Since $g(v_1(\tilde{v})) \in K(\tilde{v})$, and since the cube $K(\tilde{v})$ is defined by (5.128), inequality (5.127) holds for such g. Moreover, (5.129) would imply that g is Lipschitz and $L(g) \leq \gamma(n,k)$, as required.

To prove (5.129), we apply Corollary 1.25 from Section 1.10 to the map K regarded as a set-valued map from the pseudometric space $(\widetilde{V}, \widehat{d})$, where $\widehat{d} : (\tilde{v}, \tilde{v}') \mapsto d_w(v_1(\tilde{v}), v_1(\tilde{v}'))$. According to this result K admits a C-Lipschitz selection with $C := \gamma(n,k)$ if its restriction to every two-point subset of $\widetilde{V}$ does.

We now prove the last claim. Let $K(\tilde{v}), K(\tilde{v}')$ be distinct cubes. By Lemma 5.57 and (5.122), their centers satisfy

$$\begin{aligned} \left| \tilde{f}(\tilde{v}) - \tilde{f}(\tilde{v}') \right| &\leq \gamma(n,k) \tilde{d}(\tilde{v}, \tilde{v}') \\ &\leq \gamma(n,k) \bigl(r_{i(v)} + r_{i(v')} + d_w(v_1(\tilde{v}), v_1(\tilde{v}'))\bigr). \end{aligned}$$

We will derive from here that there exist points $x \in K(\tilde{v}), x' \in K(\tilde{v}')$ such that

$$|x - x'| \leq \gamma(n,k) d_w(v_1(\tilde{v}), v_1(\tilde{v}')). \tag{5.131}$$

These points would determine the required selection of $K|_{\{\tilde{v},\tilde{v}'\}}$ and then Corollary 1.25 will give (5.129). We derive (5.131) from

Lemma 5.59. *Let K_i be a closed n-dimensional cube in $\mathbb{R}^n$ of radius $r_i \in [0, +\infty]$ and of center c^i, $i = 1, 2$. Assume that for some constant $0 \leq \lambda \leq \infty$,*

$$|c^1 - c^2| \leq r_1 + r_2 + \lambda.$$

Then there exist points $x^i \in K_i$, $i = 1, 2$, such that

$$|x^1 - x^2| \leq \lambda. \tag{5.132}$$

Proof. The result is trivial if one of the parameters is $+\infty$ or $K_1 \cap K_2 \neq \emptyset$. So, assume that K_i are bounded and disjoint. If $Q_\lambda(0) \subset \mathbb{R}^n$ is the closed cube of radius λ centered at 0, then $K_1 + Q_\lambda(0)$ is the closed cube of radius $r_1 + \lambda$ centered at c^1. Assuming that $(K_1 + Q_\lambda(0)) \cap K_2 = \emptyset$, we clearly have $|c^1 - c^2| > r_1 + r_2 + \lambda$, a contradiction. Thus there exists $x^2 \in (K_1 + Q_\lambda(0)) \cap K_2$. Also, there exist $x^1 \in K_1$ and $v \in Q_\lambda(0)$ such that $x^2 = x^1 + v$. This implies $|x^2 - x^1| = |v| \leq \lambda$, as required. □

The proof of Proposition 5.58 is complete. □

C. Finally, we define the required compact cubes $Q(v)$, $v \in V$, and the related set-valued map $G : V \to \mathcal{C}(\mathbb{R}^n)$ (into compact subsets of $\mathbb{R}^n$) and prove, for G, the Lipschitz condition (5.118). To this end, we set

$$d(v) := d(g(v), F(v)), \tag{5.133}$$

where we recall that the distance is measured in the ℓ_∞^n-norm, and then define, for $v \in V$, the cube $Q(v)$ by

$$Q(v) := Q_{2d(v)}(g(v)). \tag{5.134}$$

In turn, the map $G : V \to \mathcal{C}(\mathbb{R}^n)$ is given by

$$G(v) := F(v) \cap Q(v). \tag{5.135}$$

Proposition 5.60. *For every pair $v, v' \in V$ joined by an edge,*

$$|d(v) - d(v')| \leq \theta d_w(v, v') \tag{5.136}$$

and, moreover,

$$G(v) \subset F(v') + \theta Q_{d_w(v,v')}(0). \tag{5.137}$$

Here $\theta = \theta(n, k) > 1$ may be taken to be equal to $12\sqrt{n}\gamma(n, k)$.

5.3. Lipschitz selections

Proof. Let $h(v) \in F(v)$ be a point closest to $g(v)$:

$$|g(v) - h(v)| = d(v).$$

By (5.129), $g(v) = \tilde{g}(\tilde{v})$ for every point $\tilde{v} \in \tilde{V}$ whose first coordinate $pr_1(\tilde{v}) = v$. Then (5.127) implies, for these $\tilde{v}$,

$$d(v) \leq |\tilde{g}(\tilde{v}) - \tilde{f}(\tilde{v})| \leq \gamma(n,k) r_{i(\tilde{v})} =: cr_{i(\tilde{v})}, \tag{5.138}$$

where hereafter $c := \gamma(n,k)$. Setting

$$\rho(v) := \inf\{r_{i(\tilde{v})} \,;\, pr_1(\tilde{v}) = v\}, \tag{5.139}$$

we then derive that

$$d(v) \leq c\rho(v).$$

By the definition of $Q(v)$, see (5.134), and $h(v)$, this implies embeddings

$$Q(v) \subset Q_{2c\rho(v)}(g(v)) \subset Q_{3c\rho(v)}(h(v)).$$

In turn, this leads to the embedding

$$G(v) \subset F(v) \cap Q_{3c\rho(v)}(h(v)). \tag{5.140}$$

Next we show that the right-hand side of (5.140) is contained in the extended polytope $5\sqrt{n}cP(v,v')$, the $5\sqrt{n}c$-homothety of $P(v,v')$ with respect to its center $x^1 = x^1(v,v')$, see (5.120).

Lemma 5.61. *For every pair* $v, v' \in V$ *joined by an edge,*

$$F(v) \cap Q_{3c\rho(v)}(h(v)) \subset 5\sqrt{n}cP(v,v'). \tag{5.141}$$

Proof. We assume without loss of generality that $x^1(v,v') = 0$. For a triple $\tilde{v} := (v, v', i) \in \tilde{V}$, we have, by (5.138),

$$|h(v) - \tilde{f}(\tilde{v})| \leq 2cr_i. \tag{5.142}$$

Therefore, for every point y belonging to the cube in (5.141), we have

$$|y - \tilde{f}(\tilde{v})| \leq 5cr_i.$$

Since $\tilde{f}(\tilde{v}) \in \tilde{F}(\tilde{v}) := L_i$, see (5.123), the inequality yields

$$Q_{3c\rho(v)}(h(v)) \subset \bigcap_{i \in J(v,v')} Q_{5cr_i}(\tilde{f}(\tilde{v})) \subset \bigcap_{i \in J(v,v')} (L_i + Q_{5cr_i}(0)).$$

Since L_i is an affine subspace of the affine space $F(v)$ passing through $x^1(v,v') = 0 \in F(v)$, both of these spaces are linear subspaces of the (vector) space $\mathbb{R}^n$.

Therefore the right-hand side of the above embedding is the $5c$-homothety of the set $\bigcap_{i \in J(v,v')} (L_i + Q_{r_i}(0))$ with respect to the origin. This homothety also maps $F(v)$ onto itself and therefore

$$Q_{3c\rho(v)}(h(v)) \cap F(v) \subset 5c\left(F(v) \cap \left[\bigcap_{i \in J(v,v')} (L_i + Q_{r_i}(0))\right]\right)$$

$$\subset 5\sqrt{n}c\left(F(v) \cap \left[\bigcap_{i \in J(v,v')} (L_i + \overline{B}_{r_i}(0))\right]\right) =: 5\sqrt{n}cP(v,v'),$$

as required. □

The lemma and (5.140) yield

$$G(v) \subset 5\sqrt{n}cP(v,v'). \tag{5.143}$$

Now we use the definition of $P(v,v')$, see (5.120), to conclude that

$$P(v,v') \subset F(v) \cap \left[(F(v') + x^1(v,v') - x^2(v,v')) + 2Q_{d_w(v,v')}(0)\right].$$

Since the affine subspace in the brackets contains $x^1(v,v') = 0$, the $5\sqrt{n}c$-homothety of the right-hand side with respect to 0 equals

$$\left((F(v') - x^2(v,v')) + 10\sqrt{n}c\, Q_{d_w(v,v')}(0)\right) \cap F(v).$$

Together with (5.143) and (5.119) this implies that

$$G(v) \subset F(v') + \theta Q_{d_w(v.v')}(0),$$

where $\theta := 12\sqrt{n}\gamma(n,k) \geq 10\sqrt{n}c + 1$.

To prove the remaining inequality (5.136) we use the embedding just established to find a point z in $F(v')$ such that

$$|h(v) - z| \leq (10\sqrt{n}c + 1)d_w(v,v').$$

This leads to the estimate

$$d(v') := d(g(v'), F(v')) \leq |g(v') - z|$$
$$\leq |g(v') - g(v)| + |g(v) - h(v)| + |h(v) - z|$$
$$\leq |g(v') - g(v)| + d(v) + (10\sqrt{n}c + 1)d_w(v,v').$$

Due to Proposition 5.58, the first term of the right-hand side is at most $\gamma(n,k)d_w(v,v') =: cd_w(v,v')$. Therefore

$$d(v') \leq d(v) + (11\sqrt{n}c + 1)d_w(v,v'),$$

whence

$$|d(v) - d(v')| \leq \theta d_w(v,v').$$

The proof of Proposition 5.60 is complete. □

5.3. Lipschitz selections

We use this proposition to estimate the Lipschitz constant of the set-valued map G. To this end, we need a property of the parallel bodies of convex sets. Let us recall that the parallel body $[S]_r$ of a subset $S \subset \mathbb{R}^n$, $r > 0$, is defined in ℓ_∞-norm by

$$[S]_r := S + Q_r(0).$$

Lemma 5.62. *Let $C \subset \mathbb{R}^n$ be convex. Assume that for some $r > 0$ the distance from a point x to C is at most r. Then*

$$[C]_s \cap [Q_{2r}(x)]_s \subset [C \cap Q_{2r}(x)]_{7s}. \tag{5.144}$$

Proof. We may assume that $x = 0$. Hence, we must prove that

$$(C + Q_s(0)) \cap Q_{2r+s}(0) \subset (C \cap Q_{2r}(0)) + Q_{7s}(0).$$

To this end, we must show for a point y from the left-hand side that it belongs to the set from the right. By the choice of y, there exists a point z from C such that $|y - z| \leq s$. If, in addition, $|z| \leq 2r$, then it belongs to $C \cap Q_{2r}(0)$, and therefore $y = z + (y - z) \in C \cap Q_{2r}(0) + Q_s(0)$, as required.

In the remaining case, $z = y + (z - y)$ satisfies

$$|z| > 2r \quad \text{and} \quad |z| \leq |y| + |y - z| \leq 2r + 2s, \tag{5.145}$$

since $y \in Q_{r+2s}(0)$.

By the assumption, there exists a point $z' \in C \cap Q_r(0)$. Therefore for every $\lambda \in (0, 1)$, the point $z_\lambda := (1 - \lambda)z + \lambda z'$ belongs to C and satisfies:

$$|z_\lambda| > 2r \ \text{if} \ \lambda = 0 \ \text{and} \ |z_\lambda| \leq r \ \text{if} \ \lambda = 1.$$

Hence, there exists $\lambda \in (0, 1)$ such that $|z_\lambda| = 2r$. For this λ the point z_λ belongs to $Q_{2r}(0)$, and therefore $z_\lambda \in C \cap Q_{2r}(0)$.

We next show that

$$|y - z_\lambda| \leq 7s.$$

Since $y = z_\lambda + (y - z_\lambda)$, this would imply that $y \in (C \cap Q_{2r}(0)) + Q_{7s}(0)$, as required.

By the triangle inequality and (5.145),

$$2r = |z_\lambda| \leq \lambda|z'| + (1 - \lambda)|z| \leq \lambda r + (1 - \lambda)(2r + 2s),$$

whence $\lambda \leq \frac{2s}{r+2s}$. The last estimate and (5.145) yield

$$|y - z_\lambda| \leq |y - z| + |z - z_\lambda| \leq s + \lambda(|z| + |z'|) \leq s + \frac{2s}{r + 2s}(3r + 2s) < 7s.$$

This completes the proof. $\square$

We apply this lemma to the convex set $C := F(v)$ and the point $x := g(v)$ with
$$r := d(v) \ (:= d(g(v), F(v))) \quad \text{and} \quad s := 25\sqrt{n}\gamma(n,k)d_w(v,v'),$$
where $v' \in V$ is joined by an edge with v.

Under this choice $Q_{2r}(x)$ equals $Q(v)$, and the lemma yields
$$\bigl(F(v) + Q_s(0)\bigr) \cap \bigl(Q(v) + Q_s(0)\bigr) \subset \bigl(F(v) \cap Q(v)\bigr) + Q_{7s}(0). \tag{5.146}$$

On the other hand, (5.137) implies that
$$G(v') \subset F(v) + \theta Q_{d_w(v,v')}(0) \subset F(v) + Q_s(0)$$
(recall that $\theta := 12\sqrt{n}\gamma(n,k)$).

Further, $L(g) \leq \gamma(n,k)$, see Proposition 5.58. This, together with (5.136) and the definition of s, yield
$$G(v') \subset Q(v') \subset Q(v) + [\gamma(n,k)d_w(v,v') + 2\theta d_w(v,v')]Q_1(0) \subset Q(v) + Q_s(0).$$

Combining this with the previous embedding, we conclude that $G(v')$ is contained in the left-hand side of (5.146). Hence, for every pair $v, v' \in V$ joined by an edge,
$$G(v') \subset G(v) + Q_{7s}(0).$$

Because of symmetry in v and v', this implies that
$$d_{\mathcal{H}}\bigl(G(v), G(v')\bigr) \leq 7s = C\sqrt{n}\gamma(n,k)d_w(v,v'),$$
where C is a numerical constant.

Now let $s_n : \mathcal{C}(\mathbb{R}^n) \to \mathbb{R}^n$ be the Steiner selector for ℓ_∞^n, see (5.92). Then $s_n(G)$ is a point of G and, therefore, of $F \supset G$. Further, Corollary 5.48 implies that for some constant $a(n) > 1$ and all v, v',
$$\bigl|s_n(G(v)) - s_n(G(v'))\bigr| \leq a(n)d_{\mathcal{H}}\bigl(G(v), G(v')\bigr).$$

Together with the previous inequality this estimates the Lipschitz constant of the map $s_n \circ G : (V, d_w) \to \ell_\infty^n$ by $Ca(n)\sqrt{n}\gamma(n,k)$.

Theorem 5.56 is proved. □

As an immediate consequence of Theorem 5.56 we get

Corollary 5.63. *Let F map a pseudometric space $(\mathcal{M}, d)$ into $\mathrm{Aff}_k(\mathbb{R}^n)$. If the restriction of F to every subset of cardinality at most 2^k admits a 1-Lipschitz selection, then F admits a $\gamma(n,k)$-Lipschitz selection.*

Proof. $\mathcal{M}$ is the vertex set of a complete graph whose edges are determined by pairs $m, m' \in \mathcal{M}$ with $m \neq m'$. The corresponding pseudometric d_w equals $d(m, m')$ at an edge $[m, m']$. Then F is defined on the vertex set of this metric graph and satisfies the assumptions of Theorem 5.56. The result follows. □

5.4 Simultaneous Lipschitz selections

5.4.1 The problem

We study a selection problem for a *family* of set-valued maps from a metric space $\mathcal{M} := (M, d)$ into the set $\mathrm{Aff}(\mathbb{R}^n)$. The family consists of all maps $f + F$, where F sends $\mathcal{M}$ into the set $\mathrm{Lin}_k(\mathbb{R}^n)$ of k-dimensional subspaces of $\mathbb{R}^n$ and $f : \mathcal{M} \to \mathbb{R}^n$ is such that $f + F$ has a Lipschitz selection, say s. Geometrically this means that distance between two affine subspaces $f(m_i) + F(m_i)$, $i = 1, 2$, is bounded by $L(s)d(m_1, m_2)$ for all $m_1, m_2 \in \mathcal{M}$, where distances in $\mathbb{R}^n$ are measured in the ℓ_∞^n-norm denoted hereafter by $\|\cdot\|$, and

$$L(s) := |s|_{\mathrm{Lip}(\mathcal{M}, \mathbb{R}^n)} := \sup_{m' \neq m} \frac{\|f(m') - f(m)\|}{d(m', m)}.$$

By $\Sigma_F(\mathcal{M}; \mathbb{R}^n)$ we denote the linear space of all such f and define its seminorm by

$$|f|_{\Sigma_F(\mathcal{M}; \mathbb{R}^n)} := \inf\{|s|_{\mathrm{Lip}(\mathcal{M}, \mathbb{R}^n)} \,;\, s \in f + F\}. \tag{5.147}$$

Problem 5.64. *Find a linear operator* $T_F : \Sigma_F(\mathcal{M}; \mathbb{R}^n) \to \mathrm{Lip}(\mathcal{M}, \mathbb{R}^n)$ *of norm bounded by some constant* $c > 0$ *depending only on* $\mathcal{M}$ *and* n *such that for every* f *from its domain*

$$T_F f \in f + F. \tag{5.148}$$

In other words, we are asking for existence of a simultaneous selection for the family $\{f + F\}$ that linearly depends on f and has uniformly bounded Lipschitz constants.

The positive answer may be obtained only for a specific class of metric spaces. To explain the situation we extend the problem assuming that F assigns linear subspaces whose dimension may vary. It will be shown later that this problem is equivalent to the previous.

Proposition 5.65. *If the extended problem has the positive solution for a metric space* $\mathcal{M}$*, then the linear Lipschitz extension constant*

$$\lambda(\mathcal{M}) := \sup_{\mathcal{M}' \subset \mathcal{M}} \inf\{\|T_{\mathcal{M}'}\|\} < \infty.$$

Here $T_{\mathcal{M}'}$ runs over the space $\mathrm{Ext}(\mathcal{M}', \mathrm{Lip}(\mathcal{M}))$ of all linear bounded extension operators from $\mathrm{Lip}\,\mathcal{M}' := \mathrm{Lip}(\mathcal{M}', \mathbb{R})$ into $\mathrm{Lip}\,\mathcal{M}$.

Proof. Let $\mathcal{M}' \subset \mathcal{M}$ and $F_{\mathcal{M}'} : \mathcal{M} \to \mathrm{Aff}(\mathbb{R}) = \{\{0\}, \mathbb{R}\}$ be a set-valued map given by

$$F_{\mathcal{M}'}(m) := \begin{cases} \{0\} & \text{if } m \in \mathcal{M}', \\ \mathbb{R} & \text{if } m \in \mathcal{M} \setminus \mathcal{M}'. \end{cases}$$

If $s \in f + F_{\mathcal{M}'}$, then $s = f$ on $\mathcal{M}'$ and is arbitrary on $\mathcal{M} \setminus \mathcal{M}'$. Hence, every Lipschitz selection of $f + F_{\mathcal{M}'}$ is an extension of $f|_{\mathcal{M}'}$ to a function from Lip $\mathcal{M}$.

Now let $\mathcal{M}$ have the desired property, in particular, for every $F_{\mathcal{M}'}$ there exists a linear operator $T_{\mathcal{M}'} : \Sigma_{\mathcal{F}_{\mathcal{M}'}}(\mathcal{M}; \mathbb{R}) \to$ Lip $\mathcal{M}$ such that $T_{\mathcal{M}'} f \in f + F_{\mathcal{M}'}$ and $\|T_{\mathcal{M}'}\|$ is bounded by a constant $c > 0$ independent of $\mathcal{M}'$. Then by definition

$$\lambda(\mathcal{M}) \leq \sup_{\mathcal{M}'} \|T_{\mathcal{M}'}\| \leq c. \qquad \square$$

It will be much more complicated to show that the condition of this proposition is also sufficient. This result is a special (but, in fact, basic) case of a general simultaneous Lipschitz selection theorem presented in the Yu. Brudnyi and Shvartsman paper [BSh-1999], see Thm. 4.15 there.

5.4.2 Formulation of the main theorem

Let $F : (\mathcal{M}, d) \to \text{Lin}_k(\mathbb{R}^n)$, $0 \leq k \leq n$, and the seminormed space $\Sigma_F(\mathcal{M}; \mathbb{R}^n)$ is introduced by (5.147).

Theorem 5.66. *Assume that $\lambda(\mathcal{M}) < \infty$. Then there exists a linear operator $T_F : \Sigma_F(\mathcal{M}; \mathbb{R}^n) \to \text{Lip}(\mathcal{M}, \mathbb{R}^n)$ such that*

$$\|T_F\| \leq c(n) \lambda(\mathcal{M})^{2k} \quad \text{and} \quad T_F f \in f + F. \qquad (5.149)$$

We reformulate this using a characteristic of a metric space given by

$$\chi_{k,n}(\mathcal{M}) := \sup\{\inf_F \|T_F\|\}, \qquad (5.150)$$

where T_F runs over all linear operators from $\Sigma_F(\mathcal{M}; \mathbb{R}^n)$ into Lip$(\mathcal{M}, \mathbb{R}^n)$ satisfying condition (5.148), and F runs over all set-valued maps from $\mathcal{M}$ into $\text{Lin}_n(\mathbb{R}^n)$.

Example 5.67. (a) $\text{Lin}_0(\mathbb{R}^n) = \{\{0\}\}$ and therefore $T_F = Id_{\text{Lip}(\mathcal{M}, \mathbb{R}^n)}$ for a (unique) $F := \{0\}$. Hence,

$$\chi_{0,n}(\mathcal{M}) = 1.$$

(b) $\text{Lin}_n(\mathbb{R}^n) = \{\mathbb{R}^n\}$ and we may take $T_F = 0$ for $F : \mathcal{M} \to \{\mathbb{R}^n\}$. Hence,

$$\chi_{n,n}(\mathcal{M}) = 0.$$

We therefore may discard trivial cases $k = 0, n$ and reformulate the main theorem as follows:

For every $0 < k < n$,

$$\chi_{k,n}(\mathcal{M}) \leq c(n) \lambda(\mathcal{M})^{2k}. \qquad (5.151)$$

5.4.3 Auxiliary results

We prove (5.151) by induction on k with $\lambda(\mathcal{M})$ replaced by some equivalent characteristic denoted by $\kappa(\mathcal{M})$ that will be now introduced.

Let $w : \mathcal{M} \to \mathbb{R}_+ \cup \{+\infty\}$ be a weight assumed to be finite at some point. We define a weighted space $\ell^w_\infty(\mathcal{M})$ using a pseudonorm given for $f : \mathcal{M} \to \mathbb{R}_+$ by

$$\|f\|_{\ell^w_\infty(\mathcal{M})} := \inf\{\lambda > 0 \,;\, |f(m)| \leq \lambda w(m) \quad \text{for all} \quad m\}. \tag{5.152}$$

Since $w \neq +\infty$, this equals zero only if $f = 0$, but it may be $+\infty$ if the weight is zero at some point.

Using the standard convention on operations with $+\infty$ and an additional one: $0 \cdot (+\infty) = 0$, we easily see that (5.152) defines a functional (which may assign value $+\infty$) satisfying all of the axioms of norm.

The aforementioned replacement $\kappa(\mathcal{M})$ relates to the so-called K-linearity constants of the family of pairs (couples) $\{\ell^w_\infty(\mathcal{M}), \text{Lip } \mathcal{M}\}$, see [BK-1991, subsec. 3.9.1] for details. For its introduction we set

$$\Sigma_w(\mathcal{M}) := \ell^w_\infty(\mathcal{M}) + \text{Lip } \mathcal{M} \tag{5.153}$$

and equip this linear space with a (pseudo) seminorm given by

$$|f|_{\Sigma_w(\mathcal{M})} := \inf_{f = f_0 + f_1} \{\|f_0\|_{\ell^w_\infty(\mathcal{M})} + \|f_1\|_{\text{Lip } \mathcal{M}}\}.$$

As we will see every couple $(\ell^w_\infty(\mathcal{M}), \text{Lip } \mathcal{M})$ is K-*linearizable* meaning that there exist linear bounded operators

$$T_w : \Sigma_w(\mathcal{M}) \to \text{Lip } \mathcal{M}, \quad S_w : \Sigma_w(\mathcal{M}) \to \ell^w_\infty(\mathcal{M})$$

such that

$$T_w + S_w = Id_{\Sigma_w(\mathcal{M})}.$$

Then the K-*linearity constant* of this couple is defined by

$$\kappa(\ell^w_\infty(\mathcal{M}), \text{Lip } \mathcal{M}) := \inf\{\|T_w\| + \|S_w\|\}.$$

Finally, we introduce the required characteristic $\kappa(\mathcal{M})$ by setting

$$\kappa(\mathcal{M}) := \sup_w \kappa(\ell^w_\infty(\mathcal{M}), \text{Lip } \mathcal{M}). \tag{5.154}$$

Proposition 5.68. *For some numerical constant* $c > 0$

$$\lambda(\mathcal{M}) \leq \kappa(\mathcal{M}) \leq c\lambda(\mathcal{M})^2.$$

This result will be proved in subsection 5.4.5; now we only mark its relation to the basic notion (5.150). Specifically, we describe similarity between two seemingly distinct characteristics, $\chi_{1,n}(\mathcal{M})$ and $\kappa(\mathcal{M})$. The former deals with Lipschitz selections of set-valued maps from $\mathcal{M}$ into straight lines of $\mathbb{R}^n$ passing through 0. However, the latter, as will be shown, is a quantitative characteristic of a similar problem but dealing with set-valued maps into symmetric intervals of a *single* straight line, say, $\mathbb{R}$. The value of such a map F at $m \in \mathcal{M}$ has a form $[-w(m), w(m)]$, where $w : \mathcal{M} \to \mathbb{R}_+ \cup \{+\infty\}$. Then the norm of $f : \mathcal{M} \to \mathbb{R}$ in the weighted space $\ell_\infty^w(\mathcal{M})$ may be seen as a measure of deviation from being a selection of F. In fact, if $\|f\|_{\ell_\infty^w(\mathcal{M})} \leq 1$, then $f \in F$ and if $\|f\|_{\ell_\infty^w(\mathcal{M})} := \gamma > 1$, then $f \in \gamma F := [-\gamma w, \gamma w]$.

An analog of the space of Lipschitz selections $\Sigma_F(\mathcal{M}; \mathbb{R})$ for the second case is naturally identified with the space $\Sigma_w(\mathcal{M})$ given by (5.153). In turn, the linear operators T_w, S_w there give a solution to the corresponding selection problem with T_w controlling the Lipschitz constants of selected maps f and with S_w measuring their deviation from being selections of F. The optimal choice of these operators is characterized by the K-linearity constant of $(\ell_\infty^w, \mathrm{Lip}\,\mathcal{M})$. It will follow from a result presented below, see Proposition 5.70, that the supremum of these constants, i.e., $\kappa(\mathcal{M})$, bounds $\chi_{1,n}(\mathcal{M})$; in general this result states that $\chi_{k,n}(\mathcal{M})$ is bounded by $\kappa(\mathcal{M})^k$. The proof exploits the "doubling trick" passing from $(\mathcal{M}, d)$ to a (pseudo) metric space $\widetilde{\mathcal{M}}$ of triples (m', m, i), where $(m', m) \in \mathcal{M} \times \mathcal{M}$ and i runs over some finite index set $J(m', m)$, cf. (5.122) in the proof of Theorem 5.56. To return to the initial situation we then should estimate the constant $\lambda(\widetilde{\mathcal{M}})$ by that of $\mathcal{M}$. This will follow from a more general result presented now. In its formulation, given a set $\widetilde{\mathcal{M}}$ and two arbitrary maps

$$\varphi : \widetilde{\mathcal{M}} \to \mathcal{M} \quad \text{and} \quad r : \widetilde{\mathcal{M}} \to \mathbb{R}_+ \cup \{+\infty\}$$

we associate with $(\mathcal{M}, d)$ a pseudometric space $(\widetilde{\mathcal{M}}, \tilde{d})$, where the pseudometric $\tilde{d} : \widetilde{\mathcal{M}} \times \widetilde{\mathcal{M}} \to \mathbb{R}_+ \cup \{+\infty\}$ is defined for $\tilde{m}' \neq \tilde{m}$ by

$$\tilde{d}(\tilde{m}', \tilde{m}) := d(\varphi(\tilde{m}'), \varphi(\tilde{m})) + r(\tilde{m}') + r(\tilde{m})$$

and for $\tilde{m}' = \tilde{m}$ by zero.

This clearly meets the metric axioms but may assign 0 for $\tilde{m}' \neq \tilde{m}$ (if, e.g., $\varphi(\tilde{m}') = \varphi(\tilde{m})$ and $r = 0$) or $+\infty$.

Further, we define the couple $(\ell_\infty^w(\widetilde{\mathcal{M}}), \mathrm{Lip}\,\widetilde{\mathcal{M}})$ and the related constant $\kappa(\widetilde{\mathcal{M}})$ similarly to that for $\mathcal{M}$ and (5.154).

Proposition 5.69. *If $\kappa(\mathcal{M}) < \infty$, then there exist linear operators T_w, S_w mapping $\Sigma_w(\widetilde{\mathcal{M}})$ into $\mathrm{Lip}\,\widetilde{\mathcal{M}}$ and $\ell_\infty^w(\widetilde{\mathcal{M}})$, respectively, such that*

$$T_w + S_w = \mathrm{Id}_{\Sigma_w(\widetilde{\mathcal{M}})}$$

and, moreover, for some numerical constant $c > 0$,

$$\|T_w\| + \|S_w\| \leq c\kappa(\mathcal{M}).$$

5.4. Simultaneous Lipschitz selections 497

This will be proved in subsection 5.4.6 in the equivalent form asserting that

$$\kappa(\widetilde{\mathcal{M}}) \leq c\kappa(\mathcal{M}).$$

5.4.4 Proof of Theorem 5.66

In view of Proposition 5.68 the desired result will immediately follow from the next

Proposition 5.70. *For every $0 \leq k \leq n-1$ it is true that*

$$\chi_{k,n}(\mathcal{M}) \leq c(n,k)\kappa(\mathcal{M})^k. \tag{5.155}$$

Proof. (By induction on k) The result is trivial for $k = 0$, since

$$1 = \chi_{0,n}(\mathcal{M}) = \kappa(\mathcal{M})^0.$$

Assuming now that (5.155) holds for $0 \leq k < n-1$ we will prove it for $k+1$.

In other words, given a set-valued map $F : \mathcal{M} \to \mathrm{Lin}_{k+1}(\mathbb{R}^n)$ on a metric space $(\mathcal{M}, d)$ satisfying $\lambda(\mathcal{M}) < \infty$, we should find a linear operator

$$T_F : \Sigma_F(\mathcal{M}; \mathbb{R}^n) \to \mathrm{Lip}(\mathcal{M}, \mathbb{R}^n) \tag{5.156}$$

such that $T_F f \in f + F$ and, moreover,

$$\|T_F\| \leq c(n, k+1)\kappa(\mathcal{M})^{k+1}. \tag{5.157}$$

We consider the case of $F = const$ separately to emphasize some peculiarities of the general situation. So, now all affine spaces $f(m) + F(m)$, $m \in \mathcal{M}$, are parallel to a fixed $(k+1)$-dimensional subspace of $\mathbb{R}^n$, say L. Let $L^\perp$ be the orthogonal complement of L and Pr be the orthogonal projection of $\mathbb{R}^n$ onto $L^\perp$. Then $Pr(f(m))$ belongs to the affine space $f(m) + L := f(m) + F(m)$. Moreover, by the definition of $\Sigma_F(\mathcal{M}; \mathbb{R}^n)$ the ℓ_∞-distance between $f(m_1) + L$ and $f(m_2) + L$ is at most $2|f|_{\Sigma_F(\mathcal{M};\mathbb{R}^n)} d(m_1, m_2)$. Since the Euclidean distance between $f(m_1) + L$ and $f(m_2) + L$ equals

$$\|Pr(f(m_1)) - Pr(f(m_2))\|_{\ell_2^n} \leq \sqrt{n}\, \mathrm{dist}_{\ell_\infty^n}(f(m_1) + L, f(m_2) + L)$$
$$\leq 2\sqrt{n}|f|_{\Sigma_F(\mathcal{M};\mathbb{R}^n)} d(m_1, m_2),$$

the linear operator Pr maps $\Sigma_F(\mathcal{M}; \mathbb{R}^n)$ into $\mathrm{Lip}(\mathcal{M}, \mathbb{R}^n)$ and its norm is bounded by $2\sqrt{n}$. As $\kappa(\mathcal{M}) \geq 1$, Pr satisfies the required conditions (5.156) and (5.157).

Now, let $F \neq const$, hence affine subspaces $f(m) + F(m)$ change directions as m varies. In the forthcoming derivation, we use the orthogonal projections onto these subspaces to estimate qualitatively the variations of their directions; the properties required below are presented by

Lemma 5.71. *Let L_1, L_2 be linear subspaces of $\mathbb{R}^n$ and $L := L_1 + L_2$. Then there exist linear maps $a, b : \mathbb{R}^n \times \mathbb{R}^n \to \mathbb{R}^n$ such that, for every pair (x, y),*

$$a(x, y) \in x + L_1, \qquad b(x, y) \in y + L_2, \qquad (5.158)$$

and, moreover,

$$a(x,y) - b(x,y) \in L \quad \text{and} \quad \|a(x,y) - b(x,y)\|_{\ell_2^n} = d_{\ell_2^n}(x + L_1, y + L_2).$$

Proof. We may represent $L := L_1 + L_2$ as the direct sum of two subspaces, say, $E_i \subset L_i$, $i = 1, 2$. Then every $z \in L$ can be uniquely written as $a_1(z) + a_2(z)$, where each $a_i(z) \in E_i$ linearly depends on z. Further, we define a function $h : \mathbb{R}^n \times \mathbb{R}^n \to \mathbb{R}^n$ by setting

$$h(x, y) := Pr_L(x - y),$$

where Pr_L is the orthogonal projection of $\mathbb{R}^n$ onto L. Since Pr_L is a linear operator, h is linear in (x, y).

Finally, we define $a, b : \mathbb{R}^n \times \mathbb{R}^n \to \mathbb{R}^n$ by setting

$$a(x, y) := x - a_1[h(x, y)], \qquad b(x, y) := y + a_2[h(x, y)]. \qquad (5.159)$$

Clearly, (5.158) holds, and, moreover,

$$a(x, y) - b(x, y) := (x - y) - h(x, y) = (x - y) - Pr_L(x - y),$$

i.e., the right-hand side is a vector of L.

Using optimality of Pr_L we then have

$$\|a(x,y) - b(x,y)\|_{\ell_2^n} = \inf_{z \in L} \|x - y - z\|_{\ell_2^n} = \inf_{z_i \in L_i} \|(x + z_1) - (y + z_2)\|_{\ell_2^n}$$
$$=: d_{\ell_2^n}(x + L_1, y + L_2).$$

This completes the proof. $\qquad \square$

We apply the lemma to affine subspaces

$$L_f(m_i) := f(m_i) + F(m_i), \quad i = 1, 2, \quad \text{where} \quad f \in \Sigma_F(\mathcal{M}; \mathbb{R}^n).$$

Setting for brevity

$$a := a(f(m_1), f(m_2)), \qquad b := b(f(m_1), f(m_2)), \qquad (5.160)$$

we then obtain two functions from $\Sigma_F(\mathcal{M}; \mathbb{R}^n) \times \mathcal{M} \times \mathcal{M}$ into $\mathbb{R}^n$ linearly depending on f. Moreover, in virtue of the lemma, these functions measure the Euclidean distance between two L_f-spaces by

$$\|a - b\|_{\ell_2^n} = d_{\ell_2^n}(L_f(m_1), L_f(m_2)).$$

Due to the definition of $\Sigma_F(\mathcal{M};\mathbb{R}^n)$, see (5.147), there exists a map $g \in \text{Lip}(\mathcal{M},\mathbb{R}^n)$ such that $g \in f + F =: L_f$ and $|g|_{\text{Lip}(\mathcal{M},\mathbb{R}^n)} \leq 2|f|_{\Sigma_F(\mathcal{M};\mathbb{R}^n)}$. Therefore, for ℓ_∞^n-norm of $a - b$ we have

$$\|a - b\| \leq \|a - b\|_{\ell_2^n} = d_{\ell_2^n}(L_f(m_1), L_f(m_2)) \leq \sqrt{n}\|g(m_1) - g(m_2)\|$$
$$\leq 2\sqrt{n}|g|_{\Sigma_F(\mathcal{M};\mathbb{R}^n)} d(m_1, m_2).$$

Hence, the map $a - b$ is Lipschitz and

$$|a - b|_{\text{Lip}(\mathcal{M},\mathbb{R}^n)} \leq 2\sqrt{n}|g|_{\Sigma_F(\mathcal{M};\mathbb{R}^n)}. \tag{5.161}$$

Now given $(\mathcal{M}, d)$ and F we introduce new objects denoted by $(\widetilde{\mathcal{M}}, \widetilde{d})$ and $\widetilde{F}$. Unlike F, the set-valued map $\widetilde{F}$ will act from $\widetilde{\mathcal{M}}$ into the set of linear subspaces of $\mathbb{R}^n$ of dimension k and therefore the induction conjecture can be applied to these settings. This yields the corresponding linear operator $T_{\widetilde{F}}$ satisfying (5.156) and (5.157) which then will be exploited to return to the initial settings with the desired operator T_F at the final stage.

To realize this outline we begin with an auxiliary set-valued map $F_f : \mathcal{M} \times \mathcal{M} \to \text{Aff}_{k+1}(\mathbb{R}^n)$ given by

$$F_f(m, m') := F(m') + a(f(m), f(m')). \tag{5.162}$$

Due to Lemma 5.71, $f(m') - b(f(m), f(m')) \in F(m')$ and therefore

$$F_f(m, m') = L_f(m') + (a - b)(f(m), f(m')). \tag{5.163}$$

Further, we define the layer

$$\Lambda_f(m, m') := F_f(m, m') + Q_{d(m,m')},$$

where hereafter Q_r stands for the closed cube $Q_r(0) \subset \mathbb{R}^n$, and then introduce the second auxiliary set-valued map from $\mathcal{M} \times \mathcal{M}$ into the set $\mathcal{C}_0(\mathbb{R}^n)$ of centrally symmetric closed convex subsets of $\mathbb{R}^n$. Specifically, we set for $m, m' \in \mathcal{M}$,

$$G_f(m, m') := L_f(m) \cap \Lambda_f(m, m'). \tag{5.164}$$

The next representation of G_f (the doubling trick) plays an essential role in construction of the desired objects $\widetilde{F}$ and $\widetilde{\mathcal{M}}$.

Lemma 5.72. *For every $m, m' \in \mathcal{M}$ there exist finite sets of numbers $r_i \in \mathbb{R}_+ \cup \{+\infty\}$ and k-dimensional linear subspaces $L_i \subset F(m)$, $i \in J(m', m)$, such that the following holds:*

$$G_f(m, m') := L_f(m) \bigcap \left[\bigcap_{i \in J(m,m')} (L_i + Q_{r_i} + a(f(m), f(m'))) \right]. \tag{5.165}$$

Proof. First, let $F(m) \neq F(m')$. Then the set

$$\widehat{G}_f(m, m') := F(m) \cap (F(m') + Q_{d(m,m')})$$

is a nonempty convex closed centrally symmetric polytope in $\mathbb{R}^n$. Since the difference $f(m) - a(f(m), f(m'))$ belongs to $F(m)$, see (5.158) and (5.159), we get

$$L_f(m) := f(m) + F(m) = a(f(m), f(m')) + F(m).$$

Therefore (5.163) may be rewritten as

$$F_f(m, m') = (F(m) + a) \cap (F(m') + a + Q_{d(m,m')})$$

with $a := a(f(m), f(m'))$. This yields

$$G_f(m, m') = \widehat{G}_f(m, m') + a(f(m), f(m')). \tag{5.166}$$

Now, let L_i, $i \in J(m, m')$, be the set of k-dimensional linear subspaces of $F(m)$ consisting of all parallels to the facets (faces of dimension k) of the $(k+1)$-dimensional polytope $\widehat{G}_f(m, m')$. Let then $L_i + Q_{r_i}$ be the minimal layer of the form $L_i + Q_r$, $0 < r < \infty$, containing $\widehat{G}_f(m, m')$. Then, we have

$$\widehat{G}_f(m, m') = F(m) \cap \left[\bigcap_{i \in J(m,m')} (L_i + Q_{r_i}) \right].$$

Inserting this into (5.166) we get the required equality (5.165) for this case.

Now let $F(m) = F(m')$. Then by definition

$$G_f(m, m') := (F(m) + f(m)) \cap (F(m') + f(m')) + a(f(m), f(m')) + Q_{d(m,m')}$$
$$= (F(m) + f(m)) \cap (F(m) + f(m) + Q_{d(m,m')}) = F(m) + f(m) (=: L_f(m)).$$

Let L_1 be any fixed k-dimensional linear subspace of $F(m)$ and $r_1 = +\infty$. Then $Q_{r_1} = \mathbb{R}^n$ and

$$G_f(m, m') = (F(m) + f(m)) \cap \mathbb{R}^n = L_f(m) \cap (L_1 + Q_{r_1} + a(f(m), f(m'))),$$

i.e., (5.165) holds for this case too. $\square$

Now we use the representation of the lemma to define the desired pseudometric space $(\widetilde{\mathcal{M}}, \widetilde{d})$ and the set-valued map $\widetilde{F} : \widetilde{\mathcal{M}} \to \mathrm{Aff}_k(\mathbb{R}^n)$.

Specifically, we set

$$\widetilde{\mathcal{M}} := \{(m_1, m_2, i) \,;\, m_1, m_2 \in \mathcal{M}, \, i \in J(m_1, m_2)\}, \tag{5.167}$$

and for points $\xi := (m_1, m_2, i)$, $\xi' := (m_1', m_2', i')$ of $\widetilde{\mathcal{M}}$ define

$$\widetilde{d}(\xi, \xi') := d(m_1, m_1') + r_i + r_i'. \tag{5.168}$$

5.4. Simultaneous Lipschitz selections

Extending this expression to pairs (ξ, ξ) by zero we define a function satisfying the metric axioms but assigning $+\infty$ at (ξ, ξ') if $F(m_1(\xi)) = F(m_1(\xi'))$.

Hereafter we use the notation

$$m_j(\xi) := m_j, \ j = 1, 2, \text{ and } i(\xi) := i \quad \text{whenever} \quad \xi = (m_1, m_2, i).$$

In this notation, the set-valued map $\widetilde{F} : \widetilde{\mathcal{M}} \to \mathrm{Aff}_k(\mathbb{R}^n)$ is defined by

$$\widetilde{F}(\xi) := L_{i(\xi)}. \tag{5.169}$$

Since $\dim \widetilde{F} = k$, we may apply the induction hypothesis resulting in

Lemma 5.73. *There exists a linear operator*

$$T_{\widetilde{F}} : \Sigma_{\widetilde{F}}(\widetilde{\mathcal{M}}; \mathbb{R}^n) \to \mathrm{Lip}(\widetilde{\mathcal{M}}, \mathbb{R}^n)$$

such that

$$T_{\widetilde{F}} f \in f + \widetilde{F} \quad \text{and} \quad \|T_{\widetilde{F}}\| \leq c(n, k) \kappa(\mathcal{M})^k. \tag{5.170}$$

Proof. Due to (5.155) the result holds with $\kappa(\widetilde{\mathcal{M}})$ in place of $\kappa(\mathcal{M})$. On the other hand, the space $(\widetilde{\mathcal{M}}, \widetilde{d})$ meets the conditions of Proposition 5.69 with maps $\varphi : \widetilde{\mathcal{M}} \to \mathcal{M}$ and $r : \widetilde{\mathcal{M}} \to \mathbb{R}_+ \cup \{+\infty\}$ given at $\xi := (m, m', i)$ by

$$\varphi(\xi) := \varphi(m) \quad \text{and} \quad r(\xi) := r_i.$$

The proposition then implies $\kappa(\widetilde{\mathcal{M}}) \leq c\kappa(\mathcal{M})$ with $c > 0$ being a numerical constant. $\square$

To exploit such obtained $T_{\widetilde{F}}$ for construction of the desired operator $T_F : \Sigma_F(\mathcal{M}; \mathbb{R}^n) \to \mathrm{Lip}(\mathcal{M}, \mathbb{R}^n)$, we will transfer $T_{\widetilde{F}}$ to a new domain, i.e., $\Sigma_F(\mathcal{M}; \mathbb{R}^n)$, and then map its image from $\mathrm{Lip}(\widetilde{\mathcal{M}}, \mathbb{R}^n)$ into $\mathrm{Lip}(\mathcal{M}, \mathbb{R}^n)$ linearly.

The corresponding transfer operator R is given at $f \in \Sigma_F(\mathcal{M}; \mathbb{R}^n)$ by

$$(Rf)(\xi) := a(f(m_1(\xi)), f(m_2(\xi))), \quad \xi \in \widetilde{\mathcal{M}}, \tag{5.171}$$

see (5.160).

Lemma 5.74. *R is a linear operator from $\Sigma_F(\mathcal{M}; \mathbb{R}^n)$ into $\Sigma_{\widetilde{F}}(\widetilde{\mathcal{M}}; \mathbb{R}^n)$ and*

$$\|R\| \leq c(n).$$

Proof. Since $f \in \Sigma_F(\mathcal{M}; \mathbb{R}^n)$, there exists by definition a function $g \in \mathrm{Lip}(\mathcal{M}, \mathbb{R}^n)$ such that $g \in f + F$ and

$$|g|_{\mathrm{Lip}(\mathcal{M}, \mathbb{R}^n)} \leq 2|f|_{\Sigma_F(\mathcal{M}; \mathbb{R}^n)}.$$

By $h(\xi)$ we denote a point of the affine space $\widetilde{F}(\xi) + (Rf)(\xi)$ nearest to $g(m(\xi))$ (in the ℓ_∞^n-norm). This defines a function $h : \widetilde{\mathcal{M}} \to \mathbb{R}^n$ whose Lipschitz constant is now estimated to obtain the desired bound of $\|Rf\|_{\Sigma_F(\mathcal{M};\mathbb{R}^n)}$.

To this end we use the second function in (5.160) to set
$$\widehat{g}(m, m') := g(m') + (a - b)(f(m), f(m')).$$

Since $g(m') \in L_f(m')$, we get by (5.163) $\widehat{g}(m, m') \in F_f(m, m')$. Moreover, due to (5.161),
$$\|g(m) - \widehat{g}(m, m')\| \leq \|g(m) - g(m')\| + \|(a-b)(f(m), f(m'))\| \leq c(n)\lambda(f)d(m, m'),$$
where we set $\lambda(f) := |f|_{\Sigma_F(\mathcal{M};\mathbb{R}^n)}$, $c(n) := 2(\sqrt{n} + 1)$. These two relations imply that
$$g(m) \in F_f(m, m') + c(n)\lambda(f)Q_{d(m,m')}$$
which, together with (5.163)–(5.165), gives for any $i \in J(m, m')$
$$g(m) \in L_i + a(f(m), f(m')) + Q_{r_i}.$$

Due to (5.169) and (5.171) this can be rewritten as
$$g(m(\xi)) \in \widetilde{F}(\xi) + (Rf)(\xi) + Q_{r_{i(\xi)}} \text{ with } \xi := (m, m', i)$$
which, in turn, yields
$$\|g(m(\xi)) - h(\xi)\| := d\big(g(m(\xi)), \widetilde{F}(\xi) + (Rf)(\xi)\big) \leq c(n)\lambda(f)r_{i(\xi)}.$$

Now the required Lipschitz constant is estimated as follows:
$$\|h(\xi) - h(\xi')\| \leq \|h(\xi) - g(m(\xi))\| + \|g(m(\xi)) - g(m(\xi'))\| + \|h(\xi') - g(m(\xi'))\|$$
$$\leq c(n)\lambda(f)[r_{i(\xi)} + d(m(\xi), m(\xi')) + r_{i(\xi')}] =: c(n)\lambda(f)\widetilde{d}(\xi, \xi').$$

Hence, $h \in \text{Lip}(\widetilde{\mathcal{M}}, \mathbb{R}^n)$ and its seminorm is at most $c(n)\lambda(f)$; moreover, $h \in \widetilde{F} + Rf$ by definition. This immediately implies
$$|Rf|_{\Sigma_{\widetilde{F}}(\widetilde{\mathcal{M}};\mathbb{R}^n)} \leq |h|_{\text{Lip}(\widetilde{\mathcal{M}},\mathbb{R}^n)} \leq c(n)\lambda(f) =: c(n)|f|_{\Sigma_F(\mathcal{M};\mathbb{R}^n)}.$$

Hence, $\|R\|$ is bounded by $c(n)$ as required. □

At the next stage of our construction of T_F we should find a linear operator, say T, that maps $\text{Lip}(\widetilde{\mathcal{M}}, \mathbb{R}^n)$ into $\text{Lip}(\mathcal{M}, \mathbb{R}^n)$ with norm bounded by some $c(n) > 0$. It will appear (as a by-product) within the proof of an n-dimensional version of Proposition 5.69 concerning a couple of spaces defined on a pseudometric space $\widehat{\mathcal{M}} := (\widehat{\mathcal{M}}, \widehat{d})$ with the underlying set $\widehat{\mathcal{M}} := \widetilde{\mathcal{M}}$ and the pseudometric $\widehat{d}$ given at (ξ, ξ') by
$$\widehat{d}(\xi, \xi') := d(m_1(\xi), m_1(\xi')). \tag{5.172}$$

5.4. Simultaneous Lipschitz selections

The first space of the couple is $\operatorname{Lip}(\widehat{\mathcal{M}}, \mathbb{R}^n)$. Since (5.172) depends only on the first coordinates of ξ, ξ', every function $f \in \operatorname{Lip}(\widehat{\mathcal{M}}, \mathbb{R}^n)$ is a constant on each slice $\{\xi = (m_1(\xi), m_2(\xi), i(\xi)) \in \widetilde{\mathcal{M}}\,;\, m_1(\xi) = m\}$. Hence, f can and will be regarded as a function on the metric space $(\mathcal{M}, d)$. This identification gives rise to the linear surjection

$$I : \operatorname{Lip}(\widehat{\mathcal{M}}, \mathbb{R}^n) \to \operatorname{Lip}(\mathcal{M}, \mathbb{R}^n) \tag{5.173}$$

of norm $\|I\| = 1$.

The second space of the couple, an analog of ℓ_∞^w, is defined on $\widehat{\mathcal{M}}$ using a set-valued weight $W : \widehat{\mathcal{M}} \to \{Q_r\,;\, r \in \mathbb{R}_+ \cup \{+\infty\}\}$ given at $\xi \in \widetilde{\mathcal{M}}\,(=: \widehat{\mathcal{M}})$ by

$$W(\xi) := Q_{r_{i(\xi)}}. \tag{5.174}$$

Then the space in question denoted by $\ell_\infty^W(\widehat{\mathcal{M}}; \mathbb{R}^n)$ is defined by finiteness of a functional given at $f : \widehat{\mathcal{M}} \to \mathbb{R}^n$ by

$$\|f\|_{\ell_\infty^W(\widehat{\mathcal{M}}; \mathbb{R}^n)} := \inf\{\lambda > 0\,;\, f(\xi) \in \lambda W(\xi) \text{ for all } \xi \in \widetilde{\mathcal{M}}\}. \tag{5.175}$$

Since the case of $F = const$ has been already considered, some r_i in (5.174) are finite, and therefore (5.175) is zero only for $f = 0$. Moreover, all $r_i > 0$ so that (5.175) is finite, i.e., it defines a *norm*.

Using Proposition 5.69 we show that the couple $(\operatorname{Lip}(\widehat{\mathcal{M}}, \mathbb{R}^n), \ell_\infty^W(\widehat{\mathcal{M}}; \mathbb{R}^n))$ is K-linearizable and estimate its K-linearity constant. But beforehand we reveal a relation between the space $\operatorname{Lip}((\widetilde{\mathcal{M}}, \widetilde{d}), \mathbb{R}^n)$ and the sum of the couple

$$\Sigma_W(\widehat{\mathcal{M}}; \mathbb{R}^n) := \operatorname{Lip}(\widehat{\mathcal{M}}, \mathbb{R}^n) + \ell_\infty^W(\widehat{\mathcal{M}}; \mathbb{R}^n).$$

Lemma 5.75. *As linear spaces*

$$\operatorname{Lip}(\widetilde{\mathcal{M}}, \mathbb{R}^n) = \Sigma_W(\widehat{\mathcal{M}}; \mathbb{R}^n)$$

and distortion of the identity map is at most 4.

Proof. Let first $f \in \operatorname{Lip}(\widetilde{\mathcal{M}}) := \operatorname{Lip}((\widetilde{\mathcal{M}}, \widetilde{d}), \mathbb{R}^n)$. If $f = f_0 + f_1$, where

$$f_0 \in \operatorname{Lip}(\widehat{\mathcal{M}}) := \operatorname{Lip}((\widehat{\mathcal{M}}, \widehat{d}), \mathbb{R}^n) \text{ and } f_1 \in \ell_\infty^W := \ell_\infty^W(\widehat{\mathcal{M}}; \mathbb{R}^n),$$

then by definition

$$|f|_{\operatorname{Lip}(\widetilde{\mathcal{M}})} := \sup_{\xi \neq \xi'} \frac{\|f(\xi) - f(\xi')\|}{\widehat{d}(\xi, \xi') + r_{i(\xi)} + r_{i(\xi')}} \leq \sup_{\xi \neq \xi'} \frac{\|f_0(\xi) - f_0(\xi')\|}{\widehat{d}(\xi, \xi')} + 2 \sup_\xi \frac{\|f_1(\xi)\|}{r_{i(\xi)}}$$

$$\leq 2(|f_0|_{\operatorname{Lip}(\widehat{\mathcal{M}})} + |f_1|_{\ell_\infty^W}).$$

Taking infimum over all such decompositions of f we get

$$|f|_{\operatorname{Lip}(\widetilde{\mathcal{M}})} \leq 2|f|_{\Sigma_W(\widehat{\mathcal{M}}; \mathbb{R}^n)}. \tag{5.176}$$

Conversely, let $f \in \mathrm{Lip}(\widetilde{\mathcal{M}})$ and $g = (g_i)_{1 \le i \le n} : \widetilde{\mathcal{M}} \to \mathbb{R}^n$ be given by

$$g_i(\xi) := \inf_{\xi'}\{f_i(\xi') + |f|_{\mathrm{Lip}(\widetilde{\mathcal{M}})}(r_{i(\xi')} + \widehat{d}(\xi,\xi'))\}, \; 1 \le i \le n.$$

Let us first show that

$$|g|_{\mathrm{Lip}(\widehat{\mathcal{M}})} \le |f|_{\mathrm{Lip}(\widetilde{\mathcal{M}})}.$$

Actually, if for definiteness $g_i(\xi) - g_i(\xi') \ge 0$ and the infimum is, up to $\varepsilon > 0$, attained for $g_i(\xi')$ at ξ^o, then

$$g_i(\xi) - g_i(\xi') \le f_i(\xi^o) + |f|_{\mathrm{Lip}(\widetilde{\mathcal{M}})}(d_{i(\xi^o)} + \widehat{d}(\xi^o,\xi')) + \varepsilon \le |f|_{\mathrm{Lip}(\widetilde{\mathcal{M}})}\widehat{d}(\xi,\xi') + \varepsilon$$

and the result is proved.

Further, we show that

$$|f - g|_{\ell_\infty^W} \le |f|_{\mathrm{Lip}(\widetilde{\mathcal{M}})}.$$

By the definition of g_i we have

$$|f_i - g_i|(\xi) \le \inf_{\xi'}\bigl(|f_i(\xi) - f_i(\xi')| - |f|_{\mathrm{Lip}(\widetilde{\mathcal{M}})}(r_{i(\xi')} + \widehat{d}(\xi,\xi'))\bigr)$$

$$\le |f|_{\mathrm{Lip}(\widetilde{\mathcal{M}})} \inf_{\xi'}\bigl(\widehat{d}(\xi,\xi') + r_{i(\xi)} + r_{i(\xi')} - r_{i(\xi')} - \widehat{d}(\xi,\xi')\bigr)$$

$$= |f|_{\mathrm{Lip}(\widetilde{\mathcal{M}})} r_{i(\xi)}.$$

Then this implies the second required inequality

$$|f - g|_{\ell_\infty^W} := \max_{1 \le i \le n} \sup_{\xi} \frac{|f_i - g_i|(\xi)}{r_{i(\xi)}} \le |f|_{\mathrm{Lip}(\widetilde{\mathcal{M}})}.$$

Combining the inequalities now proved we finally have

$$|f|_{\Sigma_W} \le |g|_{\mathrm{Lip}(\widehat{M})} + |f - g|_{\ell_\infty^W} \le 2|f|_{\mathrm{Lip}(\widetilde{\mathcal{M}})}.$$

Together with (5.176) this proves the lemma. $\square$

Now we prove for the couple introduced the aforementioned analog of Proposition 5.69.

Lemma 5.76. *There exist linear operators*

$$T : \Sigma_W(\widehat{\mathcal{M}}; \mathbb{R}^n) \to \mathrm{Lip}(\widehat{\mathcal{M}}, \mathbb{R}^n), \quad S : \Sigma_W(\widehat{\mathcal{M}}; \mathbb{R}^n) \to \ell_\infty^W(\widehat{\mathcal{M}}; \mathbb{R}^n)$$

such that

$$T + S = \mathrm{Id}_{\Sigma_W(\widehat{\mathcal{M}};\mathbb{R}^n)} \tag{5.177}$$

and, moreover,

$$\|T\| + \|S\| \le c(n)\kappa(\mathcal{M}). \tag{5.178}$$

5.4. Simultaneous Lipschitz selections

Proof. The weight W is the n-th degree of a map $\widehat{W} : \widehat{\mathcal{M}} \to \mathrm{Aff}(\mathbb{R})$ given by

$$\widehat{W}(\xi) := \{t \in \mathbb{R}\,;\, |t| \leq r_{i(\xi)}\}.$$

Applying Proposition 5.69 to this map we find linear operator

$$\widehat{T} : \Sigma_{\widehat{W}}(\widehat{\mathcal{M}}; \mathbb{R}) \to \mathrm{Lip}(\widehat{\mathcal{M}}, \mathbb{R}), \qquad \widehat{S} : \Sigma_{\widehat{W}}(\widehat{\mathcal{M}}; \mathbb{R}) \to \ell_\infty^{\widehat{W}}(\widehat{\mathcal{M}}; \mathbb{R})$$

such that

$$\widehat{T} + \widehat{S} = \mathit{Id}_{\Sigma_{\widehat{W}}(\widehat{\mathcal{M}};\mathbb{R})}$$

and, moreover, for some numerical constant $c > 0$,

$$\|\widehat{T}\| + \|\widehat{S}\| \leq c\kappa(\widehat{\mathcal{M}}).$$

Using these we define new linear operators T and S, acting from $\Sigma_W(\widehat{\mathcal{M}}; \mathbb{R}^n)$ into $\mathrm{Lip}(\widehat{\mathcal{M}}, \mathbb{R}^n)$ and $\ell_\infty^W(\widehat{\mathcal{M}}; \mathbb{R}^n)$, respectively. Specifically, for $f := (f_i)_{1 \leq i \leq n} : \widehat{\mathcal{M}} \to \mathbb{R}^n$ we set

$$Tf := (\widehat{T}f_i)_{1 \leq i \leq n} \qquad Sf := (\widehat{S}f_i)_{1 \leq i \leq n}.$$

The operators so obtained clearly satisfy (5.177) and (5.178), but the latter with $\kappa(\widehat{\mathcal{M}})$ in place of $\kappa(\mathcal{M})$. Applying then again Proposition 5.69 with the maps $\varphi : \widehat{\mathcal{M}} \to \mathcal{M}$ given by $\varphi(\xi) := m_1(\xi)$ and $r := 0$, we obtain for some numerical constant $c > 0$,

$$\kappa(\widehat{\mathcal{M}}) \leq c\kappa(\mathcal{M}).$$

This completes the proof. $\square$

Combining Lemmas 5.75 and 5.76 we get

Corollary 5.77. *The linear operator T maps $\mathrm{Lip}(\widetilde{\mathcal{M}}, \mathbb{R}^n)$ into $\mathrm{Lip}(\widehat{\mathcal{M}}, \mathbb{R}^n)$ and its norm satisfies*

$$\|T\| \leq c(n)\kappa(\mathcal{M}).$$

Further, we compose all of the operators introduced which are presented in the diagram:

$$\Sigma_F(\mathcal{M}; \mathbb{R}^n) \xrightarrow{R} \Sigma_{\widetilde{F}}(\widetilde{\mathcal{M}}; \mathbb{R}^n) \xrightarrow{T_{\widetilde{F}}} \mathrm{Lip}(\widetilde{\mathcal{M}}, \mathbb{R}^n) \xrightarrow{T} \mathrm{Lip}(\widehat{\mathcal{M}}, \mathbb{R}^n).$$

The operator so obtained is denoted by U, i.e.,

$$U := TT_{\widetilde{F}}R. \tag{5.179}$$

Using the identification of the last space in the diagram with $\mathrm{Lip}(\mathcal{M}, \mathbb{R}^n)$, see (5.173), we regard every function Uf as an element of the latter space; hence,

U maps $\Sigma_F(\mathcal{M}; \mathbb{R}^n)$ into $\text{Lip}(\mathcal{M}, \mathbb{R}^n)$. Moreover, due to Lemmas 5.73, 5.74 and Corollary 5.77,

$$\|U\| \leq c(n)\kappa(\mathcal{M})c(n,k)\kappa(\mathcal{M})^k c(n) = c(k+1,n)\kappa(\mathcal{M})^{k+1}. \tag{5.180}$$

Unfortunately, U cannot be used as the desired operator T_F, since Uf is not a selection of $L_f := f + F$, but U may be transformed into the T_F in the following way.

Let $Pr(\cdot; L)$ denote the orthogonal projection of $\mathbb{R}^n$ onto an affine subspace $L \subset \mathbb{R}^n$. Then we define the required operator by setting

$$T_F f := Pr(Uf; L_f). \tag{5.181}$$

Clearly, $T_F f$ is a selection of $L_f := f + F$, as required. To show that T_F is linear we use a relation between the orthogonal projection onto the parallel subspaces $L_f(m)$ and $F(m)$ to write

$$T_F f = Pr(Uf; F) - Pr(f; F) + f.$$

Since $F(m)$ is a linear subspace, the orthogonal projector is a linear map, hence T_F is linear.

It remains to show that for every f from its domain $\Sigma_F(\mathcal{M}; \mathbb{R}^n)$ and $m, m' \in \mathcal{M}$,

$$\|(T_F f)(m) - (T_F f)(m')\| \leq c(k+1,n)\mu d(m,m'), \tag{5.182}$$

where we set for brevity

$$\mu := \kappa(\mathcal{M})^{k+1} |f|_{\Sigma_F(\mathcal{M}; \mathbb{R}^n)}. \tag{5.183}$$

In the subsequent derivation we also set for brevity

$$u := Uf, \quad g := T_F f \quad \text{and} \quad h := (T_{\widetilde{F}} R)f. \tag{5.184}$$

Due to (5.181) the left-hand side of (5.182) is bounded by the sum $J_1 + J_2 + J_3$, where we set

$$J_1 := \|Pr(u(m), L_f(m)) - Pr(u(m), g(m) + F(m'))\|,$$
$$J_2 := \|Pr(u(m), g(m) + F(m')) - Pr(u(m'), g(m) + F(m'))\|,$$
$$J_3 := \|Pr(u(m'), g(m) + F(m')) - Pr(u(m'), L_f(m'))\|.$$

Hence, the desired inequality (5.182) will straightforwardly follow from

Claim. $J_i \leq c(k,n)\mu d(m,m')$.

Its proof is rather technical and requires several auxiliary results.

5.4. Simultaneous Lipschitz selections

Lemma 5.78. *For $\xi := (m, m', i)$ the next inequalities*

$$\|u(m) - h(\xi)\| \leq c(k+1, n)\mu r_{i(\xi)}, \tag{5.185}$$

$$\|g(m) - h(\xi)\| \leq \sqrt{n} c(k+1, n)\mu r_{i(\xi)} \tag{5.186}$$

are true; here $c(k+1, n)$ is a constant in (5.180).

Proof. By the definition of U, see (5.179), and the equality $T + S = \mathrm{Id}$ of Lemma 5.76 we have

$$\|u(m) - h(\xi)\| := \|u(m(\xi)) - h(\xi)\| = \|Th - h\|(\xi) = \|Sh\|(\xi) := \|(ST_{\widetilde{F}}R)f\|(\xi).$$

By the definition of the norm of $\ell_\infty^W(\widehat{\mathcal{M}}; \mathbb{R}^n)$, see (5.175), the right-hand side is at most $\|(ST_{\widetilde{F}}R)f\|_{\ell_\infty^W(\widetilde{\mathcal{M}}; \mathbb{R}^n)} r_{i(\xi)}$ which, in turn, is bounded by

$$c(k+1, n)\kappa(\mathcal{M})^{k+1} r_{i(\xi)} |f|_{\Sigma_F(\widehat{\mathcal{M}}; \mathbb{R}^n)} := c(k+1, n)\mu r_{i(\xi)}.$$

This proves (5.185).

To prove (5.186) we first note that due to Lemma 5.73 the map $h := T_{\widetilde{F}}(Rf)$ is a selection of $\widetilde{F} + Rf$, so that by the definition of $\widetilde{F}$ and R, see (5.169) and (5.171), we have

$$h(\xi) \in \widetilde{F}(\xi) + (Rf)(\xi) := L_{i(\xi)} + a(f(m), f(m')) \subset F(m) + a(f(m), f(m')).$$

Applying now Lemma 5.71 with $L_i := L_f(m)$, $i = 1, 2$, and $x := f(m)$, $y := f(m')$ we conclude that $a(f(m), f(m'))$ belongs to $f(m) + F(m)$; hence we get

$$h(\xi) \in F(m) + f(m) + F(m) = L_f(m).$$

This, (5.181) and (5.184) then imply that

$$\|g(m) - h(\xi)\| \leq \|g(m) - h(\xi)\|_{\ell_2^n} := \|Pr(u(m), L_f(m)) - h(\xi)\|_{\ell_2^n}$$
$$\leq \|u(m) - h(\xi)\|_{\ell_2^n} \leq \sqrt{n}\|u(m) - h(\xi)\|.$$

Together with (5.184) this proves the second inequality. $\square$

To formulate a consequence of this result we set

$$r(m) := \mu \inf\{r_{i(\xi)} \,;\, m_1(\xi) = m\} \text{ and } K_m := Q_{r(m)}(g(m)), \tag{5.187}$$

where we recall $\mu := \kappa(\mathcal{M})^{k+1} |f|_{\Sigma_F(\mathcal{M}; \mathbb{R}^n)}$.

Corollary 5.79. *The point $u(m) \in \widetilde{c} K_m$, where hereafter*

$$\widetilde{c} := (1 + \sqrt{n}) c(k+1, n).$$

Proof. Using (5.185) and (5.186) we estimate the distance from the center $g(m)$ of K_m to $u(m)$ as

$$\|u(m) - g(m)\| \leq \|u(m) - h(\xi)\| + \|h(\xi) - g(m)\| \leq \tilde{c}\mu r_{i(\xi)}.$$

This implies the desired estimate

$$\|u(m) - g(m)\| \leq \tilde{c}\mu \inf\{r_{i(\xi)}\,;\, m_1(\xi) = m\} := \tilde{c}r(m). \qquad \square$$

Lemma 5.80. *Let $\rho := \tilde{c}\mu d(m,m')$. Then*

$$K_m \cap L_f(m) \subset F(m') + a(f(m), f(m')) + Q_\rho. \tag{5.188}$$

Proof. Let $x \in K_m$; by definition $\|x - g(m)\| \leq r(m) \leq \mu r_{i(\xi)}$ and therefore

$$\|x - h(\xi)\| \leq \|x - g(m)\| + \|g(m) - h(\xi)\| \leq \tilde{c}\mu r_{i(\xi)}.$$

Since, moreover, $h(\xi) \in L_{i(\xi)} + a$, where $a = a(f(m), f(m'))$, we get from this inequality

$$K_m \subset h(\xi) + \tilde{c}\mu Q_{r_{i(\xi)}} \subset a + L_{i(\xi)} + \tilde{c}\mu Q_{r_{i(\xi)}}.$$

This holds for every index $i \in J(m,m')$ and therefore

$$K_m - a \subset \bigcap_{i \in J(m,m')} \{L_i + \tilde{c}\mu Q_{r_i}\} = \tilde{c}\mu \left(\bigcap_{i \in J(m,m')} \{L_i + Q_{r_i}\} \right).$$

Since $a + F(m) = f(m) + F(m) =: L_f(m)$, this, (5.165) and (5.166) imply

$$K_m \cap L_f(m) = K_m \cap \{F(m) + a\} \subset \tilde{c}\mu \left[(\cap_{i \in J(m,m')}(L_i + Q_{r_i})) \cap F(m) \right] + a$$
$$= \tilde{c}\mu[F(m) \cap \{F(m') + Q_{d(m,m')}\}] + a$$
$$\subset \{F(m') + a\} + Q_{\tilde{c}\mu d(m,m')}. \qquad \square$$

Lemma 5.81. *There exists a point $x(m,m')$ of $F(m') + a(f(m), f(m'))$ such that*

$$\|x(m,m') - g(m)\| \leq \tilde{c}\mu d(m,m')\,(=: \rho). \tag{5.189}$$

Proof. By the embedding of the previous lemma

$$g(m) \in K_m \cap \{f(m) + F(m)\} \subset F(m') + a(f(m), f(m')) + Q_\rho.$$

Hence, there exists the required point $x(m,m')$ satisfying (5.189). $\qquad \square$

Now we are prepared to prove our CLAIM. We begin with the estimate of

$$J_2 := \|Pr(u(m), g(m) + F(m')) - Pr(u(m'), g(m) + F(m'))\|.$$

5.4. Simultaneous Lipschitz selections

Since the metric projection onto an affine subspace is 1-Lipschitz, we have

$$J_2 \leq \|u(m) - u(m')\|_{\ell_2^n} \leq \sqrt{n}\|u(m) - u(m')\| := \sqrt{n}\|(Uf)(m) - (Uf)(m')\|$$
$$\leq \sqrt{n}|Uf|_{\mathrm{Lip}(\mathcal{M},\mathbb{R}^n)}d(m,m') \leq \sqrt{n}\|U\|\|f\|_{\Sigma_F(\mathcal{M};\mathbb{R}^n)}d(m,m').$$

Moreover, $\|U\| \leq c(k+1,n)\kappa(\mathcal{M})^{k+1}$ by (5.180) and the desired result for J_2 follows.

To estimate the term

$$J_3 := \|Pr(u(m'), g(m) + F(m')) - Pr(u(m'), L_f(m'))\|$$

we show that the affine subspace $L_f(m') := f(m') + F(m')$ is the shift of that of $g(m') + F(m')$ by a factor v whose Euclidean norm satisfies

$$\|v\|_{\ell_2^n} \leq \widehat{c}(k,n)|f|_{\Sigma_F(\mathcal{M};\mathbb{R}^n)}d(m,m'). \tag{5.190}$$

Since $J_3 \leq \|v\|_{\ell_2^n}$, this would give the desired estimate for J_3.

To find v we first note that by (5.159),

$$L_f(m') = F(m') + b =: F(m') + b(f(m), f(m'));$$

a similar abbreviation will be used for $a(f(m), f(m'))$.

Then $x(m,m') \in F(m') + a = L_f(m') + (b-a)$ and therefore

$$L_f(m') = F(m') + b = \{F(m') + g(m)\} + (x(m,m') - g(m) + b - a).$$

The vector in the parenthesis denoted by v is the required shift. Actually, due to (5.189) and inequality (5.161) estimating the Lipschitz constant of $b - a$ we have

$$\|v\|_{\ell_2^n} \leq \sqrt{n}\|v\| \leq \sqrt{n}(\|x(m,m') - g(m)\| + \|b-a\|)$$
$$\leq \sqrt{n}(\widetilde{c}\mu d(m,m') + |b-a|_{\mathrm{Lip}(\mathcal{M},\mathbb{R}^n)}d(m,m'))$$
$$\leq \sqrt{n}(\widetilde{c}\kappa(\mathcal{M})^{k+1} + 2\sqrt{n})|f|_{\Sigma_F(\mathcal{M};\mathbb{R}^n)}d(m,m')$$

and (5.190) follows.

It remains to estimate

$$J_1 := \|Pr(u(m), L_f(m)) - Pr(u(m), g(m) + F(m'))\|.$$

To this end we need an additional lemma in which formulation and proof $B_r(x)$ denotes a closed Euclidean ball of $\mathbb{R}^n$ and $\|\cdot\|_2$ stands for the ℓ_2^n-norm (so distances there are measured in this norm).

Lemma 5.82. *Let L_1, L_2 be linear subspaces of $\mathbb{R}^n$ and the orthogonal projection $y := Pr(x, L_1)$ belong to $L_1 \cap L_2$. Then for every ball $B := B_r(y)$ containing x,*

$$\|Pr(x, L_1) - Pr(x, L_2)\|_2 \leq d_{\mathcal{H}}(B \cap L_1, B \cap L_2). \tag{5.191}$$

Let us recall that $d_{\mathcal{H}}$ is the Hausdorff metric given by
$$d_{\mathcal{H}}(S_0, S_1) := \inf\{r > 0;\ S_i + B_r(0) \supset S_{1-i},\ i = 0, 1\}.$$

Proof. We set
$$x' := Pr(x, L_2), \qquad x'' := Pr(x', L_1).$$
Then x'' is contained in the ball $B_\rho(y)$ of radius $\rho := \|y - x'\|_2$ (as $y \in L_1 \cap L_2$). This, minimality of orthogonal projection and the definition of $d_{\mathcal{H}}$ then yield
$$\|x' - x''\|_2 = d(x', B_\rho(y) \cap L_1) \leq d_{\mathcal{H}}(B_\rho(y) \cap L_1, B_\rho(y) \cap L_2)$$
$$\leq d_{\mathcal{H}}(B_r(y) \cap L_1, B_r(y) \cap L_2).$$

Hence, we get in the notation introduced
$$\|x' - x''\|_2 \leq \frac{\|y - x'\|_2}{r} d_{\mathcal{H}}(B \cap L_1, B \cap L_2). \tag{5.192}$$

Now we will show that
$$\|y - x'\|_2^2 \leq r\|x' - x''\|_2. \tag{5.193}$$

Multiplying (5.192) by $\|y - x'\|_2$, using (5.193) and recalling that $\|y - x'\|_2$ is the left-hand side of (5.191) we obtain the desired result.

To prove (5.193) we denote by ℓ a 1-dimensional subspace of the affine hull $\mathrm{aff}\{x, y, x'\}$ which is orthogonal to $x - y$ (briefly, $(x - y) \perp \ell$). But $x - y := x - Pr(x, L_1)$ is orthogonal to L_1 and therefore $(x - y) \perp \mathrm{aff}(\ell \cup L_1)$. Hence, the vectors $x - y$ and $x' - Pr(x', \ell)$ lie in the plane $\mathrm{aff}\{x, y, x'\}$ and perpendicular to ℓ. In particular, they are parallel and therefore $(x' - Pr(x', \ell)) \perp \mathrm{aff}(\ell \cup L_1)$ and so
$$\|x' - Pr(x', \ell)\|_2 = d(x', \ell) = d(x', \mathrm{aff}(\ell \cup L_1)) \leq d(x', L_1) = \|x' - x''\|_2.$$

Further, $\Delta_1 := \mathrm{conv}\{y, x', Pr(x', \ell)\}$ and $\Delta_2 := \mathrm{conv}\{x, y, x'\}$ are triangles with the common side $[x, y]$ and parallel sides $x' - Pr(x', \ell)$ and $x - y$. Due to their similarity
$$\frac{\|y - x'\|_2}{\|x' - Pr(x', \ell)\|_2} = \frac{\|y - x\|_2}{\|y - x'\|_2},$$
and we conclude that
$$\|y - x'\|_2^2 = \|y - x\|_2 \cdot \|x' - Pr(x', \ell)\|_2 \leq r\|x' - Pr(x', \ell)\|_2.$$

Since the norm in the right-hand side is, clearly, less than $r\|x' - Pr(x', L_1)\|_2 := r\|x' - x''\|_2$, (5.193) follows. □

Now we apply this lemma to the case
$$L_1 := L_f(m),\ L_2 := g(m) + F(m'),\ x := u(m)\ \text{and}\ y := Pr(u(m), L_f(m)).$$

5.4. Simultaneous Lipschitz selections

Since by (5.184) $Pr(u(m), L_f(m)) =: g(m)$, and $g(m) := (T_F f)(m) \in f(m) + F(m) := L_f(m)$, i.e., the point $g(m) \in L_f(m) \cap \{g(m) + F(m')\}$, the lemma yields

$$J_1 \leq d_{\mathcal{H}}(B \cap L_f(m), B \cap \{y(m) + F(m')\}) \qquad (5.194)$$

for $B := B_r(g(m))$ and r satisfying

$$\|g(m) - u(m)\|_2 \leq r.$$

By Corollary 5.79 the norm in the left-hand side is at most $\sqrt{n}\tilde{c}r(m)$ and therefore we can take, in (5.194),

$$r := \sqrt{n}\tilde{c}r(m) \text{ and } B := B_r(g(m)).$$

To estimate the Hausdorff distance in (5.194) for the chosen B we first show that

$$B \cap L_f(m) \subset B \cap \{g(m) + F(m')\} + B_{2\sqrt{n}f}(0) \qquad (5.195)$$

and from here then derive the similar embedding with interchanging $L_f(m)$ and $g(m) + F(m')$. This clearly would estimate the right-hand side of (5.194) by $2\sqrt{n}r$, i.e., we have obtained the required inequality

$$J_1 \leq 2\sqrt{n}r := 2\sqrt{n}(1 + \sqrt{n})c(k+1, n)|f|_{\Sigma_F(\mathcal{M}; \mathbb{R}^n)} d(m, m'). \qquad (5.196)$$

By Lemma 5.81 we have, for $\rho := \tilde{c}\mu d(m, m')$ and $a := a(f(m), f(m'))$,

$$F(m') + a = F(m') + x(m, m') = \{F(m') + g(m)\} + (x(m, m') - g(m))$$
$$\subset \{F(m') + g(m)\} + Q_\rho.$$

This and (5.188) then imply

$$K_m \cap L_f(m) \subset F(m') + a + Q_\rho \subset \{F(m') + g(m)\} + Q_{2\rho}.$$

The sets in both sides have the center of symmetry $g(m)$. Dilating with respect to the center with the factor $\sqrt{n}\tilde{c}$ we then obtain

$$(\sqrt{n}\tilde{c}K_m) \cap L_f(m) \subset \{F(m') + g(m)\} + Q_{2\sqrt{n}\tilde{c}\rho}.$$

Since K_m and B have the same center $g(m)$ and, moreover, $r_B = \sqrt{n}\tilde{c}r_{K_m}$ $(:= \sqrt{n}\tilde{c}r(m))$, this embedding implies

$$B \cap L_f(m) \subset \{F(m') + g(m)\} + \sqrt{n}Q_{2\sqrt{n}\rho} \subset \{F(M') + g(m)\} + B_{2n\rho}(0).$$

Moreover, for every x from the left-hand side its point nearest to $F(m') + g(m)$ is contained in B. Therefore the previous embedding leads to

$$B \cap L_f(m) \subset B \cap \{F(m') + g(m)\} + B_{2n\rho}(0).$$

This coincides with (5.195), since $2n\rho := 2\sqrt{n}r$.

Finally, since the affine subspaces $L_f(m)$ and $F(m') + g(m)$ are of the same dimension, we can rotate $L_f(m)$ about $g(m)$ to transform $B \cap L_f(m)$ into $B \cap \{F(m') + g(m)\}$. This clearly leads to the second embedding:

$$B \cap \{F(m') + g(m)\} \subset B \cap L_f(m) + B_{2n\rho}(0)$$

and proves (5.196) and our CLAIM; hence, the remaining property of the operator T_F, see (5.182), has also been proved. This completes the proof of the induction hypothesis (5.155) and therefore establishes Theorem 5.66 up to the derivations of Propositions 5.68 and 5.69 presented below. □

5.4.5 Proof of Proposition 5.68

We divide the proof in two parts showing in part A the basic inequality of the proposition under additional restrictions and then in part B derive from here the general result.

Part A. We assume that a *weight* w defined on a metric space $\mathcal{M} := (\mathcal{M}, d)$ is *finite*. Then the seminorm

$$\|f\|_{\ell_\infty^w(\mathcal{M})} := \inf\{\lambda > 0; |f(m)| \leq \lambda w(m) \text{ for all } m \in \mathcal{M}\}$$

is nondegenerate but equals $+\infty$ if $f \neq 0$ on the zero set of w (i.e., $\ell_\infty^w(\mathcal{M})$ is a pseudometric space).

Further, due to Lemma 5.75 for $n = 1$ the sum

$$\Sigma_w(\mathcal{M}) := \mathrm{Lip}(\mathcal{M}) + \ell_\infty^w(\mathcal{M})$$

coincides as a linear space with the Lipschitz space $\mathrm{Lip}(\mathcal{M}_w)$ and

$$\frac{1}{2}|f|_{\Sigma_w(\mathcal{M})} \leq |f|_{\mathrm{Lip}(\mathcal{M}_w)} + 2|f|_{\Sigma_w(\mathcal{M})}. \tag{5.197}$$

Here $\mathcal{M}_w := (\mathcal{M}, d_w)$, where a pseudometric d_w is given for $m \neq m'$ by

$$d_w(m, m') := d(m, m') + w(m) + w(m') \tag{5.198}$$

and for $m \neq m'$ by 0. Since $w < \infty$, this, in fact, is a metric.

Our second restriction asserts:

The metric spaces $\mathcal{M}$ and $\mathcal{M}_w$ are complete.

Under the restrictions imposed we should prove that $\kappa(\mathcal{M}) \leq O(1)\lambda(\mathcal{M})^2$; hereafter $O(1)$ stands for a numerical constant. In fact, we will prove a stronger result with $\lambda(\mathcal{M})$ in place of its square. Due to the definition of $\kappa(\mathcal{M})$, see (5.154), we should find linear operators T_w, S_w mapping $\Sigma_w(\mathcal{M})$ into $\mathrm{Lip}(\mathcal{M})$ and ℓ_∞^w, respectively, such that

$$T_w + S_w = \mathrm{Id}_{\Sigma_w(\mathcal{M})} \text{ and } \|T_w\| + \|S_w\| \leq O(1)\lambda(\mathcal{M}). \tag{5.199}$$

5.4. Simultaneous Lipschitz selections

The optimal constant there is called the K-linearity constant and is denoted by $\kappa(\mathrm{Lip}(\mathcal{M}), \ell_\infty^w(\mathcal{M}))$ (it is introduced by (5.154)). The basic assumption on $\mathcal{M}$, finiteness of the characteristic

$$\lambda(\mathcal{M}) := \sup_{S \subset \mathcal{M}} \inf_T \{\|T\|\}, \tag{5.200}$$

where T runs over all linear extension operators from $\mathrm{Lip}(S)$ into $\mathrm{Lip}(\mathcal{M})$, is used in the proof as follows.

We first introduce a metric subspace $\Omega := (\Omega, d_w|_{\Omega \times \Omega})$ of the space $\mathcal{M}_w$ whose metric is equivalent to $d|_{\Omega \times \Omega}$. This implies that the restriction operator $R_\Omega : f \to f|_\Omega$ maps $\mathrm{Lip}(\mathcal{M}_w)$ into $\mathrm{Lip}(\Omega) := \mathrm{Lip}\left((\Omega, d|_{\Omega \times \Omega})\right)$; moreover, its norm will be proved to be at most 2.

Further, using finiteness of $\lambda(\mathcal{M})$ we find a linear extension operator $E_\Omega : \mathrm{Lip}(\Omega) \to \mathrm{Lip}(\mathcal{M})$ whose norm is bounded by $2\lambda(\mathcal{M})$. Then the desired operator T_w is obtained by composing the operators presented in the following diagram:

$$\Sigma_w(\mathcal{M}) \xrightarrow{I} \mathrm{Lip}(\mathcal{M}_w) \xrightarrow{R_\Omega} \mathrm{Lip}(\Omega) \xrightarrow{E_\Omega} \mathrm{Lip}(\mathcal{M}). \tag{5.201}$$

The norm of T_w is, hence, bounded by

$$\|T_w\| \leq \|I\| \cdot \|R_\Omega\| \cdot \|E_\Omega\| \leq 32\lambda(\mathcal{M}).$$

The norm of the second operator $S_w := Id_{\Sigma_w(\mathcal{M})} - T_w$ is also estimated from the properties of the subspace Ω; they are described in

Lemma 5.83. *Assume that $\mathcal{M}$ and $\mathcal{M}_w$ are complete and $w < \infty$. There exists a metric subspace Ω of $\mathcal{M}_w$ such that*

(a) *For every point $m \in \mathcal{M}$ there exists a point $\omega \in \Omega$ such that*

$$d_w(\omega, m) \leq 24w(m). \tag{5.202}$$

(b) *For every pair $\omega \neq \omega'$ from Ω,*

$$d_w(\omega, \omega') \geq 8(w(\omega) + w(\omega')). \tag{5.203}$$

Proof. We decompose $\mathcal{M}$ into slices $\mathcal{M}_j$, $j \in \mathbb{Z} \cup \{+\infty\}$, setting for $j < \infty$,

$$\mathcal{M}_j := \{m \in \mathcal{M} \,;\, 2^{-j-1} < w(m) \leq 2^{-j}\} \tag{5.204}$$

and for $j = \infty$

$$\mathcal{M}_\infty := \{m \in \mathcal{M} \,;\, w(m) = 0\}. \tag{5.205}$$

Then

$$\mathcal{M} := \bigsqcup_{j \in \mathbb{Z} \cup \{+\infty\}} \mathcal{M}_j. \tag{5.206}$$

Regarding $\mathcal{M}_j \neq \emptyset$ with $j < \infty$ as a metric subspace of $\mathcal{M}_w := (\mathcal{M}, d_w)$ we denote by N_j a maximal ϵ_j-separated subset (ϵ_j-net) of $\mathcal{M}_j$, where $\epsilon_j := 2^{-j+2}$. Due to its definition N_j is a subset of $\mathcal{M}_j$ satisfying the conditions:

(a) If card $N_j > 1$, then for every m, m' from N_j,
$$d_w(m, m') \geq \epsilon_j.$$

(b) For every $m \in \mathcal{M}_j$ there exists a point $\widehat{m} \in N_j$ such that
$$d_w(m, \widehat{m}) < \epsilon_j.$$

These properties and (5.204) yield for $m, m' \in N_j$, $j < \infty$, the inequality
$$d_w(m, m') \geq 2(w(m) + w(m')). \tag{5.207}$$

Moreover, for the points $m, \widehat{m}$ from (b) we get the inequality
$$d_w(m, \widehat{m}) < 8w(m). \tag{5.208}$$

Further, we throw out points of N_j that are ϵ_j-closed to the ϵ_j-nets N_i with $i > j$. Specifically, we set
$$\widehat{N}_j := \{m \in N_j \,;\, d_w(m, \sqcup_{i>j} N_i) > \epsilon_j\}.$$

We also set for convenience $\widehat{N}_\infty := \mathcal{M}_\infty$. Using this disjoint sequence we finally define the desired subspace $\Omega \subset \mathcal{M}_w$ by setting
$$\Omega := \bigsqcup_{j \in \mathbb{Z} \cup \{\infty\}} \widehat{N}_j. \tag{5.209}$$

It is clear that $\Omega \neq \emptyset$ if card $\mathcal{M} < \infty$. Let us show that this holds also for card $\mathcal{M} = \infty$.

Actually, if $\Omega = \emptyset$ in this case, then there exists $j_0 \in \mathbb{Z}$ such that $N_{j_0} \neq \emptyset$ but $\widehat{N}_{j_0} = \emptyset$. Further, given a point $m_0 \in N_{j_0}$ there exist $j_1 > j_0$ and $m_1 \in N_{j_1}$ such that $d_w(m_0, m_1) \leq \epsilon_{j_0}$ (otherwise, card $\mathcal{M} < \infty$, since in this case $\mathcal{M} = \sqcup_{j \in \mathbb{Z}} N_j$). But $\widehat{N}_{j_1} = \emptyset$ and, hence, there exist $j_2 > j_1$ and $m_2 \in N_{j_2}$ such that $d_w(m_1, m_2) \leq \epsilon_{j_1}$. Proceeding this way we finally obtain a sequence $\{m_i\}_{i \geq 1}$ satisfying for $i < j$,
$$d_w(m_i, m_j) \leq \sum_{k=i}^{j-1} \epsilon_{j_k} \leq 2^{-(j_i - 3)}.$$

In particular, $\{m_i\}$ is a Cauchy sequence in the complete metric space $\mathcal{M}_w$ and therefore for some $m \in \mathcal{M}$,
$$w(m) \leq \lim_{i \to \infty} d_w(m, m_i) = 0.$$

5.4. Simultaneous Lipschitz selections

This means that $m \in \mathcal{M}_\infty \subset \Omega = \emptyset$, a contradiction.

Now we prove part (b), i.e., inequality (5.203). Let $\omega \neq \omega' \in \Omega$ and, for definiteness, let

$$w(\omega') \leq w(\omega). \tag{5.210}$$

If both of the points belong to some $\widehat{N}_j$, then (5.203) immediately follows from (5.207). Otherwise, $\omega \in \widehat{N}_i$, $\omega' \in \widehat{N}_j$ and $i < j \leq \infty$ by (5.210) and the definition of $\{\widehat{N}_j\}$. By the same reason $d_w(\omega, \omega') \geq \epsilon_i := 4 \cdot 2^{-i}$. On the other hand, $\widehat{N}_j \subset \mathcal{M}_j$ so that $w(\omega') \leq 2^{-j} \leq 2^{-i-1} < w(\omega) < 2^{-i}$. Combining with the previous inequality we get

$$d_w(\omega, \omega') \geq 8w(\omega) \geq 4(w(\omega) + w(\omega')).$$

To prove the remaining inequality (5.202) we fix $m \in \mathcal{M}$ and assume, first, that $m \in \Omega$. Then the desired result immediately follows from (5.208), if $m \notin \widehat{N}_\infty$, and is trivial if $m \in \widehat{N}_\infty$ with $\omega := m$, since $w(m) = 0$ in this case.

If now $m \notin \Omega$, then $m \in \mathcal{M}_{j_0} \setminus \Omega$ for some $j_0 \in \mathbb{Z}$. Since N_{j_0} is a maximal ϵ_{j_0}-net in $\mathcal{M}_{j_0}$, there exists a point $m_0 \in N_{j_0}$ such that $d_w(m, m_0) < \epsilon_{j_0}$. Together with (5.208), this yields

$$d_w(m, m_0) < \epsilon_{j_0} \leq 8w(m). \tag{5.211}$$

Hence, this gives the desired inequality (5.202), if $m_0 \in \Omega$. Otherwise, $m_0 \in N_{j_0} \setminus \Omega$ and, by the definitions of $\widehat{N}_j$ and Ω there exist $j_1 > j_0$ and $m_1 \in N_{j_1}$ such that $d_w(m_1, m_0) < \epsilon_{j_0}$. Proceeding this way we either obtain a point, say $m_\ell \in N_{j_\ell}$, $\ell \geq 1$, belonging to Ω or an infinite sequence $\{m_k\}_{k \geq 0}$. In the former case, the sequence $\{m_k\}_{-1 \leq k \leq \ell}$ where $m_{-1} := m$ satisfies

$$d_w(m, m_\ell) \leq \sum_{k=0}^{\ell} d_w(m_{k-1}, m_k) < \epsilon_{j_0} + \sum_{k=0}^{\ell} \epsilon_{j_k}.$$

This and (5.211) yield

$$d_w(m, m_\ell) \leq 8w(m) + \sum_{i \geq j_0} 2^{-i+2} \leq 8(w(m) + 2^{-j_0}).$$

Moreover, $m \in \mathcal{M}_{j_0}$, i.e., $w(m) \geq 2^{-j_0-1}$. Inserting this into the previous inequality we get the required result:

$$d_w(m, m_\ell) \leq 8\left(w(m) + \frac{1}{2}w(m)\right) = 12w(m).$$

If now $\{m_k\}_{k \geq 1}$ is infinite, then $d_w(m_k, m_{k+1}) \leq \epsilon_{j_k} \to 0$ as $k \to \infty$ and there exists a point in $\mathcal{M}$, say m_∞, such that $w(m_\infty) \leq \lim_{k \to \infty} d_w(m_\infty, m_k) = 0$. Hence, $m_\infty \in \mathcal{M}_\infty \subset \Omega$ and as above

$$d_w(m, m_\infty) \leq 8w(m) + \sum_{j \geq j_0} 2^{-j+2} \leq 8(w(m) + 2^{-j_0}) \leq 24w(m).$$

This proves (5.202) with $m := m_\infty$ and gives the lemma. □

Now we apply (5.203) to prove that the restriction operator $R_\Omega : f \mapsto f|_\Omega$ maps $\mathrm{Lip}(\mathcal{M}_w)$ into $\mathrm{Lip}(\Omega)$ with $\|R_\Omega\| < 2$.
To this end we fix $m, m' \in \Omega$ and derive from (5.202) that

$$d_w(m, m') := d(m, m') + (w(m) + w(m')) \leq d(m, m') + \frac{1}{8} d_w(m, m').$$

This immediately yields $d_w(m, m') \leq \frac{8}{7} d(m, m')$ which, in turn, implies for $f \in \mathrm{Lip}(\mathcal{M}_w)$, $m, m' \in \Omega$ and $\epsilon > 0$,

$$|(R_\Omega f)(m) - (R_\Omega f)(m')| := |f(m) - f(m')| \leq (1+\epsilon)|f|_{\mathrm{Lip}(\mathcal{M}_w)} d_w(m, m')$$
$$\leq \frac{8}{7}(1+\epsilon)|f|_{\mathrm{Lip}(\mathcal{M}_w)} d(m, m').$$

Hence, $R_\Omega : \mathrm{Lip}(\mathcal{M}_w) \to \mathrm{Lip}(\Omega)$ and its norm is at most $\frac{8}{7}(1+\epsilon) < 2$.

The result now proved together with the one established above, see diagram (5.201), implies that the linear operator

$$T_w := E_\Omega R_\Omega I$$

maps $\Sigma_w(\mathcal{M})$ into $\mathrm{Lip}(\mathcal{M})$ and its norm is at most $8\lambda(\mathcal{M})$.

It remains to show that the operator $S_w := Id_{\Sigma_w(\mathcal{M})} - T_w$ maps $\Sigma_w(\mathcal{M})$ into $\ell_\infty^w(\mathcal{M})$ and its norm is bounded by a numerical constant.

To this end, given a fixed $m \in \mathcal{M}$ we find using (5.202) a point $\omega \in \Omega$ such that $d_w(m, \omega) \leq 24 w(m)$. Further, since $\omega \in \Omega$ and E_Ω is an extension operator,

$$(T_w f)(\omega) := (E_\Omega R_\Omega I f)(\omega) = E_\Omega(f|_\Omega)(\omega) = f(\omega)$$

and therefore for $f \in \Sigma_w(\mathcal{M})$ we have

$$|S_w f|(m) \leq |f(m) - f(m')| + |(T_w f)(m) - (T_w f)(\omega)|$$
$$\leq |f|_{\mathrm{Lip}(\mathcal{M}_w)} d_w(m, \omega) + 8\lambda(\mathcal{M}) |f|_{\Sigma_w(\mathcal{M})} d_w(m, \omega).$$

This, (5.202) and (5.197) yield

$$|S_w f|(m) \leq |f|_{\Sigma_w(\mathcal{M})} d_w(m, \omega)(2 + 8\lambda(\mathcal{M})) \leq 10\lambda(\mathcal{M}) \cdot 24 w(m) |f|_{\Sigma_w(\mathcal{M})}.$$

Hence, we get $\|S_w\| \leq 240\lambda(\mathcal{M})$. Together with the just proved inequality for $\|T_w\|$ we obtain

$$\|T_w\| + \|S_w\| \leq 272\lambda(\mathcal{M}). \tag{5.212}$$

Thus Proposition 5.68 has been proved under the assumptions of Lemma 5.83.

5.4. Simultaneous Lipschitz selections

Part B. Now we derive from the result of part A the next one where we discard the assumption on completeness of $\mathcal{M}$ and $\mathcal{M}_w$ but retain the condition $w < \infty$.

Given a metric space $\mathcal{M}$ by $\mathcal{M}^{cmp}$ we denote its completion and consider a relation between this functor and that transforming $\mathcal{M}$ into $\mathcal{M}_w$. To this end we extend the weight $w : \mathcal{M} \to \mathbb{R}_+$ to $\mathcal{M}^{cmp}$ by setting

$$w^*(m) := \begin{cases} w(m) & \text{if } m \in \mathcal{M}, \\ 0 & \text{if } \in \mathcal{M}^{cmp} \setminus \mathcal{M}. \end{cases}$$

Hereafter, we may and will consider $\mathcal{M}$ as a dense subspace of $\mathcal{M}^{cmp}$. Further, we set for brevity

$$(\mathcal{M}, d)^{cmp} := (\mathcal{M}^*, d^*) \text{ and } (\mathcal{M}_w, d_w)^{cmp} := (\mathcal{M}_w^*, d_w^*).$$

Lemma 5.84. *The following isometric embedding is true:*

$$\mathcal{M}_w^* \subset (\mathcal{M}^*)_{w^*}. \tag{5.213}$$

Proof. The spaces $(\mathcal{M}, d)$ and $(\mathcal{M}_w, d_w)$ have the common underlying set $\mathcal{M}$ and, moreover, $d \leq d_w$. Hence, every Cauchy sequence in $\mathcal{M}_w$ is one in $\mathcal{M}$ and therefore $\mathcal{M}_w^* \subset \mathcal{M}^*$.

Further, if $\{m_i\}_{i \geq 1}$ is a Cauchy sequence in $\mathcal{M}_w$, then $w(m_i) \to 0$ as $i \to \infty$. Therefore, we have for $m, m' \in \mathcal{M}_w^*$,

$$d_w^*(m, m') = d^*(m, m') + w^*(m) + w^*(m'),$$

i.e., embedding (5.213) is isometric. $\square$

Now we construct the desired operator $T_w : \Sigma_w(\mathcal{M}) \to \mathrm{Lip}(\mathcal{M})$. By density of $\mathcal{M}_w$ in $\mathcal{M}_w^*$ the continuity extension operator denoted by E maps $\mathrm{Lip}(\mathcal{M}_w)$ into $\mathrm{Lip}(\mathcal{M}_w^*)$ with $\|E\| = 1$. Further, due to Lemma 5.84 the restriction operator $R : f \mapsto f|_{(\mathcal{M}^*)_{w^*}}$ maps $\mathrm{Lip}(\mathcal{M}_w^*)$ into $\mathrm{Lip}((\mathcal{M}^*)_{w^*})$ with $\|R\| = 1$. Finally, the restriction operator $R' : f \mapsto f|_\mathcal{M}$ maps $\mathrm{Lip}(\mathcal{M}^*)$ into $\mathrm{Lip}(\mathcal{M})$ and $\ell_\infty^{w^*}(\mathcal{M}^*)$ into $\ell_\infty^w(\mathcal{M})$; in both cases its norm is 1.

Now we apply the result proved in part A to the pair of the complete metric spaces $\mathcal{M}^*$ and $(\mathcal{M}^*)_{w^*}$ and the weight $w^* : \mathcal{M}^* \to \mathbb{R}_+$. According to this result there exist linear operators T_{w^*}, S_{w^*} mapping $\Sigma_{w^*}(\mathcal{M}^*)$ into $\mathrm{Lip}(\mathcal{M}^*)$ and $\ell_\infty^{w^*}(\mathcal{M}^*)$, respectively, such that $\|T_{w^*}\| \leq 32\lambda(\mathcal{M}^*)$, and $\|S_{w^*}\| \leq 240\lambda(\mathcal{M}^*)$, see (5.212). We then define the desired operator T_w by composing the linear operators presented in the diagram

$$\Sigma_w(\mathcal{M}) \xrightarrow{I} \mathrm{Lip}(\mathcal{M}_w) \xrightarrow{E} \mathrm{Lip}(\mathcal{M}_w^*) \xrightarrow{R} \mathrm{Lip}((\mathcal{M}^*)_{w^*}) \xrightarrow{T_{w^*}} \mathrm{Lip}(\mathcal{M}^*) \xrightarrow{R'} \mathrm{Lip}(\mathcal{M}).$$

The norm of T_w is then bounded by $\|I\| \cdot \|T_{w^*}\| \leq 32\lambda(\mathcal{M}^*)$.

Further, for the second operator S_w we have

$$\|S_w\| := \|Id_{\Sigma_w(\mathcal{M})} - T_w\| = \|R'(Id_{\Sigma_{w^*}(\mathcal{M}^*)} - T_{w^*})EI\| := \|R'S_{w^*}EI\| \leq 240\lambda(\mathcal{M}^*).$$

So we have found the required operators T_w, S_w but with the bound $O(1)\lambda(\mathcal{M}^*)$ in place of $O(1)\lambda(\mathcal{M})$ for $\|T_w\| + \|S_w\|$. To complete the proof it suffices to use the next result that will be proved after Corollary 7.19 of Volume II.

Lemma 5.85. *If S is a dense subspace of a metric space $\mathcal{M}$, then*

$$\lambda(S) = \lambda(\mathcal{M}).$$

In particular, $\lambda(\mathcal{M}) = \lambda(\mathcal{M}^*)$ and the desired result now follows under the restriction $w < \infty$.

It remains to prove the proposition for weights assigning $+\infty$.

Let $w : \mathcal{M} \to \mathbb{R}_+ \cup \{+\infty\}$ be such a weight and

$$M_0 := \{m \in \mathcal{M}\,;\, 0 \le w(m) < \infty\}.$$

Since $\lambda(\mathcal{M}) < \infty$, there exists a linear extension operator E_0 such that

$$E_0 : \mathrm{Lip}(M_0) \to \mathrm{Lip}(\mathcal{M}) \text{ and } \|E_0\| \le 2\lambda(M_0).$$

Further, we apply the result proved above to the metric space

$$\mathcal{M}_0 := (M_0, d|_{M_0 \times M_0})$$

and the weight $w_0 := w|_{M_0}$. It implies that there exists a linear operator $T_{w_0} : \Sigma_{w_0}(\mathcal{M}_0) \to \mathrm{Lip}(\mathcal{M}_0)$ with $\|T_{w_0}\| \le 32\lambda(\mathcal{M}_0) \le 32\lambda(\mathcal{M})$.

Now let $R_0 : f \mapsto f|_{M_0}$ be the restriction operator. Then R_0 maps $\Sigma_w(\mathcal{M})$ into $\Sigma_{w_0}(\mathcal{M}_0)$ with $\|R_0\| = 1$ and therefore $T_w := E_0 T_{w_0} R_0$ maps $\Sigma_w(\mathcal{M})$ into $\mathrm{Lip}(\mathcal{M})$ and its norm satisfies

$$\|T_w\| \le \|E_0\| \cdot \|R_0\| \cdot \|T_{w_0}\| \le 32\lambda(\mathcal{M})^2.$$

Finally, we set

$$S_w := \mathrm{Id}_{\Sigma_w(\mathcal{M})} - T_w = E_0(\mathrm{Id}_{\Sigma_{w_0}(\mathcal{M}_0)} - T_{w_0})R_0 := E_0 S_{w_0} R_0.$$

It is readily seen that S_w maps $\Sigma_w(\mathcal{M})$ into $\ell_\infty^w(\mathcal{M})$ and its norm is bounded by $\|E_0\| \cdot \|S_{w_0}\| \cdot \|R_0\| \le 240\lambda(\mathcal{M})^2$.

As the final result we, hence, have

$$\kappa(\mathcal{M}) \le \sup_w\{\|T_w\| + \|S_w\|\} \le 272\lambda(\mathcal{M})^2,$$

i.e., the required right inequality of Proposition 5.68 has been proved.

It remains to prove the left inequality of this proposition, i.e., to show that

$$\lambda(\mathcal{M}) \le \kappa(\mathcal{M}). \tag{5.214}$$

5.4. Simultaneous Lipschitz selections

To this end, given $\mathcal{M}' \subset \mathcal{M}$ we define a set-valued map $F_{\mathcal{M}'} : \mathcal{M} \to \text{Aff}(\mathbb{R}) := \{\{0\}, \mathbb{R}\}$ by

$$F_{\mathcal{M}'}(m) := \begin{cases} \{0\} & \text{if } m \in \mathcal{M}', \\ \mathbb{R} & \text{if } m \in \mathcal{M} \setminus \mathcal{M}'. \end{cases}$$

The corresponding to $F_{\mathcal{M}'}$ weight $w_{\mathcal{M}'}$ equals 0 on $\mathcal{M}'$ and $+\infty$ on $\mathcal{M} \setminus \mathcal{M}'$. Therefore, the restriction of $f \in \Sigma_w(\mathcal{M})$ to $\mathcal{M}'$ belongs to $\text{Lip}(\mathcal{M}')$ while $f|_{\mathcal{M} \setminus \mathcal{M}'}$ is an extension of $f|_{\mathcal{M}'}$. We conclude that any Lipschitz selection of $f + F_{\mathcal{M}'}$ for this f is an extension of $f|_{\mathcal{M}'} \in \text{Lip}(\mathcal{M}')$ to a function from $\text{Lip}(\mathcal{M})$.

Now, if $\kappa(\mathcal{M}) < \infty$, then there exists a linear operator $T_{\mathcal{M}'} : \Sigma_{w_{\mathcal{M}'}}(\mathcal{M}) \to \text{Lip}(\mathcal{M})$ such that $T_{\mathcal{M}'}f$ is a Lipschitz selection of $f + F$ and $\|T_{\mathcal{M}'}\| \leq (1+\epsilon)\kappa(\mathcal{M})$ for a fixed $\epsilon > 0$. Then $T_{\mathcal{M}'}$ is a linear extension operator from $\text{Lip}(\mathcal{M}')$ to $\text{Lip}(\mathcal{M})$ of norm at most $(1+\epsilon)\kappa(\mathcal{M})$. Due to the definition of λ, see (5.200),

$$\lambda(\mathcal{M}) \leq \sup_{\mathcal{M}' \subset \mathcal{M}} \|T_{\mathcal{M}'}\| \leq (1+\epsilon)\kappa(\mathcal{M}).$$

This proves (5.214) and completes the proof of Proposition 5.68.

5.4.6 Proof of Proposition 5.69

Let $(\mathcal{M}, d)$ be a metric space with K-linearity constant $\kappa(\mathcal{M}) < \infty$. Given a set $\widetilde{\mathcal{M}}$ and two maps $\varphi : \widetilde{\mathcal{M}} \to \mathcal{M}$ and $r : \widetilde{\mathcal{M}} \to \mathbb{R}_+ \cup \{+\infty\}$ a pseudometric $\tilde{d}$ on $\widetilde{\mathcal{M}}$ is defined for $\widetilde{m} \neq \widetilde{m}'$ by

$$\tilde{d}(\widetilde{m}, \widetilde{m}') := d(\varphi(\widetilde{m}), \varphi(\widetilde{m}')) + r(\widetilde{m}) + r(\widetilde{m}') \tag{5.215}$$

and for $\widetilde{m} = \widetilde{m}'$ by zero.

The pseudometric space $(\widetilde{\mathcal{M}}, \tilde{d})$ is equipped with a weight $w : \widetilde{\mathcal{M}} \to \mathbb{R}_+ \cup \{+\infty\}$ and we should find an associated to $(\text{Lip}(\widetilde{\mathcal{M}}), \ell_\infty^w(\widetilde{\mathcal{M}}))$ pair of linear operators (T_w, S_w) satisfying

$$T_w + S_w = \text{Id}_{\Sigma_w(\widetilde{\mathcal{M}})} \quad \text{and} \quad \|T_w\| + \|S_w\| \leq O(1)\kappa(\mathcal{M}).$$

We begin the proof with the case $r = 0$, so that $\tilde{d} = d \circ \varphi$, i.e., we set for $\widetilde{m}, \widetilde{m}' \in \widetilde{\mathcal{M}}$,

$$\tilde{d}(\widetilde{m}, \widetilde{m}') := d(\varphi(\widetilde{m}), \varphi(\widetilde{m}')). \tag{5.216}$$

To exploit the assumption $\kappa(\mathcal{M}) < \infty$ we equip the space $\mathcal{M}$ with a weight $\widehat{w}$ related to w by

$$\widehat{w}(m) := \inf\{w(\widetilde{m}) \,;\, \widetilde{m} \in \varphi^{-1}(m)\}. \tag{5.217}$$

Then there exists a pair $(T_{\widehat{w}}, S_{\widehat{w}})$ of linear operators associated to $(\mathrm{Lip}(\mathcal{M}), \ell_\infty^{\widehat{w}}(\mathcal{M}))$ such that

$$\|T_{\widehat{w}}\| + \|S_{\widehat{w}}\| \leq 2\kappa(\mathcal{M}).$$

To pass from here to the initial situation we use a "transfer" operator $R : \Sigma_w(\widetilde{\mathcal{M}}) \to \Sigma_{\widehat{w}}(\mathcal{M})$ with $\|R\| \leq 4$. We first define $(Rf)(m)$ for $m \in \widehat{w}^{-1}(0) := \{m \in \mathcal{M}\,;\, w(\widetilde{m}) = 0\}$. Then there exists a sequence $\{\widetilde{m}_i\}_{i\in\mathbb{N}} \subset \varphi^{-1}(m)$ such that

$$\lim_{i\to\infty} w(\widetilde{m}_i) = \widehat{w}(m) = 0. \tag{5.218}$$

To define $(Rf)(m)$ we first present $f \in \Sigma_w(\widetilde{\mathcal{M}})$ as $f = f_0 + f_1$ with

$$\|f_0\|_{\mathrm{Lip}(\widetilde{\mathcal{M}})} + \|f_1\|_{\ell_\infty^w(\widetilde{\mathcal{M}})} \leq 2|f|_{\Sigma_w(\widetilde{\mathcal{M}})}. \tag{5.219}$$

Then we will derive from here for $\widetilde{m}, \widetilde{m}' \in \widetilde{\mathcal{M}}$ the inequality

$$|f(\widetilde{m}) - f(\widetilde{m}')| \leq 2|f|_{\Sigma_w(\widetilde{\mathcal{M}})}((d\circ\varphi)(\widetilde{m},\widetilde{m}') + w(\widetilde{m}) + w(\widetilde{m}')). \tag{5.220}$$

Assuming for a while that it is true we apply the inequality to $\widetilde{m}_i, \widetilde{m}_j$, $i < j$. Then (5.219) and (5.218) imply that

$$\lim_{i,j\to\infty} |f(\widetilde{m}_i) - f(\widetilde{m}_j)| = 0,$$

i.e., there exists $\lim_{i\to\infty} f(\widetilde{m}_i)$ that is clearly independent of the choice of $\{\widetilde{m}_i\} \subset \varphi^{-1}(m)$. We then set for $w(m) = 0$,

$$(Rf)(m) := \lim_{i\to\infty} f(\widetilde{m}_i), \quad \{\widetilde{m}_i\} \subset \varphi^{-1}(m). \tag{5.221}$$

Now let $\widehat{w}(m) > 0$ and let $\widetilde{m} := \widetilde{m}(m)$ be such that

$$\widetilde{m} \in \varphi^{-1}(m) \text{ and } w(\widetilde{m}) \leq 2\widehat{w}(m).$$

We then set for this m,

$$(Rf)(m) := f(\widetilde{m}(m)).$$

Hence, in this case $(Rf)(m)$ is also given by (5.221) with $\widetilde{m}_i := \widetilde{m}(m)$ for all $i \geq 1$; in both cases

$$\lim_{j\to\infty} w(\widetilde{m}_j) \leq 2\widehat{w}(m). \tag{5.222}$$

Due to (5.222) and (5.220) the linear operator R satisfies, for $m, m' \in \mathcal{M}$,

$$|(Rf)(m) - (Rf)(m')| \leq 2|f|_{\Sigma_w(\widetilde{\mathcal{M}})} \lim_{i\to\infty}[(d\circ\varphi)(\widetilde{m}_i,\widetilde{m}_i') + w(\widetilde{m}_i) + w(\widetilde{m}_i')]$$

$$\leq 2|f|_{\Sigma_w(\widetilde{\mathcal{M}})} \cdot [d(m,m') + 2\widehat{w}(m) + 2\widehat{w}(m')] = 2|f|_{\Sigma_w(\widetilde{\mathcal{M}})} \cdot d_{\widehat{w}}(m, m').$$

5.4. Simultaneous Lipschitz selections

Hence, R maps $\Sigma_w(\widetilde{\mathcal{M}})$ into $\text{Lip}(\mathcal{M}_{\widehat{w}})$ and its norm is bounded by 2. Moreover, by Lemma 5.75 with $n = 1$,

$$\text{Lip}(\mathcal{M}_{\widehat{w}}) = \Sigma_{\widehat{w}}(\mathcal{M}) \text{ and } |f|_{\text{Lip}(M_{\widehat{w}})} \leq 2|f|_{\Sigma_{\widehat{w}}(\mathcal{M})}.$$

Hence, R maps $\Sigma_w(\widetilde{\mathcal{M}})$ into $\Sigma_{\widehat{w}}(\mathcal{M})$ with norm bounded by 4, as required.

It remains to check (5.220). If $f = f_0 + f_1$ and (5.219) holds, then

$$|f(\widetilde{m}) - f(\widetilde{m}')| \leq |f_0(\widetilde{m}) - f_1(\widetilde{m}')| + |f_0(\widetilde{m})| + |f_1(\widetilde{m}')|$$
$$\leq 2|f|_{\Sigma_w(\widetilde{\mathcal{M}})} \cdot [\widetilde{d}(\widetilde{m}, \widetilde{m}') + w(\widetilde{m}) + w(\widetilde{m}')].$$

Hence, in general, we get

$$|f(\widetilde{m}) - f(\widetilde{m}')| \leq 2|f|_{\Sigma_w(\widetilde{\mathcal{M}})} \cdot [(d \circ \varphi)(\widetilde{m}, \widetilde{m}') + (r + w)(\widetilde{m}) + (r + w)(\widetilde{m}')], \quad (5.223)$$

which for $r = 0$ implies (5.220).

Now we use finiteness of $\kappa(\mathcal{M})$ to find linear operators $T_{\widehat{w}}, S_{\widehat{w}}$ mapping $\Sigma_{\widehat{w}}(\mathcal{M})$ into $\text{Lip}(\mathcal{M})$ and $\ell_\infty^{\widehat{w}}(\mathcal{M})$, respectively, such that

$$T_{\widehat{w}} + S_{\widehat{w}} = Id_{\Sigma_{\widehat{w}}(\mathcal{M})} \text{ and } \|T_{\widehat{w}}\| + \|S_{\widehat{w}}\| \leq 2\kappa(\mathcal{M}). \quad (5.224)$$

The desired operators T_w, S_w will then be defined on $\Sigma_w(\widetilde{\mathcal{M}})$ by

$$T_w := \varphi^* T_{\widehat{w}} R, \qquad S_w := \varphi^* S_{\widehat{w}} R,$$

where we set $\varphi^*(f) := f \circ \varphi$. Then we have for $f \in \Sigma_w(\widetilde{\mathcal{M}})$,

$$|(T_w f)(\widetilde{m}) - (T_w f)(\widetilde{m}')| := |(T_{\widehat{w}} R)(f \circ \varphi)(\widetilde{m}) - (T_{\widehat{w}} R)(f \circ \varphi)(\widetilde{m}')|$$
$$\leq 2\kappa(\mathcal{M})|(Rf)(\varphi(\widetilde{m})) - (Rf)(\varphi(\widetilde{m}'))| \leq 8\kappa(\mathcal{M})d(\varphi(\widetilde{m}), \varphi(\widetilde{m}'))$$
$$:= 8\kappa(\mathcal{M})\widetilde{d}(\widetilde{m}, \widetilde{m}').$$

Hence, $T_w : \Sigma_w(\widetilde{\mathcal{M}}) \to \text{Lip}(\widetilde{\mathcal{M}}, \widetilde{d})$ and $\|T_w\| \leq 8\kappa(\mathcal{M})$.

A similar argument shows that $S_w : \Sigma_w(\widetilde{\mathcal{M}}) \to \ell_\infty^w(\widetilde{\mathcal{M}})$ and $\|S_w\| \leq 8\kappa(\mathcal{M})$.

Thus Proposition 5.69 has been proved for $r = 0$. To prove the result for $r \neq 0$ we, as in the previous subsection, define $\widetilde{\mathcal{M}}_w := (\widetilde{\mathcal{M}}, \widetilde{d}_w)$ which is now a pseudometric space as $\widetilde{d}_w$ is now a pseudometric given for $\widetilde{m} \neq \widetilde{m}'$ by

$$\widetilde{d}_w(\widetilde{m}, \widetilde{m}') := \widetilde{d}(\widetilde{m}, \widetilde{m}') + w(\widetilde{m}) + w(\widetilde{m}')$$
$$:= [(d \circ \varphi)(\widetilde{m}, \widetilde{m}') + r(\widetilde{m}) + r(\widetilde{m}')] + w(m) + w(m'),$$

see (5.215). Setting $v := r + w$ we rewrite the previous as

$$\widetilde{d}_w(\widetilde{m}, \widetilde{m}') = (d \circ \varphi)(\widetilde{m}, \widetilde{m}') + v(\widetilde{m}) + v(\widetilde{m}') := (d \circ \varphi)_v(\widetilde{m}, \widetilde{m}').$$

This equality and Lemma 5.75 imply that
$$\mathrm{Lip}(\widetilde{\mathcal{M}}_w) = \mathrm{Lip}(\widetilde{\mathcal{M}}, (d\circ\varphi)_v) \subset \Sigma_v(\widetilde{\mathcal{M}}, d\circ\varphi)$$
with the embedding constant 2.

Further, by (5.223)
$$\Sigma_w(\widetilde{\mathcal{M}}) \subset \mathrm{Lip}(\widetilde{\mathcal{M}}, (d\circ\varphi)_v)$$
with the same embedding constant.

Combining these results we get
$$|f|_{\Sigma_v(\widetilde{\mathcal{M}}, d\circ\varphi)} \leq 4|f|_{\Sigma_w(\widetilde{\mathcal{M}})}. \tag{5.225}$$

Further, we apply the result already proved to the pseudometric space $(\widetilde{\mathcal{M}}, d\circ\varphi)$ and the weight v. It gives linear operators T_v, S_v associated to $(\mathrm{Lip}(\widetilde{\mathcal{M}}), \ell^v_\infty(\widetilde{\mathcal{M}}))$ such that
$$\|T_v\| + \|S_v\| \leq 16\kappa(\mathcal{M}). \tag{5.226}$$

Now we introduce the desired operators T_w, S_w by setting
$$T_w := T_v + \frac{r}{v}S_v, \quad S_w := \frac{w}{v}S_v.$$

In this definition and in subsequent computations, we conventionally set for $0 \leq c < \infty$ and $0 < d \leq \infty$,
$$\frac{0}{0} := 0, \quad \frac{+\infty}{+\infty} := +\infty, \quad \frac{c}{\infty} := 0, \quad \frac{d}{0} := +\infty.$$

By (5.226) and (5.225) we now obtain for $\widetilde{m} \in \widetilde{\mathcal{M}}$,
$$|(S_w f)(\widetilde{m})| := \frac{w}{v}(\widetilde{m})|(S_v f)(\widetilde{m})| \leq 8\kappa(\mathcal{M})\frac{w}{v}(\widetilde{m})v(\widetilde{m})|f|_{\Sigma_v(\widetilde{\mathcal{M}}, d\circ\varphi)}$$
$$\leq 32\kappa(\mathcal{M})|f|_{\Sigma_w(\widetilde{\mathcal{M}})}w(\widetilde{m}),$$

that is to say,
$$|S_w f|_{\ell^w_\infty(\widetilde{\mathcal{M}})} \leq 32\kappa(\mathcal{M})|f|_{\Sigma_w(\widetilde{\mathcal{M}})}.$$

For T_w we get by definition
$$|(Tf)(\widetilde{m}) - (Tf)(\widetilde{m}')|$$
$$\leq |(T_v f)(\widetilde{m}) - (T_v f)(\widetilde{m}')| + \frac{r}{v}(\widetilde{m})|(S_v f)(\widetilde{m})| + \frac{r}{v}(\widetilde{m}')|(S_v f)(\widetilde{m}')|.$$

Applying then (5.226) and (5.215) we bound the right-hand side by
$$8\kappa(\mathcal{M})|f|_{\Sigma_v(\widetilde{\mathcal{M}}, d\circ\varphi)}\left(d(\varphi(\widetilde{m}), \varphi(\widetilde{m}')) + \left(\frac{rv}{v}\right)(\widetilde{m}) + \left(\frac{rv}{v}\right)(\widetilde{m}')\right)$$
$$\leq 32\kappa(\mathcal{M})\widetilde{d}(\widetilde{m}, \widetilde{m}')|f|_{\Sigma_w(\widetilde{\mathcal{M}})}.$$

Hence, we also prove that

$$|T_w f|_{\mathrm{Lip}(\widetilde{\mathcal{M}})} \leq 32\kappa(\mathcal{M})|f|_{\Sigma_w(\widetilde{\mathcal{M}})}.$$

It remains to show that the pair (T_w, S_w) defined on $\Sigma_w(\widetilde{\mathcal{M}}, d \circ \varphi)$ is in fact associated to the couple $(\mathrm{Lip}(\widetilde{\mathcal{M}}), \ell_\infty^w(\widetilde{\mathcal{M}}))$. To this end we need only show that

$$T_w + S_w = Id_{\Sigma_w(\widetilde{\mathcal{M}})}. \qquad (5.227)$$

Inequality (5.226) implies that $T_w + S_w$ maps $\Sigma_w(\widetilde{\mathcal{M}}, d \circ \varphi)$ into $\Sigma_w(\widetilde{\mathcal{M}})$ while (5.225) gives the embedding $\Sigma_w(\widetilde{\mathcal{M}}) \subset \Sigma_v(\widetilde{\mathcal{M}}, d \circ \varphi)$. Hence, $T_w + S_w$ maps $\Sigma_w(\widetilde{\mathcal{M}})$ into itself. Since, moreover, $(T_w + S_w)f = f$ for $f \in \Sigma_w(\widetilde{\mathcal{M}}) \subset \Sigma_v(\widetilde{\mathcal{M}}, d \circ \varphi)$, equality (5.227) holds.

Proposition 5.69 has been proved.

Comments

Bourgain's Theorem 5.1 was the first essential result of the quickly developing area studying low-distortion embeddings of finite metric spaces into Euclidean spaces. Let us note that Bourgain's embedding required an exponential number of dimensions to embed the metric. This drawback was overcome in the paper of Linial, London and Rabinovich [LLR-1995], where a new approach gives an embedding $O(\log^2 n)$ dimensions with $O(\log n)$ distortion.

At the initial stage the driving force of the investigations was a program formulated and partially carried out by Bourgain, Johnson, Lindenstrauss, V. Milman and their collaborators and followers, see, in particular, [JL-1984], [Bou-1985], [Bou-1986], [BMW-1986]. The aim of the program is to develop the structure theory for general metric spaces analogous to the local theory of Banach spaces. For instance, according to Bourgain's theorem, $\log n$ would play the role of dimension for n-point subspaces of a metric space. Unfortunately, such analogies cannot be extended too far. Actually, accepting this concept of dimension would result in the corresponding analog of John's ellipsoid theorem [Jo-1948] asserting that an n-point metric space can be embedded in ℓ_n^2 with distortion $O(\sqrt{\log n})$. This result is not true.

However, as has frequently occurred in mathematics, results of pure theoretical interest may become powerful tools in applied fields. This is the case for the area discussed which now plays an important role in Computer Science. Its results become essential algorithmic tools serving for the visualization of finite metric space structure, finding clusters, small separators, etc., see, e.g., the above mentioned paper [LLR-1995] and Matoušek's book [Mat-2004].

The nonembedding result formulated at the beginning of subsection 5.1.2 is a consequence of the general Bourgain theorem [Bou-1985]. It asserts that a Banach space X is not superreflexive if and only if every dyadic metric tree $\mathcal{T}_k$

(on 2^k-vertices) admits a C-bi-Lipschitz embedding in X with a constant $C \geq 1$ independent of k. We refer the reader to the classical James paper [Ja-1972] for the definition of superreflexivity, noting only that the space L_p possesses this property whenever $1 < p < \infty$.

Uniform boundedness of degree in Theorem 5.4 is a necessary condition. In fact, $\mathbb{H}^n$ is a metric space of bounded geometry and therefore any metric space admitting a bi-Lipschitz embedding into $\mathbb{H}^n$ should possess this property.

As it has been noted, the snowflake map $(\mathcal{M}, d) \mapsto (\mathcal{M}, d^p)$, $0 < p < 1$, used in Assouad's Theorem 5.10, essentially worsens the geometric structure of a space. In particular, it increases the Hausdorff dimension and turns all rectifiable curves into nonrectifiable. This naturally leads to the question of characterization of metric spaces admitting a bi-Lipschitz embedding into $\mathbb{R}^n$. It is a matter of definition to check that such a space should be doubling. The examples presented in subsection 5.1.3 demonstrate that this condition is far from being sufficient. Under some additional restrictions on a space this condition becomes sufficient. To formulate this result, due to Lang and Plaut [LaPl-2001], we need the following concept.

Let $(\mathcal{M}, d)$ be a metric space. A *geodesic bicombing* (henceforth called a "bicombing") on a subset $S \subset \mathcal{M}$ is a family of curves $\{\gamma_{mm'} : [0, 1] \to \mathcal{M}\,;\, m, m' \in S\}$, where $\gamma_{mm'}$ is a geodesic parametrized proportionally to the arc length which joins m and m'.

A bicombing is said to be *weakly convex* if for some constant $C \geq 1$ and all m, m', m'' in S and all $0 \leq t \leq 1$,

$$d\big(\gamma_{mm'}(t), \gamma_{mm''}(t)\big) \leq Ct d(m', m'').$$

Theorem ([LaPl-2001]). *Let $(\mathcal{M}, d)$ be a metric space with a weakly convex bicombing on $\mathcal{M}$. Assume that for every triple $m, m' \in \mathcal{M}$, $t \in [0, 1]$ there exists a point $m'' \in \mathcal{M}$ such that $m' = \gamma_{mm''}(t)$. Then $(\mathcal{M}, d)$ admits a bi-Lipschitz embedding into some $\mathbb{R}^n$ if and only if it is doubling.*

The assumptions of the theorem hold, e.g., for Hadamard manifolds, C-convex subsets of normed spaces and $\mathbb{R}$-trees (in particular, combinatorial metric trees). Hence, these spaces admit bi-Lipschitz embeddings in suitable Euclidean spaces if and only if they are doubling.

Theorem 3.99, whose proof is presented in the final part of subsection 5.1.4, is apparently known, see, e.g., Lemmas 2.1 and 2.2 and Remarks 5.3 and 5.6 from [LaPl-2001] for the proof of a similar result.

The Bonk–Schramm Theorem 5.36 is one of the basic results of the new field of Geometric Analysis known by the name of *Coarse Geometry*. For the proofs of the results surveyed in Section 5.2 we refer the reader to the papers and books mentioned there and to the books by Roe [Ro-2003] and by Buyalo and Schroeder [BuSch-2007]. The latter book contains the streamlined proof of the Bonk–Schramm theorem.

A version of Shvartsman's result [Shv-1986] presented as Theorem 5.56 is taken from the Yu. Brudnyi and Shvartsman paper [BSh-2001b] devoted to the solution of the Finiteness Problem for the trace spaces of $C^{1,\omega}(\mathbb{R}^n)$. A general result of this type with k-dimensional subspaces in an infinite-dimensional Banach space is due to Shvartsman [Shv-2001].

Variants of Theorem 5.46 have been discovered by several authors; in the form presented here it was due to Przeslawski and Ribinski [PR-1992].

Theorem 5.66 is a special case of the Yu. Brudnyi and Shvartsman general result [BSh-1999, Thm. 4.16] dealing with families of set-valued maps $\{f + F\}_{f \in \text{Lip}(\mathcal{M}, \mathbb{R}^n)}$ where F maps a metric space $\mathcal{M}$ into the set of centrally-symmetric convex subsets in $\mathbb{R}^n$.

Bibliography

[Al-1928]   P. Aleksandroff, *Über den allgemeinen Dimensionsbegriff und seine Beziehungen zur elementaren geometrischen Anschauung*, Math. Ann. **98** (1928), 617–635.

[Ale-1943]   A. Aleksandrov, *Additive set functions in abstract spaces*, Mat. Sb. **13** (1943), 169–238.

[Ale-1951]   A. D. Aleksandrov, *A theorem on triangles in a metric space and some of its applications*, Trudy Mat. Inst. Steklova **38** (1951), 5–23.

[ABN-1986]   A. D. Aleksandrov, V. N. Berestovski and I. G. Nikolaev, *Generalized Riemannian manifolds*, Uspekhi Mat. Nauk **41:3** (1986), 3–44; English translation in: Russian Math. Surveys **41:3** (1986), 1–54.

[Ant-1921]   L. Antoine, *Sur l'homéomorphisme de deux figures et de leurs voisinages*, J. Math. Pures Appl. **4** (1921), 221–325.

[Aro-1935]   N. Aronszajn, *Neuer Beweis der Streckenverbundenheit vollständiger konvexer Räume*, Ergebnisse eines Math. Kolloquiums (Wien) **6** (1935), 45–56.

[AP-1956]   N. Aronszajn and P. Panitchpakdi, *Extension of uniformly continuous transformations and hyperconvex metric spaces*, Pacific J. Math. **6** (1956), 405–439.

[As-1980]   P. Assouad, *Pseudodistances, facteurs et dimension métrique*, Séminaire d'Analyse Harmonique 1979–1980, Publ. Math. Orsey **80**, No. 7 (1980), 1–33.

[As-1982]   P. Assouad, *Sur la distance de Nagata*, C. R. Acad Sci. Paris Sér. I Math. **294**, No. 1 (1982), 31–34.

[As-1983]   P. Assouad, *Plongement Lipschitziens dans $\mathbb{R}^n$*, Bull. Soc. Math. de France **111** (1983), 429–448.

[AF-1990]   J.-P. Aubin and H. Frankowska, *Set-valued Analysis*, Birkhäuser, 1990.

[Bach-2001]   M. Bachir, *A non-convex analogue to Fenchel duality*, J. Funct. Anal. **181** (2001), 300–312.

[B-1992]   K. Ball, *Markov chains, Riesz transforms and Lipschitz maps*, GAFA **2** (1992), 137–172.

[BCL-1994]   K. Ball, E. Carlen and E. Liel, *Sharp uniform convexity and smoothness inequalities for trace norms*, Invent. Math. **115** (1994), 463–482.

[BaBo-2000]   Z. M. Balogh and M. Bonk, *Gromov hyperbolicity and the Kobayashi metric on strictly pseudoconvex domains*, Comment. Math. Helv. **75** (2000), 504–533.

[Ban-1932]   S. Banach, *Théorie des Opérations Linéaires*, Monographie Matematyczne 1, Warszawa, 1932.

[BLMN-2005]   Y. Bartal, N. Linial, M. Mendel and A. Naor, *On metric Ramsey type phenomena*, Ann. of Math. **162**, No. 2 (2005), 643–709.

[BG-1952]   R. G. Bartle and L. M. Graves, *Mappings between function spaces*, Trans. Amer. Math. Soc. **72** (1952), 400–413.

[Ba-1972]   H. Bass, *The degree of polynomial growth of finitely generated nilpotent groups*, Proc. London Math. Soc. **25** (1972), 603–614.

[BL-2000]   Y. Benyamini and J. Lindenstrauss, *Geometric Nonlinear Functional Analysis*, vol. 1, AMS Colloquium Publ. 48, Providence, RI, 2000.

[BLo-1976]   J. Bergh and J. Löfström, *Interpolation Spaces. An Introduction*, Springer, 1976.

[Ber-1960]   D. Berman, *Linear polynomial operators on groups*, Izv. Vuzov (matem.) **4** (1960), 17–28.

[Ber-1912]   S. Bernstein, *Sur les recherches récentes relatives à la meilleure approximation des fonctions continues par des polynômes*. In: Proc. 5th Intern. Math. Congress, Vol. 1, 1912, 256–266.

[Ber-1940]   S. N. Bernstein, *On the question of local polynomial approximation of functions*, Dokl. AN SSSR **26** (1940), 839–842; Reprinted in: Collected Papers, Vol. II, Izdat. AN SSSR, Moskwa, 1954 (in Russian).

[Bes-1946]   A. Besicovitch, *A general form of the covering principle and relative differentiation of additive functions II*, Proc. Cambridge Phil. Soc. **42** (1946), 1–10.

[Bes-1952]   A. Besicovich, *On the existence of subsets of finite measure of sets of infinite measure*, Indag. Math. **14** (1952), 339–344.

Bibliography

[BeSch-1961] A. Besicovich and I. Schoenberg, *On Jordan arcs and Lipschitz classes of functions defined on them*, Acta Math. **106** (1961), 113–136.

[BA-1956] A. Beurling and L. Ahlfors, *The boundary correspondance under quasiconformal mappings*, Acta Math. **96** (1956), 125–142.

[BM-1991] E. Bierstone and P. Milman, *Geometric and differential properties of subanalytic sets*, Bull. AMS **25** (1991), 385–393.

[BMP-2003] E. Bierstone, P. Milman, and W. Pawlucki, *Differentiable functions defined on closed sets. A problem of Whitney*, Invent. Math. **151** (2003), 329–352.

[BM-2007] E. Bierstone and P. Milman, C^m *norms of finite sets and* C^m *extension criteria*, Duke Math. J. **137** (2007), 1–18.

[BS-1983] E. Bierstone and G. Schwartz, *Continuous linear division and extension of* C^∞-*functions*, Duke Math. J. **50** (1983), 233–271.

[Bil-1968] P. Billingsley, *Convergence of Probability Measures*, John Wiley & Sons, Inc., 1968.

[BI-1985] P. Binew and K. Ivanov, *On a representation of mixed finite differences*, SEDDICA Bulg. Math. Publ., Vol. **11** (1985), 259–268.

[Bin-1953] R. H. Bing, *A convex metric with unique segments*, Proc. Amer. Math. Soc. **4** (1953), 167–174.

[Blan-1951] W. Blankenship, *Generalization of a construction of Antoine*, Ann. Math. **53** (1951), 276–297.

[Bom-1977] J. Boman, *On comparison theorems for generalized moduli of continuity*. In: Lect. Notes in Math., Vol. 571, Springer, 1977, 38–52.

[BHK-2001] M. Bonk, J. Heinonen and P. Koskela, *Uniformizing Gromov hyperbolic spaces*, Astérisque **270** (2001).

[BSch-2000] M. Bonk and O. Schramm, *Embeddings of Gromov hyperbolic spaces*, GAFA **10** (2000), 206–266.

[deB-2001] C. de Boor, *A Practical Guide to Splines*, Applied Math. Sciences **27**, Springer, 2001.

[Bo-1963] A. Borel, *Compact Clifford–Klein forms of symmetric spaces*, Topology **2** (1963), 111–122.

[EBo-1895] E. Borel, *Sur quelques points de la théorie des fonctiones*, Ann. Sci. École Normal Superior **3** (1895), 9–55.

[Bor-1931]	K. Borsuk, *Sur les rétractes*, Fund. Math. **17** (1931), 152–170.
[Bor-1933a]	K. Borsuk, *Über Isomorphie der Funktionalräume*, Bull. Intern. Acad. Polon. Sér. A (1933), 1–10.
[Bor-1933b]	K. Borsuk, *Drei Sätze über die n-dimensionale euklidische Sphäre*, Fund. Math. **20** (1933), 177–190.
[Bo-1953]	N. Bourbaki, Elements de Mathématique. Livre V, Espaces Vectoriels Topologiques, Hermann, Paris, 1953.
[Bou-1985]	J. Bourgain, *On Lipschitz embedding of finite metric spaces in Hilbert space*, Israel J. Math. **52** (1985), 46–52.
[Bou-1986]	J. Bourgain, *The metric interpretation of superreflexivity in Banach spaces*, Israel J. Math. **56** (1986), 222–230.
[BFM-1986]	J. Bourgain, T. Figiel and V. Milman, *On Hilbertian subsets of finite metric spaces*, Israel J. Math. **55**, No. 2 (1986), 147–152.
[BMW-1986]	J. Bourgain, V. Milman and H. Wolfson, *On type of metric spaces*, Trans. AMS **294** (1986), 295–317.
[BrH-1970]	J. H. Bramble and S. R. Hilbert, *Estimation of linear functionals on Sobolev spaces with application to Fourier transforms and spline interpolation*, SIAM J. Numer. Analysis **7** (1970), 112–124.
[BH-1999]	M. R. Bridson and A. Haefliger, *Metric Spaces of Non-Positive Curvature*, Springer, 1999.
[Bro-1967]	M. Brodskii, *Admissible uniform estimates from below for dilations in mappings of a cube into a cube of higher dimension*, Mat. Sbornik **73** (1967), 8–20 (in Russian).
[Brom-1982]	S. Bromberg, *An extension theorem in the class C^1*, Bol. Soc. Mat. Mexicana, Ser. **27** (1982), 35–44.
[Bro-1908]	L. E. J. Brouwer, *Differences and Derivatives*, Kengl. Akad. Wetenschaft Amsterdam, **38** 1908.
[Bro-1912]	L. E. J. Brouwer, *Beweis der Invarianz des n-dimensionalen Gebiets*, Math. Ann. **71** (1912), 305–313.
[Bro-1913]	L. E. J. Brouwer, *Über den natürlichen Dimensionbegriff*, J. Reine. Angew. Math. **142** (1913), 146–152.
[Bro-1918]	L. E. J. Brouwer, *Über die Erweiterung des Definitionsbereichs einer stetigen Funktionen*, Math. Ann. **79** (1918), 209–211.
[BB-2006]	A. Brudnyi and Yu. Brudnyi, *Extensions of Lipschitz functions defined on metric subspaces of homogeneous type*, Revista Mat. Compultensa **19** (2006), 347–359.

Bibliography

[BB-2007a]   A. Brudnyi and Yu. Brudnyi, *Remez type inequalities and Morrey–Campanato spaces on Ahlfors regular sets*, Contemporary Mathematics **445** (2007), 19–44.

[BB-2007b]   A. Brudnyi and Yu. Brudnyi, *Metric spaces with linear extensions preserving Lipschitz condition*, Amer. Math. J. **129** (2007), 217–314.

[BB-2007c]   A. Brudnyi and Yu. Brudnyi, *A universal Lipschitz extension property of Gromov hyperbolic spaces*, Revista Math. Iberoamericana **23** (2007), 861–896.

[BB-2008]   A. Brudnyi and Yu. Brudnyi, *Linear and nonlinear extensions of Lipschitz functions from subsets of metric spaces*, St. Petersburg J. Math. **19**, no. 3 (2008), 397–406.

[Br-1964]   Yu. Brudnyi, *On a theorem of local best approximation*, Uchen. Zap., Kazan. Univ. **124** (1964), 43–49 (in Russian).

[Br-1965a]   Yu. Brudnyi, *On the local best approximation of functions by polynomials*, Dokl. Akad. Nauk SSSR **161** (1965), 746–749 (in Russian).

[Br-1965b]   Yu. Brudnyi, *Study of properties of multivariate functions by polynomial and quasipolynomial approximation*, Uspekhi Mat. Nauk **20**, No. 5 (1965), 276–279 (in Russian).

[Br-1967]   Yu. Brudnyi, *Criteria for existence of derivatives in L_p*, Matem. Sbornik **73** (1967), 42–64 (in Russian).

[Br-1970a]   Yu. Brudnyi, *A multidimensional analog of a theorem of Whitney*, Mat. Sbornik **82** (124) (1970), 175–191; English translation in: Math. USSR Sbornik **11** (1970), 157–170.

[Br-1970b]   Yu. Brudnyi, *On an extension theorem*, Funk. Anal. Pril. **4** (1970), 96–97; English translation in: Funct. Anal. Appl. **4** (1970), 252–253.

[Br-1971]   Yu. Brudnyi, *Spaces defined by local polynomial approximations*, Trudy Mosk. Mat. Ob-va **24** (1971), 69–132; English translation in: Translations of the Moscow Math. Soc. **24** (1971), 73–139.

[Br-1976]   Yu. Brudnyi, *An extension theorem for a family of function spaces*, Zapiski Nauch. Seminarov LOMI AN SSSR **VI** (1976), 170–173.

[Br-1977]   Yu. Brudnyi, *Piecewise polynomial approximation, embedding theorems and rational approximation*, Lecture Notes in Math., Vol. 554, Springer, 1977, 73–98.

[Br-1980]   Yu. Brudnyi, *Extension of functions preserving order of decay for moduli of continuity*. In: Studies in Function Theory of Several Real Variables, Yaroslavl' State Univ., Yaroslavl', 1980, 33–53 (in Russian).

[Br-1994]   Yu. Brudnyi, *Adaptive approximation of functions with singularities*, Trudy Mosk. Mat. Ob-va **55** (1994), 149–242; English translation in: Translations of the Moscow Math. Soc. (1994), 123–186.

[Br-2002]   Yu. Brudnyi, *Local approximation of multivariate functions*. In: Approximation Theory, a volume dedicated to B. Sendov, DARBA, Sofia, 2002, 65–83.

[Br-2004]   Yu. Brudnyi, *Nonlinear N-term approximation of refinable functions*, St. Petersburg Math. J. **16** (2005), 143–179.

[Br-2008]   Yu. Brudnyi, *Sobolev spaces and their relatives: local polynomial approximation approach*. In: Sobolev Spaces in Mathematics, Vol. 2, Springer, 2008.

[BrG-1973]   Yu. Brudnyi and M. Ganzburg, *On an extremal problems for polynomials in n variables*, Izv. Akad. Nauk SSSR **37** (1973), 344–355; English translation in: Math. USSR–Izv. **7** (1973), 345–356.

[BGo-1961]   Yu. Brudnyi and I. Gopengauz, *Approximation by piecewise polynomial functions*, Dokl. Akad. Nauk SSSR **141** (1961), 1283–1286 (in Russian).

[BGo-1963]   Yu. Brudnyi and I. Gopengauz, *Approximation by piecewise polynomial functions*, Izv. Akad. Nauk SSSR **27** (1963), 723–746 (in Russian).

[BKa-2000]   Yu. Brudnyi and N. Kalton, *Polynomial approximation on convex subsets of* $\mathbb{R}^n$, Constr. Appr. **16** (2000), 161–199.

[BKo-1970]   Yu. Brudnyi and B. Kotlyar, *On a problem of combinatorial geometry*, Sib. Mat. J. **XI**, No. 5 (1970), 1171–1173; English translation in: Siberian Math. J. **11** (1970).

[BK-1981a]   Yu. Brudnyi and N. Krugljak, *Real interpolation functors*, Dokl. AN SSSR **256** (1981), 14–17; English translation in: Soviet Doklady **23** (1981), 5–8.

[BK-1981b]   Yu. Brudnyi and N. Krugljak, *Real Interpolation Functors*, Yaroslavl, 1981, 221 pps. Deposited in VINITI 13 May 1981.

[BK-1991]   Yu. Brudnyi and N. Krugljak, *Interpolation Functors and Interpolation Spaces*, North Holland, 1991.

Bibliography

[BSha-1971] Yu. Brudnyi and V. Shalashov, *Lipschitz function spaces*, Dokl. AN SSSR **197**, No. 1 (1971), 18–20; English translation in: Soviet Math. Dokl. **12** (1971), 383–386.

[BSha-1973] Yu. Brudnyi and V. Shalashov, *Lipschitz spaces of functions*. In: Metric Theory of Functions and Maps, Vol. 4, Ukrainian Acad. Sci., Donetzk, 1973, 1–60 (in Russian).

[BSht-1996] Yu. Brudnyi and A. Shteinberg, *Calderón couples of Banach spaces*, J. Funct. Analysis **131** (1996), 459–498.

[BSht-1997] Yu. Brudnyi and A. Shteinberg, *Calderón couples of finite dimensional Banach spaces*, Israel J. Math. **101** (1997), 289–322.

[BSh-1982] Yu. Brudnyi and P. Shvartsman, *Description of the trace of a general Lipschitz space to an arbitrary compact*. In: Studies in Theory of Functions of Several Real Variables, Yaroslavl' State Univ., Yaroslavl', 1982, 16–24 (in Russian).

[BSh-1985] Yu. Brudnyi and P. Shvartsman, *A linear extension operator for a space of smooth functions defined on closed subsets of* $\mathbb{R}^n$, Dokl. AN SSSR **280** (1985), 268–272; English translation in: Soviet Math. Doklady **31** (1985), 48–51.

[BSh-1994] Yu. Brudnyi and P. Shvartsman, *Generalizations of Whitney's extension theorem*, Intern. Math. Research Notes **3** (1994), 129–139.

[BSh-1998] Yu. Brudnyi and P. Shvartsman, *The trace of jet space* $J^k\Lambda^\omega$ *to an arbitrary closed subset of* $\mathbb{R}^n$, Trans. Amer. Math. Soc. **350** (1998), 1519–1553.

[BSh-1999] Yu. Brudnyi and P. Shvartsman, *The Whitney problem of existence of a linear extension operator*, J. of Geom. Analysis **7**, No. 4 (1999), 515–574.

[BSh-2001a] Yu. Brudnyi and P. Shvartsman, *The Whitney Extension Problem*, 2001, Preprint, 1–65.

[BSh-2001b] Yu. Brudnyi and P. Shvartsman, *Whitney extension problem for multivariate* $C^{1,\omega}$-*functions*, Trans. Amer. Math. Soc. **353** (2001), 2487–2512.

[BSh-2002] Yu. Brudnyi and P. Shvartsman, *Stability of the Lipshcitz extension property under metric transforms*, GAFA **12** (2002), 73–79.

[BBI-2001] D. Burago, Yu. Burago and S. Ivanov, *Course of Metric Geometry*, GSM, Vol. 33, AMS, Providence, 2001.

[BGP-1992]   Yu. Burago, M. Gromov and G. Perel'man, *A. D. Aleksandrov spaces with curvature bounded below*, Uspekhi Mat. Nauk **47:2** (1992), 1–58; English translation in: Russian Math. Surveys **47** (1992), 3–51.

[BuSch-2007]   S. Buyalo and V. Schroeder, *Elements of Asymptotic Geometry*, EMS Monographs in Mathematics, EMS Publishing House, 2007.

[BZ-1988]   Yu. Burago and V. Zalgaller, *Geometric Inequalities*, Springer, 1988.

[BuGo-1979]   V. Burenkov and M. Goldman, *On extensions of functions from L_p*, Trudy Mat. Inst. AN SSSR **150** (1979), 31–61 (in Russian).

[Bu-1952]   H. Burkill, *Cesaro–Perron almost periodic functions*, Proc. London Math. Soc. **3**, No. 2 (1952), 150–174.

[Bu-1955]   H. Buseman, *The Geometry of Geodesics*, Academic Press, 1955.

[BuSch-2001]   S. Buyalo and V. Schroeder, *Extensions of Lipschitz maps into 3-manifolds*, Asian J. Math. **5**, no. 4 (2001), 685–704.

[Cal-1964]   A. Calderón, *Intermediate spaces and interpolation, the complex method*, Studia Math. **24**, No. 2 (1964), 113–190.

[Cal-1972]   A. Calderón, *Estimates for singular integral operators in terms of maximal functions*, Studia Math. **44** (1972), 167–186.

[CKR-2004]   G. Calinescu, H. Karloff and Y. Rabani, *Approximation algorithms for the 0-extension problem*, SIAM J. Comput. **34**, No. 2 (2004), 358–372 (electronic version).

[Cam-1964]   S. Campanato, *Proprietà di una famiglia di spazi funzionali*, Ann. Scuola Norm. Super. Pisa **18** (1964), 137–160.

[Ca-1918]   C. Carathéodory, *Vorlesungen über reelle Funktionen*, Leipzig und Berlin, 1918.

[CW-2001]   A. Carbery and J. Wright, *Distributional and L^p norm inequalities for polynomials over convex bodies in $\mathbb{R}^n$*, Math. Research Letters **8** (2001), 233–248.

[Cay-1878]   A. Cayley, *On the theory of groups*, Proc. London Math. Soc. **9** (1878), 126–133.

[Cha-1993]   J. Chavel, *Riemannian Geometry: A Modern Introduction*, Cambridge Univ. Press, 1993.

[Ch-1999]   J. Cheeger, *Differentiability of Lipschitz functions on metric measure spaces*, GAFA **9** (1999), 428–517.

Bibliography

[CE-1975] J. Cheeger and D. Ebin, *Comparison Theorems in Riemannian Geometry*, North-Holland, 1975.

[ChK-2006a] J. Cheeger and B. Kleiner, *On the differentiability of Lipschitz maps from metric measure spaces to Banach spaces*. Inspired by S. S. Chern, 129–152, Nankai Tracts Math., **11**, World Sci. Publ., Hackensack, NJ, 2006.

[ChK-2006b] J. Cheeger and B. Kleiner, *Differentiating maps into L_1, and the geometry of BV functions*, Preprint, 2006, 46 pp.

[Chr-1984] M. Christ, *The extension problem for certain function spaces involving fractional order of differentiability*, Ark. Mat. **22** (1984), 63–81.

[CK-1995] P. Callahan and S. Kosaraju, *A decomposition of multi-dimensional point-sets with applications to k-nearest-neighbors and n-body potential fields*, J. Association for Computer Machinery **42** (1995), 67–90.

[CW-1971] R. R. Coifman and G. Weiss, *Analyse Harmonique Non-Commutative sur Certains Espaces Homogènes*, Lecture Notes in Math., vol. 242, Springer, 1971.

[CDP-1991] M. Coornaert, T. Delzant and A. Papadopoulos, *Géométrie et Théorie des Groupes*, Lecture Notes in Math., vol. 1441, Springer, 1991.

[CG-1990] L. Corwin and F. P. Greenleaf, *Representations of Nilpotent Lie Groups and Their Applications. Part I: Basic Theory and Examples*, Cambridge Univ. Press, 1990.

[Cw-1984] M. Cwikel, *K-divisibility of the K-functional of Carderón couples*, Ark. Mat. **22** (1984), 39–62.

[Cw-2003] M. Cwikel, *The K-divisibility constant for couples of Banach lattices*, J. Appr. Theory **124** (2003), 124–136.

[CwJM-1990] M. Cwikel, B. Jawerth and M. Milman, *On the fundamental lemma of Interpolation Theory*, J. Appr. Theory **60** (1990), 70–82.

[Cy-1981] J. Cygan, *Subadditivity of homogeneous norms on certain nilpotent Lie groups*, Proc. Amer. Math. Soc. **83** (1981), 69–70.

[Da-1979] B. Dahlberg, *Regularity properties of Riesz potentials*, Indiana Univ. Math. J. **28** (1979), 257–268.

[DaSe-1997] G. David and S. Semmes, *Fractured Fractals and Broken Dreams: Self-Similar Geometry through Metric and Measure*, Clarendon Press, Oxford, 1997.

[D-1998] M. W. Davis, *Buildings are CAT(0)*. In: Geometry and Cohomology in Group Theory (P. H. Knopholler et al., eds.), LMS Lecture Notes Series **252**, Cambridge Univ. Pless, 1998, 108–123.

[Dav-1970] R. O. Davies, *Increasing sequences of sets and Hausdorff measure*, Proc. London Math. Soc. (3) **20** (1970), 222–236.

[DGK-1963] L. Danzer, B. Grünbaum and V. Klee, *Helly's theorem and its relatives*, Convexity, Proc. Sympos. Pure Math, vol. VII, AMS, Providence, RI, 1963.

[Dau-1968] I. Daugavet, *Some applications of the generalized Marcinkiewicz–Berman identity*, Vestnik Leningrad Univ., Ser. Mat., Mech., Astron. **19** (1968), 59–64 (in Russian).

[Da-1955] M. M. Day, *Normed Linear Spaces*, Ergebnisse der Mathematik und ihre Grenzgebiete, vol. 21, Springer, 1973 (third edition).

[DeVL-1993] R. A. DeVore and G. G. Lorentz, *Constructive Approximation*, Springer, 1993.

[DeVSh-1984] R. A. DeVore and R. Sharpley, *Maximal functions measuring smoothness*, Memoirs AMS **47** No. 223 (1984), Providence RI.

[Di-1948] J. Dixmier, *Sur un theorèmé de Banach*, Duke Math. J. **15** (1948), 1054–1071.

[Dol-1993] V. Dol'nikov, *Generalized transversals of families of sets in $\mathbb{R}^n$ and connections between the Helly and Borsuk theorems*, Mat. Sbornik **184** (1993), 111–132 (in Russian); English translation in: Russian Acad. Sci. Sb. Mat. **79** No. 1 (1994), 93–107.

[Dor-1986] J. Dorronsoro, *Poisson integrals and regular functions*, Trans. Amer. Math. Soc. **297** (1986), 669–685.

[DH-1956] C. H. Dowker and W. Hurewicz, *Dimension of metric spaces*, Fund. Math. **43** (1956), 83–87.

[Dud-1989] R. Dudley, *Real Analysis and Probability*, Chapman and Hall, 1989.

[DR-1962] R. Dudley and B. Randol, *Implications of pointwise bounds on polynomials*, Duke Math. J. **29** (1962), 455-458.

[DS-1958] N. Dunford and J. T. Schwartz, *Linear Operators*, Interscience Publ. Inc., NY, 1958.

[Du-1951] J. Dugundji, *An extension of Tietze's theorem*, Pacific J. Math. **1** (1951), 353–367.

[Dv-1961]	A. Dvoretzky, *Some results on convex bodies and Banach spaces*, Proc. Sympos. on Linear Spaces, Jerusalem, 1961, 123–160.
[Dy-1983]	E. M. Dyn'kin, *Free interpolation by functions with derivatives in H^1*, Zap. Nauchn. Sem LOMI **126** (1983), 77–83 (in Russian); English translation in: J. Soviet Math. **27** (1984), 2475–2481.
[Dy-1984]	E. M. Dyn'kin, *Homogeneous measures on subsets of $\mathbb{R}^n$*, Lecture Notes in Math., vol. 1043, Springer, 1984, 698–699.
[DShe-1983]	V. K. Dziadik and I. A. Shevchuk, *Extension of functions which are traces of functions with prescribed modulus of smoothness to arbitrary subsets of the real line*, Izv. AN SSSR, Ser. Matem. **47** (1983), 248–267; English translation in: Math. USSR – Izv. **22** (1984), 227–245.
[Eck-1993]	J. Eckhoff, *Helly, Radon and Carathéodory type theorems*. In: Handbook of Convex Geometry, North-Holland, 1993, 389–448.
[Ei-1936]	M. Eidelheit, *Zur Theorie der Systeme linearer Gleichungen*, Studia Math. **6** (1936), 139–148.
[En-1978]	R. Engelking, *Dimension Theory*, North-Holland, 1978.
[EO-1938]	S. Eilenberg and E. Otto, *Quelques propriétés caractéristiques de la théorie de la dimension*, Fund. Math. **31** (1938), 149–153.
[Er-1940]	P. Erdős, *The dimension of the rational points in Hilbert space*, Ann. Math. **41** (1940), 734–736.
[Fal-1999]	K. Falconer, *Fractal Geometry. Mathematical Foundations and Applications*, John Wiley, 1999.
[Fe-1969]	H. Federer, *Geometric Measure Theory*, Stringer, 1969.
[F-2003]	Ch. Fefferman, *Whitney's extension problem for certain function spaces*, Preprint, 2003, Nov., 1–78.
[F-2005a]	Ch. Fefferman, *Interpolation and extrapolation of smooth functions by linear operators*, Revista Mat. Iberoamericana **21** (2005), 313–348.
[F-2005b]	Ch. Fefferman, *A sharp form of Whitney's extension theorem*, Ann. Math. **161** (2005), 509–577.
[F-2006a]	Ch. Fefferman, *Whitney's extension problem for C^m*, Ann. Math. **164** (2006), 313–359.
[F-2007a]	Ch. Fefferman, *C^m extensions by linear operators*, Ann. Math. **166**, No. 12 (2007), 779–835.

[F-2007b]	Ch. Fefferman, *The structure of linear extension operators for C^m*, Revista Mat. Iberoamericana **23** (2007), 269–280.
[F-2009]	Ch. Fefferman, *Fitting a C^m-smooth function to data III*, Ann. Math. **170** (2009), No. 1, 427–441.
[F-2009a]	Ch. Fefferman, *Whitney's extension problems and interpolation of data*, Bull. AMS **46** (2009), 207–220.
[F-2009b]	Ch. Fefferman, *Extension of $C^{m,\omega}$-smooth functions by linear operators*, Revista Mat. Iberoamericana **25** (2009), 1–48.
[Fr-1937]	A. H. Frink, *Distance functions and the metrization problem*, Bull. AMS **43** (1937), 133–142.
[FK-2007]	Ch. Fefferman and B. Klartag, *Fitting a C^m-smooth function to data II*, Revista Math. Iberoamericana **25**, no. 5 (2009), 49–273.
[FK-2009]	Ch. Fefferman and B. Klartag, *Fitting a C^m-smooth function to data I*, Ann. Math. **169** (2009), 315–346.
[FS-1982]	G. B. Folland and E. M. Stein, *Hardy Spaces on Homogeneous Groups*, Princeton Univ. Press, 1982.
[Ga-1957]	E. Gagliardo, Caratterizzazioni delle trace sulla frontiera relative ad alcune classi di funzioni in n variabili, Rend. Sem. Mat. Univ. Padova **27** (1957), 284–305.
[Ga-1958]	E. Gagliardo, Proprietá di alcune classi di funzioni in piú variabili, Ricerche di Mat. Napoli **7** (1958), 102–137.
[GM-2005]	J. Garnett and O. Marshall, *Harmonic Measures*, New Math. Monographs 2, Cambridge Univ. Press, Cambridge, 2005.
[Ge-1982]	F. W. Gehring, *Characteristic Properties of Quasidisks*, Les Presses de L'Universitè de Montreal, 1982.
[GO-1979]	F. W. Gehring and B. G. Osgood, *Uniform domains and the quasi-hyperbolic metric*, J. D'Analyse Math. **36** (1979), 50–74.
[Gel-1938]	I. Gelfand, *Abstracte Funktionen und lineare Operatoren*, Mat. Sb. **46** (1938), 235–286.
[GH-1990]	E. Ghys and P. De la Harpe (eds.), *Sur les Groupes Hyperboliques d'apres Mikhail Gromov*, Progress in Math. **83**, Birkhaüser, 1990.
[GKS-2002]	J. Gilewicz, Yu. Kryakin and J. Shevchuk, *Boundedness by 3 of the Whitney Interpolation Constant*, J. Appr. Theory **119**, No. 2 (2002), 971–290.
[Gl-1958]	G. Glaeser, *Étude de quelques algebres Tayloriennes*, J. d'Analyse Math. **6** (1958), 1–125.

Bibliography

[GK-2003] G. Godefroy and N. Kalton, *Lipschitz-free Banach spaces*, Studia Math. **159** (2003), 121–141.

[GV-1980] V. M. Gol'dshtein and S. K. Vodop'yanov, *Prolongement des fonctions de classe $\mathcal{L}_p^1$ et applications quasi conformes*, C. R. Acad. Sci. Ser. A–B **290**, No. 10 (1980), A453–A456.

[GV-1981] V. M. Gol'dshtein and S. K. Vodop'yanov, *Prolongement des fonctions différentiables hors de domaines plans*, C. R. Acad. Sci. Ser. I Math. **293**, No. 12 (1981), 581–584.

[GR-2000] J. E. Goodman and J. O'Rourke (Eds.), *Handbook of Discrete and Computational Geometry*, CRC, 2000.

[Gr-1981] M. Gromov, *Groups of polynomial growth and expanding maps*, IHES Publ. Math. **53** (1981), 53–78.

[Gr-1987] M. Gromov, *Monotonicity of the volumes of intersections of balls*, Lecture Notes in Math., vol. 1267, Springer, 1987, 1–4.

[Gr-1987a] M. Gromov, *Hyperbolic groups*. In: S. M. Gersten (Ed.), Essays in Group Theory, Springer, 1987, 75–263.

[Gr-2000] M. Gromov, *Metric Structures for Riemannian and Non-Riemannian Spaces* (with appendices by M. Katz, P. Pansu and S. Semmes), Birkhäuser, 2000.

[Gro-1953] A. Grothendieck, *Sur les applications linéaires faiblement compactes d'espace du type $C(K)$*, Canad. J. Math. **5** (1953), 129–173.

[GP-1988] M. Gromov and J. Piatetski–Shapiro, *Non-arithmetic groups on Lobachevski spaces*, IHES Sci. Publ. Math. **66** (1988), 93–103.

[G-1960] B. Grünbaum, *Projection constants*, Trans. Amer. Math. Soc. **95** (1960), 451–465.

[G-1963] B. Grünbaum, *Measures of symmetries for convex sets*. In: Proc. Symposia Pure Math. VII, Providence, RI, 1963, 233–270.

[GKL-2003] A. Gupta, R. Krauthgamer and J. Lee, *Bounded geometries, fractals and low-distortion embeddings*. In: Proc. of the 44th Annual Symp. on Foundations of Computer Science, 2003.

[Hal-1950] P. Halmos, *Measure Theory*, Van Nostrand, 1950.

[Had-1957] H. Hadwiger, *Vorlesungen über Inhalt, Oberfläche und Isoperimetrie*, Springer, 1957.

[HaK-2000] P. Hajlasz and P. Koskela, *Sobolev Met Poincaré*, Memoirs AMS, 145, Providence, 2000.

[Han-1987] L. Hanin, *A trace of functions with high order derivatives from Zygmund space to arbitrary closed subsets of* $\mathbb{R}$, Studies in Theory of Functions of several variables, Yaroslavl' State Univ., Yaroslavl' (1987), 128–144.

[Han-1997] L. Hanin, *Duality for general Lipschitz classes and applications*, Proc. London Math. Soc. **74** (1997), 134–156.

[Har-1969] F. Harary, *Graph Theory*, Addison–Wesley, 1969.

[Hau-1919] F. Hausdorff, *Über halbstetige Funktionen und deren Verallgemeinerung*, Math. Z. **5** (1919), 292–309.

[Hei-2001] J. Heinonen, *Lectures on Analysis of Metric Spaces*, Springer, 2001.

[Hei-2007] J. Heinonen, *Nonsmooth calculus*, Bull. AMS **44** (2007), 163–232.

[HK-1996] J. Heinonen and P. Koskela, *From local to global quasiconformal structures*, Proc. Nat. Acad. Sci. USA **93** (1996), 554–556.

[Hel-1978] S. Helgason, *Differential Geometry, Lie Groups and Symmetric Spaces*, Academic Press, 1978.

[He-1923] E. Helly, *Über Mengen konvexer Körper mit gemeinschaftlichen Punkten*, Jahresbericht Deutsch. Math. Verein **32** (1923), 175–176.

[He-1930] E. Helly, *Über Systeme abgeschlossener Mengen mit gemeinschaftlichen Punkten*, Monatsh. Math. und Phys. **37** (1930), 281–302.

[Hes-1941] M. Hestens, *Extension of the range of a differentiable function*, Duke Math. J. **8** (1941), 183–192.

[HR-1963] E. Hewitt and K. A. Ross, *Harmonic Analysis*, vol. I, Springer, 1963.

[Hor-1983] L. Hörmander, *The Analysis of Linear Partial Differential Equations*, Vol. 1, Springer, 1983.

[Ho-1995] J. D. Howroyd, *On dimension and on the existence of sets of finite Hausdorff positive measure*, Proc. London Math. Soc. (3) **70** (1995), 581–604.

[Hu-1927b] W. Hurewicz, *Über das Verhältnis separabler Räume zu kompacten Räumen*, Proc. Akad. Amsterdam **30** (1927), 425–430.

[Hu-1927b]	W. Hurewicz, *Normalbereiche und Dimensiontheorie*, Math. Ann. **96** (1927), 736–764.
[Hu-1935]	W. Hurewicz, *Über Abbildungen topologischer Räume auf die n-dimensional Sphäre*, Fund Math. **24** (1935), 144–150.
[HW-1941]	W. Hurewicz and H. Wallman, *Dimension Theory*, Princeton Univ. Press, Princeton, NJ, 1941.
[Hut-1981]	J. Hutchinson, *Fractals and self-similarity*, Indiana Univ. Math. J. **30** (1981), 713–743.
[IYa-1998]	S. A. Igonin and V. V. Yanishevski, *On an extreme problem on the plane*. In: Questions of Group Theory and Homological Algebra, Yaroslavl State University, Yaroslavl, 1998, 129–132 (in Russian).
[IT-1974]	A. D. Ioffe and V. M. Tihomirov, *The Theory of Extreme Problems*, Nauka, Moskva, 1974; English translation in: Theory of Extremal Problems, North Holland, 1979.
[Ism-1966]	R. S. Ismagilov, *Unitary representations of the Lorentz group in a space with indefinite metric*, Izv. AN SSSR, Ser. Mat. **30** (1966), 497–522.
[Ja-1972]	R. James, *Superreflexive spaces with bases*, Pacific J. Math. **41** (1972), 409–419.
[Jo-1948]	F. John, *Extreme problem with inequalities as subsidiary conditions*. In: Studies and Essays Presented to R. Courant, Interscience, N. Y., 1948, 187–204.
[Jo-1961]	F. John, *Rotation and strain*, Comm. Pure Appl. Math. **14** (1961), 391–413.
[JN-1961]	F. John and L. Nirenberg, *On functions of bounded mean oscillation*, Comm. Pure Appl. Math. **14** (1961), 415–426.
[JL-1984]	W. Johnson and J. Lindenstrauss, *Extensions of Lipschitz maps into a Hilbert space*. In: Contemp. Mathematics **26**, 1984, 189–206.
[JSch-1977]	H. Johnen and K. Scherer, *On the equivalence of the K-functional and moduli of continuity and some applications*. In: Lect. Notes in Math. **571** (1977), Springer, 119–140.
[Jon-1980]	P. W. Jones, *Extensions theorems for BMO*, Indiana Math. J. **29** (1980), 41–66.
[Jon-1981]	P. W. Jones, *Quasiconformal mappings and extendability of functions in Sobolev spaces*, Acta Math. **147** (1981), 71-88.

[Jons-1995] A. Jonsson, *Measures satisfying a refined doubling condition and absolute continuity*, Proc. AMS **123** (1995), 2441–2446.

[JW-1984] A. Jonsson and H. Wallin, *Function Spaces on Subsets of $\mathbb{R}^n$*, Harwood Academic Publishers, 1984.

[JLS-1986] W. Johnson, J. Lindenstrauss and G. Schechtman, *Extension of Lipschitz maps into Banach spaces*, Israel J. Math. **54** (1986), 129–138.

[KS-1971] M. Kadets and M. Snobar, *Certain functionals on the Minkowski compactum*, Mat. Zametki **10** (1971), 453–457; English translation in: Math. Notes **10** (1971).

[KST-1999] H. König, C. Schütt, and N. Tomczak-Jaegermann, *Projection constants of symmetric spaces and variants of Khintchine's inequality*, J. Reine Angew. Math. **511** (1999), 1–42.

[Ka-1940] S. Kakutani, *Simultaneous extension of continuous functions considered as a positive linear operation*, Japanese J. Math. **17** (1940), 187–190.

[Kal-2007] N. Kalton, *Extending Lipschitz maps into $C(K)$-spaces*, Israel J. Math. **162** (2007), 275–315.

[KR-1958] L. Kantorovich and G. Rubinshtein, *On the space of completely additive functions*, Vestnik LGU, Ser. Mat., Med. and Astronomy **7** (1958), 52–59 (in Russian).

[Kar-1994] R. Karidi, *Geometry of balls in nilpotent Lie groups*, Duke Math. J. **74** (1994), 301–317.

[Kat-1958] M. Katětov, *On the relations between the metric and topological dimensions*, Czech. Math. J. **8** (1958), 163–166 (in Russian, English summary).

[Kel-1957] J. L. Kelley, *General Topology*, Van Nostrand Co., Princeton, 1957.

[Kell-1970] J. B. Kelly, *Metric inequalities and symmetric differences*. In: Inequalities–II, O. Shisha (ed.), Academic Press, 1970, 193–212.

[Ke-1973] J. D. Kelly, *A method of constructing measures appropriate for the study of Cartesian products*, Proc. London Math. Soc. (3) **26** (1973), 521–546.

[Ki-1934] M. D. Kirszbraun, *Über die zusammenziehenden und Lipschitzschen Tranformationen*, Fund. Math. **22** (1934), 77–108.

[Kir-1976] W. A. Kirk, *Caristi's fixed point theorem and metric convexity*, Colloq. Math. **36**, No. 1 (1976), 81–86.

[Kl-1952] V. L. Klee, *Invariant metrics in groups (solution of a problem of Banach)*, Proc. Amer. Math. Soc. **6** (1952), 484–487.

[KK-1994] V. Klee and P. Kleinschmidt, *Polyhedral complexes and their relatives*. In: R. Graham, M. Gröschel and L. Lovász (Eds.), Handbook of Combinatorics, North Holland, 1994.

[KPR-1993] P. Klein, S. Plotkin and S. Rao, *Excluded minors, network decomposition, and multicommodity flow*. In: Annual ACM Symp. on Theory of Computing, 1993, 682–690.

[KKM-1929] B. Knaster, K. Kuratowski and S. Mazurkiewicz, *Ein Beweis des Fixpunktsatzes für n-dimensionale Simplexe*, Fund. Math. **14** (1929), 132–137.

[Kn-1955] M. Kneser, *Einige Bemerkungen über das Minkowskische Flächenmass*, Archiv der Math. **6** (1955), 382–390.

[Kob-2005] S. Kobayashi, *Hyperbolic Manifolds and Holomorphic Mappings*, 2nd edition, World Scientific, 2005.

[Ko-1932] A. N. Kolmogorov, *Beiträge zur Masstheorie*, Math. Ann. **107** (1932), 351–356.

[KT-1959] A. N. Kolmogorov and V. M. Tichomirov, ε-*entropy and* ε-*capacity of sets in function spaces*, Uspekhi Mat. Nauk **14**, No. 2 (1959), 3–86; English translation in: AMS Translations **17** (1961).

[KL-1997] A. Korányi and H. H. Lehman (Eds.), *Harmonic Functions on Trees and Buildings*, Contemporary Mathematics **206** (1997).

[KLMN-2005] R. Krauthgamer, J. Lee, M. Mendel and A. Naor, *Measured descent: a new embedding method for finite metrics*, GAFA **15** (2005), 839–858.

[KM-1997] A. Kriegl and P. Michor, *The Convenient Setting of Global Analysis*, Math. Surveys and Monographs, Vol. 53, AMS, Providence, 1997.

[Kry-2002] Yu. Kryakin, *Whitney's constants and Sendov's conjecture*, Math. Balcanica, New Series **16**, Fasc. 1–4 (2002), 237–247.

[Kur-1968] K. Kuratowski, *Topology*, vol. II, Academic Press, 1968.

[KP-1997] K. Kurdyka and W. Pawlucki, *Subanalytic version of Whitney's extension theorem*, Studia Math. **124** (1997), 269–280.

[Laa-2000] T. Laakso, *Ahlfors Q-regular spaces with arbitrary $Q > 1$ admitting weak Poincaré inequality*, GAFA **10** (2000), 111–123.

[La-1966]   S. Lang, *Introduction to Diophantine Approximations*, Addison–Wesley Publ. Company, 1966.

[L-1999]   U. Lang, *Extendability of large-scale Lipschitz maps*, Trans. Amer. Math. Soc. **351** (1999), 3975–3988.

[LPSch-2000]   U. Lang, B. Pavlovic' and V. Schroeder, *Extensions of Lipschitz maps into Hadamard spaces*, GAFA **10** (2000), 1527–1553.

[LaPl-2001]   U. Lang and C. Plaut, *Bilipschitz embeddings of metric spaces into space forms*, Geom. Dedicata **87** (2001), 285–307.

[LSch-1997]   U. Lang and V. Schroeder, *Kirszbraun's theorem and metric spaces of bounded curvature*, GAFA **7**, No. 3 (1997), 535–560.

[LSchl-2005]   U. Lang and T. Schlichenmaier, *Nagata dimension, quasisymmetric embeddings and Lipschitz extensions*, Intern. Math. Research Notes **58** (2005), 3625–3655.

[Le-1907]   H. Lebesgue, *Sur le probléme de Dirichlet*, Rendiconti del circ. Math. di Palermo, XXIV (1907), 1–32. Reprinted in: Oevres Scientifiques, vol. IV, L'Enseignment Math. (1973), 91–122.

[Le-1911]   H. Lebesgue, *Sur la non applicabilité de deux domaines appartenant respectivement à des espaces, de n et $n+p$ dimensions*, Math. Ann. **70** (1911), 166–168. Reprinted in: Oevres Scientifiques, vol. IV, 170–172.

[Lul-2010]   G. K. Luli, $C^{m,\omega}$ *extension by bounded-depth linear operators*, Adv. Math. **224** (2010), 1927–2021.

[LN-2005]   J. Lee and A. Naor, *Extending Lipschitz functions via random metric partitions*, Invent. Math. **160** (2005), 59–95.

[LV-1973]   O. Lehto and K. I. Virtanen, *Quasiconformal Mappings in the Plane*, 2nd edition, Springer, 1973.

[Lich-1929]   L. Lichtenstein, *Eine elementare Bemerkung zur reellen Analysis*, Math. Z. **30** (1929), 794–795.

[LLR-1995]   N. Linial, E. London and Y. Rabinovich, *The geometry of graphs and some of its algorithmic applications*, Combinatorica **15** (1995), 215–245.

[Li-1964]   J. Lindenstrauss, *On nonlinear projections in Banach spaces*, Michigan Math. J. **11** (1964), 263–287.

[LT-1971]   J. Lindenstrauss and L. Tzafriri, *On the complemented subspaces problem*, Israel J. Math. **9** (1971), 263–269.

[LT-1977]   J. Lindenstrauss and L. Tzafriri, *Classical Banach Spaces I*, Springer, 1977.

[LT-1979]	J. Lindenstrauss and L. Tzafriri, *Classical Banach Spaces II*, Springer, 1979.
[LP-1964]	J. L. Lions and J. Peetre, *Sur une classe d'espaces d'interpolation*, Publ. Math. Inst. Haute Etudes Sci. **19** (1964), 5–68.
[Lo-1966]	G. G. Lorentz, *Metric theory and approximation*, Bull. AMS **72** (1966), 903–937.
[Lu-1998]	J. Luukkainen, *Assouad dimension: antifractal metrization, porous sets, and homogeneous measures*, J. of Korean Math. Soc. **35** No. 1 (1998), 23–76.
[LuuM-L-1994]	J. Luukkainen and H. Movahedi-Lankrani, *Minimal bi-Lipschitz embedding dimension of ultrametric spaces*, Fund. Math. **144** No. 2 (1994) 181–193.
[LS-1998]	J. Luukkainen and E. Saksman, *Every complete doubling metric space carries a doubling measure*, Proc. Amer. Math. Soc. **126** (1998), 531–354.
[LyT-1969]	Yu. Lyubich and V. Tkachenko, *The reconstruction of C^∞ functions from the values of their derivatives at zero*, Teorija Funkt., Funkt. Analiz i Prilož. **9** (1969), 131–141 (in Russian).
[McSh-1934]	E. MacShane, *Extension of range of functions*, Bull. AMS **40** (1934), 837–842.
[Mag-2005]	V. Magnani, *Differentiability from representation formula and the Sobolev–Poincaré inequality*, Studia Math. **168** (2005), 251–272.
[Mal-1966]	B. Malgrange, *Ideals of Differentiable Functions*, Oxford Univ. Press, 1966.
[Mar-1927]	A. Marchaud, *Sur le dérivées et sur les différences des fonctions de variables réelles*, J. Math. Pures Appl. **6** (1927), 337–425.
[Ma-1938]	J. Marcinkiewicz, *Sur quelques integrals du type de Dini*, Ann. Soc. Polon. Math. **17** (1938), 42–50.
[MZ-1936]	J. Marcinkiewicz and A. Zygmund, *On the differentiability of functions and summability of trigonometric series*, Fund. Math. **26** (1936), 1–43.
[MP-1984]	M. Marcus and J. Pisier, *Characterizations of almost surely continuous p-stable random Fourier series and strongly stationary processes*, Acta Math. **152** (1984), 245–301.
[MRV-1971]	O. Martio, S. Rickman and J. Väisälä, *Topology and metric properties of quasiregular mappings*, Ann. Acad. Sci. Fenn., Ser. AI Math. **488** (1971), 1–31.

[MS-1979]   O. Martio and J. Sarvas, *Injectivity theorems in plane and space*, Ann. Acad. Sci. Fenn., Ser. AI Math. **4** (1979), 383–401.

[Ma-1995]   P. Mattila, *Geometry of Sets and Measures in Euclidean Spaces*, Cambridge Univ. Press, 1995.

[Mat-1990] J. Matoušek, *Extensions of Lipschitz maps on metric trees*, Univ. Carol. **31** (1990), 99–104.

[Mat-2004] J. Matoušek, *Lectures on Discrete Geometry*, GTM, vol. 212, Springer, 2004.

[Mau-1974] B. Maurey, *Une théorème de prolongement*, C.R. Acad. Sci. Paris **279** (1974), 329–332.

[Maz-1981] V. G. Maz'ya, *Extensions of functions from Sobolev spaces*, Zap. Nauch. Sem. LOMI **113** (1981), 231–236; English translation in: J. Soviet Math. **22** (1983), 1851–1855.

[Maz-1985] V. G. Maz'ya, *Sobolev Spaces*, Springer, 1985.

[McM-1987] C. McMullen, *Area and Hausdorff dimension of Julia sets of entire functions*, Trans. AMS **300** (1987), 329–342.

[MT-1989]   R. Meise and B. Taylor, *Linear extension operators for ultradifferentiable functions of Beurling type on compact sets*, Amer. J. Math. **111** (1989), 309–337.

[MN-2007]   M. Mendel and A. Naor, *Ramsey partitions and proximity data structures*, J. European Math. Soc. **9** (2007), 253–275.

[Men-1926] K. Menger, *Über die Dimension von Punktmengen, II*, Monatsh. für Math. und Phys. (1926), 137–161.

[Men-1928] K. Menger, *Untersuchungen über allgemeine Metrik*, Math. Ann. **100** (1928), 75–163.

[Mer-1966] J. Merrien, *Prolongateurs de fonctions différentiables d'une variable réelle*, J. Math. Pures Appl. (9) **45** (1966), 291–309.

[Mey-1970] W. Meyer, *Characterization of the Steiner point*, Pacific J. Math. **35** (1970), 717–725.

[Me-1964]   N. Meyers, *Mean oscillation over cubes and Hölder continuity*, Proc. Amer. Math. Soc. **15** (1964), 717–721.

[Mi-1956]   E. Michael, *Continuous selections. I*, Ann. Math. **63**, No. 2 (1956), 361–382.

[Mic-1949] E. J. Mickle, *On the extension of a transformation*, Bull. Amer. Math. Soc. **55** (1949), 160–164.

Bibliography

[Mil-2000] J. Milnor, *Dynamics in One Complex Variable. Introductory Lectures*, Vieweg, 2000.

[Min-1970] G. J. Minti, *On the extension of Lipschitz, Lipschitz–Hölder and monotone functions*, Bull. AMS **76** (1970), 334–339.

[Mitch-1985] J. Mitchell, *On Carnot–Carathéodory metrics*, J. Diff. Geometry **21** (1985), 35–45.

[Mit-1961] B. S. Mitiagin, *Approximative dimension and bases in nuclear spaces*, Uspekhi Mat. Nauk **16**, No. 4 (1961), 63–132; English translation in: Russian Math. Surveys **16**, No. 4 (1961), 59–128.

[Mon-2002] R. Montgomery, *A Tour of Sub-Riemannian Geometries. Their Geodesics and Applications*, Math. Surveys and Monographs of AMS, vol. 91, 2002.

[Mor-1946] P. Moran, *Additive functions of intervals and Hausdorff measure*, Proc. Camb. Phil. Soc. **42** (1946), 15–23.

[Mo-1938] C. B. Morrey, *On the solution of quasi-linear elliptic partial differential equations*, Trans. Amer. Math. Soc. **43** (1938), 126–166.

[Mun-1984] J. R. Munkres, *Elements of Algebraic Topology*, Addison–Wesley, 1984.

[Mun-2000] J. Munkres, *Topology*, Engelwood Cliffs, 2000.

[Nag-1958] J. I. Nagata, *Note on dimension theory for metric spaces*, Fund. Math. **45** (1958), 143–181.

[Nag-1965] J. I. Nagata, *Modern Dimension Theory*, North-Holland Publ. Company, Amsterdam, 1965.

[N-2001] A. Naor, *A phase transition phenomenon between the isometric and isomorphic extension problems for Hölder functions between L_p spaces*, Mathematika **48** (2001), 253–271.

[NPSS-2006] A. Naor, Y. Peres, D. Schramm and S. Sheffield, *Markov chains in smooth Banach spaces and Gromov hyperbolic metric spaces*, Duke Math. J. **134** (2006), 165–197.

[Na-1966] J. Nash, *Analyticity of the solutions of implicit function problems with analytic data*, Ann. Math. **84** (1966), 345–355.

[NSch-1941] J. von Neumann and I. Schoenberg, *Fourier integrals and metric geometry*, Trans. AMS **50** (1941), 226–251.

[Ne-1984] M. Nevsky, *Approximation of functions in Orlicz classes*. In: Studies in Theory of Functions of Several Variables, Yaroslavl' State Univ., Yaroslavl', 1984, 83-101 (in Russian).

[Nik-1980] I. G. Nikolaev, *Parallel translation and smoothness of the metric of spaces of bounded curvature*, Dokl. Akad. Nauk SSSR **250** (1980), 1056–1058; English translation in: Soviet Math. Doklady **21** (1980), 263–265.

[No-1931] P. S. Novikov, *Sur les fonctions implicites mesurables B*, Fund. Math. **XVII** (1931), 8–25.

[Ol'-1992] A. Yu. Ol'shanskii, *Almost every group is hyperbolic*, International J. Algebra Comput. **2** (1992), 1–17.

[OS-1978] P. Oswald and E. Storozhenko, *Jackson's theorem in $L_p(\mathbb{R}^n)$*, Sib. Mat. Zh. **19** (1978), 888–901; English translation in: Siberian Math. J. **19** (1978).

[Pa-1989] P. Pansu, *Métriques de Carnot–Carathéodory et quasi-isométries des espaces symétriques de rang un*, Ann. Math. **129** (1989), 1–60.

[PP-1988] W. Pawlucki and W. Pleśniak, *Extensions of C^∞ functions from sets with polynomial cusps*, Studia Math. **88** (1988), 279–287.

[Peet-1963] J. Peetre, *Nouvelles propriétés d'espaces d'interpolation*, C. R. Acad. Sci. **256** (1963), 1424–1426.

[Peet-1979] J. Peetre, *A counterexample connected with Gagliardo's trace theorem*, Comment. Math., Special Issue **2** (1979), 277–282.

[PSp-1972] J. Peetre and G. Sparr, *Interpolation of normed Abelian groups*, Ann. Mat. Pura Appl. **92** (1972), 217–262.

[Pe-1960] A. Pełczyński, *Projections in certain Banach spaces*, Studia Math. **19** (1960), 209–228.

[Pe-1968] A. Pełczyński, *Linear extensions, linear averaging and their applications to linear topological classification of spaces of continuous functions*, Dissertationes Math. **58** (1968), 1–92.

[PW-2002] A. Pełczyński and M. Wojciechowski, *Sobolev spaces in several variables in L^1-type norms are not isomorphic to Banach lattices*, Ark. Mat. **40** (2002), 363–382.

[Per-1994] Y. Peres, *The packing measure of self-affine carpets*, Math. Proc. Cambridge Phil Soc. **115** (1994), 437–450.

[Ph-1940] R. S. Phillips, *On linear transformations*, Trans. Amer. Math. Soc. **48** (1940), 516–541.

[Pl-1990] W. Pleśniak, *Markov's inequality and the existence of an extension operator for C^∞ functions*, J. Appr. Theory **61** (1990), 106–117.

[PSch-1932]	L. Pontriagin and L. Schnirelman, *Sur une propriété métrique de la dimension*, Ann. Math. **33** (1932), 156–162.
[Pos-1971]	E. Posicelski, *Lipschitz maps in the space of convex bodies*, Optimization **4** (21) (1971), 83–90 (in Russian).
[Pos-1973]	E. Posicelski, *Characterization of Steiner points*, Mat. Zametki **14** (1973), 243–247 (in Russian).
[PRR-1967]	E. Posner, E. Rodemich and H. Rumsey, Jr., *Epsilon entropy of stochastic processes*, Ann. Math. Statistics **38** (1967), 1000–1020.
[Pr-1987]	D. Preiss, *Geometry of measures in $\mathbb{R}^n$: Distribution, rectifiability, and densities*, Ann. Math. **125** (1987), 537–643.
[PU-1989]	F. Przytycki and M. Urbański, *On the Hausdorff dimension of some fractal sets*, Studia Math. **93** (1989), 155–186.
[PR-1992]	K. Przesławski and L. Rybinski, *Concept of lower semicontinuity and continuous selections for convex-valued multifunctions*, J. Appr. Theory **68** (1992), 262–282.
[PY-1989]	K. Przesławski and D. Yost, *Continuity properties of selectors and Michael's theorem*, Michigan Math. J. **36** (1989), 113–134.
[Ra-1919]	H. Rademacher, *Über partialle und totale Differenzierbarkeit I*, Math. Ann. **79** (1919), 340–359.
[RSch-1950]	H. Rademacher and J. I. Schoenberg, *Helly's theorem on convex domains and Tchebycheff's approximation problem*, Can. J. Math. **2** (1950), 245–256.
[Rai-1939]	D. Raikov, *On local best approximation*, Dokl. AN SSSR **24** (1939), 652–655 (in Russian).
[Ram-1930]	F. Ramsey, *On a problem of formal logic*, Proc. London Math. Soc. **30** (1930), 264–286.
[Rat-1994]	J. Ratcliffe, *Foundations of Hyperbolic Manifolds*, GTM, Springer, 1994.
[Rem-1936]	E. Remez, *Sur une proprieté extrémale des polynómes de Tchebychef*, Comm. Inst. Scie. Kharkov **13** (1936), 93–95.
[Rem-1957]	E. Ya. Remez, *General Computational Methods of Chebyshev Approximation*, Izd. Akad. Nauk Ukr. SSR, Kiev, 1957 (in Russian).
[R-1910]	F. Riesz, *Untersuchungen über Systeme integrierbarer Funktionen*, Math. Ann. **69** (1910), 449–497.

[Ri-1961]	W. Rinow, *Die Innere Geometrie der Metrischen Räume*, Springer, 1961.
[Rie-2006]	M. Rieffel, *Lipschitz extension constants equal projection constants*, Contemporary Math. **414** (2006), 147–162.
[Rob-1942]	C. V. Robinson, *Spherical theorems of Helly type and congruence indices of spherical cups*, Amer. J. Math. **64** (1942), 260–272.
[Rock-1970]	R. Rockafellar, *Convex Analysis*, Princeton Math. Ser., 28, Princeton, 1970.
[Ro-2003]	J. Roe, *Lectures on Coarse Geometry*, Univ. Lecture Series, vol. 31, AMS, Providence, RI, 2003.
[Rut-1965]	D. Rutovitz, *Some parameters associated with finite-dimensional Banach spaces*, J. London Math. Soc. **40** (1965), 241–255.
[MRu-1968]	M. E. Rudin, *A new proof that metric spaces are paracompact*, Proc. AMS **20** (1969), 603.
[Ru-1987]	W. Rudin, *Real and Complex Analysis*, Mc-Graw Hill, 1987.
[Rush-1992]	T. Rushing, *Hausdorff dimension of wild fractals*, Trans. AMS **334** (1992), 597–613.
[Ruz-1925]	S. Ruziewicz, *Contribution à l'étude des ensembles de distances de points*, Fund. Math. **7** (1925), 141–143.
[Sa-1999]	E. Saksman, *Remarks on the nonexistence of doubling measures*, Ann. Acad. Sci. Fenn. Ser. AI Math. (1999), 153–163.
[Sch-2005]	T. Schlichenmaier, *A quasi-simmetrically invariant notion and absolute Lipschitz retracts*, ETH, Diss. ETH, No. 16216 (2005), 74 pps.
[Sch-1993]	R. Schneider, *Convex Bodies: the Brunn–Minkowski Theory*, Cambridge Univ. Press, Cambridge, 1993.
[Sem-1996]	S. Semmes, *On the nonexistence of bi-Lipschitz parameterizations and geometric problems about A_∞-weights*, Revista Mat. Iberoamericana **12** (1996), 337–410.
[Sem-1999]	S. Semmes, *Metric spaces and mappings seen in many scales*, in the book [Gr-2000], Appendix B.
[Sem-2007]	S. Semmes, *Some Novel Types of Fractal Geometry*, Oxford Univ. Press, 2007.
[Sen-1987]	B. Sendov, *On the theorem and constants of H. Whitney*, Constr. Appr. **3** (1987), 1–11.

Bibliography

[Shal-1991] P. B. Shalen, *Dendrology and its applications*. In: Group Theory from Geometric View Point (E. Ghys et al., eds.), World Scientific, 1991, 543–617.

[Shan-1948] C. E. Shannon, *A mathematical theory of communication, Parts I,II*, Bell Syst. Techn. J. **27** (1948), 379–423 and 623–656.

[Sha-1971] H. Shapiro, *Topics in Approximation Theory*, Lect. Notes in Math., **183**, Springer, 1971.

[Sh-1966] G. Sheppard, *The Steiner point of a convex polytope*, Canad. J. Math. **18** (1966), 1294–1300.

[She-1992] I. A. Shevchuk, *Approximation by Polynomials and Traces of Functions Continuous on Closed Intervals*, Kiev, "Naukova Dumka", 1992 (in Russian).

[Shn-1938] L. G. Shnirelman, *On uniform approximation*, Izv. Akad. Nauk SSSR **2** (1938), 53–59 (in Russian).

[Shv-1978] P. Shvartsman, *Extension theorem for a class of spaces defined by local approximation*. In: Studies in the Theory of Functions of Several Variables, Yaroslavl' State Univ., Yaroslavl', 1978, 215–242 (in Russian).

[Shv-1982] P. A. Shvartsman, *On traces of functions of two variables satisfying Zygmund's conditions*. In: Investigations in Theory of Functions of Several Variables, Yaroslavl' State Univ., 1982, pp. 145–168 (in Russian).

[Shv-1985] P. A. Shvartsman, *The K-functional of a couple of Lipschitz spaces*. In: Geometric Problems in the Theory of Functions and Sets, Kalinin State Univ., 1985, 139–147.

[Shv-1986] P. Shvartsman, *Extension theorems preserving local polynomial approximation*, Preprint, Yaroslavl', 1986, 1–173 (in Russian).

[Shv-1986a] P. Shvartsman, *Lipschitz selections of multivalued maps* In: Studies in the Theory of Functions of Several Variables, Yaroslavl' State Univ., Yaroslavl', 1986, 121–132 (in Russian).

[Shv-1987] P. A. Shvartsman, *On the traces of functions of the Zygmund class*, Sib. Math. J. **28** (5) (1987), 203–215; English tanslation in: Siberian Math. J. **28** (1987), 853–863.

[Shv-1992] P. Shvartsman, *K-functionals of weighted Lipschitz spaces and Lipschitz selections of multivalued mappings*. In: Interpolation Spaces and Related Topics, Math. Conf. Proceedings **5**, Jerusalem, 1992, 245–268.

[Shv-2001] P. Shvartsman, *On Lipschitz selections of affine-set valued mappings*, GAFA **11** (2001), 840–868.

[Shv-2002] P. Shvartsman, *Lipschitz selections of set-valued mappings and Helly's theorem*, J. Geom. Analysis **12** (2002), 289–324.

[Shv-2004] P. Shvartsman, *Barycentric selectors and a Steiner-type point of a convex body in a Banach space*, J. Funct. Analysis **210** (2004), 1–42.

[Shv-2008] P. Shvartsman, *The Whitney extension problem and Lipschitz selections of set-valued mappings in jet-spaces*, Trans. Amer. Math. Soc. **360** (2008), 5529–5550.

[Shv-2009] P. Shvartsman, *The Whitney extension problem for Zygmund spaces and Lipschitz selections in hyperbolic jet-spaces*, arXiv: 0905.2602 [math.FA] (2009), 1–43.

[Sie-1928] W. Sierpiński, *Sur les ensembles completes d'un espace (D)*, Fund. Math. **11** (1928), 203–205.

[Sim-1983] L. Simon, *Lectures on Geometric Measure Theory*, Proc. of Center of Math. Analysis, Austral. Nat. University, 3, 1983.

[STh-1967] J. M. Singer and J. A. Thorpe, *Lectures on Elementary Topology and Geometry*, Springer, 1967.

[Sit-1953] K. Sitnikov, *An example of two-dimensional set in three-dimensional Euclidean space allowing arbitrarily small deformations into a one-dimensional polyhedron and a certain new characterization of the dimension of sets in Euclidean spaces*, Dokl. Akad. Nauk SSSR **88** (1953), 21–24 (in Russian).

[Sit-1955] K. Sitnikov, *Combinatorial topology of nonclosed sets II*, Mat. Sbornik **37** (1955), 385–434.

[So-1941] A. Sobczyk, *Projections in Minkowski and Banach spaces*, Duke Math. J. **8** (1941), 78–106.

[Sob-1938] S. L. Sobolev, *On a theorem of functional analysis*, Matem. Sb. **4**, No. 3 (1938), 471–497 (in Russian); English translation in: AMS Translations (2) **34** (1963), 39–68.

[Sob-1950] S. L. Sobolev, *Some applications of functional analysis in mathematical physics*, Izd. LGU, Leningrad, 1950 (in Russian); English Translation in: AMS Translations **7** (1963).

[Spe-1928] E. Sperner, *Neuer Beweis für die Invarianz der Dimensionzahl und des Gebietes*, Abh. Math. Sem. Hamburg **6** (1928), 265–272.

[St-1957]	N. E. Steenrod, *Cohomology Operations, and Obstructions to Extending Continuous Functions*, Coll. Lectures, Princeton Univ. Press, Princeton, NJ, 1957.
[Ste-1970]	E. M. Stein, *Singular Integrals and Differentiability Properties of Functions*, Princeton Univ. Press, 1970.
[Ste-1993]	E. Stein, *Harmonic Analysis, Real-variable Methods, Orthogonality and Oscillatory Integrals*, Princeton, NJ, 1993.
[S-1840]	J. Steiner, *Von dem Krümmungsschwerpunkte ebener Curven*, J. Reine Angew. Math. **21** (1840), 33–63 and 101–122.
[Step-1925]	V. Stepanoff, *Sur les conditions de l'existence la différentielle totale*, Mat. Sb. **32** (1925), 511–526.
[Sto-1948]	A. H. Stone, *Paracompactness and product spaces*. Bull. AMS **54** (1948), 977–982.
[Str-1967]	R. Strichartz, *Multipliers on fractional Sobolev spaces*, J. Math. Mech. **16** (1967), 1031–1060.
[Str-1979]	J.-O. Strömberg, *Bounded mean oscillation with Orlicz norms and duality of Hardy spaces*, Indiana Univ. Math. J. **28** (1979), 511–544.
[Sul-1979]	D. Sullivan, *Hyperbolic Geometry and Homeomorphisms*. In: Geometric Topology, Acad. Press, New York, 1979, 543–545.
[Szp-1937]	E. Szpilrajn, *La dimension et la mesure*, Fund. Math. **28** (1937), 81–89.
[SzW-1968]	G. Szekeres and H. S. Wilf, *An inequality for the chromatic number of a graph*, J. Combinatorial Theory **4** (1968), 1–3.
[Tid-1979]	M. Tidten, *Fortsetzungen von C^∞-Funktionen, welche auf einer abgeschlossenen Menge in $\mathbb{R}^n$ definiert sind*, Manuscripta Math. **27** (1979), 291–312.
[Tid-1983]	M. Tidten, *Kriterien für die Existenz von Ausdehnungsoperatoren zu $\mathcal{E}(K)$ für kompakte Teilmengen K von $\mathbb{R}$*, Arch. Math. (Basel) **40** (1983), 73–81.
[Ti-1915]	H. Tietze, *Über Functionen, die auf einer abgeschlossenen Menge stetig sind*, J. für Mat. **145** (1915), 9–14.
[MTim-1969]	M. F. Timan, *Difference properties of functions of several variables*, Izv. AN SSSR, Ser. Matem. **33** (1969), 667–678 (in Russian).
[Tim-1963]	A. F. Timan, *Theory of Approximation of Functions of Real Variable*, Pergamon Press, Oxford, 1963.

[Tim-1975] A. Timan, *On the isometric mapping of some ultrametric spaces into L_p*, Trudy Mat. Inst. **134** (1975), 314–326.

[TV-1983] A. Timan and I. Vestfrid, *Any separable ultrametric space can be isometrically embedded in ℓ_2* Funk. Anal. i Prilozh. **17** (1983), 85–86; English translation in: Func. Anal. Appl. **17** (1983), 70–71.

[Tr-1995] C. Tricot, *Curves and Fractal Dimensions*, Springer, 1995.

[Tri-1992] H. Triebel, *Theory of Function Spaces, II*, Birkhäuser, 1992.

[TB-2004] R. Trigub and E. Belinsky, *Fourier Analysis and Approximation of Functions*, Kluwer Acad. Publ., 2004.

[Ts-1999] I. Tsarkov, *Extension of Hilbert-valued Lipschitz mappings*, Moscow Univ. Math. Bull. **54**, No. 6 (1999), 7–14.

[Tu-1989] P. Tukia, *Hausdorff dimension and quasisymmetric mappings*, Math. Scand. **65** (1989), 152–160.

[Us-2004] V. Uspenskij, *The Urusohn Universal space is homeomorphic to a Hilbert space*, Topology and its Appl. **139** (2004), 145–149.

[Ur-1925] P. Urysohn, *Über die Mächtigkeit der zusammenhängenden Mengen*, Math. Ann. **94** (1925), 262–295.

[Ur-1927] P. Urysohn, *Sur un espace métrique universal*, Bull. de Sci. Mathématique, serie 2, **51** (1927), 1–38.

[Vai-2005] J. Väisälä, *Gromov Hyperbolic Groups*, Expo. Math. **23**, No. 3 (2005), 187–231.

[Va-1944] F. A. Valentine, *Contractions in non-Euclidean spaces*, Bull. Amer. Math. Soc. **50** (1944), 710–713.

[Va-1945] F. A. Valentine, *A Lipschitz condition preserving extension for a vector function*, Amer. J. Math. **67** (1945), 83–93.

[V-1998] A. M. Vershik, *The universal Urysohn space, Gromov metric triples and random metrics on the natural numbers*, Uspekhi Mat. Nauk **53** (1998), 57–64 (in Russian); English translation in: Russian Math. Surveys **53** (1998), 921–928.

[VG-1994] A. M. Vershik and V. Ya. Gershkovich, *Nonholonomic dynamic systems. Geometry of distributions and variational problems.* In: Dynamical Systems VII, Springer, 1994, 1–81. Russian original 1987.

[Vit-1985] R. Vitale, *The Steiner point in infinite dimensions*, Israel J. Math. **52** (1985), 245–250.

[VGr-1996]	S. K. Vodop'yanov and A. V. Greshnov, *Extending differentiable functions and quasiconformal mappings on Carnot groups*, Doklady Ak. Nauk **348**, No. 1 (1996), 15–18; English translation in: Doklady Math. **53**, No. 3 (1996), 331–333.
[Vogt-1983]	D. Vogt, *Sequence space representations of spaces of test functions and distributions.* In: Lect. Notes in Pure and Appl. Math. **83**, Marcel Dekker, 1983, 405–443.
[VK-1987]	A. L. Vol'berg and S. V. Koniagin, *On measures with the doubling condition*, Izv. Akad. Nauk SSSR, Ser. Mat. **51** (1987), 666–675; English translation in: Math. USSR–Izv. **30** (1988), 629–638.
[Vop-1959]	P. Vopenka, *Remarks on the dimension of metric spaces*, Czech. Math. J. **9** (1959), 519–522.
[W-1999]	N. Weaver, *Lipschitz Algebras*, Scientific World, 1999.
[We-1952]	A. Weil, *Sur les théorèmes de de Rham*, Comment. Math. Helv. **26** (1952), 119–145.
[WW-1975]	J. Wells and L. Williams, *Embeddings and Extensions in Analysis*, Springer, 1975.
[Whi-2000]	A. T. White, *Graph and map colorings.* In: K. H. Rosen (Ed.), Hand book of Discrete and Combinatorial Mathematics, CRC Press, 2000, 557–566.
[Wh-1934a]	H. Whitney, *Analytic extensions of differentiable functions defined in closed sets*, Trans. Amer. Math. Soc. **36** (1934), 63–89.
[Wh-1934b]	H. Whitney, *Differentiable functions defined in closed sets, I*, Trans. Amer. Math. Soc. **36** (1934), 369–387.
[Wh-1934c]	H. Whitney, *Derivatives, difference quotients and Taylor's formula*, Bull. Amer. Math. Soc. **40** (1934), 89–94.
[Wh-1934d]	H. Whitney, *Functions differentiable on boundaries of regions*, Ann. Math. **35**, No. 3 (1934), 482–485.
[Wh-1957]	H. Whitney, *On functions with bounded n-th differences*, J. Math. Pure Appl. **9**, No. 3 (1957), 67–95.
[Wh-1957a]	H. Whitney, *Geometric Integration Theory*, Princeton Univ. Press, Princeton, NJ, 1957.
[Wh-1959]	H. Whitney, *On boundded functions with bounded n-th difference*, Proc. Amer. Math. Soc. **10** (1959), 480–481.
[Wo-1980]	Th. Wolff, *Restrictions of A_p weights*, Dept. of Math. CALTEX, 1980, 1–5.

[Woj-1991]   P. Wojtaszczyk, *Banach Spaces for Analysts*, Studies in Adv. Math. **25**, Cambridge Univ. Press, 1991.

[Wu-1998]   J.-M. Wu, *Hausdorff dimension and doubling measures on metric spaces*, Proc. AMS **126** (1998), 1453–1459.

[Zi-1989]   W. P. Ziemer, *Weakly Differentiable Functions*, Springer, 1989.

[Zo-1998]   N. Zobin, *Whitney's problem on extendability of functions and an intrinsic metric*, Adv. Math. **133**, No. 1 (1998), 96–132.

[Zo-1999]   N. Zobin, *Extension of smooth functions from finitely connected planar domains*, J. Geometric Analysis **9**, No. 3 (1999), 489–509.

[Z-1945]   A. Zygmund, *Smooth functions*, Duke Math. J. **12** (1945), 47–76.

Index

F_σ-set, 204
G_δ-set, 204
K-functional, 151
ε-
    capacity, 216
    chain, 235
    dense subset, 236
    entropy, 313
    lattice, 235
    net, 236
    separated subset, 235
μ-measurable map, 253
k-
    jet, 86
    majorant, 89
    modulus of continuity, 87
    oscillation, 110
    roughly geodesic ray, 454
p-
    adic numbers, 275
    Jacobian, 257
    rectifiable subset of $\mathbb{R}^n$, 332
    set, 332
    sum of a family of metric spaces, 237

Antoine Necklace, 338

Banach
    couple, 151
    indicatrix, 257
    limit, 251
barycenter, 69
    of a bounded convex body, 471
    of the face, 342

barycentric selector, 471
barycentric subdivision, 69
bi-Lipschitz embedding, 21
bi-Lipschitz homeomorphism, 21
boundary at infinity, 455
Brownian motion, 340

Calderón couple, 398
Chebyshev radius, 248
child of a vertex, 299
chromatic number of a graph, 292
coloring of a graph, 292
complex
    metric polyhedral, 342
    polyhedral, 68
    simplicial, 69
cover, 213
    locally finite, 213
    open (closed), 213
covering number, 216
curvature, 284
curve, 211
    k-roughly geodesic, 454
    geodesic, 211
    Jordan, 258
    rectifiable, 226
    uniform, 289

degree of a vertex, 293
difference
    k-th, 87
    divided, 102
    mixed, 103
dimension
    Assouad, 334

fractal (homogeneous), 309
Hausdorff, 329
metric, 326
Minkowski, 223
Nagata, 341
packing, 414
topological (covering), 61
distance
　Gromov–Hausdorff, 242
　Hausdorff, 23
　hyperbolic, 38
　spherical, 79
distortion, 21
domain
　orderly convex, 174
　quasiconvex, 136
　special Lipschitz, 149
　uniform, 155
doubling
　constant, 222
　constant of a measure, 263
　function, 222

family of measures
　K-uniform, 266
　coherent, 267
finiteness constant
　local, 131
　strong, 132
　uniform, 131
function
　Bochner integrable, 254
　Borel measurable, 253
　characteristic (indicator), 254
　consistency, 266
　dilation, 265

geodesic bicombing, 524
geodesic triangle, 284
graph
　Cayley, 306
　combinatorial, 292
　metric, 293
　simple, 292

spatially colored cover, 300
group
　acting cocompactly, 304
　acting freely, 304
　acting properly, 304
　Carnot, 307
　finitely generated, 303
　free, 307
　Heisenberg, 224
　hyperbolic, 306
　linear, 301
　metric, 301
　of polynomial growth, 304
　uniform lattice, 307

Helly index, 32

isometry, 205

Lipschitz
　absolute retract, 48
　constant, 21
　extension constant, 22
　map, 21
　retract, 48
　selection, 23
　simultaneous extension, 53
local polynomial (best) approximation, 110
loop, 292

manifold
　"pinched" Hadamard, 357
　Lipschitz, 389
　Riemannian, 280
　sub-Riemannian, 291
Markov type inequality, 109
measure, 252
　(C, s)-homogeneous, 357
　σ-finite, 253
　Borel, 252
　counting, 257
　doubling, 263
　Hausdorff, 256

Index

Minkowski p-area, 259
packing, 414
Radon, 333
metric
    Carnot-Carathéodory, 309
    inner, 229
    quasihyperbolic, 289
    word, 303
metric space, 6
    (C, s)-homogeneous, 357
    C-quasiconvex, 232
    δ-hyperbolic, 286
    k-visual, 456
    absolute neighborhood retract, 18
    absolute retract, 18
    Ahlfors regular, 333
    arcwise connected, 212
    cocompact, 288
    complete, 204
    convex, 210
    doubling, 222
    generalized hyperbolic, 285
    geodesic, 211
    Gromov hyperbolic, 286
    homogeneous, 205
    length, 230
    locally doubling, 226
    of bounded geometry, 278
    of bounded growth in some scale, 458
    of homogeneous type, 264
    of pointwise homogeneous type, 267
    Polish, 205
    proper, 219
    retract, 17
    transitive, 205
    uniformly proper, 222
    with the ball intersection property, 32
    with the finite intersection property, 37
modulus of continuity, 29

nerve, 70
normal topology, 204

order (multiplicity) of cover, 95

paracompact, 213
parent of a vertex, 299
partition of unity
    continuous, 215
    smooth, 96
polynomial
    Chebyshev, 188
    Taylor, 85
projection constant, 49
pseudometric, 201

quasiball, 266
quasimetric, 202

Radon–Nikodym property, 387
rearrangement, 193
reduced remainder, 93
refinement, 213
rough similarity, 288

set
    Cantor, 328
    countably p-rectifiable, 333
    Fatou, 341
    Julia, 341
    self-similar, 336
Sierpiński gasket, 328
similitude, 336
space
    $C^k(G)$, 85
    $C^k\Lambda^{s,\omega}(\mathbb{R}^n)$, 90
    $C_b^k(G)$, 85
    $C_b^k\Lambda^{s,\omega}(\mathbb{R}^n)$, 89
    $C_u^k(G)$, 86
    $C_b^{k,\omega}(G)$, 85
    $C_b(\mathcal{M}, X)$ of bounded continuous maps $f : \mathcal{M} \to X$, 7
    $C_u(\mathcal{M}, X)$ of bounded uniformly continuous maps $f : \mathcal{M} \to X$, 7

$J^k F$ of all k-jets with values in $\mathbb{R}^N$, 86
$J^k \Lambda^{s,\omega}(\mathbb{R}^n)$, 90
$\Lambda^{k,\omega}(\mathbb{R}^n)$ of Lipschitz functions of order k, 89
$\operatorname{Lip}(\mathcal{M}, \widetilde{\mathcal{M}})$ of Lipschitz maps $f : \mathcal{M} \to \widetilde{\mathcal{M}}$, 21
$\dot{C}^{k,\omega}(G)$, 85
$\dot{\Lambda}^{k,\omega}(\mathbb{R}^n)$ homogeneous of Lipschitz functions of order k, 89
Besov, 91
BMO, 11
Carnot–Carathéodory, 391
constrained in its bidual, 49
interpolation, 398
Laakso, 392
Lipschitz-free, 406
of balls, 372
Sobolev, 11
ultrametric, 274
Zygmund, 91
star, 69
Steiner point, 472
support of measure, 253

Taylor chain condition, 94
total variation of a signed measure, 266
trace space, 7
tree, 299
triangulation, 73

Urysohn k-width, 217

Whitney constants, 186

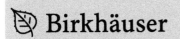

Monographs in Mathematics

The foundations of this outstanding book series were laid in 1944. Until the end of the 1970s, a total of 77 volumes appeared, including works of such distinguished mathematicians as Carathéodory, Nevanlinna and Shafarevich, to name a few. The series came to its name and present appearance in the 1980s. In keeping its wellestablished tradition, only monographs of excellent quality are published in this collection. Comprehensive, in-depth treatments of areas of current interest are presented to a readership ranging from graduate students to professional mathematicians. Concrete examples and applications both within and beyond the immediate domain of mathematics illustrate the import and consequences of the theory under discussion.

Managing Editors
Amann, H., Universität Zürich, Switzerland
Bourguignon, J.-P., IHES, France
Grove, K., University of Maryland, USA
Lions, P.-L., Paris-Dauphine, France

■ **Vol. 101: Volpert, V.**, Elliptic Partial Differential Equations. Volume 1: Fredholm Theory of Elliptic Problems in Unbounded Domains
2011. 656 pages. Hardcover.
ISBN 978-3-0346-0536-6

The theory of elliptic partial differential equations has undergone an important development over the last two centuries. Together with electrostatics, heat and mass diffusion, hydrodynamics and many other applications, it has become one of the most richly enhanced fields of mathematics. This monograph undertakes a systematic presentation of the theory of general elliptic operators. The author discusses a priori estimates, normal solvability, the Fredholm property, the index of an elliptic operator, operators with a parameter, and nonlinear Fredholm operators. Particular attention is paid to elliptic problems in unbounded domains which have not yet been sufficiently treated in the literature and which require some special approaches. The book also contains an analysis of non-Fredholm operators and discrete operators as well as extensive historical and bibliographical comments. The selected topics and the author's level of discourse will make this book a most useful resource for researchers and graduate students working in the broad field of partial differential equations and applications.

■ **Vol. 96: Arendt, W., Batty, C.J.K., Hieber M. and Neubrander, F.**, Vector-valued Laplace Transforms and Cauchy Problems, 2nd edition
2011. 552 pages. Hardcover.
ISBN 978-3-0348-0086-0

This monograph gives a systematic account of the theory of vector-valued Laplace transforms, ranging from representation theory to Tauberian theorems. In parallel, the theory of linear Cauchy problems and semigroups of operators is developed completely in the spirit of Laplace transforms. Existence and uniqueness, regularity, approximation and above all asymptotic behaviour of solutions are studied. Diverse applications to partial differential equations are given. The book contains an introduction to the Bochner integral and several appendices on background material. It is addressed to students and researchers interested in evolution equations, Laplace and Fourier transforms, and functional analysis.

This authoritative work is likely to become a standard reference on both the Laplace transform and its applications to the abstract Cauchy problem. ... The book is an excellent textbook as well. Proofs are always transparent and complete. All this makes the text very accessible and self-contained.

J. van Neerven, *Nieuw Archief voor Wiskunde*, No. 3, 2003